## Basic Differentiation Rules

**1.** $\dfrac{d}{dx}[cu] = cu'$

**2.** $\dfrac{d}{dx}[u \pm v] = u' \pm v'$

**3.** $\dfrac{d}{dx}[uv] = uv' + vu'$

**4.** $\dfrac{d}{dx}\left[\dfrac{u}{v}\right] = \dfrac{vu' - uv'}{v^2}$

**5.** $\dfrac{d}{dx}[c] = 0$

**6.** $\dfrac{d}{dx}[u^n] = nu^{n-1}u'$

**7.** $\dfrac{d}{dx}[x] = 1$

**8.** $\dfrac{d}{dx}[|u|] = \dfrac{u}{|u|}(u'), \quad u \neq 0$

**9.** $\dfrac{d}{dx}[\ln u] = \dfrac{u'}{u}$

**10.** $\dfrac{d}{dx}[e^u] = e^u u'$

**11.** $\dfrac{d}{dx}[\log_a u] = \dfrac{u'}{(\ln a)u}$

**12.** $\dfrac{d}{dx}[a^u] = (\ln a)a^u u'$

**13.** $\dfrac{d}{dx}[\sin u] = (\cos u)u'$

**14.** $\dfrac{d}{dx}[\cos u] = -(\sin u)u'$

**15.** $\dfrac{d}{dx}[\tan u] = (\sec^2 u)u'$

**16.** $\dfrac{d}{dx}[\cot u] = -(\csc^2 u)u'$

**17.** $\dfrac{d}{dx}[\sec u] = (\sec u \tan u)u'$

**18.** $\dfrac{d}{dx}[\csc u] = -(\csc u \cot u)u'$

**19.** $\dfrac{d}{dx}[\arcsin u] = \dfrac{u'}{\sqrt{1 - u^2}}$

**20.** $\dfrac{d}{dx}[\arccos u] = \dfrac{-u'}{\sqrt{1 - u^2}}$

**21.** $\dfrac{d}{dx}[\arctan u] = \dfrac{u'}{1 + u^2}$

**22.** $\dfrac{d}{dx}[\operatorname{arccot} u] = \dfrac{-u'}{1 + u^2}$

**23.** $\dfrac{d}{dx}[\operatorname{arcsec} u] = \dfrac{u'}{|u|\sqrt{u^2 - 1}}$

**24.** $\dfrac{d}{dx}[\operatorname{arccsc} u] = \dfrac{-u'}{|u|\sqrt{u^2 - 1}}$

**25.** $\dfrac{d}{dx}[\sinh u] = (\cosh u)u'$

**26.** $\dfrac{d}{dx}[\cosh u] = (\sinh u)u'$

**27.** $\dfrac{d}{dx}[\tanh u] = (\operatorname{sech}^2 u)u'$

**28.** $\dfrac{d}{dx}[\coth u] = -(\operatorname{csch}^2 u)u'$

**29.** $\dfrac{d}{dx}[\operatorname{sech} u] = -(\operatorname{sech} u \tanh u)u'$

**30.** $\dfrac{d}{dx}[\operatorname{csch} u] = -(\operatorname{csch} u \coth u)u'$

**31.** $\dfrac{d}{dx}[\sinh^{-1} u] = \dfrac{u'}{\sqrt{u^2 + 1}}$

**32.** $\dfrac{d}{dx}[\cosh^{-1} u] = \dfrac{u'}{\sqrt{u^2 - 1}}$

**33.** $\dfrac{d}{dx}[\tanh^{-1} u] = \dfrac{u'}{1 - u^2}$

**34.** $\dfrac{d}{dx}[\coth^{-1} u] = \dfrac{u'}{1 - u^2}$

**35.** $\dfrac{d}{dx}[\operatorname{sech}^{-1} u] = \dfrac{-u'}{u\sqrt{1 - u^2}}$

**36.** $\dfrac{d}{dx}[\operatorname{csch}^{-1} u] = \dfrac{-u'}{|u|\sqrt{1 + u^2}}$

## Basic Integration Formulas

**1.** $\displaystyle\int kf(u)\, du = k\int f(u)\, du$

**2.** $\displaystyle\int [f(u) \pm g(u)]\, du = \int f(u)\, du \pm \int g(u)\, du$

**3.** $\displaystyle\int du = u + C$

**4.** $\displaystyle\int u^n\, du = \dfrac{u^{n+1}}{n+1} + C, \quad n \neq -1$

**5.** $\displaystyle\int \dfrac{du}{u} = \ln|u| + C$

**6.** $\displaystyle\int e^u\, du = e^u + C$

**7.** $\displaystyle\int a^u\, du = \left(\dfrac{1}{\ln a}\right)a^u + C$

**8.** $\displaystyle\int \sin u\, du = -\cos u + C$

**9.** $\displaystyle\int \cos u\, du = \sin u + C$

**10.** $\displaystyle\int \tan u\, du = -\ln|\cos u| + C$

**11.** $\displaystyle\int \cot u\, du = \ln|\sin u| + C$

**12.** $\displaystyle\int \sec u\, du = \ln|\sec u + \tan u| + C$

**13.** $\displaystyle\int \csc u\, du = -\ln|\csc u + \cot u| + C$

**14.** $\displaystyle\int \sec^2 u\, du = \tan u + C$

**15.** $\displaystyle\int \csc^2 u\, du = -\cot u + C$

**16.** $\displaystyle\int \sec u \tan u\, du = \sec u + C$

**17.** $\displaystyle\int \csc u \cot u\, du = -\csc u + C$

**18.** $\displaystyle\int \dfrac{du}{\sqrt{a^2 - u^2}} = \arcsin \dfrac{u}{a} + C$

**19.** $\displaystyle\int \dfrac{du}{a^2 + u^2} = \dfrac{1}{a}\arctan \dfrac{u}{a} + C$

**20.** $\displaystyle\int \dfrac{du}{u\sqrt{u^2 - a^2}} = \dfrac{1}{a}\operatorname{arcsec} \dfrac{|u|}{a} + C$

# TRIGONOMETRY

## Definition of the Six Trigonometric Functions

*Right triangle definitions, where $0 < \theta < \pi/2$.*

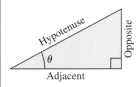

$$\sin \theta = \frac{\text{opp}}{\text{hyp}} \quad \csc \theta = \frac{\text{hyp}}{\text{opp}}$$

$$\cos \theta = \frac{\text{adj}}{\text{hyp}} \quad \sec \theta = \frac{\text{hyp}}{\text{adj}}$$

$$\tan \theta = \frac{\text{opp}}{\text{adj}} \quad \cot \theta = \frac{\text{adj}}{\text{opp}}$$

*Circular function definitions, where $\theta$ is any angle.*

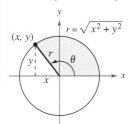

$$\sin \theta = \frac{y}{r} \quad \csc \theta = \frac{r}{y}$$

$$\cos \theta = \frac{x}{r} \quad \sec \theta = \frac{r}{x}$$

$$\tan \theta = \frac{y}{x} \quad \cot \theta = \frac{x}{y}$$

*Unit circle diagram with angles in degrees and radians and coordinate points:*

$\left(-\frac{1}{2}, \frac{\sqrt{3}}{2}\right)$   $(0,1)$   $\left(\frac{1}{2}, \frac{\sqrt{3}}{2}\right)$

$\left(-\frac{\sqrt{2}}{2}, \frac{\sqrt{2}}{2}\right)$   $\frac{\pi}{2}$   $90°$   $\frac{2\pi}{3}$   $120°$   $\frac{\pi}{3}$   $60°$   $\frac{\pi}{4}$   $\left(\frac{\sqrt{2}}{2}, \frac{\sqrt{2}}{2}\right)$

$\left(-\frac{\sqrt{3}}{2}, \frac{1}{2}\right)$   $\frac{3\pi}{4}$   $135°$   $\frac{5\pi}{6}$   $150°$   $45°$   $\frac{\pi}{6}$   $30°$   $\left(\frac{\sqrt{3}}{2}, \frac{1}{2}\right)$

$(-1,0)$   $\pi$   $180°$   $0°$   $0$   $360°$   $2\pi$   $(1,0)$

$\left(-\frac{\sqrt{3}}{2}, -\frac{1}{2}\right)$   $\frac{7\pi}{6}$   $210°$   $225°$   $\frac{5\pi}{4}$   $240°$   $\frac{4\pi}{3}$   $330°$   $\frac{11\pi}{6}$   $\left(\frac{\sqrt{3}}{2}, -\frac{1}{2}\right)$

$\left(-\frac{\sqrt{2}}{2}, -\frac{\sqrt{2}}{2}\right)$   $315°$   $300°$   $\frac{5\pi}{3}$   $\frac{7\pi}{4}$   $270°$   $\frac{3\pi}{2}$   $\left(\frac{\sqrt{2}}{2}, -\frac{\sqrt{2}}{2}\right)$

$\left(-\frac{1}{2}, -\frac{\sqrt{3}}{2}\right)$   $(0,-1)$   $\left(\frac{1}{2}, -\frac{\sqrt{3}}{2}\right)$

## Reciprocal Identities

$$\sin x = \frac{1}{\csc x} \quad \sec x = \frac{1}{\cos x} \quad \tan x = \frac{1}{\cot x}$$

$$\csc x = \frac{1}{\sin x} \quad \cos x = \frac{1}{\sec x} \quad \cot x = \frac{1}{\tan x}$$

## Quotient Identities

$$\tan x = \frac{\sin x}{\cos x} \quad \cot x = \frac{\cos x}{\sin x}$$

## Pythagorean Identities

$$\sin^2 x + \cos^2 x = 1$$
$$1 + \tan^2 x = \sec^2 x \qquad 1 + \cot^2 x = \csc^2 x$$

## Cofunction Identities

$$\sin\left(\frac{\pi}{2} - x\right) = \cos x \qquad \cos\left(\frac{\pi}{2} - x\right) = \sin x$$

$$\csc\left(\frac{\pi}{2} - x\right) = \sec x \qquad \tan\left(\frac{\pi}{2} - x\right) = \cot x$$

$$\sec\left(\frac{\pi}{2} - x\right) = \csc x \qquad \cot\left(\frac{\pi}{2} - x\right) = \tan x$$

## Even/Odd Identities

$$\sin(-x) = -\sin x \qquad \cos(-x) = \cos x$$
$$\csc(-x) = -\csc x \qquad \tan(-x) = -\tan x$$
$$\sec(-x) = \sec x \qquad \cot(-x) = -\cot x$$

## Sum and Difference Formulas

$$\sin(u \pm v) = \sin u \cos v \pm \cos u \sin v$$
$$\cos(u \pm v) = \cos u \cos v \mp \sin u \sin v$$
$$\tan(u \pm v) = \frac{\tan u \pm \tan v}{1 \mp \tan u \tan v}$$

## Double-Angle Formulas

$$\sin 2u = 2 \sin u \cos u$$
$$\cos 2u = \cos^2 u - \sin^2 u = 2 \cos^2 u - 1 = 1 - 2 \sin^2 u$$
$$\tan 2u = \frac{2 \tan u}{1 - \tan^2 u}$$

## Power-Reducing Formulas

$$\sin^2 u = \frac{1 - \cos 2u}{2}$$

$$\cos^2 u = \frac{1 + \cos 2u}{2}$$

$$\tan^2 u = \frac{1 - \cos 2u}{1 + \cos 2u}$$

## Sum-to-Product Formulas

$$\sin u + \sin v = 2 \sin\left(\frac{u + v}{2}\right) \cos\left(\frac{u - v}{2}\right)$$

$$\sin u - \sin v = 2 \cos\left(\frac{u + v}{2}\right) \sin\left(\frac{u - v}{2}\right)$$

$$\cos u + \cos v = 2 \cos\left(\frac{u + v}{2}\right) \cos\left(\frac{u - v}{2}\right)$$

$$\cos u - \cos v = -2 \sin\left(\frac{u + v}{2}\right) \sin\left(\frac{u - v}{2}\right)$$

## Product-to-Sum Formulas

$$\sin u \sin v = \frac{1}{2}[\cos(u - v) - \cos(u + v)]$$

$$\cos u \cos v = \frac{1}{2}[\cos(u - v) + \cos(u + v)]$$

$$\sin u \cos v = \frac{1}{2}[\sin(u + v) + \sin(u - v)]$$

$$\cos u \sin v = \frac{1}{2}[\sin(u + v) - \sin(u - v)]$$

# MULTIVARIABLE
# CALCULUS
### with CalcChat® and CalcView®

**11e**

**Ron Larson**
The Pennsylvania State University
The Behrend College

**Bruce Edwards**
University of Florida

CENGAGE
Learning·

Australia • Brazil • Mexico • Singapore • United Kingdom • United States

*Multivariable Calculus*, **Eleventh Edition**
**Ron Larson, Bruce Edwards**

Product Director: Terry Boyle

Product Manager: Gary Whalen

Senior Content Developer: Stacy Green

Associate Content Developer: Samantha Lugtu

Product Assistant: Katharine Werring

Media Developer: Lynh Pham

Marketing Manager: Ryan Ahern

Content Project Manager: Jennifer Risden

Manufacturing Planner: Doug Bertke

Production Service: Larson Texts, Inc.

Photo Researcher: Lumina Datamatics

Text Researcher: Lumina Datamatics

Illustrator: Larson Texts, Inc.

Text Designer: Larson Texts, Inc.

Compositor: Larson Texts, Inc.

Cover Designer: Larson Texts, Inc.

Cover photograph by Caryn B. Davis | carynbdavis.com

Cover background: iStockphoto.com/briddy_

Umbilic Torus by Helaman Ferguson, donated to Stony Brook University

The cover image is the Umbilic Torus statue created in 2012 by the famed sculptor and mathematician Dr. Helaman Ferguson. This statue weighs 10 tons and has a height of 24 feet. It is located at Stony Brook University in Stony Brook, New York.

For product information and technology assistance, contact us at **Cengage Learning Customer & Sales Support, 1-800-354-9706.**
For permission to use material from this text or product, submit all requests online at **www.cengage.com/permissions.** Further permissions questions can be emailed to **permissionrequest@cengage.com.**

Library of Congress Control Number: 2016944975

Student Edition:
ISBN: 978-1-337-27537-8

Loose-leaf Edition:
ISBN: 978-1-337-27559-0

**Cengage Learning**
20 Channel Center Street
Boston, MA 02210
USA

Cengage Learning is a leading provider of customized learning solutions with employees residing in nearly 40 different countries and sales in more than 125 countries around the world. Find your local representative at **www.cengage.com.**

Cengage Learning products are represented in Canada by Nelson Education, Ltd.

To learn more about Cengage Learning Solutions, visit **www.cengage.com.** Purchase any of our products at your local college store or at our preferred online store **www.cengagebrain.com.**

QR Code is a registered trademark of Denso Wave Incorporated

Printed in the United States of America
Print Number: 02    Print Year: 2017

# Contents

# Appendices

\*Available at the text-specific website *www.cengagebrain.com*

# Preface

Welcome to *Calculus*, Eleventh Edition. We are excited to offer you a new edition with even more resources that will help you understand and master calculus. This textbook includes features and resources that continue to make *Calculus* a valuable learning tool for students and a trustworthy teaching tool for instructors.

*Calculus* provides the clear instruction, precise mathematics, and thorough coverage that you expect for your course. Additionally, this new edition provides you with **free** access to three companion websites:

- **CalcView.com**—video solutions to selected exercises
- **CalcChat.com**—worked-out solutions to odd-numbered exercises and access to online tutors
- **LarsonCalculus.com**—companion website with resources to supplement your learning

These websites will help enhance and reinforce your understanding of the material presented in this text and prepare you for future mathematics courses. CalcView® and CalcChat® are also available as free mobile apps.

## Features

### NEW CalcView®

The website *CalcView.com* contains video solutions of selected exercises. Watch instructors progress step-by-step through solutions, providing guidance to help you solve the exercises. The CalcView mobile app is available for free at the Apple® App Store® or Google Play™ store. The app features an embedded QR Code® reader that can be used to scan the on-page codes and go directly to the videos. You can also access the videos at CalcView.com.

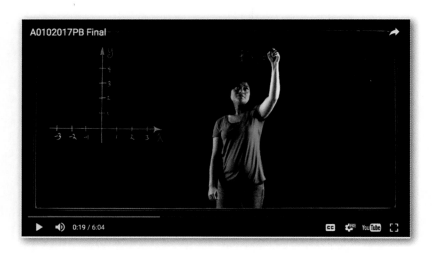

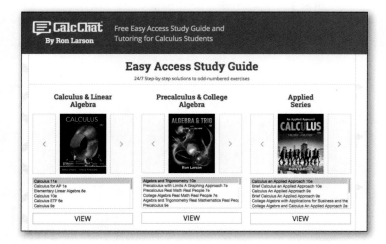

### UPDATED CalcChat®

In each exercise set, be sure to notice the reference to *CalcChat.com*. This website provides free step-by-step solutions to all odd-numbered exercises in many of our textbooks. Additionally, you can chat with a tutor, at no charge, during the hours posted at the site. For over 14 years, hundreds of thousands of students have visited this site for help. The CalcChat mobile app is also available as a free download at the Apple® App Store® or Google Play™ store and features an embedded QR Code® reader.

---

App Store is a service mark of Apple Inc. Google Play is a trademark of Google Inc.
QR Code is a registered trademark of Denso Wave Incorporated.

## REVISED LarsonCalculus.com

All companion website features have been updated based on this revision. Watch videos explaining concepts or proofs from the book, explore examples, view three-dimensional graphs, download articles from math journals, and much more.

## NEW Conceptual Exercises

The *Concept Check* exercises and *Exploring Concepts* exercises appear in each section. These exercises will help you develop a deeper and clearer knowledge of calculus. Work through these exercises to build and strengthen your understanding of the calculus concepts and to prepare you for the rest of the section exercises.

## REVISED Exercise Sets

The exercise sets have been carefully and extensively examined to ensure they are rigorous and relevant and to include topics our users have suggested. The exercises are organized and titled so you can better see the connections between examples and exercises. Multi-step, real-life exercises reinforce problem-solving skills and mastery of concepts by giving you the opportunity to apply the concepts in real-life situations.

## REVISED Section Projects

Projects appear in selected sections and encourage you to explore applications related to the topics you are studying. We have added new projects, revised others, and kept some of our favorites. All of these projects provide an interesting and engaging way for you and other students to work and investigate ideas collaboratively.

## How Do You See It? Exercise

The How Do You See It? exercise in each section presents a problem that you will solve by visual inspection using the concepts learned in the lesson. This exercise is excellent for classroom discussion or test preparation.

## Applications

Carefully chosen applied exercises and examples are included throughout to address the question, "When will I use this?" These applications are pulled from diverse sources, such as current events, world data, industry trends, and more, and relate to a wide range of interests. Understanding where calculus is (or can be) used promotes fuller understanding of the material.

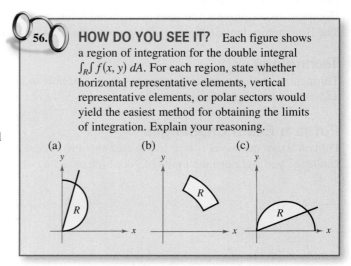

**56. HOW DO YOU SEE IT?** Each figure shows a region of integration for the double integral $\int_R \int f(x, y)\, dA$. For each region, state whether horizontal representative elements, vertical representative elements, or polar sectors would yield the easiest method for obtaining the limits of integration. Explain your reasoning.

## Chapter Opener

Each Chapter Opener highlights real-life applications used in the examples and exercises.

## Section Objectives

A bulleted list of learning objectives provides you with the opportunity to preview what will be presented in the upcoming section.

## Theorems

Theorems provide the conceptual framework for calculus. Theorems are clearly stated and separated from the rest of the text by boxes for quick visual reference. Key proofs often follow the theorem and can be found at *LarsonCalculus.com.*

## Definitions

As with theorems, definitions are clearly stated using precise, formal wording and are separated from the text by boxes for quick visual reference.

## Explorations

Explorations provide unique challenges to study concepts that have not yet been formally covered in the text. They allow you to learn by discovery and introduce topics related to ones presently being studied. Exploring topics in this way encourages you to think outside the box.

## Remarks

These hints and tips reinforce or expand upon concepts, help you learn how to study mathematics, caution you about common errors, address special cases, or show alternative or additional steps to a solution of an example.

## Historical Notes and Biographies

Historical Notes provide you with background information on the foundations of calculus. The Biographies introduce you to the people who created and contributed to calculus.

## Technology

Throughout the book, technology boxes show you how to use technology to solve problems and explore concepts of calculus. These tips also point out some pitfalls of using technology.

## Putnam Exam Challenges

Putnam Exam questions appear in selected sections. These actual Putnam Exam questions will challenge you and push the limits of your understanding of calculus.

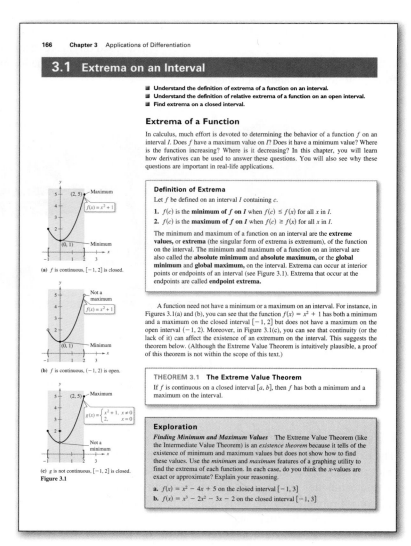

**166**  **Chapter 3**  Applications of Differentiation

### 3.1  Extrema on an Interval

■ Understand the definition of extrema of a function on an interval.
■ Understand the definition of relative extrema of a function on an open interval.
■ Find extrema on a closed interval.

#### Extrema of a Function

In calculus, much effort is devoted to determining the behavior of a function $f$ on an interval $I$. Does $f$ have a maximum value on $I$? Does it have a minimum value? Where is the function increasing? Where is it decreasing? In this chapter, you will learn how derivatives can be used to answer these questions. You will also see why these questions are important in real-life applications.

**Definition of Extrema**

Let $f$ be defined on an interval $I$ containing $c$.

1. $f(c)$ is the **minimum of $f$ on $I$** when $f(c) \le f(x)$ for all $x$ in $I$.
2. $f(c)$ is the **maximum of $f$ on $I$** when $f(c) \ge f(x)$ for all $x$ in $I$.

The minimum and maximum of a function on an interval are the **extreme values**, or **extrema** (the singular form of extrema is extremum), of the function on the interval. The minimum and maximum of a function on an interval are also called the **absolute minimum** and **absolute maximum**, or the **global minimum** and **global maximum**, on the interval. Extrema can occur at interior points or endpoints of an interval (see Figure 3.1). Extrema that occur at the endpoints are called **endpoint extrema**.

A function need not have a minimum or a maximum on an interval. For instance, in Figures 3.1(a) and (b), you can see that the function $f(x) = x^2 + 1$ has both a minimum and a maximum on the closed interval $[-1, 2]$ but does not have a maximum on the open interval $(-1, 2)$. Moreover, in Figure 3.1(c), you can see that continuity (or the lack of it) can affect the existence of an extremum on the interval. This suggests the theorem below. (Although the Extreme Value Theorem is intuitively plausible, a proof of this theorem is not within the scope of this text.)

**THEOREM 3.1  The Extreme Value Theorem**

If $f$ is continuous on a closed interval $[a, b]$, then $f$ has both a minimum and a maximum on the interval.

**Exploration**

*Finding Minimum and Maximum Values*  The Extreme Value Theorem (like the Intermediate Value Theorem) is an *existence theorem* because it tells of the existence of minimum and maximum values but does not show how to find these values. Use the *minimum* and *maximum* features of a graphing utility to find the extrema of each function. In each case, do you think the $x$-values are exact or approximate? Explain your reasoning.

a. $f(x) = x^2 - 4x + 5$ on the closed interval $[-1, 3]$
b. $f(x) = x^3 - 2x^2 - 3x - 2$ on the closed interval $[-1, 3]$

# Student Resources

**Student Solutions Manual for Multivariable Calculus**
ISBN-13: 978-1-337-27539-2

Need a leg up on your homework or help to prepare for an exam? The *Student Solutions Manual* contains worked-out solutions for all odd-numbered exercises. This manual is a great resource to help you understand how to solve those tough problems.

**CengageBrain.com**
To access additional course materials, please visit *www.cengagebrain.com*. At the *CengageBrain.com* home page, search for the ISBN of your title (from the back cover of your book) using the search box at the top of the page. This will take you to the product page where these resources can be found.

**MindTap for Mathematics**
MindTap® provides you with the tools you need to better manage your limited time—you can complete assignments whenever and wherever you are ready to learn with course material specifically customized for you by your instructor and streamlined in one proven, easy-to-use interface. With an array of tools and apps—from note taking to flashcards—you'll get a true understanding of course concepts, helping you to achieve better grades and setting the groundwork for your future courses. This access code entitles you to 3 terms of usage.

**Enhanced WebAssign®**
Enhanced WebAssign (assigned by the instructor) provides you with instant feedback on homework assignments. This online homework system is easy to use and includes helpful links to textbook sections, video examples, and problem-specific tutorials.

**Complete Solutions Manual for Multivariable Calculus**
ISBN-13: 978-1-337-27542-2

The *Complete Solutions Manual* contains worked-out solutions to all exercises in the text. It is posted on the instructor companion website.

**Instructor's Resource Guide   (on instructor companion site)**
This robust manual contains an abundance of instructor resources keyed to the textbook at the section and chapter level, including section objectives, teaching tips, and chapter projects.

**Cengage Learning Testing Powered by Cognero   (login.cengage.com)**
CLT is a flexible online system that allows you to author, edit, and manage test bank content; create multiple test versions in an instant; and deliver tests from your LMS, your classroom, or wherever you want. This is available online via *www.cengage.com/login.*

**Instructor Companion Site**
Everything you need for your course in one place! This collection of book-specific lecture and class tools is available online via *www.cengage.com/login.* Access and download PowerPoint® presentations, images, instructor's manual, and more.

**Test Bank   (on instructor companion site)**
The Test Bank contains text-specific multiple-choice and free-response test forms.

**MindTap for Mathematics**
MindTap® is the digital learning solution that helps you engage and transform today's students into critical thinkers. Through paths of dynamic assignments and applications that you can personalize, real-time course analytics, and an accessible reader, MindTap helps you turn cookie cutter into cutting edge, apathy into engagement, and memorizers into higher-level thinkers.

**Enhanced WebAssign®**
Exclusively from Cengage Learning, Enhanced WebAssign combines the exceptional mathematics content that you know and love with the most powerful online homework solution, WebAssign. Enhanced WebAssign engages students with immediate feedback, rich tutorial content, and interactive, fully customizable e-books (YouBook), helping students to develop a deeper conceptual understanding of their subject matter. Quick Prep and Just In Time exercises provide opportunities for students to review prerequisite skills and content, both at the start of the course and at the beginning of each section. Flexible assignment options give instructors the ability to release assignments conditionally on the basis of students' prerequisite assignment scores. Visit us at **www.cengage.com/ewa** to learn more.

# Acknowledgments

We would like to thank the many people who have helped us at various stages of *Calculus* over the last 43 years. Their encouragement, criticisms, and suggestions have been invaluable.

## Reviewers

Stan Adamski, *Owens Community College;* Tilak de Alwis; Darry Andrews; Alexander Arhangelskii, *Ohio University;* Seth G. Armstrong, *Southern Utah University;* Jim Ball, *Indiana State University;* Denis Bell, *University of Northern Florida;* Marcelle Bessman, *Jacksonville University;* Abraham Biggs, *Broward Community College;* Jesse Blosser, *Eastern Mennonite School;* Linda A. Bolte, *Eastern Washington University;* James Braselton, *Georgia Southern University;* Harvey Braverman, *Middlesex County College;* Mark Brittenham, *University of Nebraska;* Tim Chappell, *Penn Valley Community College;* Mingxiang Chen, *North Carolina A&T State University;* Oiyin Pauline Chow, *Harrisburg Area Community College;* Julie M. Clark, *Hollins University;* P.S. Crooke, *Vanderbilt University;* Jim Dotzler, *Nassau Community College;* Murray Eisenberg, *University of Massachusetts at Amherst;* Donna Flint, *South Dakota State University;* Michael Frantz, *University of La Verne;* David French, *Tidewater Community College;* Sudhir Goel, *Valdosta State University;* Arek Goetz, *San Francisco State University;* Donna J. Gorton, *Butler County Community College;* John Gosselin, *University of Georgia;* Arran Hamm; Shahryar Heydari, *Piedmont College;* Guy Hogan, *Norfolk State University;* Marcia Kleinz, *Atlantic Cape Community College;* Ashok Kumar, *Valdosta State University;* Kevin J. Leith, *Albuquerque Community College;* Maxine Lifshitz, *Friends Academy;* Douglas B. Meade, *University of South Carolina;* Bill Meisel, *Florida State College at Jacksonville;* Shahrooz Moosavizadeh; Teri Murphy, *University of Oklahoma;* Darren Narayan, *Rochester Institute of Technology;* Susan A. Natale, *The Ursuline School, NY;* Martha Nega, *Georgia Perimeter College;* Sam Pearsall, *Los Angeles Pierce College;* Terence H. Perciante, *Wheaton College;* James Pommersheim, *Reed College;* Laura Ritter, *Southern Polytechnic State University;* Leland E. Rogers, *Pepperdine University;* Paul Seeburger, *Monroe Community College;* Edith A. Silver, *Mercer County Community College;* Howard Speier, *Chandler-Gilbert Community College;* Desmond Stephens, *Florida A&M University;* Jianzhong Su, *University of Texas at Arlington;* Patrick Ward, *Illinois Central College;* Chia-Lin Wu, *Richard Stockton College of New Jersey;* Diane M. Zych, *Erie Community College*

Many thanks to Robert Hostetler, The Behrend College, The Pennsylvania State University, and David Heyd, The Behrend College, The Pennsylvania State University, for their significant contributions to previous editions of this text.

We would also like to thank the staff at Larson Texts, Inc., who assisted in preparing the manuscript, rendering the art package, typesetting, and proofreading the pages and supplements.

On a personal level, we are grateful to our wives, Deanna Gilbert Larson and Consuelo Edwards, for their love, patience, and support. Also, a special note of thanks goes out to R. Scott O'Neil.

If you have suggestions for improving this text, please feel free to write to us. Over the years we have received many useful comments from both instructors and students, and we value these very much.

Ron Larson

Bruce Edwards

# 11 Vectors and the Geometry of Space

Geography *(Exercise 47, p. 807)*

Modeling Data
*(Exercise 105, p. 796)*

Work *(Exercise 62, p. 778)*

Auditorium Lights
*(Exercise 99, p. 769)*

Navigation *(Exercise 84, p. 761)*

# 11.1 Vectors in the Plane

■ Write the component form of a vector.
■ Perform vector operations and interpret the results geometrically.
■ Write a vector as a linear combination of standard unit vectors.

## Component Form of a Vector

Many quantities in geometry and physics, such as area, volume, temperature, mass, and time, can be characterized by a single real number that is scaled to appropriate units of measure. These are called **scalar quantities,** and the real number associated with each is called a **scalar.**

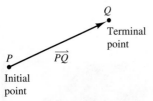

Other quantities, such as force, velocity, and acceleration, involve both magnitude and direction and cannot be characterized completely by a single real number. A **directed line segment** is used to represent such a quantity, as shown in Figure 11.1. The directed line segment $\overrightarrow{PQ}$ has **initial point** $P$ and **terminal point** $Q$, and its **length** (or **magnitude**) is denoted by $\|\overrightarrow{PQ}\|$. Directed line segments that have the same length and direction are **equivalent,** as shown in Figure 11.2. The set of all directed line segments that are equivalent to a given directed line segment $\overrightarrow{PQ}$ is a **vector in the plane** and is denoted by

$$\mathbf{v} = \overrightarrow{PQ}.$$

A directed line segment
**Figure 11.1**

In typeset material, vectors are usually denoted by lowercase, boldface letters such as **u**, **v**, and **w**. When written by hand, however, vectors are often denoted by letters with arrows above them, such as $\vec{u}$, $\vec{v}$, and $\vec{w}$.

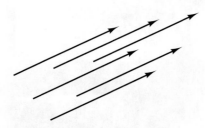

Be sure you understand that a vector represents a *set* of directed line segments (each having the same length and direction). In practice, however, it is common not to distinguish between a vector and one of its representatives.

Equivalent directed line segments
**Figure 11.2**

### EXAMPLE 1   Vector Representation: Directed Line Segments

Let **v** be represented by the directed line segment from $(0, 0)$ to $(3, 2)$, and let **u** be represented by the directed line segment from $(1, 2)$ to $(4, 4)$. Show that **v** and **u** are equivalent.

**Solution**   Let $P(0, 0)$ and $Q(3, 2)$ be the initial and terminal points of **v**, and let $R(1, 2)$ and $S(4, 4)$ be the initial and terminal points of **u**, as shown in Figure 11.3. You can use the Distance Formula to show that $\overrightarrow{PQ}$ and $\overrightarrow{RS}$ have the *same length*.

$$\|\overrightarrow{PQ}\| = \sqrt{(3 - 0)^2 + (2 - 0)^2} = \sqrt{13}$$
$$\|\overrightarrow{RS}\| = \sqrt{(4 - 1)^2 + (4 - 2)^2} = \sqrt{13}$$

Both line segments have the *same direction,* because they both are directed toward the upper right on lines having the same slope.

$$\text{Slope of } \overrightarrow{PQ} = \frac{2 - 0}{3 - 0} = \frac{2}{3}$$

and

$$\text{Slope of } \overrightarrow{RS} = \frac{4 - 2}{4 - 1} = \frac{2}{3}$$

Because $\overrightarrow{PQ}$ and $\overrightarrow{RS}$ have the same length and direction, you can conclude that the two vectors are equivalent. That is, **v** and **u** are equivalent.

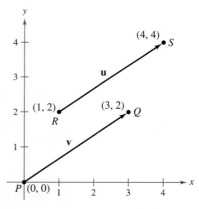

The vectors **u** and **v** are equivalent.
**Figure 11.3**

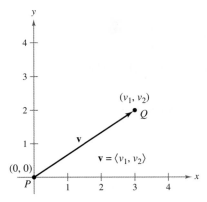

A vector in standard position
**Figure 11.4**

The directed line segment whose initial point is the origin is often the most convenient representative of a set of equivalent directed line segments such as those shown in Figure 11.3. This representation of **v** is said to be in **standard position**. A directed line segment whose initial point is the origin can be uniquely represented by the coordinates of its terminal point $Q(v_1, v_2)$, as shown in Figure 11.4. In the next definition, note the difference in the notation between the *component form* of a vector $\mathbf{v} = \langle v_1, v_2 \rangle$ and the point $(v_1, v_2)$.

---

**Definition of Component Form of a Vector in the Plane**

If **v** is a vector in the plane whose initial point is the origin and whose terminal point is $(v_1, v_2)$, then the **component form of v** is $\mathbf{v} = \langle v_1, v_2 \rangle$. The coordinates $v_1$ and $v_2$ are called the **components of v**. If both the initial point and the terminal point lie at the origin, then **v** is called the **zero vector** and is denoted by $\mathbf{0} = \langle 0, 0 \rangle$.

---

This definition implies that two vectors $\mathbf{u} = \langle u_1, u_2 \rangle$ and $\mathbf{v} = \langle v_1, v_2 \rangle$ are **equal** if and only if $u_1 = v_1$ and $u_2 = v_2$.

The procedures listed below can be used to convert directed line segments to component form or vice versa.

1. If $P(p_1, p_2)$ and $Q(q_1, q_2)$ are the initial and terminal points of a directed line segment, then the component form of the vector **v** represented by $\overrightarrow{PQ}$ is

$$\langle v_1, v_2 \rangle = \langle q_1 - p_1, q_2 - p_2 \rangle.$$

Moreover, from the Distance Formula, you can see that the **length** (or **magnitude**) **of v** is

$$\|\mathbf{v}\| = \sqrt{(q_1 - p_1)^2 + (q_2 - p_2)^2} \qquad \text{Length of a vector}$$
$$= \sqrt{v_1^2 + v_2^2}.$$

2. If $\mathbf{v} = \langle v_1, v_2 \rangle$, then **v** can be represented by the directed line segment, in standard position, from $P(0, 0)$ to $Q(v_1, v_2)$.

The length of **v** is also called the **norm of v**. If $\|\mathbf{v}\| = 1$, then **v** is a **unit vector**. Moreover, $\|\mathbf{v}\| = 0$ if and only if **v** is the zero vector **0**.

**EXAMPLE 2**   **Component Form and Length of a Vector**

Find the component form and length of the vector **v** that has initial point $(3, -7)$ and terminal point $(-2, 5)$.

**Solution**   Let $P(3, -7) = (p_1, p_2)$ and $Q(-2, 5) = (q_1, q_2)$. Then the components of $\mathbf{v} = \langle v_1, v_2 \rangle$ are

$$v_1 = q_1 - p_1 = -2 - 3 = -5$$

and

$$v_2 = q_2 - p_2 = 5 - (-7) = 12.$$

So, as shown in Figure 11.5, $\mathbf{v} = \langle -5, 12 \rangle$, and the length of **v** is

$$\|\mathbf{v}\| = \sqrt{(-5)^2 + 12^2}$$
$$= \sqrt{169}$$
$$= 13.$$

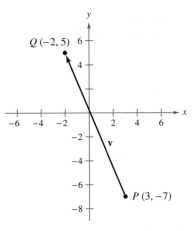

Component form of **v**:  $\mathbf{v} = \langle -5, 12 \rangle$
**Figure 11.5**

# Vector Operations

### Definitions of Vector Addition and Scalar Multiplication

Let $\mathbf{u} = \langle u_1, u_2 \rangle$ and $\mathbf{v} = \langle v_1, v_2 \rangle$ be vectors and let $c$ be a scalar.

1. The **vector sum** of $\mathbf{u}$ and $\mathbf{v}$ is the vector $\mathbf{u} + \mathbf{v} = \langle u_1 + v_1, u_2 + v_2 \rangle$.

2. The **scalar multiple** of $c$ and $\mathbf{u}$ is the vector
$$c\mathbf{u} = \langle cu_1, cu_2 \rangle.$$

3. The **negative** of $\mathbf{v}$ is the vector
$$-\mathbf{v} = (-1)\mathbf{v} = \langle -v_1, -v_2 \rangle.$$

4. The **difference** of $\mathbf{u}$ and $\mathbf{v}$ is
$$\mathbf{u} - \mathbf{v} = \mathbf{u} + (-\mathbf{v}) = \langle u_1 - v_1, u_2 - v_2 \rangle.$$

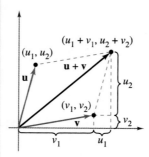

The scalar multiplication of $\mathbf{v}$
**Figure 11.6**

Geometrically, the scalar multiple of a vector $\mathbf{v}$ and a scalar $c$ is the vector that is $|c|$ times as long as $\mathbf{v}$, as shown in Figure 11.6. If $c$ is positive, then $c\mathbf{v}$ has the same direction as $\mathbf{v}$. If $c$ is negative, then $c\mathbf{v}$ has the opposite direction.

The sum of two vectors can be represented geometrically by positioning the vectors (without changing their magnitudes or directions) so that the initial point of one coincides with the terminal point of the other, as shown in Figure 11.7. The vector $\mathbf{u} + \mathbf{v}$, called the **resultant vector,** is the diagonal of a parallelogram having $\mathbf{u}$ and $\mathbf{v}$ as its adjacent sides.

**WILLIAM ROWAN HAMILTON
(1805–1865)**

Some of the earliest work with vectors was done by the Irish mathematician William Rowan Hamilton. Hamilton spent many years developing a system of vector-like quantities called *quaternions*. It was not until the latter half of the nineteenth century that the Scottish physicist James Maxwell (1831–1879) restructured Hamilton's quaternions in a form useful for representing physical quantities such as force, velocity, and acceleration. *See LarsonCalculus.com to read more of this biography.*

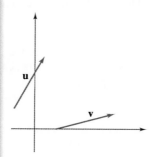 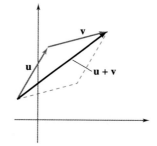 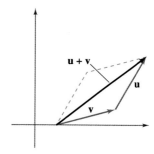

To find $\mathbf{u} + \mathbf{v}$,

(1) move the initial point of $\mathbf{v}$ to the terminal point of $\mathbf{u}$, or

(2) move the initial point of $\mathbf{u}$ to the terminal point of $\mathbf{v}$.

**Figure 11.7**

Figure 11.8 shows the equivalence of the geometric and algebraic definitions of vector addition and scalar multiplication and presents (at far right) a geometric interpretation of $\mathbf{u} - \mathbf{v}$.

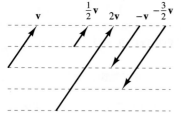

Vector addition          Scalar multiplication          Vector subtraction
**Figure 11.8**

**EXAMPLE 3** **Vector Operations**

For $\mathbf{v} = \langle -2, 5 \rangle$ and $\mathbf{w} = \langle 3, 4 \rangle$, find each of the vectors.

**a.** $\frac{1}{2}\mathbf{v}$    **b.** $\mathbf{w} - \mathbf{v}$    **c.** $\mathbf{v} + 2\mathbf{w}$

**Solution**

**a.** $\frac{1}{2}\mathbf{v} = \left\langle \frac{1}{2}(-2), \frac{1}{2}(5) \right\rangle = \left\langle -1, \frac{5}{2} \right\rangle$

**b.** $\mathbf{w} - \mathbf{v} = \langle w_1 - v_1, w_2 - v_2 \rangle$

$= \langle 3 - (-2), 4 - 5 \rangle$

$= \langle 5, -1 \rangle$

**c.** Using $2\mathbf{w} = \langle 6, 8 \rangle$, you have

$$\mathbf{v} + 2\mathbf{w} = \langle -2, 5 \rangle + \langle 6, 8 \rangle$$
$$= \langle -2 + 6, 5 + 8 \rangle$$
$$= \langle 4, 13 \rangle.$$

Vector addition and scalar multiplication share many properties of ordinary arithmetic, as shown in the next theorem.

**EMMY NOETHER (1882–1935)**

One person who contributed to our knowledge of axiomatic systems was the German mathematician Emmy Noether. Noether is generally recognized as the leading woman mathematician in recent history.

**THEOREM 11.1  Properties of Vector Operations**

Let $\mathbf{u}$, $\mathbf{v}$, and $\mathbf{w}$ be vectors in the plane, and let $c$ and $d$ be scalars.

1. $\mathbf{u} + \mathbf{v} = \mathbf{v} + \mathbf{u}$ — Commutative Property
2. $(\mathbf{u} + \mathbf{v}) + \mathbf{w} = \mathbf{u} + (\mathbf{v} + \mathbf{w})$ — Associative Property
3. $\mathbf{u} + \mathbf{0} = \mathbf{u}$ — Additive Identity Property
4. $\mathbf{u} + (-\mathbf{u}) = \mathbf{0}$ — Additive Inverse Property
5. $c(d\mathbf{u}) = (cd)\mathbf{u}$
6. $(c + d)\mathbf{u} = c\mathbf{u} + d\mathbf{u}$ — Distributive Property
7. $c(\mathbf{u} + \mathbf{v}) = c\mathbf{u} + c\mathbf{v}$ — Distributive Property
8. $1(\mathbf{u}) = \mathbf{u}$, $0(\mathbf{u}) = \mathbf{0}$

**Proof**  The proof of the *Associative Property* of vector addition uses the Associative Property of addition of real numbers.

$$(\mathbf{u} + \mathbf{v}) + \mathbf{w} = [\langle u_1, u_2 \rangle + \langle v_1, v_2 \rangle] + \langle w_1, w_2 \rangle$$
$$= \langle u_1 + v_1, u_2 + v_2 \rangle + \langle w_1, w_2 \rangle$$
$$= \langle (u_1 + v_1) + w_1, (u_2 + v_2) + w_2 \rangle$$
$$= \langle u_1 + (v_1 + w_1), u_2 + (v_2 + w_2) \rangle$$
$$= \langle u_1, u_2 \rangle + \langle v_1 + w_1, v_2 + w_2 \rangle$$
$$= \mathbf{u} + (\mathbf{v} + \mathbf{w})$$

The other properties can be proved in a similar manner.

**FOR FURTHER INFORMATION**
For more information on Emmy Noether, see the article "Emmy Noether, Greatest Woman Mathematician" by Clark Kimberling in *Mathematics Teacher*. To view this article, go to *MathArticles.com*.

Any set of vectors (with an accompanying set of scalars) that satisfies the eight properties listed in Theorem 11.1 is a **vector space.**\* The eight properties are the *vector space axioms*. So, this theorem states that the set of vectors in the plane (with the set of real numbers) forms a vector space.

---

\* For more information about vector spaces, see *Elementary Linear Algebra*, Eight Edition, by Ron Larson (Boston, Massachusetts: Cengage Learning, 2017).

---

**THEOREM 11.2   Length of a Scalar Multiple**

Let $\mathbf{v}$ be a vector and let $c$ be a scalar. Then

$\|c\mathbf{v}\| = |c|\,\|\mathbf{v}\|.$     $|c|$ is the absolute value of $c$.

---

**Proof**   Because $c\mathbf{v} = \langle cv_1, cv_2 \rangle$, it follows that

$$\|c\mathbf{v}\| = \|\langle cv_1, cv_2 \rangle\|$$
$$= \sqrt{(cv_1)^2 + (cv_2)^2}$$
$$= \sqrt{c^2 v_1^2 + c^2 v_2^2}$$
$$= \sqrt{c^2(v_1^2 + v_2^2)}$$
$$= |c|\sqrt{v_1^2 + v_2^2}$$
$$= |c|\,\|\mathbf{v}\|.$$

■

In many applications of vectors, it is useful to find a unit vector that has the same direction as a given vector. The next theorem gives a procedure for doing this.

---

**THEOREM 11.3   Unit Vector in the Direction of v**

If $\mathbf{v}$ is a nonzero vector in the plane, then the vector

$$\mathbf{u} = \frac{\mathbf{v}}{\|\mathbf{v}\|} = \frac{1}{\|\mathbf{v}\|}\mathbf{v}$$

has length 1 and the same direction as $\mathbf{v}$.

---

**Proof**   Because $1/\|\mathbf{v}\|$ is positive and $\mathbf{u} = (1/\|\mathbf{v}\|)\mathbf{v}$, you can conclude that $\mathbf{u}$ has the same direction as $\mathbf{v}$. To see that $\|\mathbf{u}\| = 1$, note that

$$\|\mathbf{u}\| = \left\|\left(\frac{1}{\|\mathbf{v}\|}\right)\mathbf{v}\right\| = \left|\frac{1}{\|\mathbf{v}\|}\right|\|\mathbf{v}\| = \frac{1}{\|\mathbf{v}\|}\|\mathbf{v}\| = 1.$$

So, $\mathbf{u}$ has length 1 and the same direction as $\mathbf{v}$.

■

In Theorem 11.3, $\mathbf{u}$ is called a **unit vector in the direction of v.** The process of multiplying $\mathbf{v}$ by $1/\|\mathbf{v}\|$ to get a unit vector is called **normalization of v.**

---

**EXAMPLE 4**   **Finding a Unit Vector**

Find a unit vector in the direction of $\mathbf{v} = \langle -2, 5 \rangle$ and verify that it has length 1.

**Solution**   From Theorem 11.3, the unit vector in the direction of $\mathbf{v}$ is

$$\frac{\mathbf{v}}{\|\mathbf{v}\|} = \frac{\langle -2, 5 \rangle}{\sqrt{(-2)^2 + (5)^2}}$$
$$= \frac{1}{\sqrt{29}}\langle -2, 5 \rangle$$
$$= \left\langle \frac{-2}{\sqrt{29}}, \frac{5}{\sqrt{29}} \right\rangle.$$

This vector has length 1, because

$$\sqrt{\left(\frac{-2}{\sqrt{29}}\right)^2 + \left(\frac{5}{\sqrt{29}}\right)^2} = \sqrt{\frac{4}{29} + \frac{25}{29}} = \sqrt{\frac{29}{29}} = 1.$$

■

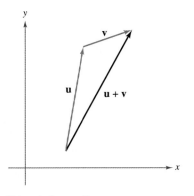

Triangle inequality
**Figure 11.9**

Generally, the length of the sum of two vectors is not equal to the sum of their lengths. To see this, consider the vectors **u** and **v** as shown in Figure 11.9. With **u** and **v** as two sides of a triangle, the length of the third side is $\|\mathbf{u} + \mathbf{v}\|$, and

$$\|\mathbf{u} + \mathbf{v}\| \le \|\mathbf{u}\| + \|\mathbf{v}\|.$$

Equality occurs only when the vectors **u** and **v** have the *same direction.* This result is called the **triangle inequality** for vectors. (You are asked to prove this in Exercise 73, Section 11.3.)

## Standard Unit Vectors

The unit vectors $\langle 1, 0 \rangle$ and $\langle 0, 1 \rangle$ are called the **standard unit vectors** in the plane and are denoted by

$$\mathbf{i} = \langle 1, 0 \rangle \quad \text{and} \quad \mathbf{j} = \langle 0, 1 \rangle \qquad \text{Standard unit vectors}$$

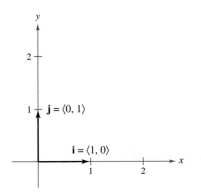

Standard unit vectors **i** and **j**
**Figure 11.10**

as shown in Figure 11.10. These vectors can be used to represent any vector uniquely, as follows.

$$\mathbf{v} = \langle v_1, v_2 \rangle = \langle v_1, 0 \rangle + \langle 0, v_2 \rangle = v_1 \langle 1, 0 \rangle + v_2 \langle 0, 1 \rangle = v_1 \mathbf{i} + v_2 \mathbf{j}$$

The vector $\mathbf{v} = v_1 \mathbf{i} + v_2 \mathbf{j}$ is called a **linear combination** of **i** and **j**. The scalars $v_1$ and $v_2$ are called the **horizontal** and **vertical components of v.**

### EXAMPLE 5  Writing a Linear Combination of Unit Vectors

Let **u** be the vector with initial point $(2, -5)$ and terminal point $(-1, 3)$, and let $\mathbf{v} = 2\mathbf{i} - \mathbf{j}$. Write each vector as a linear combination of **i** and **j**.

**a. u**

**b. w = 2u − 3v**

**Solution**

a. $\mathbf{u} = \langle q_1 - p_1, q_2 - p_2 \rangle = \langle -1 - 2, 3 - (-5) \rangle = \langle -3, 8 \rangle = -3\mathbf{i} + 8\mathbf{j}$

b. $\mathbf{w} = 2\mathbf{u} - 3\mathbf{v} = 2(-3\mathbf{i} + 8\mathbf{j}) - 3(2\mathbf{i} - \mathbf{j}) = -6\mathbf{i} + 16\mathbf{j} - 6\mathbf{i} + 3\mathbf{j} = -12\mathbf{i} + 19\mathbf{j}$

If **u** is a unit vector and $\theta$ is the angle (measured counterclockwise) from the positive $x$-axis to **u**, then the terminal point of **u** lies on the unit circle, and you have

$$\mathbf{u} = \langle \cos \theta, \sin \theta \rangle = \cos \theta \mathbf{i} + \sin \theta \mathbf{j} \qquad \text{Unit vector}$$

as shown in Figure 11.11. Moreover, it follows that any other nonzero vector **v** making an angle $\theta$ with the positive $x$-axis has the same direction as **u**, and you can write

$$\mathbf{v} = \|\mathbf{v}\| \langle \cos \theta, \sin \theta \rangle = \|\mathbf{v}\| \cos \theta \mathbf{i} + \|\mathbf{v}\| \sin \theta \mathbf{j}.$$

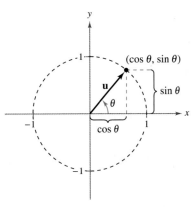

The angle $\theta$ from the positive $x$-axis to the vector **u**
**Figure 11.11**

### EXAMPLE 6  Writing a Vector of Given Magnitude and Direction

The vector **v** has a magnitude of 3 and makes an angle of $30° = \pi/6$ with the positive $x$-axis. Write **v** as a linear combination of the unit vectors **i** and **j**.

**Solution** Because the angle between **v** and the positive $x$-axis is $\theta = \pi/6$, you can write

$$\mathbf{v} = \|\mathbf{v}\| \cos \theta \mathbf{i} + \|\mathbf{v}\| \sin \theta \mathbf{j} = 3 \cos \frac{\pi}{6} \mathbf{i} + 3 \sin \frac{\pi}{6} \mathbf{j} = \frac{3\sqrt{3}}{2} \mathbf{i} + \frac{3}{2} \mathbf{j}.$$

Vectors have many applications in physics and engineering. One example is force. A vector can be used to represent force, because force has both magnitude and direction. If two or more forces are acting on an object, then the **resultant force** on the object is the vector sum of the vector forces.

### EXAMPLE 7    Finding the Resultant Force

Two tugboats are pushing an ocean liner, as shown in Figure 11.12. Each boat is exerting a force of 400 pounds. What is the resultant force on the ocean liner?

**Solution**   Using Figure 11.12, you can represent the forces exerted by the first and second tugboats as

$$\mathbf{F}_1 = 400\langle\cos 20°, \sin 20°\rangle = 400 \cos(20°)\mathbf{i} + 400 \sin(20°)\mathbf{j}$$
$$\mathbf{F}_2 = 400\langle\cos(-20°), \sin(-20°)\rangle = 400 \cos(20°)\mathbf{i} - 400 \sin(20°)\mathbf{j}.$$

The resultant force on the ocean liner is

$$\mathbf{F} = \mathbf{F}_1 + \mathbf{F}_2$$
$$= [400 \cos(20°)\mathbf{i} + 400 \sin(20°)\mathbf{j}] + [400 \cos(20°)\mathbf{i} - 400 \sin(20°)\mathbf{j}]$$
$$= 800 \cos(20°)\mathbf{i}$$
$$\approx 752\mathbf{i}.$$

So, the resultant force on the ocean liner is approximately 752 pounds in the direction of the positive $x$-axis.

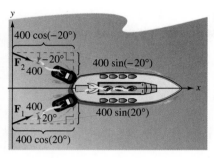

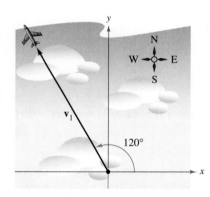

The resultant force on the ocean liner that is exerted by the two tugboats
**Figure 11.12**

In surveying and navigation, a **bearing** is a direction that measures the acute angle that a path or line of sight makes with a fixed north-south line. In air navigation, bearings are measured in degrees clockwise from north.

### EXAMPLE 8    Finding a Velocity

•••▷  *See LarsonCalculus.com for an interactive version of this type of example.*

An airplane is traveling at a fixed altitude with a negligible wind factor. The airplane is traveling at a speed of 500 miles per hour with a bearing of 330°, as shown in Figure 11.13(a). As the airplane reaches a certain point, it encounters wind with a velocity of 70 miles per hour in the direction N 45° E (45° east of north), as shown in Figure 11.13(b). What are the resultant speed and direction of the airplane?

**Solution**   Using Figure 11.13(a), represent the velocity of the airplane (alone) as

$$\mathbf{v}_1 = 500 \cos(120°)\mathbf{i} + 500 \sin(120°)\mathbf{j}.$$

The velocity of the wind is represented by the vector

$$\mathbf{v}_2 = 70 \cos(45°)\mathbf{i} + 70 \sin(45°)\mathbf{j}.$$

The resultant velocity of the airplane (in the wind) is

$$\mathbf{v} = \mathbf{v}_1 + \mathbf{v}_2$$
$$= 500 \cos(120°)\mathbf{i} + 500 \sin(120°)\mathbf{j} + 70 \cos(45°)\mathbf{i} + 70 \sin(45°)\mathbf{j}$$
$$\approx -200.5\mathbf{i} + 482.5\mathbf{j}.$$

To find the resultant speed and direction, write $\mathbf{v} = \|\mathbf{v}\|(\cos\theta\mathbf{i} + \sin\theta\mathbf{j})$. Because $\|\mathbf{v}\| \approx \sqrt{(-200.5)^2 + (482.5)^2} \approx 522.5$, you can write

$$\mathbf{v} \approx 522.5\left(\frac{-200.5}{522.5}\mathbf{i} + \frac{482.5}{522.5}\mathbf{j}\right) \approx 522.5[\cos(112.6°)\mathbf{i} + \sin(112.6°)\mathbf{j}].$$

The new speed of the airplane, as altered by the wind, is approximately 522.5 miles per hour in a path that makes an angle of 112.6° with the positive $x$-axis.

**(a)** Direction without wind

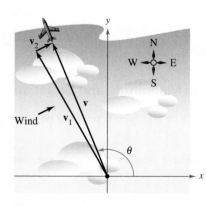

**(b)** Direction with wind
**Figure 11.13**

# 11.1 Exercises

See **CalcChat.com** for tutorial help and worked-out solutions to odd-numbered exercises.

**CONCEPT CHECK**

1. **Scalar and Vector** Describe the difference between a scalar and a vector. Give examples of each.

2. **Vector** Two points and a vector are given. Determine which point is the initial point and which point is the terminal point. Explain.

$$P(2, -1), \quad Q(-4, 6), \quad \text{and} \quad \mathbf{v} = \langle 6, -7 \rangle$$

**Sketching a Vector** In Exercises 3 and 4, (a) find the component form of the vector **v** and (b) sketch the vector with its initial point at the origin.

3.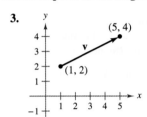

4.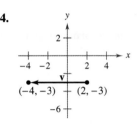

 **Equivalent Vectors** In Exercises 5–8, find the vectors u and v whose initial and terminal points are given. Show that u and v are equivalent.

| | Initial Point | Terminal Point | | Initial Point | Terminal Point |
|---|---|---|---|---|---|
| 5. | **u**: $(3, 2)$ | $(5, 6)$ | 6. | **u**: $(-4, 0)$ | $(1, 8)$ |
| | **v**: $(1, 4)$ | $(3, 8)$ | | **v**: $(2, -1)$ | $(7, 7)$ |
| 7. | **u**: $(0, 3)$ | $(6, -2)$ | 8. | **u**: $(-4, -1)$ | $(11, -4)$ |
| | **v**: $(3, 10)$ | $(9, 5)$ | | **v**: $(10, 13)$ | $(25, 10)$ |

 **Writing a Vector in Different Forms** In Exercises 9–16, the initial and terminal points of a vector v are given. (a) Sketch the given directed line segment. (b) Write the vector in component form. (c) Write the vector as the linear combination of the standard unit vectors i and j. (d) Sketch the vector with its initial point at the origin.

| | Initial Point | Terminal Point | | Initial Point | Terminal Point |
|---|---|---|---|---|---|
| 9. | $(2, 0)$ | $(5, 5)$ | 10. | $(4, -6)$ | $(3, 6)$ |
| 11. | $(8, 3)$ | $(6, -1)$ | 12. | $(0, -4)$ | $(-5, -1)$ |
| 13. | $(6, 2)$ | $(6, 6)$ | 14. | $(7, -1)$ | $(-3, -1)$ |
| 15. | $\left(\frac{3}{2}, \frac{4}{3}\right)$ | $\left(\frac{1}{2}, 3\right)$ | 16. | $(0.12, 0.60)$ | $(0.84, 1.25)$ |

**Finding a Terminal Point** In Exercises 17 and 18, the vector v and its initial point are given. Find the terminal point.

17. $\mathbf{v} = \langle -1, 3 \rangle$; Initial point: $(4, 2)$

18. $\mathbf{v} = \langle 4, -9 \rangle$; Initial point: $(5, 3)$

 **Finding a Magnitude of a Vector** In Exercises 19–24, find the magnitude of v.

19. $\mathbf{v} = 4\mathbf{i}$

20. $\mathbf{v} = -9\mathbf{j}$

21. $\mathbf{v} = \langle 8, 15 \rangle$

22. $\mathbf{v} = \langle -24, 7 \rangle$

23. $\mathbf{v} = -\mathbf{i} - 5\mathbf{j}$

24. $\mathbf{v} = 3\mathbf{i} + 3\mathbf{j}$

**Sketching Scalar Multiples** In Exercises 25 and 26, sketch each scalar multiple of v.

25. $\mathbf{v} = \langle 3, 5 \rangle$   (a) $2\mathbf{v}$   (b) $-3\mathbf{v}$   (c) $\frac{7}{2}\mathbf{v}$   (d) $\frac{2}{3}\mathbf{v}$

26. $\mathbf{v} = \langle -2, 3 \rangle$   (a) $4\mathbf{v}$   (b) $-\frac{1}{2}\mathbf{v}$   (c) $0\mathbf{v}$   (d) $-6\mathbf{v}$

 **Using Vector Operations** In Exercises 27 and 28, find (a) $\frac{2}{3}\mathbf{u}$, (b) $3\mathbf{v}$, (c) $\mathbf{v} - \mathbf{u}$, and (d) $2\mathbf{u} + 5\mathbf{v}$.

27. $\mathbf{u} = \langle 4, 9 \rangle$, $\mathbf{v} = \langle 2, -5 \rangle$    28. $\mathbf{u} = \langle -3, -8 \rangle$, $\mathbf{v} = \langle 8, 7 \rangle$

**Sketching a Vector** In Exercises 29–34, use the figure to sketch a graph of the vector. To print an enlarged copy of the graph, go to *MathGraphs.com*.

29. $-\mathbf{u}$

30. $2\mathbf{u}$

31. $-\mathbf{v}$

32. $\frac{1}{2}\mathbf{v}$

33. $\mathbf{u} - \mathbf{v}$

34. $\mathbf{u} + 2\mathbf{v}$

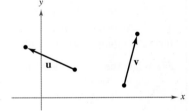

 **Finding a Unit Vector** In Exercises 35–38, find the unit vector in the direction of v and verify that it has length 1.

35. $\mathbf{v} = \langle 3, 12 \rangle$

36. $\mathbf{v} = \langle -5, 15 \rangle$

37. $\mathbf{v} = \left\langle \frac{3}{2}, \frac{5}{2} \right\rangle$

38. $\mathbf{v} = \langle -6.2, 3.4 \rangle$

**Finding Magnitudes** In Exercises 39–42, find the following.

(a) $\|\mathbf{u}\|$    (b) $\|\mathbf{v}\|$    (c) $\|\mathbf{u} + \mathbf{v}\|$

(d) $\left\| \dfrac{\mathbf{u}}{\|\mathbf{u}\|} \right\|$    (e) $\left\| \dfrac{\mathbf{v}}{\|\mathbf{v}\|} \right\|$    (f) $\left\| \dfrac{\mathbf{u} + \mathbf{v}}{\|\mathbf{u} + \mathbf{v}\|} \right\|$

39. $\mathbf{u} = \langle 1, -1 \rangle$, $\mathbf{v} = \langle -1, 2 \rangle$

40. $\mathbf{u} = \langle 0, 1 \rangle$, $\mathbf{v} = \langle 3, -3 \rangle$

41. $\mathbf{u} = \left\langle 1, \frac{1}{2} \right\rangle$, $\mathbf{v} = \langle 2, 3 \rangle$

42. $\mathbf{u} = \langle 2, -4 \rangle$, $\mathbf{v} = \langle 5, 5 \rangle$

**Using the Triangle Inequality** In Exercises 43 and 44, sketch a graph of u, v, and u + v. Then demonstrate the triangle inequality using the vectors u and v.

43. $\mathbf{u} = \langle 2, 1 \rangle$, $\mathbf{v} = \langle 5, 4 \rangle$

44. $\mathbf{u} = \langle -3, 2 \rangle$, $\mathbf{v} = \langle 1, -2 \rangle$

**Finding a Vector** In Exercises 45–48, find the vector **v** with the given magnitude and the same direction as **u**.

| Magnitude | Direction |
|---|---|
| **45.** $\|\mathbf{v}\| = 6$ | $\mathbf{u} = \langle 0, 3 \rangle$ |
| **46.** $\|\mathbf{v}\| = 4$ | $\mathbf{u} = \langle 1, 1 \rangle$ |
| **47.** $\|\mathbf{v}\| = 5$ | $\mathbf{u} = \langle -1, 2 \rangle$ |
| **48.** $\|\mathbf{v}\| = 2$ | $\mathbf{u} = \langle \sqrt{3}, 3 \rangle$ |

**Finding a Vector** In Exercises 49–52, find the component form of **v** given its magnitude and the angle it makes with the positive *x*-axis.

**49.** $\|\mathbf{v}\| = 3, \quad \theta = 0°$  **50.** $\|\mathbf{v}\| = 5, \quad \theta = 120°$

**51.** $\|\mathbf{v}\| = 2, \quad \theta = 150°$  **52.** $\|\mathbf{v}\| = 4, \quad \theta = 3.5°$

**Finding a Vector** In Exercises 53–56, find the component form of **u** + **v** given the lengths of **u** and **v** and the angles that **u** and **v** make with the positive *x*-axis.

**53.** $\|\mathbf{u}\| = 1, \quad \theta_u = 0°$  **54.** $\|\mathbf{u}\| = 4, \quad \theta_u = 0°$
$\|\mathbf{v}\| = 3, \quad \theta_v = 45°$    $\|\mathbf{v}\| = 2, \quad \theta_v = 60°$

**55.** $\|\mathbf{u}\| = 2, \quad \theta_u = 4$  **56.** $\|\mathbf{u}\| = 5, \quad \theta_u = -0.5$
$\|\mathbf{v}\| = 1, \quad \theta_v = 2$    $\|\mathbf{v}\| = 5, \quad \theta_v = 0.5$

**EXPLORING CONCEPTS**

**Think About It** In Exercises 57 and 58, consider two forces of equal magnitude acting on a point.

**57.** When the magnitude of the resultant is the sum of the magnitudes of the two forces, make a conjecture about the angle between the forces.

**58.** When the resultant of the forces is 0, make a conjecture about the angle between the forces.

**59. Triangle** Consider a triangle with vertices $X$, $Y$, and $Z$. What is $\overrightarrow{XY} + \overrightarrow{YZ} + \overrightarrow{ZX}$? Explain.

**60. HOW DO YOU SEE IT?** Use the figure to determine whether each statement is true or false. Justify your answer.

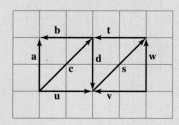

(a) $\mathbf{a} = -\mathbf{d}$  (b) $\mathbf{c} = \mathbf{s}$
(c) $\mathbf{a} + \mathbf{u} = \mathbf{c}$  (d) $\mathbf{v} + \mathbf{w} = -\mathbf{s}$
(e) $\mathbf{a} + \mathbf{d} = \mathbf{0}$  (f) $\mathbf{u} - \mathbf{v} = -2(\mathbf{b} + \mathbf{t})$

**Finding Values** In Exercises 61–66, find $a$ and $b$ such that $\mathbf{v} = a\mathbf{u} + b\mathbf{w}$, where $\mathbf{u} = \langle 1, 2 \rangle$ and $\mathbf{w} = \langle 1, -1 \rangle$.

**61.** $\mathbf{v} = \langle 4, 5 \rangle$  **62.** $\mathbf{v} = \langle -7, -2 \rangle$
**63.** $\mathbf{v} = \langle -6, 0 \rangle$  **64.** $\mathbf{v} = \langle 0, 6 \rangle$
**65.** $\mathbf{v} = \langle 1, -3 \rangle$  **66.** $\mathbf{v} = \langle -1, 8 \rangle$

**Finding Unit Vectors** In Exercises 67–72, find a unit vector (a) parallel to and (b) perpendicular to the graph of $f$ at the given point. Then sketch the graph of $f$ and sketch the vectors at the given point.

**67.** $f(x) = x^2, \quad (3, 9)$  **68.** $f(x) = -x^2 + 5, \quad (1, 4)$
**69.** $f(x) = x^3, \quad (1, 1)$  **70.** $f(x) = x^3, \quad (-2, -8)$
**71.** $f(x) = \sqrt{25 - x^2}, \quad (3, 4)$

**72.** $f(x) = \tan x, \quad \left( \dfrac{\pi}{4}, 1 \right)$

**Finding a Vector** In Exercises 73 and 74, find the component form of **v** given the magnitudes of **u** and **u** + **v** and the angles that **u** and **u** + **v** make with the positive *x*-axis.

**73.** $\|\mathbf{u}\| = 1, \quad \theta = 45°$  **74.** $\|\mathbf{u}\| = 4, \quad \theta = 30°$
$\|\mathbf{u} + \mathbf{v}\| = \sqrt{2}, \quad \theta = 90°$    $\|\mathbf{u} + \mathbf{v}\| = 6, \quad \theta = 120°$

**75. Resultant Force** Forces with magnitudes of 500 pounds and 200 pounds act on a machine part at angles of 30° and $-45°$, respectively, with the *x*-axis (see figure). Find the direction and magnitude of the resultant force.

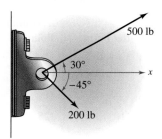

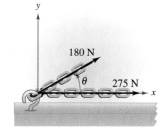

Figure for 75       Figure for 76

**76. Numerical and Graphical Analysis** Forces with magnitudes of 180 newtons and 275 newtons act on a hook (see figure). The angle between the two forces is $\theta$ degrees.

(a) When $\theta = 30°$, find the direction and magnitude of the resultant force.

(b) Write the magnitude $M$ and direction $\alpha$ of the resultant force as functions of $\theta$, where $0° \le \theta \le 180°$.

(c) Use a graphing utility to complete the table.

| $\theta$ | 0° | 30° | 60° | 90° | 120° | 150° | 180° |
|---|---|---|---|---|---|---|---|
| $M$ | | | | | | | |
| $\alpha$ | | | | | | | |

(d) Use a graphing utility to graph the two functions $M$ and $\alpha$.

(e) Explain why one of the functions decreases for increasing values of $\theta$, whereas the other does not.

**77. Resultant Force** Three forces with magnitudes of 75 pounds, 100 pounds, and 125 pounds act on an object at angles of 30°, 45°, and 120°, respectively, with the positive *x*-axis. Find the direction and magnitude of the resultant force.

**78. Resultant Force** Three forces with magnitudes of 400 newtons, 280 newtons, and 350 newtons act on an object at angles of −30°, 45°, and 135°, respectively, with the positive *x*-axis. Find the direction and magnitude of the resultant force.

**Cable Tension** In Exercises 79 and 80, determine the tension in the cable supporting the given load.

**79.** 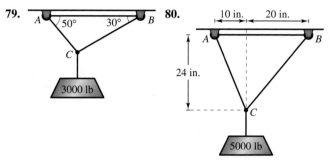 **80.**

**81. Projectile Motion** A gun with a muzzle velocity of 1200 feet per second is fired at an angle of 6° above the horizontal. Find the vertical and horizontal components of the velocity.

**82. Shared Load** To carry a 100-pound cylindrical weight, two workers lift on the ends of short ropes tied to an eyelet on the top center of the cylinder. One rope makes a 20° angle away from the vertical and the other makes a 30° angle (see figure).

(a) Find each rope's tension when the resultant force is vertical.

(b) Find the vertical component of each worker's force.

**83. Navigation** A plane is flying with a bearing of 302°. Its speed with respect to the air is 900 kilometers per hour. The wind at the plane's altitude is from the southwest at 100 kilometers per hour (see figure). What is the true direction of the plane, and what is its speed with respect to the ground?

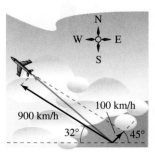

A plane flies at a constant groundspeed of 400 miles per hour due east and encounters a 50-mile-per-hour wind from the northwest. Find the airspeed and compass direction that will allow the plane to maintain its groundspeed and eastward direction.

**True or False?** In Exercises 85–94, determine whether the statement is true or false. If it is false, explain why or give an example that shows it is false.

**85.** The weight of a car is a scalar.

**86.** The mass of a book is a scalar.

**87.** The temperature of your blood is a scalar.

**88.** The velocity of a bicycle is a vector.

**89.** If **u** and **v** have the same magnitude and direction, then **u** and **v** are equivalent.

**90.** If **u** is a unit vector in the direction of **v**, then $\mathbf{v} = \|\mathbf{v}\|\mathbf{u}$.

**91.** If $\mathbf{u} = a\mathbf{i} + b\mathbf{j}$ is a unit vector, then $a^2 + b^2 = 1$.

**92.** If $\mathbf{v} = a\mathbf{i} + b\mathbf{j} = \mathbf{0}$, then $a = -b$.

**93.** If $a = b$, then $\|a\mathbf{i} + b\mathbf{j}\| = \sqrt{2}\,a$.

**94.** If **u** and **v** have the same magnitude but opposite directions, then $\mathbf{u} + \mathbf{v} = \mathbf{0}$.

**95. Proof** Prove that

$$\mathbf{u} = (\cos\theta)\mathbf{i} - (\sin\theta)\mathbf{j} \quad \text{and} \quad \mathbf{v} = (\sin\theta)\mathbf{i} + (\cos\theta)\mathbf{j}$$

are unit vectors for any angle $\theta$.

**96. Geometry** Using vectors, prove that the line segment joining the midpoints of two sides of a triangle is parallel to, and one-half the length of, the third side.

**97. Geometry** Using vectors, prove that the diagonals of a parallelogram bisect each other.

**98. Proof** Prove that the vector $\mathbf{w} = \|\mathbf{u}\|\mathbf{v} + \|\mathbf{v}\|\mathbf{u}$ bisects the angle between **u** and **v**.

**99. Using a Vector** Consider the vector $\mathbf{u} = \langle x, y \rangle$. Describe the set of all points $(x, y)$ such that $\|\mathbf{u}\| = 5$.

---

**PUTNAM EXAM CHALLENGE**

**100.** A coast artillery gun can fire at any angle of elevation between 0° and 90° in a fixed vertical plane. If air resistance is neglected and the muzzle velocity is constant $(= v_0)$, determine the set $H$ of points in the plane and above the horizontal which can be hit.

# 11.2 Space Coordinates and Vectors in Space

■ Understand the three-dimensional rectangular coordinate system.
■ Analyze vectors in space.

## Coordinates in Space

Up to this point in the text, you have been primarily concerned with the two-dimensional coordinate system. Much of the remaining part of your study of calculus will involve the three-dimensional coordinate system.

Before extending the concept of a vector to three dimensions, you must be able to identify points in the **three-dimensional coordinate system.** You can construct this system by passing a $z$-axis perpendicular to both the $x$- and $y$-axes at the origin, as shown in Figure 11.14. Taken as pairs, the axes determine three **coordinate planes:** the **xy-plane,** the **xz-plane,** and the **yz-plane.** These three coordinate planes separate three-space into eight **octants.** The first octant is the one for which all three coordinates are positive. In this three-dimensional system, a point $P$ in space is determined by an ordered triple $(x, y, z)$, where $x$, $y$, and $z$ are as follows.

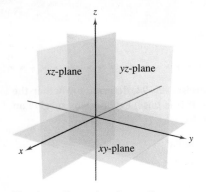

The three-dimensional coordinate system
**Figure 11.14**

$x$ = directed distance from $yz$-plane to $P$

$y$ = directed distance from $xz$-plane to $P$

$z$ = directed distance from $xy$-plane to $P$

Several points are shown in Figure 11.15.

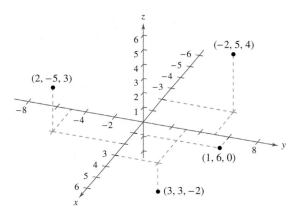

Points in the three-dimensional coordinate system are represented by ordered triples.
**Figure 11.15**

**·· REMARK** The three-dimensional rotatable graphs that are available at *LarsonCalculus.com* can help you visualize points or objects in a three-dimensional coordinate system.

A three-dimensional coordinate system can have either a **right-handed** or a **left-handed** orientation. To determine the orientation of a system, imagine that you are standing at the origin, with your arms pointing in the direction of the positive $x$- and $y$-axes and with the positive $z$-axis pointing up, as shown in Figure 11.16. The system is right-handed or left-handed depending on which hand points along the $x$-axis. In this text, you will work exclusively with the right-handed system.

Right-handed system    Left-handed system
**Figure 11.16**

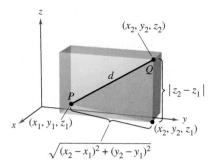

The distance between two points in space

**Figure 11.17**

Many of the formulas established for the two-dimensional coordinate system can be extended to three dimensions. For example, to find the distance between two points in space, you can use the Pythagorean Theorem twice, as shown in Figure 11.17. By doing this, you will obtain the formula for the distance between the points $(x_1, y_1, z_1)$ and $(x_2, y_2, z_2)$.

$$d = \sqrt{(x_2 - x_1)^2 + (y_2 - y_1)^2 + (z_2 - z_1)^2}$$   Distance Formula

**EXAMPLE 1**   **Finding the Distance Between Two Points in Space**

Find the distance between the points $(2, -1, 3)$ and $(1, 0, -2)$.

**Solution**

$$\begin{aligned}
d &= \sqrt{(1 - 2)^2 + (0 + 1)^2 + (-2 - 3)^2} \quad \text{Distance Formula} \\
&= \sqrt{1 + 1 + 25} \\
&= \sqrt{27} \\
&= 3\sqrt{3}
\end{aligned}$$

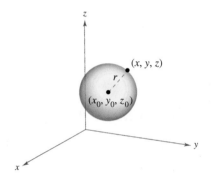

**Figure 11.18**

A **sphere** with center at $(x_0, y_0, z_0)$ and radius $r$ is defined to be the set of all points $(x, y, z)$ such that the distance between $(x, y, z)$ and $(x_0, y_0, z_0)$ is $r$. You can use the Distance Formula to find the **standard equation of a sphere** of radius $r$, centered at $(x_0, y_0, z_0)$. If $(x, y, z)$ is an arbitrary point on the sphere, then the equation of the sphere is

$$(x - x_0)^2 + (y - y_0)^2 + (z - z_0)^2 = r^2$$   Equation of sphere

as shown in Figure 11.18. Moreover, the midpoint of the line segment joining the points $(x_1, y_1, z_1)$ and $(x_2, y_2, z_2)$ has coordinates

$$\left( \frac{x_1 + x_2}{2}, \frac{y_1 + y_2}{2}, \frac{z_1 + z_2}{2} \right).$$   Midpoint Formula

**EXAMPLE 2**   **Finding the Equation of a Sphere**

Find the standard equation of the sphere that has the points

$$(5, -2, 3) \quad \text{and} \quad (0, 4, -3)$$

as endpoints of a diameter.

**Solution**   Using the Midpoint Formula, the center of the sphere is

$$\left( \frac{5 + 0}{2}, \frac{-2 + 4}{2}, \frac{3 - 3}{2} \right) = \left( \frac{5}{2}, 1, 0 \right).$$   Midpoint Formula

By the Distance Formula, the radius is

$$r = \sqrt{\left( 0 - \frac{5}{2} \right)^2 + (4 - 1)^2 + (-3 - 0)^2} = \sqrt{\frac{97}{4}} = \frac{\sqrt{97}}{2}.$$

Therefore, the standard equation of the sphere is

$$\left( x - \frac{5}{2} \right)^2 + (y - 1)^2 + z^2 = \frac{97}{4}.$$   Equation of sphere

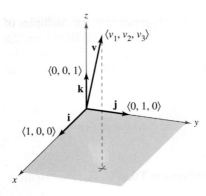

The standard unit vectors in space
**Figure 11.19**

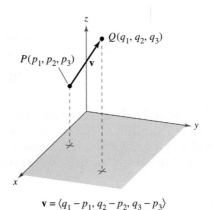

$$\mathbf{v} = \langle q_1 - p_1, q_2 - p_2, q_3 - p_3 \rangle$$

**Figure 11.20**

## Vectors in Space

In space, vectors are denoted by ordered triples $\mathbf{v} = \langle v_1, v_2, v_3 \rangle$. The **zero vector** is denoted by $\mathbf{0} = \langle 0, 0, 0 \rangle$. Using the unit vectors

$$\mathbf{i} = \langle 1, 0, 0 \rangle, \quad \mathbf{j} = \langle 0, 1, 0 \rangle, \quad \text{and} \quad \mathbf{k} = \langle 0, 0, 1 \rangle$$

the **standard unit vector notation** for $\mathbf{v}$ is

$$\mathbf{v} = v_1 \mathbf{i} + v_2 \mathbf{j} + v_3 \mathbf{k}$$

as shown in Figure 11.19. If $\mathbf{v}$ is represented by the directed line segment from $P(p_1, p_2, p_3)$ to $Q(q_1, q_2, q_3)$, as shown in Figure 11.20, then the component form of $\mathbf{v}$ is written by subtracting the coordinates of the initial point from the coordinates of the terminal point, as follows.

$$\mathbf{v} = \langle v_1, v_2, v_3 \rangle = \langle q_1 - p_1, q_2 - p_2, q_3 - p_3 \rangle$$

---

**Vectors in Space**

Let $\mathbf{u} = \langle u_1, u_2, u_3 \rangle$ and $\mathbf{v} = \langle v_1, v_2, v_3 \rangle$ be vectors in space and let $c$ be a scalar.

1. *Equality of Vectors:* $\mathbf{u} = \mathbf{v}$ if and only if $u_1 = v_1, u_2 = v_2,$ and $u_3 = v_3$.
2. *Component Form:* If $\mathbf{v}$ is represented by the directed line segment from $P(p_1, p_2, p_3)$ to $Q(q_1, q_2, q_3)$, then
$$\mathbf{v} = \langle v_1, v_2, v_3 \rangle = \langle q_1 - p_1, q_2 - p_2, q_3 - p_3 \rangle.$$
3. *Length:* $\|\mathbf{v}\| = \sqrt{v_1^2 + v_2^2 + v_3^2}$
4. *Unit Vector in the Direction of* $\mathbf{v}$: $\dfrac{\mathbf{v}}{\|\mathbf{v}\|} = \left(\dfrac{1}{\|\mathbf{v}\|}\right)\langle v_1, v_2, v_3 \rangle, \quad \mathbf{v} \neq \mathbf{0}$
5. *Vector Addition:* $\mathbf{v} + \mathbf{u} = \langle v_1 + u_1, v_2 + u_2, v_3 + u_3 \rangle$
6. *Scalar Multiplication:* $c\mathbf{v} = \langle cv_1, cv_2, cv_3 \rangle$

---

Note that the properties of vector operations listed in Theorem 11.1 (see Section 11.1) are also valid for vectors in space.

### EXAMPLE 3   Finding the Component Form of a Vector in Space

⋯▷ *See LarsonCalculus.com for an interactive version of this type of example.*

Find the component form and magnitude of the vector $\mathbf{v}$ having initial point $(-2, 3, 1)$ and terminal point $(0, -4, 4)$. Then find a unit vector in the direction of $\mathbf{v}$.

**Solution**   The component form of $\mathbf{v}$ is

$$\mathbf{v} = \langle q_1 - p_1, q_2 - p_2, q_3 - p_3 \rangle = \langle 0 - (-2), -4 - 3, 4 - 1 \rangle = \langle 2, -7, 3 \rangle$$

which implies that its magnitude is

$$\|\mathbf{v}\| = \sqrt{2^2 + (-7)^2 + 3^2} = \sqrt{62}.$$

The unit vector in the direction of $\mathbf{v}$ is

$$\begin{aligned}
\mathbf{u} &= \frac{\mathbf{v}}{\|\mathbf{v}\|} \\
&= \frac{1}{\sqrt{62}}\langle 2, -7, 3 \rangle \\
&= \left\langle \frac{2}{\sqrt{62}}, \frac{-7}{\sqrt{62}}, \frac{3}{\sqrt{62}} \right\rangle.
\end{aligned}$$

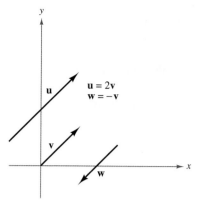

Parallel vectors
**Figure 11.21**

Recall from the definition of scalar multiplication that positive scalar multiples of a nonzero vector **v** have the same direction as **v**, whereas negative multiples have the direction opposite of **v**. In general, two nonzero vectors **u** and **v** are **parallel** when there is some scalar $c$ such that $\mathbf{u} = c\mathbf{v}$. For example, in Figure 11.21, the vectors **u**, **v**, and **w** are parallel because

$$\mathbf{u} = 2\mathbf{v} \quad \text{and} \quad \mathbf{w} = -\mathbf{v}.$$

---

**Definition of Parallel Vectors**

Two nonzero vectors **u** and **v** are **parallel** when there is some scalar $c$ such that $\mathbf{u} = c\mathbf{v}$.

---

**EXAMPLE 4**  **Parallel Vectors**

Vector **w** has initial point $(2, -1, 3)$ and terminal point $(-4, 7, 5)$. Which of the following vectors is parallel to **w**?

**a.** $\mathbf{u} = \langle 3, -4, -1 \rangle$

**b.** $\mathbf{v} = \langle 12, -16, 4 \rangle$

**Solution**  Begin by writing **w** in component form.

$$\mathbf{w} = \langle -4 - 2, 7 - (-1), 5 - 3 \rangle = \langle -6, 8, 2 \rangle$$

**a.** Because $\mathbf{u} = \langle 3, -4, -1 \rangle = -\frac{1}{2}\langle -6, 8, 2 \rangle = -\frac{1}{2}\mathbf{w}$, you can conclude that **u** is parallel to **w**.

**b.** In this case, you want to find a scalar $c$ such that

$$\langle 12, -16, 4 \rangle = c\langle -6, 8, 2 \rangle.$$

To find $c$, equate the corresponding components and solve as shown.

$$12 = -6c \quad \Longrightarrow \quad c = -2$$
$$-16 = 8c \quad \Longrightarrow \quad c = -2$$
$$4 = 2c \quad \Longrightarrow \quad c = 2$$

Note that $c = -2$ for the first two components and $c = 2$ for the third component. This means that the equation $\langle 12, -16, 4 \rangle = c\langle -6, 8, 2 \rangle$ has no solution, and the vectors are not parallel.

**EXAMPLE 5**  **Using Vectors to Determine Collinear Points**

Determine whether the points

$$P(1, -2, 3), \quad Q(2, 1, 0), \quad \text{and} \quad R(4, 7, -6)$$

are collinear.

**Solution**  The component forms of $\overrightarrow{PQ}$ and $\overrightarrow{PR}$ are

$$\overrightarrow{PQ} = \langle 2 - 1, 1 - (-2), 0 - 3 \rangle = \langle 1, 3, -3 \rangle$$

and

$$\overrightarrow{PR} = \langle 4 - 1, 7 - (-2), -6 - 3 \rangle = \langle 3, 9, -9 \rangle.$$

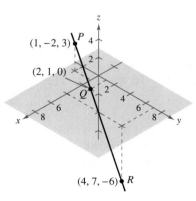

The points $P$, $Q$, and $R$ lie on the same line.
**Figure 11.22**

These two vectors have a common initial point. So, $P$, $Q$, and $R$ lie on the same line if and only if $\overrightarrow{PQ}$ and $\overrightarrow{PR}$ are parallel—which they are because $\overrightarrow{PR} = 3\,\overrightarrow{PQ}$, as shown in Figure 11.22.

EXAMPLE 6    **Standard Unit Vector Notation**

a. Write the vector $\mathbf{v} = 4\mathbf{i} - 5\mathbf{k}$ in component form.

b. Find the terminal point of the vector $\mathbf{v} = 7\mathbf{i} - \mathbf{j} + 3\mathbf{k}$, given that the initial point is $P(-2, 3, 5)$.

c. Find the magnitude of the vector $\mathbf{v} = -6\mathbf{i} + 2\mathbf{j} - 3\mathbf{k}$. Then find a unit vector in the direction of $\mathbf{v}$.

**Solution**

a. Because $\mathbf{j}$ is missing, its component is 0 and

$$\mathbf{v} = 4\mathbf{i} - 5\mathbf{k} = \langle 4, 0, -5 \rangle.$$

b. You need to find $Q(q_1, q_2, q_3)$ such that

$$\mathbf{v} = \overrightarrow{PQ} = 7\mathbf{i} - \mathbf{j} + 3\mathbf{k}.$$

This implies that $q_1 - (-2) = 7$, $q_2 - 3 = -1$, and $q_3 - 5 = 3$. The solution of these three equations is $q_1 = 5$, $q_2 = 2$, and $q_3 = 8$. Therefore, $Q$ is $(5, 2, 8)$.

c. Note that $v_1 = -6$, $v_2 = 2$, and $v_3 = -3$. So, the magnitude of $\mathbf{v}$ is

$$\|\mathbf{v}\| = \sqrt{(-6)^2 + 2^2 + (-3)^2} = \sqrt{49} = 7.$$

The unit vector in the direction of $\mathbf{v}$ is

$$\tfrac{1}{7}(-6\mathbf{i} + 2\mathbf{j} - 3\mathbf{k}) = -\tfrac{6}{7}\mathbf{i} + \tfrac{2}{7}\mathbf{j} - \tfrac{3}{7}\mathbf{k}.$$

EXAMPLE 7    **Measuring Force**

A television camera weighing 120 pounds is supported by a tripod, as shown in Figure 11.23. Represent the force exerted on each leg of the tripod as a vector.

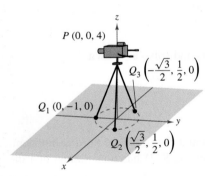

**Figure 11.23**

**Solution**    Let the vectors $\mathbf{F}_1$, $\mathbf{F}_2$, and $\mathbf{F}_3$ represent the forces exerted on the three legs. From Figure 11.23, you can determine the directions of $\mathbf{F}_1$, $\mathbf{F}_2$, and $\mathbf{F}_3$ to be as follows.

$$\mathbf{F}_1 = \overrightarrow{PQ_1} = \langle 0 - 0, -1 - 0, 0 - 4 \rangle = \langle 0, -1, -4 \rangle$$

$$\mathbf{F}_2 = \overrightarrow{PQ_2} = \left\langle \frac{\sqrt{3}}{2} - 0, \frac{1}{2} - 0, 0 - 4 \right\rangle = \left\langle \frac{\sqrt{3}}{2}, \frac{1}{2}, -4 \right\rangle$$

$$\mathbf{F}_3 = \overrightarrow{PQ_3} = \left\langle -\frac{\sqrt{3}}{2} - 0, \frac{1}{2} - 0, 0 - 4 \right\rangle = \left\langle -\frac{\sqrt{3}}{2}, \frac{1}{2}, -4 \right\rangle.$$

Because all three legs have the same length and the total force is distributed equally among the three legs, you know that $\|\mathbf{F}_1\| = \|\mathbf{F}_2\| = \|\mathbf{F}_3\|$. So, there exists a constant $c$ such that

$$\mathbf{F}_1 = c\langle 0, -1, -4 \rangle, \quad \mathbf{F}_2 = c\left\langle \frac{\sqrt{3}}{2}, \frac{1}{2}, -4 \right\rangle, \quad \text{and} \quad \mathbf{F}_3 = c\left\langle -\frac{\sqrt{3}}{2}, \frac{1}{2}, -4 \right\rangle.$$

Let the total force exerted by the object be given by $\mathbf{F} = \langle 0, 0, -120 \rangle$. Then, using the fact that

$$\mathbf{F} = \mathbf{F}_1 + \mathbf{F}_2 + \mathbf{F}_3$$

you can conclude that $\mathbf{F}_1$, $\mathbf{F}_2$, and $\mathbf{F}_3$ all have a vertical component of $-40$. This implies that $c(-4) = -40$ and $c = 10$. Therefore, the forces exerted on the legs can be represented by

$$\mathbf{F}_1 = \langle 0, -10, -40 \rangle,$$
$$\mathbf{F}_2 = \langle 5\sqrt{3}, 5, -40 \rangle,$$

and

$$\mathbf{F}_3 = \langle -5\sqrt{3}, 5, -40 \rangle.$$

# 11.2 Exercises

See **CalcChat.com** for tutorial help and worked-out solutions to odd-numbered exercises.

**CONCEPT CHECK**

1. **Describing Coordinates** A point in the three-dimensional coordinate system has coordinates $(x_0, y_0, z_0)$. Describe what each coordinate measures.

2. **Coordinates in Space** What is the $y$-coordinate of any point in the $xz$-plane?

3. **Comparing Graphs** Describe the graph of $x = 4$ on (a) the number line, (b) the two-dimensional coordinate system, and (c) the three-dimensional coordinate system.

4. **Parallel Vectors** Explain how to determine whether two nonzero vectors **u** and **v** are parallel.

**Plotting Points** In Exercises 5–8, plot the points in the same three-dimensional coordinate system.

5. (a) $(2, 1, 3)$    (b) $(-1, 2, 1)$

6. (a) $(3, -2, 5)$    (b) $\left(\frac{3}{2}, 4, -2\right)$

7. (a) $(5, -2, 2)$    (b) $(5, -2, -2)$

8. (a) $(0, 4, -5)$    (b) $(4, 0, 5)$

 **Finding Coordinates of a Point** In Exercises 9–12, find the coordinates of the point.

9. The point is located three units behind the $yz$-plane, four units to the right of the $xz$-plane, and five units above the $xy$-plane.

10. The point is located seven units in front of the $yz$-plane, two units to the left of the $xz$-plane, and one unit below the $xy$-plane.

11. The point is located on the $x$-axis, 12 units in front of the $yz$-plane.

12. The point is located in the $yz$-plane, three units to the right of the $xz$-plane, and two units above the $xy$-plane.

**Using the Three-Dimensional Coordinate System** In Exercises 13–24, determine the location of a point $(x, y, z)$ that satisfies the condition(s).

13. $z = 1$    14. $y = 6$

15. $x = -3$    16. $z = -5$

17. $y < 0$    18. $x > 0$

19. $|y| \le 3$    20. $|x| > 4$

21. $xy > 0, \quad z = -3$    22. $xy < 0, \quad z = 4$

23. $xyz < 0$    24. $xyz > 0$

 **Finding the Distance Between Two Points in Space** In Exercises 25–28, find the distance between the points.

25. $(4, 1, 5), (8, 2, 6)$    26. $(-1, 1, 1), (-3, 5, -3)$

27. $(0, 2, 4), (3, 2, 8)$    28. $(-3, 7, 1), (-5, 8, -4)$

**Classifying a Triangle** In Exercises 29–32, find the lengths of the sides of the triangle with the indicated vertices, and determine whether the triangle is a right triangle, an isosceles triangle, or neither.

29. $(0, 0, 4), (2, 6, 7), (6, 4, -8)$

30. $(3, 4, 1), (0, 6, 2), (3, 5, 6)$

31. $(-1, 0, -2), (-1, 5, 2), (-3, -1, 1)$

32. $(4, -1, -1), (2, 0, -4), (3, 5, -1)$

 **Finding the Midpoint** In Exercises 33–36, find the coordinates of the midpoint of the line segment joining the points.

33. $(4, 0, -6), (8, 8, 20)$

34. $(7, 2, 2), (-5, -2, -3)$

35. $(3, 4, 6), (1, 8, 0)$

36. $(5, -9, 7), (-2, 3, 3)$

 **Finding the Equation of a Sphere** In Exercises 37–42, find the standard equation of the sphere with the given characteristics.

37. Center: $(7, 1, -2)$; Radius: 1

38. Center: $(-1, -5, 8)$; Radius: 5

39. Endpoints of a diameter: $(2, 1, 3), (1, 3, -1)$

40. Endpoints of a diameter: $(-2, 4, -5), (-4, 0, 3)$

41. Center: $(-7, 7, 6)$, tangent to the $xy$-plane

42. Center: $(-4, 0, 0)$, tangent to the $yz$-plane

**Finding the Equation of a Sphere** In Exercises 43–46, complete the square to write the equation of the sphere in standard form. Find the center and radius.

43. $x^2 + y^2 + z^2 - 2x + 6y + 8z + 1 = 0$

44. $x^2 + y^2 + z^2 + 9x - 2y + 10z + 19 = 0$

45. $9x^2 + 9y^2 + 9z^2 - 6x + 18y + 1 = 0$

46. $4x^2 + 4y^2 + 4z^2 - 24x - 4y + 8z - 23 = 0$

**Finding the Component Form of a Vector in Space** In Exercises 47 and 48, (a) find the component form of the vector v, (b) write the vector using standard unit vector notation, and (c) sketch the vector with its initial point at the origin.

47.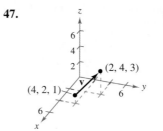

48.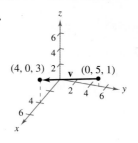

**Writing a Vector in Different Forms** In Exercises 49 and 50, the initial and terminal points of a vector v are given. **(a)** Sketch the directed line segment. **(b)** Find the component form of the vector. **(c)** Write the vector using standard unit vector notation. **(d)** Sketch the vector with its initial point at the origin.

**49.** Initial point: $(-1, 2, 3)$

Terminal point: $(3, 3, 4)$

**50.** Initial point: $(2, -1, -2)$

Terminal point: $(-4, 3, 7)$

 **Finding the Component Form of a Vector in Space** In Exercises 51–54, find the component form and magnitude of the vector v with the given initial and terminal points. Then find a unit vector in the direction of v.

**51.** Initial point: $(3, 2, 0)$      **52.** Initial point: $(1, -2, 4)$

Terminal point: $(4, 1, 6)$      Terminal point: $(2, 4, -2)$

**53.** Initial point: $(4, 2, 0)$      **54.** Initial point: $(1, -2, 0)$

Terminal point: $(0, 5, 2)$      Terminal point: $(1, -2, -3)$

**Finding a Terminal Point** In Exercises 55 and 56, the vector v and its initial point are given. Find the terminal point.

**55.** $\mathbf{v} = \langle 3, -5, 6 \rangle$

Initial point: $(0, 6, 2)$

**56.** $\mathbf{v} = \langle 1, -\frac{2}{3}, \frac{1}{2} \rangle$

Initial point: $\left(0, 2, \frac{5}{2}\right)$

**Finding Scalar Multiples** In Exercises 57 and 58, find each scalar multiple of v and sketch its graph.

**57.** $\mathbf{v} = \langle 1, 2, 2 \rangle$

(a) $2\mathbf{v}$     (b) $-\mathbf{v}$

(c) $\frac{3}{2}\mathbf{v}$     (d) $0\mathbf{v}$

**58.** $\mathbf{v} = \langle 2, -2, 1 \rangle$

(a) $-\mathbf{v}$     (b) $2\mathbf{v}$

(c) $\frac{1}{2}\mathbf{v}$     (d) $\frac{5}{2}\mathbf{v}$

**Finding a Vector** In Exercises 59–62, find the vector z, given that $\mathbf{u} = \langle 1, 2, 3 \rangle$, $\mathbf{v} = \langle 2, 2, -1 \rangle$, and $\mathbf{w} = \langle 4, 0, -4 \rangle$.

**59.** $\mathbf{z} = \mathbf{u} - \mathbf{v} + \mathbf{w}$      **60.** $\mathbf{z} = 5\mathbf{u} - 3\mathbf{v} - \frac{1}{2}\mathbf{w}$

**61.** $\frac{1}{3}\mathbf{z} - 3\mathbf{u} = \mathbf{w}$      **62.** $2\mathbf{u} + \mathbf{v} - \mathbf{w} + 3\mathbf{z} = 0$

 **Parallel Vectors** In Exercises 63–66, determine which of the vectors is/are parallel to z. Use a graphing utility to confirm your results.

**63.** $\mathbf{z} = \langle 3, 2, -5 \rangle$      **64.** $\mathbf{z} = \frac{1}{2}\mathbf{i} - \frac{2}{3}\mathbf{j} + \frac{3}{4}\mathbf{k}$

(a) $\langle -6, -4, 10 \rangle$      (a) $6\mathbf{i} - 4\mathbf{j} + 9\mathbf{k}$

(b) $\langle 2, \frac{4}{3}, -\frac{10}{3} \rangle$      (b) $-\mathbf{i} + \frac{4}{3}\mathbf{j} - \frac{3}{2}\mathbf{k}$

(c) $\langle 6, 4, 10 \rangle$      (c) $12\mathbf{i} + 9\mathbf{k}$

(d) $\langle 1, -4, 2 \rangle$      (d) $\frac{3}{4}\mathbf{i} - \mathbf{j} + \frac{9}{8}\mathbf{k}$

**65.** z has initial point $(1, -1, 3)$ and terminal point $(-2, 3, 5)$.

(a) $-6\mathbf{i} + 8\mathbf{j} + 4\mathbf{k}$      (b) $4\mathbf{j} + 2\mathbf{k}$

**66.** z has initial point $(5, 4, 1)$ and terminal point $(-2, -4, 4)$.

(a) $\langle 7, 6, 2 \rangle$      (b) $\langle 14, 16, -6 \rangle$

 **Using Vectors to Determine Collinear Points** In Exercises 67–70, use vectors to determine whether the points are collinear.

**67.** $(0, -2, -5), (3, 4, 4), (2, 2, 1)$

**68.** $(4, -2, 7), (-2, 0, 3), (7, -3, 9)$

**69.** $(1, 2, 4), (2, 5, 0), (0, 1, 5)$

**70.** $(0, 0, 0), (1, 3, -2), (2, -6, 4)$

**Verifying a Parallelogram** In Exercises 71 and 72, use vectors to show that the points form the vertices of a parallelogram.

**71.** $(2, 9, 1), (3, 11, 4), (0, 10, 2), (1, 12, 5)$

**72.** $(1, 1, 3), (9, -1, -2), (11, 2, -9), (3, 4, -4)$

**Finding the Magnitude** In Exercises 73–78, find the magnitude of v.

**73.** $\mathbf{v} = \langle -1, 0, 1 \rangle$      **74.** $\mathbf{v} = \langle -5, -3, -4 \rangle$

**75.** $\mathbf{v} = 3\mathbf{j} - 5\mathbf{k}$      **76.** $\mathbf{v} = 2\mathbf{i} + 5\mathbf{j} - \mathbf{k}$

**77.** $\mathbf{v} = \mathbf{i} - 2\mathbf{j} - 3\mathbf{k}$      **78.** $\mathbf{v} = -4\mathbf{i} + 3\mathbf{j} + 7\mathbf{k}$

 **Finding Unit Vectors** In Exercises 79–82, find a unit vector **(a)** in the direction of v and **(b)** in the direction opposite of v.

**79.** $\mathbf{v} = \langle 2, -1, 2 \rangle$      **80.** $\mathbf{v} = \langle 6, 0, 8 \rangle$

**81.** $\mathbf{v} = 4\mathbf{i} - 5\mathbf{j} + 3\mathbf{k}$      **82.** $\mathbf{v} = 5\mathbf{i} + 3\mathbf{j} - \mathbf{k}$

**Finding a Vector** In Exercises 83–86, find the vector v with the given magnitude and the same direction as u.

| Magnitude | Direction |
|---|---|
| **83.** $\|\mathbf{v}\| = 10$ | $\mathbf{u} = \langle 0, 3, 3 \rangle$ |
| **84.** $\|\mathbf{v}\| = 3$ | $\mathbf{u} = \langle 1, 1, 1 \rangle$ |
| **85.** $\|\mathbf{v}\| = \frac{3}{2}$ | $\mathbf{u} = \langle 2, -2, 1 \rangle$ |
| **86.** $\|\mathbf{v}\| = 7$ | $\mathbf{u} = \langle -4, 6, 2 \rangle$ |

**Sketching a Vector** In Exercises 87 and 88, sketch the vector v and write its component form.

**87.** v lies in the $yz$-plane, has magnitude 2, and makes an angle of $30°$ with the positive $y$-axis.

**88.** v lies in the $xz$-plane, has magnitude 5, and makes an angle of $45°$ with the positive $z$-axis.

**Finding a Point Using Vectors** In Exercises 89 and 90, use vectors to find the point that lies two-thirds of the way from P to Q.

**89.** $P(4, 3, 0), Q(1, -3, 3)$

**90.** $P(1, 2, 5), Q(6, 8, 2)$

## EXPLORING CONCEPTS

**91. Writing** The initial and terminal points of the vector **v** are $(x_1, y_1, z_1)$ and $(x, y, z)$. Describe the set of all points $(x, y, z)$ such that $\|\mathbf{v}\| = 4$.

**92. Writing** Let $\mathbf{r} = \langle x, y, z \rangle$ and $\mathbf{r}_0 = \langle 1, 1, 1 \rangle$. Describe the set of all points $(x, y, z)$ such that $\|\mathbf{r} - \mathbf{r}_0\| = 2$.

**93. Writing** Let $\mathbf{r} = \langle x, y, z \rangle$. Describe the set of all points $(x, y, z)$ such that $\|\mathbf{r}\| > 1$.

**94. HOW DO YOU SEE IT?** Determine $(x, y, z)$ for each figure. Then find the component form of the vector from the point on the $x$-axis to the point $(x, y, z)$.

(a) 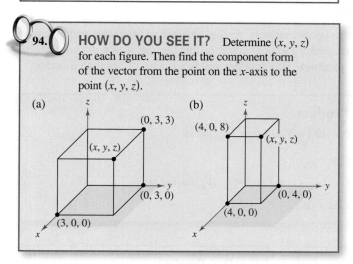 (b)

**95. Using Vectors** Consider two nonzero vectors **u** and **v**, and let $s$ and $t$ be real numbers. Describe the geometric figure generated by connecting the terminal points of the three vectors $t\mathbf{v}$, $\mathbf{u} + t\mathbf{v}$, and $s\mathbf{u} + t\mathbf{v}$.

**96. Using Vectors** Let $\mathbf{u} = \mathbf{i} + \mathbf{j}$, $\mathbf{v} = \mathbf{j} + \mathbf{k}$, and $\mathbf{w} = a\mathbf{u} + b\mathbf{v}$.

(a) Sketch **u** and **v**.

(b) If $\mathbf{w} = \mathbf{0}$, show that $a$ and $b$ must both be zero.

(c) Find $a$ and $b$ such that $\mathbf{w} = \mathbf{i} + 2\mathbf{j} + \mathbf{k}$.

(d) Show that no choice of $a$ and $b$ yields $\mathbf{w} = \mathbf{i} + 2\mathbf{j} + 3\mathbf{k}$.

**97. Diagonal of a Cube** Find the component form of the unit vector **v** in the direction of the diagonal of the cube shown in the figure.

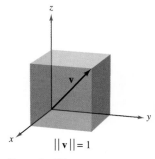

Figure for 97

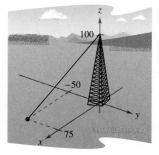

Figure for 98

**98. Tower Guy Wire** The guy wire supporting a 100-foot tower has a tension of 550 pounds. Using the distance shown in the figure, write the component form of the vector **F** representing the tension in the wire.

**99. Auditorium Lights**

The lights in an auditorium are 24-pound discs of radius 18 inches. Each disc is supported by three equally spaced cables that are $L$ inches long (see figure).

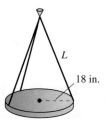 

(a) Write the tension $T$ in each cable as a function of $L$. Determine the domain of the function.

(b) Use a graphing utility and the function in part (a) to complete the table.

| $L$ | 20 | 25 | 30 | 35 | 40 | 45 | 50 |
|-----|----|----|----|----|----|----|----|
| $T$ |    |    |    |    |    |    |    |

(c) Use a graphing utility to graph the function in part (a). Determine the asymptotes of the graph.

(d) Confirm the asymptotes of the graph in part (c) analytically.

(e) Determine the minimum length of each cable when a cable is designed to carry a maximum load of 10 pounds.

**100. Think About It** Suppose the length of each cable in Exercise 99 has a fixed length $L = a$ and the radius of each disc is $r_0$ inches. Make a conjecture about the limit $\lim\limits_{r_0 \to a^-} T$ and give a reason for your answer.

**101. Load Supports** Find the tension in each of the supporting cables in the figure when the weight of the crate is 500 newtons.

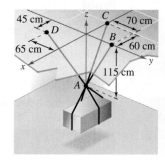

Figure for 101

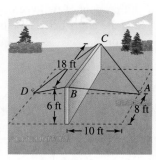

Figure for 102

**102. Construction** A precast concrete wall is temporarily kept in its vertical position by ropes (see figure). Find the total force exerted on the pin at position $A$. The tensions in $AB$ and $AC$ are 420 pounds and 650 pounds, respectively.

**103. Geometry** Write an equation whose graph consists of the set of points $P(x, y, z)$ that are twice as far from $A(0, -1, 1)$ as from $B(1, 2, 0)$. Describe the geometric figure represented by the equation.

# 11.3 The Dot Product of Two Vectors

■ Use properties of the dot product of two vectors.
■ Find the angle between two vectors using the dot product.
■ Find the direction cosines of a vector in space.
■ Find the projection of a vector onto another vector.
■ Use vectors to find the work done by a constant force.

## The Dot Product

So far, you have studied two operations with vectors—vector addition and multiplication by a scalar—each of which yields another vector. In this section, you will study a third vector operation, the **dot product.** This product yields a scalar, rather than a vector.

**··REMARK** Because the dot product of two vectors yields a scalar, it is also called the *scalar product* (or *inner product*) of the two vectors.

> ### Definition of Dot Product
> The **dot product** of $\mathbf{u} = \langle u_1, u_2 \rangle$ and $\mathbf{v} = \langle v_1, v_2 \rangle$ is
> $$\mathbf{u} \cdot \mathbf{v} = u_1 v_1 + u_2 v_2.$$
> The **dot product** of $\mathbf{u} = \langle u_1, u_2, u_3 \rangle$ and $\mathbf{v} = \langle v_1, v_2, v_3 \rangle$ is
> $$\mathbf{u} \cdot \mathbf{v} = u_1 v_1 + u_2 v_2 + u_3 v_3.$$

### Exploration

*Interpreting a Dot Product*
Several vectors are shown below on the unit circle. Find the dot products of several pairs of vectors. Then find the angle between each pair that you used. Make a conjecture about the relationship between the dot product of two vectors and the angle between the vectors.

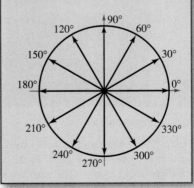

> ### THEOREM 11.4 Properties of the Dot Product
> Let $\mathbf{u}$, $\mathbf{v}$, and $\mathbf{w}$ be vectors in the plane or in space and let $c$ be a scalar.
>
> 1. $\mathbf{u} \cdot \mathbf{v} = \mathbf{v} \cdot \mathbf{u}$    Commutative Property
> 2. $\mathbf{u} \cdot (\mathbf{v} + \mathbf{w}) = \mathbf{u} \cdot \mathbf{v} + \mathbf{u} \cdot \mathbf{w}$    Distributive Property
> 3. $c(\mathbf{u} \cdot \mathbf{v}) = c\mathbf{u} \cdot \mathbf{v} = \mathbf{u} \cdot c\mathbf{v}$    Associative Property
> 4. $\mathbf{0} \cdot \mathbf{v} = 0$
> 5. $\mathbf{v} \cdot \mathbf{v} = \|\mathbf{v}\|^2$

**Proof**   To prove the first property, let $\mathbf{u} = \langle u_1, u_2, u_3 \rangle$ and $\mathbf{v} = \langle v_1, v_2, v_3 \rangle$. Then
$$\mathbf{u} \cdot \mathbf{v} = u_1 v_1 + u_2 v_2 + u_3 v_3 = v_1 u_1 + v_2 u_2 + v_3 u_3 = \mathbf{v} \cdot \mathbf{u}.$$
For the fifth property, let $\mathbf{v} = \langle v_1, v_2, v_3 \rangle$. Then
$$\mathbf{v} \cdot \mathbf{v} = v_1^2 + v_2^2 + v_3^2 = \left( \sqrt{v_1^2 + v_2^2 + v_3^2} \right)^2 = \|\mathbf{v}\|^2.$$
Proofs of the other properties are left to you. ■

### EXAMPLE 1   Finding Dot Products

Let $\mathbf{u} = \langle 2, -2 \rangle$, $\mathbf{v} = \langle 5, 8 \rangle$, and $\mathbf{w} = \langle -4, 3 \rangle$.

**a.** $\mathbf{u} \cdot \mathbf{v} = \langle 2, -2 \rangle \cdot \langle 5, 8 \rangle = 2(5) + (-2)(8) = -6$
**b.** $(\mathbf{u} \cdot \mathbf{v})\mathbf{w} = -6\langle -4, 3 \rangle = \langle 24, -18 \rangle$
**c.** $\mathbf{u} \cdot (2\mathbf{v}) = 2(\mathbf{u} \cdot \mathbf{v}) = 2(-6) = -12$
**d.** $\|\mathbf{w}\|^2 = \mathbf{w} \cdot \mathbf{w} = \langle -4, 3 \rangle \cdot \langle -4, 3 \rangle = (-4)(-4) + (3)(3) = 25$

Notice that the result of part (b) is a *vector* quantity, whereas the results of the other three parts are *scalar* quantities. ■

## Angle Between Two Vectors

The **angle between two nonzero vectors** is the angle $\theta$, $0 \le \theta \le \pi$, between their respective standard position vectors, as shown in Figure 11.24. The next theorem shows how to find this angle using the dot product. (Note that the angle between the zero vector and another vector is not defined here.)

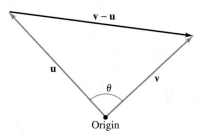

The angle between two vectors
**Figure 11.24**

---

**THEOREM 11.5 Angle Between Two Vectors**

If $\theta$ is the angle between two nonzero vectors $\mathbf{u}$ and $\mathbf{v}$, where $0 \le \theta \le \pi$, then

$$\cos \theta = \frac{\mathbf{u} \cdot \mathbf{v}}{\|\mathbf{u}\| \|\mathbf{v}\|}.$$

---

**Proof** Consider the triangle determined by vectors $\mathbf{u}$, $\mathbf{v}$, and $\mathbf{v} - \mathbf{u}$, as shown in Figure 11.24. By the Law of Cosines, you can write

$$\|\mathbf{v} - \mathbf{u}\|^2 = \|\mathbf{u}\|^2 + \|\mathbf{v}\|^2 - 2\|\mathbf{u}\| \|\mathbf{v}\| \cos \theta.$$

Using the properties of the dot product, the left side can be rewritten as

$$\begin{aligned} \|\mathbf{v} - \mathbf{u}\|^2 &= (\mathbf{v} - \mathbf{u}) \cdot (\mathbf{v} - \mathbf{u}) \\ &= (\mathbf{v} - \mathbf{u}) \cdot \mathbf{v} - (\mathbf{v} - \mathbf{u}) \cdot \mathbf{u} \\ &= \mathbf{v} \cdot \mathbf{v} - \mathbf{u} \cdot \mathbf{v} - \mathbf{v} \cdot \mathbf{u} + \mathbf{u} \cdot \mathbf{u} \\ &= \|\mathbf{v}\|^2 - 2\mathbf{u} \cdot \mathbf{v} + \|\mathbf{u}\|^2 \end{aligned}$$

and substitution back into the Law of Cosines yields

$$\begin{aligned} \|\mathbf{v}\|^2 - 2\mathbf{u} \cdot \mathbf{v} + \|\mathbf{u}\|^2 &= \|\mathbf{u}\|^2 + \|\mathbf{v}\|^2 - 2\|\mathbf{u}\| \|\mathbf{v}\| \cos \theta \\ -2\mathbf{u} \cdot \mathbf{v} &= -2\|\mathbf{u}\| \|\mathbf{v}\| \cos \theta \\ \cos \theta &= \frac{\mathbf{u} \cdot \mathbf{v}}{\|\mathbf{u}\| \|\mathbf{v}\|}. \end{aligned}$$

∎

Note in Theorem 11.5 that because $\|\mathbf{u}\|$ and $\|\mathbf{v}\|$ are always positive, $\mathbf{u} \cdot \mathbf{v}$ and $\cos \theta$ will always have the same sign. Figure 11.25 shows the possible orientations of two vectors.

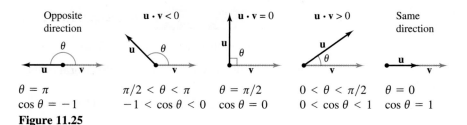

**Figure 11.25**

From Theorem 11.5, you can see that two nonzero vectors meet at a right angle if and only if their dot product is zero. Two such vectors are said to be **orthogonal.**

---

**Definition of Orthogonal Vectors**

The vectors **u** and **v** are orthogonal when $\mathbf{u} \cdot \mathbf{v} = 0$.

---

• • • • • • **REMARK** The terms "perpendicular," "orthogonal," and "normal" all mean essentially the same thing—meeting at right angles. It is common, however, to say that two vectors are *orthogonal*, two lines or planes are *perpendicular*, and a vector is *normal* to a line or plane.

From this definition, it follows that the zero vector is orthogonal to every vector **u**, because $\mathbf{0} \cdot \mathbf{u} = 0$. Moreover, for $0 \le \theta \le \pi$, you know that $\cos \theta = 0$ if and only if $\theta = \pi/2$. So, you can use Theorem 11.5 to conclude that two *nonzero* vectors are orthogonal if and only if the angle between them is $\pi/2$.

**EXAMPLE 2** Finding the Angle Between Two Vectors

• • • ▷ *See LarsonCalculus.com for an interactive version of this type of example.*

For $\mathbf{u} = \langle 3, -1, 2 \rangle$, $\mathbf{v} = \langle -4, 0, 2 \rangle$, $\mathbf{w} = \langle 1, -1, -2 \rangle$, and $\mathbf{z} = \langle 2, 0, -1 \rangle$, find the angle between each pair of vectors.

**a.** **u** and **v**    **b.** **u** and **w**    **c.** **v** and **z**

**Solution**

**a.** $\cos \theta = \dfrac{\mathbf{u} \cdot \mathbf{v}}{\|\mathbf{u}\| \|\mathbf{v}\|} = \dfrac{-12 + 0 + 4}{\sqrt{14}\sqrt{20}} = \dfrac{-8}{2\sqrt{14}\sqrt{5}} = \dfrac{-4}{\sqrt{70}}$

Because $\mathbf{u} \cdot \mathbf{v} < 0$, $\theta = \arccos \dfrac{-4}{\sqrt{70}} \approx 2.069$ radians.

**REMARK** The angle between **u** and **v** in Example 3(a) can also be written as approximately 118.561°.

**b.** $\cos \theta = \dfrac{\mathbf{u} \cdot \mathbf{w}}{\|\mathbf{u}\| \|\mathbf{w}\|} = \dfrac{3 + 1 - 4}{\sqrt{14}\sqrt{6}} = \dfrac{0}{\sqrt{84}} = 0$

Because $\mathbf{u} \cdot \mathbf{w} = 0$, **u** and **w** are *orthogonal*. So, $\theta = \pi/2$.

**c.** $\cos \theta = \dfrac{\mathbf{v} \cdot \mathbf{z}}{\|\mathbf{v}\| \|\mathbf{z}\|} = \dfrac{-8 + 0 - 2}{\sqrt{20}\sqrt{5}} = \dfrac{-10}{\sqrt{100}} = -1$

Consequently, $\theta = \pi$. Note that **v** and **z** are parallel, with $\mathbf{v} = -2\mathbf{z}$. ∎

When the angle between two vectors is known, rewriting Theorem 11.5 in the form

$$\mathbf{u} \cdot \mathbf{v} = \|\mathbf{u}\| \|\mathbf{v}\| \cos \theta \qquad \text{Alternative form of dot product}$$

produces an alternative way to calculate the dot product.

**EXAMPLE 3** Alternative Form of the Dot Product

Given that $\|\mathbf{u}\| = 10$, $\|\mathbf{v}\| = 7$, and the angle between **u** and **v** is $\pi/4$, find $\mathbf{u} \cdot \mathbf{v}$.

**Solution** Use the alternative form of the dot product as shown.

$$\mathbf{u} \cdot \mathbf{v} = \|\mathbf{u}\| \|\mathbf{v}\| \cos \theta = (10)(7) \cos \frac{\pi}{4} = 35\sqrt{2} \qquad ∎$$

## Direction Cosines

For a vector in the plane, you have seen that it is convenient to measure direction in terms of the angle, measured counterclockwise, *from* the positive $x$-axis *to* the vector. In space, it is more convenient to measure direction in terms of the angles *between* the nonzero vector $\mathbf{v}$ and the three unit vectors $\mathbf{i}$, $\mathbf{j}$, and $\mathbf{k}$, as shown in Figure 11.26. The angles $\alpha$, $\beta$, and $\gamma$ are the **direction angles of v,** and $\cos\alpha$, $\cos\beta$, and $\cos\gamma$ are the **direction cosines of v.** Because

$$\mathbf{v} \cdot \mathbf{i} = \|\mathbf{v}\|\|\mathbf{i}\| \cos\alpha = \|\mathbf{v}\| \cos\alpha$$

and

$$\mathbf{v} \cdot \mathbf{i} = \langle v_1, v_2, v_3 \rangle \cdot \langle 1, 0, 0 \rangle = v_1$$

it follows that $\cos\alpha = v_1/\|\mathbf{v}\|$. By similar reasoning with the unit vectors $\mathbf{j}$ and $\mathbf{k}$, you have

$$\cos\alpha = \frac{v_1}{\|\mathbf{v}\|} \qquad \text{$\alpha$ is the angle between $\mathbf{v}$ and $\mathbf{i}$.}$$

$$\cos\beta = \frac{v_2}{\|\mathbf{v}\|} \qquad \text{$\beta$ is the angle between $\mathbf{v}$ and $\mathbf{j}$.}$$

$$\cos\gamma = \frac{v_3}{\|\mathbf{v}\|}. \qquad \text{$\gamma$ is the angle between $\mathbf{v}$ and $\mathbf{k}$.}$$

Consequently, any nonzero vector $\mathbf{v}$ in space has the normalized form

$$\frac{\mathbf{v}}{\|\mathbf{v}\|} = \frac{v_1}{\|\mathbf{v}\|}\mathbf{i} + \frac{v_2}{\|\mathbf{v}\|}\mathbf{j} + \frac{v_3}{\|\mathbf{v}\|}\mathbf{k} = \cos\alpha\,\mathbf{i} + \cos\beta\,\mathbf{j} + \cos\gamma\,\mathbf{k}$$

and because $\mathbf{v}/\|\mathbf{v}\|$ is a unit vector, it follows that

$$\cos^2\alpha + \cos^2\beta + \cos^2\gamma = 1.$$

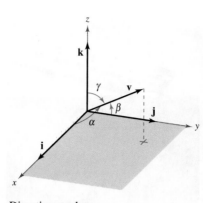

Direction angles
**Figure 11.26**

> **·· REMARK**  Recall that $\alpha$, $\beta$, and $\gamma$ are the Greek letters alpha, beta, and gamma, respectively.

**EXAMPLE 4**   **Finding Direction Angles**

Find the direction cosines and angles for the vector $\mathbf{v} = 2\mathbf{i} + 3\mathbf{j} + 4\mathbf{k}$, and show that $\cos^2\alpha + \cos^2\beta + \cos^2\gamma = 1$.

**Solution**   Because $\|\mathbf{v}\| = \sqrt{2^2 + 3^2 + 4^2} = \sqrt{29}$, you can write the following.

$$\cos\alpha = \frac{v_1}{\|\mathbf{v}\|} = \frac{2}{\sqrt{29}} \implies \alpha \approx 68.2° \qquad \text{Angle between $\mathbf{v}$ and $\mathbf{i}$}$$

$$\cos\beta = \frac{v_2}{\|\mathbf{v}\|} = \frac{3}{\sqrt{29}} \implies \beta \approx 56.1° \qquad \text{Angle between $\mathbf{v}$ and $\mathbf{j}$}$$

$$\cos\gamma = \frac{v_3}{\|\mathbf{v}\|} = \frac{4}{\sqrt{29}} \implies \gamma \approx 42.0° \qquad \text{Angle between $\mathbf{v}$ and $\mathbf{k}$}$$

Furthermore, the sum of the squares of the direction cosines is

$$\cos^2\alpha + \cos^2\beta + \cos^2\gamma = \frac{4}{29} + \frac{9}{29} + \frac{16}{29}$$

$$= \frac{29}{29}$$

$$= 1.$$

See Figure 11.27.

$\alpha$ = angle between $\mathbf{v}$ and $\mathbf{i}$
$\beta$ = angle between $\mathbf{v}$ and $\mathbf{j}$
$\gamma$ = angle between $\mathbf{v}$ and $\mathbf{k}$

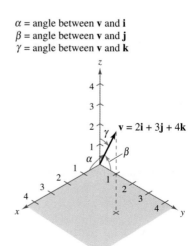

The direction angles of $\mathbf{v}$
**Figure 11.27**

## Projections and Vector Components

You have already seen applications in which two vectors are added to produce a resultant vector. Many applications in physics and engineering pose the reverse problem—decomposing a vector into the sum of two **vector components.** The following physical example enables you to see the usefulness of this procedure.

Consider a boat on an inclined ramp, as shown in Figure 11.28. The force $\mathbf{F}$ due to gravity pulls the boat *down* the ramp and *against* the ramp. These two forces, $\mathbf{w}_1$ and $\mathbf{w}_2$, are orthogonal—they are called the vector components of $\mathbf{F}$.

$$\mathbf{F} = \mathbf{w}_1 + \mathbf{w}_2 \qquad \text{Vector components of } \mathbf{F}$$

The forces $\mathbf{w}_1$ and $\mathbf{w}_2$ help you analyze the effect of gravity on the boat. For example, $\mathbf{w}_1$ indicates the force necessary to keep the boat from rolling down the ramp, whereas $\mathbf{w}_2$ indicates the force that the tires must withstand.

The force due to gravity pulls the boat against the ramp and down the ramp.

**Figure 11.28**

---

**Definitions of Projection and Vector Components**

Let $\mathbf{u}$ and $\mathbf{v}$ be nonzero vectors. Moreover, let

$$\mathbf{u} = \mathbf{w}_1 + \mathbf{w}_2$$

where $\mathbf{w}_1$ is parallel to $\mathbf{v}$ and $\mathbf{w}_2$ is orthogonal to $\mathbf{v}$, as shown in Figure 11.29.

1. $\mathbf{w}_1$ is called the **projection of u onto v** or the **vector component of u along v,** and is denoted by $\mathbf{w}_1 = \text{proj}_{\mathbf{v}}\mathbf{u}$.

2. $\mathbf{w}_2 = \mathbf{u} - \mathbf{w}_1$ is called the **vector component of u orthogonal to v.**

---

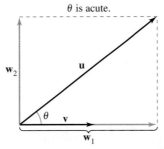
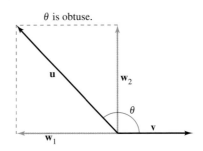

$\mathbf{w}_1 = \text{proj}_{\mathbf{v}}\mathbf{u} = $ projection of $\mathbf{u}$ onto $\mathbf{v} = $ vector component of $\mathbf{u}$ along $\mathbf{v}$
$\mathbf{w}_2 = $ vector component of $\mathbf{u}$ orthogonal to $\mathbf{v}$

**Figure 11.29**

---

**EXAMPLE 5** **Finding a Vector Component of u Orthogonal to v**

Find the vector component of $\mathbf{u} = \langle 5, 10 \rangle$ that is orthogonal to $\mathbf{v} = \langle 4, 3 \rangle$, given that

$$\mathbf{w}_1 = \text{proj}_{\mathbf{v}}\mathbf{u} = \langle 8, 6 \rangle$$

and

$$\mathbf{u} = \langle 5, 10 \rangle = \mathbf{w}_1 + \mathbf{w}_2.$$

**Solution** Because $\mathbf{u} = \mathbf{w}_1 + \mathbf{w}_2$, where $\mathbf{w}_1$ is parallel to $\mathbf{v}$, it follows that $\mathbf{w}_2$ is the vector component of $\mathbf{u}$ orthogonal to $\mathbf{v}$. So, you have

$$\begin{aligned} \mathbf{w}_2 &= \mathbf{u} - \mathbf{w}_1 \\ &= \langle 5, 10 \rangle - \langle 8, 6 \rangle \\ &= \langle -3, 4 \rangle. \end{aligned}$$

Check to see that $\mathbf{w}_2$ is orthogonal to $\mathbf{v}$, as shown in Figure 11.30.

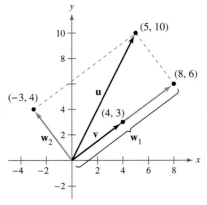

$\mathbf{u} = \mathbf{w}_1 + \mathbf{w}_2$

**Figure 11.30**

From Example 5, you can see that it is easy to find the vector component $\mathbf{w}_2$ once you have found the projection, $\mathbf{w}_1$, of $\mathbf{u}$ onto $\mathbf{v}$. To find this projection, use the dot product in the next theorem, which you will prove in Exercise 74.

> **THEOREM 11.6    Projection Using the Dot Product**
>
> If $\mathbf{u}$ and $\mathbf{v}$ are nonzero vectors, then the projection of $\mathbf{u}$ onto $\mathbf{v}$ is
>
> $$\operatorname{proj}_{\mathbf{v}}\mathbf{u} = \left(\frac{\mathbf{u} \cdot \mathbf{v}}{\|\mathbf{v}\|^2}\right)\mathbf{v}.$$

•• **REMARK**  Note the distinction between the terms "component" and "vector component." For example, using the standard unit vectors with $\mathbf{u} = u_1\mathbf{i} + u_2\mathbf{j}$, $u_1$ is the *component* of $\mathbf{u}$ in the direction of $\mathbf{i}$, and $u_1\mathbf{i}$ is the *vector component* in the direction of $\mathbf{i}$.

▷ The projection of $\mathbf{u}$ onto $\mathbf{v}$ can be written as a scalar multiple of a unit vector in the direction of $\mathbf{v}$. That is,

$$\left(\frac{\mathbf{u} \cdot \mathbf{v}}{\|\mathbf{v}\|^2}\right)\mathbf{v} = \left(\frac{\mathbf{u} \cdot \mathbf{v}}{\|\mathbf{v}\|}\right)\frac{\mathbf{v}}{\|\mathbf{v}\|} = (k)\frac{\mathbf{v}}{\|\mathbf{v}\|}.$$

The scalar $k$ is called the **component of u in the direction of v.** So,

$$k = \frac{\mathbf{u} \cdot \mathbf{v}}{\|\mathbf{v}\|} = \|\mathbf{u}\| \cos \theta.$$

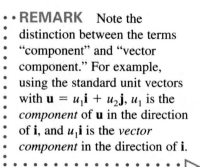

$$\mathbf{u} = 3\mathbf{i} - 5\mathbf{j} + 2\mathbf{k}$$
$$\mathbf{v} = 7\mathbf{i} + \mathbf{j} - 2\mathbf{k}$$

$$\mathbf{u} = \mathbf{w}_1 + \mathbf{w}_2$$
**Figure 11.31**

**EXAMPLE 6    Decomposing a Vector into Vector Components**

Find the projection of $\mathbf{u}$ onto $\mathbf{v}$ and the vector component of $\mathbf{u}$ orthogonal to $\mathbf{v}$ for

$$\mathbf{u} = 3\mathbf{i} - 5\mathbf{j} + 2\mathbf{k} \quad \text{and} \quad \mathbf{v} = 7\mathbf{i} + \mathbf{j} - 2\mathbf{k}.$$

**Solution**    The projection of $\mathbf{u}$ onto $\mathbf{v}$ is

$$\mathbf{w}_1 = \operatorname{proj}_{\mathbf{v}}\mathbf{u} = \left(\frac{\mathbf{u} \cdot \mathbf{v}}{\|\mathbf{v}\|^2}\right)\mathbf{v} = \left(\frac{12}{54}\right)(7\mathbf{i} + \mathbf{j} - 2\mathbf{k}) = \frac{14}{9}\mathbf{i} + \frac{2}{9}\mathbf{j} - \frac{4}{9}\mathbf{k}.$$

The vector component of $\mathbf{u}$ orthogonal to $\mathbf{v}$ is the vector

$$\mathbf{w}_2 = \mathbf{u} - \mathbf{w}_1 = (3\mathbf{i} - 5\mathbf{j} + 2\mathbf{k}) - \left(\frac{14}{9}\mathbf{i} + \frac{2}{9}\mathbf{j} - \frac{4}{9}\mathbf{k}\right) = \frac{13}{9}\mathbf{i} - \frac{47}{9}\mathbf{j} + \frac{22}{9}\mathbf{k}.$$

See Figure 11.31.

**EXAMPLE 7    Finding a Force**

A 600-pound boat sits on a ramp inclined at 30°, as shown in Figure 11.32. What force is required to keep the boat from rolling down the ramp?

**Solution**    Because the force due to gravity is vertical and downward, you can represent the gravitational force by the vector $\mathbf{F} = -600\mathbf{j}$. To find the force required to keep the boat from rolling down the ramp, project $\mathbf{F}$ onto a unit vector $\mathbf{v}$ in the direction of the ramp, as follows.

$$\mathbf{v} = \cos 30°\mathbf{i} + \sin 30°\mathbf{j} = \frac{\sqrt{3}}{2}\mathbf{i} + \frac{1}{2}\mathbf{j} \qquad \text{Unit vector along ramp}$$

Therefore, the projection of $\mathbf{F}$ onto $\mathbf{v}$ is

$$\mathbf{w}_1 = \operatorname{proj}_{\mathbf{v}}\mathbf{F} = \left(\frac{\mathbf{F} \cdot \mathbf{v}}{\|\mathbf{v}\|^2}\right)\mathbf{v} = (\mathbf{F} \cdot \mathbf{v})\mathbf{v} = (-600)\left(\frac{1}{2}\right)\mathbf{v} = -300\left(\frac{\sqrt{3}}{2}\mathbf{i} + \frac{1}{2}\mathbf{j}\right).$$

The magnitude of this force is 300, so a force of 300 pounds is required to keep the boat from rolling down the ramp. ∎

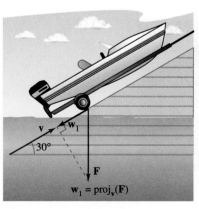

$$\mathbf{w}_1 = \operatorname{proj}_{\mathbf{v}}(\mathbf{F})$$

**Figure 11.32**

## Work

The work $W$ done by the constant force $\mathbf{F}$ acting along the line of motion of an object is given by

$$W = (\text{magnitude of force})(\text{distance}) = \|\mathbf{F}\| \|\overrightarrow{PQ}\|$$

as shown in Figure 11.33(a). When the constant force $\mathbf{F}$ is not directed along the line of motion, you can see from Figure 11.33(b) that the work $W$ done by the force is

$$W = \|\text{proj}_{\overrightarrow{PQ}}\mathbf{F}\| \|\overrightarrow{PQ}\| = (\cos\theta)\|\mathbf{F}\| \|\overrightarrow{PQ}\| = \mathbf{F} \cdot \overrightarrow{PQ}.$$

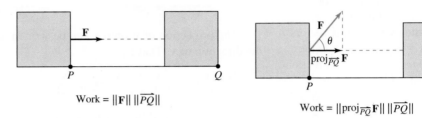

Work $= \|\mathbf{F}\| \|\overrightarrow{PQ}\|$

Work $= \|\text{proj}_{\overrightarrow{PQ}}\mathbf{F}\| \|\overrightarrow{PQ}\|$

**(a)** Force acts along the line of motion.

**(b)** Force acts at angle $\theta$ with the line of motion.

**Figure 11.33**

This notion of work is summarized in the next definition.

---

### Definition of Work

The work $W$ done by a constant force $\mathbf{F}$ as its point of application moves along the vector $\overrightarrow{PQ}$ is one of the following.

1. $W = \|\text{proj}_{\overrightarrow{PQ}}\mathbf{F}\| \|\overrightarrow{PQ}\|$    Projection form

2. $W = \mathbf{F} \cdot \overrightarrow{PQ}$    Dot product form

---

### EXAMPLE 8    Finding Work

To close a sliding door, a person pulls on a rope with a constant force of 50 pounds at a constant angle of 60°, as shown in Figure 11.34. Find the work done in moving the door 12 feet to its closed position.

**Figure 11.34**

**Solution**    Using a projection, you can calculate the work as follows.

$$W = \|\text{proj}_{\overrightarrow{PQ}}\mathbf{F}\| \|\overrightarrow{PQ}\| = \cos(60°)\|\mathbf{F}\| \|\overrightarrow{PQ}\| = \frac{1}{2}(50)(12) = 300 \text{ foot-pounds} \quad \blacksquare$$

# 11.3 Exercises

See CalcChat.com for tutorial help and worked-out solutions to odd-numbered exercises.

**CONCEPT CHECK**

**1. Dot Product** What can you say about the relative position of two nonzero vectors if their dot product is zero?

**2. Direction Cosines** Consider the vector

$$\mathbf{v} = \langle v_1, v_2, v_3 \rangle.$$

What is the meaning of $\arccos \dfrac{v_2}{\|\mathbf{v}\|} = 30°$?

 **Finding Dot Products** In Exercises 3–10, find (a) $\mathbf{u} \cdot \mathbf{v}$, (b) $\mathbf{u} \cdot \mathbf{u}$, (c) $\|\mathbf{v}\|^2$, (d) $(\mathbf{u} \cdot \mathbf{v})\mathbf{v}$, and (e) $\mathbf{u} \cdot (3\mathbf{v})$.

3. $\mathbf{u} = \langle 3, 4 \rangle$, $\mathbf{v} = \langle -1, 5 \rangle$    4. $\mathbf{u} = \langle 4, 10 \rangle$, $\mathbf{v} = \langle -2, 3 \rangle$

5. $\mathbf{u} = \langle 6, -4 \rangle$, $\mathbf{v} = \langle -3, 2 \rangle$  6. $\mathbf{u} = \langle -7, -1 \rangle$, $\mathbf{v} = \langle -4, -1 \rangle$

7. $\mathbf{u} = \langle 2, -3, 4 \rangle$, $\mathbf{v} = \langle 0, 6, 5 \rangle$

8. $\mathbf{u} = \langle -5, 0, 5 \rangle$, $\mathbf{v} = \langle -1, 2, 1 \rangle$

9. $\mathbf{u} = 2\mathbf{i} - \mathbf{j} + \mathbf{k}$    10. $\mathbf{u} = 2\mathbf{i} + \mathbf{j} - 2\mathbf{k}$
   $\mathbf{v} = \mathbf{i} - \mathbf{k}$         $\mathbf{v} = \mathbf{i} - 3\mathbf{j} + 2\mathbf{k}$

 **Finding the Angle Between Two Vectors** In Exercises 11–18, find the angle $\theta$ between the vectors (a) in radians and (b) in degrees.

11. $\mathbf{u} = \langle 1, 1 \rangle$, $\mathbf{v} = \langle 2, -2 \rangle$

12. $\mathbf{u} = \langle 3, 1 \rangle$, $\mathbf{v} = \langle 2, -1 \rangle$

13. $\mathbf{u} = 3\mathbf{i} + \mathbf{j}$, $\mathbf{v} = -2\mathbf{i} + 4\mathbf{j}$

14. $\mathbf{u} = \cos\left(\dfrac{\pi}{6}\right)\mathbf{i} + \sin\left(\dfrac{\pi}{6}\right)\mathbf{j}$, $\mathbf{v} = \cos\left(\dfrac{3\pi}{4}\right)\mathbf{i} + \sin\left(\dfrac{3\pi}{4}\right)\mathbf{j}$

15. $\mathbf{u} = \langle 1, 1, 1 \rangle$, $\mathbf{v} = \langle 2, 1, -1 \rangle$

16. $\mathbf{u} = 3\mathbf{i} + 2\mathbf{j} + \mathbf{k}$, $\mathbf{v} = 2\mathbf{i} - 3\mathbf{j}$

17. $\mathbf{u} = 3\mathbf{i} + 4\mathbf{j}$, $\mathbf{v} = -2\mathbf{j} + 3\mathbf{k}$

18. $\mathbf{u} = 2\mathbf{i} - 3\mathbf{j} + \mathbf{k}$, $\mathbf{v} = \mathbf{i} - 2\mathbf{j} + \mathbf{k}$

 **Alternative Form of Dot Product** In Exercises 19 and 20, use the alternative form of the dot product to find $\mathbf{u} \cdot \mathbf{v}$.

19. $\|\mathbf{u}\| = 8$, $\|\mathbf{v}\| = 5$, and the angle between $\mathbf{u}$ and $\mathbf{v}$ is $\pi/3$.

20. $\|\mathbf{u}\| = 40$, $\|\mathbf{v}\| = 25$, and the angle between $\mathbf{u}$ and $\mathbf{v}$ is $5\pi/6$.

**Comparing Vectors** In Exercises 21–26, determine whether $\mathbf{u}$ and $\mathbf{v}$ are orthogonal, parallel, or neither.

21. $\mathbf{u} = \langle 4, 3 \rangle$          22. $\mathbf{u} = -\frac{1}{3}(\mathbf{i} - 2\mathbf{j})$
    $\mathbf{v} = \langle \frac{1}{2}, -\frac{2}{3} \rangle$        $\mathbf{v} = 2\mathbf{i} - 4\mathbf{j}$

23. $\mathbf{u} = \mathbf{j} + 6\mathbf{k}$        24. $\mathbf{u} = -2\mathbf{i} + 3\mathbf{j} - \mathbf{k}$
    $\mathbf{v} = \mathbf{i} - 2\mathbf{j} - \mathbf{k}$       $\mathbf{v} = 2\mathbf{i} + \mathbf{j} - \mathbf{k}$

25. $\mathbf{u} = \langle 2, -3, 1 \rangle$       26. $\mathbf{u} = \langle \cos\theta, \sin\theta, -1 \rangle$
    $\mathbf{v} = \langle -1, -1, -1 \rangle$      $\mathbf{v} = \langle \sin\theta, -\cos\theta, 0 \rangle$

**Classifying a Triangle** In Exercises 27–30, the vertices of a triangle are given. Determine whether the triangle is an acute triangle, an obtuse triangle, or a right triangle. Explain your reasoning.

27. $(1, 2, 0), (0, 0, 0), (-2, 1, 0)$

28. $(-3, 0, 0), (0, 0, 0), (1, 2, 3)$

29. $(2, 0, 1), (0, 1, 2), (-0.5, 1.5, 0)$

30. $(2, -7, 3), (-1, 5, 8), (4, 6, -1)$

 **Finding Direction Angles** In Exercises 31–36, find the direction cosines and angles of $\mathbf{u}$ and show that $\cos^2\alpha + \cos^2\beta + \cos^2\gamma = 1$.

31. $\mathbf{u} = \mathbf{i} + 2\mathbf{j} + 2\mathbf{k}$        32. $\mathbf{u} = 5\mathbf{i} + 3\mathbf{j} - \mathbf{k}$

33. $\mathbf{u} = 7\mathbf{i} + \mathbf{j} - \mathbf{k}$        34. $\mathbf{u} = -4\mathbf{i} + 3\mathbf{j} + 5\mathbf{k}$

35. $\mathbf{u} = \langle 0, 6, -4 \rangle$        36. $\mathbf{u} = \langle -1, 5, 2 \rangle$

 **Finding the Projection of u onto v** In Exercises 37–44, (a) find the projection of $\mathbf{u}$ onto $\mathbf{v}$ and (b) find the vector component of $\mathbf{u}$ orthogonal to $\mathbf{v}$.

37. $\mathbf{u} = \langle 6, 7 \rangle$, $\mathbf{v} = \langle 1, 4 \rangle$    38. $\mathbf{u} = \langle 9, 7 \rangle$, $\mathbf{v} = \langle 1, 3 \rangle$

39. $\mathbf{u} = 2\mathbf{i} + 3\mathbf{j}$, $\mathbf{v} = 5\mathbf{i} + \mathbf{j}$

40. $\mathbf{u} = 2\mathbf{i} - 3\mathbf{j}$, $\mathbf{v} = 3\mathbf{i} + 2\mathbf{j}$

41. $\mathbf{u} = \langle 0, 3, 3 \rangle$, $\mathbf{v} = \langle -1, 1, 1 \rangle$

42. $\mathbf{u} = \langle 8, 2, 0 \rangle$, $\mathbf{v} = \langle 2, 1, -1 \rangle$

43. $\mathbf{u} = -9\mathbf{i} - 2\mathbf{j} - 4\mathbf{k}$, $\mathbf{v} = 4\mathbf{j} + 4\mathbf{k}$

44. $\mathbf{u} = 5\mathbf{i} - \mathbf{j} - \mathbf{k}$, $\mathbf{v} = -\mathbf{i} + 5\mathbf{j} + 8\mathbf{k}$

**EXPLORING CONCEPTS**

**45. Using Vectors** Explain why $\mathbf{u} + \mathbf{v} \cdot \mathbf{w}$ is not defined, where $\mathbf{u}$, $\mathbf{v}$, and $\mathbf{w}$ are nonzero vectors.

**46. Projection** What can be said about the vectors $\mathbf{u}$ and $\mathbf{v}$ when the projection of $\mathbf{u}$ onto $\mathbf{v}$ equals $\mathbf{u}$?

**47. Projection** When the projection of $\mathbf{u}$ onto $\mathbf{v}$ has the same magnitude as the projection of $\mathbf{v}$ onto $\mathbf{u}$, can you conclude that $\|\mathbf{u}\| = \|\mathbf{v}\|$? Explain.

**48. HOW DO YOU SEE IT?** What is known about $\theta$, the angle between two nonzero vectors $\mathbf{u}$ and $\mathbf{v}$, when

(a) $\mathbf{u} \cdot \mathbf{v} = 0$?  (b) $\mathbf{u} \cdot \mathbf{v} > 0$?  (c) $\mathbf{u} \cdot \mathbf{v} < 0$?

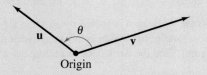

Origin

**49. Revenue** The vector $\mathbf{u} = \langle 3240, 1450, 2235 \rangle$ gives the numbers of hamburgers, chicken sandwiches, and cheeseburgers, respectively, sold at a fast-food restaurant in one week. The vector $\mathbf{v} = \langle 2.25, 2.95, 2.65 \rangle$ gives the prices (in dollars) per unit for the three food items. Find the dot product $\mathbf{u} \cdot \mathbf{v}$ and explain what information it gives.

**50. Revenue** Repeat Exercise 49 after decreasing the prices by 2%. Identify the vector operation used to decrease the prices by 2%.

**Orthogonal Vectors** **In Exercises 51–54, find two vectors in opposite directions that are orthogonal to the vector u. (The answers are not unique.)**

**51.** $\mathbf{u} = -\frac{1}{4}\mathbf{i} + \frac{3}{2}\mathbf{j}$        **52.** $\mathbf{u} = 9\mathbf{i} - 4\mathbf{j}$

**53.** $\mathbf{u} = \langle 3, 1, -2 \rangle$        **54.** $\mathbf{u} = \langle 4, -3, 6 \rangle$

**55. Finding an Angle** Find the angle between a cube's diagonal and one of its edges.

**56. Finding an Angle** Find the angle between the diagonal of a cube and the diagonal of one of its sides.

**57. Braking Load** A 48,000-pound truck is parked on a 10° slope (see figure). Assume the only force to overcome is that due to gravity. Find (a) the force required to keep the truck from rolling down the hill and (b) the force perpendicular to the hill.

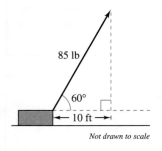

Weight = 48,000 lb

**58. Braking Load** A 5400-pound sport utility vehicle is parked on an 18° slope. Assume the only force to overcome is that due to gravity. Find (a) the force required to keep the vehicle from rolling down the hill and (b) the force perpendicular to the hill.

**59. Work** An object is pulled 10 feet across a floor using a force of 85 pounds. The direction of the force is 60° above the horizontal (see figure). Find the work done.

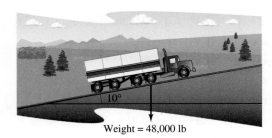

85 lb

60°

10 ft

*Not drawn to scale*

Figure for 59        Figure for 60

**60. Work** A wagon is pulled by exerting a force of 65 pounds on a handle that makes a 20° angle with the horizontal (see figure). Find the work done in pulling the wagon 50 feet.

**61. Work** A car is towed using a force of 1600 newtons. The chain used to pull the car makes a 25° angle with the horizontal. Find the work done in towing the car 2 kilometers.

**62. Work**

A pallet truck is pulled by exerting a force of 400 newtons on a handle that makes a 60° angle with the horizontal. Find the work done in pulling the truck 40 meters.

**True or False?** **In Exercises 63 and 64, determine whether the statement is true or false. If it is false, explain why or give an example that shows it is false.**

**63.** If $\mathbf{u} \cdot \mathbf{v} = \mathbf{u} \cdot \mathbf{w}$ and $\mathbf{u} \neq \mathbf{0}$, then $\mathbf{v} = \mathbf{w}$.

**64.** If $\mathbf{u}$ and $\mathbf{v}$ are orthogonal to $\mathbf{w}$, then $\mathbf{u} + \mathbf{v}$ is orthogonal to $\mathbf{w}$.

**Using Points of Intersection** **In Exercises 65–68, (a) find all points of intersection of the graphs of the two equations, (b) find the unit tangent vectors to each curve at their points of intersection, and (c) find the angles ($0° \leq \theta \leq 90°$) between the curves at their points of intersection.**

**65.** $y = x^2, \quad y = x^{1/3}$        **66.** $y = x^3, \quad y = x^{1/3}$

**67.** $y = 1 - x^2, \quad y = x^2 - 1$        **68.** $(y + 1)^2 = x, \quad y = x^3 - 1$

**69. Proof** Use vectors to prove that the diagonals of a rhombus are perpendicular.

**70. Proof** Use vectors to prove that a parallelogram is a rectangle if and only if its diagonals are equal in length.

**71. Bond Angle** Consider a regular tetrahedron with vertices $(0, 0, 0)$, $(k, k, 0)$, $(k, 0, k)$, and $(0, k, k)$, where $k$ is a positive real number.

(a) Sketch the graph of the tetrahedron.

(b) Find the length of each edge.

(c) Find the angle between any two edges.

(d) Find the angle between the line segments from the centroid $(k/2, k/2, k/2)$ to two vertices. This is the bond angle for a molecule, such as $CH_4$ (methane) or $PbCl_4$ (lead tetrachloride), where the structure of the molecule is a tetrahedron.

**72. Proof** Consider the vectors $\mathbf{u} = \langle \cos \alpha, \sin \alpha, 0 \rangle$ and $\mathbf{v} = \langle \cos \beta, \sin \beta, 0 \rangle$, where $\alpha > \beta$. Find the dot product of the vectors and use the result to prove the identity

$$\cos(\alpha - \beta) = \cos \alpha \cos \beta + \sin \alpha \sin \beta.$$

**73. Proof** Prove the triangle inequality $\|\mathbf{u} + \mathbf{v}\| \leq \|\mathbf{u}\| + \|\mathbf{v}\|$.

**74. Proof** Prove Theorem 11.6.

**75. Proof** Prove the **Cauchy-Schwarz Inequality,**

$$|\mathbf{u} \cdot \mathbf{v}| \leq \|\mathbf{u}\|\|\mathbf{v}\|.$$

# 11.4 The Cross Product of Two Vectors in Space

■ Find the cross product of two vectors in space.
■ Use the triple scalar product of three vectors in space.

## The Cross Product

Many applications in physics, engineering, and geometry involve finding a vector in space that is orthogonal to two given vectors. In this section, you will study a product that will yield such a vector. It is called the **cross product,** and it is most conveniently defined and calculated using the standard unit vector form. Because the cross product yields a vector, it is also called the **vector product.**

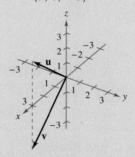
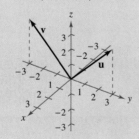
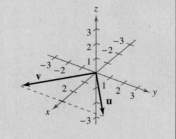
---

### Definition of Cross Product of Two Vectors in Space

Let

$$\mathbf{u} = u_1\mathbf{i} + u_2\mathbf{j} + u_3\mathbf{k} \quad \text{and} \quad \mathbf{v} = v_1\mathbf{i} + v_2\mathbf{j} + v_3\mathbf{k}$$

be vectors in space. The **cross product** of **u** and **v** is the vector

$$\mathbf{u} \times \mathbf{v} = (u_2v_3 - u_3v_2)\mathbf{i} - (u_1v_3 - u_3v_1)\mathbf{j} + (u_1v_2 - u_2v_1)\mathbf{k}.$$

---

It is important to note that this definition applies only to three-dimensional vectors. The cross product is not defined for two-dimensional vectors.

A convenient way to calculate $\mathbf{u} \times \mathbf{v}$ is to use the *determinant form* with cofactor expansion shown below. (This $3 \times 3$ determinant form is used simply to help remember the formula for the cross product. The corresponding array is technically not a matrix because its entries are not all numbers.)

$$\mathbf{u} \times \mathbf{v} = \begin{vmatrix} \mathbf{i} & \mathbf{j} & \mathbf{k} \\ u_1 & u_2 & u_3 \\ v_1 & v_2 & v_3 \end{vmatrix} \quad \begin{array}{l} \longleftarrow \text{Put "u" in Row 2.} \\ \longleftarrow \text{Put "v" in Row 3.} \end{array}$$

$$= \begin{vmatrix} \mathbf{i} & \mathbf{j} & \mathbf{k} \\ u_1 & u_2 & u_3 \\ v_1 & v_2 & v_3 \end{vmatrix} \mathbf{i} - \begin{vmatrix} \mathbf{i} & \mathbf{j} & \mathbf{k} \\ u_1 & u_2 & u_3 \\ v_1 & v_2 & v_3 \end{vmatrix} \mathbf{j} + \begin{vmatrix} \mathbf{i} & \mathbf{j} & \mathbf{k} \\ u_1 & u_2 & u_3 \\ v_1 & v_2 & v_3 \end{vmatrix} \mathbf{k}$$

$$= \begin{vmatrix} u_2 & u_3 \\ v_2 & v_3 \end{vmatrix} \mathbf{i} - \begin{vmatrix} u_1 & u_3 \\ v_1 & v_3 \end{vmatrix} \mathbf{j} + \begin{vmatrix} u_1 & u_2 \\ v_1 & v_2 \end{vmatrix} \mathbf{k}$$

$$= (u_2v_3 - u_3v_2)\mathbf{i} - (u_1v_3 - u_3v_1)\mathbf{j} + (u_1v_2 - u_2v_1)\mathbf{k}$$

Note the minus sign in front of the **j**-component. Each of the three $2 \times 2$ determinants can be evaluated by using the diagonal pattern

$$\begin{vmatrix} a & b \\ c & d \end{vmatrix} = ad - bc.$$

Here are a couple of examples.

$$\begin{vmatrix} 2 & 4 \\ 3 & -1 \end{vmatrix} = (2)(-1) - (4)(3) = -2 - 12 = -14$$

and

$$\begin{vmatrix} 4 & 0 \\ -6 & 3 \end{vmatrix} = (4)(3) - (0)(-6) = 12$$

**EXAMPLE 1**  **Finding the Cross Product**

For $\mathbf{u} = \mathbf{i} - 2\mathbf{j} + \mathbf{k}$ and $\mathbf{v} = 3\mathbf{i} + \mathbf{j} - 2\mathbf{k}$, find each of the following.

**a.** $\mathbf{u} \times \mathbf{v}$   **b.** $\mathbf{v} \times \mathbf{u}$   **c.** $\mathbf{v} \times \mathbf{v}$

**Solution**

**a.** $\mathbf{u} \times \mathbf{v} = \begin{vmatrix} \mathbf{i} & \mathbf{j} & \mathbf{k} \\ 1 & -2 & 1 \\ 3 & 1 & -2 \end{vmatrix}$

$= \begin{vmatrix} -2 & 1 \\ 1 & -2 \end{vmatrix} \mathbf{i} - \begin{vmatrix} 1 & 1 \\ 3 & -2 \end{vmatrix} \mathbf{j} + \begin{vmatrix} 1 & -2 \\ 3 & 1 \end{vmatrix} \mathbf{k}$

$= (4 - 1)\mathbf{i} - (-2 - 3)\mathbf{j} + (1 + 6)\mathbf{k}$

$= 3\mathbf{i} + 5\mathbf{j} + 7\mathbf{k}$

**b.** $\mathbf{v} \times \mathbf{u} = \begin{vmatrix} \mathbf{i} & \mathbf{j} & \mathbf{k} \\ 3 & 1 & -2 \\ 1 & -2 & 1 \end{vmatrix}$

$= \begin{vmatrix} 1 & -2 \\ -2 & 1 \end{vmatrix} \mathbf{i} - \begin{vmatrix} 3 & -2 \\ 1 & 1 \end{vmatrix} \mathbf{j} + \begin{vmatrix} 3 & 1 \\ 1 & -2 \end{vmatrix} \mathbf{k}$

$= (1 - 4)\mathbf{i} - (3 + 2)\mathbf{j} + (-6 - 1)\mathbf{k}$

$= -3\mathbf{i} - 5\mathbf{j} - 7\mathbf{k}$

> **· · REMARK** Note that this result is the negative of that in part (a).
> · · · · · · · · · · · · · · · ▷

**c.** $\mathbf{v} \times \mathbf{v} = \begin{vmatrix} \mathbf{i} & \mathbf{j} & \mathbf{k} \\ 3 & 1 & -2 \\ 3 & 1 & -2 \end{vmatrix} = \mathbf{0}$

The results obtained in Example 1 suggest some interesting *algebraic* properties of the cross product. For instance, $\mathbf{u} \times \mathbf{v} = -(\mathbf{v} \times \mathbf{u})$, and $\mathbf{v} \times \mathbf{v} = \mathbf{0}$. These properties, and several others, are summarized in the next theorem.

---

**THEOREM 11.7  Algebraic Properties of the Cross Product**

Let $\mathbf{u}$, $\mathbf{v}$, and $\mathbf{w}$ be vectors in space, and let $c$ be a scalar.

**1.** $\mathbf{u} \times \mathbf{v} = -(\mathbf{v} \times \mathbf{u})$

**2.** $\mathbf{u} \times (\mathbf{v} + \mathbf{w}) = (\mathbf{u} \times \mathbf{v}) + (\mathbf{u} \times \mathbf{w})$

**3.** $c(\mathbf{u} \times \mathbf{v}) = (c\mathbf{u}) \times \mathbf{v} = \mathbf{u} \times (c\mathbf{v})$

**4.** $\mathbf{u} \times \mathbf{0} = \mathbf{0} \times \mathbf{u} = \mathbf{0}$

**5.** $\mathbf{u} \times \mathbf{u} = \mathbf{0}$

**6.** $\mathbf{u} \cdot (\mathbf{v} \times \mathbf{w}) = (\mathbf{u} \times \mathbf{v}) \cdot \mathbf{w}$

---

**Proof**  To prove Property 1, let $\mathbf{u} = u_1\mathbf{i} + u_2\mathbf{j} + u_3\mathbf{k}$ and $\mathbf{v} = v_1\mathbf{i} + v_2\mathbf{j} + v_3\mathbf{k}$. Then

$$\mathbf{u} \times \mathbf{v} = (u_2v_3 - u_3v_2)\mathbf{i} - (u_1v_3 - u_3v_1)\mathbf{j} + (u_1v_2 - u_2v_1)\mathbf{k}$$

and

$$\mathbf{v} \times \mathbf{u} = (v_2u_3 - v_3u_2)\mathbf{i} - (v_1u_3 - v_3u_1)\mathbf{j} + (v_1u_2 - v_2u_1)\mathbf{k}$$

which implies that $\mathbf{u} \times \mathbf{v} = -(\mathbf{v} \times \mathbf{u})$. Proofs of Properties 2, 3, 5, and 6 are left as exercises (see Exercises 47–50).

Note that Property 1 of Theorem 11.7 indicates that the cross product is *not commutative*. In particular, this property indicates that the vectors $\mathbf{u} \times \mathbf{v}$ and $\mathbf{v} \times \mathbf{u}$ have equal lengths but opposite directions. The next theorem lists some other *geometric* properties of the cross product of two vectors.

---

**THEOREM 11.8  Geometric Properties of the Cross Product**

Let $\mathbf{u}$ and $\mathbf{v}$ be nonzero vectors in space, and let $\theta$ be the angle between $\mathbf{u}$ and $\mathbf{v}$.

1. $\mathbf{u} \times \mathbf{v}$ is orthogonal to both $\mathbf{u}$ and $\mathbf{v}$.
2. $\|\mathbf{u} \times \mathbf{v}\| = \|\mathbf{u}\|\,\|\mathbf{v}\| \sin \theta$
3. $\mathbf{u} \times \mathbf{v} = \mathbf{0}$ if and only if $\mathbf{u}$ and $\mathbf{v}$ are scalar multiples of each other.
4. $\|\mathbf{u} \times \mathbf{v}\|$ = area of parallelogram having $\mathbf{u}$ and $\mathbf{v}$ as adjacent sides.

---

·· **REMARK**  It follows from Properties 1 and 2 in Theorem 11.8 that if $\mathbf{n}$ is a unit vector orthogonal to both $\mathbf{u}$ and $\mathbf{v}$, then

$$\mathbf{u} \times \mathbf{v} = \pm(\|\mathbf{u}\|\,\|\mathbf{v}\| \sin \theta)\mathbf{n}.$$

**Proof**  To prove Property 2, note because $\cos \theta = (\mathbf{u} \cdot \mathbf{v})/(\|\mathbf{u}\|\,\|\mathbf{v}\|)$, it follows that

$$\|\mathbf{u}\|\,\|\mathbf{v}\| \sin \theta = \|\mathbf{u}\|\,\|\mathbf{v}\| \sqrt{1 - \cos^2 \theta}$$

$$= \|\mathbf{u}\|\,\|\mathbf{v}\| \sqrt{1 - \frac{(\mathbf{u} \cdot \mathbf{v})^2}{\|\mathbf{u}\|^2\,\|\mathbf{v}\|^2}}$$

$$= \sqrt{\|\mathbf{u}\|^2\,\|\mathbf{v}\|^2 - (\mathbf{u} \cdot \mathbf{v})^2}$$

$$= \sqrt{(u_1^2 + u_2^2 + u_3^2)(v_1^2 + v_2^2 + v_3^2) - (u_1 v_1 + u_2 v_2 + u_3 v_3)^2}$$

$$= \sqrt{(u_2 v_3 - u_3 v_2)^2 + (u_1 v_3 - u_3 v_1)^2 + (u_1 v_2 - u_2 v_1)^2}$$

$$= \|\mathbf{u} \times \mathbf{v}\|.$$

To prove Property 4, refer to Figure 11.35, which is a parallelogram having $\mathbf{v}$ and $\mathbf{u}$ as adjacent sides. Because the height of the parallelogram is $\|\mathbf{v}\| \sin \theta$, the area is

$$\text{Area} = (\text{base})(\text{height})$$
$$= \|\mathbf{u}\|\,\|\mathbf{v}\| \sin \theta$$
$$= \|\mathbf{u} \times \mathbf{v}\|.$$

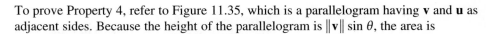

Proofs of Properties 1 and 3 are left as exercises (see Exercises 51 and 52).

Both $\mathbf{u} \times \mathbf{v}$ and $\mathbf{v} \times \mathbf{u}$ are perpendicular to the plane determined by $\mathbf{u}$ and $\mathbf{v}$. One way to remember the orientations of the vectors $\mathbf{u}$, $\mathbf{v}$, and $\mathbf{u} \times \mathbf{v}$ is to compare them with the unit vectors $\mathbf{i}$, $\mathbf{j}$, and $\mathbf{k} = \mathbf{i} \times \mathbf{j}$, as shown in Figure 11.36. The three vectors $\mathbf{u}$, $\mathbf{v}$, and $\mathbf{u} \times \mathbf{v}$ form a *right-handed system*, whereas the three vectors $\mathbf{u}$, $\mathbf{v}$, and $\mathbf{v} \times \mathbf{u}$ form a *left-handed system*.

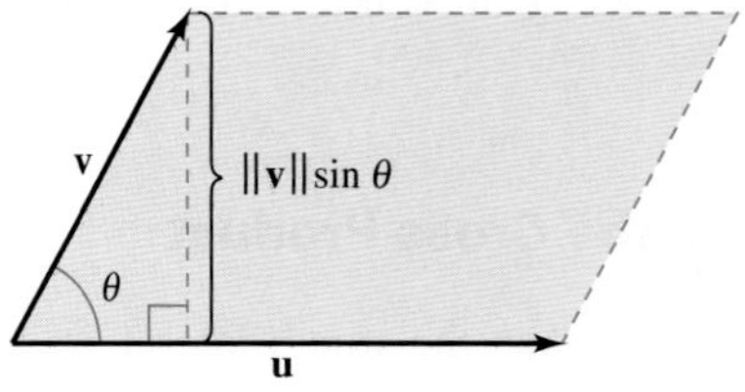

The vectors $\mathbf{u}$ and $\mathbf{v}$ form adjacent sides of a parallelogram.
**Figure 11.35**

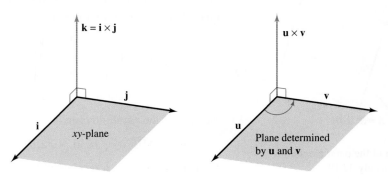

Right-handed systems
**Figure 11.36**

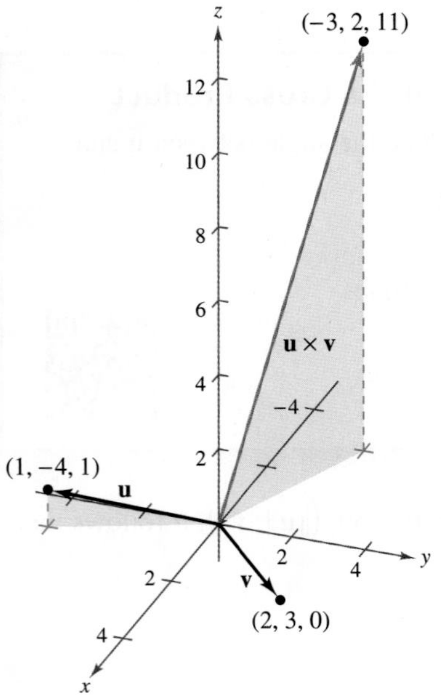

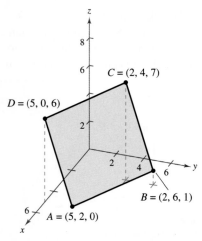

The vector $\mathbf{u} \times \mathbf{v}$ is orthogonal to both $\mathbf{u}$ and $\mathbf{v}$.

**Figure 11.37**

---

**EXAMPLE 2** **Using the Cross Product**

$\cdots\triangleright$ *See LarsonCalculus.com for an interactive version of this type of example.*

Find a unit vector that is orthogonal to both

$$\mathbf{u} = \mathbf{i} - 4\mathbf{j} + \mathbf{k}$$

and

$$\mathbf{v} = 2\mathbf{i} + 3\mathbf{j}.$$

**Solution** The cross product $\mathbf{u} \times \mathbf{v}$, as shown in Figure 11.37, is orthogonal to both $\mathbf{u}$ and $\mathbf{v}$.

$$\mathbf{u} \times \mathbf{v} = \begin{vmatrix} \mathbf{i} & \mathbf{j} & \mathbf{k} \\ 1 & -4 & 1 \\ 2 & 3 & 0 \end{vmatrix} \qquad \text{Cross product}$$

$$= -3\mathbf{i} + 2\mathbf{j} + 11\mathbf{k}$$

Because

$$\|\mathbf{u} \times \mathbf{v}\| = \sqrt{(-3)^2 + 2^2 + 11^2} = \sqrt{134}$$

a unit vector orthogonal to both $\mathbf{u}$ and $\mathbf{v}$ is

$$\frac{\mathbf{u} \times \mathbf{v}}{\|\mathbf{u} \times \mathbf{v}\|} = -\frac{3}{\sqrt{134}}\mathbf{i} + \frac{2}{\sqrt{134}}\mathbf{j} + \frac{11}{\sqrt{134}}\mathbf{k}.$$

In Example 2, note that you could have used the cross product $\mathbf{v} \times \mathbf{u}$ to form a unit vector that is orthogonal to both $\mathbf{u}$ and $\mathbf{v}$. With that choice, you would have obtained the negative of the unit vector found in the example.

---

**EXAMPLE 3** **Geometric Application of the Cross Product**

The vertices of a quadrilateral are listed below. Show that the quadrilateral is a parallelogram and find its area.

$$A = (5, 2, 0) \qquad B = (2, 6, 1)$$
$$C = (2, 4, 7) \qquad D = (5, 0, 6)$$

**Solution** From Figure 11.38, you can see that the sides of the quadrilateral correspond to the following four vectors.

$$\overrightarrow{AB} = -3\mathbf{i} + 4\mathbf{j} + \mathbf{k} \qquad \overrightarrow{CD} = 3\mathbf{i} - 4\mathbf{j} - \mathbf{k} = -\overrightarrow{AB}$$
$$\overrightarrow{AD} = 0\mathbf{i} - 2\mathbf{j} + 6\mathbf{k} \qquad \overrightarrow{CB} = 0\mathbf{i} + 2\mathbf{j} - 6\mathbf{k} = -\overrightarrow{AD}$$

So, $\overrightarrow{AB}$ is parallel to $\overrightarrow{CD}$ and $\overrightarrow{AD}$ is parallel to $\overrightarrow{CB}$, and you can conclude that the quadrilateral is a parallelogram with $\overrightarrow{AB}$ and $\overrightarrow{AD}$ as adjacent sides. Moreover, because

$$\overrightarrow{AB} \times \overrightarrow{AD} = \begin{vmatrix} \mathbf{i} & \mathbf{j} & \mathbf{k} \\ -3 & 4 & 1 \\ 0 & -2 & 6 \end{vmatrix} \qquad \text{Cross product}$$

$$= 26\mathbf{i} + 18\mathbf{j} + 6\mathbf{k}$$

the area of the parallelogram is

$$\|\overrightarrow{AB} \times \overrightarrow{AD}\| = \sqrt{1036} \approx 32.19.$$

Is the parallelogram a rectangle? You can determine whether it is by finding the angle between the vectors $\overrightarrow{AB}$ and $\overrightarrow{AD}$.

The area of the parallelogram is approximately 32.19.

**Figure 11.38**

In physics, the cross product can be used to measure **torque**—the **moment M of a force F about a point *P*,** as shown in Figure 11.39. If the point of application of the force is *Q*, then the moment of **F** about *P* is

$$\mathbf{M} = \overrightarrow{PQ} \times \mathbf{F}.$$    Moment of **F** about *P*

The magnitude of the moment **M** measures the tendency of the vector $\overrightarrow{PQ}$ to rotate counterclockwise (using the right-hand rule) about an axis directed along the vector **M**.

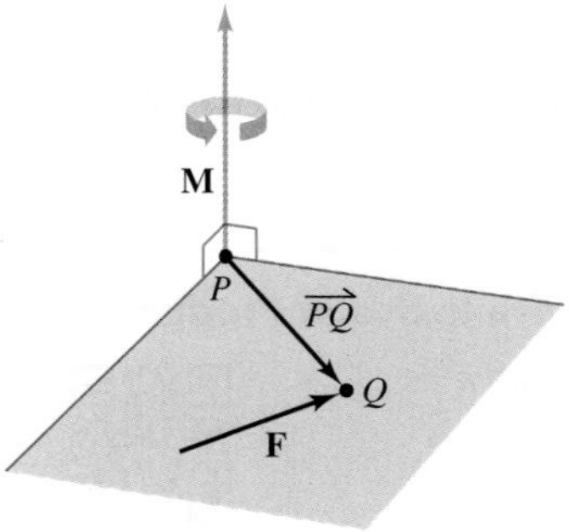

## EXAMPLE 4   An Application of the Cross Product

A vertical force of 50 pounds is applied to the end of a one-foot lever that is attached to an axle at point *P*, as shown in Figure 11.40. Find the moment of this force about the point *P* when $\theta = 60°$.

**Solution**   Represent the 50-pound force as

$$\mathbf{F} = -50\mathbf{k}$$

and the lever as

$$\overrightarrow{PQ} = \cos(60°)\mathbf{j} + \sin(60°)\mathbf{k} = \frac{1}{2}\mathbf{j} + \frac{\sqrt{3}}{2}\mathbf{k}.$$

The moment of **F** about *P* is

$$\mathbf{M} = \overrightarrow{PQ} \times \mathbf{F} = \begin{vmatrix} \mathbf{i} & \mathbf{j} & \mathbf{k} \\ 0 & \dfrac{1}{2} & \dfrac{\sqrt{3}}{2} \\ 0 & 0 & -50 \end{vmatrix} = -25\mathbf{i}.$$    Moment of **F** about *P*

The magnitude of this moment is 25 foot-pounds.    ∎

In Example 4, note that the moment (the tendency of the lever to rotate about its axle) is dependent on the angle $\theta$. When $\theta = \pi/2$, the moment is 0. The moment is greatest when $\theta = 0$.

## The Triple Scalar Product

For vectors **u**, **v**, and **w** in space, the dot product of **u** and $\mathbf{v} \times \mathbf{w}$

$$\mathbf{u} \cdot (\mathbf{v} \times \mathbf{w})$$

is called the **triple scalar product,** as defined in Theorem 11.9. The proof of this theorem is left as an exercise (see Exercise 55).

---

### THEOREM 11.9   The Triple Scalar Product

For $\mathbf{u} = u_1\mathbf{i} + u_2\mathbf{j} + u_3\mathbf{k}$, $\mathbf{v} = v_1\mathbf{i} + v_2\mathbf{j} + v_3\mathbf{k}$, and $\mathbf{w} = w_1\mathbf{i} + w_2\mathbf{j} + w_3\mathbf{k}$, the triple scalar product is

$$\mathbf{u} \cdot (\mathbf{v} \times \mathbf{w}) = \begin{vmatrix} u_1 & u_2 & u_3 \\ v_1 & v_2 & v_3 \\ w_1 & w_2 & w_3 \end{vmatrix}.$$

---

Note that the value of a determinant is multiplied by $-1$ when two rows are interchanged. After two such interchanges, the value of the determinant will be unchanged. So, the following triple scalar products are equivalent.

$$\mathbf{u} \cdot (\mathbf{v} \times \mathbf{w}) = \mathbf{v} \cdot (\mathbf{w} \times \mathbf{u}) = \mathbf{w} \cdot (\mathbf{u} \times \mathbf{v})$$

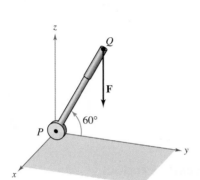

The moment of **F** about *P*
**Figure 11.39**

A vertical force of 50 pounds is applied at point *Q*.
**Figure 11.40**

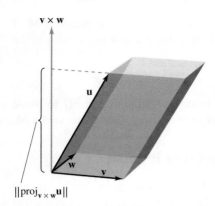

$\|\text{proj}_{\mathbf{v}\times\mathbf{w}}\mathbf{u}\|$

Area of base $= \|\mathbf{v} \times \mathbf{w}\|$
Volume of parallelepiped $= |\mathbf{u} \cdot (\mathbf{v} \times \mathbf{w})|$
**Figure 11.41**

If the vectors **u**, **v**, and **w** do not lie in the same plane, then the triple scalar product **u** · (**v** × **w**) can be used to determine the volume of the parallelepiped (a polyhedron, all of whose faces are parallelograms) with **u**, **v**, and **w** as adjacent edges, as shown in Figure 11.41. This is established in the next theorem.

---

### THEOREM 11.10 Geometric Property of the Triple Scalar Product

The volume $V$ of a parallelepiped with vectors **u**, **v**, and **w** as adjacent edges is

$$V = |\mathbf{u} \cdot (\mathbf{v} \times \mathbf{w})|.$$

---

**Proof** In Figure 11.41, note that the area of the base is $\|\mathbf{v} \times \mathbf{w}\|$ and the height of the parallelepiped is $\|\text{proj}_{\mathbf{v}\times\mathbf{w}}\mathbf{u}\|$. Therefore, the volume is

$$V = (\text{height})(\text{area of base})$$
$$= \|\text{proj}_{\mathbf{v}\times\mathbf{w}}\mathbf{u}\| \, \|\mathbf{v} \times \mathbf{w}\|$$
$$= \left| \frac{\mathbf{u} \cdot (\mathbf{v} \times \mathbf{w})}{\|\mathbf{v} \times \mathbf{w}\|} \right| \|\mathbf{v} \times \mathbf{w}\|$$
$$= |\mathbf{u} \cdot (\mathbf{v} \times \mathbf{w})|.$$

---

**EXAMPLE 5** **Volume by the Triple Scalar Product**

Find the volume of the parallelepiped shown in Figure 11.42 having

$$\mathbf{u} = 3\mathbf{i} - 5\mathbf{j} + \mathbf{k}$$
$$\mathbf{v} = 2\mathbf{j} - 2\mathbf{k}$$

and

$$\mathbf{w} = 3\mathbf{i} + \mathbf{j} + \mathbf{k}$$

as adjacent edges.

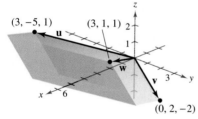

The parallelepiped has a volume of 36.
**Figure 11.42**

**Solution** By Theorem 11.10, you have

$$V = |\mathbf{u} \cdot (\mathbf{v} \times \mathbf{w})| \qquad \text{Triple scalar product}$$
$$= \begin{vmatrix} 3 & -5 & 1 \\ 0 & 2 & -2 \\ 3 & 1 & 1 \end{vmatrix}$$
$$= 3\begin{vmatrix} 2 & -2 \\ 1 & 1 \end{vmatrix} - (-5)\begin{vmatrix} 0 & -2 \\ 3 & 1 \end{vmatrix} + (1)\begin{vmatrix} 0 & 2 \\ 3 & 1 \end{vmatrix}$$
$$= 3(4) + 5(6) + 1(-6)$$
$$= 36.$$

A natural consequence of Theorem 11.10 is that the volume of the parallelepiped is 0 if and only if the three vectors are coplanar. That is, when the vectors $\mathbf{u} = \langle u_1, u_2, u_3 \rangle$, $\mathbf{v} = \langle v_1, v_2, v_3 \rangle$, and $\mathbf{w} = \langle w_1, w_2, w_3 \rangle$ have the same initial point, they lie in the same plane if and only if

$$\mathbf{u} \cdot (\mathbf{v} \times \mathbf{w}) = \begin{vmatrix} u_1 & u_2 & u_3 \\ v_1 & v_2 & v_3 \\ w_1 & w_2 & w_3 \end{vmatrix} = 0.$$

# 11.4 Exercises

See CalcChat.com for tutorial help and worked-out solutions to odd-numbered exercises.

**CONCEPT CHECK**

1. **Vectors** Explain what $\mathbf{u} \times \mathbf{v}$ represents geometrically.

2. **Area** Explain how to find the area of a parallelogram using vectors.

**Cross Product of Unit Vectors** In Exercises 3–6, find the cross product of the unit vectors and sketch your result.

3. $\mathbf{j} \times \mathbf{i}$

4. $\mathbf{j} \times \mathbf{k}$

5. $\mathbf{i} \times \mathbf{k}$

6. $\mathbf{k} \times \mathbf{i}$

 **Finding Cross Products** In Exercises 7–10, find (a) $\mathbf{u} \times \mathbf{v}$, (b) $\mathbf{v} \times \mathbf{u}$, and (c) $\mathbf{v} \times \mathbf{v}$.

7. $\mathbf{u} = -2\mathbf{i} + 4\mathbf{j}$
   $\mathbf{v} = 3\mathbf{i} + 2\mathbf{j} + 5\mathbf{k}$

8. $\mathbf{u} = 3\mathbf{i} + 5\mathbf{k}$
   $\mathbf{v} = 2\mathbf{i} + 3\mathbf{j} - 2\mathbf{k}$

9. $\mathbf{u} = \langle 7, 3, 2 \rangle$
   $\mathbf{v} = \langle 1, -1, 5 \rangle$

10. $\mathbf{u} = \langle 2, 1, -9 \rangle$
    $\mathbf{v} = \langle -6, -2, -1 \rangle$

**Finding a Cross Product** In Exercises 11–14, find $\mathbf{u} \times \mathbf{v}$ and show that it is orthogonal to both $\mathbf{u}$ and $\mathbf{v}$.

11. $\mathbf{u} = \langle 4, -1, 0 \rangle$
    $\mathbf{v} = \langle -6, 3, 0 \rangle$

12. $\mathbf{u} = \langle -5, 2, 2 \rangle$
    $\mathbf{v} = \langle 0, 1, 8 \rangle$

13. $\mathbf{u} = \mathbf{i} + \mathbf{j} + \mathbf{k}$
    $\mathbf{v} = 2\mathbf{i} + \mathbf{j} - \mathbf{k}$

14. $\mathbf{u} = \mathbf{i} + 6\mathbf{j}$
    $\mathbf{v} = -2\mathbf{i} + \mathbf{j} + \mathbf{k}$

 **Finding a Unit Vector** In Exercises 15–18, find a unit vector that is orthogonal to both $\mathbf{u}$ and $\mathbf{v}$.

15. $\mathbf{u} = \langle 4, -3, 1 \rangle$
    $\mathbf{v} = \langle 2, 5, 3 \rangle$

16. $\mathbf{u} = \langle -8, -6, 4 \rangle$
    $\mathbf{v} = \langle 10, -12, -2 \rangle$

17. $\mathbf{u} = -3\mathbf{i} + 2\mathbf{j} - 5\mathbf{k}$
    $\mathbf{v} = \mathbf{i} - \mathbf{j} + 4\mathbf{k}$

18. $\mathbf{u} = 2\mathbf{k}$
    $\mathbf{v} = 4\mathbf{i} + 6\mathbf{k}$

**Area** In Exercises 19–22, find the area of the parallelogram that has the given vectors as adjacent sides. Use a computer algebra system or a graphing utility to verify your result.

19. $\mathbf{u} = \mathbf{j}$
    $\mathbf{v} = \mathbf{j} + \mathbf{k}$

20. $\mathbf{u} = \mathbf{i} + \mathbf{j} + \mathbf{k}$
    $\mathbf{v} = \mathbf{j} + \mathbf{k}$

21. $\mathbf{u} = \langle 3, 2, -1 \rangle$
    $\mathbf{v} = \langle 1, 2, 3 \rangle$

22. $\mathbf{u} = \langle 2, -1, 0 \rangle$
    $\mathbf{v} = \langle -1, 2, 0 \rangle$

 **Area** In Exercises 23 and 24, verify that the points are the vertices of a parallelogram, and find its area.

23. $A(0, 3, 2)$, $B(1, 5, 5)$, $C(6, 9, 5)$, $D(5, 7, 2)$

24. $A(2, -3, 1)$, $B(6, 5, -1)$, $C(7, 2, 2)$, $D(3, -6, 4)$

**Area** In Exercises 25 and 26, find the area of the triangle with the given vertices. $\left( Hint:\ \frac{1}{2}\|\mathbf{u} \times \mathbf{v}\| \text{ is the area of the triangle having } \mathbf{u} \text{ and } \mathbf{v} \text{ as adjacent sides.}\right)$

25. $A(0, 0, 0)$, $B(1, 0, 3)$, $C(-3, 2, 0)$

26. $A(2, -3, 4)$, $B(0, 1, 2)$, $C(-1, 2, 0)$

27. **Torque** The brakes on a bicycle are applied using a downward force of 20 pounds on the pedal when the crank makes a $40°$ angle with the horizontal (see figure). The crank is 6 inches in length. Find the torque at $P$.

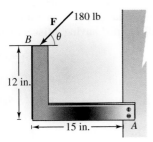

Figure for 27          Figure for 28

28. **Torque** Both the magnitude and the direction of the force on a crankshaft change as the crankshaft rotates. Find the torque on the crankshaft using the position and data shown in the figure.

29. **Optimization** A force of 180 pounds acts on the bracket shown in the figure.

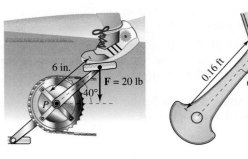

(a) Determine the vector $\overrightarrow{AB}$ and the vector $\mathbf{F}$ representing the force. ($\mathbf{F}$ will be in terms of $\theta$.)

(b) Find the magnitude of the moment about $A$ by evaluating $\|\overrightarrow{AB} \times \mathbf{F}\|$.

(c) Use the result of part (b) to determine the magnitude of the moment when $\theta = 30°$.

(d) Use the result of part (b) to determine the angle $\theta$ when the magnitude of the moment is maximum. At that angle, what is the relationship between the vectors $\mathbf{F}$ and $\overrightarrow{AB}$? Is it what you expected? Why or why not?

(e) Use a graphing utility to graph the function for the magnitude of the moment about $A$ for $0° \leq \theta \leq 180°$. Find the zero of the function in the given domain. Interpret the meaning of the zero in the context of the problem.

**30. Optimization** A force of 56 pounds acts on the pipe wrench shown in the figure.

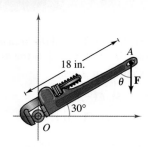

18 in.

$A$

$\theta$ **F**

30°

$O$

 (a) Find the magnitude of the moment about $O$ by evaluating $\|\overrightarrow{OA} \times \mathbf{F}\|$. Use a graphing utility to graph the resulting function of $\theta$.

(b) Use the result of part (a) to determine the magnitude of the moment when $\theta = 45°$.

(c) Use the result of part (a) to determine the angle $\theta$ when the magnitude of the moment is maximum. Is the answer what you expected? Why or why not?

**Finding a Triple Scalar Product** In Exercises 31–34, find $\mathbf{u} \cdot (\mathbf{v} \times \mathbf{w})$.

31. $\mathbf{u} = \mathbf{i}$
    $\mathbf{v} = \mathbf{j}$
    $\mathbf{w} = \mathbf{k}$

32. $\mathbf{u} = \langle 1, 1, 1 \rangle$
    $\mathbf{v} = \langle 2, 1, 0 \rangle$
    $\mathbf{w} = \langle 0, 0, 1 \rangle$

33. $\mathbf{u} = \langle 2, 0, 1 \rangle$
    $\mathbf{v} = \langle 0, 3, 0 \rangle$
    $\mathbf{w} = \langle 0, 0, 1 \rangle$

34. $\mathbf{u} = \langle 2, 0, 0 \rangle$
    $\mathbf{v} = \langle 1, 1, 1 \rangle$
    $\mathbf{w} = \langle 0, 2, 2 \rangle$

**Volume** In Exercises 35 and 36, use the triple scalar product to find the volume of the parallelepiped having adjacent edges u, v, and w.

35. $\mathbf{u} = \mathbf{i} + \mathbf{j}$
    $\mathbf{v} = \mathbf{j} + \mathbf{k}$
    $\mathbf{w} = \mathbf{i} + \mathbf{k}$

36. $\mathbf{u} = \langle 1, 3, 1 \rangle$
    $\mathbf{v} = \langle 0, 6, 6 \rangle$
    $\mathbf{w} = \langle -4, 0, -4 \rangle$

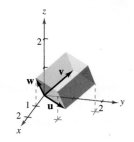

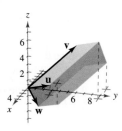

**Volume** In Exercises 37 and 38, find the volume of the parallelepiped with the given vertices.

37. $(0, 0, 0)$, $(3, 0, 0)$, $(0, 5, 1)$, $(2, 0, 5)$, $(3, 5, 1)$, $(5, 0, 5)$, $(2, 5, 6)$, $(5, 5, 6)$

38. $(0, 0, 0)$, $(0, 4, 0)$, $(-3, 0, 0)$, $(-1, 1, 5)$, $(-3, 4, 0)$, $(-1, 5, 5)$, $(-4, 1, 5)$, $(-4, 5, 5)$

**EXPLORING CONCEPTS**

**39. Comparing Dot Products** Identify the dot products that are equal. Explain your reasoning. (Assume **u**, **v**, and **w** are nonzero vectors.)

(a) $\mathbf{u} \cdot (\mathbf{v} \times \mathbf{w})$
(b) $(\mathbf{v} \times \mathbf{w}) \cdot \mathbf{u}$
(c) $(\mathbf{u} \times \mathbf{v}) \cdot \mathbf{w}$
(d) $(\mathbf{u} \times -\mathbf{w}) \cdot \mathbf{v}$
(e) $\mathbf{u} \cdot (\mathbf{w} \times \mathbf{v})$
(f) $\mathbf{w} \cdot (\mathbf{v} \times \mathbf{u})$
(g) $(-\mathbf{u} \times \mathbf{v}) \cdot \mathbf{w}$
(h) $(\mathbf{w} \times \mathbf{u}) \cdot \mathbf{v}$

**40. Using Dot and Cross Products** When $\mathbf{u} \times \mathbf{v} = 0$ and $\mathbf{u} \cdot \mathbf{v} = 0$, what can you conclude about **u** and **v**?

**41. Cross Product** Two nonzero vectors lie in the $yz$-plane. Where does the cross product of the vectors lie? Explain.

**42.** **HOW DO YOU SEE IT?** The vertices of a triangle in space are $(x_1, y_1, z_1)$, $(x_2, y_2, z_2)$, and $(x_3, y_3, z_3)$. Explain how to find a vector perpendicular to the triangle.

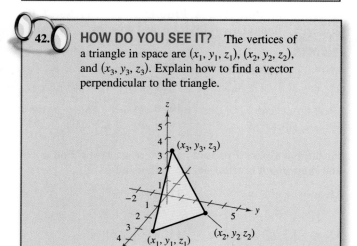

**True or False?** In Exercises 43–46, determine whether the statement is true or false. If it is false, explain why or give an example that shows it is false.

43. It is possible to find the cross product of two vectors in a two-dimensional coordinate system.

44. The cross product of two nonzero vectors is a nonzero vector.

45. If $\mathbf{u} \neq 0$ and $\mathbf{u} \times \mathbf{v} = \mathbf{u} \times \mathbf{w}$, then $\mathbf{v} = \mathbf{w}$.

46. If $\mathbf{u} \neq 0$, $\mathbf{u} \cdot \mathbf{v} = \mathbf{u} \cdot \mathbf{w}$, and $\mathbf{u} \times \mathbf{v} = \mathbf{u} \times \mathbf{w}$, then $\mathbf{v} = \mathbf{w}$.

**Proof** In Exercises 47–52, prove the property of the cross product.

47. $\mathbf{u} \times (\mathbf{v} + \mathbf{w}) = (\mathbf{u} \times \mathbf{v}) + (\mathbf{u} \times \mathbf{w})$

48. $c(\mathbf{u} \times \mathbf{v}) = (c\mathbf{u}) \times \mathbf{v} = \mathbf{u} \times (c\mathbf{v})$

49. $\mathbf{u} \times \mathbf{u} = 0$

50. $\mathbf{u} \cdot (\mathbf{v} \times \mathbf{w}) = (\mathbf{u} \times \mathbf{v}) \cdot \mathbf{w}$

51. $\mathbf{u} \times \mathbf{v}$ is orthogonal to both **u** and **v**.

52. $\mathbf{u} \times \mathbf{v} = 0$ if and only if **u** and **v** are scalar multiples of each other.

**53. Proof** Prove that $\|\mathbf{u} \times \mathbf{v}\| = \|\mathbf{u}\| \|\mathbf{v}\|$ if **u** and **v** are orthogonal.

**54. Proof** Prove that $\mathbf{u} \times (\mathbf{v} \times \mathbf{w}) = (\mathbf{u} \cdot \mathbf{w})\mathbf{v} - (\mathbf{u} \cdot \mathbf{v})\mathbf{w}$.

**55. Proof** Prove Theorem 11.9.

# 11.5 Lines and Planes in Space

■ Write a set of parametric equations for a line in space.
■ Write a linear equation to represent a plane in space.
■ Sketch the plane given by a linear equation.
■ Find the distances between points, planes, and lines in space.

## Lines in Space

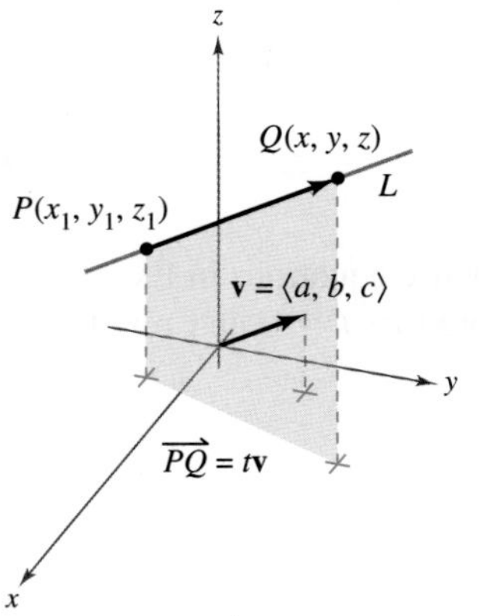

In the plane, *slope* is used to determine the equation of a line. In space, it is more convenient to use *vectors* to determine the equation of a line.

In Figure 11.43, consider the line $L$ through the point $P(x_1, y_1, z_1)$ and parallel to the vector $\mathbf{v} = \langle a, b, c \rangle$. The vector $\mathbf{v}$ is a **direction vector** for the line $L$, and $a$, $b$, and $c$ are **direction numbers.** One way of describing the line $L$ is to say that it consists of all points $Q(x, y, z)$ for which the vector $\overrightarrow{PQ}$ is parallel to $\mathbf{v}$. This means that $\overrightarrow{PQ}$ is a scalar multiple of $\mathbf{v}$ and you can write $\overrightarrow{PQ} = t\mathbf{v}$, where $t$ is a scalar (a real number).

$$\overrightarrow{PQ} = \langle x - x_1, y - y_1, z - z_1 \rangle = \langle at, bt, ct \rangle = t\mathbf{v}$$

By equating corresponding components, you can obtain **parametric equations** of a line in space.

Line $L$ and its direction vector $\mathbf{v}$
**Figure 11.43**

### THEOREM 11.11 Parametric Equations of a Line in Space

A line $L$ parallel to the vector $\mathbf{v} = \langle a, b, c \rangle$ and passing through the point $P(x_1, y_1, z_1)$ is represented by the **parametric equations**

$$x = x_1 + at, \quad y = y_1 + bt, \quad \text{and} \quad z = z_1 + ct.$$

If the direction numbers $a$, $b$, and $c$ are all nonzero, then you can eliminate the parameter $t$ in the parametric equations to obtain **symmetric equations** of the line.

$$\frac{x - x_1}{a} = \frac{y - y_1}{b} = \frac{z - z_1}{c} \qquad \text{Symmetric equations}$$

## EXAMPLE 1 Finding Parametric and Symmetric Equations

Find parametric and symmetric equations of the line $L$ that passes through the point $(1, -2, 4)$ and is parallel to $\mathbf{v} = \langle 2, 4, -4 \rangle$, as shown in Figure 11.44.

**Solution** To find a set of parametric equations of the line, use the coordinates $x_1 = 1$, $y_1 = -2$, and $z_1 = 4$ and direction numbers $a = 2$, $b = 4$, and $c = -4$.

$$x = 1 + 2t, \quad y = -2 + 4t, \quad z = 4 - 4t \qquad \text{Parametric equations}$$

Because $a$, $b$, and $c$ are all nonzero, a set of symmetric equations is

$$\frac{x - 1}{2} = \frac{y + 2}{4} = \frac{z - 4}{-4}. \qquad \text{Symmetric equations}$$

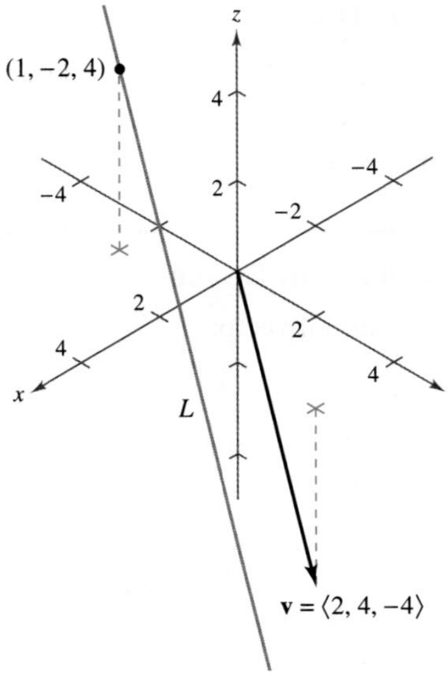

The vector $\mathbf{v}$ is parallel to the line $L$.
**Figure 11.44**

Neither parametric equations nor symmetric equations of a given line are unique. For instance, in Example 1, by letting $t = 1$ in the parametric equations, you would obtain the point $(3, 2, 0)$. Using this point with the direction numbers $a = 2$, $b = 4$, and $c = -4$ would produce a different set of parametric equations

$$x = 3 + 2t, \quad y = 2 + 4t, \quad \text{and} \quad z = -4t.$$

| EXAMPLE 2 | **Parametric Equations of a Line Through Two Points** |

·····▷ *See LarsonCalculus.com for an interactive version of this type of example.*

Find a set of parametric equations of the line that passes through the points

$$(-2, 1, 0) \quad \text{and} \quad (1, 3, 5).$$

**Solution**  Begin by using the points $P(-2, 1, 0)$ and $Q(1, 3, 5)$ to find a direction vector for the line passing through $P$ and $Q$.

$$\mathbf{v} = \overrightarrow{PQ} = \langle 1 - (-2), 3 - 1, 5 - 0 \rangle = \langle 3, 2, 5 \rangle = \langle a, b, c \rangle$$

Using the direction numbers $a = 3$, $b = 2$, and $c = 5$ with the point $P(-2, 1, 0)$, you obtain the parametric equations

····▷      $x = -2 + 3t, \quad y = 1 + 2t, \quad \text{and} \quad z = 5t.$      ■

····· **REMARK**  As $t$ varies over all real numbers, the parametric equations in Example 2 determine the points $(x, y, z)$ on the line. In particular, note that $t = 0$ and $t = 1$ give the original points $(-2, 1, 0)$ and $(1, 3, 5)$.

## Planes in Space

You have seen how an equation of a line in space can be obtained from a point on the line and a vector *parallel* to it. You will now see that an equation of a plane in space can be obtained from a point in the plane and a vector *normal* (perpendicular) to the plane.

Consider the plane containing the point $P(x_1, y_1, z_1)$ having a nonzero normal vector

$$\mathbf{n} = \langle a, b, c \rangle$$

as shown in Figure 11.45. This plane consists of all points $Q(x, y, z)$ for which vector $\overrightarrow{PQ}$ is orthogonal to $\mathbf{n}$. Using the dot product, you can write the following.

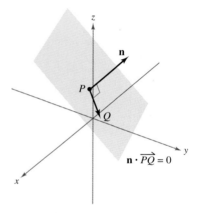

The normal vector $\mathbf{n}$ is orthogonal to each vector $\overrightarrow{PQ}$ in the plane.
**Figure 11.45**

$$\mathbf{n} \cdot \overrightarrow{PQ} = 0$$
$$\langle a, b, c \rangle \cdot \langle x - x_1, y - y_1, z - z_1 \rangle = 0$$
$$a(x - x_1) + b(y - y_1) + c(z - z_1) = 0$$

The third equation of the plane is said to be in **standard form.**

---

**THEOREM 11.12  Standard Equation of a Plane in Space**

The plane containing the point $(x_1, y_1, z_1)$ and having normal vector

$$\mathbf{n} = \langle a, b, c \rangle$$

can be represented by the **standard form** of the equation of a plane

$$a(x - x_1) + b(y - y_1) + c(z - z_1) = 0.$$

---

By regrouping terms in the standard form of the equation of a plane, you obtain the **general form.**

$$ax + by + cz + d = 0 \qquad \text{General form of equation of plane}$$

Given the general form of the equation of a plane, it is easy to find a normal vector to the plane. Simply use the coefficients of $x$, $y$, and $z$ and write $\mathbf{n} = \langle a, b, c \rangle$.

**EXAMPLE 3** **Finding an Equation of a Plane in Three-Space**

Find an equation (in standard form and in general form) of the plane containing the points

$$(2, 1, 1), \quad (1, 4, 1), \quad \text{and} \quad (-2, 0, 4).$$

**Solution** To apply Theorem 11.12, you need a point in the plane and a vector that is normal to the plane. There are three choices for the point, but no normal vector is given. To obtain a normal vector, use the cross product of vectors $\mathbf{u}$ and $\mathbf{v}$ extending from the point $(2, 1, 1)$ to the points $(1, 4, 1)$ and $(-2, 0, 4)$, as shown in Figure 11.46. The component forms of $\mathbf{u}$ and $\mathbf{v}$ are

$$\mathbf{u} = \langle 1 - 2, 4 - 1, 1 - 1 \rangle = \langle -1, 3, 0 \rangle$$

and

$$\mathbf{v} = \langle -2 - 2, 0 - 1, 4 - 1 \rangle = \langle -4, -1, 3 \rangle.$$

So, it follows that a vector normal to the given plane is

$$\mathbf{n} = \mathbf{u} \times \mathbf{v}$$
$$= \begin{vmatrix} \mathbf{i} & \mathbf{j} & \mathbf{k} \\ -1 & 3 & 0 \\ -4 & -1 & 3 \end{vmatrix}$$
$$= 9\mathbf{i} + 3\mathbf{j} + 13\mathbf{k}$$
$$= \langle a, b, c \rangle.$$

Using the direction numbers for $\mathbf{n}$ and the point $(x_1, y_1, z_1) = (2, 1, 1)$, you can determine an equation of the plane in standard form to be

$$a(x - x_1) + b(y - y_1) + c(z - z_1) = 0$$
$$9(x - 2) + 3(y - 1) + 13(z - 1) = 0. \qquad \text{Standard form}$$

By regrouping terms, the general form is

$$9x - 18 + 3y - 3 + 13z - 13 = 0$$
$$9x + 3y + 13z - 34 = 0. \qquad \text{General form}$$

**REMARK** In Example 3, check to see that each of the three original points satisfies the equation $9x + 3y + 13z - 34 = 0$.

Two distinct planes in three-space either are parallel or intersect in a line. For two planes that intersect, you can determine the angle $(0 \le \theta \le \pi/2)$ between them from the angle between their normal vectors, as shown in Figure 11.47. Specifically, if vectors $\mathbf{n}_1$ and $\mathbf{n}_2$ are normal to two intersecting planes, then the angle $\theta$ between the normal vectors is equal to the angle between the two planes and is

$$\cos \theta = \frac{|\mathbf{n}_1 \cdot \mathbf{n}_2|}{\|\mathbf{n}_1\| \|\mathbf{n}_2\|}. \qquad \text{Angle between two planes}$$

Consequently, two planes with normal vectors $\mathbf{n}_1$ and $\mathbf{n}_2$ are

1. *perpendicular* when $\mathbf{n}_1 \cdot \mathbf{n}_2 = 0$.
2. *parallel* when $\mathbf{n}_1$ is a scalar multiple of $\mathbf{n}_2$.

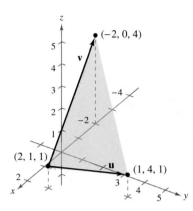

A plane determined by $\mathbf{u}$ and $\mathbf{v}$
**Figure 11.46**

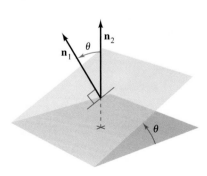

The angle $\theta$ between two planes
**Figure 11.47**

## EXAMPLE 4  Finding the Line of Intersection of Two Planes

Find the angle between the two planes $x - 2y + z = 0$ and $2x + 3y - 2z = 0$. Then find parametric equations of their line of intersection (see Figure 11.48).

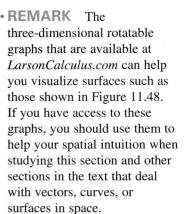

•• **REMARK**   The three-dimensional rotatable graphs that are available at *LarsonCalculus.com* can help you visualize surfaces such as those shown in Figure 11.48. If you have access to these graphs, you should use them to help your spatial intuition when studying this section and other sections in the text that deal with vectors, curves, or surfaces in space.

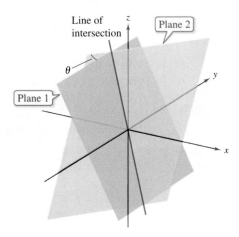

**Figure 11.48**

**Solution**   Normal vectors for the planes are $\mathbf{n}_1 = \langle 1, -2, 1 \rangle$ and $\mathbf{n}_2 = \langle 2, 3, -2 \rangle$. Consequently, the angle between the two planes is determined as follows.

$$\cos \theta = \frac{|\mathbf{n}_1 \cdot \mathbf{n}_2|}{\|\mathbf{n}_1\| \|\mathbf{n}_2\|} = \frac{|-6|}{\sqrt{6}\sqrt{17}} = \frac{6}{\sqrt{102}} \approx 0.59409$$

This implies that the angle between the two planes is $\theta \approx 53.55°$. You can find the line of intersection of the two planes by simultaneously solving the two linear equations representing the planes. One way to do this is to multiply the first equation by $-2$ and add the result to the second equation.

$$
\begin{array}{lll}
x - 2y + z = 0 & \Rightarrow & -2x + 4y - 2z = 0 \quad & \text{Multiply Equation 1 by } -2. \\
2x + 3y - 2z = 0 & \Rightarrow & \underline{\phantom{-}2x + 3y - 2z = 0} \quad & \text{Write Equation 2.} \\
& & 7y - 4z = 0 & \text{Add equations.} \\
& & y = \dfrac{4z}{7} & \text{Solve for } y.
\end{array}
$$

Substituting $y = 4z/7$ back into one of the original equations, you can determine that $x = z/7$. Finally, by letting $t = z/7$, you obtain the parametric equations

$$x = t, \quad y = 4t, \quad \text{and} \quad z = 7t \qquad \text{Line of intersection}$$

which indicate that 1, 4, and 7 are direction numbers for the line of intersection. ∎

Note that the direction numbers in Example 4 can be obtained from the cross product of the two normal vectors as follows.

$$
\mathbf{n}_1 \times \mathbf{n}_2 = \begin{vmatrix} \mathbf{i} & \mathbf{j} & \mathbf{k} \\ 1 & -2 & 1 \\ 2 & 3 & -2 \end{vmatrix}
$$

$$
= \begin{vmatrix} -2 & 1 \\ 3 & -2 \end{vmatrix} \mathbf{i} - \begin{vmatrix} 1 & 1 \\ 2 & -2 \end{vmatrix} \mathbf{j} + \begin{vmatrix} 1 & -2 \\ 2 & 3 \end{vmatrix} \mathbf{k}
$$

$$
= \mathbf{i} + 4\mathbf{j} + 7\mathbf{k}
$$

This means that the line of intersection of the two planes is parallel to the cross product of their normal vectors.

# Sketching Planes in Space

If a plane in space intersects one of the coordinate planes, then the line of intersection is called the **trace** of the given plane in the coordinate plane. To sketch a plane in space, it is helpful to find its points of intersection with the coordinate axes and its traces in the coordinate planes. For example, consider the plane

$$3x + 2y + 4z = 12. \qquad \text{Equation of plane}$$

You can find the $xy$-trace by letting $z = 0$ and sketching the line

$$3x + 2y = 12 \qquad xy\text{-trace}$$

in the $xy$-plane. This line intersects the $x$-axis at $(4, 0, 0)$ and the $y$-axis at $(0, 6, 0)$. In Figure 11.49, this process is continued by finding the $yz$-trace and the $xz$-trace and then shading the triangular region lying in the first octant.

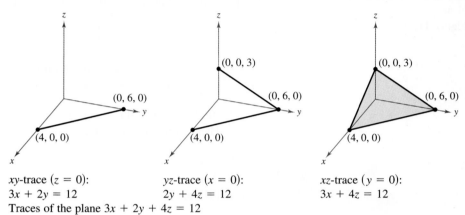

$xy$-trace $(z = 0)$:
$3x + 2y = 12$
Traces of the plane $3x + 2y + 4z = 12$
**Figure 11.49**

$yz$-trace $(x = 0)$:
$2y + 4z = 12$

$xz$-trace $(y = 0)$:
$3x + 4z = 12$

If an equation of a plane has a missing variable, such as

$$2x + z = 1$$

then the plane must be *parallel to the axis* represented by the missing variable, as shown in Figure 11.50. If two variables are missing from an equation of a plane, such as

$$ax + d = 0$$

then it is *parallel to the coordinate plane* represented by the missing variables, as shown in Figure 11.51.

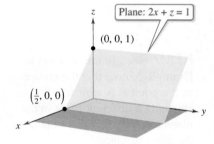

Plane $2x + z = 1$ is parallel to the $y$-axis.
**Figure 11.50**

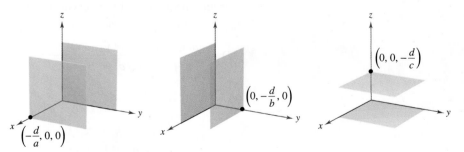

Plane $ax + d = 0$ is parallel to the $yz$-plane.
**Figure 11.51**

Plane $by + d = 0$ is parallel to the $xz$-plane.

Plane $cz + d = 0$ is parallel to the $xy$-plane.

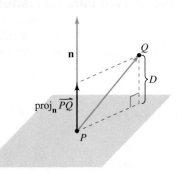

$$D = \|\text{proj}_\mathbf{n}\,\overrightarrow{PQ}\|$$

The distance between a point and a plane

**Figure 11.52**

## Distances Between Points, Planes, and Lines

Consider two types of problems involving distance in space: (1) finding the distance between a point and a plane and (2) finding the distance between a point and a line. The solutions of these problems illustrate the versatility and usefulness of vectors in coordinate geometry: the first problem uses the *dot product* of two vectors, and the second problem uses the *cross product*.

The distance $D$ between a point $Q$ and a plane is the length of the shortest line segment connecting $Q$ to the plane, as shown in Figure 11.52. For *any* point $P$ in the plane, you can find this distance by projecting the vector $\overrightarrow{PQ}$ onto the normal vector $\mathbf{n}$. The length of this projection is the desired distance.

---

**THEOREM 11.13  Distance Between a Point and a Plane**

The distance between a plane and a point $Q$ (not in the plane) is

$$D = \|\text{proj}_\mathbf{n}\overrightarrow{PQ}\| = \frac{|\overrightarrow{PQ} \cdot \mathbf{n}|}{\|\mathbf{n}\|}$$

where $P$ is a point in the plane and $\mathbf{n}$ is normal to the plane.

---

To find a point in the plane $ax + by + cz + d = 0$, where $a \neq 0$, let $y = 0$ and $z = 0$. Then, from the equation $ax + d = 0$, you can conclude that the point

$$\left(-\frac{d}{a}, 0, 0\right)$$

lies in the plane.

---

**EXAMPLE 5**  **Finding the Distance Between a Point and a Plane**

Find the distance between the point $Q(1, 5, -4)$ and the plane $3x - y + 2z = 6$.

**Solution**  You know that $\mathbf{n} = \langle 3, -1, 2 \rangle$ is normal to the plane. To find a point in the plane, let $y = 0$ and $z = 0$, and obtain the point $P(2, 0, 0)$. The vector from $P$ to $Q$ is

$$\overrightarrow{PQ} = \langle 1 - 2, 5 - 0, -4 - 0 \rangle$$
$$= \langle -1, 5, -4 \rangle.$$

Using the Distance Formula given in Theorem 11.13 produces

$$D = \frac{|\overrightarrow{PQ} \cdot \mathbf{n}|}{\|\mathbf{n}\|} = \frac{|\langle -1, 5, -4 \rangle \cdot \langle 3, -1, 2 \rangle|}{\sqrt{9 + 1 + 4}} = \frac{|-3 - 5 - 8|}{\sqrt{14}} = \frac{16}{\sqrt{14}} \approx 4.28.$$

**• • REMARK**  In the solution to Example 5, note that the choice of the point $P$ is arbitrary. Try choosing a different point in the plane to verify that you obtain the same distance.

From Theorem 11.13, you can determine that the distance between the point $Q(x_0, y_0, z_0)$ and the plane $ax + by + cz + d = 0$ is

$$D = \frac{|a(x_0 - x_1) + b(y_0 - y_1) + c(z_0 - z_1)|}{\sqrt{a^2 + b^2 + c^2}}$$

or

$$D = \frac{|ax_0 + by_0 + cz_0 + d|}{\sqrt{a^2 + b^2 + c^2}} \qquad \text{Distance between a point and a plane}$$

where $P(x_1, y_1, z_1)$ is a point in the plane and $d = -(ax_1 + by_1 + cz_1)$.

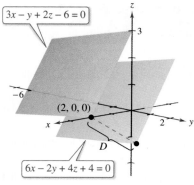

$3x - y + 2z - 6 = 0$

$(2, 0, 0)$

$D$

$6x - 2y + 4z + 4 = 0$

The distance between the parallel planes is approximately 2.14.
**Figure 11.53**

**EXAMPLE 6** **Finding the Distance Between Two Parallel Planes**

Two parallel planes, $3x - y + 2z - 6 = 0$ and $6x - 2y + 4z + 4 = 0$, are shown in Figure 11.53. To find the distance between the planes, choose a point in the first plane, such as $(x_0, y_0, z_0) = (2, 0, 0)$. Then, from the second plane, you can determine that $a = 6$, $b = -2$, $c = 4$, and $d = 4$ and conclude that the distance is

$$D = \frac{|ax_0 + by_0 + cz_0 + d|}{\sqrt{a^2 + b^2 + c^2}}$$

$$= \frac{|6(2) + (-2)(0) + (4)(0) + 4|}{\sqrt{6^2 + (-2)^2 + 4^2}}$$

$$= \frac{16}{\sqrt{56}} = \frac{8}{\sqrt{14}} \approx 2.14.$$

The formula for the distance between a point and a line in space resembles that for the distance between a point and a plane—except that you replace the dot product with the length of the cross product and the normal vector $\mathbf{n}$ with a direction vector for the line.

**THEOREM 11.14  Distance Between a Point and a Line in Space**

The distance between a point $Q$ and a line in space is

$$D = \frac{\|\overrightarrow{PQ} \times \mathbf{u}\|}{\|\mathbf{u}\|}$$

where $\mathbf{u}$ is a direction vector for the line and $P$ is a point on the line.

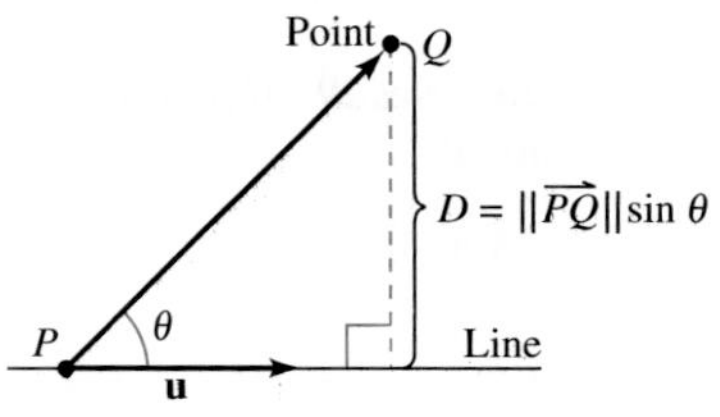

Point $Q$

$D = \|\overrightarrow{PQ}\| \sin \theta$

$P$ $\theta$

$\mathbf{u}$ Line

The distance between a point and a line
**Figure 11.54**

**Proof** In Figure 11.54, let $D$ be the distance between the point $Q$ and the line. Then $D = \|\overrightarrow{PQ}\| \sin \theta$, where $\theta$ is the angle between $\mathbf{u}$ and $\overrightarrow{PQ}$. By Property 2 of Theorem 11.8, you have $\|\mathbf{u}\| \|\overrightarrow{PQ}\| \sin \theta = \|\mathbf{u} \times \overrightarrow{PQ}\| = \|\overrightarrow{PQ} \times \mathbf{u}\|$. Consequently,

$$D = \|\overrightarrow{PQ}\| \sin \theta = \frac{\|\overrightarrow{PQ} \times \mathbf{u}\|}{\|\mathbf{u}\|}.$$

**EXAMPLE 7** **Finding the Distance Between a Point and a Line**

Find the distance between the point $Q(3, -1, 4)$ and the line

$$x = -2 + 3t, \quad y = -2t, \quad \text{and} \quad z = 1 + 4t.$$

**Solution** Using the direction numbers 3, $-2$, and 4, a direction vector for the line is $\mathbf{u} = \langle 3, -2, 4 \rangle$. To find a point on the line, let $t = 0$ and obtain $P = (-2, 0, 1)$. So,

$$\overrightarrow{PQ} = \langle 3 - (-2), -1 - 0, 4 - 1 \rangle = \langle 5, -1, 3 \rangle$$

and you can form the cross product

$$\overrightarrow{PQ} \times \mathbf{u} = \begin{vmatrix} \mathbf{i} & \mathbf{j} & \mathbf{k} \\ 5 & -1 & 3 \\ 3 & -2 & 4 \end{vmatrix} = 2\mathbf{i} - 11\mathbf{j} - 7\mathbf{k} = \langle 2, -11, -7 \rangle.$$

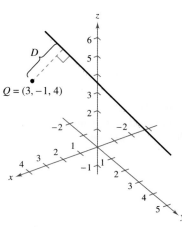

$Q = (3, -1, 4)$

The distance between the point $Q$ and the line is $\sqrt{6} \approx 2.45$.
**Figure 11.55**

Finally, using Theorem 11.14, you can find the distance to be

$$D = \frac{\|\overrightarrow{PQ} \times \mathbf{u}\|}{\|\mathbf{u}\|} = \frac{\sqrt{174}}{\sqrt{29}} = \sqrt{6} \approx 2.45.$$  See Figure 11.55.

# 11.5 Exercises

See **CalcChat.com** for tutorial help and worked-out solutions to odd-numbered exercises.

## CONCEPT CHECK

1. **Parametric and Symmetric Equations** Give the parametric equations and the symmetric equations of a line in space. Describe what is required to find these equations.

2. **Normal Vector** The equation of a plane in space is $2(x - 1) + 4(y - 3) - (z + 5) = 0$. What is the normal vector to this plane?

3. **Plane** Write an equation of a plane in space that is parallel to the x-axis.

4. **Parallel Planes** Explain how to find the distance between two parallel planes.

**Checking Points on a Line** In Exercises 5 and 6, determine whether each point lies on the line.

5. $x = -2 + t, y = 3t, z = 4 + t$
   (a) $(0, 6, 6)$    (b) $(2, 3, 5)$    (c) $(-4, -6, 2)$

6. $\dfrac{x - 3}{2} = \dfrac{y - 7}{8} = z + 2$
   (a) $(7, 23, 0)$    (b) $(1, -1, -3)$    (c) $(-7, 47, -7)$

 **Finding Parametric and Symmetric Equations** In Exercises 7–12, find sets of (a) parametric equations and (b) symmetric equations of the line that passes through the given point and is parallel to the given vector or line. (For each line, write the direction numbers as integers.)

| Point | Parallel to |
|-------|-------------|
| 7. $(0, 0, 0)$ | $\mathbf{v} = \langle 3, 1, 5 \rangle$ |
| 8. $(0, 0, 0)$ | $\mathbf{v} = \langle -2, \frac{5}{2}, 1 \rangle$ |
| 9. $(-2, 0, 3)$ | $\mathbf{v} = 2\mathbf{i} + 4\mathbf{j} - 2\mathbf{k}$ |
| 10. $(-3, 0, 2)$ | $\mathbf{v} = 6\mathbf{j} + 3\mathbf{k}$ |
| 11. $(1, 0, 1)$ | $x = 3 + 3t, y = 5 - 2t, z = -7 + t$ |
| 12. $(-3, 5, 4)$ | $\dfrac{x - 1}{3} = \dfrac{y + 1}{-2} = z - 3$ |

 **Finding Parametric and Symmetric Equations** In Exercises 13–16, find sets of (a) parametric equations and (b) symmetric equations of the line that passes through the two points (if possible). (For each line, write the direction numbers as integers.)

13. $(5, -3, -2), \left(-\frac{2}{3}, \frac{2}{3}, 1\right)$    14. $(0, 4, 3), (-1, 2, 5)$

15. $(7, -2, 6), (-3, 0, 6)$    16. $(0, 0, 25), (10, 10, 0)$

**Finding Parametric Equations** In Exercises 17–24, find a set of parametric equations of the line with the given characteristics.

17. The line passes through the point $(2, 3, 4)$ and is parallel to the xz-plane and the yz-plane.

18. The line passes through the point $(-4, 5, 2)$ and is parallel to the xy-plane and the yz-plane.

19. The line passes through the point $(2, 3, 4)$ and is perpendicular to the plane given by $3x + 2y - z = 6$.

20. The line passes through the point $(-4, 5, 2)$ and is perpendicular to the plane given by $-x + 2y + z = 5$.

21. The line passes through the point $(5, -3, -4)$ and is parallel to $\mathbf{v} = \langle 2, -1, 3 \rangle$.

22. The line passes through the point $(-1, 4, -3)$ and is parallel to $\mathbf{v} = 5\mathbf{i} - \mathbf{j}$.

23. The line passes through the point $(2, 1, 2)$ and is parallel to the line $x = -t, y = 1 + t, z = -2 + t$.

24. The line passes through the point $(-6, 0, 8)$ and is parallel to the line $x = 5 - 2t, y = -4 + 2t, z = 0$.

**Using Parametric and Symmetric Equations** In Exercises 25–28, find the coordinates of a point $P$ on the line and a vector $\mathbf{v}$ parallel to the line.

25. $x = 3 - t, \quad y = -1 + 2t, \quad z = -2$

26. $x = 4t, \quad y = 5 - t, \quad z = 4 + 3t$

27. $\dfrac{x - 7}{4} = \dfrac{y + 6}{2} = z + 2$    28. $\dfrac{x + 3}{5} = \dfrac{y}{8} = \dfrac{z - 3}{6}$

**Determining Parallel Lines** In Exercises 29–32, determine whether the lines are parallel or identical.

29. $x = 6 - 3t, \quad y = -2 + 2t, \quad z = 5 + 4t$
    $x = 6t, \quad y = 2 - 4t, \quad z = 13 - 8t$

30. $x = 1 + 2t, \quad y = -1 - t, \quad z = 3t$
    $x = 5 + 2t, \quad y = 1 - t, \quad z = 8 + 3t$

31. $\dfrac{x - 8}{4} = \dfrac{y + 5}{-2} = \dfrac{z + 9}{3}$
    $\dfrac{x + 4}{-8} = \dfrac{y - 1}{4} = \dfrac{z + 18}{-6}$

32. $\dfrac{x - 1}{4} = \dfrac{y - 1}{2} = \dfrac{z + 3}{4}$
    $\dfrac{x + 2}{1} = \dfrac{y - 1}{0.5} = \dfrac{z - 3}{1}$

**Finding a Point of Intersection** In Exercises 33–36, determine whether the lines intersect, and if so, find the point of intersection and the angle between the lines.

33. $x = 4t + 2, \quad y = 3, \quad z = -t + 1$
    $x = 2s + 2, \quad y = 2s + 3, \quad z = s + 1$

34. $x = -3t + 1, \quad y = 4t + 1, \quad z = 2t + 4$
    $x = 3s + 1, \quad y = 2s + 4, \quad z = -s + 1$

35. $\dfrac{x}{3} = \dfrac{y - 2}{-1} = z + 1, \quad \dfrac{x - 1}{4} = y + 2 = \dfrac{z + 3}{-3}$

36. $\dfrac{x - 2}{-3} = \dfrac{y - 2}{6} = z - 3, \quad \dfrac{x - 3}{2} = y + 5 = \dfrac{z + 2}{4}$

**Checking Points in a Plane**   In Exercises 37 and 38, determine whether each point lies in the plane.

37. $x + 2y - 4z - 1 = 0$

   (a) $(-7, 2, -1)$   (b) $(5, 2, 2)$   (c) $(-6, 1, -1)$

38. $2x + y + 3z - 6 = 0$

   (a) $(3, 6, -2)$   (b) $(-1, 5, -1)$   (c) $(2, 1, 0)$

 **Finding an Equation of a Plane**   In Exercises 39–44, find an equation of the plane that passes through the given point and is perpendicular to the given vector or line.

| Point | Perpendicular to |
|---|---|
| 39. $(1, 3, -7)$ | $\mathbf{n} = \mathbf{j}$ |
| 40. $(0, -1, 4)$ | $\mathbf{n} = \mathbf{k}$ |
| 41. $(3, 2, 2)$ | $\mathbf{n} = 2\mathbf{i} + 3\mathbf{j} - \mathbf{k}$ |
| 42. $(0, 0, 0)$ | $\mathbf{n} = -3\mathbf{i} + 2\mathbf{k}$ |
| 43. $(-1, 4, 0)$ | $x = -1 + 2t,\ y = 5 - t,\ z = 3 - 2t$ |
| 44. $(3, 2, 2)$ | $\dfrac{x - 1}{4} = y + 2 = \dfrac{z + 3}{-3}$ |

 **Finding an Equation of a Plane**   In Exercises 45–56, find an equation of the plane with the given characteristics.

45. The plane passes through $(0, 0, 0)$, $(2, 0, 3)$, and $(-3, -1, 5)$.

46. The plane passes through $(3, -1, 2)$, $(2, 1, 5)$, and $(1, -2, -2)$.

47. The plane passes through $(1, 2, 3)$, $(3, 2, 1)$, and $(-1, -2, 2)$.

48. The plane passes through the point $(1, 2, 3)$ and is parallel to the $yz$-plane.

49. The plane passes through the point $(1, 2, 3)$ and is parallel to the $xy$-plane.

50. The plane contains the $y$-axis and makes an angle of $\pi/6$ with the positive $x$-axis.

51. The plane contains the lines given by
$$\frac{x - 1}{-2} = y - 4 = z \quad \text{and} \quad \frac{x - 2}{-3} = \frac{y - 1}{4} = \frac{z - 2}{-1}.$$

52. The plane passes through the point $(2, 2, 1)$ and contains the line given by
$$\frac{x}{2} = \frac{y - 4}{-1} = z.$$

53. The plane passes through the points $(2, 2, 1)$ and $(-1, 1, -1)$ and is perpendicular to the plane
$$2x - 3y + z = 3.$$

54. The plane passes through the points $(3, 2, 1)$ and $(3, 1, -5)$ and is perpendicular to the plane
$$6x + 7y + 2z = 10.$$

55. The plane passes through the points $(1, -2, -1)$ and $(2, 5, 6)$ and is parallel to the $x$-axis.

56. The plane passes through the points $(4, 2, 1)$ and $(-3, 5, 7)$ and is parallel to the $z$-axis.

**Finding an Equation of a Plane**   In Exercises 57–60, find an equation of the plane that contains all the points that are equidistant from the given points.

57. $(2, 2, 0)$,   $(0, 2, 2)$

58. $(1, 0, 2)$,   $(2, 0, 1)$

59. $(-3, 1, 2)$,   $(6, -2, 4)$

60. $(-5, 1, -3)$,   $(2, -1, 6)$

**Parallel Planes**   In Exercises 61–64, determine whether the planes are parallel or identical.

61. $-5x + 2y - 8z = 6$
   $15x - 6y + 24z = 17$

62. $2x - y + 3z = 8$
   $8x - 4y + 12z = 5$

63. $3x - 2y + 5z = 10$
   $75x - 50y + 125z = 250$

64. $-x + 4y - z = 6$
   $-\frac{5}{2}x + 10y - \frac{5}{2}z = 15$

 **Intersection of Planes**   In Exercises 65–68, (a) find the angle between the two planes and (b) find a set of parametric equations for the line of intersection of the planes.

65. $3x + 2y - z = 7$
   $x - 4y + 2z = 0$

66. $-2x + y + z = 2$
   $6x - 3y + 2z = 4$

67. $3x - y + z = 7$
   $4x + 6y + 3z = 2$

68. $6x - 3y + z = 5$
   $-x + y + 5z = 5$

**Comparing Planes**   In Exercises 69–74, determine whether the planes are parallel, orthogonal, or neither. If they are neither parallel nor orthogonal, find the angle between the planes.

69. $5x - 3y + z = 4$
   $x + 4y + 7z = 1$

70. $3x + y - 4z = 3$
   $-9x - 3y + 12z = 4$

71. $x - 3y + 6z = 4$
   $5x + y - z = 4$

72. $3x + 2y - z = 7$
   $x - 4y + 2z = 0$

73. $x - 5y - z = 1$
   $5x - 25y - 5z = -3$

74. $2x - z = 1$
   $4x + y + 8z = 10$

 **Sketching a Graph of a Plane**   In Exercises 75–82, sketch a graph of the plane and label any intercepts.

75. $y = -2$

76. $z = 1$

77. $x + z = 6$

78. $2x + y = 8$

79. $4x + 2y + 6z = 12$

80. $3x + 6y + 2z = 6$

81. $2x - y + 3z = 4$

82. $2x - y + z = 4$

**Intersection of a Plane and a Line**   In Exercises 83–86, find the point(s) of intersection (if any) of the plane and the line. Also, determine whether the line lies in the plane.

83. $x + 3y - z = 6$,   $\dfrac{x + 7}{2} = y - 4 = \dfrac{z + 1}{5}$

84. $2x + 3y = -5$,   $\dfrac{x - 1}{4} = \dfrac{y}{2} = \dfrac{z - 3}{6}$

85. $2x + 3y = 10, \quad \dfrac{x-1}{3} = \dfrac{y+1}{-2} = z - 3$

86. $5x + 3y = 17, \quad \dfrac{x-4}{2} = \dfrac{y+1}{-3} = \dfrac{z+2}{5}$

 **Finding the Distance Between a Point and a Plane** In Exercises 87–90, find the distance between the point and the plane.

87. $(0, 0, 0)$

$2x + 3y + z = 12$

88. $(0, 0, 0)$

$5x + y - z = 9$

89. $(2, 8, 4)$

$2x + y + z = 5$

90. $(1, 3, -1)$

$3x - 4y + 5z = 6$

 **Finding the Distance Between Two Parallel Planes** In Exercises 91–94, verify that the two planes are parallel and find the distance between the planes.

91. $x - 3y + 4z = 10$

$x - 3y + 4z = 6$

92. $2x + 7y + z = 13$

$2x + 7y + z = 9$

93. $-3x + 6y + 7z = 1$

$6x - 12y - 14z = 25$

94. $-x + 6y + 2z = 3$

$-\frac{1}{2}x + 3y + z = 4$

 **Finding the Distance Between a Point and a Line** In Exercises 95–98, find the distance between the point and the line given by the set of parametric equations.

95. $(1, 5, -2);$  $x = 4t - 2,$  $y = 3,$  $z = -t + 1$

96. $(1, -2, 4);$  $x = 2t,$  $y = t - 3,$  $z = 2t + 2$

97. $(-2, 1, 3);$  $x = 1 - t,$  $y = 2 + t,$  $z = -2t$

98. $(4, -1, 5);$  $x = 3,$  $y = 1 + 3t,$  $z = 1 + t$

**Finding the Distance Between Two Parallel Lines** In Exercises 99 and 100, verify that the two lines are parallel and find the distance between the lines.

99. $L_1$: $x = 2 - t,$  $y = 3 + 2t,$  $z = 4 + t$

$L_2$: $x = 3t,$  $y = 1 - 6t,$  $z = 4 - 3t$

100. $L_1$: $x = 3 + 6t,$  $y = -2 + 9t,$  $z = 1 - 12t$

$L_2$: $x = -1 + 4t,$  $y = 3 + 6t,$  $z = -8t$

**EXPLORING CONCEPTS**

101. **Planes** Consider a line and a point not on the line. How many planes contain the line and the point? Explain.

102. **Planes** How many planes are orthogonal to a given plane in space? Explain.

103. **Think About It** Do two distinct lines in space determine a unique plane? Explain.

104.  **HOW DO YOU SEE IT?** Match the general equation with its graph. Then state what axis or plane the equation is parallel to.

(a) $ax + by + d = 0$

(b) $ax + d = 0$

(c) $cz + d = 0$

(d) $ax + cz + d = 0$

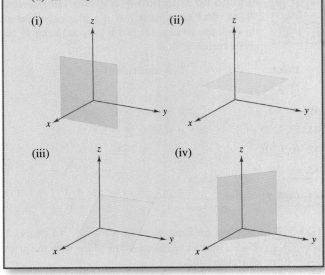

(i) (ii) (iii) (iv)

**105. Modeling Data**

Personal consumption expenditures (in billions of dollars) for several types of recreation from 2009 through 2014 are shown in the table, where $x$ is the expenditures on amusement parks and campgrounds, $y$ is the expenditures on live entertainment (excluding sports), and $z$ is the expenditures on spectator sports. *(Source: U.S. Bureau of Economic Analysis)*

| Year | 2009 | 2010 | 2011 | 2012 | 2013 | 2014 |
|------|------|------|------|------|------|------|
| $x$ | 37.2 | 38.8 | 41.3 | 44.6 | 47.0 | 50.3 |
| $y$ | 25.2 | 26.3 | 28.3 | 28.5 | 28.0 | 30.0 |
| $z$ | 18.8 | 19.2 | 20.4 | 20.6 | 21.6 | 22.4 |

A model for the data is given by

$0.23x + 0.14y - z = -6.85.$

(a) Complete a fourth row in the table using the model to approximate $z$ for the given values of $x$ and $y$. Compare the approximations with the actual values of $z$.

(b) According to this model, increases in expenditures on recreation types $x$ and $y$ would correspond to what kind of change in expenditures on recreation type $z$?

**106. Mechanical Design** The figure shows a chute at the top of a grain elevator of a combine that funnels the grain into a bin. Find the angle between two adjacent sides.

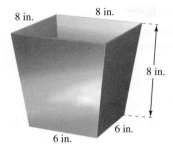

8 in.    8 in.

8 in.

6 in.

6 in.

**107. Distance** Two insects are crawling along different lines in three-space. At time $t$ (in minutes), the first insect is at the point $(x, y, z)$ on the line $x = 6 + t$, $y = 8 - t$, $z = 3 + t$. Also, at time $t$, the second insect is at the point $(x, y, z)$ on the line $x = 1 + t$, $y = 2 + t$, $z = 2t$. Assume that distances are given in inches.

(a) Find the distance between the two insects at time $t = 0$.

(b) Use a graphing utility to graph the distance between the insects from $t = 0$ to $t = 10$.

(c) Using the graph from part (b), what can you conclude about the distance between the insects?

(d) How close to each other do the insects get?

**108. Finding an Equation of a Sphere** Find the standard equation of the sphere with center $(-3, 2, 4)$ that is tangent to the plane given by $2x + 4y - 3z = 8$.

**109. Finding a Point of Intersection** Find the point of intersection of the plane $3x - y + 4z = 7$ and the line through $(5, 4, -3)$ that is perpendicular to this plane.

**110. Finding the Distance Between a Plane and a Line** Show that the plane $2x - y - 3z = 4$ is parallel to the line $x = -2 + 2t$, $y = -1 + 4t$, $z = 4$, and find the distance between them.

**111. Finding Parametric Equations** Find a set of parametric equations for the line passing through the point $(0, 1, 4)$ that is perpendicular to $\mathbf{u} = \langle 2, -5, 1 \rangle$ and $\mathbf{v} = \langle -3, 1, 4 \rangle$.

**112. Finding Parametric Equations** Find a set of parametric equations for the line passing through the point $(1, 0, 2)$ that is parallel to the plane given by $x + y + z = 5$ and perpendicular to the line $x = t$, $y = 1 + t$, $z = 1 + t$.

**True or False?** In Exercises 113–118, determine whether the statement is true or false. If it is false, explain why or give an example that shows it is false.

**113.** If $\mathbf{v} = a_1 \mathbf{i} + b_1 \mathbf{j} + c_1 \mathbf{k}$ is any vector in the plane given by $a_2 x + b_2 y + c_2 z + d_2 = 0$, then $a_1 a_2 + b_1 b_2 + c_1 c_2 = 0$.

**114.** Two lines in space are either intersecting or parallel.

**115.** Two planes in space are either intersecting or parallel.

**116.** If two lines $L_1$ and $L_2$ are each parallel to a plane, then $L_1$ and $L_2$ are parallel.

**117.** If two planes $P_1$ and $P_2$ are each perpendicular to a third plane in space, then $P_1$ and $P_2$ are parallel.

**118.** A plane and a line in space are either intersecting or parallel.

## SECTION PROJECT

## Distances in Space

You have learned two distance formulas in this section—one for the distance between a point and a plane, and one for the distance between a point and a line. In this project, you will study a third distance problem—the distance between two skew lines. Two lines in space are *skew* if they are neither parallel nor intersecting (see figure).

(a) Consider the following two lines in space.

$L_1$: $x = 4 + 5t$, $y = 5 + 5t$, $z = 1 - 4t$

$L_2$: $x = 4 + s$, $y = -6 + 8s$, $z = 7 - 3s$

(i) Show that these lines are not parallel.

(ii) Show that these lines do not intersect and therefore are skew lines.

(iii) Show that the two lines lie in parallel planes.

(iv) Find the distance between the parallel planes from part (iii). This is the distance between the original skew lines.

(b) Use the procedure in part (a) to find the distance between the lines.

$L_1$: $x = 2t$, $y = 4t$, $z = 6t$

$L_2$: $x = 1 - s$, $y = 4 + s$, $z = -1 + s$

(c) Use the procedure in part (a) to find the distance between the lines.

$L_1$: $x = 3t$, $y = 2 - t$, $z = -1 + t$

$L_2$: $x = 1 + 4s$, $y = -2 + s$, $z = -3 - 3s$

(d) Develop a formula for finding the distance between the skew lines.

$L_1$: $x = x_1 + a_1 t$, $y = y_1 + b_1 t$, $z = z_1 + c_1 t$

$L_2$: $x = x_2 + a_2 s$, $y = y_2 + b_2 s$, $z = z_2 + c_2 s$

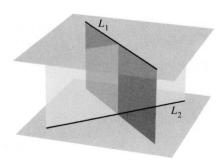

$L_1$

$L_2$

# 11.6 Surfaces in Space

■ Recognize and write equations of cylindrical surfaces.
■ Recognize and write equations of quadric surfaces.
■ Recognize and write equations of surfaces of revolution.

## Cylindrical Surfaces

The first five sections of this chapter contained the vector portion of the preliminary work necessary to study vector calculus and the calculus of space. In this and the next section, you will study surfaces in space and alternative coordinate systems for space. You have already studied two special types of surfaces.

1. Spheres: $(x - x_0)^2 + (y - y_0)^2 + (z - z_0)^2 = r^2$     Section 11.2
2. Planes: $ax + by + cz + d = 0$     Section 11.5

A third type of surface in space is a **cylindrical surface,** or simply a **cylinder.** To define a cylinder, consider the familiar right circular cylinder shown in Figure 11.56. The cylinder was generated by a vertical line moving around the circle $x^2 + y^2 = a^2$ in the $xy$-plane. This circle is a **generating curve** for the cylinder, as indicated in the next definition.

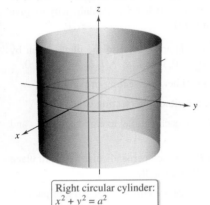

Right circular cylinder:
$x^2 + y^2 = a^2$

Rulings are parallel to $z$-axis
**Figure 11.56**

---

**Definition of a Cylinder**

Let $C$ be a curve in a plane and let $L$ be a line not in a parallel plane. The set of all lines parallel to $L$ and intersecting $C$ is a **cylinder.** The curve $C$ is the **generating curve** (or **directrix**) of the cylinder, and the parallel lines are **rulings.**

---

Without loss of generality, you can assume that $C$ lies in one of the three coordinate planes. Moreover, this text restricts the discussion to *right* cylinders— cylinders whose rulings are perpendicular to the coordinate plane containing $C$, as shown in Figure 11.57. Note that the rulings intersect $C$ and are parallel to the line $L$.

For the right circular cylinder shown in Figure 11.56, the equation of the generating curve in the $xy$-plane is

$$x^2 + y^2 = a^2.$$

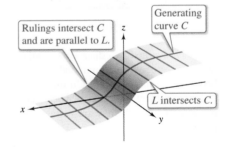

Right cylinder: A cylinder whose rulings are perpendicular to the coordinate plane containing $C$
**Figure 11.57**

To find an equation of the cylinder, note that you can generate any one of the rulings by fixing the values of $x$ and $y$ and then allowing $z$ to take on all real values. In this sense, the value of $z$ is arbitrary and is, therefore, not included in the equation. In other words, the equation of this cylinder is simply the equation of its generating curve.

$$x^2 + y^2 = a^2$$     Equation of cylinder in space

---

**Equations of Cylinders**

The equation of a cylinder whose rulings are parallel to one of the coordinate axes contains only the variables corresponding to the other two axes.

---

EXAMPLE 1 **Sketching a Cylinder**

Sketch the surface represented by each equation.

**a.** $z = y^2$     **b.** $z = \sin x, \quad 0 \le x \le 2\pi$

**Solution**

**a.** The graph is a cylinder whose generating curve, $z = y^2$, is a parabola in the $yz$-plane. The rulings of the cylinder are parallel to the $x$-axis, as shown in Figure 11.58(a).

**b.** The graph is a cylinder generated by the sine curve in the $xz$-plane. The rulings are parallel to the $y$-axis, as shown in Figure 11.58(b).

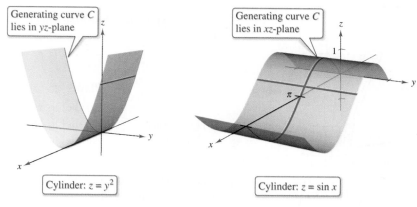

Generating curve $C$ lies in $yz$-plane

Generating curve $C$ lies in $xz$-plane

Cylinder: $z = y^2$     Cylinder: $z = \sin x$

**(a)** Rulings are parallel to $x$-axis.     **(b)** Rulings are parallel to $y$-axis.

**Figure 11.58**

## Quadric Surfaces

The fourth basic type of surface in space is a **quadric surface.** Quadric surfaces are the three-dimensional analogs of conic sections.

---

**Quadric Surface**

The equation of a **quadric surface** in space is a second-degree equation in three variables. The **general form** of the equation is

$$Ax^2 + By^2 + Cz^2 + Dxy + Exz + Fyz + Gx + Hy + Iz + J = 0.$$

There are six basic types of quadric surfaces: **ellipsoid, hyperboloid of one sheet, hyperboloid of two sheets, elliptic cone, elliptic paraboloid,** and **hyperbolic paraboloid.**

---

The intersection of a surface with a plane is called the **trace of the surface** in the plane. To visualize a surface in space, it is helpful to determine its traces in some well-chosen planes. The traces of quadric surfaces are conics. These traces, together with the **standard form** of the equation of each quadric surface, are shown in the table on the next two pages.

In the table on the next two pages, only one of several orientations of each quadric surface is shown. When the surface is oriented along a different axis, its standard equation will change accordingly, as illustrated in Examples 2 and 3. The fact that the two types of paraboloids have one variable raised to the first power can be helpful in classifying quadric surfaces. The other four types of basic quadric surfaces have equations that are of *second degree* in all three variables.

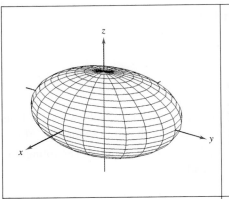

### Ellipsoid

$$\frac{x^2}{a^2} + \frac{y^2}{b^2} + \frac{z^2}{c^2} = 1$$

| Trace | Plane |
|---|---|
| Ellipse | Parallel to $xy$-plane |
| Ellipse | Parallel to $xz$-plane |
| Ellipse | Parallel to $yz$-plane |

The surface is a sphere when $a = b = c \neq 0$.

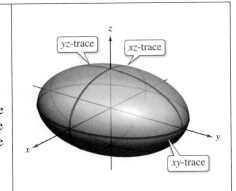

### Hyperboloid of One Sheet

$$\frac{x^2}{a^2} + \frac{y^2}{b^2} - \frac{z^2}{c^2} = 1$$

| Trace | Plane |
|---|---|
| Ellipse | Parallel to $xy$-plane |
| Hyperbola | Parallel to $xz$-plane |
| Hyperbola | Parallel to $yz$-plane |

The axis of the hyperboloid corresponds to the variable whose coefficient is negative.

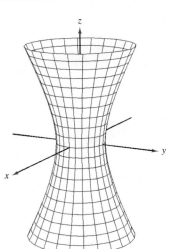

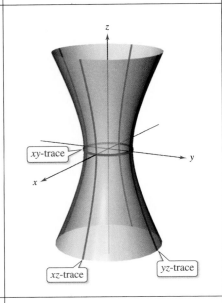

### Hyperboloid of Two Sheets

$$\frac{z^2}{c^2} - \frac{x^2}{a^2} - \frac{y^2}{b^2} = 1$$

| Trace | Plane |
|---|---|
| Ellipse | Parallel to $xy$-plane |
| Hyperbola | Parallel to $xz$-plane |
| Hyperbola | Parallel to $yz$-plane |

The axis of the hyperboloid corresponds to the variable whose coefficient is positive. There is no trace in the coordinate plane perpendicular to this axis.

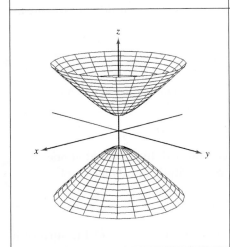

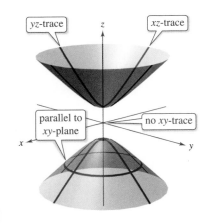

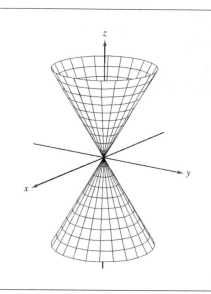

### Elliptic Cone

$$\frac{x^2}{a^2} + \frac{y^2}{b^2} - \frac{z^2}{c^2} = 0$$

| Trace | Plane |
|---|---|
| Ellipse | Parallel to $xy$-plane |
| Hyperbola | Parallel to $xz$-plane |
| Hyperbola | Parallel to $yz$-plane |

The axis of the cone corresponds to the variable whose coefficient is negative. The traces in the coordinate planes parallel to this axis are intersecting lines.

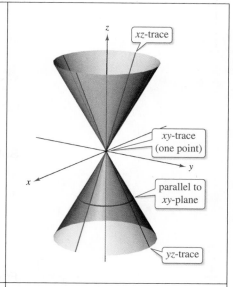

### Elliptic Paraboloid

$$z = \frac{x^2}{a^2} + \frac{y^2}{b^2}$$

| Trace | Plane |
|---|---|
| Ellipse | Parallel to $xy$-plane |
| Parabola | Parallel to $xz$-plane |
| Parabola | Parallel to $yz$-plane |

The axis of the paraboloid corresponds to the variable raised to the first power.

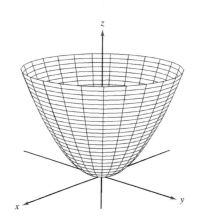

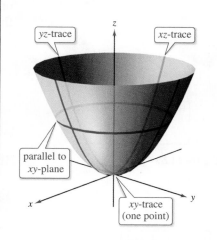

### Hyperbolic Paraboloid

$$z = \frac{y^2}{b^2} - \frac{x^2}{a^2}$$

| Trace | Plane |
|---|---|
| Hyperbola | Parallel to $xy$-plane |
| Parabola | Parallel to $xz$-plane |
| Parabola | Parallel to $yz$-plane |

The axis of the paraboloid corresponds to the variable raised to the first power.

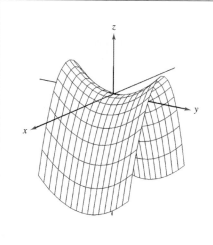

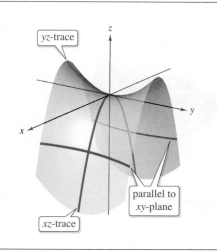

To classify a quadric surface, begin by writing the equation of the surface in standard form. Then, determine several traces taken in the coordinate planes *or* taken in planes that are parallel to the coordinate planes.

**EXAMPLE 2**     **Sketching a Quadric Surface**

Classify and sketch the surface

$$4x^2 - 3y^2 + 12z^2 + 12 = 0.$$

**Solution**     Begin by writing the equation in standard form.

| | |
|---|---|
| $4x^2 - 3y^2 + 12z^2 + 12 = 0$ | Write original equation. |
| $\dfrac{x^2}{-3} + \dfrac{y^2}{4} - z^2 - 1 = 0$ | Divide by $-12$. |
| $\dfrac{y^2}{4} - \dfrac{x^2}{3} - z^2 = 1$ | Standard form |

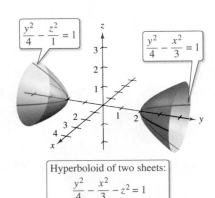

$\dfrac{y^2}{4} - \dfrac{z^2}{1} = 1$

$\dfrac{y^2}{4} - \dfrac{x^2}{3} = 1$

Hyperboloid of two sheets:
$$\dfrac{y^2}{4} - \dfrac{x^2}{3} - z^2 = 1$$

**Figure 11.59**

From the table on pages 800 and 801, you can conclude that the surface is a hyperboloid of two sheets with the $y$-axis as its axis. To sketch the graph of this surface, it helps to find the traces in the coordinate planes.

| | | |
|---|---|---|
| $xy$-trace ($z = 0$): | $\dfrac{y^2}{4} - \dfrac{x^2}{3} = 1$ | Hyperbola |
| $xz$-trace ($y = 0$): | $\dfrac{x^2}{3} + \dfrac{z^2}{1} = -1$ | No trace |
| $yz$-trace ($x = 0$): | $\dfrac{y^2}{4} - \dfrac{z^2}{1} = 1$ | Hyperbola |

The graph is shown in Figure 11.59.

**EXAMPLE 3**     **Sketching a Quadric Surface**

Classify and sketch the surface

$$x - y^2 - 4z^2 = 0.$$

**Solution**     Because $x$ is raised only to the first power, the surface is a paraboloid. The axis of the paraboloid is the $x$-axis. In standard form, the equation is

$$x = y^2 + 4z^2.$$     Standard form

Some convenient traces are listed below.

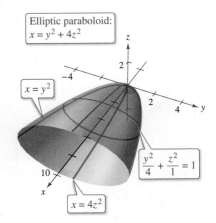

Elliptic paraboloid:
$x = y^2 + 4z^2$

$x = y^2$

$\dfrac{y^2}{4} + \dfrac{z^2}{1} = 1$

$x = 4z^2$

**Figure 11.60**

| | | |
|---|---|---|
| $xy$-trace ($z = 0$): | $x = y^2$ | Parabola |
| $xz$-trace ($y = 0$): | $x = 4z^2$ | Parabola |
| parallel to $yz$-plane ($x = 4$): | $\dfrac{y^2}{4} + \dfrac{z^2}{1} = 1$ | Ellipse |

The surface is an *elliptic* paraboloid, as shown in Figure 11.60.     ■

Some second-degree equations in $x$, $y$, and $z$ do not represent any of the basic types of quadric surfaces. For example, the graph of

$$x^2 + y^2 + z^2 = 0$$     Single point

is a single point, and the graph of

$$x^2 + y^2 = 1$$     Right circular cylinder

is a right circular cylinder.

For a quadric surface not centered at the origin, you can form the standard equation by completing the square, as demonstrated in Example 4.

**EXAMPLE 4** **A Quadric Surface Not Centered at the Origin**

•••▷ *See LarsonCalculus.com for an interactive version of this type of example.*

Classify and sketch the surface

$$x^2 + 2y^2 + z^2 - 4x + 4y - 2z + 3 = 0.$$

**Solution** Begin by grouping terms and factoring where possible.

$$x^2 - 4x + 2(y^2 + 2y) + z^2 - 2z = -3$$

Next, complete the square for each variable and write the equation in standard form.

$$(x^2 - 4x +\quad) + 2(y^2 + 2y +\quad) + (z^2 - 2z +\quad) = -3$$
$$(x^2 - 4x + 4) + 2(y^2 + 2y + 1) + (z^2 - 2z + 1) = -3 + 4 + 2 + 1$$
$$(x - 2)^2 + 2(y + 1)^2 + (z - 1)^2 = 4$$
$$\frac{(x - 2)^2}{4} + \frac{(y + 1)^2}{2} + \frac{(z - 1)^2}{4} = 1$$

From this equation, you can see that the quadric surface is an ellipsoid that is centered at $(2, -1, 1)$. Its graph is shown in Figure 11.61. ▪

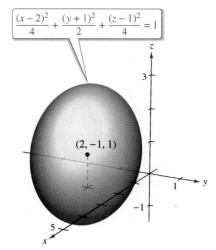

$$\frac{(x - 2)^2}{4} + \frac{(y + 1)^2}{2} + \frac{(z - 1)^2}{4} = 1$$

$(2, -1, 1)$

An ellipsoid centered at $(2, -1, 1)$
**Figure 11.61**

▷ **TECHNOLOGY** A 3-D graphing utility can help you visualize a surface in space.* Such a graphing utility may create a three-dimensional graph by sketching several traces of the surface and then applying a "hidden-line" routine that blocks out portions of the surface that lie behind other portions of the surface. Two examples of figures that were generated by *Mathematica* are shown below.

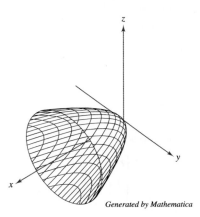

*Generated by Mathematica*

Elliptic paraboloid
$$x = \frac{y^2}{2} + \frac{z^2}{2}$$

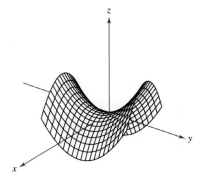

*Generated by Mathematica*

Hyperbolic paraboloid
$$z = \frac{y^2}{16} - \frac{x^2}{16}$$

Using a graphing utility to graph a surface in space requires practice. For one thing, you must know enough about the surface to be able to specify a *viewing window* that gives a representative view of the surface. Also, you can often improve the view of a surface by rotating the axes. For instance, note that the elliptic paraboloid in the figure is seen from a line of sight that is "higher" than the line of sight used to view the hyperbolic paraboloid.

* Some 3-D graphing utilities require surfaces to be entered with parametric equations. For a discussion of this technique, see Section 15.5.

## Surfaces of Revolution

The fifth special type of surface you will study is a **surface of revolution.** In Section 7.4, you studied a method for finding the *area* of such a surface. You will now look at a procedure for finding its *equation.* Consider the graph of the **radius function**

$$y = r(z) \qquad \text{Generating curve}$$

in the $yz$-plane. When this graph is revolved about the $z$-axis, it forms a surface of revolution, as shown in Figure 11.62. The trace of the surface in the plane $z = z_0$ is a circle whose radius is $r(z_0)$ and whose equation is

$$x^2 + y^2 = [r(z_0)]^2. \qquad \text{Circular trace in plane: } z = z_0$$

Replacing $z_0$ with $z$ produces an equation that is valid for all values of $z$. In a similar manner, you can obtain equations for surfaces of revolution for the other two axes, and the results are summarized as follows.

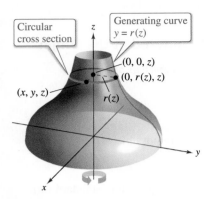

Circular cross section

$z$

Generating curve $y = r(z)$

$(0, 0, z)$

$(0, r(z), z)$

$(x, y, z)$

$r(z)$

$x$

**Figure 11.62**

---

### Surface of Revolution

If the graph of a radius function $r$ is revolved about one of the coordinate axes, then the equation of the resulting surface of revolution has one of the forms listed below.

1. Revolved about the $x$-axis: $y^2 + z^2 = [r(x)]^2$
2. Revolved about the $y$-axis: $x^2 + z^2 = [r(y)]^2$
3. Revolved about the $z$-axis: $x^2 + y^2 = [r(z)]^2$

---

**EXAMPLE 5** **Finding an Equation for a Surface of Revolution**

Find an equation for the surface of revolution formed by revolving (a) the graph of $y = 1/z$ about the $z$-axis and (b) the graph of $9x^2 = y^3$ about the $y$-axis.

**Solution**

a. An equation for the surface of revolution formed by revolving the graph of

$$y = \frac{1}{z} \qquad \text{Radius function}$$

about the $z$-axis is

$$x^2 + y^2 = [r(z)]^2 \qquad \text{Revolved about the } z\text{-axis}$$

$$x^2 + y^2 = \left(\frac{1}{z}\right)^2. \qquad \text{Substitute } 1/z \text{ for } r(z).$$

b. To find an equation for the surface formed by revolving the graph of $9x^2 = y^3$ about the $y$-axis, solve for $x$ in terms of $y$ to obtain

$$x = \frac{1}{3}y^{3/2} = r(y). \qquad \text{Radius function}$$

So, the equation for this surface is

$$x^2 + z^2 = [r(y)]^2 \qquad \text{Revolved about the } y\text{-axis}$$

$$x^2 + z^2 = \left(\frac{1}{3}y^{3/2}\right)^2 \qquad \text{Substitute } \tfrac{1}{3}y^{3/2} \text{ for } r(y).$$

$$x^2 + z^2 = \frac{1}{9}y^3. \qquad \text{Equation of surface}$$

The graph is shown in Figure 11.63.

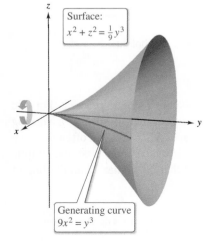

$z$

Surface: $x^2 + z^2 = \frac{1}{9}y^3$

$y$

$x$

Generating curve $9x^2 = y^3$

**Figure 11.63**

The generating curve for a surface of revolution is not unique. For instance, the surface

$$x^2 + z^2 = e^{-2y}$$

can be formed by revolving either the graph of

$$x = e^{-y}$$

about the $y$-axis or the graph of

$$z = e^{-y}$$

about the $y$-axis, as shown in Figure 11.64.

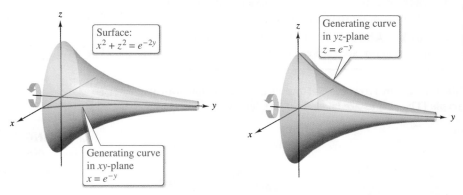

**Figure 11.64**

### EXAMPLE 6    Finding a Generating Curve

Find a generating curve and the axis of revolution for the surface

$$x^2 + 3y^2 + z^2 = 9.$$

**Solution**    The equation has one of the forms listed below.

$$x^2 + y^2 = [r(z)]^2 \qquad \text{Revolved about } z\text{-axis}$$
$$y^2 + z^2 = [r(x)]^2 \qquad \text{Revolved about } x\text{-axis}$$
$$x^2 + z^2 = [r(y)]^2 \qquad \text{Revolved about } y\text{-axis}$$

Because the coefficients of $x^2$ and $z^2$ are equal, you should choose the third form and write

$$x^2 + z^2 = 9 - 3y^2.$$

The $y$-axis is the axis of revolution. You can choose a generating curve from either of the traces

$$x^2 = 9 - 3y^2 \qquad \text{Trace in } xy\text{-plane}$$

or

$$z^2 = 9 - 3y^2. \qquad \text{Trace in } yz\text{-plane}$$

For instance, using the first trace, the generating curve is the semiellipse

$$x = \sqrt{9 - 3y^2}. \qquad \text{Generating curve}$$

The graph of this surface is shown in Figure 11.65.

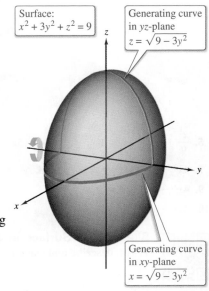

**Figure 11.65**

## 11.6 Exercises

See **CalcChat.com** for tutorial help and worked-out solutions to odd-numbered exercises.

### CONCEPT CHECK

**1. Quadric Surfaces** How are quadric surfaces and conic sections related?

**2. Classifying an Equation** What does the equation $z = x^2$ represent in the $xz$-plane? What does it represent in three-space?

**3. Trace of a Surface** What is meant by the trace of a surface? How do you find a trace?

**4. Think About It** Does every second-degree equation in $x$, $y$, and $z$ represent a quadric surface? Explain.

**Matching** In Exercises 5–10, match the equation with its graph. [The graphs are labeled (a), (b), (c), (d), (e), and (f).]

**(a)**

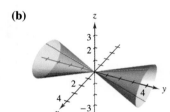

**(b)**

**(c)**

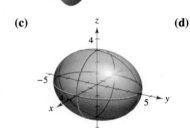

**(d)**

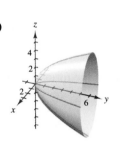

**(e)**

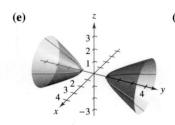

**(f)**

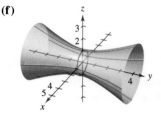

**5.** $\dfrac{x^2}{9} + \dfrac{y^2}{16} + \dfrac{z^2}{9} = 1$

**6.** $15x^2 - 4y^2 + 15z^2 = -4$

**7.** $4x^2 - y^2 + 4z^2 = 4$

**8.** $y^2 = 4x^2 + 9z^2$

**9.** $4x^2 - 4y + z^2 = 0$

**10.** $4x^2 - y^2 + 4z = 0$

**Sketching a Surface in Space** In Exercises 11–14, describe and sketch the surface.

**11.** $y^2 + z^2 = 9$

**12.** $y^2 + z = 6$

**13.** $4x^2 + y^2 = 4$

**14.** $y^2 - z^2 = 25$

**Sketching a Quadric Surface** In Exercises 15–26, classify and sketch the quadric surface. Use a computer algebra system or a graphing utility to confirm your sketch.

**15.** $4x^2 - y^2 - z^2 = 1$

**16.** $\dfrac{x^2}{16} + \dfrac{y^2}{25} + \dfrac{z^2}{25} = 1$

**17.** $16x^2 - y^2 + 16z^2 = 4$

**18.** $z = x^2 + 4y^2$

**19.** $x^2 + \dfrac{y^2}{4} + z^2 = 1$

**20.** $z^2 - x^2 - \dfrac{y^2}{4} = 1$

**21.** $z^2 = x^2 + \dfrac{y^2}{9}$

**22.** $3z = -y^2 + x^2$

**23.** $x^2 - y^2 + z = 0$

**24.** $x^2 = 2y^2 + 2z^2$

**25.** $x^2 - y + z^2 = 0$

**26.** $-8x^2 + 18y^2 + 18z^2 = 2$

### EXPLORING CONCEPTS

**27. Hyperboloid** Explain how to determine whether a quadric surface is a hyperboloid of one sheet or a hyperboloid of two sheets.

**28. Ellipsoid** Is every trace of an ellipsoid an ellipse? Explain.

**29. Quadric Surface** Is there a quadric surface whose traces are all parabolas? Explain.

**30. HOW DO YOU SEE IT?** The four figures below are graphs of the quadric surface $z = x^2 + y^2$. Match each of the four graphs with the point in space from which the paraboloid is viewed.

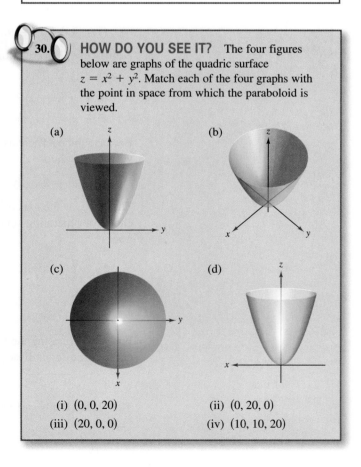

**(a)**   **(b)**

**(c)**   **(d)**

(i)  $(0, 0, 20)$      (ii)  $(0, 20, 0)$

(iii) $(20, 0, 0)$     (iv)  $(10, 10, 20)$

**Finding an Equation for a Surface of Revolution** In Exercises 31–36, find an equation for the surface of revolution formed by revolving the curve in the indicated coordinate plane about the given axis.

| Equation of Curve | Coordinate Plane | Axis of Revolution |
|---|---|---|
| **31.** $z = 5y$ | $yz$-plane | $y$-axis |
| **32.** $z^2 = 9y$ | $yz$-plane | $y$-axis |
| **33.** $y^3 = 8z$ | $yz$-plane | $z$-axis |
| **34.** $z = \ln x$ | $xz$-plane | $z$-axis |
| **35.** $xy = 2$ | $xy$-plane | $x$-axis |
| **36.** $2z = \sqrt{4 - x^2}$ | $xz$-plane | $x$-axis |

**Finding a Generating Curve** In Exercises 37–40, find an equation of a generating curve given the equation of its surface of revolution.

**37.** $x^2 + y^2 - 2z = 0$

**38.** $x^2 + z^2 = \cos^2 y$

**39.** $8x^2 + y^2 + z^2 = 5$

**40.** $6x^2 + 2y^2 + 2z^2 = 1$

**Finding the Volume of a Solid** In Exercises 41 and 42, use the shell method to find the volume of the solid below the surface of revolution and above the $xy$-plane.

**41.** The curve $z = 4x - x^2$ in the $xz$-plane is revolved about the $z$-axis.

**42.** The curve

$$z = \sin y, \quad 0 \le y \le \pi$$

in the $yz$-plane is revolved about the $z$-axis.

**Analyzing a Trace** In Exercises 43 and 44, analyze the trace when the surface

$$z = \tfrac{1}{2}x^2 + \tfrac{1}{4}y^2$$

is intersected by the indicated planes.

**43.** Find the lengths of the major and minor axes and the coordinates of the foci of the ellipse generated when the surface is intersected by the planes given by

(a) $z = 2$   and   (b) $z = 8$.

**44.** Find the coordinates of the focus of the parabola formed when the surface is intersected by the planes given by

(a) $y = 4$   and   (b) $x = 2$.

**Finding an Equation of a Surface** In Exercises 45 and 46, find an equation of the surface satisfying the conditions, and identify the surface.

**45.** The set of all points equidistant from the point $(0, 2, 0)$ and the plane $y = -2$

**46.** The set of all points equidistant from the point $(0, 0, 4)$ and the $xy$-plane

**47. Geography**

Because of the forces caused by its rotation, Earth is an oblate ellipsoid rather than a sphere. The equatorial radius is 3963 miles and the polar radius is 3950 miles. Find an equation of the ellipsoid. (Assume that the center of Earth is at the origin and that the trace formed by the plane $z = 0$ corresponds to the equator.)

**48. Machine Design** The top of a rubber bushing designed to absorb vibrations in an automobile is the surface of revolution generated by revolving the curve

$$z = \frac{1}{2}y^2 + 1$$

for $0 \le y \le 2$ in the $yz$-plane about the $z$-axis.

(a) Find an equation for the surface of revolution.

(b) All measurements are in centimeters and the bushing is set on the $xy$-plane. Use the shell method to find its volume.

(c) The bushing has a hole of diameter 1 centimeter through its center and parallel to the axis of revolution. Find the volume of the rubber bushing.

**49. Using a Hyperbolic Paraboloid** Determine the intersection of the hyperbolic paraboloid

$$z = \frac{y^2}{b^2} - \frac{x^2}{a^2}$$

with the plane $bx + ay - z = 0$. (Assume $a, b > 0$.)

**50. Intersection of Surfaces** Explain why the curve of intersection of the surfaces $x^2 + 3y^2 - 2z^2 + 2y = 4$ and $2x^2 + 6y^2 - 4z^2 - 3x = 2$ lies in a plane.

**51. Think About It** Three types of classic *topological surfaces* are shown below. The sphere and torus have both an "inside" and an "outside." Does the Klein bottle have both an "inside" and an "outside?" Explain.

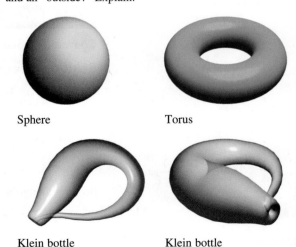

Sphere          Torus

Klein bottle      Klein bottle

# 11.7  Cylindrical and Spherical Coordinates

■ Use cylindrical coordinates to represent surfaces in space.
■ Use spherical coordinates to represent surfaces in space.

## Cylindrical Coordinates

You have already seen that some two-dimensional graphs are easier to represent in polar coordinates than in rectangular coordinates. A similar situation exists for surfaces in space. In this section, you will study two alternative space-coordinate systems. The first, the **cylindrical coordinate system,** is an extension of polar coordinates in the plane to three-dimensional space.

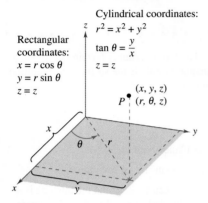

Cylindrical coordinates:
$$r^2 = x^2 + y^2$$
$$\tan \theta = \frac{y}{x}$$
$$z = z$$

Rectangular coordinates:
$$x = r \cos \theta$$
$$y = r \sin \theta$$
$$z = z$$

**Figure 11.66**

### The Cylindrical Coordinate System

In a **cylindrical coordinate system,** a point $P$ in space is represented by an ordered triple $(r, \theta, z)$.

1. $(r, \theta)$ is a polar representation of the projection of $P$ in the $xy$-plane.
2. $z$ is the directed distance from $(r, \theta)$ to $P$.

To convert from rectangular to cylindrical coordinates (or vice versa), use the conversion guidelines for polar coordinates listed below and illustrated in Figure 11.66.

*Cylindrical to rectangular:*

$$x = r \cos \theta, \quad y = r \sin \theta, \quad z = z$$

*Rectangular to cylindrical:*

$$r^2 = x^2 + y^2, \quad \tan \theta = \frac{y}{x}, \quad z = z$$

The point $(0, 0, 0)$ is called the **pole.** Moreover, because the representation of a point in the polar coordinate system is not unique, it follows that the representation in the cylindrical coordinate system is also not unique.

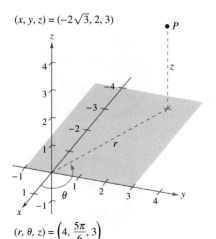

$(x, y, z) = (-2\sqrt{3}, 2, 3)$

$(r, \theta, z) = \left(4, \dfrac{5\pi}{6}, 3\right)$

**Figure 11.67**

### EXAMPLE 1    Cylindrical-to-Rectangular Conversion

Convert the point $(r, \theta, z) = (4, 5\pi/6, 3)$ to rectangular coordinates.

**Solution**    Using the cylindrical-to-rectangular conversion equations produces

$$x = 4 \cos \frac{5\pi}{6} = 4\left(-\frac{\sqrt{3}}{2}\right) = -2\sqrt{3}$$

$$y = 4 \sin \frac{5\pi}{6} = 4\left(\frac{1}{2}\right) = 2$$

$$z = 3.$$

So, in rectangular coordinates, the point is

$$(x, y, z) = \left(-2\sqrt{3}, 2, 3\right)$$

as shown in Figure 11.67.

**EXAMPLE 2** **Rectangular-to-Cylindrical Conversion**

Convert the point

$$(x, y, z) = \left(1, \sqrt{3}, 2\right)$$

to cylindrical coordinates.

**Solution**   Use the rectangular-to-cylindrical conversion equations.

$$r = \pm\sqrt{1 + 3} = \pm 2$$

$$\tan \theta = \sqrt{3} \implies \theta = \arctan \sqrt{3} + n\pi = \frac{\pi}{3} + n\pi$$

$$z = 2$$

You have two choices for $r$ and infinitely many choices for $\theta$. As shown in Figure 11.68, two convenient representations of the point are

$$\left(2, \frac{\pi}{3}, 2\right) \qquad \text{$r > 0$ and $\theta$ in Quadrant I}$$

and

$$\left(-2, \frac{4\pi}{3}, 2\right). \qquad \text{$r < 0$ and $\theta$ in Quadrant III}$$

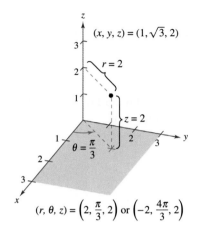

$(r, \theta, z) = \left(2, \frac{\pi}{3}, 2\right)$ or $\left(-2, \frac{4\pi}{3}, 2\right)$

**Figure 11.68**

Cylindrical coordinates are especially convenient for representing cylindrical surfaces and surfaces of revolution with the $z$-axis as the axis of symmetry, as shown in Figure 11.69.

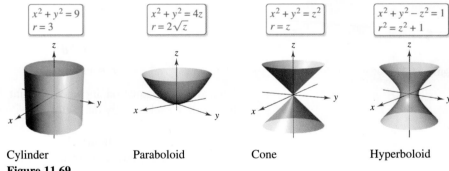

| $x^2 + y^2 = 9$ $r = 3$ | $x^2 + y^2 = 4z$ $r = 2\sqrt{z}$ | $x^2 + y^2 = z^2$ $r = z$ | $x^2 + y^2 - z^2 = 1$ $r^2 = z^2 + 1$ |

Cylinder          Paraboloid          Cone          Hyperboloid
**Figure 11.69**

Vertical planes containing the $z$-axis and horizontal planes also have simple cylindrical coordinate equations, as shown in Figure 11.70.

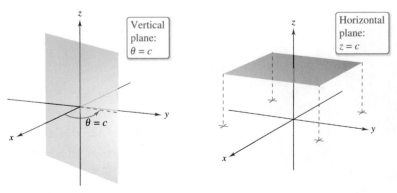

Vertical plane: $\theta = c$

Horizontal plane: $z = c$

**Figure 11.70**

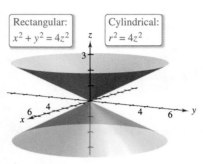

Figure 11.71

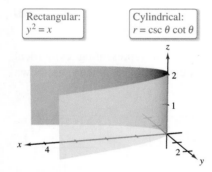

Figure 11.72

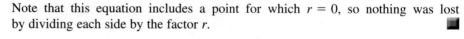

Figure 11.73

**EXAMPLE 3** **Rectangular-to-Cylindrical Conversion**

Find an equation in cylindrical coordinates for the surface represented by each rectangular equation.

**a.** $x^2 + y^2 = 4z^2$

**b.** $y^2 = x$

**Solution**

**a.** From Section 11.6, you know that the graph of

$$x^2 + y^2 = 4z^2$$

is an elliptic cone with its axis along the $z$-axis, as shown in Figure 11.71. When you replace $x^2 + y^2$ with $r^2$, the equation in cylindrical coordinates is

$$x^2 + y^2 = 4z^2 \qquad \text{Rectangular equation}$$
$$r^2 = 4z^2. \qquad \text{Cylindrical equation}$$

**b.** The graph of the surface

$$y^2 = x$$

is a parabolic cylinder with rulings parallel to the $z$-axis, as shown in Figure 11.72. To obtain the equation in cylindrical coordinates, replace $y^2$ with $r^2 \sin^2 \theta$ and $x$ with $r \cos \theta$, as shown.

$$y^2 = x \qquad \text{Rectangular equation}$$
$$r^2 \sin^2 \theta = r \cos \theta \qquad \text{Substitute } r \sin \theta \text{ for } y \text{ and } r \cos \theta \text{ for } x.$$
$$r(r \sin^2 \theta - \cos \theta) = 0 \qquad \text{Collect terms and factor.}$$
$$r \sin^2 \theta - \cos \theta = 0 \qquad \text{Divide each side by } r.$$
$$r = \frac{\cos \theta}{\sin^2 \theta} \qquad \text{Solve for } r.$$
$$r = \csc \theta \cot \theta \qquad \text{Cylindrical equation}$$

Note that this equation includes a point for which $r = 0$, so nothing was lost by dividing each side by the factor $r$. ∎

Converting from cylindrical coordinates to rectangular coordinates is less straightforward than converting from rectangular coordinates to cylindrical coordinates, as demonstrated in Example 4.

**EXAMPLE 4** **Cylindrical-to-Rectangular Conversion**

Find an equation in rectangular coordinates for the surface represented by the cylindrical equation

$$r^2 \cos 2\theta + z^2 + 1 = 0.$$

**Solution**

$$r^2 \cos 2\theta + z^2 + 1 = 0 \qquad \text{Cylindrical equation}$$
$$r^2(\cos^2 \theta - \sin^2 \theta) + z^2 + 1 = 0 \qquad \text{Trigonometric identity}$$
$$r^2 \cos^2 \theta - r^2 \sin^2 \theta + z^2 = -1$$
$$x^2 - y^2 + z^2 = -1 \qquad \text{Replace } r \cos \theta \text{ with } x \text{ and } r \sin \theta \text{ with } y.$$
$$y^2 - x^2 - z^2 = 1 \qquad \text{Rectangular equation}$$

This is a hyperboloid of two sheets whose axis lies along the $y$-axis, as shown in Figure 11.73. ∎

# Spherical Coordinates

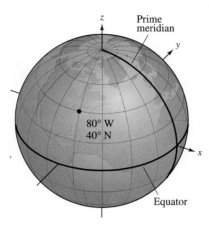

**Figure 11.74**

In the **spherical coordinate system,** each point is represented by an ordered triple: the first coordinate is a distance, and the second and third coordinates are angles. This system is similar to the latitude-longitude system used to identify points on the surface of Earth. For example, the point on the surface of Earth whose latitude is 40° North (of the equator) and whose longitude is 80° West (of the prime meridian) is shown in Figure 11.74. Assuming that Earth is spherical and has a radius of 4000 miles, you would label this point as

$$(4000, \ -80°, \ 50°).$$

Radius     80° clockwise from     50° down from
prime meridian     North Pole

---

## The Spherical Coordinate System

In a **spherical coordinate system,** a point $P$ in space is represented by an ordered triple $(\rho, \theta, \phi)$, where $\rho$ is the lowercase Greek letter rho and $\phi$ is the lowercase Greek letter phi.

1. $\rho$ is the distance between $P$ and the origin, $\rho \geq 0$.
2. $\theta$ is the same angle used in cylindrical coordinates for $r \geq 0$.
3. $\phi$ is the angle *between* the positive $z$-axis and the line segment $\overrightarrow{OP}$, $0 \leq \phi \leq \pi$.

Note that the first and third coordinates, $\rho$ and $\phi$, are nonnegative.

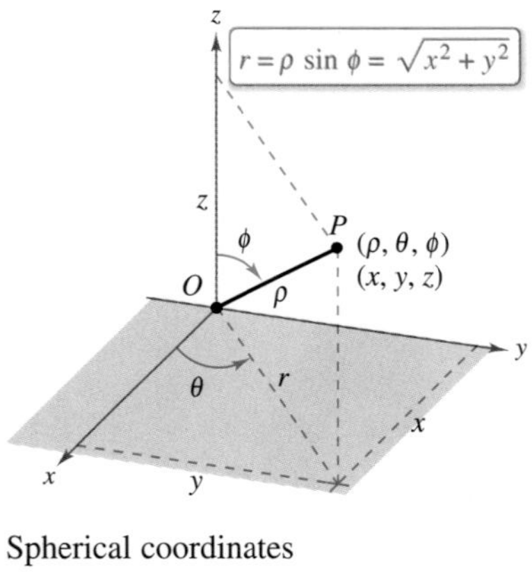

---

The relationship between rectangular and spherical coordinates is illustrated in Figure 11.75. To convert from one system to the other, use the conversion guidelines listed below.

Spherical coordinates
**Figure 11.75**

*Spherical to rectangular:*

$$x = \rho \sin \phi \cos \theta, \quad y = \rho \sin \phi \sin \theta, \quad z = \rho \cos \phi$$

*Rectangular to spherical:*

$$\rho^2 = x^2 + y^2 + z^2, \quad \tan \theta = \frac{y}{x}, \quad \phi = \arccos \frac{z}{\sqrt{x^2 + y^2 + z^2}}$$

To change coordinates between the cylindrical and spherical systems, use the conversion guidelines listed below.

*Spherical to cylindrical $(r \geq 0)$:*

$$r^2 = \rho^2 \sin^2 \phi, \quad \theta = \theta, \quad z = \rho \cos \phi$$

*Cylindrical to spherical $(r \geq 0)$:*

$$\rho = \sqrt{r^2 + z^2}, \quad \theta = \theta, \quad \phi = \arccos \frac{z}{\sqrt{r^2 + z^2}}$$

The spherical coordinate system is useful primarily for surfaces in space that have a *point* or *center* of symmetry. For example, Figure 11.76 shows three surfaces with simple spherical equations.

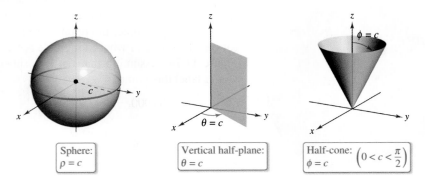

| Sphere: $\rho = c$ | Vertical half-plane: $\theta = c$ | Half-cone: $\phi = c$ $\left(0 < c < \dfrac{\pi}{2}\right)$ |

**Figure 11.76**

### EXAMPLE 5   Rectangular-to-Spherical Conversion

•••▷ *See LarsonCalculus.com for an interactive version of this type of example.*

Find an equation in spherical coordinates for the surface represented by each rectangular equation.

**a.** Cone: $x^2 + y^2 = z^2$    **b.** Sphere: $x^2 + y^2 + z^2 - 4z = 0$

**Solution**

**a.** Use the spherical-to-rectangular equations

$$x = \rho \sin \phi \cos \theta, \quad y = \rho \sin \phi \sin \theta, \quad \text{and} \quad z = \rho \cos \phi$$

and substitute in the rectangular equation as shown.

$$x^2 + y^2 = z^2$$
$$\rho^2 \sin^2 \phi \cos^2 \theta + \rho^2 \sin^2 \phi \sin^2 \theta = \rho^2 \cos^2 \phi$$
$$\rho^2 \sin^2 \phi(\cos^2 \theta + \sin^2 \theta) = \rho^2 \cos^2 \phi$$
$$\rho^2 \sin^2 \phi = \rho^2 \cos^2 \phi$$
$$\frac{\sin^2 \phi}{\cos^2 \phi} = 1 \qquad\qquad \rho \geq 0$$
$$\tan^2 \phi = 1$$
$$\tan \phi = \pm 1$$

| Rectangular: $x^2 + y^2 + z^2 - 4z = 0$ | Spherical: $\rho = 4 \cos \phi$ |

So, you can conclude that

$$\phi = \frac{\pi}{4} \quad \text{or} \quad \phi = \frac{3\pi}{4}.$$

The equation $\phi = \pi/4$ represents the *upper* half-cone, and the equation $\phi = 3\pi/4$ represents the *lower* half-cone.

**b.** Because $\rho^2 = x^2 + y^2 + z^2$ and $z = \rho \cos \phi$, the rectangular equation has the following spherical form.

$$\rho^2 - 4\rho \cos \phi = 0 \quad\implies\quad \rho(\rho - 4 \cos \phi) = 0$$

Temporarily discarding the possibility that $\rho = 0$, you have the spherical equation

$$\rho - 4 \cos \phi = 0 \quad \text{or} \quad \rho = 4 \cos \phi.$$

Note that the solution set for this equation includes a point for which $\rho = 0$, so nothing is lost by discarding the factor $\rho$. The sphere represented by the equation $\rho = 4 \cos \phi$ is shown in Figure 11.77.

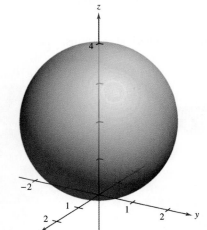

**Figure 11.77**

# 11.7 Exercises

See **CalcChat.com** for tutorial help and worked-out solutions to odd-numbered exercises.

**CONCEPT CHECK**

**1. Cylindrical Coordinates** Describe the cylindrical coordinate system in your own words.

**2. Spherical Coordinates** Describe the position of the point $(2, 0°, 30°)$ given in spherical coordinates.

 **Cylindrical-to-Rectangular Conversion** In Exercises 3–8, convert the point from cylindrical coordinates to rectangular coordinates.

**3.** $(-7, 0, 5)$

**4.** $(2, -\pi, -4)$

**5.** $\left(3, \dfrac{\pi}{4}, 1\right)$

**6.** $\left(6, -\dfrac{3\pi}{2}, 2\right)$

**7.** $\left(4, \dfrac{7\pi}{6}, -3\right)$

**8.** $\left(-\dfrac{2}{3}, \dfrac{4\pi}{3}, 8\right)$

 **Rectangular-to-Cylindrical Conversion** In Exercises 9–14, convert the point from rectangular coordinates to cylindrical coordinates.

**9.** $(0, 5, 1)$

**10.** $(6, 2\sqrt{3}, -1)$

**11.** $(2, -2, -4)$

**12.** $(3, -3, 7)$

**13.** $(1, \sqrt{3}, 4)$

**14.** $(2\sqrt{3}, -2, 6)$

 **Rectangular-to-Cylindrical Conversion** In Exercises 15–22, find an equation in cylindrical coordinates for the surface represented by the rectangular equation.

**15.** $z = 4$

**16.** $x = 9$

**17.** $x^2 + y^2 - 2z^2 = 5$

**18.** $z = x^2 + y^2 - 11$

**19.** $y = x^2$

**20.** $x^2 + y^2 = 8x$

**21.** $y^2 = 10 - z^2$

**22.** $x^2 + y^2 + z^2 - 3z = 0$

 **Cylindrical-to-Rectangular Conversion** In Exercises 23–30, find an equation in rectangular coordinates for the surface represented by the cylindrical equation, and sketch its graph.

**23.** $r = 3$

**24.** $z = -2$

**25.** $\theta = \dfrac{\pi}{6}$

**26.** $r = \dfrac{1}{2}z$

**27.** $r^2 + z^2 = 5$

**28.** $z = r^2 \cos^2 \theta$

**29.** $r = 4 \sin \theta$

**30.** $r = 2 \cos \theta$

**Rectangular-to-Spherical Conversion** In Exercises 31–36, convert the point from rectangular coordinates to spherical coordinates.

**31.** $(4, 0, 0)$

**32.** $(-4, 0, 0)$

**33.** $(-2, 2\sqrt{3}, 4)$

**34.** $(-5, -5, \sqrt{2})$

**35.** $(\sqrt{3}, 1, 2\sqrt{3})$

**36.** $(-1, 2, 1)$

**Spherical-to-Rectangular Conversion** In Exercises 37–42, convert the point from spherical coordinates to rectangular coordinates.

**37.** $\left(4, \dfrac{\pi}{6}, \dfrac{\pi}{4}\right)$

**38.** $\left(6, \pi, \dfrac{\pi}{2}\right)$

**39.** $\left(12, -\dfrac{\pi}{4}, 0\right)$

**40.** $\left(9, \dfrac{\pi}{4}, \pi\right)$

**41.** $\left(5, \dfrac{\pi}{4}, \dfrac{\pi}{12}\right)$

**42.** $\left(7, \dfrac{3\pi}{4}, \dfrac{\pi}{9}\right)$

 **Rectangular-to-Spherical Conversion** In Exercises 43–50, find an equation in spherical coordinates for the surface represented by the rectangular equation.

**43.** $y = 2$

**44.** $z = 6$

**45.** $x^2 + y^2 + z^2 = 49$

**46.** $x^2 + y^2 - 3z^2 = 0$

**47.** $x^2 + y^2 = 16$

**48.** $x = 13$

**49.** $x^2 + y^2 = 2z^2$

**50.** $x^2 + y^2 + z^2 - 9z = 0$

 **Spherical-to-Rectangular Conversion** In Exercises 51–58, find an equation in rectangular coordinates for the surface represented by the spherical equation, and sketch its graph.

**51.** $\rho = 1$

**52.** $\theta = \dfrac{3\pi}{4}$

**53.** $\phi = \dfrac{\pi}{6}$

**54.** $\phi = \dfrac{\pi}{2}$

**55.** $\rho = 4 \cos \phi$

**56.** $\rho = 2 \sec \phi$

**57.** $\rho = \csc \phi$

**58.** $\rho = 4 \csc \phi \sec \theta$

**Cylindrical-to-Spherical Conversion** In Exercises 59–64, convert the point from cylindrical coordinates to spherical coordinates.

**59.** $\left(4, \dfrac{\pi}{4}, 0\right)$

**60.** $\left(3, -\dfrac{\pi}{4}, 0\right)$

**61.** $\left(6, \dfrac{\pi}{2}, -6\right)$

**62.** $\left(-4, \dfrac{\pi}{3}, 4\right)$

**63.** $(12, \pi, 5)$

**64.** $\left(4, \dfrac{\pi}{2}, 3\right)$

**Spherical-to-Cylindrical Conversion** In Exercises 65–70, convert the point from spherical coordinates to cylindrical coordinates.

**65.** $\left(10, \dfrac{\pi}{6}, \dfrac{\pi}{2}\right)$

**66.** $\left(4, \dfrac{\pi}{18}, \dfrac{\pi}{2}\right)$

**67.** $\left(6, -\dfrac{\pi}{6}, \dfrac{\pi}{3}\right)$

**68.** $\left(5, -\dfrac{5\pi}{6}, \pi\right)$

**69.** $\left(8, \dfrac{7\pi}{6}, \dfrac{\pi}{6}\right)$

**70.** $\left(7, \dfrac{\pi}{4}, \dfrac{3\pi}{4}\right)$

**Matching** In Exercises 71–76, match the equation (written in terms of cylindrical or spherical coordinates) with its graph. [The graphs are labeled (a), (b), (c), (d), (e), and (f).]

(a)

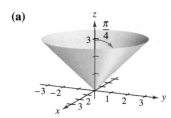

(b)

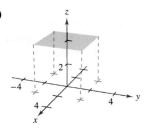

(c)

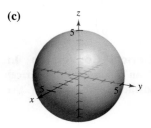

(d)

(e)

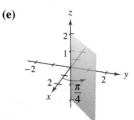

(f)

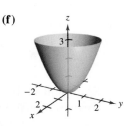

**71.** $r = 5$

**72.** $\theta = \dfrac{\pi}{4}$

**73.** $\rho = 5$

**74.** $\phi = \dfrac{\pi}{4}$

**75.** $r^2 = z$

**76.** $\rho = 4 \sec \phi$

**77. Spherical Coordinates** Explain why in spherical coordinates the graph of $\theta = c$ is a half-plane and not an entire plane.

**78. HOW DO YOU SEE IT?** Identify the surface graphed and match the graph with its rectangular equation. Then find an equation in cylindrical coordinates for the equation given in rectangular coordinates.

(a)

(b)

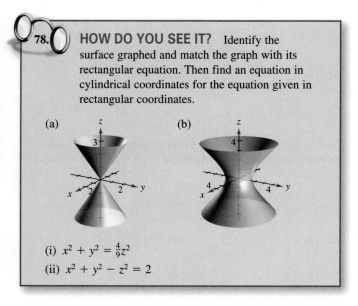

(i) $x^2 + y^2 = \frac{4}{9}z^2$

(ii) $x^2 + y^2 - z^2 = 2$

**Converting a Rectangular Equation** In Exercises 79–86, convert the rectangular equation to an equation in (a) cylindrical coordinates and (b) spherical coordinates.

**79.** $x^2 + y^2 + z^2 = 27$

**80.** $4(x^2 + y^2) = z^2$

**81.** $x^2 + y^2 + z^2 - 2z = 0$

**82.** $x^2 + y^2 = z$

**83.** $x^2 + y^2 = 4y$

**84.** $x^2 + y^2 = 45$

**85.** $x^2 - y^2 = 9$

**86.** $y = 4$

**Sketching a Solid** In Exercises 87–90, sketch the solid that has the given description in cylindrical coordinates.

**87.** $0 \le \theta \le \pi/2, 0 \le r \le 2, 0 \le z \le 4$

**88.** $-\pi/2 \le \theta \le \pi/2, 0 \le r \le 3, 0 \le z \le r \cos \theta$

**89.** $0 \le \theta \le 2\pi, 0 \le r \le a, r \le z \le a$

**90.** $0 \le \theta \le 2\pi, 2 \le r \le 4, z^2 \le -r^2 + 6r - 8$

**Sketching a Solid** In Exercises 91–94, sketch the solid that has the given description in spherical coordinates.

**91.** $0 \le \theta \le 2\pi, 0 \le \phi \le \pi/6, 0 \le \rho \le a \sec \phi$

**92.** $0 \le \theta \le 2\pi, \pi/4 \le \phi \le \pi/2, 0 \le \rho \le 1$

**93.** $0 \le \theta \le \pi/2, 0 \le \phi \le \pi/2, 0 \le \rho \le 2$

**94.** $0 \le \theta \le \pi, 0 \le \phi \le \pi/2, 1 \le \rho \le 3$

**EXPLORING CONCEPTS**

**Think About It** In Exercises 95–100, find inequalities that describe the solid and state the coordinate system used. Position the solid on the coordinate system such that the inequalities are as simple as possible.

**95.** A cube with each edge 10 centimeters long

**96.** A cylindrical shell 8 meters long with an inside diameter of 0.75 meter and an outside diameter of 1.25 meters

**97.** A spherical shell with inside and outside radii of 4 inches and 6 inches, respectively

**98.** The solid that remains after a hole 1 inch in diameter is drilled through the center of a sphere 6 inches in diameter

**99.** The solid inside both $x^2 + y^2 + z^2 = 9$ and $\left(x - \frac{3}{2}\right)^2 + y^2 = \frac{9}{4}$

**100.** The solid between the spheres $x^2 + y^2 + z^2 = 4$ and $x^2 + y^2 + z^2 = 9$, and inside the cone $z^2 = x^2 + y^2$

**True or False?** In Exercises 101 and 102, determine whether the statement is true or false. If it is false, explain why or give an example that shows it is false.

**101.** The cylindrical coordinates of a point $(x, y, z)$ are unique.

**102.** The spherical coordinates of a point $(x, y, z)$ are unique.

**103. Intersection of Surfaces** Identify the curve of intersection of the surfaces (in cylindrical coordinates) $z = \sin \theta$ and $r = 1$.

**104. Intersection of Surfaces** Identify the curve of intersection of the surfaces (in spherical coordinates) $\rho = 2 \sec \phi$ and $\rho = 4$.

**Writing Vectors in Different Forms** In Exercises 1 and 2, let $\mathbf{u} = \overrightarrow{PQ}$ and $\mathbf{v} = \overrightarrow{PR}$ and (a) write $\mathbf{u}$ and $\mathbf{v}$ in component form, (b) write $\mathbf{u}$ and $\mathbf{v}$ as the linear combination of the standard unit vectors $\mathbf{i}$ and $\mathbf{j}$, (c) find the magnitudes of $\mathbf{u}$ and $\mathbf{v}$, and (d) find $-3\mathbf{u} + \mathbf{v}$.

1. $P = (1, 2), Q = (4, 1), R = (5, 4)$

2. $P = (-2, -1), Q = (5, -1), R = (2, 4)$

**Finding a Vector** In Exercises 3 and 4, find the component form of $\mathbf{v}$ given its magnitude and the angle it makes with the positive $x$-axis.

3. $\|\mathbf{v}\| = 8, \theta = 60°$      4. $\|\mathbf{v}\| = \frac{1}{2}, \theta = 225°$

5. **Finding Coordinates of a Point** Find the coordinates of the point located in the $xy$-plane, four units to the right of the $xz$-plane, and five units behind the $yz$-plane.

6. **Using the Three-Dimensional Coordinate System** Determine the location of a point $(x, y, z)$ that satisfies the condition $y = 3$.

**Finding the Distance Between Two Points in Space** In Exercises 7 and 8, find the distance between the points.

7. $(1, 6, 3), (-2, 3, 5)$

8. $(-2, 1, -5), (4, -1, -1)$

**Finding the Equation of a Sphere** In Exercises 9 and 10, find the standard equation of the sphere with the given characteristics.

9. Center: $(3, -2, 6)$; Radius: $4$

10. Endpoints of a diameter: $(0, 0, 4), (4, 6, 0)$

**Finding the Equation of a Sphere** In Exercises 11 and 12, complete the square to write the equation of the sphere in standard form. Find the center and radius.

11. $x^2 + y^2 + z^2 - 4x - 6y + 4 = 0$

12. $x^2 + y^2 + z^2 - 10x + 6y - 4z + 34 = 0$

**Writing a Vector in Different Forms** In Exercises 13 and 14, the initial and terminal points of a vector are given. (a) Sketch the directed line segment. (b) Find the component form of the vector. (c) Write the vector using standard unit vector notation. (d) Sketch the vector with its initial point at the origin.

13. Initial point: $(2, -1, 3)$      14. Initial point: $(6, 2, 0)$

   Terminal point: $(4, 4, -7)$      Terminal point: $(3, -3, 8)$

**Finding a Vector** In Exercises 15 and 16, find the vector $\mathbf{z}$, given that $\mathbf{u} = \langle 5, -2, 3\rangle$, $\mathbf{v} = \langle 0, 2, 1\rangle$, and $\mathbf{w} = \langle -6, -6, 2\rangle$.

15. $\mathbf{z} = -\mathbf{u} + 3\mathbf{v} + \frac{1}{2}\mathbf{w}$

16. $\mathbf{u} - \mathbf{v} + \mathbf{w} - 2\mathbf{z} = \mathbf{0}$

**Using Vectors to Determine Collinear Points** In Exercises 17 and 18, use vectors to determine whether the points are collinear.

17. $(3, 4, -1), (-1, 6, 9), (5, 3, -6)$

18. $(5, -4, 7), (8, -5, 5), (11, 6, 3)$

19. **Finding a Unit Vector** Find a unit vector in the direction of $\mathbf{u} = \langle 2, 3, 5\rangle$.

20. **Finding a Vector** Find the vector $\mathbf{v}$ of magnitude 8 in the direction $\langle 6, -3, 2\rangle$.

**Finding Dot Products** In Exercises 21 and 22, let $\mathbf{u} = \overrightarrow{PQ}$ and $\mathbf{v} = \overrightarrow{PR}$, and find (a) the component forms of $\mathbf{u}$ and $\mathbf{v}$, (b) $\mathbf{u} \cdot \mathbf{v}$, and (c) $\mathbf{v} \cdot \mathbf{v}$.

21. $P = (5, 0, 0), Q = (4, 4, 0), R = (2, 0, 6)$

22. $P = (2, -1, 3), Q = (0, 5, 1), R = (5, 5, 0)$

**Finding the Angle Between Two Vectors** In Exercises 23 and 24, find the angle $\theta$ between the vectors (a) in radians and (b) in degrees.

23. $\mathbf{u} = 5[\cos(3\pi/4)\mathbf{i} + \sin(3\pi/4)\mathbf{j}]$
    $\mathbf{v} = 2[\cos(2\pi/3)\mathbf{i} + \sin(2\pi/3)\mathbf{j}]$

24. $\mathbf{u} = \langle 1, 0, -3\rangle, \quad \mathbf{v} = \langle 2, -2, 1\rangle$

**Comparing Vectors** In Exercises 25 and 26, determine whether $\mathbf{u}$ and $\mathbf{v}$ are orthogonal, parallel, or neither.

25. $\mathbf{u} = \langle 7, -2, 3\rangle$      26. $\mathbf{u} = \langle -3, 0, 9\rangle$
    $\mathbf{v} = \langle -1, 4, 5\rangle$          $\mathbf{v} = \langle 1, 0, -3\rangle$

**Finding the Projection of u onto v** In Exercises 27 and 28, (a) find the projection of $\mathbf{u}$ onto $\mathbf{v}$, and (b) find the vector component of $\mathbf{u}$ orthogonal to $\mathbf{v}$.

27. $\mathbf{u} = 4\mathbf{i} + 2\mathbf{j}, \quad \mathbf{v} = 3\mathbf{i} + 4\mathbf{j}$

28. $\mathbf{u} = \langle 1, -1, 1\rangle, \quad \mathbf{v} = \langle 2, 0, 2\rangle$

29. **Orthogonal Vectors** Find two vectors in opposite directions that are orthogonal to the vector $\mathbf{u} = \langle 5, 6, -3\rangle$.

30. **Work** An object is pulled 8 feet across a floor using a force of 75 pounds. The direction of the force is $30°$ above the horizontal. Find the work done.

**Finding Cross Products** In Exercises 31 and 32, find (a) $\mathbf{u} \times \mathbf{v}$, (b) $\mathbf{v} \times \mathbf{u}$, and (c) $\mathbf{v} \times \mathbf{v}$.

31. $\mathbf{u} = 4\mathbf{i} + 3\mathbf{j} + 6\mathbf{k}$      32. $\mathbf{u} = \langle 0, 2, 1\rangle$
    $\mathbf{v} = 5\mathbf{i} + 2\mathbf{j} + \mathbf{k}$          $\mathbf{v} = \langle 1, -3, 4\rangle$

33. **Finding a Unit Vector** Find a unit vector that is orthogonal to both $\mathbf{u} = \langle 2, -10, 8\rangle$ and $\mathbf{v} = \langle 4, 6, -8\rangle$.

34. **Area** Find the area of the parallelogram that has the vectors $\mathbf{u} = \langle 3, -1, 5\rangle$ and $\mathbf{v} = \langle 2, -4, 1\rangle$ as adjacent sides.

**35. Torque** A vertical force of 40 pounds acts on a wrench, as shown in the figure. Find the torque at $P$.

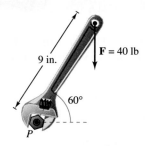

9 in.

$F = 40$ lb

60°

$P$

**36. Volume** Use the triple scalar product to find the volume of the parallelepiped having adjacent edges $\mathbf{u} = 2\mathbf{i} + \mathbf{j}$, $\mathbf{v} = 2\mathbf{j} + \mathbf{k}$, and $\mathbf{w} = -\mathbf{j} + 2\mathbf{k}$.

**Finding Parametric and Symmetric Equations** In Exercises 37 and 38, find sets of (a) parametric equations and (b) symmetric equations of the line that passes through the two points. (For each line, write the direction numbers as integers.)

**37.** $(3, 0, 2)$, $(9, 11, 6)$  **38.** $(-1, 4, 3)$, $(8, 10, 5)$

**Finding Parametric Equations** In Exercises 39 and 40, find a set of parametric equations of the line with the given characteristics.

**39.** The line passes through the point $(-6, -8, 2)$ and is perpendicular to the $xz$-plane.

**40.** The line passes through the point $(1, 2, 3)$ and is parallel to the line given by $x = y = z$.

**Finding an Equation of a Plane** In Exercises 41–44, find an equation of the plane with the given characteristics.

**41.** The plane passes through $(-3, -4, 2)$, $(-3, 4, 1)$, and $(1, 1, -2)$.

**42.** The plane passes through the point $(-2, 3, 1)$ and is perpendicular to $\mathbf{n} = 3\mathbf{i} - \mathbf{j} + \mathbf{k}$.

**43.** The plane contains the lines given by

$$\frac{x - 1}{-2} = y = z + 1$$

and

$$\frac{x + 1}{-2} = y - 1 = z - 2.$$

**44.** The plane passes through the points $(5, 1, 3)$ and $(2, -2, 1)$ and is perpendicular to the plane $2x + y - z = 4$.

**45. Distance** Find the distance between the point $(1, 0, 2)$ and the plane $2x - 3y + 6z = 6$.

**46. Distance** Find the distance between the point $(3, -2, 4)$ and the plane $2x - 5y + z = 10$.

**47. Distance** Find the distance between the planes $5x - 3y + z = 2$ and $5x - 3y + z = -3$.

**48. Distance** Find the distance between the point $(-5, 1, 3)$ and the line given by $x = 1 + t$, $y = 3 - 2t$, and $z = 5 - t$.

**Sketching a Surface in Space** In Exercises 49–58, describe and sketch the surface.

**49.** $x + 2y + 3z = 6$  **50.** $y = z^2$

**51.** $y = \frac{1}{2}z$  **52.** $y = \cos z$

**53.** $\dfrac{x^2}{16} + \dfrac{y^2}{9} + z^2 = 1$  **54.** $16x^2 + 16y^2 - 9z^2 = 0$

**55.** $\dfrac{x^2}{16} - \dfrac{y^2}{9} + z^2 = -1$  **56.** $\dfrac{x^2}{25} + \dfrac{y^2}{4} - \dfrac{z^2}{100} = 1$

**57.** $x^2 + z^2 = 4$  **58.** $y^2 + z^2 = 16$

**59. Surface of Revolution** Find an equation for the surface of revolution formed by revolving the curve $z^2 = 2y$ in the $yz$-plane about the $y$-axis.

**60. Surface of Revolution** Find an equation for the surface of revolution formed by revolving the curve $2x + 3z = 1$ in the $xz$-plane about the $x$-axis.

**Converting Rectangular Coordinates** In Exercises 61 and 62, convert the point from rectangular coordinates to (a) cylindrical coordinates and (b) spherical coordinates.

**61.** $\left(-\sqrt{3}, 3, -5\right)$  **62.** $(8, 8, 1)$

**Cylindrical-to-Rectangular Conversion** In Exercises 63 and 64, convert the point from cylindrical coordinates to rectangular coordinates.

**63.** $(5, \pi, 1)$  **64.** $\left(-2, \dfrac{\pi}{3}, 3\right)$

**Spherical-to-Rectangular Conversion** In Exercises 65 and 66, convert the point from spherical coordinates to rectangular coordinates.

**65.** $\left(4, \pi, \dfrac{\pi}{4}\right)$  **66.** $\left(8, -\dfrac{\pi}{6}, \dfrac{\pi}{3}\right)$

**Converting a Rectangular Equation** In Exercises 67 and 68, convert the rectangular equation to an equation in (a) cylindrical coordinates and (b) spherical coordinates.

**67.** $x^2 - y^2 = 2z$  **68.** $x^2 + y^2 + z^2 = 16$

**Cylindrical-to-Rectangular Conversion** In Exercises 69 and 70, find an equation in rectangular coordinates for the surface represented by the cylindrical equation, and sketch its graph.

**69.** $z = r^2 \sin^2 \theta + 3r \cos \theta$

**70.** $r = -5z$

**Spherical-to-Rectangular Conversion** In Exercises 71 and 72, find an equation in rectangular coordinates for the surface represented by the spherical equation, and sketch its graph.

**71.** $\phi = \dfrac{\pi}{4}$

**72.** $\rho = 9 \sec \theta$

# P.S. Problem Solving

See **CalcChat.com** for tutorial help and worked-out solutions to odd-numbered exercises.

**1. Proof**   Using vectors, prove the Law of Sines: If **a**, **b**, and **c** are the three sides of the triangle shown in the figure, then

$$\frac{\sin A}{\|\mathbf{a}\|} = \frac{\sin B}{\|\mathbf{b}\|} = \frac{\sin C}{\|\mathbf{c}\|}.$$

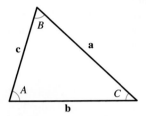

**2. Using an Equation**   Consider the function

$$f(x) = \int_0^x \sqrt{t^4 + 1}\, dt.$$

(a) Use a graphing utility to graph the function on the interval $-2 \le x \le 2$.

(b) Find a unit vector parallel to the graph of $f$ at the point $(0, 0)$.

(c) Find a unit vector perpendicular to the graph of $f$ at the point $(0, 0)$.

(d) Find the parametric equations of the tangent line to the graph of $f$ at the point $(0, 0)$.

**3. Proof**   Using vectors, prove that the line segments joining the midpoints of the sides of a parallelogram form a parallelogram (see figure).

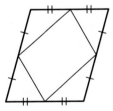

**4. Proof**   Using vectors, prove that the diagonals of a rhombus are perpendicular (see figure).

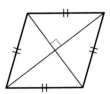

**5. Distance**

(a) Find the shortest distance between the point $Q(2, 0, 0)$ and the line determined by the points $P_1(0, 0, 1)$ and $P_2(0, 1, 2)$.

(b) Find the shortest distance between the point $Q(2, 0, 0)$ and the line segment joining the points $P_1(0, 0, 1)$ and $P_2(0, 1, 2)$.

**6. Orthogonal Vectors**   Let $P_0$ be a point in the plane with normal vector **n**. Describe the set of points $P$ in the plane for which $\left(\mathbf{n} + \overrightarrow{PP_0}\right)$ is orthogonal to $\left(\mathbf{n} - \overrightarrow{PP_0}\right)$.

**7. Volume**

(a) Find the volume of the solid bounded below by the paraboloid

$$z = x^2 + y^2$$

and above by the plane $z = 1$.

(b) Find the volume of the solid bounded below by the elliptic paraboloid

$$z = \frac{x^2}{a^2} + \frac{y^2}{b^2}$$

and above by the plane $z = k$, where $k > 0$.

(c) Show that the volume of the solid in part (b) is equal to one-half the product of the area of the base times the altitude, as shown in the figure.

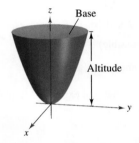

**8. Volume**

(a) Use the disk method to find the volume of the sphere $x^2 + y^2 + z^2 = r^2$.

(b) Find the volume of the ellipsoid $\dfrac{x^2}{a^2} + \dfrac{y^2}{b^2} + \dfrac{z^2}{c^2} = 1$.

**9. Proof**   Prove the following property of the cross product.

$$(\mathbf{u} \times \mathbf{v}) \times (\mathbf{w} \times \mathbf{z}) = [(\mathbf{u} \times \mathbf{v}) \cdot \mathbf{z}]\mathbf{w} - [(\mathbf{u} \times \mathbf{v}) \cdot \mathbf{w}]\mathbf{z}$$

**10. Using Parametric Equations**   Consider the line given by the parametric equations

$$x = -t + 3, \quad y = \tfrac{1}{2}t + 1, \quad z = 2t - 1$$

and the point $(4, 3, s)$ for any real number $s$.

(a) Write the distance between the point and the line as a function of $s$.

(b) Use a graphing utility to graph the function in part (a). Use the graph to find the value of $s$ such that the distance between the point and the line is minimum.

(c) Use the *zoom* feature of a graphing utility to zoom out several times on the graph in part (b). Does it appear that the graph has slant asymptotes? Explain. If it appears to have slant asymptotes, find them.

**11. Sketching Graphs** Sketch the graph of each equation given in spherical coordinates.

(a) $\rho = 2 \sin \phi$    (b) $\rho = 2 \cos \phi$

**12. Sketching Graphs** Sketch the graph of each equation given in cylindrical coordinates.

(a) $r = 2 \cos \theta$    (b) $z = r^2 \cos 2\theta$

**13. Tetherball** A tetherball weighing 1 pound is pulled outward from the pole by a horizontal force **u** until the rope makes an angle of $\theta$ degrees with the pole (see figure).

(a) Determine the resulting tension in the rope and the magnitude of **u** when $\theta = 30°$.

(b) Write the tension $T$ in the rope and the magnitude of **u** as functions of $\theta$. Determine the domains of the functions.

(c) Use a graphing utility to complete the table.

| $\theta$ | 0° | 10° | 20° | 30° | 40° | 50° | 60° |
|---|---|---|---|---|---|---|---|
| $T$ | | | | | | | |
| $\|\mathbf{u}\|$ | | | | | | | |

(d) Use a graphing utility to graph the two functions for $0° \le \theta \le 60°$.

(e) Compare $T$ and $\|\mathbf{u}\|$ as $\theta$ increases.

(f) Find (if possible)

$$\lim_{\theta \to \pi/2^-} T \quad \text{and} \quad \lim_{\theta \to \pi/2^-} \|\mathbf{u}\|.$$

Are the results what you expected? Explain.

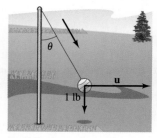 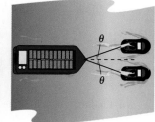

Figure for 13                    Figure for 14

**14. Towing** A loaded barge is being towed by two tugboats, and the magnitude of the resultant is 6000 pounds directed along the axis of the barge (see figure). Each towline makes an angle of $\theta$ degrees with the axis of the barge.

(a) Find the tension in the towlines when $\theta = 20°$.

(b) Write the tension $T$ of each line as a function of $\theta$. Determine the domain of the function.

(c) Use a graphing utility to complete the table.

| $\theta$ | 10° | 20° | 30° | 40° | 50° | 60° |
|---|---|---|---|---|---|---|
| $T$ | | | | | | |

(d) Use a graphing utility to graph the tension function.

(e) Explain why the tension increases as $\theta$ increases.

**15. Proof** Consider the vectors

$$\mathbf{u} = \langle \cos \alpha, \sin \alpha, 0 \rangle \quad \text{and} \quad \mathbf{v} = \langle \cos \beta, \sin \beta, 0 \rangle$$

where $\alpha > \beta$. Find the cross product of the vectors and use the result to prove the identity

$$\sin(\alpha - \beta) = \sin \alpha \cos \beta - \cos \alpha \sin \beta.$$

**16. Latitude-Longitude System** Los Angeles is located at 34.05° North latitude and 118.24° West longitude, and Rio de Janeiro, Brazil, is located at 22.90° South latitude and 43.23° West longitude (see figure). Assume that Earth is spherical and has a radius of 4000 miles.

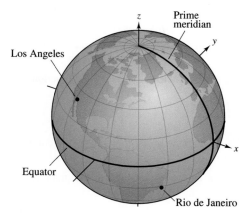

(a) Find the spherical coordinates for the location of each city.

(b) Find the rectangular coordinates for the location of each city.

(c) Find the angle (in radians) between the vectors from the center of Earth to the two cities.

(d) Find the great-circle distance $s$ between the cities. (*Hint:* $s = r\theta$)

(e) Repeat parts (a)–(d) for the cities of Boston, located at 42.36° North latitude and 71.06° West longitude, and Honolulu, located at 21.31° North latitude and 157.86° West longitude.

**17. Distance Between a Point and a Plane** Consider the plane that passes through the points $P$, $R$, and $S$. Show that the distance from a point $Q$ to this plane is

$$\text{Distance} = \frac{|\mathbf{u} \cdot (\mathbf{v} \times \mathbf{w})|}{\|\mathbf{u} \times \mathbf{v}\|}$$

where $\mathbf{u} = \overrightarrow{PR}$, $\mathbf{v} = \overrightarrow{PS}$, and $\mathbf{w} = \overrightarrow{PQ}$.

**18. Distance Between Parallel Planes** Show that the distance between the parallel planes

$$ax + by + cz + d_1 = 0 \quad \text{and} \quad ax + by + cz + d_2 = 0$$

is

$$\text{Distance} = \frac{|d_1 - d_2|}{\sqrt{a^2 + b^2 + c^2}}.$$

**19. Intersection of Planes** Show that the curve of intersection of the plane $z = 2y$ and the cylinder $x^2 + y^2 = 1$ is an ellipse.

# 12 Vector-Valued Functions

Speed *(Exercise 66, p. 865)*

**Air Traffic Control**
*(Exercise 61, p. 854)*

Football *(Exercise 34, p. 843)*

**Shot-Put Throw**
*(Exercise 44, p. 843)*

Staircase *(Exercise 81, p. 827)*

# 12.1 Vector-Valued Functions

■ Analyze and sketch a space curve given by a vector-valued function.
■ Extend the concepts of limits and continuity to vector-valued functions.

## Space Curves and Vector-Valued Functions

In Section 10.2, a *plane curve* was defined as the set of ordered pairs $(f(t), g(t))$ together with their defining parametric equations $x = f(t)$ and $y = g(t)$, where $f$ and $g$ are continuous functions of $t$ on an interval $I$. This definition can be extended naturally to three-dimensional space. A **space curve** $C$ is the set of all ordered triples $(f(t), g(t), h(t))$ together with their defining parametric equations

$$x = f(t), \quad y = g(t), \quad \text{and} \quad z = h(t)$$

where $f$, $g$, and $h$ are continuous functions of $t$ on an interval $I$.

Before looking at examples of space curves, a new type of function, called a **vector-valued function,** is introduced. This type of function maps real numbers to vectors.

---

**Definition of Vector-Valued Function**

A function of the form

$$\mathbf{r}(t) = f(t)\mathbf{i} + g(t)\mathbf{j} \qquad \text{Plane}$$

or

$$\mathbf{r}(t) = f(t)\mathbf{i} + g(t)\mathbf{j} + h(t)\mathbf{k} \qquad \text{Space}$$

is a **vector-valued function,** where the **component functions** $f$, $g$, and $h$ are real-valued functions of the parameter $t$. Vector-valued functions are sometimes denoted as

$$\mathbf{r}(t) = \langle f(t), g(t) \rangle \qquad \text{Plane}$$

or

$$\mathbf{r}(t) = \langle f(t), g(t), h(t) \rangle. \qquad \text{Space}$$

---

Technically, a curve in a plane or in space consists of a collection of points and the defining parametric equations. Two different curves can have the same graph. For instance, each of the curves

$$\mathbf{r}(t) = \sin t\,\mathbf{i} + \cos t\,\mathbf{j} \quad \text{and} \quad \mathbf{r}(t) = \sin t^2\mathbf{i} + \cos t^2\mathbf{j}$$

has the unit circle as its graph, but these equations do not represent the same curve—because the circle is traced out in different ways on the graphs.

Be sure you see the distinction between the vector-valued function $\mathbf{r}$ and the real-valued functions $f$, $g$, and $h$. All are functions of the real variable $t$, but $\mathbf{r}(t)$ is a vector, whereas $f(t)$, $g(t)$, and $h(t)$ are real numbers (for each specific value of $t$). Real-valued functions are sometimes called **scalar functions** to distinguish them from vector-valued functions.

Vector-valued functions serve dual roles in the representation of curves. By letting the parameter $t$ represent time, you can use a vector-valued function to represent *motion* along a curve. Or, in the more general case, you can use a vector-valued function to *trace the graph* of a curve. In either case, the terminal point of the position vector $\mathbf{r}(t)$ coincides with the point $(x, y)$ or $(x, y, z)$ on the curve given by the parametric equations, as shown in Figure 12.1. The arrowhead on the curve indicates the curve's *orientation* by pointing in the direction of increasing values of $t$.

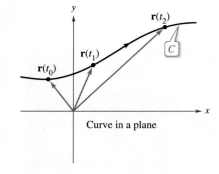

Curve in a plane

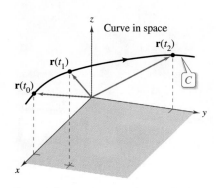

Curve $C$ is traced out by the terminal point of position vector $\mathbf{r}(t)$.

**Figure 12.1**

Unless stated otherwise, the **domain** of a vector-valued function **r** is considered to be the intersection of the domains of the component functions $f$, $g$, and $h$. For instance, the domain of $\mathbf{r}(t) = \ln t\,\mathbf{i} + \sqrt{1 - t}\,\mathbf{j} + t\mathbf{k}$ is the interval $(0, 1]$.

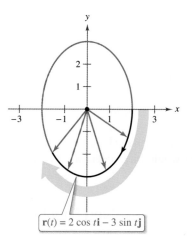

The ellipse is traced clockwise as $t$ increases from 0 to $2\pi$.
**Figure 12.2**

### EXAMPLE 1   Sketching a Plane Curve

Sketch the plane curve represented by the vector-valued function

$$\mathbf{r}(t) = 2\cos t\,\mathbf{i} - 3\sin t\,\mathbf{j}, \quad 0 \le t \le 2\pi. \qquad \text{Vector-valued function}$$

**Solution**   From the position vector $\mathbf{r}(t)$, you can write the parametric equations

$$x = 2\cos t \quad \text{and} \quad y = -3\sin t.$$

Solving for $\cos t$ and $\sin t$ and using the identity $\cos^2 t + \sin^2 t = 1$, you get the rectangular equation

$$\frac{x^2}{2^2} + \frac{y^2}{3^2} = 1. \qquad \text{Rectangular equation}$$

The graph of this rectangular equation is the ellipse shown in Figure 12.2. The curve has a *clockwise* orientation. That is, as $t$ increases from 0 to $2\pi$, the position vector $\mathbf{r}(t)$ moves clockwise, and its terminal point traces the ellipse.

### EXAMPLE 2   Sketching a Space Curve

⋯▷ *See LarsonCalculus.com for an interactive version of this type of example.*

Sketch the space curve represented by the vector-valued function

$$\mathbf{r}(t) = 4\cos t\,\mathbf{i} + 4\sin t\,\mathbf{j} + t\mathbf{k}, \quad 0 \le t \le 4\pi. \qquad \text{Vector-valued function}$$

**Solution**   From the first two parametric equations

$$x = 4\cos t \quad \text{and} \quad y = 4\sin t$$

you can obtain

$$x^2 + y^2 = 16. \qquad \text{Rectangular equation}$$

This means that the curve lies on a right circular cylinder of radius 4, centered about the $z$-axis. To locate the curve on this cylinder, you can use the third parametric equation

$$z = t.$$

In Figure 12.3, note that as $t$ increases from 0 to $4\pi$, the point $(x, y, z)$ spirals up the cylinder to produce a **helix.** A real-life example of a helix is shown in the drawing at the left.

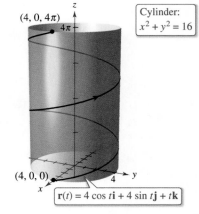

As $t$ increases from 0 to $4\pi$, two spirals on the helix are traced out.
**Figure 12.3**

In 1953, Francis Crick and James D. Watson discovered the double helix structure of DNA.

In Examples 1 and 2, you were given a vector-valued function and were asked to sketch the corresponding curve. The next two examples address the reverse problem—finding a vector-valued function to represent a given graph. Of course, when the graph is described parametrically, representation by a vector-valued function is straightforward. For instance, to represent the line in space given by $x = 2 + t$, $y = 3t$, and $z = 4 - t$, you can simply use the vector-valued function

$$\mathbf{r}(t) = (2 + t)\mathbf{i} + 3t\mathbf{j} + (4 - t)\mathbf{k}.$$

When a set of parametric equations for the graph is not given, the problem of representing the graph by a vector-valued function boils down to finding a set of parametric equations.

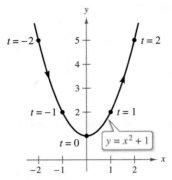

There are many ways to parametrize this graph. One way is to let $x = t$.
**Figure 12.4**

**EXAMPLE 3** **Representing a Graph: Vector-Valued Function**

Represent the parabola

$$y = x^2 + 1$$

by a vector-valued function.

**Solution** Although there are many ways to choose the parameter $t$, a natural choice is to let $x = t$. Then $y = t^2 + 1$ and you have

$$\mathbf{r}(t) = t\mathbf{i} + (t^2 + 1)\mathbf{j}. \qquad \text{Vector-valued function}$$

Note in Figure 12.4 the orientation produced by this particular choice of parameter. Had you chosen $x = -t$ as the parameter, the curve would have been oriented in the opposite direction.

**EXAMPLE 4** **Representing a Graph: Vector-Valued Function**

Sketch the space curve $C$ represented by the intersection of the semiellipsoid

$$\frac{x^2}{12} + \frac{y^2}{24} + \frac{z^2}{4} = 1, \quad z \geq 0$$

and the parabolic cylinder $y = x^2$. Then find a vector-valued function to represent the graph.

**Solution** The intersection of the two surfaces is shown in Figure 12.5. As in Example 3, a natural choice of parameter is $x = t$. For this choice, you can use the given equation $y = x^2$ to obtain $y = t^2$. Then it follows that

$$\frac{z^2}{4} = 1 - \frac{x^2}{12} - \frac{y^2}{24} = 1 - \frac{t^2}{12} - \frac{t^4}{24} = \frac{24 - 2t^2 - t^4}{24} = \frac{(6 + t^2)(4 - t^2)}{24}.$$

Because the curve lies above the $xy$-plane, you should choose the positive square root for $z$ and obtain the parametric equations

$$x = t, \quad y = t^2, \quad \text{and} \quad z = \sqrt{\frac{(6 + t^2)(4 - t^2)}{6}}.$$

The resulting vector-valued function is

$$\mathbf{r}(t) = t\mathbf{i} + t^2\mathbf{j} + \sqrt{\frac{(6 + t^2)(4 - t^2)}{6}}\mathbf{k}, \quad -2 \leq t \leq 2. \qquad \text{Vector-valued function}$$

(Note that the **k**-component of $\mathbf{r}(t)$ implies $-2 \leq t \leq 2$.) From the points $(-2, 4, 0)$ and $(2, 4, 0)$ shown in Figure 12.5, you can see that the curve is traced as $t$ increases from $-2$ to $2$.

• **REMARK** Curves in space can be specified in various ways. For instance, the curve in Example 4 is described as the intersection of two surfaces in space.

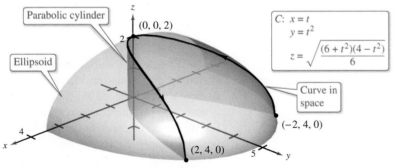

The curve $C$ is the intersection of the semiellipsoid and the parabolic cylinder.
**Figure 12.5**

## Limits and Continuity

Many techniques and definitions used in the calculus of real-valued functions can be applied to vector-valued functions. For instance, you can add and subtract vector-valued functions, multiply a vector-valued function by a scalar, take the limit of a vector-valued function, differentiate a vector-valued function, and so on. The basic approach is to capitalize on the linearity of vector operations by extending the definitions on a component-by-component basis. For example, to add two vector-valued functions (in the plane), you can write

$$\mathbf{r}_1(t) + \mathbf{r}_2(t) = [f_1(t)\mathbf{i} + g_1(t)\mathbf{j}] + [f_2(t)\mathbf{i} + g_2(t)\mathbf{j}] \qquad \text{Sum}$$

$$= [f_1(t) + f_2(t)]\mathbf{i} + [g_1(t) + g_2(t)]\mathbf{j}.$$

To subtract two vector-valued functions, you can write

$$\mathbf{r}_1(t) - \mathbf{r}_2(t) = [f_1(t)\mathbf{i} + g_1(t)\mathbf{j}] - [f_2(t)\mathbf{i} + g_2(t)\mathbf{j}] \qquad \text{Difference}$$

$$= [f_1(t) - f_2(t)]\mathbf{i} + [g_1(t) - g_2(t)]\mathbf{j}.$$

Similarly, to multiply a vector-valued function by a scalar, you can write

$$c\mathbf{r}(t) = c[f_1(t)\mathbf{i} + g_1(t)\mathbf{j}] \qquad \text{Scalar multiplication}$$

$$= cf_1(t)\mathbf{i} + cg_1(t)\mathbf{j}.$$

To divide a vector-valued function by a scalar, you can write

$$\frac{\mathbf{r}(t)}{c} = \frac{[f_1(t)\mathbf{i} + g_1(t)\mathbf{j}]}{c}, \quad c \neq 0 \qquad \text{Scalar division}$$

$$= \frac{f_1(t)}{c}\mathbf{i} + \frac{g_1(t)}{c}\mathbf{j}.$$

This component-by-component extension of operations with real-valued functions to vector-valued functions is further illustrated in the definition of the limit of a vector-valued function.

---

**Definition of the Limit of a Vector-Valued Function**

1. If $\mathbf{r}$ is a vector-valued function such that $\mathbf{r}(t) = f(t)\mathbf{i} + g(t)\mathbf{j}$, then

$$\lim_{t \to a} \mathbf{r}(t) = \left[\lim_{t \to a} f(t)\right]\mathbf{i} + \left[\lim_{t \to a} g(t)\right]\mathbf{j} \qquad \text{Plane}$$

provided $f$ and $g$ have limits as $t \to a$.

2. If $\mathbf{r}$ is a vector-valued function such that $\mathbf{r}(t) = f(t)\mathbf{i} + g(t)\mathbf{j} + h(t)\mathbf{k}$, then

$$\lim_{t \to a} \mathbf{r}(t) = \left[\lim_{t \to a} f(t)\right]\mathbf{i} + \left[\lim_{t \to a} g(t)\right]\mathbf{j} + \left[\lim_{t \to a} h(t)\right]\mathbf{k} \qquad \text{Space}$$

provided $f$, $g$, and $h$ have limits as $t \to a$.

---

If $\mathbf{r}(t)$ approaches the vector $\mathbf{L}$ as $t \to a$, then the length of the vector $\mathbf{r}(t) - \mathbf{L}$ approaches 0. That is,

$$\|\mathbf{r}(t) - \mathbf{L}\| \to 0 \quad \text{as} \quad t \to a.$$

This is illustrated graphically in Figure 12.6. With this definition of the limit of a vector-valued function, you can develop vector versions of most of the limit theorems given in Chapter 1. For example, the limit of the sum of two vector-valued functions is the sum of their individual limits. Also, you can use the orientation of the curve $\mathbf{r}(t)$ to define one-sided limits of vector-valued functions. The next definition extends the notion of continuity to vector-valued functions.

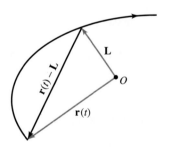

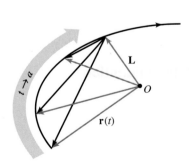

As $t$ approaches $a$, $\mathbf{r}(t)$ approaches the limit $\mathbf{L}$. For the limit $\mathbf{L}$ to exist, it is not necessary that $\mathbf{r}(a)$ be defined or that $\mathbf{r}(a)$ be equal to $\mathbf{L}$.

**Figure 12.6**

> ## Definition of Continuity of a Vector-Valued Function
>
> A vector-valued function **r** is **continuous at the point** given by $t = a$ when the limit of **r**$(t)$ exists as $t \to a$ and
>
> $$\lim_{t \to a} \mathbf{r}(t) = \mathbf{r}(a).$$
>
> A vector-valued function **r** is **continuous on an interval** $I$ when it is continuous at every point in the interval.

From this definition, it follows that a vector-valued function is continuous at $t = a$ if and only if each of its component functions is continuous at $t = a$.

### EXAMPLE 5    Continuity of a Vector-Valued Function

Discuss the continuity of the vector-valued function

$$\mathbf{r}(t) = t\mathbf{i} + a\mathbf{j} + (a^2 - t^2)\mathbf{k} \qquad a \text{ is a constant.}$$

at $t = 0$.

**Solution**    As $t$ approaches 0, the limit is

$$\lim_{t \to 0} \mathbf{r}(t) = \left[\lim_{t \to 0} t\right]\mathbf{i} + \left[\lim_{t \to 0} a\right]\mathbf{j} + \left[\lim_{t \to 0} (a^2 - t^2)\right]\mathbf{k}$$

$$= 0\mathbf{i} + a\mathbf{j} + a^2\mathbf{k}$$

$$= a\mathbf{j} + a^2\mathbf{k}.$$

Because

$$\mathbf{r}(0) = (0)\mathbf{i} + (a)\mathbf{j} + (a^2)\mathbf{k}$$

$$= a\mathbf{j} + a^2\mathbf{k}$$

you can conclude that **r** is continuous at $t = 0$. By similar reasoning, you can conclude that the vector-valued function **r** is continuous at all real-number values of $t$. ∎

For each value of $a$, the curve represented by the vector-valued function in Example 5

$$\mathbf{r}(t) = t\mathbf{i} + a\mathbf{j} + (a^2 - t^2)\mathbf{k} \qquad a \text{ is a constant.}$$

is a parabola. You can think of each parabola as the intersection of the vertical plane $y = a$ and the hyperbolic paraboloid

$$y^2 - x^2 = z$$

as shown in Figure 12.7.

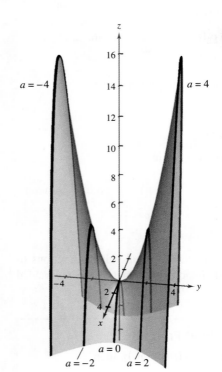

For each value of $a$, the curve represented by the vector-valued function $\mathbf{r}(t) = t\mathbf{i} + a\mathbf{j} + (a^2 - t^2)\mathbf{k}$ is a parabola.
**Figure 12.7**

▷ **TECHNOLOGY** Almost any type of three-dimensional sketch is difficult to do by hand, but sketching curves in space is especially difficult. The problem is trying to create the illusion of three dimensions. Graphing utilities use a variety of techniques to add "three-dimensionality" to graphs of space curves. One way is to show the curve on a surface, as in Figure 12.7.

### EXAMPLE 6    Continuity of a Vector-Valued Function

Determine the interval(s) on which the vector-valued function

$$\mathbf{r}(t) = t\mathbf{i} + \sqrt{t + 1}\,\mathbf{j} + (t^2 + 1)\mathbf{k}$$

is continuous.

**Solution**    The component functions are

$$f(t) = t, \quad g(t) = \sqrt{t + 1}, \quad \text{and} \quad h(t) = (t^2 + 1).$$

Both $f$ and $h$ are continuous for all real-number values of $t$. The function $g$, however, is continuous only for $t \geq -1$. So, **r** is continuous on the interval $[-1, \infty)$. ∎

# 12.1  Exercises

See **CalcChat.com** for tutorial help and worked-out solutions to odd-numbered exercises.

## CONCEPT CHECK

1. **Vector-Valued Function**  Describe how you can use a vector-valued function to represent a curve.

2. **Continuity of a Vector-Valued Function**  Describe what it means for a vector-valued function $\mathbf{r}(t)$ to be continuous at a point.

  **Finding the Domain**  In Exercises 3–10, find the domain of the vector-valued function.

3. $\mathbf{r}(t) = \dfrac{1}{t+1}\mathbf{i} + \dfrac{t}{2}\mathbf{j} - 3t\mathbf{k}$

4. $\mathbf{r}(t) = \sqrt{4 - t^2}\,\mathbf{i} + t^2\mathbf{j} - 6t\mathbf{k}$

5. $\mathbf{r}(t) = \ln t\,\mathbf{i} - e^t\mathbf{j} - t\mathbf{k}$

6. $\mathbf{r}(t) = \sin t\,\mathbf{i} + 4 \cos t\,\mathbf{j} + t\mathbf{k}$

7. $\mathbf{r}(t) = \mathbf{F}(t) + \mathbf{G}(t)$, where

   $\mathbf{F}(t) = \cos t\,\mathbf{i} - \sin t\,\mathbf{j} + \sqrt{t}\,\mathbf{k}$,  $\mathbf{G}(t) = \cos t\,\mathbf{i} + \sin t\,\mathbf{j}$

8. $\mathbf{r}(t) = \mathbf{F}(t) - \mathbf{G}(t)$, where

   $\mathbf{F}(t) = \ln t\,\mathbf{i} + 5t\mathbf{j} - 3t^2\mathbf{k}$,  $\mathbf{G}(t) = \mathbf{i} + 4t\mathbf{j} - 3t^2\mathbf{k}$

9. $\mathbf{r}(t) = \mathbf{F}(t) \times \mathbf{G}(t)$, where

   $\mathbf{F}(t) = \sin t\,\mathbf{i} + \cos t\,\mathbf{j}$,  $\mathbf{G}(t) = \sin t\,\mathbf{j} + \cos t\,\mathbf{k}$

10. $\mathbf{r}(t) = \mathbf{F}(t) \times \mathbf{G}(t)$, where

    $\mathbf{F}(t) = t^3\mathbf{i} - t\mathbf{j} + t\mathbf{k}$,  $\mathbf{G}(t) = \sqrt[3]{t}\,\mathbf{i} + \dfrac{1}{t+1}\mathbf{j} + (t+2)\mathbf{k}$

  **Evaluating a Function**  In Exercises 11 and 12, evaluate the vector-valued function at each given value of $t$.

11. $\mathbf{r}(t) = \frac{1}{2}t^2\mathbf{i} - (t-1)\mathbf{j}$

    (a) $\mathbf{r}(1)$    (b) $\mathbf{r}(0)$    (c) $\mathbf{r}(s+1)$

    (d) $\mathbf{r}(2 + \Delta t) - \mathbf{r}(2)$

12. $\mathbf{r}(t) = \cos t\,\mathbf{i} + 2 \sin t\,\mathbf{j}$

    (a) $\mathbf{r}(0)$    (b) $\mathbf{r}(\pi/4)$    (c) $\mathbf{r}(\theta - \pi)$

    (d) $\mathbf{r}(\pi/6 + \Delta t) - \mathbf{r}(\pi/6)$

**Writing a Vector-Valued Function**  In Exercises 13–16, represent the line segment from $P$ to $Q$ by a vector-valued function and by a set of parametric equations.

13. $P(0, 0, 0)$, $Q(5, 2, 2)$     14. $P(0, 2, -1)$, $Q(4, 7, 2)$

15. $P(-3, -6, -1)$, $Q(-1, -9, -8)$

16. $P(1, -6, 8)$, $Q(-3, -2, 5)$

**Think About It**  In Exercises 17 and 18, find $\mathbf{r}(t) \cdot \mathbf{u}(t)$. Is the result a vector-valued function? Explain.

17. $\mathbf{r}(t) = (3t - 1)\mathbf{i} + \frac{1}{4}t^3\mathbf{j} + 4\mathbf{k}$,  $\mathbf{u}(t) = t^2\mathbf{i} - 8\mathbf{j} + t^3\mathbf{k}$

18. $\mathbf{r}(t) = \langle 3 \cos t, 2 \sin t, t - 2 \rangle$,  $\mathbf{u}(t) = \langle 4 \sin t, -6 \cos t, t^2 \rangle$

**Matching**  In Exercises 19–22, match the equation with its graph. [The graphs are labeled (a), (b), (c), and (d).]

(a)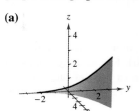

(b)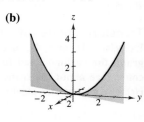

(c)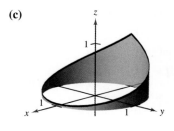

(d)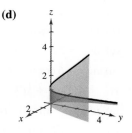

19. $\mathbf{r}(t) = t\mathbf{i} + 2t\mathbf{j} + t^2\mathbf{k}$,  $-2 \le t \le 2$

20. $\mathbf{r}(t) = \cos(\pi t)\mathbf{i} + \sin(\pi t)\mathbf{j} + t^2\mathbf{k}$,  $-1 \le t \le 1$

21. $\mathbf{r}(t) = t\mathbf{i} + t^2\mathbf{j} + e^{0.75t}\mathbf{k}$,  $-2 \le t \le 2$

22. $\mathbf{r}(t) = t\mathbf{i} + \ln t\,\mathbf{j} + \dfrac{2t}{3}\mathbf{k}$,  $0.1 \le t \le 5$

  **Sketching a Plane Curve**  In Exercises 23–30, sketch the plane curve represented by the vector-valued function and give the orientation of the curve.

23. $\mathbf{r}(t) = \dfrac{t}{4}\mathbf{i} + (t-1)\mathbf{j}$     24. $\mathbf{r}(t) = (5 - t)\mathbf{i} + \sqrt{t}\,\mathbf{j}$

25. $\mathbf{r}(t) = t^3\mathbf{i} + t^2\mathbf{j}$

26. $\mathbf{r}(t) = (t^2 + t)\mathbf{i} + (t^2 - t)\mathbf{j}$

27. $\mathbf{r}(\theta) = \cos \theta\,\mathbf{i} + 3 \sin \theta\,\mathbf{j}$

28. $\mathbf{r}(t) = 2 \cos t\,\mathbf{i} + 2 \sin t\,\mathbf{j}$

29. $\mathbf{r}(\theta) = 3 \sec \theta\,\mathbf{i} + 2 \tan \theta\,\mathbf{j}$

30. $\mathbf{r}(t) = 2 \cos^3 t\,\mathbf{i} + 2 \sin^3 t\,\mathbf{j}$

**Sketching a Space Curve**  In Exercises 31–38, sketch the space curve represented by the vector-valued function and give the orientation of the curve.

31. $\mathbf{r}(t) = (-t + 1)\mathbf{i} + (4t + 2)\mathbf{j} + (2t + 3)\mathbf{k}$

32. $\mathbf{r}(t) = t\mathbf{i} + (2t - 5)\mathbf{j} + 3t\mathbf{k}$

33. $\mathbf{r}(t) = 2 \cos t\,\mathbf{i} + 2 \sin t\,\mathbf{j} + t\mathbf{k}$

34. $\mathbf{r}(t) = t\mathbf{i} + 3 \cos t\,\mathbf{j} + 3 \sin t\,\mathbf{k}$

35. $\mathbf{r}(t) = 2 \sin t\,\mathbf{i} + 2 \cos t\,\mathbf{j} + e^{-t}\mathbf{k}$

36. $\mathbf{r}(t) = t^2\mathbf{i} + 2t\mathbf{j} + \frac{3}{2}t\mathbf{k}$

37. $\mathbf{r}(t) = \left\langle t, t^2, \frac{2}{3}t^3 \right\rangle$

38. $\mathbf{r}(t) = \langle \cos t + t \sin t, \sin t - t \cos t, t \rangle$

**Identifying a Common Curve** In Exercises 39 and 40, use a computer algebra system to graph the vector-valued function and identify the common curve.

**39.** $\mathbf{r}(t) = -\frac{1}{2}t^2\mathbf{i} + t\mathbf{j} - \frac{\sqrt{3}}{2}t^2\mathbf{k}$

**40.** $\mathbf{r}(t) = -\sqrt{2}\sin t\mathbf{i} + 2\cos t\mathbf{j} + \sqrt{2}\sin t\mathbf{k}$

**Transformations of Vector-Valued Functions** In Exercises 41 and 42, use a computer algebra system to graph the vector-valued function $\mathbf{r}(t)$. For each $\mathbf{u}(t)$, make a conjecture about the transformation (if any) of the graph of $\mathbf{r}(t)$. Use a computer algebra system to verify your conjecture.

**41.** $\mathbf{r}(t) = 2\cos t\mathbf{i} + 2\sin t\mathbf{j} + \frac{1}{2}t\mathbf{k}$

(a) $\mathbf{u}(t) = 2(\cos t - 1)\mathbf{i} + 2\sin t\mathbf{j} + \frac{1}{2}t\mathbf{k}$

(b) $\mathbf{u}(t) = 2\cos t\mathbf{i} + 2\sin t\mathbf{j} + 2t\mathbf{k}$

(c) $\mathbf{u}(t) = 2\cos(-t)\mathbf{i} + 2\sin(-t)\mathbf{j} + \frac{1}{2}(-t)\mathbf{k}$

(d) $\mathbf{u}(t) = 6\cos t\mathbf{i} + 6\sin t\mathbf{j} + \frac{1}{2}t\mathbf{k}$

**42.** $\mathbf{r}(t) = t\mathbf{i} + t^2\mathbf{j} + \frac{1}{2}t^3\mathbf{k}$

(a) $\mathbf{u}(t) = (-t)\mathbf{i} + (-t)^2\mathbf{j} + \frac{1}{2}(-t)^3\mathbf{k}$

(b) $\mathbf{u}(t) = t^2\mathbf{i} + t\mathbf{j} + \frac{1}{2}t^3\mathbf{k}$

(c) $\mathbf{u}(t) = t\mathbf{i} + t^2\mathbf{j} + (\frac{1}{2}t^3 + 4)\mathbf{k}$

(d) $\mathbf{u}(t) = t\mathbf{i} + t^2\mathbf{j} + \frac{1}{8}t^3\mathbf{k}$

**Writing a Transformation** In Exercises 43–46, consider the vector-valued function $\mathbf{r}(t) = 3t^2\mathbf{i} + (t-1)\mathbf{j} + t\mathbf{k}$. Write a vector-valued function $\mathbf{u}(t)$ that is the specified transformation of $\mathbf{r}$.

**43.** A vertical translation two units upward

**44.** A horizontal translation one unit in the direction of the positive $x$-axis

**45.** The $y$-value increases by a factor of two

**46.** The $z$-value increases by a factor of three

 **Representing a Graph by a Vector-Valued Function** In Exercises 47–54, represent the plane curve by a vector-valued function. (There are many correct answers.)

**47.** $y = x + 5$

**48.** $2x - 3y + 5 = 0$

**49.** $y = (x - 2)^2$

**50.** $y = 4 - x^2$

**51.** $x^2 + y^2 = 25$

**52.** $(x - 2)^2 + y^2 = 4$

**53.** $\frac{x^2}{16} - \frac{y^2}{4} = 1$

**54.** $\frac{x^2}{9} + \frac{y^2}{16} = 1$

**Representing a Graph by a Vector-Valued Function** In Exercises 55–62, sketch the space curve represented by the intersection of the surfaces. Then represent the curve by a vector-valued function using the given parameter.

| Surfaces | Parameter |
|---|---|
| **55.** $z = x^2 + y^2$, $x + y = 0$ | $x = t$ |
| **56.** $z = x^2 + y^2$, $z = 4$ | $x = 2\cos t$ |
| **57.** $x^2 + y^2 = 4$, $z = x^2$ | $x = 2\sin t$ |
| **58.** $4x^2 + 4y^2 + z^2 = 16$, $x = z^2$ | $z = t$ |
| **59.** $x^2 + y^2 + z^2 = 4$, $x + z = 2$ | $x = 1 + \sin t$ |
| **60.** $x^2 + y^2 + z^2 = 10$, $x + y = 4$ | $x = 2 + \sin t$ |
| **61.** $x^2 + z^2 = 4$, $y^2 + z^2 = 4$ | $x = t$ (first octant) |
| **62.** $x^2 + y^2 + z^2 = 16$, $xy = 4$ | $x = t$ (first octant) |

**63. Sketching a Curve** Show that the vector-valued function $\mathbf{r}(t) = t\mathbf{i} + 2t\cos t\mathbf{j} + 2t\sin t\mathbf{k}$ lies on the cone $4x^2 = y^2 + z^2$. Sketch the curve.

**64. Sketching a Curve** Show that the vector-valued function $\mathbf{r}(t) = e^{-t}\cos t\mathbf{i} + e^{-t}\sin t\mathbf{j} + e^{-t}\mathbf{k}$ lies on the cone $z^2 = x^2 + y^2$. Sketch the curve.

 **Finding a Limit** In Exercises 65–70, find the limit (if it exists).

**65.** $\lim\limits_{t \to \pi} (t\mathbf{i} + \cos t\mathbf{j} + \sin t\mathbf{k})$

**66.** $\lim\limits_{t \to 2} \left(3t\mathbf{i} + \frac{2}{t^2 - 1}\mathbf{j} + \frac{1}{t}\mathbf{k}\right)$

**67.** $\lim\limits_{t \to 0} \left(t^2\mathbf{i} + 3t\mathbf{j} + \frac{1 - \cos t}{t}\mathbf{k}\right)$

**68.** $\lim\limits_{t \to 1} \left(\sqrt{t}\mathbf{i} + \frac{\ln t}{t^2 - 1}\mathbf{j} + \frac{1}{t - 1}\mathbf{k}\right)$

**69.** $\lim\limits_{t \to 0} \left(e^t\mathbf{i} + \frac{\sin t}{t}\mathbf{j} + e^{-t}\mathbf{k}\right)$

**70.** $\lim\limits_{t \to \infty} \left(e^{-t}\mathbf{i} + \frac{1}{t}\mathbf{j} + t^{1/t}\mathbf{k}\right)$

**Continuity of a Vector-Valued Function** In Exercises 71–76, determine the interval(s) on which the vector-valued function is continuous.

**71.** $\mathbf{r}(t) = \frac{1}{2t + 1}\mathbf{i} + \frac{1}{t}\mathbf{j}$

**72.** $\mathbf{r}(t) = \sqrt{t}\mathbf{i} + \sqrt{t - 1}\mathbf{j}$

**73.** $\mathbf{r}(t) = t\mathbf{i} + \arcsin t\mathbf{j} + (t - 1)\mathbf{k}$

**74.** $\mathbf{r}(t) = 2e^{-t}\mathbf{i} + e^{-t}\mathbf{j} + \ln(t - 1)\mathbf{k}$

**75.** $\mathbf{r}(t) = \langle e^{-t}, t^2, \tan t \rangle$

**76.** $\mathbf{r}(t) = \langle 8, \sqrt{t}, \sqrt[3]{t} \rangle$

**EXPLORING CONCEPTS**

**77. Think About It** Consider first-degree polynomial functions $f(t)$, $g(t)$, and $h(t)$. Determine whether the curve represented by $\mathbf{r}(t) = f(t)\mathbf{i} + g(t)\mathbf{j} + h(t)\mathbf{k}$ is a line. Explain.

**78. Think About It** The curve represented by $\mathbf{r}(t) = f(t)\mathbf{i} + g(t)\mathbf{j} + h(t)\mathbf{k}$ is a line. Are $f$, $g$, and $h$ first-degree polynomial functions of $t$? Explain.

**79. Continuity of a Vector-Valued Function** Give an example of a vector-valued function that is defined but not continuous at $t = 3$.

**80. Comparing Functions** Which of the following vector-valued functions represent the same graph?

(a) $\mathbf{r}(t) = (-3\cos t + 1)\mathbf{i} + (5\sin t + 2)\mathbf{j} + 4\mathbf{k}$

(b) $\mathbf{r}(t) = 4\mathbf{i} + (-3\cos t + 1)\mathbf{j} + (5\sin t + 2)\mathbf{k}$

(c) $\mathbf{r}(t) = (3\cos t - 1)\mathbf{i} + (-5\sin t - 2)\mathbf{j} + 4\mathbf{k}$

(d) $\mathbf{r}(t) = (-3\cos 2t + 1)\mathbf{i} + (5\sin 2t + 2)\mathbf{j} + 4\mathbf{k}$

**· · 81. Staircase · · · · · · · · · · · · · · · · · · · · ·**

The outer bottom edge of a staircase is in the shape of a helix of radius 1 meter. The staircase has a height of 4 meters and makes two complete revolutions from top to bottom. Find a vector-valued function for the staircase. Use a computer algebra system to graph your function. (There are many correct answers.)

**82. HOW DO YOU SEE IT?** The four figures below are graphs of the vector-valued function

$$\mathbf{r}(t) = 4\cos t\,\mathbf{i} + 4\sin t\,\mathbf{j} + \frac{t}{4}\mathbf{k}.$$

Match each of the four graphs with the point in space from which the helix is viewed.

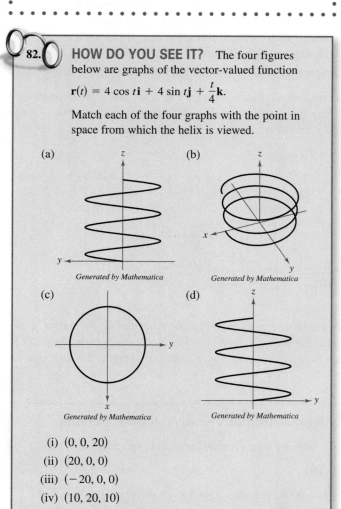

(a) *Generated by Mathematica*

(b) *Generated by Mathematica*

(c) *Generated by Mathematica*

(d) *Generated by Mathematica*

(i) $(0, 0, 20)$

(ii) $(20, 0, 0)$

(iii) $(-20, 0, 0)$

(iv) $(10, 20, 10)$

**83. Proof** Let $\mathbf{r}(t)$ and $\mathbf{u}(t)$ be vector-valued functions whose limits exist as $t \to c$. Prove that

$$\lim_{t \to c}\left[\mathbf{r}(t) \times \mathbf{u}(t)\right] = \lim_{t \to c}\mathbf{r}(t) \times \lim_{t \to c}\mathbf{u}(t).$$

**84. Proof** Let $\mathbf{r}(t)$ and $\mathbf{u}(t)$ be vector-valued functions whose limits exist as $t \to c$. Prove that

$$\lim_{t \to c}\left[\mathbf{r}(t) \cdot \mathbf{u}(t)\right] = \lim_{t \to c}\mathbf{r}(t) \cdot \lim_{t \to c}\mathbf{u}(t).$$

**85. Proof** Prove that if $\mathbf{r}$ is a vector-valued function that is continuous at $c$, then $\|\mathbf{r}\|$ is continuous at $c$.

**86. Verifying a Converse** Verify that the converse of Exercise 85 is not true by finding a vector-valued function $\mathbf{r}$ such that $\|\mathbf{r}\|$ is continuous at $c$ but $\mathbf{r}$ is not continuous at $c$.

**Think About It** In Exercises 87 and 88, two particles travel along the space curves $\mathbf{r}(t)$ and $\mathbf{u}(t)$.

**87.** If $\mathbf{r}(t)$ and $\mathbf{u}(t)$ intersect, will the particles collide?

**88.** If the particles collide, do their paths $\mathbf{r}(t)$ and $\mathbf{u}(t)$ intersect?

**Particle Motion** In Exercises 89 and 90, two particles travel along the space curves $\mathbf{r}(t)$ and $\mathbf{u}(t)$. Do the particles collide? Do their paths intersect?

**89.** $\mathbf{r}(t) = t^2\mathbf{i} + (9t - 20)\mathbf{j} + t^2\mathbf{k}$

$\mathbf{u}(t) = (3t + 4)\mathbf{i} + t^2\mathbf{j} + (5t - 4)\mathbf{k}$

**90.** $\mathbf{r}(t) = t\mathbf{i} + t^2\mathbf{j} + t^3\mathbf{k}$

$\mathbf{u}(t) = (-2t + 3)\mathbf{i} + 8t\mathbf{j} + (12t + 2)\mathbf{k}$

**SECTION PROJECT**

# Witch of Agnesi

In Section 3.5, you studied a famous curve called the **Witch of Agnesi**. In this project, you will take a closer look at this function.

Consider a circle of radius $a$ centered on the $y$-axis at $(0, a)$. Let $A$ be a point on the horizontal line $y = 2a$, let $O$ be the origin, and let $B$ be the point where the segment $OA$ intersects the circle. A point $P$ is on the Witch of Agnesi when $P$ lies on the horizontal line through $B$ and on the vertical line through $A$.

(a) Show that the point $A$ is traced out by the vector-valued function

$$\mathbf{r}_A(\theta) = 2a\cot\theta\,\mathbf{i} + 2a\mathbf{j}, \quad 0 < \theta < \pi$$

where $\theta$ is the angle that $OA$ makes with the positive $x$-axis.

(b) Show that the point $B$ is traced out by the vector-valued function $\mathbf{r}_B(\theta) = a\sin 2\theta\,\mathbf{i} + a(1 - \cos 2\theta)\mathbf{j}, \; 0 < \theta < \pi$.

(c) Combine the results of parts (a) and (b) to find the vector-valued function $\mathbf{r}(\theta)$ for the Witch of Agnesi. Use a graphing utility to graph this curve for $a = 1$.

(d) Describe the limits $\lim\limits_{\theta \to 0^+}\mathbf{r}(\theta)$ and $\lim\limits_{\theta \to \pi^-}\mathbf{r}(\theta)$.

(e) Eliminate the parameter $\theta$ and determine the rectangular equation of the Witch of Agnesi. Use a graphing utility to graph this function for $a = 1$ and compare your graph with that obtained in part (c).

# 12.2 Differentiation and Integration of Vector-Valued Functions

■ Differentiate a vector-valued function.
■ Integrate a vector-valued function.

## Differentiation of Vector-Valued Functions

In Sections 12.3–12.5, you will study several important applications involving the calculus of vector-valued functions. In preparation for that study, this section is devoted to the mechanics of differentiation and integration of vector-valued functions.

The definition of the derivative of a vector-valued function parallels the definition for real-valued functions.

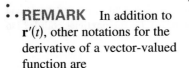

**·· REMARK** In addition to $\mathbf{r}'(t)$, other notations for the derivative of a vector-valued function are

$$\frac{d}{dt}[\mathbf{r}(t)], \quad \frac{d\mathbf{r}}{dt}, \quad \text{and} \quad D_t[\mathbf{r}(t)].$$

> **Definition of the Derivative of a Vector-Valued Function**
>
> The **derivative of a vector-valued function r** is
>
> $$\mathbf{r}'(t) = \lim_{\Delta t \to 0} \frac{\mathbf{r}(t + \Delta t) - \mathbf{r}(t)}{\Delta t}$$
>
> for all $t$ for which the limit exists. If $\mathbf{r}'(t)$ exists, then **r** is **differentiable at $t$.** If $\mathbf{r}'(t)$ exists for all $t$ in an open interval $I$, then **r** is **differentiable on the interval $I$.** Differentiability of vector-valued functions can be extended to closed intervals by considering one-sided limits.

Differentiation of vector-valued functions can be done on a *component-by-component basis*. To see why this is true, consider the function $\mathbf{r}(t) = f(t)\mathbf{i} + g(t)\mathbf{j}$. Applying the definition of the derivative produces the following.

$$\mathbf{r}'(t) = \lim_{\Delta t \to 0} \frac{\mathbf{r}(t + \Delta t) - \mathbf{r}(t)}{\Delta t}$$

$$= \lim_{\Delta t \to 0} \frac{f(t + \Delta t)\mathbf{i} + g(t + \Delta t)\mathbf{j} - f(t)\mathbf{i} - g(t)\mathbf{j}}{\Delta t}$$

$$= \lim_{\Delta t \to 0} \left\{ \left[ \frac{f(t + \Delta t) - f(t)}{\Delta t} \right]\mathbf{i} + \left[ \frac{g(t + \Delta t) - g(t)}{\Delta t} \right]\mathbf{j} \right\}$$

$$= \left\{ \lim_{\Delta t \to 0} \left[ \frac{f(t + \Delta t) - f(t)}{\Delta t} \right] \right\}\mathbf{i} + \left\{ \lim_{\Delta t \to 0} \left[ \frac{g(t + \Delta t) - g(t)}{\Delta t} \right] \right\}\mathbf{j}$$

$$= f'(t)\mathbf{i} + g'(t)\mathbf{j}$$

This important result is listed in the theorem shown below. Note that the derivative of the vector-valued function **r** is itself a vector-valued function. You can see from Figure 12.8 that $\mathbf{r}'(t)$ is a vector tangent to the curve given by $\mathbf{r}(t)$ and pointing in the direction of increasing $t$-values.

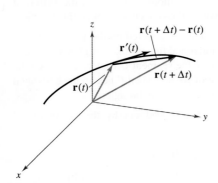

**Figure 12.8**

> **THEOREM 12.1 Differentiation of Vector-Valued Functions**
>
> **1.** If $\mathbf{r}(t) = f(t)\mathbf{i} + g(t)\mathbf{j}$, where $f$ and $g$ are differentiable functions of $t$, then
>
> $$\mathbf{r}'(t) = f'(t)\mathbf{i} + g'(t)\mathbf{j}. \qquad \text{Plane}$$
>
> **2.** If $\mathbf{r}(t) = f(t)\mathbf{i} + g(t)\mathbf{j} + h(t)\mathbf{k}$, where $f$, $g$, and $h$ are differentiable functions of $t$, then
>
> $$\mathbf{r}'(t) = f'(t)\mathbf{i} + g'(t)\mathbf{j} + h'(t)\mathbf{k}. \qquad \text{Space}$$

**EXAMPLE 1**    **Differentiation of a Vector-Valued Function**

⋮ ... ▷ *See LarsonCalculus.com for an interactive version of this type of example.*

For the vector-valued function

$$\mathbf{r}(t) = t\mathbf{i} + (t^2 + 2)\mathbf{j}$$

find $\mathbf{r}'(t)$. Then sketch the plane curve represented by $\mathbf{r}(t)$ and the graphs of $\mathbf{r}(1)$ and $\mathbf{r}'(1)$.

**Solution**    Differentiate on a component-by-component basis to obtain

$$\mathbf{r}'(t) = \mathbf{i} + 2t\mathbf{j}.$$    Derivative

From the position vector $\mathbf{r}(t)$, you can write the parametric equations $x = t$ and $y = t^2 + 2$. The corresponding rectangular equation is $y = x^2 + 2$. When $t = 1$,

$$\mathbf{r}(1) = \mathbf{i} + 3\mathbf{j}$$

and

$$\mathbf{r}'(1) = \mathbf{i} + 2\mathbf{j}.$$

In Figure 12.9, $\mathbf{r}(1)$ is drawn starting at the origin, and $\mathbf{r}'(1)$ is drawn starting at the terminal point of $\mathbf{r}(1)$. Note that at $(1, 3)$, the vector $\mathbf{r}'(1)$ is tangent to the curve given by $\mathbf{r}(t)$ and is pointing in the direction of increasing $t$-values. ∎

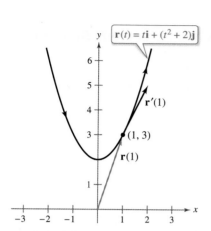

$\mathbf{r}(t) = t\mathbf{i} + (t^2 + 2)\mathbf{j}$

**Figure 12.9**

Higher-order derivatives of vector-valued functions are obtained by successive differentiation of each component function.

**EXAMPLE 2**    **Higher-Order Differentiation**

For the vector-valued function

$$\mathbf{r}(t) = \cos t\,\mathbf{i} + \sin t\,\mathbf{j} + 2t\mathbf{k}$$

find each of the following.

**a.** $\mathbf{r}'(t)$
**b.** $\mathbf{r}''(t)$
**c.** $\mathbf{r}'(t) \cdot \mathbf{r}''(t)$
**d.** $\mathbf{r}'(t) \times \mathbf{r}''(t)$

**Solution**

**a.** $\mathbf{r}'(t) = -\sin t\,\mathbf{i} + \cos t\,\mathbf{j} + 2\mathbf{k}$     First derivative

**b.** $\mathbf{r}''(t) = -\cos t\,\mathbf{i} - \sin t\,\mathbf{j} + 0\mathbf{k}$
$\qquad = -\cos t\,\mathbf{i} - \sin t\,\mathbf{j}$     Second derivative

**c.** $\mathbf{r}'(t) \cdot \mathbf{r}''(t) = \sin t \cos t - \sin t \cos t = 0$     Dot product

**d.** $\mathbf{r}'(t) \times \mathbf{r}''(t) = \begin{vmatrix} \mathbf{i} & \mathbf{j} & \mathbf{k} \\ -\sin t & \cos t & 2 \\ -\cos t & -\sin t & 0 \end{vmatrix}$     Cross product

$\qquad = \begin{vmatrix} \cos t & 2 \\ -\sin t & 0 \end{vmatrix}\mathbf{i} - \begin{vmatrix} -\sin t & 2 \\ -\cos t & 0 \end{vmatrix}\mathbf{j} + \begin{vmatrix} -\sin t & \cos t \\ -\cos t & -\sin t \end{vmatrix}\mathbf{k}$

$\qquad = 2\sin t\,\mathbf{i} - 2\cos t\,\mathbf{j} + \mathbf{k}$ ∎

In Example 2(c), note that the dot product is a real-valued function, not a vector-valued function.

The parametrization of the curve represented by the vector-valued function

$$\mathbf{r}(t) = f(t)\mathbf{i} + g(t)\mathbf{j} + h(t)\mathbf{k}$$

is **smooth on an open interval** $I$ when $f'$, $g'$, and $h'$ are continuous on $I$ and $\mathbf{r}'(t) \neq \mathbf{0}$ for any value of $t$ in the interval $I$.

---

**EXAMPLE 3**   **Finding Intervals on Which a Curve Is Smooth**

Find the intervals on which the epicycloid $C$ given by

$$\mathbf{r}(t) = (5 \cos t - \cos 5t)\mathbf{i} + (5 \sin t - \sin 5t)\mathbf{j}, \quad 0 \le t \le 2\pi$$

is smooth.

**Solution**   The derivative of $\mathbf{r}$ is

$$\mathbf{r}'(t) = (-5 \sin t + 5 \sin 5t)\mathbf{i} + (5 \cos t - 5 \cos 5t)\mathbf{j}.$$

In the interval $[0, 2\pi]$, the only values of $t$ for which

$$\mathbf{r}'(t) = 0\mathbf{i} + 0\mathbf{j}$$

are $t = 0$, $\pi/2$, $\pi$, $3\pi/2$, and $2\pi$. Therefore, you can conclude that $C$ is smooth on the intervals

$$\left(0, \frac{\pi}{2}\right), \quad \left(\frac{\pi}{2}, \pi\right), \quad \left(\pi, \frac{3\pi}{2}\right), \quad \text{and} \quad \left(\frac{3\pi}{2}, 2\pi\right)$$

as shown in Figure 12.10.

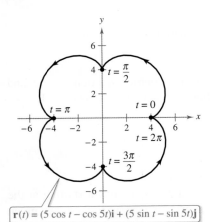

$\mathbf{r}(t) = (5 \cos t - \cos 5t)\mathbf{i} + (5 \sin t - \sin 5t)\mathbf{j}$

The epicycloid is not smooth at the points where it intersects the axes.
**Figure 12.10**

In Figure 12.10, note that the curve is not smooth at points at which the curve makes abrupt changes in direction. Such points are called *cusps* or **nodes.**

Most of the differentiation rules in Chapter 2 have counterparts for vector-valued functions, and several of these are listed in the next theorem. Note that the theorem contains three versions of "product rules." Property 3 gives the derivative of the product of a real-valued function $w$ and a vector-valued function $\mathbf{r}$, Property 4 gives the derivative of the dot product of two vector-valued functions, and Property 5 gives the derivative of the cross product of two vector-valued functions (in space).

---

**THEOREM 12.2   Properties of the Derivative**

Let $\mathbf{r}$ and $\mathbf{u}$ be differentiable vector-valued functions of $t$, let $w$ be a differentiable real-valued function of $t$, and let $c$ be a scalar.

1. $\dfrac{d}{dt}[c\mathbf{r}(t)] = c\mathbf{r}'(t)$

2. $\dfrac{d}{dt}[\mathbf{r}(t) \pm \mathbf{u}(t)] = \mathbf{r}'(t) \pm \mathbf{u}'(t)$

3. $\dfrac{d}{dt}[w(t)\mathbf{r}(t)] = w(t)\mathbf{r}'(t) + w'(t)\mathbf{r}(t)$

4. $\dfrac{d}{dt}[\mathbf{r}(t) \cdot \mathbf{u}(t)] = \mathbf{r}(t) \cdot \mathbf{u}'(t) + \mathbf{r}'(t) \cdot \mathbf{u}(t)$

5. $\dfrac{d}{dt}[\mathbf{r}(t) \times \mathbf{u}(t)] = \mathbf{r}(t) \times \mathbf{u}'(t) + \mathbf{r}'(t) \times \mathbf{u}(t)$

6. $\dfrac{d}{dt}[\mathbf{r}(w(t))] = \mathbf{r}'(w(t))w'(t)$

7. If $\mathbf{r}(t) \cdot \mathbf{r}(t) = c$, then $\mathbf{r}(t) \cdot \mathbf{r}'(t) = 0$.

**• • REMARK**   Note that Property 5 applies only to three-dimensional vector-valued functions because the cross product is not defined for two-dimensional vectors.

**Proof** To prove Property 4, let

$$\mathbf{r}(t) = f_1(t)\mathbf{i} + g_1(t)\mathbf{j} \quad \text{and} \quad \mathbf{u}(t) = f_2(t)\mathbf{i} + g_2(t)\mathbf{j}$$

where $f_1$, $f_2$, $g_1$, and $g_2$ are differentiable functions of $t$. Then

$$\mathbf{r}(t) \cdot \mathbf{u}(t) = f_1(t)f_2(t) + g_1(t)g_2(t)$$

and it follows that

$$\frac{d}{dt}[\mathbf{r}(t) \cdot \mathbf{u}(t)] = f_1(t)f_2'(t) + f_1'(t)f_2(t) + g_1(t)g_2'(t) + g_1'(t)g_2(t)$$

$$= [f_1(t)f_2'(t) + g_1(t)g_2'(t)] + [f_1'(t)f_2(t) + g_1'(t)g_2(t)]$$

$$= \mathbf{r}(t) \cdot \mathbf{u}'(t) + \mathbf{r}'(t) \cdot \mathbf{u}(t).$$

Proofs of the other properties are left as exercises (see Exercises 61–65 and Exercise 68).

---

### Exploration

Let $\mathbf{r}(t) = \cos t\mathbf{i} + \sin t\mathbf{j}$. Sketch the graph of $\mathbf{r}(t)$. Explain why the graph is a circle of radius 1 centered at the origin. Calculate $\mathbf{r}(\pi/4)$ and $\mathbf{r}'(\pi/4)$. Position the vector $\mathbf{r}'(\pi/4)$ so that its initial point is at the terminal point of $\mathbf{r}(\pi/4)$. What do you observe? Show that $\mathbf{r}(t) \cdot \mathbf{r}(t)$ is constant and that $\mathbf{r}(t) \cdot \mathbf{r}'(t) = 0$ for all $t$. How does this example relate to Property 7 of Theorem 12.2?

---

**EXAMPLE 4** **Using Properties of the Derivative**

For $\mathbf{r}(t) = \dfrac{1}{t}\mathbf{i} - \mathbf{j} + \ln t\mathbf{k}$ and $\mathbf{u}(t) = t^2\mathbf{i} - 2t\mathbf{j} + \mathbf{k}$, find each derivative.

**a.** $\dfrac{d}{dt}[\mathbf{r}(t) \cdot \mathbf{u}(t)]$

**b.** $\dfrac{d}{dt}[\mathbf{u}(t) \times \mathbf{u}'(t)]$

**Solution**

**a.** Because $\mathbf{r}'(t) = -\dfrac{1}{t^2}\mathbf{i} + \dfrac{1}{t}\mathbf{k}$ and $\mathbf{u}'(t) = 2t\mathbf{i} - 2\mathbf{j}$, you have

$$\frac{d}{dt}[\mathbf{r}(t) \cdot \mathbf{u}(t)]$$

$$= \mathbf{r}(t) \cdot \mathbf{u}'(t) + \mathbf{r}'(t) \cdot \mathbf{u}(t)$$

$$= \left(\frac{1}{t}\mathbf{i} - \mathbf{j} + \ln t\mathbf{k}\right) \cdot (2t\mathbf{i} - 2\mathbf{j}) + \left(-\frac{1}{t^2}\mathbf{i} + \frac{1}{t}\mathbf{k}\right) \cdot (t^2\mathbf{i} - 2t\mathbf{j} + \mathbf{k})$$

$$= 2 + 2 + (-1) + \frac{1}{t}$$

$$= 3 + \frac{1}{t}.$$

**b.** Because $\mathbf{u}'(t) = 2t\mathbf{i} - 2\mathbf{j}$ and $\mathbf{u}''(t) = 2\mathbf{i}$, you have

$$\frac{d}{dt}[\mathbf{u}(t) \times \mathbf{u}'(t)] = [\mathbf{u}(t) \times \mathbf{u}''(t)] + [\mathbf{u}'(t) \times \mathbf{u}'(t)]$$

$$= \begin{vmatrix} \mathbf{i} & \mathbf{j} & \mathbf{k} \\ t^2 & -2t & 1 \\ 2 & 0 & 0 \end{vmatrix} + \mathbf{0}$$

$$= \begin{vmatrix} -2t & 1 \\ 0 & 0 \end{vmatrix}\mathbf{i} - \begin{vmatrix} t^2 & 1 \\ 2 & 0 \end{vmatrix}\mathbf{j} + \begin{vmatrix} t^2 & -2t \\ 2 & 0 \end{vmatrix}\mathbf{k}$$

$$= 0\mathbf{i} - (-2)\mathbf{j} + 4t\mathbf{k}$$

$$= 2\mathbf{j} + 4t\mathbf{k}.$$

Try reworking parts (a) and (b) in Example 4 by first forming the dot and cross products and then differentiating to see that you obtain the same results.

## Integration of Vector-Valued Functions

The next definition is a consequence of the definition of the derivative of a vector-valued function.

---

**Definition of Integration of Vector-Valued Functions**

1. If $\mathbf{r}(t) = f(t)\mathbf{i} + g(t)\mathbf{j}$, where $f$ and $g$ are continuous on $[a, b]$, then the **indefinite integral (antiderivative)** of $\mathbf{r}$ is

$$\int \mathbf{r}(t)\, dt = \left[\int f(t)\, dt\right]\mathbf{i} + \left[\int g(t)\, dt\right]\mathbf{j} \qquad \text{Plane}$$

and its **definite integral** over the interval $a \le t \le b$ is

$$\int_a^b \mathbf{r}(t)\, dt = \left[\int_a^b f(t)\, dt\right]\mathbf{i} + \left[\int_a^b g(t)\, dt\right]\mathbf{j}.$$

2. If $\mathbf{r}(t) = f(t)\mathbf{i} + g(t)\mathbf{j} + h(t)\mathbf{k}$, where $f$, $g$, and $h$ are continuous on $[a, b]$, then the **indefinite integral (antiderivative)** of $\mathbf{r}$ is

$$\int \mathbf{r}(t)\, dt = \left[\int f(t)\, dt\right]\mathbf{i} + \left[\int g(t)\, dt\right]\mathbf{j} + \left[\int h(t)\, dt\right]\mathbf{k} \qquad \text{Space}$$

and its **definite integral** over the interval $a \le t \le b$ is

$$\int_a^b \mathbf{r}(t)\, dt = \left[\int_a^b f(t)\, dt\right]\mathbf{i} + \left[\int_a^b g(t)\, dt\right]\mathbf{j} + \left[\int_a^b h(t)\, dt\right]\mathbf{k}.$$

---

The antiderivative of a vector-valued function is a family of vector-valued functions all differing by a constant vector $\mathbf{C}$. For instance, if $\mathbf{r}(t)$ is a three-dimensional vector-valued function, then for the indefinite integral $\int \mathbf{r}(t)\, dt$, you obtain three constants of integration

$$\int f(t)\, dt = F(t) + C_1, \quad \int g(t)\, dt = G(t) + C_2, \quad \int h(t)\, dt = H(t) + C_3$$

where $F'(t) = f(t)$, $G'(t) = g(t)$, and $H'(t) = h(t)$. These three *scalar* constants produce one *vector* constant of integration

$$\int \mathbf{r}(t)\, dt = [F(t) + C_1]\mathbf{i} + [G(t) + C_2]\mathbf{j} + [H(t) + C_3]\mathbf{k}$$

$$= [F(t)\mathbf{i} + G(t)\mathbf{j} + H(t)\mathbf{k}] + [C_1\mathbf{i} + C_2\mathbf{j} + C_3\mathbf{k}]$$

$$= \mathbf{R}(t) + \mathbf{C}$$

where $\mathbf{R}'(t) = \mathbf{r}(t)$.

---

**EXAMPLE 5**   **Integrating a Vector-Valued Function**

Find the indefinite integral

$$\int (t\mathbf{i} + 3\mathbf{j})\, dt.$$

**Solution**   Integrating on a component-by-component basis produces

$$\int (t\mathbf{i} + 3\mathbf{j})\, dt = \frac{t^2}{2}\mathbf{i} + 3t\mathbf{j} + \mathbf{C}.$$

Example 6 shows how to evaluate the definite integral of a vector-valued function.

**EXAMPLE 6**    **Definite Integral of a Vector-Valued Function**

Evaluate the integral

$$\int_0^1 \mathbf{r}(t)\, dt = \int_0^1 \left( \sqrt[3]{t}\,\mathbf{i} + \frac{1}{t+1}\mathbf{j} + e^{-t}\mathbf{k} \right) dt.$$

**Solution**

$$\int_0^1 \mathbf{r}(t)\, dt = \left( \int_0^1 t^{1/3}\, dt \right)\mathbf{i} + \left( \int_0^1 \frac{1}{t+1}\, dt \right)\mathbf{j} + \left( \int_0^1 e^{-t}\, dt \right)\mathbf{k}$$

$$= \left[ \left( \frac{3}{4} \right)t^{4/3} \right]_0^1 \mathbf{i} + \left[ \ln|t+1| \right]_0^1 \mathbf{j} + \left[ -e^{-t} \right]_0^1 \mathbf{k}$$

$$= \frac{3}{4}\mathbf{i} + (\ln 2)\mathbf{j} + \left( 1 - \frac{1}{e} \right)\mathbf{k} \qquad \blacksquare$$

As with real-valued functions, you can narrow the family of antiderivatives of a vector-valued function $\mathbf{r}'$ down to a single antiderivative by imposing an initial condition on the vector-valued function $\mathbf{r}$. This is demonstrated in the next example.

**EXAMPLE 7**    **The Antiderivative of a Vector-Valued Function**

Find the antiderivative of

$$\mathbf{r}'(t) = \cos 2t\,\mathbf{i} - 2 \sin t\,\mathbf{j} + \frac{1}{1+t^2}\mathbf{k}$$

that satisfies the initial condition

$$\mathbf{r}(0) = 3\mathbf{i} - 2\mathbf{j} + \mathbf{k}.$$

**Solution**

$$\mathbf{r}(t) = \int \mathbf{r}'(t)\, dt$$

$$= \left( \int \cos 2t\, dt \right)\mathbf{i} + \left( \int -2 \sin t\, dt \right)\mathbf{j} + \left( \int \frac{1}{1+t^2}\, dt \right)\mathbf{k}$$

$$= \left( \frac{1}{2}\sin 2t + C_1 \right)\mathbf{i} + (2 \cos t + C_2)\mathbf{j} + (\arctan t + C_3)\mathbf{k}$$

Letting $t = 0$, you can write

$$\mathbf{r}(0) = (0 + C_1)\mathbf{i} + (2 + C_2)\mathbf{j} + (0 + C_3)\mathbf{k}.$$

Using the fact that $\mathbf{r}(0) = 3\mathbf{i} - 2\mathbf{j} + \mathbf{k}$, you have

$$(0 + C_1)\mathbf{i} + (2 + C_2)\mathbf{j} + (0 + C_3)\mathbf{k} = 3\mathbf{i} - 2\mathbf{j} + \mathbf{k}.$$

Equating corresponding components produces

$$C_1 = 3, \quad 2 + C_2 = -2, \quad \text{and} \quad C_3 = 1.$$

So, the antiderivative that satisfies the initial condition is

$$\mathbf{r}(t) = \left( \frac{1}{2}\sin 2t + 3 \right)\mathbf{i} + (2 \cos t - 4)\mathbf{j} + (\arctan t + 1)\mathbf{k}. \qquad \blacksquare$$

## 12.2 Exercises

See **CalcChat.com** for tutorial help and worked-out solutions to odd-numbered exercises.

**CONCEPT CHECK**

1. **Derivative** Describe the relationship between the graph of $\mathbf{r}'(t_0)$ and the curve represented by $\mathbf{r}(t)$.

2. **Integration** Explain why the family of vector-valued functions that are the antiderivatives of a vector-valued function differ by a constant vector.

 **Differentiation of Vector-Valued Functions** In Exercises 3–10, find $\mathbf{r}'(t)$, $\mathbf{r}(t_0)$, and $\mathbf{r}'(t_0)$ for the given value of $t_0$. Then sketch the curve represented by the vector-valued function and sketch the vectors $\mathbf{r}(t_0)$ and $\mathbf{r}'(t_0)$.

3. $\mathbf{r}(t) = (1 - t^2)\mathbf{i} + t\mathbf{j}, \quad t_0 = 3$

4. $\mathbf{r}(t) = (1 + t)\mathbf{i} + t^3\mathbf{j}, \quad t_0 = 1$

5. $\mathbf{r}(t) = \cos t\mathbf{i} + \sin t\mathbf{j}, \quad t_0 = \dfrac{\pi}{2}$

6. $\mathbf{r}(t) = 3\sin t\mathbf{i} + 4\cos t\mathbf{j}, \quad t_0 = \dfrac{\pi}{2}$

7. $\mathbf{r}(t) = \langle e^t, e^{2t} \rangle, \quad t_0 = 0$

8. $\mathbf{r}(t) = \langle e^{-t}, e^t \rangle, \quad t_0 = 0$

9. $\mathbf{r}(t) = 2\cos t\mathbf{i} + 2\sin t\mathbf{j} + t\mathbf{k}, \quad t_0 = \dfrac{3\pi}{2}$

10. $\mathbf{r}(t) = t\mathbf{i} + t^2\mathbf{j} + \tfrac{3}{2}\mathbf{k}, \quad t_0 = 2$

**Finding a Derivative** In Exercises 11–18, find $\mathbf{r}'(t)$.

11. $\mathbf{r}(t) = t^4\mathbf{i} - 5t\mathbf{j}$

12. $\mathbf{r}(t) = \sqrt{t}\mathbf{i} + (1 - t^3)\mathbf{j}$

13. $\mathbf{r}(t) = 3\cos^3 t\mathbf{i} + 2\sin^3 t\mathbf{j} + \mathbf{k}$

14. $\mathbf{r}(t) = 4\sqrt{t}\mathbf{i} + t^2\sqrt{t}\mathbf{j} + \ln t^2\mathbf{k}$

15. $\mathbf{r}(t) = e^{-t}\mathbf{i} + 4\mathbf{j} + 5te^t\mathbf{k}$

16. $\mathbf{r}(t) = \langle t^3, \cos 3t, \sin 3t \rangle$

17. $\mathbf{r}(t) = \langle t\sin t, t\cos t, t \rangle$

18. $\mathbf{r}(t) = \langle \arcsin t, \arccos t, 0 \rangle$

**Higher-Order Differentiation** In Exercises 19–22, find (a) $\mathbf{r}'(t)$, (b) $\mathbf{r}''(t)$, and (c) $\mathbf{r}'(t) \cdot \mathbf{r}''(t)$.

19. $\mathbf{r}(t) = t^3\mathbf{i} + \tfrac{1}{2}t^2\mathbf{j}$

20. $\mathbf{r}(t) = (t^2 + t)\mathbf{i} + (t^2 - t)\mathbf{j}$

21. $\mathbf{r}(t) = 4\cos t\mathbf{i} + 4\sin t\mathbf{j}$

22. $\mathbf{r}(t) = 8\cos t\mathbf{i} + 3\sin t\mathbf{j}$

 **Higher-Order Differentiation** In Exercises 23–26, find (a) $\mathbf{r}'(t)$, (b) $\mathbf{r}''(t)$, (c) $\mathbf{r}'(t) \cdot \mathbf{r}''(t)$, and (d) $\mathbf{r}'(t) \times \mathbf{r}''(t)$.

23. $\mathbf{r}(t) = \tfrac{1}{2}t^2\mathbf{i} - t\mathbf{j} + \tfrac{1}{6}t^3\mathbf{k}$

24. $\mathbf{r}(t) = t^3\mathbf{i} + (2t^2 + 3)\mathbf{j} + (3t - 5)\mathbf{k}$

25. $\mathbf{r}(t) = \langle \cos t + t\sin t, \sin t - t\cos t, t \rangle$

26. $\mathbf{r}(t) = \langle e^{-t}, t^2, \tan t \rangle$

 **Finding Intervals on Which a Curve Is Smooth** In Exercises 27–34, find the open interval(s) on which the curve given by the vector-valued function is smooth.

27. $\mathbf{r}(t) = t^2\mathbf{i} + t^3\mathbf{j}$        28. $\mathbf{r}(t) = 5t^5\mathbf{i} - t^4\mathbf{j}$

29. $\mathbf{r}(\theta) = 2\cos^3\theta\mathbf{i} + 3\sin^3\theta\mathbf{j}, \quad 0 \le \theta \le 2\pi$

30. $\mathbf{r}(\theta) = (\theta + \sin\theta)\mathbf{i} + (1 - \cos\theta)\mathbf{j}, \quad 0 \le \theta \le 2\pi$

31. $\mathbf{r}(t) = \dfrac{2t}{8 + t^3}\mathbf{i} + \dfrac{2t^2}{8 + t^3}\mathbf{j}$

32. $\mathbf{r}(t) = e^t\mathbf{i} - e^{-t}\mathbf{j} + 3t\mathbf{k}$

33. $\mathbf{r}(t) = t\mathbf{i} - 3t\mathbf{j} + \tan t\mathbf{k}$

34. $\mathbf{r}(t) = \sqrt{t}\mathbf{i} + (t^2 - 1)\mathbf{j} + \tfrac{1}{4}t\mathbf{k}$

 **Using Properties of the Derivative** In Exercises 35 and 36, use the properties of the derivative to find the following.

(a) $\mathbf{r}'(t)$        (b) $\dfrac{d}{dt}[3\mathbf{r}(t) - \mathbf{u}(t)]$        (c) $\dfrac{d}{dt}[(5t)\mathbf{u}(t)]$

(d) $\dfrac{d}{dt}[\mathbf{r}(t) \cdot \mathbf{u}(t)]$        (e) $\dfrac{d}{dt}[\mathbf{r}(t) \times \mathbf{u}(t)]$        (f) $\dfrac{d}{dt}[\mathbf{r}(2t)]$

35. $\mathbf{r}(t) = t\mathbf{i} + 3t\mathbf{j} + t^2\mathbf{k}, \quad \mathbf{u}(t) = 4t\mathbf{i} + t^2\mathbf{j} + t^3\mathbf{k}$

36. $\mathbf{r}(t) = \langle t, 2\sin t, 2\cos t \rangle, \quad \mathbf{u}(t) = \left\langle \dfrac{1}{t}, 2\sin t, 2\cos t \right\rangle$

**Using Two Methods** In Exercises 37 and 38, find (a) $\dfrac{d}{dt}[\mathbf{r}(t) \cdot \mathbf{u}(t)]$ and (b) $\dfrac{d}{dt}[\mathbf{r}(t) \times \mathbf{u}(t)]$ in two different ways.

(i) Find the product first, then differentiate.
(ii) Apply the properties of Theorem 12.2.

37. $\mathbf{r}(t) = t\mathbf{i} + 2t^2\mathbf{j} + t^3\mathbf{k}, \quad \mathbf{u}(t) = t^4\mathbf{k}$

38. $\mathbf{r}(t) = \cos t\mathbf{i} + \sin t\mathbf{j} + t\mathbf{k}, \quad \mathbf{u}(t) = \mathbf{j} + t\mathbf{k}$

 **Finding an Indefinite Integral** In Exercises 39–46, find the indefinite integral.

39. $\displaystyle\int (2t\mathbf{i} + \mathbf{j} + 9\mathbf{k})\,dt$        40. $\displaystyle\int (4t^3\mathbf{i} + 6t\mathbf{j} - 4\sqrt{t}\mathbf{k})\,dt$

41. $\displaystyle\int \left(\dfrac{1}{t}\mathbf{i} + \mathbf{j} - t^{3/2}\mathbf{k}\right)dt$        42. $\displaystyle\int \left(\ln t\mathbf{i} + \dfrac{1}{t}\mathbf{j} + \mathbf{k}\right)dt$

43. $\displaystyle\int (\mathbf{i} + 4t^3\mathbf{j} + 5^t\mathbf{k})\,dt$

44. $\displaystyle\int \left(\sec^2 t\mathbf{i} + \dfrac{1}{1 + t^2}\mathbf{j}\right)dt$

45. $\displaystyle\int (e^t\mathbf{i} + \mathbf{j} + t\cos t\mathbf{k})\,dt$

46. $\displaystyle\int (e^{-t}\sin t\mathbf{i} + \cot t\mathbf{j})\,dt$

**Evaluating a Definite Integral** In Exercises 47–52, evaluate the definite integral.

47. $\displaystyle\int_0^1 (8t\mathbf{i} + t\mathbf{j} - \mathbf{k})\, dt$

48. $\displaystyle\int_{-1}^1 \left(t\mathbf{i} + t^3\mathbf{j} + \sqrt[3]{t}\,\mathbf{k}\right) dt$

49. $\displaystyle\int_0^{\pi/2} \left[(5\cos t)\mathbf{i} + (6\sin t)\mathbf{j} + \mathbf{k}\right] dt$

50. $\displaystyle\int_0^{\pi/4} \left[(\sec t\tan t)\mathbf{i} + (\tan t)\mathbf{j} + (2\sin t\cos t)\mathbf{k}\right] dt$

51. $\displaystyle\int_0^2 (t\mathbf{i} + e^t\mathbf{j} - te^t\mathbf{k})\, dt$    52. $\displaystyle\int_0^3 \|t\mathbf{i} + t^2\mathbf{j}\|\, dt$

**Finding an Antiderivative** In Exercises 53–58, find $\mathbf{r}(t)$ that satisfies the initial condition(s).

53. $\mathbf{r}'(t) = 4e^{2t}\mathbf{i} + 3e^t\mathbf{j}, \quad \mathbf{r}(0) = 2\mathbf{i}$

54. $\mathbf{r}'(t) = 3t^2\mathbf{j} + 6\sqrt{t}\,\mathbf{k}, \quad \mathbf{r}(0) = \mathbf{i} + 2\mathbf{j}$

55. $\mathbf{r}''(t) = -32\mathbf{j}, \quad \mathbf{r}'(0) = 600\sqrt{3}\,\mathbf{i} + 600\mathbf{j}, \quad \mathbf{r}(0) = \mathbf{0}$

56. $\mathbf{r}''(t) = -4\cos t\mathbf{j} - 3\sin t\mathbf{k}, \quad \mathbf{r}'(0) = 3\mathbf{k}, \quad \mathbf{r}(0) = 4\mathbf{j}$

57. $\mathbf{r}'(t) = te^{-t^2}\mathbf{i} - e^{-t}\mathbf{j} + \mathbf{k}, \quad \mathbf{r}(0) = \frac{1}{2}\mathbf{i} - \mathbf{j} + \mathbf{k}$

58. $\mathbf{r}'(t) = \dfrac{1}{1+t^2}\mathbf{i} + \dfrac{1}{t^2}\mathbf{j} + \dfrac{1}{t}\mathbf{k}, \quad \mathbf{r}(1) = 2\mathbf{i}$

---

**EXPLORING CONCEPTS**

59. **Using a Derivative** The three components of the derivative of the vector-valued function $\mathbf{u}$ are positive at $t = t_0$. Describe the behavior of $\mathbf{u}$ at $t = t_0$.

60. **Think About It** Find two vector-valued functions $\mathbf{f}(t)$ and $\mathbf{g}(t)$ such that

$$\int_a^b \left[\mathbf{f}(t) \cdot \mathbf{g}(t)\right] dt \neq \left[\int_a^b \mathbf{f}(t)\, dt\right] \cdot \left[\int_a^b \mathbf{g}(t)\, dt\right].$$

---

**Proof** In Exercises 61–68, prove the property. In each case, assume $\mathbf{r}$, $\mathbf{u}$, and $\mathbf{v}$ are differentiable vector-valued functions of $t$ in space, $w$ is a differentiable real-valued function of $t$, and $c$ is a scalar.

61. $\dfrac{d}{dt}[c\mathbf{r}(t)] = c\mathbf{r}'(t)$

62. $\dfrac{d}{dt}[\mathbf{r}(t) \pm \mathbf{u}(t)] = \mathbf{r}'(t) \pm \mathbf{u}'(t)$

63. $\dfrac{d}{dt}[w(t)\mathbf{r}(t)] = w(t)\mathbf{r}'(t) + w'(t)\mathbf{r}(t)$

64. $\dfrac{d}{dt}[\mathbf{r}(t) \times \mathbf{u}(t)] = \mathbf{r}(t) \times \mathbf{u}'(t) + \mathbf{r}'(t) \times \mathbf{u}(t)$

65. $\dfrac{d}{dt}[\mathbf{r}(w(t))] = \mathbf{r}'(w(t))w'(t)$

66. $\dfrac{d}{dt}[\mathbf{r}(t) \times \mathbf{r}'(t)] = \mathbf{r}(t) \times \mathbf{r}''(t)$

67. $\dfrac{d}{dt}\{\mathbf{r}(t) \cdot [\mathbf{u}(t) \times \mathbf{v}(t)]\} = \mathbf{r}'(t) \cdot [\mathbf{u}(t) \times \mathbf{v}(t)] +$
$\mathbf{r}(t) \cdot [\mathbf{u}'(t) \times \mathbf{v}(t)] +$
$\mathbf{r}(t) \cdot [\mathbf{u}(t) \times \mathbf{v}'(t)]$

68. If $\mathbf{r}(t) \cdot \mathbf{r}(t)$ is a constant, then $\mathbf{r}(t) \cdot \mathbf{r}'(t) = 0$.

69. **Particle Motion** A particle moves in the $xy$-plane along the curve represented by the vector-valued function $\mathbf{r}(t) = (t - \sin t)\mathbf{i} + (1 - \cos t)\mathbf{j}$.

   (a) Use a graphing utility to graph $\mathbf{r}$. Describe the curve.

   (b) Find the minimum and maximum values of $\|\mathbf{r}'\|$ and $\|\mathbf{r}''\|$.

70. **Particle Motion** A particle moves in the $yz$-plane along the curve represented by the vector-valued function $\mathbf{r}(t) = (2\cos t)\mathbf{j} + (3\sin t)\mathbf{k}$.

   (a) Describe the curve.

   (b) Find the minimum and maximum values of $\|\mathbf{r}'\|$ and $\|\mathbf{r}''\|$.

71. **Perpendicular Vectors** Consider the vector-valued function $\mathbf{r}(t) = (e^t\sin t)\mathbf{i} + (e^t\cos t)\mathbf{j}$. Show that $\mathbf{r}(t)$ and $\mathbf{r}''(t)$ are always perpendicular to each other.

---

72.  **HOW DO YOU SEE IT?** The graph shows a vector-valued function $\mathbf{r}(t)$ for $0 \le t \le 2\pi$ and its derivative $\mathbf{r}'(t)$ for several values of $t$.

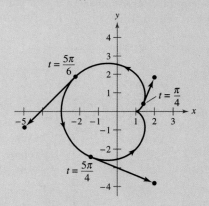

   (a) For each derivative shown in the graph, determine whether each component is positive or negative.

   (b) Is the curve smooth on the interval $[0, 2\pi]$? Explain.

---

**True or False?** In Exercises 73–76, determine whether the statement is true or false. If it is false, explain why or give an example that shows it is false.

73. If a particle moves along a sphere centered at the origin, then its derivative vector is always tangent to the sphere.

74. The definite integral of a vector-valued function is a real number.

75. $\dfrac{d}{dt}\left[\|\mathbf{r}(t)\|\right] = \|\mathbf{r}'(t)\|$

76. If $\mathbf{r}$ and $\mathbf{u}$ are differentiable vector-valued functions of $t$, then

$$\frac{d}{dt}[\mathbf{r}(t) \cdot \mathbf{u}(t)] = \mathbf{r}'(t) \cdot \mathbf{u}'(t).$$

# 12.3 Velocity and Acceleration

- Describe the velocity and acceleration associated with a vector-valued function.
- Use a vector-valued function to analyze projectile motion.

## Velocity and Acceleration

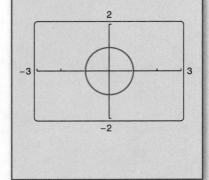
You are now ready to combine your study of parametric equations, curves, vectors, and vector-valued functions to form a model for motion along a curve. You will begin by looking at the motion of an object in the plane. (The motion of an object in space can be developed similarly.)

As an object moves along a curve in the plane, the coordinates $x$ and $y$ of its center of mass are each functions of time $t$. Rather than using the letters $f$ and $g$ to represent these two functions, it is convenient to write $x = x(t)$ and $y = y(t)$. So, the position vector $\mathbf{r}(t)$ takes the form

$$\mathbf{r}(t) = x(t)\mathbf{i} + y(t)\mathbf{j}. \qquad \text{Position vector}$$

The beauty of this vector model for representing motion is that you can use the first and second derivatives of the vector-valued function $\mathbf{r}$ to find the object's velocity and acceleration. (Recall from the preceding chapter that velocity and acceleration are both vector quantities having magnitude and direction.) To find the velocity and acceleration vectors at a given time $t$, consider a point $Q(x(t + \Delta t), y(t + \Delta t))$ that is approaching the point $P(x(t), y(t))$ along the curve $C$ given by $\mathbf{r}(t) = x(t)\mathbf{i} + y(t)\mathbf{j}$, as shown in Figure 12.11. As $\Delta t \to 0$, the direction of the vector $\overrightarrow{PQ}$ (denoted by $\Delta \mathbf{r}$) approaches the *direction of motion* at time $t$.

$$\Delta \mathbf{r} = \mathbf{r}(t + \Delta t) - \mathbf{r}(t)$$

$$\frac{\Delta \mathbf{r}}{\Delta t} = \frac{\mathbf{r}(t + \Delta t) - \mathbf{r}(t)}{\Delta t}$$

$$\lim_{\Delta t \to 0} \frac{\Delta \mathbf{r}}{\Delta t} = \lim_{\Delta t \to 0} \frac{\mathbf{r}(t + \Delta t) - \mathbf{r}(t)}{\Delta t}$$

When this limit exists, it is defined as the **velocity vector** or **tangent vector** to the curve at point $P$. Note that this is the same limit used to define $\mathbf{r}'(t)$. So, the direction of $\mathbf{r}'(t)$ gives the direction of motion at time $t$. Moreover, the magnitude of the vector $\mathbf{r}'(t)$

$$\|\mathbf{r}'(t)\| = \|x'(t)\mathbf{i} + y'(t)\mathbf{j}\| = \sqrt{[x'(t)]^2 + [y'(t)]^2}$$

gives the **speed** of the object at time $t$.

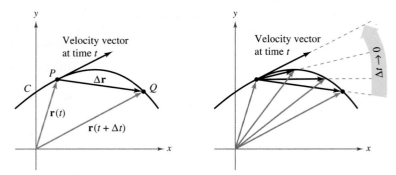

As $\Delta t \to 0$, $\dfrac{\Delta \mathbf{r}}{\Delta t}$ approaches the velocity vector.

**Figure 12.11**

Similar to how $\mathbf{r}'(t)$ is used to find velocity, you can use $\mathbf{r}''(t)$ to find acceleration, as indicated in the definitions at the top of the next page.

**Definitions of Velocity and Acceleration**

If $x$ and $y$ are twice-differentiable functions of $t$, and $\mathbf{r}$ is a vector-valued function given by $\mathbf{r}(t) = x(t)\mathbf{i} + y(t)\mathbf{j}$, then the velocity vector, acceleration vector, and speed at time $t$ are as follows.

$$\textbf{Velocity} = \mathbf{v}(t) \quad = \mathbf{r}'(t) \quad = x'(t)\mathbf{i} + y'(t)\mathbf{j}$$
$$\textbf{Acceleration} = \mathbf{a}(t) \quad = \mathbf{r}''(t) \quad = x''(t)\mathbf{i} + y''(t)\mathbf{j}$$
$$\textbf{Speed} = \|\mathbf{v}(t)\| = \|\mathbf{r}'(t)\| = \sqrt{[x'(t)]^2 + [y'(t)]^2}$$

For motion along a space curve, the definitions are similar. That is, for

$$\mathbf{r}(t) = x(t)\mathbf{i} + y(t)\mathbf{j} + z(t)\mathbf{k}$$

you have the following.

$$\textbf{Velocity} = \mathbf{v}(t) \quad = \mathbf{r}'(t) \quad = x'(t)\mathbf{i} + y'(t)\mathbf{j} + z'(t)\mathbf{k}$$
$$\textbf{Acceleration} = \mathbf{a}(t) \quad = \mathbf{r}''(t) \quad = x''(t)\mathbf{i} + y''(t)\mathbf{j} + z''(t)\mathbf{k}$$
$$\textbf{Speed} = \|\mathbf{v}(t)\| = \|\mathbf{r}'(t)\| = \sqrt{[x'(t)]^2 + [y'(t)]^2 + [z'(t)]^2}$$

---

**EXAMPLE 1**   **Velocity and Acceleration Along a Plane Curve**

Find the (a) velocity vector, (b) speed, and (c) acceleration vector for the particle that moves along the plane curve $C$ described by

$$\mathbf{r}(t) = 2\sin\frac{t}{2}\mathbf{i} + 2\cos\frac{t}{2}\mathbf{j}. \qquad \text{Position vector}$$

**Solution**

a. $\mathbf{v}(t) = \mathbf{r}'(t) = \cos\dfrac{t}{2}\mathbf{i} - \sin\dfrac{t}{2}\mathbf{j}$      Velocity vector

b. $\|\mathbf{r}'(t)\| = \sqrt{\cos^2\dfrac{t}{2} + \sin^2\dfrac{t}{2}} = 1$      Speed (at any time)

c. $\mathbf{a}(t) = \mathbf{r}''(t) = -\dfrac{1}{2}\sin\dfrac{t}{2}\mathbf{i} - \dfrac{1}{2}\cos\dfrac{t}{2}\mathbf{j}$      Acceleration vector

▷ • • **REMARK** In Example 1, note that the velocity and acceleration vectors are orthogonal at any point in time (see Figure 12.12). This is characteristic of motion at a constant speed. (See Exercise 59.)

The parametric equations for the curve in Example 1 are

$$x = 2\sin\frac{t}{2} \quad \text{and} \quad y = 2\cos\frac{t}{2}.$$

By eliminating the parameter $t$, you obtain the rectangular equation

$$x^2 + y^2 = 4. \qquad \text{Rectangular equation}$$

So, the curve is a circle of radius 2 centered at the origin, as shown in Figure 12.12. Because the velocity vector

$$\mathbf{v}(t) = \cos\frac{t}{2}\mathbf{i} - \sin\frac{t}{2}\mathbf{j}$$

has a constant magnitude but a changing direction as $t$ increases, the particle moves around the circle at a constant speed.

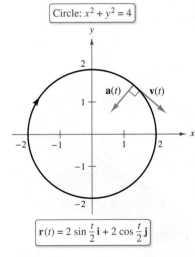

Circle: $x^2 + y^2 = 4$

$\mathbf{r}(t) = 2\sin\dfrac{t}{2}\mathbf{i} + 2\cos\dfrac{t}{2}\mathbf{j}$

The particle moves around the circle at a constant speed.
**Figure 12.12**

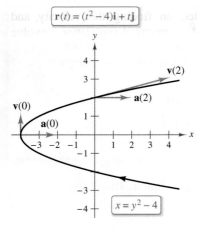

$$\mathbf{r}(t) = (t^2 - 4)\mathbf{i} + t\mathbf{j}$$

$$x = y^2 - 4$$

At each point on the curve, the acceleration vector points to the right.
**Figure 12.13**

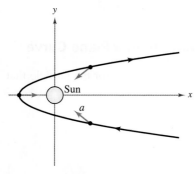

At each point in the comet's orbit, the acceleration vector points toward the sun.
**Figure 12.14**

**EXAMPLE 2**   **Velocity and Acceleration Vectors in the Plane**

Sketch the path of an object moving along the plane curve given by

$$\mathbf{r}(t) = (t^2 - 4)\mathbf{i} + t\mathbf{j} \qquad \text{Position vector}$$

and find the velocity and acceleration vectors when $t = 0$ and $t = 2$.

**Solution**   Using the parametric equations $x = t^2 - 4$ and $y = t$, you can determine that the curve is a parabola given by

$$x = y^2 - 4 \qquad \text{Rectangular equation}$$

as shown in Figure 12.13. The velocity vector (at any time) is

$$\mathbf{v}(t) = \mathbf{r}'(t) = 2t\mathbf{i} + \mathbf{j} \qquad \text{Velocity vector}$$

and the acceleration vector (at any time) is

$$\mathbf{a}(t) = \mathbf{r}''(t) = 2\mathbf{i}. \qquad \text{Acceleration vector}$$

When $t = 0$, the velocity and acceleration vectors are

$$\mathbf{v}(0) = 2(0)\mathbf{i} + \mathbf{j} = \mathbf{j} \quad \text{and} \quad \mathbf{a}(0) = 2\mathbf{i}.$$

When $t = 2$, the velocity and acceleration vectors are

$$\mathbf{v}(2) = 2(2)\mathbf{i} + \mathbf{j} = 4\mathbf{i} + \mathbf{j} \quad \text{and} \quad \mathbf{a}(2) = 2\mathbf{i}. \qquad ■$$

For the object moving along the path shown in Figure 12.13, note that the acceleration vector is constant (it has a magnitude of 2 and points to the right). This implies that the speed of the object is decreasing as the object moves toward the vertex of the parabola, and the speed is increasing as the object moves away from the vertex of the parabola.

This type of motion is *not* characteristic of comets that travel on parabolic paths through our solar system. For such comets, the acceleration vector always points to the origin (the sun), which implies that the comet's speed increases as it approaches the vertex of the path and decreases as it moves away from the vertex. (See Figure 12.14.)

**EXAMPLE 3**   **Velocity and Acceleration Vectors in Space**

$\vdots\cdots\triangleright$ *See LarsonCalculus.com for an interactive version of this type of example.*

Sketch the path of an object moving along the space curve $C$ given by

$$\mathbf{r}(t) = t\mathbf{i} + t^3\mathbf{j} + 3t\mathbf{k}, \quad t \geq 0 \qquad \text{Position vector}$$

and find the velocity and acceleration vectors when $t = 1$.

**Solution**   Using the parametric equations $x = t$ and $y = t^3$, you can determine that the path of the object lies on the cubic cylinder given by

$$y = x^3. \qquad \text{Rectangular equation}$$

Moreover, because $z = 3t$, the object starts at $(0, 0, 0)$ and moves upward as $t$ increases, as shown in Figure 12.15. Because $\mathbf{r}(t) = t\mathbf{i} + t^3\mathbf{j} + 3t\mathbf{k}$, you have

$$\mathbf{v}(t) = \mathbf{r}'(t) = \mathbf{i} + 3t^2\mathbf{j} + 3\mathbf{k} \qquad \text{Velocity vector}$$

and

$$\mathbf{a}(t) = \mathbf{r}''(t) = 6t\mathbf{j}. \qquad \text{Acceleration vector}$$

When $t = 1$, the velocity and acceleration vectors are

$$\mathbf{v}(1) = \mathbf{r}'(1) = \mathbf{i} + 3\mathbf{j} + 3\mathbf{k} \quad \text{and} \quad \mathbf{a}(1) = \mathbf{r}''(1) = 6\mathbf{j}. \qquad ■$$

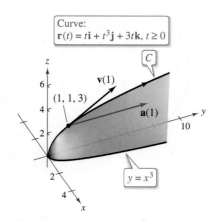

Curve:
$$\mathbf{r}(t) = t\mathbf{i} + t^3\mathbf{j} + 3t\mathbf{k}, t \geq 0$$

$$y = x^3$$

**Figure 12.15**

So far in this section, you have concentrated on finding the velocity and acceleration by differentiating the position vector. Many practical applications involve the reverse problem—finding the position vector for a given velocity or acceleration. This is demonstrated in the next example.

## EXAMPLE 4  Finding a Position Vector by Integration

An object starts from rest at the point $(1, 2, 0)$ and moves with an acceleration of

$$\mathbf{a}(t) = \mathbf{j} + 2\mathbf{k} \qquad \text{Acceleration vector}$$

where $\|\mathbf{a}(t)\|$ is measured in feet per second per second. Find the location of the object after $t = 2$ seconds.

**Solution**  From the description of the object's motion, you can deduce the following *initial conditions*. Because the object starts from rest, you have

$$\mathbf{v}(0) = \mathbf{0}.$$

Moreover, because the object starts at the point $(x, y, z) = (1, 2, 0)$, you have

$$\mathbf{r}(0) = x(0)\mathbf{i} + y(0)\mathbf{j} + z(0)\mathbf{k} = 1\mathbf{i} + 2\mathbf{j} + 0\mathbf{k} = \mathbf{i} + 2\mathbf{j}.$$

To find the position vector, you should integrate twice, each time using one of the initial conditions to solve for the constant of integration. The velocity vector is

$$\mathbf{v}(t) = \int \mathbf{a}(t)\, dt$$

$$= \int (\mathbf{j} + 2\mathbf{k})\, dt$$

$$= t\mathbf{j} + 2t\mathbf{k} + \mathbf{C}$$

where $\mathbf{C} = C_1\mathbf{i} + C_2\mathbf{j} + C_3\mathbf{k}$. Letting $t = 0$ and applying the initial condition $\mathbf{v}(0) = \mathbf{0}$, you obtain

$$\mathbf{v}(0) = C_1\mathbf{i} + C_2\mathbf{j} + C_3\mathbf{k} = \mathbf{0} \quad \Longrightarrow \quad C_1 = C_2 = C_3 = 0.$$

So, the *velocity* at any time $t$ is

$$\mathbf{v}(t) = t\mathbf{j} + 2t\mathbf{k}. \qquad \text{Velocity vector}$$

Integrating once more produces

$$\mathbf{r}(t) = \int \mathbf{v}(t)\, dt$$

$$= \int (t\mathbf{j} + 2t\mathbf{k})\, dt$$

$$= \frac{t^2}{2}\mathbf{j} + t^2\mathbf{k} + \mathbf{C}$$

where $\mathbf{C} = C_4\mathbf{i} + C_5\mathbf{j} + C_6\mathbf{k}$. Letting $t = 0$ and applying the initial condition $\mathbf{r}(0) = \mathbf{i} + 2\mathbf{j}$, you have

$$\mathbf{r}(0) = C_4\mathbf{i} + C_5\mathbf{j} + C_6\mathbf{k} = \mathbf{i} + 2\mathbf{j} \quad \Longrightarrow \quad C_4 = 1, C_5 = 2, C_6 = 0.$$

So, the *position* vector is

$$\mathbf{r}(t) = \mathbf{i} + \left(\frac{t^2}{2} + 2\right)\mathbf{j} + t^2\mathbf{k}. \qquad \text{Position vector}$$

The location of the object after $t = 2$ seconds is given by

$$\mathbf{r}(2) = \mathbf{i} + 4\mathbf{j} + 4\mathbf{k}$$

as shown in Figure 12.16.

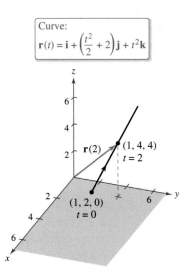

Curve:
$$\mathbf{r}(t) = \mathbf{i} + \left(\frac{t^2}{2} + 2\right)\mathbf{j} + t^2\mathbf{k}$$

The object takes 2 seconds to move from point $(1, 2, 0)$ to point $(1, 4, 4)$ along the curve.

**Figure 12.16**

## Projectile Motion

To derive the parametric equations for the path of a projectile, assume that gravity is the only force acting on the projectile after it is launched. So, the motion occurs in a vertical plane, which can be represented by the $xy$-coordinate system with the origin as a point on Earth's surface, as shown in Figure 12.17. For a projectile of mass $m$, the force due to gravity is

$$\mathbf{F} = -mg\mathbf{j}$$     Force due to gravity

where the acceleration due to gravity is $g = 32$ feet per second per second, or $9.8$ meters per second per second. By **Newton's Second Law of Motion,** this same force produces an acceleration $\mathbf{a} = \mathbf{a}(t)$ and satisfies the equation $\mathbf{F} = m\mathbf{a}$. Consequently, the acceleration of the projectile is given by $m\mathbf{a} = -mg\mathbf{j}$, which implies that

$$\mathbf{a} = -g\mathbf{j}.$$     Acceleration of projectile

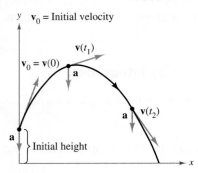

$y$   $\mathbf{v}_0 =$ Initial velocity

$\mathbf{v}(t_1)$

$\mathbf{v}_0 = \mathbf{v}(0)$   $\mathbf{a}$

$\mathbf{a}$   $\mathbf{v}(t_2)$

$\mathbf{a}$ } Initial height

$x$

**Figure 12.17**

---

**EXAMPLE 5**   **Derivation of the Position Vector for a Projectile**

A projectile of mass $m$ is launched from an initial position $\mathbf{r}_0$ with an initial velocity $\mathbf{v}_0$. Find its position vector as a function of time.

**Solution**   Begin with the acceleration $\mathbf{a}(t) = -g\mathbf{j}$ and integrate twice.

$$\mathbf{v}(t) = \int \mathbf{a}(t)\, dt = \int -g\mathbf{j}\, dt = -gt\mathbf{j} + \mathbf{C}_1$$

$$\mathbf{r}(t) = \int \mathbf{v}(t)\, dt = \int (-gt\mathbf{j} + \mathbf{C}_1)\, dt = -\frac{1}{2}gt^2\mathbf{j} + \mathbf{C}_1 t + \mathbf{C}_2$$

You can use the initial conditions $\mathbf{v}(0) = \mathbf{v}_0$ and $\mathbf{r}(0) = \mathbf{r}_0$ to solve for the constant vectors $\mathbf{C}_1$ and $\mathbf{C}_2$. Doing this produces

$$\mathbf{C}_1 = \mathbf{v}_0 \quad \text{and} \quad \mathbf{C}_2 = \mathbf{r}_0.$$

Therefore, the position vector is

$$\mathbf{r}(t) = -\frac{1}{2}gt^2\mathbf{j} + t\mathbf{v}_0 + \mathbf{r}_0.$$     Position vector     ■

---

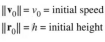

$\|\mathbf{v}_0\| = v_0 =$ initial speed
$\|\mathbf{r}_0\| = h =$ initial height

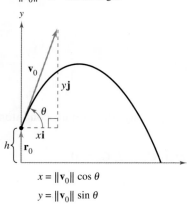

$y$

$\mathbf{v}_0$

$y\mathbf{j}$

$\theta$

$x\mathbf{i}$

$h$ { $\mathbf{r}_0$

$x$

$x = \|\mathbf{v}_0\| \cos \theta$
$y = \|\mathbf{v}_0\| \sin \theta$

**Figure 12.18**

In many projectile problems, the constant vectors $\mathbf{r}_0$ and $\mathbf{v}_0$ are not given explicitly. Often you are given the initial height $h$, the initial speed $v_0$, and the angle $\theta$ at which the projectile is launched, as shown in Figure 12.18. From the given height, you can deduce that $\mathbf{r}_0 = h\mathbf{j}$. Because the speed gives the magnitude of the initial velocity, it follows that $v_0 = \|\mathbf{v}_0\|$ and you can write

$$\mathbf{v}_0 = x\mathbf{i} + y\mathbf{j}$$
$$= \left(\|\mathbf{v}_0\| \cos \theta\right)\mathbf{i} + \left(\|\mathbf{v}_0\| \sin \theta\right)\mathbf{j}$$
$$= v_0 \cos \theta \mathbf{i} + v_0 \sin \theta \mathbf{j}.$$

So, the position vector can be written in the form

$$\mathbf{r}(t) = -\frac{1}{2}gt^2\mathbf{j} + t\mathbf{v}_0 + \mathbf{r}_0$$     Position vector

$$= -\frac{1}{2}gt^2\mathbf{j} + tv_0 \cos \theta \mathbf{i} + tv_0 \sin \theta \mathbf{j} + h\mathbf{j}$$

$$= (v_0 \cos \theta)t\mathbf{i} + \left[h + (v_0 \sin \theta)t - \frac{1}{2}gt^2\right]\mathbf{j}.$$

---

**THEOREM 12.3 Position Vector for a Projectile**

Neglecting air resistance, the path of a projectile launched from an initial height $h$ with initial speed $v_0$ and angle of elevation $\theta$ is described by the vector function

$$\mathbf{r}(t) = (v_0 \cos \theta)t\mathbf{i} + \left[h + (v_0 \sin \theta)t - \tfrac{1}{2}gt^2\right]\mathbf{j}$$

where $g$ is the acceleration due to gravity.

---

**EXAMPLE 6**  **Describing the Path of a Baseball**

A baseball is hit 3 feet above ground level at 100 feet per second and at an angle of 45° with respect to the ground, as shown in Figure 12.19. Find the maximum height reached by the baseball. Will it clear a 10-foot-high fence located 300 feet from home plate?

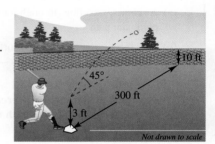

**Figure 12.19**

**Solution**  You are given

$$h = 3, \quad v_0 = 100, \quad \text{and} \quad \theta = 45°.$$

So, using Theorem 12.3 with $g = 32$ feet per second per second produces

$$\mathbf{r}(t) = \left(100 \cos \frac{\pi}{4}\right)t\mathbf{i} + \left[3 + \left(100 \sin \frac{\pi}{4}\right)t - 16t^2\right]\mathbf{j}$$

$$= \left(50\sqrt{2}\,t\right)\mathbf{i} + \left(3 + 50\sqrt{2}\,t - 16t^2\right)\mathbf{j}.$$

The velocity vector is

$$\mathbf{v}(t) = \mathbf{r}'(t) = 50\sqrt{2}\,\mathbf{i} + \left(50\sqrt{2} - 32t\right)\mathbf{j}.$$

The maximum height occurs when

$$y'(t) = 50\sqrt{2} - 32t$$

is equal to 0, which implies that

$$t = \frac{25\sqrt{2}}{16} \approx 2.21 \text{ seconds.}$$

So, the maximum height reached by the ball is

$$y = 3 + 50\sqrt{2}\left(\frac{25\sqrt{2}}{16}\right) - 16\left(\frac{25\sqrt{2}}{16}\right)^2$$

$$= \frac{649}{8}$$

$$\approx 81 \text{ feet.} \qquad \text{Maximum height when } t \approx 2.21 \text{ seconds}$$

The ball is 300 feet from where it was hit when

$$x(t) = 300 \quad \Longrightarrow \quad 50\sqrt{2}\,t = 300.$$

Solving this equation for $t$ produces $t = 3\sqrt{2} \approx 4.24$ seconds. At this time, the height of the ball is

$$y = 3 + 50\sqrt{2}\left(3\sqrt{2}\right) - 16\left(3\sqrt{2}\right)^2$$

$$= 303 - 288$$

$$= 15 \text{ feet.} \qquad \text{Height when } t \approx 4.24 \text{ seconds}$$

Therefore, the ball clears the 10-foot fence for a home run.

# 12.3 Exercises

See **CalcChat.com** for tutorial help and worked-out solutions to odd-numbered exercises.

**CONCEPT CHECK**

1. **Velocity Vector** An object moves along a curve in the plane. What information do you gain about the motion of the object from the velocity vector to the curve at time $t$?

2. **Acceleration Vectors** For each scenario, describe the direction of the acceleration vectors. Explain your reasoning.

   (a) A comet traveling through our solar system in a parabolic path

   (b) An object thrown on Earth's surface

 **Finding Velocity and Acceleration Along a Plane Curve** In Exercises 3–10, the position vector **r** describes the path of an object moving in the $xy$-plane.

(a) Find the velocity vector, speed, and acceleration vector of the object.

(b) Evaluate the velocity vector and acceleration vector of the object at the given point.

(c) Sketch a graph of the path and sketch the velocity and acceleration vectors at the given point.

| Position Vector | Point |
|---|---|
| 3. $\mathbf{r}(t) = 3t\mathbf{i} + (t-1)\mathbf{j}$ | $(3, 0)$ |
| 4. $\mathbf{r}(t) = t\mathbf{i} + (-t^2 + 4)\mathbf{j}$ | $(1, 3)$ |
| 5. $\mathbf{r}(t) = t^2\mathbf{i} + t\mathbf{j}$ | $(4, 2)$ |
| 6. $\mathbf{r}(t) = \left(\frac{1}{4}t^3 + 1\right)\mathbf{i} + t\mathbf{j}$ | $(3, 2)$ |
| 7. $\mathbf{r}(t) = 2\cos t\mathbf{i} + 2\sin t\mathbf{j}$ | $(\sqrt{2}, \sqrt{2})$ |
| 8. $\mathbf{r}(t) = 3\cos t\mathbf{i} + 2\sin t\mathbf{j}$ | $(3, 0)$ |
| 9. $\mathbf{r}(t) = \langle t - \sin t, 1 - \cos t\rangle$ | $(\pi, 2)$ |
| 10. $\mathbf{r}(t) = \langle e^{-t}, e^t\rangle$ | $(1, 1)$ |

 **Finding Velocity and Acceleration Vectors in Space** In Exercises 11–20, the position vector **r** describes the path of an object moving in space.

(a) Find the velocity vector, speed, and acceleration vector of the object.

(b) Evaluate the velocity vector and acceleration vector of the object at the given value of $t$.

| Position Vector | Time |
|---|---|
| 11. $\mathbf{r}(t) = t\mathbf{i} + 5t\mathbf{j} + 3t\mathbf{k}$ | $t = 1$ |
| 12. $\mathbf{r}(t) = 4t\mathbf{i} + 4t\mathbf{j} - 2t\mathbf{k}$ | $t = 3$ |
| 13. $\mathbf{r}(t) = t\mathbf{i} + t^2\mathbf{j} + \frac{1}{2}t^2\mathbf{k}$ | $t = 4$ |
| 14. $\mathbf{r}(t) = 3t\mathbf{i} + t\mathbf{j} + \frac{1}{4}t^2\mathbf{k}$ | $t = 2$ |
| 15. $\mathbf{r}(t) = t\mathbf{i} - t\mathbf{j} + \sqrt{9 - t^2}\,\mathbf{k}$ | $t = 0$ |
| 16. $\mathbf{r}(t) = t^2\mathbf{i} + t\mathbf{j} + 2t^{3/2}\mathbf{k}$ | $t = 4$ |

| Position Vector | Time |
|---|---|
| 17. $\mathbf{r}(t) = \langle 4t, 3\cos t, 3\sin t\rangle$ | $t = \pi$ |
| 18. $\mathbf{r}(t) = \langle 2\cos t, \sin 3t, t^2\rangle$ | $t = \dfrac{\pi}{4}$ |
| 19. $\mathbf{r}(t) = \langle e^t\cos t, e^t\sin t, e^t\rangle$ | $t = 0$ |
| 20. $\mathbf{r}(t) = \left\langle \ln t, \dfrac{1}{t^2}, t^4\right\rangle$ | $t = \sqrt{3}$ |

 **Finding a Position Vector by Integration** In Exercises 21–26, use the given acceleration vector and initial conditions to find the velocity and position vectors. Then find the position at time $t = 2$.

21. $\mathbf{a}(t) = \mathbf{i} + \mathbf{j} + \mathbf{k}$, $\mathbf{v}(0) = \mathbf{0}$, $\mathbf{r}(0) = \mathbf{0}$

22. $\mathbf{a}(t) = 2\mathbf{i} + 3\mathbf{k}$, $\mathbf{v}(0) = 4\mathbf{j}$, $\mathbf{r}(0) = \mathbf{0}$

23. $\mathbf{a}(t) = t\mathbf{j} + t\mathbf{k}$, $\mathbf{v}(1) = 5\mathbf{j}$, $\mathbf{r}(1) = \mathbf{0}$

24. $\mathbf{a}(t) = -32\mathbf{k}$, $\mathbf{v}(0) = 3\mathbf{i} - 2\mathbf{j} + \mathbf{k}$, $\mathbf{r}(0) = 5\mathbf{j} + 2\mathbf{k}$

25. $\mathbf{a}(t) = -\cos t\mathbf{i} - \sin t\mathbf{j}$, $\mathbf{v}(0) = \mathbf{j} + \mathbf{k}$, $\mathbf{r}(0) = \mathbf{i}$

26. $\mathbf{a}(t) = e^t\mathbf{i} - 8\mathbf{k}$, $\mathbf{v}(0) = 2\mathbf{i} + 3\mathbf{j} + \mathbf{k}$, $\mathbf{r}(0) = \mathbf{0}$

**Projectile Motion** In Exercises 27–40, use the model for projectile motion, assuming there is no air resistance and $g = 32$ feet per second per second.

27. A baseball is hit from a height of 2.5 feet above the ground with an initial speed of 140 feet per second and at an angle of 22° above the horizontal. Find the maximum height reached by the baseball. Determine whether it will clear a 10-foot-high fence located 375 feet from home plate.

28. Determine the maximum height and range of a projectile fired at a height of 3 feet above the ground with an initial speed of 900 feet per second and at an angle of 45° above the horizontal.

29. A baseball, hit 3 feet above the ground, leaves the bat at an angle of 45° and is caught by an outfielder 3 feet above the ground and 300 feet from home plate. What is the initial speed of the ball, and how high does it rise?

30. A baseball player at second base throws a ball 90 feet to the player at first base. The ball is released at a point 5 feet above the ground with an initial speed of 50 miles per hour and at an angle of 15° above the horizontal. At what height does the player at first base catch the ball?

31. Eliminate the parameter $t$ from the position vector for the motion of a projectile to show that the rectangular equation is

$$y = -\frac{g\sec^2\theta}{2v_0^2}x^2 + (\tan\theta)x + h.$$

32. The path of a ball is given by the rectangular equation $y = x - 0.005x^2$. Use the result of Exercise 31 to find the position vector. Then find the speed and direction of the ball at the point at which it has traveled 60 feet horizontally.

33. The Rogers Centre in Toronto, Ontario, has a center field fence that is 10 feet high and 400 feet from home plate. A ball is hit 3 feet above the ground and leaves the bat at a speed of 100 miles per hour.

    (a) The ball leaves the bat at an angle of $\theta = \theta_0$ with the horizontal. Write the vector-valued function for the path of the ball.

    (b) Use a graphing utility to graph the vector-valued function for $\theta_0 = 10°$, $\theta_0 = 15°$, $\theta_0 = 20°$, and $\theta_0 = 25°$. Use the graphs to approximate the minimum angle required for the hit to be a home run.

    (c) Determine analytically the minimum angle required for the hit to be a home run.

• • **34. Football** • • • • • • • • • • • • • • • • • • •

The quarterback of a football team releases a pass at a height of 7 feet above the playing field, and the football is caught by a receiver 30 yards directly downfield at a height of 4 feet. The pass is released at an angle of 35° with the horizontal.

(a) Find the speed of the football when it is released.

(b) Find the maximum height of the football.

(c) Find the time the receiver has to reach the proper position after the quarterback releases the football.

• • • • • • • • • • • • • • • • • • • • • • • • • • • • •

35. A bale ejector consists of two variable-speed belts at the end of a baler. Its purpose is to toss bales into a trailing wagon. In loading the back of a wagon, a bale must be thrown to a position 8 feet above and 16 feet behind the ejector.

    (a) Find the minimum initial speed of the bale and the corresponding angle at which it must be ejected from the baler.

    (b) The ejector has a fixed angle of 45°. Find the initial speed required.

36. A bomber is flying horizontally at an altitude of 30,000 feet with a speed of 540 miles per hour (see figure). When should the bomb be released for it to hit the target? (Give your answer in terms of the angle of depression from the plane to the target.) What is the speed of the bomb at the time of impact?

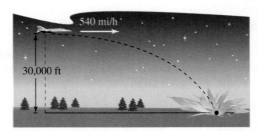

540 mi/h

30,000 ft

37. A shot fired from a gun with a muzzle speed of 1200 feet per second is to hit a target 3000 feet away. Determine the minimum angle of elevation of the gun.

38. A projectile is fired from ground level at an angle of 12° with the horizontal. The projectile is to have a range of 200 feet. Find the minimum initial speed necessary.

39. Use a graphing utility to graph the paths of a projectile for the given values of $\theta$ and $v_0$. For each case, use the graph to approximate the maximum height and range of the projectile. (Assume that the projectile is launched from ground level.)

    (a) $\theta = 10°$, $v_0 = 66$ ft/sec

    (b) $\theta = 10°$, $v_0 = 146$ ft/sec

    (c) $\theta = 45°$, $v_0 = 66$ ft/sec

    (d) $\theta = 45°$, $v_0 = 146$ ft/sec

    (e) $\theta = 60°$, $v_0 = 66$ ft/sec

    (f) $\theta = 60°$, $v_0 = 146$ ft/sec

40. Find the angles at which an object must be thrown to obtain (a) the maximum range and (b) the maximum height.

**Projectile Motion** In Exercises 41 and 42, use the model for projectile motion, assuming there is no air resistance and $g = 9.8$ meters per second per second.

41. Determine the maximum height and range of a projectile fired at a height of 1.5 meters above the ground with an initial speed of 100 meters per second and at an angle of 30° above the horizontal.

42. A projectile is fired from ground level at an angle of 8° with the horizontal. The projectile is to have a range of 50 meters. Find the minimum initial speed necessary.

43. **Shot-Put Throw** The path of a shot thrown at an angle $\theta$ is

$$\mathbf{r}(t) = (v_0 \cos \theta)t\mathbf{i} + \left[ h + (v_0 \sin \theta)t - \frac{1}{2}gt^2 \right]\mathbf{j}$$

where $v_0$ is the initial speed, $h$ is the initial height, $t$ is the time in seconds, and $g$ is the acceleration due to gravity. Verify that the shot will remain in the air for a total of

$$t = \frac{v_0 \sin \theta + \sqrt{v_0^2 \sin^2 \theta + 2gh}}{g} \text{ seconds}$$

and will travel a horizontal distance of

$$\frac{v_0^2 \cos \theta}{g} \left( \sin \theta + \sqrt{\sin^2 \theta + \frac{2gh}{v_0^2}} \right) \text{ feet.}$$

• • **44. Shot-Put Throw** • • • • • • • • • • • • •

A shot is thrown from a height of $h = 5.75$ feet with an initial speed of $v_0 = 41$ feet per second and at an angle of $\theta = 42.5°$ with the horizontal. Use the result of Exercise 43 to find the total time of travel and the total horizontal distance traveled.

• • • • • • • • • • • • • • • • • • • • • • • • • • • • •

**Cycloidal Motion** In Exercises 45 and 46, consider the motion of a point (or particle) on the circumference of a rolling circle. As the circle rolls, it generates the cycloid

$$r(t) = b(\omega t - \sin \omega t)\mathbf{i} + b(1 - \cos \omega t)\mathbf{j}$$

where $\omega$ is the constant angular speed of the circle and $b$ is the radius of the circle.

45. Find the velocity and acceleration vectors of the particle. Use the results to determine the times at which the speed of the particle will be (a) zero and (b) maximized.

46. Find the maximum speed of a point on the circumference of an automobile tire of radius 1 foot when the automobile is traveling at 60 miles per hour. Compare this speed with the speed of the automobile.

**Circular Motion** In Exercises 47–50, consider a particle moving on a circular path of radius $b$ described by

$$r(t) = b \cos \omega t\mathbf{i} + b \sin \omega t\mathbf{j}$$

where $\omega = du/dt$ is the constant angular speed.

47. Find the velocity vector and show that it is orthogonal to $\mathbf{r}(t)$.

48. (a) Show that the speed of the particle is $b\omega$.

    (b) Use a graphing utility in *parametric* mode to graph the circle for $b = 6$. Try different values of $\omega$. Does the graphing utility draw the circle faster for greater values of $\omega$?

49. Find the acceleration vector and show that its direction is always toward the center of the circle.

50. Show that the magnitude of the acceleration vector is $b\omega^2$.

**Circular Motion** In Exercises 51 and 52, use the results of Exercises 47–50.

51. A psychrometer (an instrument used to measure humidity) weighing 4 ounces is whirled horizontally using a 6-inch string (see figure). The string will break under a force of 2 pounds. Find the maximum speed the instrument can attain without breaking the string. (Use $\mathbf{F} = m\mathbf{a}$, where $m = 1/128$.)

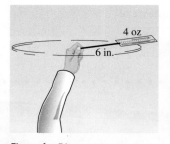
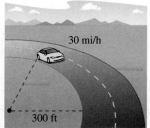

Figure for 51                Figure for 52

52. A 3400-pound automobile is negotiating a circular interchange of radius 300 feet at 30 miles per hour (see figure). Assuming the roadway is level, find the force between the tires and the road such that the car stays on the circular path and does not skid. (Use $\mathbf{F} = m\mathbf{a}$, where $m = 3400/32$.) Find the angle at which the roadway should be banked so that no lateral frictional force is exerted on the tires of the automobile.

**EXPLORING CONCEPTS**

53. **Constant Speed** Explain how a particle can be accelerating even though its speed is constant.

54. **Think About It** Consider a particle that is moving along the space curve given by $\mathbf{r}_1(t) = t^3\mathbf{i} + (3 - t)\mathbf{j} + 2t^2\mathbf{k}$. Write a vector-valued function $\mathbf{r}_2$ for a particle that moves four times as fast as the particle represented by $\mathbf{r}_1$. Explain how you found the function.

55. **Circular Motion** Consider a particle that moves around a circle. Is the velocity vector of the particle always orthogonal to the acceleration vector of the particle? Explain.

56. **Particle Motion** Consider a particle moving on an elliptical path described by $\mathbf{r}(t) = a \cos \omega t\mathbf{i} + b \sin \omega t\mathbf{j}$, where $\omega = d\theta/dt$ is the constant angular speed.

    (a) Find the velocity vector. What is the speed of the particle?

    (b) Find the acceleration vector and show that its direction is always toward the center of the ellipse.

57. **Path of an Object** When $t = 0$, an object is at the point $(0, 1)$ and has a velocity vector $\mathbf{v}(0) = -\mathbf{i}$. It moves with an acceleration of $\mathbf{a}(t) = \sin t\mathbf{i} - \cos t\mathbf{j}$. Show that the path of the object is a circle.

58. **HOW DO YOU SEE IT?** The graph shows the path of a projectile and the velocity and acceleration vectors at times $t_1$ and $t_2$. Classify the angle between the velocity vector and the acceleration vector at times $t_1$ and $t_2$. Using the vectors, is the speed increasing or decreasing at times $t_1$ and $t_2$? Explain your reasoning.

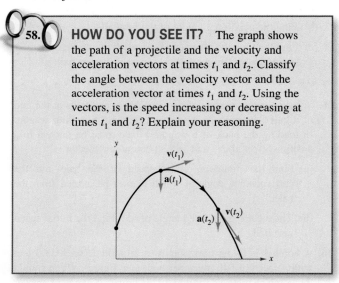

59. **Proof** Prove that when an object is traveling at a constant speed, its velocity and acceleration vectors are orthogonal.

60. **Proof** Prove that an object moving in a straight line at a constant speed has an acceleration of 0.

**True or False?** In Exercises 61–63, determine whether the statement is true or false. If it is false, explain why or give an example that shows it is false.

61. The velocity vector points in the direction of motion.

62. If a particle moves along a straight line, then the velocity and acceleration vectors are orthogonal.

63. A velocity vector of variable magnitude cannot have a constant direction.

# 12.4 Tangent Vectors and Normal Vectors

▣ Find a unit tangent vector and a principal unit normal vector at a point on a space curve.
▣ Find the tangential and normal components of acceleration.

## Tangent Vectors and Normal Vectors

In the preceding section, you learned that the velocity vector points in the direction of motion. This observation leads to the next definition, which applies to any smooth curve—not just to those for which the parameter represents time.

---

**Definition of Unit Tangent Vector**

Let $C$ be a smooth curve represented by $\mathbf{r}$ on an open interval $I$. The **unit tangent vector** $\mathbf{T}(t)$ at $t$ is defined as

$$\mathbf{T}(t) = \frac{\mathbf{r}'(t)}{\|\mathbf{r}'(t)\|}, \quad \mathbf{r}'(t) \neq \mathbf{0}.$$

---

Recall that a curve is *smooth* on an interval when $\mathbf{r}'$ is continuous and nonzero on the interval. So, "smoothness" is sufficient to guarantee that a curve has a unit tangent vector.

**EXAMPLE 1**   **Finding the Unit Tangent Vector**

Find the unit tangent vector to the curve given by

$$\mathbf{r}(t) = t\mathbf{i} + t^2\mathbf{j}$$

when $t = 1$.

**Solution**   The derivative of $\mathbf{r}(t)$ is

$$\mathbf{r}'(t) = \mathbf{i} + 2t\mathbf{j}. \qquad \text{Derivative of } \mathbf{r}(t)$$

So, the unit tangent vector is

$$\mathbf{T}(t) = \frac{\mathbf{r}'(t)}{\|\mathbf{r}'(t)\|} \qquad \text{Definition of } \mathbf{T}(t)$$

$$= \frac{1}{\sqrt{1 + 4t^2}}(\mathbf{i} + 2t\mathbf{j}). \qquad \text{Substitute for } \mathbf{r}'(t).$$

When $t = 1$, the unit tangent vector is

$$\mathbf{T}(1) = \frac{1}{\sqrt{5}}(\mathbf{i} + 2\mathbf{j})$$

as shown in Figure 12.20. ▣

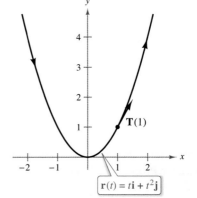

The direction of the unit tangent vector depends on the orientation of the curve.
**Figure 12.20**

In Example 1, note that the direction of the unit tangent vector depends on the orientation of the curve. For the parabola described by

$$\mathbf{r}(t) = -(t - 2)\mathbf{i} + (t - 2)^2\mathbf{j}$$

$\mathbf{T}(1)$ would still represent the unit tangent vector at the point $(1, 1)$, but it would point in the opposite direction. Try verifying this.

The **tangent line to a curve** at a point is the line that passes through the point and is parallel to the unit tangent vector. In Example 2, the unit tangent vector is used to find the tangent line at a point on a helix.

### EXAMPLE 2    Finding the Tangent Line at a Point on a Curve

Find $\mathbf{T}(t)$ and then find a set of parametric equations for the tangent line to the helix given by

$$\mathbf{r}(t) = 2 \cos t\mathbf{i} + 2 \sin t\mathbf{j} + t\mathbf{k}$$

at the point $\left( \sqrt{2}, \sqrt{2}, \dfrac{\pi}{4} \right)$.

**Solution**    The derivative of $\mathbf{r}(t)$ is

$$\mathbf{r}'(t) = -2 \sin t\mathbf{i} + 2 \cos t\mathbf{j} + \mathbf{k}$$

which implies that $\|\mathbf{r}'(t)\| = \sqrt{4 \sin^2 t + 4 \cos^2 t + 1} = \sqrt{5}$. Therefore, the unit tangent vector is

$$
\begin{aligned}
\mathbf{T}(t) &= \frac{\mathbf{r}'(t)}{\|\mathbf{r}'(t)\|} \\[2mm]
&= \frac{1}{\sqrt{5}}(-2 \sin t\mathbf{i} + 2 \cos t\mathbf{j} + \mathbf{k}). \qquad \text{Unit tangent vector}
\end{aligned}
$$

At the point $\left( \sqrt{2}, \sqrt{2}, \pi/4 \right)$, $t = \pi/4$ and the unit tangent vector is

$$
\begin{aligned}
\mathbf{T}\!\left(\frac{\pi}{4}\right) &= \frac{1}{\sqrt{5}}\left(-2\frac{\sqrt{2}}{2}\mathbf{i} + 2\frac{\sqrt{2}}{2}\mathbf{j} + \mathbf{k}\right) \\[2mm]
&= \frac{1}{\sqrt{5}}(-\sqrt{2}\mathbf{i} + \sqrt{2}\mathbf{j} + \mathbf{k}).
\end{aligned}
$$

Using the direction numbers $a = -\sqrt{2}$, $b = \sqrt{2}$, and $c = 1$, and the point $(x_1, y_1, z_1) = \left( \sqrt{2}, \sqrt{2}, \pi/4 \right)$, you can obtain the parametric equations (given with parameter $s$) listed below.

$$
\begin{aligned}
x &= x_1 + as = \sqrt{2} - \sqrt{2}s \\
y &= y_1 + bs = \sqrt{2} + \sqrt{2}s \\
z &= z_1 + cs = \frac{\pi}{4} + s
\end{aligned}
$$

This tangent line is shown in Figure 12.21. ∎

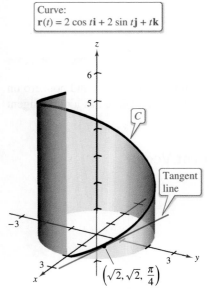

Curve:
$\mathbf{r}(t) = 2 \cos t\mathbf{i} + 2 \sin t\mathbf{j} + t\mathbf{k}$

$\left( \sqrt{2}, \sqrt{2}, \dfrac{\pi}{4} \right)$

The tangent line to a curve at a point is determined by the unit tangent vector at the point.

**Figure 12.21**

In Example 2, there are infinitely many vectors that are orthogonal to the tangent vector $\mathbf{T}(t)$. One of these is the vector $\mathbf{T}'(t)$. This follows from Property 7 of Theorem 12.2. That is,

$$\mathbf{T}(t) \cdot \mathbf{T}(t) = \|\mathbf{T}(t)\|^2 = 1 \quad \Longrightarrow \quad \mathbf{T}(t) \cdot \mathbf{T}'(t) = 0.$$

By normalizing the vector $\mathbf{T}'(t)$, you obtain a special vector called the **principal unit normal vector,** as indicated in the next definition.

---

**Definition of Principal Unit Normal Vector**

Let $C$ be a smooth curve represented by $\mathbf{r}$ on an open interval $I$. If $\mathbf{T}'(t) \neq \mathbf{0}$, then the **principal unit normal vector** at $t$ is defined as

$$\mathbf{N}(t) = \frac{\mathbf{T}'(t)}{\|\mathbf{T}'(t)\|}.$$

**EXAMPLE 3**   **Finding the Principal Unit Normal Vector**

Find $\mathbf{N}(t)$ and $\mathbf{N}(1)$ for the curve represented by $\mathbf{r}(t) = 3t\mathbf{i} + 2t^2\mathbf{j}$.

**Solution**   By differentiating, you obtain

$$\mathbf{r}'(t) = 3\mathbf{i} + 4t\mathbf{j}$$

which implies that

$$\|\mathbf{r}'(t)\| = \sqrt{9 + 16t^2}.$$

So, the unit tangent vector is

$$\begin{aligned}
\mathbf{T}(t) &= \frac{\mathbf{r}'(t)}{\|\mathbf{r}'(t)\|} \\
&= \frac{1}{\sqrt{9 + 16t^2}}(3\mathbf{i} + 4t\mathbf{j}). \qquad \text{Unit tangent vector}
\end{aligned}$$

Using Theorem 12.2, differentiate $\mathbf{T}(t)$ with respect to $t$ to obtain

$$\begin{aligned}
\mathbf{T}'(t) &= \frac{1}{\sqrt{9 + 16t^2}}(4\mathbf{j}) - \frac{16t}{(9 + 16t^2)^{3/2}}(3\mathbf{i} + 4t\mathbf{j}) \\
&= \frac{12}{(9 + 16t^2)^{3/2}}(-4t\mathbf{i} + 3\mathbf{j})
\end{aligned}$$

which implies that

$$\|\mathbf{T}'(t)\| = 12\sqrt{\frac{9 + 16t^2}{(9 + 16t^2)^3}} = \frac{12}{9 + 16t^2}.$$

Therefore, the principal unit normal vector is

$$\begin{aligned}
\mathbf{N}(t) &= \frac{\mathbf{T}'(t)}{\|\mathbf{T}'(t)\|} \\
&= \frac{1}{\sqrt{9 + 16t^2}}(-4t\mathbf{i} + 3\mathbf{j}). \qquad \text{Principal unit normal vector}
\end{aligned}$$

When $t = 1$, the principal unit normal vector is

$$\mathbf{N}(1) = \frac{1}{5}(-4\mathbf{i} + 3\mathbf{j})$$

as shown in Figure 12.22.

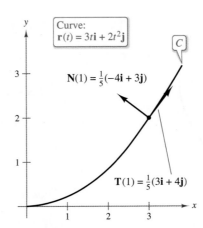

The principal unit normal vector points toward the concave side of the curve.

**Figure 12.22**

The principal unit normal vector can be difficult to evaluate algebraically. For plane curves, you can simplify the algebra by finding

$$\mathbf{T}(t) = x(t)\mathbf{i} + y(t)\mathbf{j} \qquad \text{Unit tangent vector}$$

and observing that $\mathbf{N}(t)$ must be either

$$\mathbf{N}_1(t) = y(t)\mathbf{i} - x(t)\mathbf{j} \quad \text{or} \quad \mathbf{N}_2(t) = -y(t)\mathbf{i} + x(t)\mathbf{j}.$$

Because $\sqrt{[x(t)]^2 + [y(t)]^2} = 1$, it follows that both $\mathbf{N}_1(t)$ and $\mathbf{N}_2(t)$ are unit normal vectors. The *principal* unit normal vector $\mathbf{N}$ is the one that points toward the concave side of the curve, as shown in Figure 12.22 (see Exercise 72). This also holds for curves in space. That is, for an object moving along a curve $C$ in space, the vector $\mathbf{T}(t)$ points in the direction the object is moving, whereas the vector $\mathbf{N}(t)$ is orthogonal to $\mathbf{T}(t)$ and points in the direction in which the object is turning, as shown in Figure 12.23.

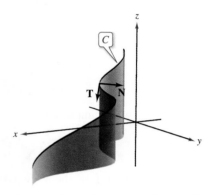

At any point on a curve, a unit normal vector is orthogonal to the unit tangent vector. The *principal* unit normal vector points in the direction in which the curve is turning.
**Figure 12.23**

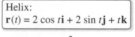

Helix:
$\mathbf{r}(t) = 2 \cos t\mathbf{i} + 2 \sin t\mathbf{j} + t\mathbf{k}$

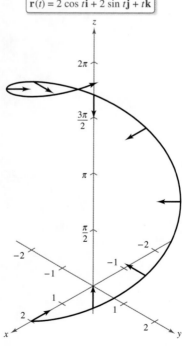

$\mathbf{N}(t)$ is horizontal and points toward the $z$-axis.
**Figure 12.24**

**EXAMPLE 4** **Finding the Principal Unit Normal Vector**

Find the principal unit normal vector for the helix $\mathbf{r}(t) = 2 \cos t\mathbf{i} + 2 \sin t\mathbf{j} + t\mathbf{k}$.

**Solution** From Example 2, you know that the unit tangent vector is

$$\mathbf{T}(t) = \frac{1}{\sqrt{5}}(-2 \sin t\mathbf{i} + 2 \cos t\mathbf{j} + \mathbf{k}). \qquad \text{Unit tangent vector}$$

So, $\mathbf{T}'(t)$ is given by

$$\mathbf{T}'(t) = \frac{1}{\sqrt{5}}(-2 \cos t\mathbf{i} - 2 \sin t\mathbf{j}).$$

Because $\|\mathbf{T}'(t)\| = 2/\sqrt{5}$, it follows that the principal unit normal vector is

$$\mathbf{N}(t) = \frac{\mathbf{T}'(t)}{\|\mathbf{T}'(t)\|}$$

$$= \frac{1}{2}(-2 \cos t\mathbf{i} - 2 \sin t\mathbf{j})$$

$$= -\cos t\mathbf{i} - \sin t\mathbf{j}. \qquad \text{Principal unit normal vector}$$

Note that this vector is horizontal and points toward the $z$-axis, as shown in Figure 12.24.

## Tangential and Normal Components of Acceleration

In the preceding section, you considered the problem of describing the motion of an object along a curve. You saw that for an object traveling at a *constant speed,* the velocity and acceleration vectors are perpendicular. This seems reasonable, because the speed would not be constant if any acceleration were acting in the direction of motion. You can verify this observation by noting that

$$\mathbf{r}''(t) \cdot \mathbf{r}'(t) = 0$$

when $\|\mathbf{r}'(t)\|$ is a constant. (See Property 7 of Theorem 12.2.)

For an object traveling at a *variable speed,* however, the velocity and acceleration vectors are not necessarily perpendicular. For instance, you saw that the acceleration vector for a projectile always points down, regardless of the direction of motion.

In general, part of the acceleration (the tangential component) acts in the line of motion, and part of it (the normal component) acts perpendicular to the line of motion. In order to determine these two components, you can use the unit vectors $\mathbf{T}(t)$ and $\mathbf{N}(t)$, which serve in much the same way as do $\mathbf{i}$ and $\mathbf{j}$ in representing vectors in the plane. The next theorem states that the acceleration vector lies in the plane determined by $\mathbf{T}(t)$ and $\mathbf{N}(t)$.

---

**THEOREM 12.4  Acceleration Vector**

If $\mathbf{r}(t)$ is the position vector for a smooth curve $C$ and $\mathbf{N}(t)$ exists, then the acceleration vector $\mathbf{a}(t)$ lies in the plane determined by $\mathbf{T}(t)$ and $\mathbf{N}(t)$.

---

**Proof**  To simplify the notation, write $\mathbf{T}$ for $\mathbf{T}(t)$, $\mathbf{T}'$ for $\mathbf{T}'(t)$, and so on. Because $\mathbf{T} = \mathbf{r}'/\|\mathbf{r}'\| = \mathbf{v}/\|\mathbf{v}\|$, it follows that

$$\mathbf{v} = \|\mathbf{v}\|\mathbf{T}.$$

By differentiating, you obtain

$$\mathbf{a} = \mathbf{v}'$$

$$= \frac{d}{dt}\big[\|\mathbf{v}\|\big]\mathbf{T} + \|\mathbf{v}\|\mathbf{T}' \qquad \text{Product Rule}$$

$$= \frac{d}{dt}\big[\|\mathbf{v}\|\big]\mathbf{T} + \|\mathbf{v}\|\mathbf{T}'\left(\frac{\|\mathbf{T}'\|}{\|\mathbf{T}'\|}\right)$$

$$= \frac{d}{dt}\big[\|\mathbf{v}\|\big]\mathbf{T} + \|\mathbf{v}\|\,\|\mathbf{T}'\|\mathbf{N}. \qquad \mathbf{N} = \mathbf{T}'/\|\mathbf{T}'\|$$

Because $\mathbf{a}$ is written as a linear combination of $\mathbf{T}$ and $\mathbf{N}$, it follows that $\mathbf{a}$ lies in the plane determined by $\mathbf{T}$ and $\mathbf{N}$. ■

The coefficients of $\mathbf{T}$ and $\mathbf{N}$ in the proof of Theorem 12.4 are called the **tangential and normal components of acceleration** and are denoted by

$$a_{\mathbf{T}} = \frac{d}{dt}\big[\|\mathbf{v}\|\big] \quad \text{and} \quad a_{\mathbf{N}} = \|\mathbf{v}\|\,\|\mathbf{T}'\|.$$

So, you can write

$$\mathbf{a}(t) = a_{\mathbf{T}}\mathbf{T}(t) + a_{\mathbf{N}}\mathbf{N}(t).$$

The next theorem lists some convenient formulas for $a_{\mathbf{T}}$ and $a_{\mathbf{N}}$.

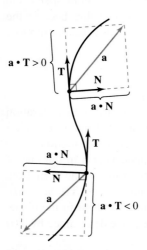

The tangential and normal components of acceleration are obtained by projecting **a** onto **T** and **N**.
**Figure 12.25**

---

## THEOREM 12.5 Tangential and Normal Components of Acceleration

If $\mathbf{r}(t)$ is the position vector for a smooth curve $C$ [for which $\mathbf{N}(t)$ exists], then the tangential and normal components of acceleration are as follows.

$$a_{\mathbf{T}} = \frac{d}{dt}\big[\|\mathbf{v}\|\big] = \mathbf{a} \cdot \mathbf{T} = \frac{\mathbf{v} \cdot \mathbf{a}}{\|\mathbf{v}\|}$$

$$a_{\mathbf{N}} = \|\mathbf{v}\|\,\|\mathbf{T}'\| = \mathbf{a} \cdot \mathbf{N} = \frac{\|\mathbf{v} \times \mathbf{a}\|}{\|\mathbf{v}\|} = \sqrt{\|\mathbf{a}\|^2 - a_{\mathbf{T}}^2}$$

Note that $a_{\mathbf{N}} \geq 0$. The normal component of acceleration is also called the **centripetal component of acceleration**.

**Proof** Note that **a** lies in the plane of **T** and **N**. So, you can use Figure 12.25 to conclude that, for any time $t$, the components of the projection of the acceleration vector onto **T** and onto **N** are given by $a_{\mathbf{T}} = \mathbf{a} \cdot \mathbf{T}$ and $a_{\mathbf{N}} = \mathbf{a} \cdot \mathbf{N}$, respectively. Moreover, because $\mathbf{a} = \mathbf{v}'$ and $\mathbf{T} = \mathbf{v}/\|\mathbf{v}\|$, you have

$$a_{\mathbf{T}} = \mathbf{a} \cdot \mathbf{T} = \mathbf{T} \cdot \mathbf{a} = \frac{\mathbf{v}}{\|\mathbf{v}\|} \cdot \mathbf{a} = \frac{\mathbf{v} \cdot \mathbf{a}}{\|\mathbf{v}\|}.$$

In Exercises 74 and 75, you are asked to prove the other parts of the theorem. ∎

---

### EXAMPLE 5 Tangential and Normal Components of Acceleration

····▷ *See LarsonCalculus.com for an interactive version of this type of example.*

Find the tangential and normal components of acceleration for the position vector given by $\mathbf{r}(t) = 3t\mathbf{i} - t\mathbf{j} + t^2\mathbf{k}$.

**Solution** Begin by finding the velocity, speed, and acceleration.

$$\mathbf{v}(t) = \mathbf{r}'(t) = 3\mathbf{i} - \mathbf{j} + 2t\mathbf{k} \qquad \text{Velocity vector}$$

$$\|\mathbf{v}(t)\| = \sqrt{9 + 1 + 4t^2} = \sqrt{10 + 4t^2} \qquad \text{Speed}$$

$$\mathbf{a}(t) = \mathbf{r}''(t) = 2\mathbf{k} \qquad \text{Acceleration vector}$$

By Theorem 12.5, the tangential component of acceleration is

$$a_{\mathbf{T}} = \frac{\mathbf{v} \cdot \mathbf{a}}{\|\mathbf{v}\|} = \frac{4t}{\sqrt{10 + 4t^2}} \qquad \text{Tangential component of acceleration}$$

and because

$$\mathbf{v} \times \mathbf{a} = \begin{vmatrix} \mathbf{i} & \mathbf{j} & \mathbf{k} \\ 3 & -1 & 2t \\ 0 & 0 & 2 \end{vmatrix} = -2\mathbf{i} - 6\mathbf{j}$$

the normal component of acceleration is

$$a_{\mathbf{N}} = \frac{\|\mathbf{v} \times \mathbf{a}\|}{\|\mathbf{v}\|} = \frac{\sqrt{4 + 36}}{\sqrt{10 + 4t^2}} = \frac{2\sqrt{10}}{\sqrt{10 + 4t^2}}. \qquad \text{Normal component of acceleration}$$

In Example 5, you could have used the alternative formula for $a_{\mathbf{N}}$ as follows.

$$a_{\mathbf{N}} = \sqrt{\|\mathbf{a}\|^2 - a_{\mathbf{T}}^2} = \sqrt{(2)^2 - \frac{16t^2}{10 + 4t^2}} = \frac{2\sqrt{10}}{\sqrt{10 + 4t^2}}$$

**EXAMPLE 6**    Finding $a_T$ and $a_N$ for a Circular Helix

Find the tangential and normal components of acceleration for the helix given by

$$\mathbf{r}(t) = b\cos t\mathbf{i} + b\sin t\mathbf{j} + ct\mathbf{k}, \quad b > 0.$$

**Solution**

$$\mathbf{v}(t) = \mathbf{r}'(t) = -b\sin t\mathbf{i} + b\cos t\mathbf{j} + c\mathbf{k} \qquad \text{Velocity vector}$$

$$\|\mathbf{v}(t)\| = \sqrt{b^2\sin^2 t + b^2\cos^2 t + c^2} = \sqrt{b^2 + c^2} \qquad \text{Speed}$$

$$\mathbf{a}(t) = \mathbf{r}''(t) = -b\cos t\mathbf{i} - b\sin t\mathbf{j} \qquad \text{Acceleration vector}$$

By Theorem 12.5, the tangential component of acceleration is

$$a_T = \frac{\mathbf{v} \cdot \mathbf{a}}{\|\mathbf{v}\|} = \frac{b^2\sin t\cos t - b^2\sin t\cos t + 0}{\sqrt{b^2 + c^2}} = 0. \qquad \begin{array}{l}\text{Tangential component}\\\text{of acceleration}\end{array}$$

Moreover, because

$$\|\mathbf{a}\| = \sqrt{b^2\cos^2 t + b^2\sin^2 t} = b$$

you can use the alternative formula for the normal component of acceleration to obtain

$$a_N = \sqrt{\|\mathbf{a}\|^2 - a_T^2} = \sqrt{b^2 - 0^2} = b. \qquad \begin{array}{l}\text{Normal component}\\\text{of acceleration}\end{array}$$

Note that the normal component of acceleration is equal to the magnitude of the acceleration. In other words, because the speed is constant, the acceleration is perpendicular to the velocity. See Figure 12.26.

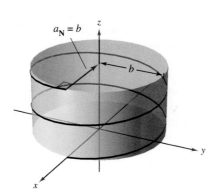

$a_N = b$

The normal component of acceleration is equal to the radius of the cylinder around which the helix is spiraling.
**Figure 12.26**

**EXAMPLE 7**    Projectile Motion

The position vector for the projectile shown in Figure 12.27 is

$$\mathbf{r}(t) = \left(50\sqrt{2}t\right)\mathbf{i} + \left(50\sqrt{2}t - 16t^2\right)\mathbf{j}. \qquad \text{Position vector}$$

Find the tangential components of acceleration when $t = 0$, $1$, and $25\sqrt{2}/16$.

**Solution**

$$\mathbf{v}(t) = 50\sqrt{2}\mathbf{i} + \left(50\sqrt{2} - 32t\right)\mathbf{j} \qquad \text{Velocity vector}$$

$$\|\mathbf{v}(t)\| = 2\sqrt{50^2 - 16(50)\sqrt{2}t + 16^2t^2} \qquad \text{Speed}$$

$$\mathbf{a}(t) = -32\mathbf{j} \qquad \text{Acceleration vector}$$

The tangential component of acceleration is

$$a_T(t) = \frac{\mathbf{v}(t) \cdot \mathbf{a}(t)}{\|\mathbf{v}(t)\|} = \frac{-32\left(50\sqrt{2} - 32t\right)}{2\sqrt{50^2 - 16(50)\sqrt{2}t + 16^2t^2}}. \qquad \begin{array}{l}\text{Tangential component}\\\text{of acceleration}\end{array}$$

At the specified times, you have

$$a_T(0) = \frac{-32\left(50\sqrt{2}\right)}{100} = -16\sqrt{2} \approx -22.6$$

$$a_T(1) = \frac{-32\left(50\sqrt{2} - 32\right)}{2\sqrt{50^2 - 16(50)\sqrt{2} + 16^2}} \approx -15.4$$

$$a_T\!\left(\frac{25\sqrt{2}}{16}\right) = \frac{-32\left(50\sqrt{2} - 50\sqrt{2}\right)}{50\sqrt{2}} = 0.$$

You can see from Figure 12.27 that at the maximum height, when $t = 25\sqrt{2}/16$, the tangential component is 0. This is reasonable because the direction of motion is horizontal at the point and the tangential component of the acceleration is equal to the horizontal component of the acceleration.

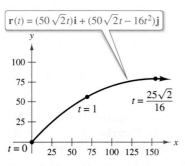

$\mathbf{r}(t) = (50\sqrt{2}t)\mathbf{i} + (50\sqrt{2}t - 16t^2)\mathbf{j}$

$t = \dfrac{25\sqrt{2}}{16}$

$t = 1$

$t = 0$

The path of a projectile
**Figure 12.27**

# 12.4 Exercises

See **CalcChat.com** for tutorial help and worked-out solutions to odd-numbered exercises.

 **Finding the Unit Tangent Vector** In Exercises 3–8, find the unit tangent vector to the curve at the specified value of the parameter.

3. $\mathbf{r}(t) = t^2\mathbf{i} + 2t\mathbf{j}, \quad t = 1$

4. $\mathbf{r}(t) = t^3\mathbf{i} + 2t^2\mathbf{j}, \quad t = 1$

5. $\mathbf{r}(t) = 5\cos t\mathbf{i} + 5\sin t\mathbf{j}, \quad t = \dfrac{\pi}{3}$

6. $\mathbf{r}(t) = 6\sin t\mathbf{i} - 2\cos t\mathbf{j}, \quad t = \dfrac{\pi}{6}$

7. $\mathbf{r}(t) = 3t\mathbf{i} - \ln t\mathbf{j}, \quad t = e$

8. $\mathbf{r}(t) = e^t\cos t\mathbf{i} + e^t\mathbf{j}, \quad t = 0$

 **Finding a Tangent Line** In Exercises 9–14, find the unit tangent vector $\mathbf{T}(t)$ and a set of parametric equations for the tangent line to the space curve at point $P$.

9. $\mathbf{r}(t) = t\mathbf{i} + t^2\mathbf{j} + t\mathbf{k}, \quad P(0, 0, 0)$

10. $\mathbf{r}(t) = t^2\mathbf{i} + t\mathbf{j} + \frac{4}{3}\mathbf{k}, \quad P\left(1, 1, \frac{4}{3}\right)$

11. $\mathbf{r}(t) = \cos t\mathbf{i} + 3\sin t\mathbf{j} + (3t - 4)\mathbf{k}, \quad P(1, 0, -4)$

12. $\mathbf{r}(t) = \left\langle t, t, \sqrt{4 - t^2} \right\rangle, \quad P\left(1, 1, \sqrt{3}\right)$

13. $\mathbf{r}(t) = \langle 2\cos t, 2\sin t, 4 \rangle, \quad P\left(\sqrt{2}, \sqrt{2}, 4\right)$

14. $\mathbf{r}(t) = \langle 2\sin t, 2\cos t, 4\sin^2 t \rangle, \quad P\left(1, \sqrt{3}, 1\right)$

 **Finding the Principal Unit Normal Vector** In Exercises 15–20, find the principal unit normal vector to the curve at the specified value of the parameter.

15. $\mathbf{r}(t) = t\mathbf{i} + \frac{1}{2}t^2\mathbf{j}, \quad t = 2$    16. $\mathbf{r}(t) = t\mathbf{i} + \dfrac{6}{t}\mathbf{j}, \quad t = 3$

17. $\mathbf{r}(t) = t\mathbf{i} + t^2\mathbf{j} + \ln t\mathbf{k}, \quad t = 1$

18. $\mathbf{r}(t) = \sqrt{2}t\mathbf{i} + e^t\mathbf{j} + e^{-t}\mathbf{k}, \quad t = 0$

19. $\mathbf{r}(t) = 6\cos t\mathbf{i} + 6\sin t\mathbf{j} + \mathbf{k}, \quad t = \dfrac{3\pi}{4}$

20. $\mathbf{r}(t) = \cos 3t\mathbf{i} + 2\sin 3t\mathbf{j} + \mathbf{k}, \quad t = \pi$

**Sketching a Graph and Vectors** In Exercises 21–24, sketch the graph of the plane curve $\mathbf{r}(t)$ and sketch the vectors $\mathbf{T}(t)$ and $\mathbf{N}(t)$ at the given value of $t$.

21. $\mathbf{r}(t) = t\mathbf{i} + \dfrac{1}{t}\mathbf{j}, \quad t = 2$

22. $\mathbf{r}(t) = t\mathbf{i} - t^3\mathbf{j}, \quad t = 1$

23. $\mathbf{r}(t) = (2t + 1)\mathbf{i} - t^2\mathbf{j}, \quad t = 2$

24. $\mathbf{r}(t) = 2\cos t\mathbf{i} + 2\sin t\mathbf{j}, \quad t = \dfrac{7\pi}{6}$

 **Finding Tangential and Normal Components of Acceleration** In Exercises 25–30, find the tangential and normal components of acceleration at the given time $t$ for the plane curve $\mathbf{r}(t)$.

25. $\mathbf{r}(t) = t\mathbf{i} + \dfrac{1}{t}\mathbf{j}, \quad t = 1$

26. $\mathbf{r}(t) = t^2\mathbf{i} + 2t\mathbf{j}, \quad t = 1$

27. $\mathbf{r}(t) = e^t\mathbf{i} + e^{-2t}\mathbf{j}, \quad t = 0$

28. $\mathbf{r}(t) = e^t\mathbf{i} + e^{-t}\mathbf{j}, \quad t = 0$

29. $\mathbf{r}(t) = e^t\cos t\mathbf{i} + e^t\sin t\mathbf{j}, \quad t = \dfrac{\pi}{2}$

30. $\mathbf{r}(t) = 4\cos 3t\mathbf{i} + 4\sin 3t\mathbf{j}, \quad t = \pi$

**Circular Motion** In Exercises 31–34, consider an object moving according to the position vector

$$\mathbf{r}(t) = a\cos \omega t\mathbf{i} + a\sin \omega t\mathbf{j}.$$

31. Find $\mathbf{T}(t)$, $\mathbf{N}(t)$, $a_{\mathbf{T}}$, and $a_{\mathbf{N}}$.

32. Determine the directions of $\mathbf{T}$ and $\mathbf{N}$ relative to the position vector $\mathbf{r}$.

33. Determine the speed of the object at any time $t$ and explain its value relative to the value of $a_{\mathbf{T}}$.

34. When the angular speed $\omega$ is halved, by what factor is $a_{\mathbf{N}}$ changed?

 **Finding Tangential and Normal Components of Acceleration** In Exercises 35–40, find the tangential and normal components of acceleration at the given time $t$ for the space curve $\mathbf{r}(t)$.

35. $\mathbf{r}(t) = t\mathbf{i} + 2t\mathbf{j} - 3t\mathbf{k}, \quad t = 1$

36. $\mathbf{r}(t) = \cos t\mathbf{i} + \sin t\mathbf{j} + 2t\mathbf{k}, \quad t = \dfrac{\pi}{3}$

37. $\mathbf{r}(t) = t\mathbf{i} + t^2\mathbf{j} + \dfrac{t^2}{2}\mathbf{k}, \quad t = 1$

38. $\mathbf{r}(t) = (2t - 1)\mathbf{i} + t^2\mathbf{j} - 4t\mathbf{k}, \quad t = 2$

39. $\mathbf{r}(t) = e^t\sin t\mathbf{i} + e^t\cos t\mathbf{j} + e^t\mathbf{k}, \quad t = 0$

40. $\mathbf{r}(t) = e^t\mathbf{i} + 2t\mathbf{j} + e^{-t}\mathbf{k}, \quad t = 0$

**43. Finding Vectors** An object moves along the path given by

$$\mathbf{r}(t) = 3t\mathbf{i} + 4t\mathbf{j}.$$

Find $\mathbf{v}(t)$, $\mathbf{a}(t)$, $\mathbf{T}(t)$, and $\mathbf{N}(t)$ (if it exists). What is the form of the path? Is the speed of the object constant or changing?

**44. HOW DO YOU SEE IT?** The figures show the paths of two particles.

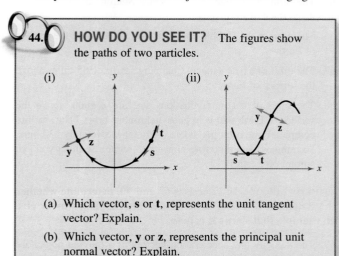

(a) Which vector, **s** or **t**, represents the unit tangent vector? Explain.

(b) Which vector, **y** or **z**, represents the principal unit normal vector? Explain.

**45. Cycloidal Motion** The figure shows the path of a particle modeled by the vector-valued function

$$\mathbf{r}(t) = \langle \pi t - \sin \pi t, 1 - \cos \pi t \rangle.$$

The figure also shows the vectors $\mathbf{v}(t)/\|\mathbf{v}(t)\|$ and $\mathbf{a}(t)/\|\mathbf{a}(t)\|$ at the indicated values of $t$.

(a) Find $a_\mathbf{T}$ and $a_\mathbf{N}$ at $t = \frac{1}{2}$, $t = 1$, and $t = \frac{3}{2}$.

(b) Determine whether the speed of the particle is increasing or decreasing at each of the indicated values of $t$. Give reasons for your answers.

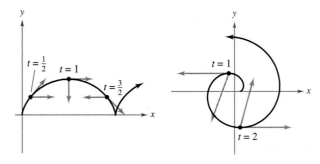

Figure for 45        Figure for 46

**46. Motion Along an Involute of a Circle** The figure shows a particle moving along a path modeled by

$$\mathbf{r}(t) = \langle \cos \pi t + \pi t \sin \pi t, \sin \pi t - \pi t \cos \pi t \rangle.$$

The figure also shows the vectors $\mathbf{v}(t)$ and $\mathbf{a}(t)$ for $t = 1$ and $t = 2$.

(a) Find $a_\mathbf{T}$ and $a_\mathbf{N}$ at $t = 1$ and $t = 2$.

(b) Determine whether the speed of the particle is increasing or decreasing at each of the indicated values of $t$. Give reasons for your answers.

**Finding a Binormal Vector** In Exercises 47–52, find the vectors **T** and **N** and the binormal vector $\mathbf{B} = \mathbf{T} \times \mathbf{N}$ for the vector-valued function $\mathbf{r}(t)$ at the given value of $t$.

**47.** $\mathbf{r}(t) = 2 \cos t\mathbf{i} + 2 \sin t\mathbf{j} + \frac{t}{2}\mathbf{k}, \quad t = \frac{\pi}{2}$

**48.** $\mathbf{r}(t) = t\mathbf{i} + t^2\mathbf{j} + \frac{t^3}{3}\mathbf{k}, \quad t = 1$

**49.** $\mathbf{r}(t) = \mathbf{i} + \sin t\mathbf{j} + \cos t\mathbf{k}, \quad t = \frac{\pi}{4}$

**50.** $\mathbf{r}(t) = 2e^t\mathbf{i} + e^t \cos t\mathbf{j} + e^t \sin t\mathbf{k}, \quad t = 0$

**51.** $\mathbf{r}(t) = 4 \sin t\mathbf{i} + 4 \cos t\mathbf{j} + 2t\mathbf{k}, \quad t = \frac{\pi}{3}$

**52.** $\mathbf{r}(t) = 3 \cos 2t\mathbf{i} + 3 \sin 2t\mathbf{j} + t\mathbf{k}, \quad t = \frac{\pi}{4}$

**Alternative Formula for the Principal Unit Normal Vector** In Exercises 53–56, use the vector-valued function $\mathbf{r}(t)$ to find the principal unit normal vector $\mathbf{N}(t)$ using the alternative formula

$$\mathbf{N} = \frac{(\mathbf{v} \cdot \mathbf{v})\mathbf{a} - (\mathbf{v} \cdot \mathbf{a})\mathbf{v}}{\|(\mathbf{v} \cdot \mathbf{v})\mathbf{a} - (\mathbf{v} \cdot \mathbf{a})\mathbf{v}\|}.$$

**53.** $\mathbf{r}(t) = 3t\mathbf{i} + 2t^2\mathbf{j}$

**54.** $\mathbf{r}(t) = 3 \cos 2t\mathbf{i} + 3 \sin 2t\mathbf{j}$

**55.** $\mathbf{r}(t) = 2t\mathbf{i} + 4t\mathbf{j} + t^2\mathbf{k}$

**56.** $\mathbf{r}(t) = 5 \cos t\mathbf{i} + 5 \sin t\mathbf{j} + 3t\mathbf{k}$

**57. Projectile Motion** Find the tangential and normal components of acceleration for a projectile fired at an angle $\theta$ with the horizontal at an initial speed of $v_0$. What are the components when the projectile is at its maximum height?

**58. Projectile Motion** Use your results from Exercise 57 to find the tangential and normal components of acceleration for a projectile fired at an angle of 45° with the horizontal at an initial speed of 150 feet per second. What are the components when the projectile is at its maximum height?

**59. Projectile Motion** A projectile is launched with an initial speed of 120 feet per second at a height of 5 feet and at an angle of 30° with the horizontal.

(a) Determine the vector-valued function for the path of the projectile.

(b) Use a graphing utility to graph the path and approximate the maximum height and range of the projectile.

(c) Find $\mathbf{v}(t)$, $\|\mathbf{v}(t)\|$, and $\mathbf{a}(t)$.

(d) Use a graphing utility to complete the table.

| $t$ | 0.5 | 1.0 | 1.5 | 2.0 | 2.5 | 3.0 |
|-------|-----|-----|-----|-----|-----|-----|
| Speed |     |     |     |     |     |     |

(e) Use a graphing utility to graph the scalar functions $a_\mathbf{T}$ and $a_\mathbf{N}$. How is the speed of the projectile changing when $a_\mathbf{T}$ and $a_\mathbf{N}$ have opposite signs?

**60. Projectile Motion** A projectile is launched with an initial speed of 220 feet per second at a height of 4 feet and at an angle of 45° with the horizontal.

(a) Determine the vector-valued function for the path of the projectile.

(b) Use a graphing utility to graph the path and approximate the maximum height and range of the projectile.

(c) Find $\mathbf{v}(t)$, $\|\mathbf{v}(t)\|$, and $\mathbf{a}(t)$.

(d) Use a graphing utility to complete the table.

| $t$ | 0.5 | 1.0 | 1.5 | 2.0 | 2.5 | 3.0 |
|-------|-----|-----|-----|-----|-----|-----|
| Speed | | | | | | |

**61. Air Traffic Control**

Because of a storm, ground controllers instruct the pilot of a plane flying at an altitude of 4 miles to make a 90° turn and climb to an altitude of 4.2 miles. The model for the path of the plane during this maneuver is

$$\mathbf{r}(t) = \langle 10 \cos 10\pi t, 10 \sin 10\pi t, 4 + 4t \rangle, \quad 0 \le t \le \tfrac{1}{20}$$

where $t$ is the time in hours and $\mathbf{r}$ is the distance in miles.

(a) Determine the speed of the plane.

(b) Calculate $a_{\mathbf{T}}$ and $a_{\mathbf{N}}$. Why is one of these equal to 0?

**62. Projectile Motion** A plane flying at an altitude of 36,000 feet at a speed of 600 miles per hour releases a bomb. Find the tangential and normal components of acceleration acting on the bomb.

**63. Centripetal Acceleration** An object is spinning at a constant speed on the end of a string, according to the position vector $\mathbf{r}(t) = a \cos \omega t \mathbf{i} + a \sin \omega t \mathbf{j}$.

(a) When the angular speed $\omega$ is doubled, how is the centripetal component of acceleration changed?

(b) When the angular speed is unchanged but the length of the string is halved, how is the centripetal component of acceleration changed?

**64. Centripetal Force** An object of mass $m$ moves at a constant speed $v$ in a circular path of radius $r$, according to the position vector $\mathbf{r}(t) = r \cos \omega t \mathbf{i} + r \sin \omega t \mathbf{j}$.

(a) The force required to produce the centripetal component of acceleration is called the *centripetal force* and is given by $F = mv^2/r$. Use $\mathbf{F} = m\mathbf{a}$ to verify the centripetal force.

(b) Newton's Law of Universal Gravitation is given by $F = GMm/d^2$, where $d$ is the distance between the centers of the two bodies of masses $M$ and $m$, and $G$ is the gravitational constant. Use this law to show that the speed required for circular motion is $v = \sqrt{GM/r}$.

**Orbital Speed** In Exercises 65–68, use the result of Exercise 64 to find the speed necessary for the given circular orbit around Earth. Let $GM = 9.56 \times 10^4$ cubic miles per second per second, and assume the radius of Earth is 4000 miles.

**65.** The orbit of the International Space Station 255 miles above the surface of Earth

**66.** The orbit of the Hubble telescope 340 miles above the surface of Earth

**67.** The orbit of a heat capacity mapping satellite 385 miles above the surface of Earth

**68.** The orbit of a communications satellite $r$ miles above the surface of Earth that is in geosynchronous orbit. [The satellite completes one orbit per sidereal day (approximately 23 hours, 56 minutes) and therefore appears to remain stationary above a point on Earth.]

**True or False?** In Exercises 69 and 70, determine whether the statement is true or false. If it is false, explain why or give an example that shows it is false.

**69.** The velocity and acceleration vectors of a moving object are always perpendicular.

**70.** If $a_{\mathbf{N}} = 0$ for a moving object, then the object is moving in a straight line.

**71. Motion of a Particle** A particle moves along a path modeled by

$$\mathbf{r}(t) = \cosh(bt)\mathbf{i} + \sinh(bt)\mathbf{j}$$

where $b$ is a positive constant.

(a) Show that the path of the particle is a hyperbola.

(b) Show that $\mathbf{a}(t) = b^2 \mathbf{r}(t)$.

**72. Proof** Prove that the principal unit normal vector $\mathbf{N}$ points toward the concave side of a plane curve.

**73. Proof** Prove that the vector $\mathbf{T}'(t)$ is 0 for an object moving in a straight line.

**74. Proof** Prove that $a_{\mathbf{N}} = \dfrac{\|\mathbf{v} \times \mathbf{a}\|}{\|\mathbf{v}\|}$.

**75. Proof** Prove that $a_{\mathbf{N}} = \sqrt{\|\mathbf{a}\|^2 - a_{\mathbf{T}}^2}$.

**PUTNAM EXAM CHALLENGE**

**76.** A particle of unit mass moves on a straight line under the action of a force which is a function $f(v)$ of the velocity $v$ of the particle, but the form of this function is not known. A motion is observed, and the distance $x$ covered in time $t$ is found to be connected with $t$ by the formula

$$x = at + bt^2 + ct^3$$

where $a$, $b$, and $c$ have numerical values determined by observation of the motion. Find the function $f(v)$ for the range of $v$ covered by the experiment.

# 12.5 Arc Length and Curvature

■ Find the arc length of a space curve.
■ Use the arc length parameter to describe a plane curve or space curve.
■ Find the curvature of a curve at a point on the curve.
■ Use a vector-valued function to find frictional force.

## Arc Length

In Section 10.3, you saw that the arc length of a smooth *plane* curve $C$ given by the parametric equations $x = x(t)$ and $y = y(t)$, $a \le t \le b$, is

$$s = \int_a^b \sqrt{[x'(t)]^2 + [y'(t)]^2} \, dt.$$

In vector form, where $C$ is given by $\mathbf{r}(t) = x(t)\mathbf{i} + y(t)\mathbf{j}$, you can rewrite this equation for arc length as

$$s = \int_a^b \|\mathbf{r}'(t)\| \, dt.$$

The formula for the arc length of a plane curve has a natural extension to a smooth curve in *space*, as stated in the next theorem.

> ### THEOREM 12.6 Arc Length of a Space Curve
>
> If $C$ is a smooth curve given by $\mathbf{r}(t) = x(t)\mathbf{i} + y(t)\mathbf{j} + z(t)\mathbf{k}$ on an interval $[a, b]$, then the arc length of $C$ on the interval is
>
> $$s = \int_a^b \sqrt{[x'(t)]^2 + [y'(t)]^2 + [z'(t)]^2} \, dt = \int_a^b \|\mathbf{r}'(t)\| \, dt.$$

### Exploration

***Arc Length Formula*** The formula for the arc length of a space curve is given in terms of the parametric equations used to represent the curve. Does this mean that the arc length of the curve depends on the parameter being used? Would you want this to be true? Explain your reasoning.

Here is a different parametric representation of the curve in Example 1.

$$\mathbf{r}(t) = t^2\mathbf{i} + \frac{4}{3}t^3\mathbf{j} + \frac{1}{2}t^4\mathbf{k}$$

Find the arc length from $t = 0$ to $t = \sqrt{2}$ and compare the result with that found in Example 1.

**EXAMPLE 1** Finding the Arc Length of a Curve in Space

⋯▷ *See LarsonCalculus.com for an interactive version of this type of example.*

Find the arc length of the curve given by

$$\mathbf{r}(t) = t\mathbf{i} + \frac{4}{3}t^{3/2}\mathbf{j} + \frac{1}{2}t^2\mathbf{k}$$

from $t = 0$ to $t = 2$, as shown in Figure 12.28.

**Solution** Using $x(t) = t$, $y(t) = \frac{4}{3}t^{3/2}$, and $z(t) = \frac{1}{2}t^2$, you obtain $x'(t) = 1$, $y'(t) = 2t^{1/2}$, and $z'(t) = t$. So, the arc length from $t = 0$ to $t = 2$ is given by

$$s = \int_0^2 \sqrt{[x'(t)]^2 + [y'(t)]^2 + [z'(t)]^2} \, dt \qquad \text{Formula for arc length}$$

$$= \int_0^2 \sqrt{1 + 4t + t^2} \, dt$$

$$= \int_0^2 \sqrt{(t+2)^2 - 3} \, dt \qquad \text{Integration tables (Appendix B), Formula 26}$$

$$= \left[ \frac{t+2}{2}\sqrt{(t+2)^2 - 3} - \frac{3}{2}\ln\left|(t+2) + \sqrt{(t+2)^2 - 3}\right| \right]_0^2$$

$$= 2\sqrt{13} - \frac{3}{2}\ln\left(4 + \sqrt{13}\right) - 1 + \frac{3}{2}\ln 3$$

$$\approx 4.816.$$

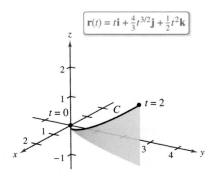

$$\mathbf{r}(t) = t\mathbf{i} + \frac{4}{3}t^{3/2}\mathbf{j} + \frac{1}{2}t^2\mathbf{k}$$

As $t$ increases from 0 to 2, the vector $\mathbf{r}(t)$ traces out a curve.
**Figure 12.28**

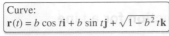

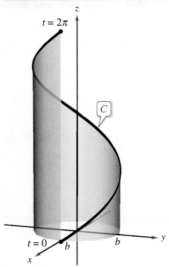

One turn of a helix
**Figure 12.29**

**EXAMPLE 2**    **Finding the Arc Length of a Helix**

Find the length of one turn of the helix given by

$$\mathbf{r}(t) = b \cos t\mathbf{i} + b \sin t\mathbf{j} + \sqrt{1 - b^2}\, t\mathbf{k}$$

as shown in Figure 12.29.

**Solution**    Begin by finding the derivative.

$$\mathbf{r}'(t) = -b \sin t\mathbf{i} + b \cos t\mathbf{j} + \sqrt{1 - b^2}\,\mathbf{k} \qquad \text{Derivative}$$

Now, using the formula for arc length, you can find the length of one turn of the helix by integrating $\|\mathbf{r}'(t)\|$ from 0 to $2\pi$.

$$
\begin{aligned}
s &= \int_0^{2\pi} \|\mathbf{r}'(t)\|\, dt && \text{Formula for arc length} \\
&= \int_0^{2\pi} \sqrt{b^2(\sin^2 t + \cos^2 t) + (1 - b^2)}\, dt \\
&= \int_0^{2\pi} dt \\
&= t\Big]_0^{2\pi} \\
&= 2\pi
\end{aligned}
$$

So, the length is $2\pi$ units.  ∎

## Arc Length Parameter

You have seen that curves can be represented by vector-valued functions in different ways, depending on the choice of parameter. For *motion* along a curve, the convenient parameter is time $t$. For studying the *geometric properties* of a curve, however, the convenient parameter is often arc length $s$.

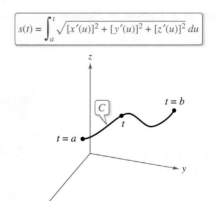

**Figure 12.30**

**Definition of Arc Length Function**

Let $C$ be a smooth curve given by $\mathbf{r}(t)$ defined on the closed interval $[a, b]$. For $a \le t \le b$, the **arc length function** is

$$s(t) = \int_a^t \|\mathbf{r}'(u)\|\, du = \int_a^t \sqrt{[x'(u)]^2 + [y'(u)]^2 + [z'(u)]^2}\, du.$$

The arc length $s$ is called the **arc length parameter**. (See Figure 12.30.)

Note that the arc length function $s$ is *nonnegative*. It measures the distance along $C$ from the initial point $(x(a), y(a), z(a))$ to the point $(x(t), y(t), z(t))$.

Using the definition of the arc length function and the Second Fundamental Theorem of Calculus, you can conclude that

$$\frac{ds}{dt} = \|\mathbf{r}'(t)\|. \qquad \text{Derivative of arc length function}$$

In differential form, you can write

$$ds = \|\mathbf{r}'(t)\|\, dt.$$

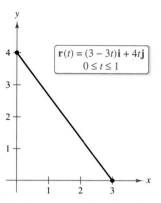

The line segment from $(3, 0)$ to $(0, 4)$ can be parametrized using the arc length parameter $s$.

**Figure 12.31**

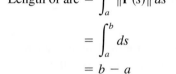  **EXAMPLE 3**    **Finding the Arc Length Function for a Line**

Find the arc length function $s(t)$ for the line segment given by

$$\mathbf{r}(t) = (3 - 3t)\mathbf{i} + 4t\mathbf{j}, \quad 0 \le t \le 1$$

and write $\mathbf{r}$ as a function of the parameter $s$. (See Figure 12.31.)

**Solution**   Because $\mathbf{r}'(t) = -3\mathbf{i} + 4\mathbf{j}$ and

$$\|\mathbf{r}'(t)\| = \sqrt{(-3)^2 + 4^2} = 5$$

you have

$$s(t) = \int_0^t \|\mathbf{r}'(u)\|\, du$$

$$= \int_0^t 5\, du$$

$$= 5t.$$

Using $s = 5t$ (or $t = s/5$), you can rewrite $\mathbf{r}$ using the arc length parameter as follows.

$$\mathbf{r}(s) = \left(3 - \frac{3}{5}s\right)\mathbf{i} + \frac{4}{5}s\mathbf{j}, \quad 0 \le s \le 5$$

One of the advantages of writing a vector-valued function in terms of the arc length parameter is that $\|\mathbf{r}'(s)\| = 1$. For instance, in Example 3, you have

$$\|\mathbf{r}'(s)\| = \sqrt{\left(-\frac{3}{5}\right)^2 + \left(\frac{4}{5}\right)^2} = 1.$$

So, for a smooth curve $C$ represented by $\mathbf{r}(s)$, where $s$ is the arc length parameter, the arc length between $a$ and $b$ is

$$\text{Length of arc} = \int_a^b \|\mathbf{r}'(s)\|\, ds$$

$$= \int_a^b ds$$

$$= b - a$$

$$= \text{length of interval.}$$

Furthermore, if $t$ is *any* parameter such that $\|\mathbf{r}'(t)\| = 1$, then $t$ must be the arc length parameter. These results are summarized in the next theorem, which is stated without proof.

---

**THEOREM 12.7   Arc Length Parameter**

If $C$ is a smooth curve given by

$$\mathbf{r}(s) = x(s)\mathbf{i} + y(s)\mathbf{j} \qquad \text{Plane curve}$$

or

$$\mathbf{r}(s) = x(s)\mathbf{i} + y(s)\mathbf{j} + z(s)\mathbf{k} \qquad \text{Space curve}$$

where $s$ is the arc length parameter, then

$$\|\mathbf{r}'(s)\| = 1.$$

Moreover, if $t$ is *any* parameter for the vector-valued function $\mathbf{r}$ such that $\|\mathbf{r}'(t)\| = 1$, then $t$ must be the arc length parameter.

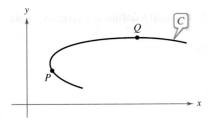

Curvature at $P$ is greater than at $Q$.
**Figure 12.32**

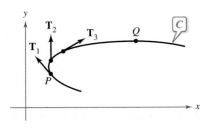

The magnitude of the rate of change of **T** with respect to the arc length is the curvature of a curve.
**Figure 12.33**

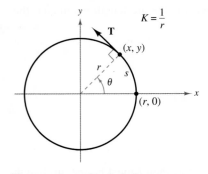

The curvature of a circle is constant.
**Figure 12.34**

# Curvature

An important use of the arc length parameter is to find **curvature**—the measure of how sharply a curve bends. For instance, in Figure 12.32, the curve bends more sharply at $P$ than at $Q$, and you can say that the curvature is greater at $P$ than at $Q$. You can calculate curvature by calculating the magnitude of the rate of change of the unit tangent vector **T** with respect to the arc length $s$, as shown in Figure 12.33.

---

### Definition of Curvature

Let $C$ be a smooth curve (in the plane *or* in space) given by $\mathbf{r}(s)$, where $s$ is the arc length parameter. The **curvature** $K$ at $s$ is

$$K = \left\| \frac{d\mathbf{T}}{ds} \right\| = \|\mathbf{T}'(s)\|.$$

---

A circle has the same curvature at any point. Moreover, the curvature and the radius of the circle are inversely related. That is, a circle with a large radius has a small curvature, and a circle with a small radius has a large curvature. This inverse relationship is made explicit in the next example.

**EXAMPLE 4**    **Finding the Curvature of a Circle**

Show that the curvature of a circle of radius $r$ is

$$K = \frac{1}{r}.$$

**Solution**    Without loss of generality, you can consider the circle to be centered at the origin. Let $(x, y)$ be any point on the circle and let $s$ be the length of the arc from $(r, 0)$ to $(x, y)$, as shown in Figure 12.34. By letting $\theta$ be the central angle of the circle, you can represent the circle by

$$\mathbf{r}(\theta) = r \cos \theta \mathbf{i} + r \sin \theta \mathbf{j}. \qquad \text{\small $\theta$ is the parameter.}$$

Using the formula for the length of a circular arc $s = r\theta$, you can rewrite $\mathbf{r}(\theta)$ in terms of the arc length parameter as follows.

$$\mathbf{r}(s) = r \cos \frac{s}{r}\mathbf{i} + r \sin \frac{s}{r}\mathbf{j} \qquad \text{\small Arc length $s$ is the parameter.}$$

So, $\mathbf{r}'(s) = -\sin \frac{s}{r}\mathbf{i} + \cos \frac{s}{r}\mathbf{j}$, and it follows that $\|\mathbf{r}'(s)\| = 1$, which implies that the unit tangent vector is

$$\mathbf{T}(s) = \frac{\mathbf{r}'(s)}{\|\mathbf{r}'(s)\|} = -\sin \frac{s}{r}\mathbf{i} + \cos \frac{s}{r}\mathbf{j}$$

and the curvature is

$$K = \|\mathbf{T}'(s)\| = \left\| -\frac{1}{r}\cos \frac{s}{r}\mathbf{i} - \frac{1}{r}\sin \frac{s}{r}\mathbf{j} \right\| = \frac{1}{r}$$

at every point on the circle. ■

Because a straight line does not curve, you would expect its curvature to be 0. Try checking this by finding the curvature of the line given by

$$\mathbf{r}(s) = \left( 3 - \frac{3}{5}s \right)\mathbf{i} + \frac{4}{5}s\mathbf{j}.$$

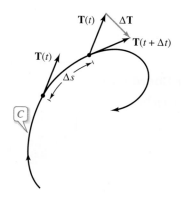

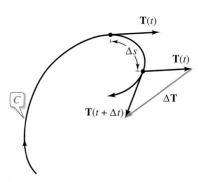

**Figure 12.35**

In Example 4, the curvature was found by applying the definition directly. This requires that the curve be written in terms of the arc length parameter $s$. The next theorem gives two other formulas for finding the curvature of a curve written in terms of an arbitrary parameter $t$. The proof of this theorem is left as an exercise [see Exercise 82, parts (a) and (b)].

---

**THEOREM 12.8   Formulas for Curvature**

If $C$ is a smooth curve given by $\mathbf{r}(t)$, then the curvature $K$ of $C$ at $t$ is

$$K = \frac{\|\mathbf{T}'(t)\|}{\|\mathbf{r}'(t)\|} = \frac{\|\mathbf{r}'(t) \times \mathbf{r}''(t)\|}{\|\mathbf{r}'(t)\|^3}.$$

---

Because $\|\mathbf{r}'(t)\| = ds/dt$, the first formula implies that curvature is the ratio of the rate of change of the unit tangent vector $\mathbf{T}$ to the rate of change of the arc length. To see that this is reasonable, let $\Delta t$ be a "small number." Then,

$$\frac{\mathbf{T}'(t)}{ds/dt} \approx \frac{[\mathbf{T}(t + \Delta t) - \mathbf{T}(t)]/\Delta t}{[s(t + \Delta t) - s(t)]/\Delta t} = \frac{\mathbf{T}(t + \Delta t) - \mathbf{T}(t)}{s(t + \Delta t) - s(t)} = \frac{\Delta \mathbf{T}}{\Delta s}.$$

In other words, for a given $\Delta s$, the greater the length of $\Delta \mathbf{T}$, the more the curve bends at $t$, as shown in Figure 12.35.

**EXAMPLE 5**   **Finding the Curvature of a Space Curve**

Find the curvature of the curve given by

$$\mathbf{r}(t) = 2t\mathbf{i} + t^2\mathbf{j} - \frac{1}{3}t^3\mathbf{k}.$$

**Solution**   It is not apparent whether this parameter represents arc length, so you should use the formula $K = \|\mathbf{T}'(t)\|/\|\mathbf{r}'(t)\|$.

$$\mathbf{r}'(t) = 2\mathbf{i} + 2t\mathbf{j} - t^2\mathbf{k}$$

$$\|\mathbf{r}'(t)\| = \sqrt{4 + 4t^2 + t^4} \qquad \text{Length of } \mathbf{r}'(t)$$

$$= t^2 + 2$$

$$\mathbf{T}(t) = \frac{\mathbf{r}'(t)}{\|\mathbf{r}'(t)\|}$$

$$= \frac{2\mathbf{i} + 2t\mathbf{j} - t^2\mathbf{k}}{t^2 + 2}$$

$$\mathbf{T}'(t) = \frac{(t^2 + 2)(2\mathbf{j} - 2t\mathbf{k}) - (2t)(2\mathbf{i} + 2t\mathbf{j} - t^2\mathbf{k})}{(t^2 + 2)^2}$$

$$= \frac{-4t\mathbf{i} + (4 - 2t^2)\mathbf{j} - 4t\mathbf{k}}{(t^2 + 2)^2}$$

$$\|\mathbf{T}'(t)\| = \frac{\sqrt{16t^2 + 16 - 16t^2 + 4t^4 + 16t^2}}{(t^2 + 2)^2}$$

$$= \frac{2(t^2 + 2)}{(t^2 + 2)^2}$$

$$= \frac{2}{t^2 + 2} \qquad \text{Length of } \mathbf{T}'(t)$$

Therefore,

$$K = \frac{\|\mathbf{T}'(t)\|}{\|\mathbf{r}'(t)\|} = \frac{2}{(t^2 + 2)^2}. \qquad \text{Curvature} \quad \blacksquare$$

The next theorem presents a formula for calculating the curvature of a plane curve given by $y = f(x)$.

---

**THEOREM 12.9   Curvature in Rectangular Coordinates**

If $C$ is the graph of a twice-differentiable function given by $y = f(x)$, then the curvature $K$ at the point $(x, y)$ is

$$K = \frac{|y''|}{[1 + (y')^2]^{3/2}}.$$

---

**Proof** By representing the curve $C$ by $\mathbf{r}(x) = x\mathbf{i} + f(x)\mathbf{j} + 0\mathbf{k}$, where $x$ is the parameter, you obtain $\mathbf{r}'(x) = \mathbf{i} + f'(x)\mathbf{j}$,

$$\|\mathbf{r}'(x)\| = \sqrt{1 + [f'(x)]^2}$$

and $\mathbf{r}''(x) = f''(x)\mathbf{j}$. Because $\mathbf{r}'(x) \times \mathbf{r}''(x) = f''(x)\mathbf{k}$, it follows that the curvature is

$$K = \frac{\|\mathbf{r}'(x) \times \mathbf{r}''(x)\|}{\|\mathbf{r}'(x)\|^3}$$

$$= \frac{|f''(x)|}{\{1 + [f'(x)]^2\}^{3/2}}$$

$$= \frac{|y''|}{[1 + (y')^2]^{3/2}}.$$ ∎

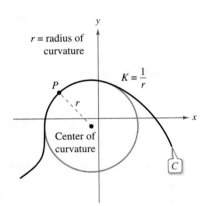

The circle of curvature

**Figure 12.36**

Let $C$ be a curve with curvature $K$ at point $P$. The circle passing through point $P$ with radius $r = 1/K$ is called the **circle of curvature** when the circle lies on the concave side of the curve and shares a common tangent line with the curve at point $P$. The radius is called the **radius of curvature** at $P$, and the center of the circle is called the **center of curvature.**

The circle of curvature gives you a nice way to estimate the curvature $K$ at a point $P$ on a curve graphically. Using a compass, you can sketch a circle that lies against the concave side of the curve at point $P$, as shown in Figure 12.36. If the circle has a radius of $r$, then you can estimate the curvature to be $K = 1/r$.

**EXAMPLE 6**   **Finding Curvature in Rectangular Coordinates**

Find the curvature of the parabola given by

$$y = x - \frac{1}{4}x^2$$

at $x = 2$. Sketch the circle of curvature at $(2, 1)$.

**Solution**   The curvature at $x = 2$ is as follows.

$$y' = 1 - \frac{x}{2} \qquad\qquad y' = 0$$

$$y'' = -\frac{1}{2} \qquad\qquad y'' = -\frac{1}{2}$$

$$K = \frac{|y''|}{[1 + (y')^2]^{3/2}} \qquad\qquad K = \frac{1}{2}$$

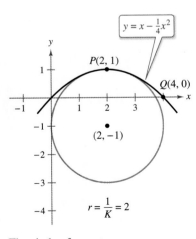

The circle of curvature

**Figure 12.37**

Because the curvature at $P(2, 1)$ is $\frac{1}{2}$, it follows that the radius of the circle of curvature at that point is 2. So, the center of curvature is $(2, -1)$, as shown in Figure 12.37. [In the figure, note that the curve has the greatest curvature at $P$. Try showing that the curvature at $Q(4, 0)$ is $1/2^{5/2} \approx 0.177$.] ∎

The amount of thrust felt by passengers in a car that is turning depends on two things—the speed of the car and the sharpness of the turn.

**Figure 12.38**

Arc length and curvature are closely related to the tangential and normal components of acceleration. The tangential component of acceleration is the rate of change of the speed, which in turn is the rate of change of the arc length. This component is negative as a moving object slows down and positive as it speeds up—regardless of whether the object is turning or traveling in a straight line. So, the tangential component is solely a function of the arc length and is independent of the curvature.

On the other hand, the normal component of acceleration is a function of *both* speed and curvature. This component measures the acceleration acting perpendicular to the direction of motion. To see why the normal component is affected by both speed and curvature, imagine that you are driving a car around a turn, as shown in Figure 12.38. When your speed is high and the turn is sharp, you feel yourself thrown against the car door. By lowering your speed *or* taking a more gentle turn, you are able to lessen this sideways thrust.

The next theorem explicitly states the relationships among speed, curvature, and the components of acceleration.

• • • • • • • • • • • • • • • ▷

**• • REMARK** Note that Theorem 12.10 gives additional formulas for $a_T$ and $a_N$.

**THEOREM 12.10  Acceleration, Speed, and Curvature**

If $\mathbf{r}(t)$ is the position vector for a smooth curve $C$, then the acceleration vector is given by

$$\mathbf{a}(t) = \frac{d^2 s}{dt^2}\mathbf{T} + K\left(\frac{ds}{dt}\right)^2\mathbf{N}$$

where $K$ is the curvature of $C$ and $ds/dt$ is the speed.

**Proof**  For the position vector $\mathbf{r}(t)$, you have

$$\mathbf{a}(t) = a_T\mathbf{T} + a_N\mathbf{N}$$

$$= \frac{d}{dt}\left[\|\mathbf{v}\|\right]\mathbf{T} + \|\mathbf{v}\|\|\mathbf{T}'\|\mathbf{N}$$

$$= \frac{d^2 s}{dt^2}\mathbf{T} + \frac{ds}{dt}\left(\|\mathbf{v}\|K\right)\mathbf{N}$$

$$= \frac{d^2 s}{dt^2}\mathbf{T} + K\left(\frac{ds}{dt}\right)^2\mathbf{N}.$$

**EXAMPLE 7**  **Tangential and Normal Components of Acceleration**

Find $a_T$ and $a_N$ for the curve given by

$$\mathbf{r}(t) = 2t\mathbf{i} + t^2\mathbf{j} - \tfrac{1}{3}t^3\mathbf{k}.$$

**Solution**  From Example 5, you know that

$$\frac{ds}{dt} = \|\mathbf{r}'(t)\| = t^2 + 2 \quad \text{and} \quad K = \frac{2}{(t^2 + 2)^2}.$$

Therefore,

$$a_T = \frac{d^2 s}{dt^2} = 2t \qquad\qquad \text{Tangential component}$$

and

$$a_N = K\left(\frac{ds}{dt}\right)^2 = \frac{2}{(t^2 + 2)^2}(t^2 + 2)^2 = 2. \qquad \text{Normal component}$$

## Application

There are many applications in physics and engineering dynamics that involve the relationships among speed, arc length, curvature, and acceleration. One such application concerns frictional force.

A moving object with mass $m$ is in contact with a stationary object. The total force required to produce an acceleration **a** along a given path is

$$\mathbf{F} = m\mathbf{a}$$
$$= m\left(\frac{d^2s}{dt^2}\right)\mathbf{T} + mK\left(\frac{ds}{dt}\right)^2\mathbf{N}$$
$$= ma_{\mathbf{T}}\mathbf{T} + ma_{\mathbf{N}}\mathbf{N}.$$

The portion of this total force that is supplied by the stationary object is called the **force of friction.** For example, when a car moving with constant speed is rounding a turn, the roadway exerts a frictional force that keeps the car from sliding off the road. If the car is not sliding, the frictional force is perpendicular to the direction of motion and has magnitude equal to the normal component of acceleration, as shown in Figure 12.39. The potential frictional force of a road around a turn can be increased by banking the roadway.

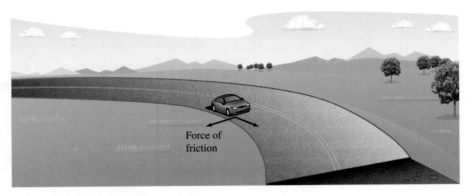

Force of friction

The force of friction is perpendicular to the direction of motion.
**Figure 12.39**

---

**EXAMPLE 8**    **Frictional Force**

A 360-kilogram go-cart is driven at a speed of 60 kilometers per hour around a circular racetrack of radius 12 meters, as shown in Figure 12.40. To keep the cart from skidding off course, what frictional force must the track surface exert on the tires?

**Solution**    The frictional force must equal the mass times the normal component of acceleration. For this circular path, you know that the curvature is

$$K = \frac{1}{12}.$$    Curvature of circular racetrack

Therefore, the frictional force is

$$ma_{\mathbf{N}} = mK\left(\frac{ds}{dt}\right)^2$$
$$= (360 \text{ kg})\left(\frac{1}{12 \text{ m}}\right)\left(\frac{60,000 \text{ m}}{3600 \text{ sec}}\right)^2$$
$$\approx 8333 \text{ (kg)(m)/sec}^2.$$

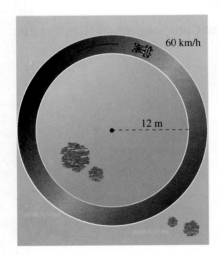

60 km/h

12 m

**Figure 12.40**

## SUMMARY OF VELOCITY, ACCELERATION, AND CURVATURE

Unless noted otherwise, let $C$ be a curve (in the plane or in space) given by the position vector

$$\mathbf{r}(t) = x(t)\mathbf{i} + y(t)\mathbf{j} \qquad\qquad \text{Curve in the plane}$$

or

$$\mathbf{r}(t) = x(t)\mathbf{i} + y(t)\mathbf{j} + z(t)\mathbf{k} \qquad\qquad \text{Curve in space}$$

where $x$, $y$, and $z$ are twice-differentiable functions of $t$.

### Velocity vector, speed, and acceleration vector

$$\mathbf{v}(t) = \mathbf{r}'(t) \qquad\qquad \text{Velocity vector}$$

$$\|\mathbf{v}(t)\| = \|\mathbf{r}'(t)\| = \frac{ds}{dt} \qquad\qquad \text{Speed}$$

$$\mathbf{a}(t) = \mathbf{r}''(t) \qquad\qquad \text{Acceleration vector}$$

$$= a_{\mathbf{T}}\mathbf{T}(t) + a_{\mathbf{N}}\mathbf{N}(t)$$

$$= \frac{d^2s}{dt^2}\mathbf{T}(t) + K\left(\frac{ds}{dt}\right)^2\mathbf{N}(t) \qquad\qquad K \text{ is curvature and } \frac{ds}{dt} \text{ is speed.}$$

### Unit tangent vector and principal unit normal vector

$$\mathbf{T}(t) = \frac{\mathbf{r}'(t)}{\|\mathbf{r}'(t)\|} \qquad\qquad \text{Unit tangent vector}$$

$$\mathbf{N}(t) = \frac{\mathbf{T}'(t)}{\|\mathbf{T}'(t)\|} \qquad\qquad \text{Principal unit normal vector}$$

### Components of acceleration

$$a_{\mathbf{T}} = \mathbf{a} \cdot \mathbf{T} = \frac{\mathbf{v} \cdot \mathbf{a}}{\|\mathbf{v}\|} = \frac{d^2s}{dt^2} \qquad\qquad \text{Tangential component of acceleration}$$

$$a_{\mathbf{N}} = \mathbf{a} \cdot \mathbf{N} \qquad\qquad \text{Normal component of acceleration}$$

$$= \frac{\|\mathbf{v} \times \mathbf{a}\|}{\|\mathbf{v}\|}$$

$$= \sqrt{\|\mathbf{a}\|^2 - a_{\mathbf{T}}^2}$$

$$= K\left(\frac{ds}{dt}\right)^2 \qquad\qquad K \text{ is curvature and } \frac{ds}{dt} \text{ is speed.}$$

### Formulas for curvature in the plane

$$K = \frac{|y''|}{[1 + (y')^2]^{3/2}} \qquad\qquad C \text{ given by } y = f(x)$$

$$K = \frac{|x'y'' - y'x''|}{[(x')^2 + (y')^2]^{3/2}} \qquad\qquad C \text{ given by } x = x(t),\ y = y(t)$$

### Formulas for curvature in the plane or in space

$$K = \|\mathbf{T}'(s)\| = \|\mathbf{r}''(s)\| \qquad\qquad s \text{ is arc length parameter.}$$

$$K = \frac{\|\mathbf{T}'(t)\|}{\|\mathbf{r}'(t)\|} = \frac{\|\mathbf{r}'(t) \times \mathbf{r}''(t)\|}{\|\mathbf{r}'(t)\|^3} \qquad\qquad t \text{ is general parameter.}$$

$$K = \frac{\mathbf{a}(t) \cdot \mathbf{N}(t)}{\|\mathbf{v}(t)\|^2}$$

Cross product formulas apply only to curves in space.

## 12.5 Exercises

See **CalcChat.com** for tutorial help and worked-out solutions to odd-numbered exercises.

### CONCEPT CHECK

**1. Curvature** Consider points $P$ and $Q$ on a curve. What does it mean for the curvature at $P$ to be less than the curvature at $Q$?

**2. Arc Length Parameter** Let $\mathbf{r}(t)$ be a space curve. How can you determine whether $t$ is the arc length parameter?

**Finding the Arc Length of a Plane Curve** In Exercises 3–8, sketch the plane curve and find its length over the given interval.

**3.** $\mathbf{r}(t) = 3t\mathbf{i} - t\mathbf{j}, \quad [0, 3]$    **4.** $\mathbf{r}(t) = t\mathbf{i} + t^2\mathbf{j}, \quad [0, 4]$

**5.** $\mathbf{r}(t) = t^3\mathbf{i} + t^2\mathbf{j}, \quad [0, 1]$

**6.** $\mathbf{r}(t) = t^2\mathbf{i} - 4t\mathbf{j}, \quad [0, 5]$

**7.** $\mathbf{r}(t) = a \cos^3 t\mathbf{i} + a \sin^3 t\mathbf{j}, \quad [0, 2\pi]$

**8.** $\mathbf{r}(t) = a \cos t\mathbf{i} + a \sin t\mathbf{j}, \quad [0, 2\pi]$

**9. Projectile Motion** The position of a baseball is represented by $\mathbf{r}(t) = 50\sqrt{2}\,t\mathbf{i} + \left(3 + 50\sqrt{2}\,t - 16t^2\right)\mathbf{j}$. Find the arc length of the trajectory of the baseball.

**10. Projectile Motion** The position of a baseball is represented by $\mathbf{r}(t) = 40\sqrt{3}\,t\mathbf{i} + \left(4 + 40t - 16t^2\right)\mathbf{j}$. Find the arc length of the trajectory of the baseball.

 **Finding the Arc Length of a Curve in Space** In Exercises 11–16, sketch the space curve and find its length over the given interval.

**11.** $\mathbf{r}(t) = -t\mathbf{i} + 4t\mathbf{j} + 3t\mathbf{k}, \quad [0, 1]$

**12.** $\mathbf{r}(t) = \mathbf{i} + t^2\mathbf{j} + t^3\mathbf{k}, \quad [0, 2]$

**13.** $\mathbf{r}(t) = \langle 4t, -\cos t, \sin t \rangle, \quad \left[0, \dfrac{3\pi}{2}\right]$

**14.** $\mathbf{r}(t) = \langle 2 \sin t, 5t, 2 \cos t \rangle, \quad [0, \pi]$

**15.** $\mathbf{r}(t) = a \cos t\mathbf{i} + a \sin t\mathbf{j} + bt\mathbf{k}, \quad [0, 2\pi]$

**16.** $\mathbf{r}(t) = \langle \cos t + t \sin t, \sin t - t \cos t, t^2 \rangle, \quad \left[0, \dfrac{\pi}{2}\right]$

**17. Investigation** Consider the graph of the vector-valued function $\mathbf{r}(t) = t\mathbf{i} + (4 - t^2)\mathbf{j} + t^3\mathbf{k}$ on the interval $[0, 2]$.

(a) Approximate the length of the curve by finding the length of the line segment connecting its endpoints.

(b) Approximate the length of the curve by summing the lengths of the line segments connecting the terminal points of the vectors $\mathbf{r}(0)$, $\mathbf{r}(0.5)$, $\mathbf{r}(1)$, $\mathbf{r}(1.5)$, and $\mathbf{r}(2)$.

(c) Describe how you could obtain a more accurate approximation by continuing the processes in parts (a) and (b).

(d) Use the integration capabilities of a graphing utility to approximate the length of the curve. Compare this result with the answers in parts (a) and (b).

**18. Investigation** Consider the helix represented by the vector-valued function $\mathbf{r}(t) = \langle 2 \cos t, 2 \sin t, t \rangle$.

(a) Write the length of the arc $s$ on the helix as a function of $t$ by evaluating the integral

$$s = \int_0^t \sqrt{[x'(u)]^2 + [y'(u)]^2 + [z'(u)]^2}\;du.$$

(b) Solve for $t$ in the relationship derived in part (a), and substitute the result into the original vector-valued function. This yields a parametrization of the curve in terms of the arc length parameter $s$.

(c) Find the coordinates of the point on the helix for arc lengths $s = \sqrt{5}$ and $s = 4$.

(d) Verify that $\|\mathbf{r}'(s)\| = 1$.

 **Finding Curvature** In Exercises 19–22, find the curvature of the curve, where $s$ is the arc length parameter.

**19.** $\mathbf{r}(s) = \left(1 + \dfrac{\sqrt{2}}{2}s\right)\mathbf{i} + \left(1 - \dfrac{\sqrt{2}}{2}s\right)\mathbf{j}$

**20.** $\mathbf{r}(s) = (3 + s)\mathbf{i} + \mathbf{j}$

**21.** $\mathbf{r}(s) = \cos \dfrac{1}{2}s\mathbf{i} + \dfrac{\sqrt{3}}{2}s\mathbf{j} + \sin \dfrac{1}{2}s\mathbf{k}$

**22.** $\mathbf{r}(s) = \cos s\mathbf{i} + \sin s\mathbf{j} + 5\mathbf{k}$

**Finding Curvature** In Exercises 23–28, find the curvature of the plane curve at the given value of the parameter.

**23.** $\mathbf{r}(t) = 4t\mathbf{i} - 2t\mathbf{j}, \quad t = 1$    **24.** $\mathbf{r}(t) = t^2\mathbf{i} + \mathbf{j}, \quad t = 2$

**25.** $\mathbf{r}(t) = t\mathbf{i} + \dfrac{1}{t}\mathbf{j}, \quad t = 1$    **26.** $\mathbf{r}(t) = t\mathbf{i} + \dfrac{1}{9}t^3\mathbf{j}, \quad t = 2$

**27.** $\mathbf{r}(t) = \langle t, \sin t \rangle, \quad t = \dfrac{\pi}{2}$

**28.** $\mathbf{r}(t) = \langle 5 \cos t, 4 \sin t \rangle, \quad t = \dfrac{\pi}{3}$

**Finding Curvature** In Exercises 29–36, find the curvature of the curve.

**29.** $\mathbf{r}(t) = 4 \cos 2\pi t\mathbf{i} + 4 \sin 2\pi t\mathbf{j}$

**30.** $\mathbf{r}(t) = 2 \cos \pi t\mathbf{i} + \sin \pi t\mathbf{j}$

**31.** $\mathbf{r}(t) = a \cos \omega t\mathbf{i} + a \sin \omega t\mathbf{j}$

**32.** $\mathbf{r}(t) = a \cos \omega t\mathbf{i} + b \sin \omega t\mathbf{j}$

**33.** $\mathbf{r}(t) = t\mathbf{i} + t^2\mathbf{j} + \dfrac{t^2}{2}\mathbf{k}$

**34.** $\mathbf{r}(t) = 2t^2\mathbf{i} + t\mathbf{j} + \dfrac{1}{2}t^2\mathbf{k}$

**35.** $\mathbf{r}(t) = 4t\mathbf{i} + 3 \cos t\mathbf{j} + 3 \sin t\mathbf{k}$

**36.** $\mathbf{r}(t) = e^{2t}\mathbf{i} + e^{2t} \cos t\mathbf{j} + e^{2t} \sin t\mathbf{k}$

**Finding Curvature**   In Exercises 37–40, find the curvature of the curve at the point $P$.

**37.** $\mathbf{r}(t) = 3t\mathbf{i} + 2t^2\mathbf{j}$,   $P(-3, 2)$

**38.** $\mathbf{r}(t) = e^t\mathbf{i} + 4t\mathbf{j}$,   $P(1, 0)$

**39.** $\mathbf{r}(t) = t\mathbf{i} + t^2\mathbf{j} + \dfrac{t^3}{4}\mathbf{k}$,   $P(2, 4, 2)$

**40.** $\mathbf{r}(t) = e^t \cos t\mathbf{i} + e^t \sin t\mathbf{j} + e^t\mathbf{k}$,   $P(1, 0, 1)$

 **Finding Curvature in Rectangular Coordinates**   In Exercises 41–48, find the curvature and radius of curvature of the plane curve at the given value of $x$.

**41.** $y = 6x$,   $x = 3$

**42.** $y = x - \dfrac{4}{x}$,   $x = 2$

**43.** $y = 5x^2 + 7$,   $x = -1$

**44.** $y = 2\sqrt{9 - x^2}$,   $x = 0$

**45.** $y = \sin 2x$,   $x = \dfrac{\pi}{4}$

**46.** $y = e^{-x/4}$,   $x = 8$

**47.** $y = x^3$,   $x = 2$

**48.** $y = x^n$,   $x = 1$,   $n \geq 2$

**Maximum Curvature**   In Exercises 49–54, (a) find the point on the curve at which the curvature is a maximum and (b) find the limit of the curvature as $x \to \infty$.

**49.** $y = (x - 1)^2 + 3$

**50.** $y = x^3$

**51.** $y = x^{2/3}$

**52.** $y = \dfrac{1}{x}$

**53.** $y = \ln x$

**54.** $y = e^x$

**Curvature**   In Exercises 55–58, find all points on the graph of the function such that the curvature is zero.

**55.** $y = 1 - x^4$

**56.** $y = (x - 2)^6 + 3x$

**57.** $y = \cos \dfrac{x}{2}$

**58.** $y = \sin x$

---

**EXPLORING CONCEPTS**

**59. Curvature**   Consider the function $f(x) = e^{cx}$. What value(s) of $c$ produce a maximum curvature at $x = 0$?

**60. Curvature**   Given a twice-differentiable function $y = f(x)$, determine its curvature at a relative extremum. Can the curvature ever be greater than it is at a relative extremum? Why or why not?

---

 **61. Investigation**   Consider the function $f(x) = x^4 - x^2$.

(a) Use a computer algebra system to find the curvature $K$ of the curve as a function of $x$.

(b) Use the result of part (a) to find the circles of curvature to the graph of $f$ when $x = 0$ and $x = 1$. Use a computer algebra system to graph the function and the two circles of curvature.

(c) Graph the function $K(x)$ and compare it with the graph of $f(x)$. For example, do the extrema of $f$ and $K$ occur at the same critical numbers? Explain your reasoning.

**62. Motion of a Particle**   A particle moves along the plane curve $C$ described by $\mathbf{r}(t) = t\mathbf{i} + t^2\mathbf{j}$.

(a) Find the length of $C$ on the interval $0 \leq t \leq 2$.

(b) Find the curvature of $C$ at $t = 0$, $t = 1$, and $t = 2$.

(c) Describe the curvature of $C$ as $t$ changes from $t = 0$ to $t = 2$.

**63. Investigation**   Find all $a$ and $b$ such that the two curves given by

$$y_1 = ax(b - x) \quad \text{and} \quad y_2 = \dfrac{x}{x + 2}$$

intersect at only one point and have a common tangent line and equal curvature at that point. Sketch a graph for each set of values of $a$ and $b$.

**64.** 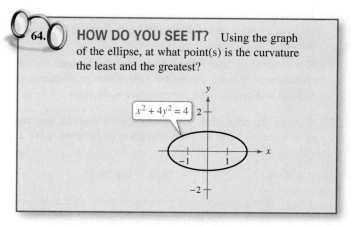 **HOW DO YOU SEE IT?**   Using the graph of the ellipse, at what point(s) is the curvature the least and the greatest?

**65. Sphere and Paraboloid**   A sphere of radius 4 is dropped into the paraboloid given by $z = x^2 + y^2$.

(a) How close will the sphere come to the vertex of the paraboloid?

(b) What is the radius of the largest sphere that will touch the vertex?

**66. Speed**

The smaller the curvature of a bend in a road, the faster a car can travel. Assume that the maximum speed around a turn is inversely proportional to the square root of the curvature. A car moving on the path $y = \frac{1}{3}x^3$, where $x$ and $y$ are measured in miles, can safely go 30 miles per hour at $\left(1, \frac{1}{3}\right)$. How fast can it go at $\left(\frac{3}{2}, \frac{9}{8}\right)$?

**67. Center of Curvature**   Let $C$ be a curve given by $y = f(x)$. Let $K$ be the curvature $(K \neq 0)$ at the point $P(x_0, y_0)$ and let

$$z = \dfrac{1 + f'(x_0)^2}{f''(x_0)}.$$

Show that the coordinates $(\alpha, \beta)$ of the center of curvature at $P$ are $(\alpha, \beta) = (x_0 - f'(x_0)z, y_0 + z)$.

**68. Center of Curvature** Use the result of Exercise 67 to find the center of curvature for the curve at the given point.

(a) $y = e^x$, $(0, 1)$     (b) $y = \dfrac{x^2}{2}$, $\left(1, \dfrac{1}{2}\right)$     (c) $y = x^2$, $(0, 0)$

**69. Curvature** A curve $C$ is given by the polar equation $r = f(\theta)$. Show that the curvature $K$ at the point $(r, \theta)$ is

$$K = \frac{|2(r')^2 - rr'' + r^2|}{[(r')^2 + r^2]^{3/2}}.$$

[*Hint:* Represent the curve by $\mathbf{r}(\theta) = r \cos \theta \mathbf{i} + r \sin \theta \mathbf{j}$.]

**70. Curvature** Use the result of Exercise 69 to find the curvature of each polar curve.

(a) $r = 1 + \sin \theta$     (b) $r = \theta$

(c) $r = a \sin \theta$     (d) $r = e^\theta$

**71. Curvature** Given the polar curve $r = e^{a\theta}$, $a > 0$, use the result of Exercise 69 to find the curvature $K$ and determine the limit of $K$ as (a) $\theta \to \infty$ and (b) $a \to \infty$.

**72. Curvature at the Pole** Show that the formula for the curvature of a polar curve $r = f(\theta)$ given in Exercise 69 reduces to $K = 2/|r'|$ for the curvature at the pole.

**Curvature at the Pole** In Exercises 73 and 74, use the result of Exercise 72 to find the curvature of the rose curve at the pole.

**73.** $r = 4 \sin 2\theta$     **74.** $r = \cos 3\theta$

**75. Proof** For a smooth curve given by the parametric equations $x = f(t)$ and $y = g(t)$, prove that the curvature is given by

$$K = \frac{|f'(t)g''(t) - g'(t)f''(t)|}{\{[f'(t)]^2 + [g'(t)]^2\}^{3/2}}.$$

**76. Horizontal Asymptotes** Use the result of Exercise 75 to find the curvature $K$ of the curve represented by the parametric equations $x(t) = t^3$ and $y(t) = \frac{1}{2}t^2$. Use a graphing utility to graph $K$ and determine any horizontal asymptotes. Interpret the asymptotes in the context of the problem.

**77. Curvature of a Cycloid** Use the result of Exercise 75 to find the curvature $K$ of the cycloid represented by the parametric equations

$$x(\theta) = a(\theta - \sin \theta) \quad \text{and} \quad y(\theta) = a(1 - \cos \theta).$$

What are the minimum and maximum values of $K$?

**78. Tangential and Normal Components of Acceleration** Use Theorem 12.10 to find $a_{\mathbf{T}}$ and $a_{\mathbf{N}}$ for each curve given by the vector-valued function.

(a) $\mathbf{r}(t) = 3t^2\mathbf{i} + (3t - t^3)\mathbf{j}$

(b) $\mathbf{r}(t) = t\mathbf{i} + t^2\mathbf{j} + \frac{1}{2}t^2\mathbf{k}$

**79. Frictional Force** A 5500-pound vehicle is driven at a speed of 30 miles per hour on a circular interchange of radius 100 feet. To keep the vehicle from skidding off course, what frictional force must the road surface exert on the tires?

**80. Frictional Force** A 6400-pound vehicle is driven at a speed of 35 miles per hour on a circular interchange of radius 250 feet. To keep the vehicle from skidding off course, what frictional force must the road surface exert on the tires?

**81. Curvature** Verify that the curvature at any point $(x, y)$ on the graph of $y = \cosh x$ is $1/y^2$.

**82. Formulas for Curvature** Use the definition of curvature in space, $K = \|\mathbf{T}'(s)\| = \|\mathbf{r}''(s)\|$, to verify each formula.

(a) $K = \dfrac{\|\mathbf{T}'(t)\|}{\|\mathbf{r}'(t)\|}$

(b) $K = \dfrac{\|\mathbf{r}'(t) \times \mathbf{r}''(t)\|}{\|\mathbf{r}'(t)\|^3}$

(c) $K = \dfrac{\mathbf{a}(t) \cdot \mathbf{N}(t)}{\|\mathbf{v}(t)\|^2}$

**True or False?** In Exercises 83–86, determine whether the statement is true or false. If it is false, explain why or give an example that shows it is false.

**83.** The arc length of a space curve depends on the parametrization.

**84.** The curvature of a plane curve at an inflection point is zero.

**85.** The curvature of a parabola is a maximum at its vertex.

**86.** The normal component of acceleration is a function of both speed and curvature.

**Kepler's Laws** In Exercises 87–94, you are asked to verify Kepler's Laws of Planetary Motion. For these exercises, assume that each planet moves in an orbit given by the vector-valued function $\mathbf{r}$. Let $r = \|\mathbf{r}\|$, let $G$ represent the universal gravitational constant, let $M$ represent the mass of the sun, and let $m$ represent the mass of the planet.

**87.** Prove that $\mathbf{r} \cdot \mathbf{r}' = r\dfrac{dr}{dt}$.

**88.** Using Newton's Second Law of Motion, $\mathbf{F} = m\mathbf{a}$, and Newton's Second Law of Gravitation

$$\mathbf{F} = -\frac{GmM}{r^3}\mathbf{r}$$

show that $\mathbf{a}$ and $\mathbf{r}$ are parallel, and that $\mathbf{r}(t) \times \mathbf{r}'(t) = \mathbf{L}$ is a constant vector. So, $\mathbf{r}(t)$ moves *in a fixed plane*, orthogonal to $\mathbf{L}$.

**89.** Prove that $\dfrac{d}{dt}\left[\dfrac{\mathbf{r}}{r}\right] = \dfrac{1}{r^3}[(\mathbf{r} \times \mathbf{r}') \times \mathbf{r}]$.

**90.** Show that $\dfrac{\mathbf{r}'}{GM} \times \mathbf{L} - \dfrac{\mathbf{r}}{r} = \mathbf{e}$ is a constant vector.

**91.** Prove Kepler's First Law: Each planet moves in an elliptical orbit with the sun as a focus.

**92.** Assume that the elliptical orbit

$$r = \frac{ed}{1 + e \cos \theta}$$

is in the $xy$-plane, with $\mathbf{L}$ along the $z$-axis. Prove that

$$\|\mathbf{L}\| = r^2\frac{d\theta}{dt}.$$

**93.** Prove Kepler's Second Law: Each ray from the sun to a planet sweeps out equal areas of the ellipse in equal times.

**94.** Prove Kepler's Third Law: The square of the period of a planet's orbit is proportional to the cube of the mean distance between the planet and the sun.

# Review Exercises    See CalcChat.com for tutorial help and worked-out solutions to odd-numbered exercises.

**Domain and Continuity**  In Exercises 1–4, (a) find the domain of **r**, and (b) determine the interval(s) on which the function is continuous.

**1.** $\mathbf{r}(t) = \tan t\,\mathbf{i} + \mathbf{j} + t\mathbf{k}$    **2.** $\mathbf{r}(t) = \sqrt{t}\,\mathbf{i} + \dfrac{1}{t-4}\mathbf{j} + \mathbf{k}$

**3.** $\mathbf{r}(t) = \sqrt{t^2 - 9}\,\mathbf{i} - \mathbf{j} + \ln(t-1)\mathbf{k}$

**4.** $\mathbf{r}(t) = (2t+1)\mathbf{i} + t^2\mathbf{j} + t\mathbf{k}$

**Evaluating a Function**  In Exercises 5 and 6, evaluate the vector-valued function at each given value of $t$.

**5.** $\mathbf{r}(t) = (2t+1)\mathbf{i} + t^2\mathbf{j} - \sqrt{t+2}\,\mathbf{k}$

   (a) $\mathbf{r}(0)$    (b) $\mathbf{r}(-2)$    (c) $\mathbf{r}(c-1)$

   (d) $\mathbf{r}(1 + \Delta t) - \mathbf{r}(1)$

**6.** $\mathbf{r}(t) = 3\cos t\,\mathbf{i} + (1 - \sin t)\mathbf{j} - t\mathbf{k}$

   (a) $\mathbf{r}(0)$    (b) $\mathbf{r}\!\left(\dfrac{\pi}{2}\right)$    (c) $\mathbf{r}(s - \pi)$

   (d) $\mathbf{r}(\pi + \Delta t) - \mathbf{r}(\pi)$

**Writing a Vector-Valued Function**  In Exercises 7 and 8, represent the line segment from $P$ to $Q$ by a vector-valued function and by a set of parametric equations.

**7.** $P(3, 0, 5),\quad Q(2, -2, 3)$

**8.** $P(-2, -3, 8),\quad Q(5, 1, -2)$

**Sketching a Curve**  In Exercises 9–12, sketch the curve represented by the vector-valued function and give the orientation of the curve.

**9.** $\mathbf{r}(t) = \langle \pi \cos t, \pi \sin t \rangle$

**10.** $\mathbf{r}(t) = \langle t + 2, t^2 - 1 \rangle$

**11.** $\mathbf{r}(t) = (t+1)\mathbf{i} + (3t-1)\mathbf{j} + 2t\mathbf{k}$

**12.** $\mathbf{r}(t) = 2\cos t\,\mathbf{i} + t\mathbf{j} + 2\sin t\,\mathbf{k}$

**Representing a Graph by a Vector-Valued Function**  In Exercises 13 and 14, represent the plane curve by a vector-valued function. (There are many correct answers.)

**13.** $3x + 4y - 12 = 0$    **14.** $y = 9 - x^2$

**Representing a Graph by a Vector-Valued Function**  In Exercises 15 and 16, sketch the space curve represented by the intersection of the surfaces. Then use the parameter $x = t$ to find a vector-valued function for the space curve.

**15.** $z = x^2 + y^2,\quad y = 2$

**16.** $x^2 + z^2 = 4,\quad x - y = 0$

**Finding a Limit**  In Exercises 17 and 18, find the limit.

**17.** $\lim\limits_{t \to 3} \left( \sqrt{3 - t}\,\mathbf{i} + \ln t\,\mathbf{j} - \dfrac{1}{t}\mathbf{k} \right)$

**18.** $\lim\limits_{t \to 0} \left( \dfrac{\sin 2t}{t}\mathbf{i} + e^{-t}\mathbf{j} + 4\mathbf{k} \right)$

**Higher-Order Differentiation**  In Exercises 19 and 20, find (a) $\mathbf{r}'(t)$, (b) $\mathbf{r}''(t)$, and (c) $\mathbf{r}'(t) \cdot \mathbf{r}''(t)$.

**19.** $\mathbf{r}(t) = (t^2 + 4t)\mathbf{i} - 3t^2\mathbf{j}$

**20.** $\mathbf{r}(t) = 5\cos t\,\mathbf{i} + 2\sin t\,\mathbf{j}$

**Higher-Order Differentiation**  In Exercises 21 and 22, find (a) $\mathbf{r}'(t)$, (b) $\mathbf{r}''(t)$, (c) $\mathbf{r}'(t) \cdot \mathbf{r}''(t)$, and (d) $\mathbf{r}'(t) \times \mathbf{r}''(t)$.

**21.** $\mathbf{r}(t) = 2t^3\mathbf{i} + 4t\mathbf{j} - t^2\mathbf{k}$

**22.** $\mathbf{r}(t) = (4t + 3)\mathbf{i} + t^2\mathbf{j} + (2t^2 + 4)\mathbf{k}$

**Finding Intervals on Which a Curve is Smooth**  In Exercises 23 and 24, find the open interval(s) on which the curve given by the vector-valued function is smooth.

**23.** $\mathbf{r}(t) = (t-1)^3\mathbf{i} + (t-1)^4\mathbf{j}$

**24.** $\mathbf{r}(t) = \dfrac{t}{t-2}\mathbf{i} + t\mathbf{j} + \sqrt{1 + t}\,\mathbf{k}$

**Using Properties of the Derivative**  In Exercises 25 and 26, use the properties of the derivative to find the following.

(a) $\mathbf{r}'(t)$    (b) $\dfrac{d}{dt}[\mathbf{u}(t) - 2\mathbf{r}(t)]$    (c) $\dfrac{d}{dt}[(3t)\mathbf{r}(t)]$

(d) $\dfrac{d}{dt}[\mathbf{r}(t) \cdot \mathbf{u}(t)]$    (e) $\dfrac{d}{dt}[\mathbf{r}(t) \times \mathbf{u}(t)]$    (f) $\dfrac{d}{dt}[\mathbf{u}(2t)]$

**25.** $\mathbf{r}(t) = 3t\mathbf{i} + (t - 1)\mathbf{j},\quad \mathbf{u}(t) = t\mathbf{i} + t^2\mathbf{j} + \frac{2}{3}t^3\mathbf{k}$

**26.** $\mathbf{r}(t) = \sin t\,\mathbf{i} + \cos t\,\mathbf{j} + t\mathbf{k},\quad \mathbf{u}(t) = \sin t\,\mathbf{i} + \cos t\,\mathbf{j} + \dfrac{1}{t}\mathbf{k}$

**Finding an Indefinite Integral**  In Exercises 27–30, find the indefinite integral.

**27.** $\displaystyle\int (t^2\mathbf{i} + 5t\mathbf{j} + 8t^3\mathbf{k})\,dt$    **28.** $\displaystyle\int (6\mathbf{i} - 2t\mathbf{j} + \ln t\,\mathbf{k})\,dt$

**29.** $\displaystyle\int \left(3\sqrt{t}\,\mathbf{i} + \dfrac{2}{t}\mathbf{j} + \mathbf{k}\right) dt$    **30.** $\displaystyle\int (\sin t\,\mathbf{i} + \mathbf{j} + e^{2t}\,\mathbf{k})\,dt$

**Evaluating a Definite Integral**  In Exercises 31–34, evaluate the definite integral.

**31.** $\displaystyle\int_{-2}^{2} (3t\mathbf{i} + 2t^2\mathbf{j} - t^3\mathbf{k})\,dt$

**32.** $\displaystyle\int_{0}^{3} (t\mathbf{i} + \sqrt{t}\,\mathbf{j} + 4t\mathbf{k})\,dt$

**33.** $\displaystyle\int_{0}^{2} (e^{t/2}\mathbf{i} - 3t^2\mathbf{j} - \mathbf{k})\,dt$

**34.** $\displaystyle\int_{0}^{\pi/3} (2\cos t\,\mathbf{i} + \sin t\,\mathbf{j} + 3\mathbf{k})\,dt$

**Finding an Antiderivative**  In Exercises 35 and 36, find $\mathbf{r}(t)$ that satisfies the initial condition(s).

**35.** $\mathbf{r}'(t) = 2t\mathbf{i} + e^t\mathbf{j} + e^{-t}\mathbf{k},\quad \mathbf{r}(0) = \mathbf{i} + 3\mathbf{j} - 5\mathbf{k}$

**36.** $\mathbf{r}'(t) = \sec t\,\mathbf{i} + \tan t\,\mathbf{j} + t^2\mathbf{k},\quad \mathbf{r}(0) = 3\mathbf{k}$

**Finding Velocity and Acceleration Vectors in Space**
In Exercises 37–40, the position vector r describes the path of an object moving in space. (a) Find the velocity vector, speed, and acceleration vector of the object. (b) Evaluate the velocity vector and acceleration vector of the object at the given value of t.

| Position Vector | Time |
|---|---|
| **37.** $\mathbf{r}(t) = 4t\mathbf{i} + t^3\mathbf{j} - t\mathbf{k}$ | $t = 1$ |
| **38.** $\mathbf{r}(t) = \sqrt{t}\,\mathbf{i} + 5t\mathbf{j} + 2t^2\mathbf{k}$ | $t = 4$ |
| **39.** $\mathbf{r}(t) = \langle \cos^3 t, \sin^3 t, 3t \rangle$ | $t = \pi$ |
| **40.** $\mathbf{r}(t) = \langle t, -\tan t, e^t \rangle$ | $t = 0$ |

**Projectile Motion**   In Exercises 41 and 42, use the model for projectile motion, assuming there is no air resistance and $g = 32$ feet per second per second.

**41.** A baseball is hit from a height of 3.5 feet above the ground with an initial speed of 120 feet per second and at an angle of 30° above the horizontal. Find the maximum height reached by the baseball. Determine whether it will clear an 8-foot-high fence located 375 feet from home plate.

**42.** Determine the maximum height and range of a projectile fired at a height of 6 feet above the ground with an initial speed of 400 feet per second and an angle of 60° above the horizontal.

**Finding the Unit Tangent Vector**   In Exercises 43 and 44, find the unit tangent vector to the curve at the specified value of the parameter.

**43.** $\mathbf{r}(t) = 6t\mathbf{i} - t^2\mathbf{j}, \quad t = 2$

**44.** $\mathbf{r}(t) = 2\sin t\mathbf{i} + 4\cos t\mathbf{j}, \quad t = \dfrac{\pi}{6}$

**Finding a Tangent Line**   In Exercises 45 and 46, find the unit tangent vector T(t) and a set of parametric equations for the tangent line to the space curve at point P.

**45.** $\mathbf{r}(t) = e^{2t}\mathbf{i} + \cos t\mathbf{j} - \sin 3t\mathbf{k}, \quad P(1, 1, 0)$

**46.** $\mathbf{r}(t) = t\mathbf{i} + t^2\mathbf{j} + \tfrac{2}{3}t^3\mathbf{k}, \quad P\!\left(2, 4, \tfrac{16}{3}\right)$

**Finding the Principal Unit Normal Vector**   In Exercises 47–50, find the principal unit normal vector to the curve at the specified value of the parameter.

**47.** $\mathbf{r}(t) = 2t\mathbf{i} + 3t^2\mathbf{j}, \quad t = 1$ **48.** $\mathbf{r}(t) = t\mathbf{i} + \ln t\mathbf{j}, \quad t = 2$

**49.** $\mathbf{r}(t) = 3\cos 2t\mathbf{i} + 3\sin 2t\mathbf{j} + 3\mathbf{k}, \quad t = \dfrac{\pi}{4}$

**50.** $\mathbf{r}(t) = 4\cos t\mathbf{i} + 4\sin t\mathbf{j} + \mathbf{k}, \quad t = \dfrac{2\pi}{3}$

**Finding Tangential and Normal Components of Acceleration**   In Exercises 51 and 52, find the tangential and normal components of acceleration at the given time t for the plane curve r(t).

**51.** $\mathbf{r}(t) = \dfrac{3}{t}\mathbf{i} - 6t\mathbf{j}, \quad t = 3$

**52.** $\mathbf{r}(t) = 3\cos 2t\mathbf{i} + 3\sin 2t\mathbf{j}, \quad t = \dfrac{\pi}{6}$

**Finding Tangential and Normal Components of Acceleration**   In Exercises 53 and 54, find the tangential and normal components of acceleration at the given time t for the space curve r(t).

**53.** $\mathbf{r}(t) = \sin t\mathbf{i} - 3t\mathbf{j} + \cos t\mathbf{k}, \quad t = \dfrac{\pi}{6}$

**54.** $\mathbf{r}(t) = \dfrac{t^3}{3}\mathbf{i} - 6t\mathbf{j} + t^2\mathbf{k}, \quad t = 2$

**Finding the Arc Length of a Plane Curve**   In Exercises 55–58, sketch the plane curve and find its length over the given interval.

**55.** $\mathbf{r}(t) = 2t\mathbf{i} - 3t\mathbf{j}, \quad [0, 5]$

**56.** $\mathbf{r}(t) = t^2\mathbf{i} + 2t\mathbf{k}, \quad [0, 3]$

**57.** $\mathbf{r}(t) = 2\sin t\mathbf{i} + \mathbf{j}, \quad \left[\dfrac{\pi}{2}, \pi\right]$

**58.** $\mathbf{r}(t) = 10\cos t\mathbf{i} + 10\sin t\mathbf{j}, \quad [0, 2\pi]$

**Finding the Arc Length of a Curve in Space**   In Exercises 59–62, sketch the space curve and find its length over the given interval.

**59.** $\mathbf{r}(t) = -3t\mathbf{i} + 2t\mathbf{j} + 4t\mathbf{k}, \quad [0, 3]$

**60.** $\mathbf{r}(t) = t\mathbf{i} + t^2\mathbf{j} + 2t\mathbf{k}, \quad [0, 2]$

**61.** $\mathbf{r}(t) = \langle 8\cos t, 8\sin t, t \rangle, \quad \left[0, \dfrac{\pi}{2}\right]$

**62.** $\mathbf{r}(t) = \langle 2(\sin t - t\cos t), 2(\cos t + t\sin t), t \rangle, \quad \left[0, \dfrac{\pi}{2}\right]$

**Finding Curvature**   In Exercises 63–66, find the curvature of the curve.

**63.** $\mathbf{r}(t) = 3t\mathbf{i} + 2t\mathbf{j}$

**64.** $\mathbf{r}(t) = 2\sqrt{t}\,\mathbf{i} + 3t\mathbf{j}$

**65.** $\mathbf{r}(t) = 2t\mathbf{i} + \tfrac{1}{2}t^2\mathbf{j} + t^2\mathbf{k}$

**66.** $\mathbf{r}(t) = 2t\mathbf{i} + 5\cos t\mathbf{j} + 5\sin t\mathbf{k}$

**Finding Curvature**   In Exercises 67 and 68, find the curvature of the curve at the point P.

**67.** $\mathbf{r}(t) = \tfrac{1}{2}t^2\mathbf{i} + t\mathbf{j} + \tfrac{1}{3}t^3\mathbf{k}, \quad P\!\left(\tfrac{1}{2}, 1, \tfrac{1}{3}\right)$

**68.** $\mathbf{r}(t) = 4\cos t\mathbf{i} + 3\sin t\mathbf{j} + t\mathbf{k}, \quad P(-4, 0, \pi)$

**Finding Curvature in Rectangular Coordinates**   In Exercises 69–72, find the curvature and radius of curvature of the plane curve at the given value of x.

**69.** $y = \tfrac{1}{2}x^2 + x, \quad x = 4$   **70.** $y = e^{-x/2}, \quad x = 0$

**71.** $y = \ln x, \quad x = 1$

**72.** $y = \tan x, \quad x = \dfrac{\pi}{4}$

**73. Frictional Force**   A 7200-pound vehicle is driven at a speed of 25 miles per hour on a circular interchange of radius 150 feet. To keep the vehicle from skidding off course, what frictional force must the road surface exert on the tires?

# P.S. Problem Solving

See **CalcChat.com** for tutorial help and worked-out solutions to odd-numbered exercises.

**1. Cornu Spiral** The **cornu spiral** is given by

$$x(t) = \int_0^t \cos\left(\frac{\pi u^2}{2}\right) du \quad \text{and} \quad y(t) = \int_0^t \sin\left(\frac{\pi u^2}{2}\right) du.$$

The spiral shown in the figure was plotted over the interval $-\pi \le t \le \pi$.

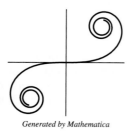

*Generated by Mathematica*

(a) Find the arc length of this curve from $t = 0$ to $t = a$.

(b) Find the curvature of the graph when $t = a$.

(c) The cornu spiral was discovered by James Bernoulli. He found that the spiral has an amazing relationship between curvature and arc length. What is this relationship?

**2. Radius of Curvature** Let $T$ be the tangent line at the point $P(x, y)$ on the graph of the curve $x^{2/3} + y^{2/3} = a^{2/3}$, $a > 0$, as shown in the figure. Show that the radius of curvature at $P$ is three times the distance from the origin to the tangent line $T$.

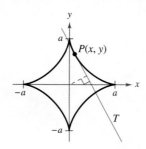

**3. Projectile Motion** A bomber is flying horizontally at an altitude of 3200 feet with a speed of 400 feet per second when it releases a bomb. A projectile is launched 5 seconds later from a cannon at a site facing the bomber and 5000 feet from the point that was directly beneath the bomber when the bomb was released, as shown in the figure. The projectile is to intercept the bomb at an altitude of 1600 feet. Determine the required initial speed and angle of inclination of the projectile. (Ignore air resistance.)

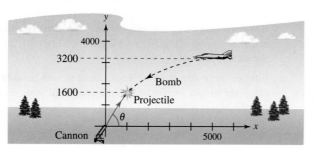

**4. Projectile Motion** Repeat Exercise 3 for the case in which the bomber is facing *away* from the launch site, as shown in the figure.

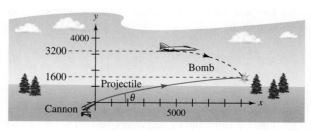

**5. Cycloid** Consider one arch of the cycloid

$$\mathbf{r}(\theta) = (\theta - \sin\theta)\mathbf{i} + (1 - \cos\theta)\mathbf{j}, \quad 0 \le \theta \le 2\pi$$

as shown in the figure. Let $s(\theta)$ be the arc length from the highest point on the arch to the point $(x(\theta), y(\theta))$, and let $\rho(\theta) = 1/K$ be the radius of curvature at the point $(x(\theta), y(\theta))$. Show that $s$ and $\rho$ are related by the equation $s^2 + \rho^2 = 16$. (This equation is called a *natural equation* for the curve.)

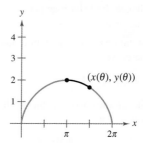

**6. Cardioid** Consider the cardioid

$$r = 1 - \cos\theta, \quad 0 \le \theta \le 2\pi$$

as shown in the figure. Let $s(\theta)$ be the arc length from the point $(2, \pi)$ on the cardioid to the point $(r, \theta)$, and let $p(\theta) = 1/K$ be the radius of curvature at the point $(r, \theta)$. Show that $s$ and $\rho$ are related by the equation $s^2 + 9\rho^2 = 16$. (This equation is called a *natural equation* for the curve.)

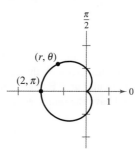

**7. Proof** If $\mathbf{r}(t)$ is a nonzero differentiable function of $t$, prove that

$$\frac{d}{dt}\big[\|\mathbf{r}(t)\|\big] = \frac{\mathbf{r}(t) \cdot \mathbf{r}'(t)}{\|\mathbf{r}(t)\|}.$$

**8. Satellite** A communications satellite moves in a circular orbit around Earth at a distance of 42,000 kilometers from the center of Earth. The angular speed

$$\frac{d\theta}{dt} = \omega = \frac{\pi}{12} \text{ radian per hour}$$

is constant.

(a) Use polar coordinates to show that the acceleration vector is given by

$$\mathbf{a} = \frac{d^2\mathbf{r}}{dt^2} = \left[\frac{d^2r}{dt^2} - r\left(\frac{d\theta}{dt}\right)^2\right]\mathbf{u}_r + \left[r\frac{d^2\theta}{dt^2} + 2\frac{dr}{dt}\frac{d\theta}{dt}\right]\mathbf{u}_\theta$$

where $\mathbf{u}_r = \cos\theta\,\mathbf{i} + \sin\theta\,\mathbf{j}$ is the unit vector in the radial direction and $\mathbf{u}_\theta = -\sin\theta\,\mathbf{i} + \cos\theta\,\mathbf{j}$.

(b) Find the radial and angular components of acceleration for the satellite.

**Binormal Vector** **In Exercises 9–11, use the binormal vector defined by the equation B = T × N.**

**9.** Find the unit tangent, principal unit normal, and binormal vectors for the helix

$$\mathbf{r}(t) = 4\cos t\,\mathbf{i} + 4\sin t\,\mathbf{j} + 3t\mathbf{k}$$

at $t = \pi/2$. Sketch the helix together with these three mutually orthogonal unit vectors.

**10.** Find the unit tangent, principal unit normal, and binormal vectors for the curve

$$\mathbf{r}(t) = \cos t\,\mathbf{i} + \sin t\,\mathbf{j} - \mathbf{k}$$

at $t = \pi/4$. Sketch the curve together with these three mutually orthogonal unit vectors.

**11.** (a) Prove that there exists a scalar $\tau$, called the **torsion**, such that $d\mathbf{B}/ds = -\tau\mathbf{N}$.

(b) Prove that $\dfrac{d\mathbf{N}}{ds} = -K\mathbf{T} + \tau\mathbf{B}$.

(The three equations $d\mathbf{T}/ds = K\mathbf{N}$, $d\mathbf{N}/ds = -K\mathbf{T} + \tau\mathbf{B}$, and $d\mathbf{B}/ds = -\tau\mathbf{N}$ are called the *Frenet-Serret formulas*.)

**12. Exit Ramp** A highway has an exit ramp that begins at the origin of a coordinate system and follows the curve

$$y = \frac{1}{32}x^{5/2}$$

to the point $(4, 1)$ (see figure). Then it follows a circular path whose curvature is that given by the curve at $(4, 1)$. What is the radius of the circular arc? Explain why the curve and the circular arc should have the same curvature at $(4, 1)$.

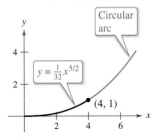

**13. Arc Length and Curvature** Consider the vector-valued function

$$\mathbf{r}(t) = \langle t\cos\pi t, t\sin\pi t\rangle, \quad 0 \le t \le 2.$$

(a) Use a graphing utility to graph the function.

(b) Find the length of the arc in part (a).

(c) Find the curvature $K$ as a function of $t$. Find the curvature at $t = 0$, $t = 1$, and $t = 2$.

(d) Use a graphing utility to graph the function $K$.

(e) Find (if possible) $\lim_{t\to\infty} K$.

(f) Using the result of part (e), make a conjecture about the graph of $\mathbf{r}$ as $t \to \infty$.

**14. Ferris Wheel** You want to toss an object to a friend who is riding a Ferris wheel (see figure). The following parametric equations give the path of the friend $\mathbf{r}_1(t)$ and the path of the object $\mathbf{r}_2(t)$. Distance is measured in meters, and time is measured in seconds.

$$\mathbf{r}_1(t) = 15\left(\sin\frac{\pi t}{10}\right)\mathbf{i} + \left(16 - 15\cos\frac{\pi t}{10}\right)\mathbf{j}$$

$$\mathbf{r}_2(t) = [22 - 8.03(t - t_0)]\mathbf{i} + [1 + 11.47(t - t_0) - 4.9(t - t_0)^2]\mathbf{j}$$

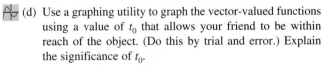

(a) Locate your friend's position on the Ferris wheel at time $t = 0$.

(b) Determine the number of revolutions per minute of the Ferris wheel.

(c) What are the speed and angle of inclination (in degrees) at which the object is thrown at time $t = t_0$?

(d) Use a graphing utility to graph the vector-valued functions using a value of $t_0$ that allows your friend to be within reach of the object. (Do this by trial and error.) Explain the significance of $t_0$.

(e) Find the approximate time your friend should be able to catch the object. Approximate the speeds of your friend and the object at that time.

# 13 Functions of Several Variables

Manufacturing
*(Example 2, p. 959)*

Ocean Floor *(Exercise 66, p. 930)*

Wind Chill *(Exercise 27, p. 910)*

Marginal Costs
*(Exercise 118, p. 902)*

Forestry *(Exercise 77, p. 882)*

# 13.1 Introduction to Functions of Several Variables

- ■ Understand the notation for a function of several variables.
- ■ Sketch the graph of a function of two variables.
- ■ Sketch level curves for a function of two variables.
- ■ Sketch level surfaces for a function of three variables.
- ■ Use computer graphics to graph a function of two variables.

## Functions of Several Variables

So far in this text, you have dealt only with functions of a single (independent) variable. Many familiar quantities, however, are functions of two or more variables. Here are three examples.

1. The work done by a force, $W = FD$, is a function of two variables.
2. The volume of a right circular cylinder, $V = \pi r^2 h$, is a function of two variables.
3. The volume of a rectangular solid, $V = lwh$, is a function of three variables.

The notation for a function of two or more variables is similar to that for a function of a single variable. Here are two examples.

$$z = f(\underbrace{x, y}_{\text{2 variables}}) = x^2 + xy \qquad \text{Function of two variables}$$

and

$$w = f(\underbrace{x, y, z}_{\text{3 variables}}) = x + 2y - 3z \qquad \text{Function of three variables}$$

**MARY FAIRFAX SOMERVILLE (1780–1872)**

Somerville was interested in the problem of creating geometric models for functions of several variables. Her most well-known book, *The Mechanics of the Heavens*, was published in 1831.
*See LarsonCalculus.com to read more of this biography.*

### Definition of a Function of Two Variables

Let $D$ be a set of ordered pairs of real numbers. If to each ordered pair $(x, y)$ in $D$ there corresponds a unique real number $f(x, y)$, then $f$ is a **function of x and y.** The set $D$ is the **domain** of $f$, and the corresponding set of values for $f(x, y)$ is the **range** of $f$. For the function

$$z = f(x, y)$$

$x$ and $y$ are called the **independent variables** and $z$ is called the **dependent variable.**

Similar definitions can be given for functions of three, four, or $n$ variables, where the domains consist of ordered triples $(x_1, x_2, x_3)$, quadruples $(x_1, x_2, x_3, x_4)$, and $n$-tuples $(x_1, x_2, \ldots, x_n)$. In all cases, the range is a set of real numbers. In this chapter, you will study only functions of two or three variables.

As with functions of one variable, the most common way to describe a function of several variables is with an *equation*, and unless it is otherwise restricted, you can assume that the domain is the set of all points for which the equation is defined. For instance, the domain of the function

$$f(x, y) = x^2 + y^2$$

is the entire $xy$-plane. Similarly, the domain of

$$f(x, y) = \ln xy$$

is the set of all points $(x, y)$ in the plane for which $xy > 0$. This consists of all points in the first and third quadrants.

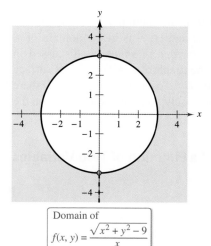

Domain of
$$f(x, y) = \frac{\sqrt{x^2 + y^2 - 9}}{x}$$

**Figure 13.1**

---

**EXAMPLE 1**    **Domains of Functions of Several Variables**

Find the domain of each function.

**a.** $f(x, y) = \dfrac{\sqrt{x^2 + y^2 - 9}}{x}$    **b.** $g(x, y, z) = \dfrac{x}{\sqrt{9 - x^2 - y^2 - z^2}}$

**Solution**

**a.** The function $f$ is defined for all points $(x, y)$ such that $x \neq 0$ and

$$x^2 + y^2 \geq 9.$$

So, the domain is the set of all points lying on or outside the circle $x^2 + y^2 = 9$ *except* those points on the $y$-axis, as shown in Figure 13.1.

**b.** The function $g$ is defined for all points $(x, y, z)$ such that

$$x^2 + y^2 + z^2 < 9.$$

Consequently, the domain is the set of all points $(x, y, z)$ lying inside a sphere of radius 3 that is centered at the origin.    ∎

Functions of several variables can be combined in the same ways as functions of single variables. For instance, you can form the sum, difference, product, and quotient of two functions of two variables as follows.

$$(f \pm g)(x, y) = f(x, y) \pm g(x, y) \qquad \text{Sum or difference}$$
$$(fg)(x, y) = f(x, y)g(x, y) \qquad \text{Product}$$
$$\frac{f}{g}(x, y) = \frac{f(x, y)}{g(x, y)}, \quad g(x, y) \neq 0 \qquad \text{Quotient}$$

You cannot form the composite of two functions of several variables. You can, however, form the **composite** function $(g \circ h)(x, y)$, where $g$ is a function of a single variable and $h$ is a function of two variables.

$$(g \circ h)(x, y) = g(h(x, y)) \qquad \text{Composition}$$

The domain of this composite function consists of all $(x, y)$ in the domain of $h$ such that $h(x, y)$ is in the domain of $g$. For example, the function

$$f(x, y) = \sqrt{16 - 4x^2 - y^2}$$

can be viewed as the composite of the function of two variables given by

$$h(x, y) = 16 - 4x^2 - y^2$$

and the function of a single variable given by

$$g(u) = \sqrt{u}.$$

The domain of this function is the set of all points lying on or inside the ellipse $4x^2 + y^2 = 16$.

A function that can be written as a sum of functions of the form $cx^m y^n$ (where $c$ is a real number and $m$ and $n$ are nonnegative integers) is called a **polynomial function** of two variables. For instance, the functions

$$f(x, y) = x^2 + y^2 - 2xy + x + 2 \quad \text{and} \quad g(x, y) = 3xy^2 + x - 2$$

are polynomial functions of two variables. A **rational function** is the quotient of two polynomial functions. Similar terminology is used for functions of more than two variables.

# The Graph of a Function of Two Variables

As with functions of a single variable, you can learn a lot about the behavior of a function of two variables by sketching its graph. The **graph** of a function $f$ of two variables is the set of all points $(x, y, z)$ for which $z = f(x, y)$ and $(x, y)$ is in the domain of $f$. This graph can be interpreted geometrically as a *surface in space,* as discussed in Sections 11.5 and 11.6. In Figure 13.2, note that the graph of $z = f(x, y)$ is a surface whose projection onto the $xy$-plane is $D$, the domain of $f$. To each point $(x, y)$ in $D$ there corresponds a point $(x, y, z)$ on the surface, and, conversely, to each point $(x, y, z)$ on the surface there corresponds a point $(x, y)$ in $D$.

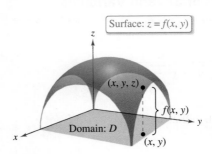

Surface: $z = f(x, y)$

Domain: $D$

**Figure 13.2**

**EXAMPLE 2** **Describing the Graph of a Function of Two Variables**

Consider the function given by

$$f(x, y) = \sqrt{16 - 4x^2 - y^2}.$$

**a.** Find the domain and range of the function.

**b.** Describe the graph of $f$.

**Solution**

**a.** The domain $D$ implied by the equation of $f$ is the set of all points $(x, y)$ such that

$$16 - 4x^2 - y^2 \geq 0.$$

So, $D$ is the set of all points lying on or inside the ellipse

$$\frac{x^2}{4} + \frac{y^2}{16} = 1. \qquad \text{Ellipse in the } xy\text{-plane}$$

The range of $f$ is all values $z = f(x, y)$ such that $0 \leq z \leq \sqrt{16}$, or

$$0 \leq z \leq 4. \qquad \text{Range of } f$$

**b.** A point $(x, y, z)$ is on the graph of $f$ if and only if

$$z = \sqrt{16 - 4x^2 - y^2}$$
$$z^2 = 16 - 4x^2 - y^2$$
$$4x^2 + y^2 + z^2 = 16$$
$$\frac{x^2}{4} + \frac{y^2}{16} + \frac{z^2}{16} = 1, \quad 0 \leq z \leq 4.$$

From Section 11.6, you know that the graph of $f$ is the upper half of an ellipsoid, as shown in Figure 13.3. ∎

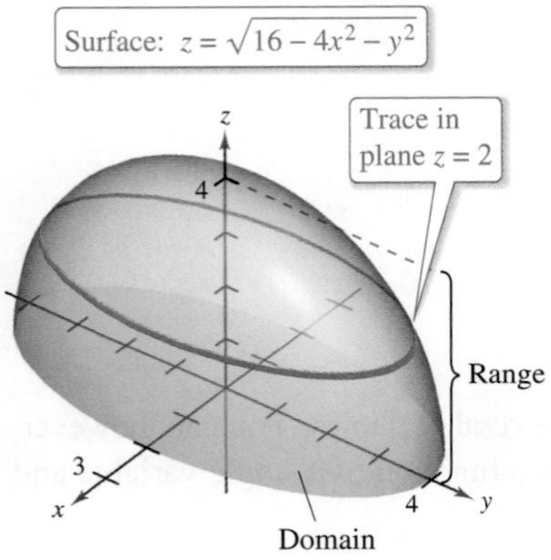

Surface: $z = \sqrt{16 - 4x^2 - y^2}$

Trace in plane $z = 2$

Range

Domain

The graph of
$f(x, y) = \sqrt{16 - 4x^2 - y^2}$ is the upper half of an ellipsoid.
**Figure 13.3**

To sketch a surface in space *by hand,* it helps to use traces in planes parallel to the coordinate planes, as shown in Figure 13.3. For example, to find the trace of the surface in the plane $z = 2$, substitute $z = 2$ in the equation $z = \sqrt{16 - 4x^2 - y^2}$ and obtain

$$2 = \sqrt{16 - 4x^2 - y^2} \quad \Longrightarrow \quad \frac{x^2}{3} + \frac{y^2}{12} = 1.$$

So, the trace is an ellipse centered at the point $(0, 0, 2)$ with major and minor axes of lengths $4\sqrt{3}$ and $2\sqrt{3}$.

Traces are also used with most three-dimensional graphing utilities. For instance, Figure 13.4 shows a computer-generated version of the surface given in Example 2. For this graph, the computer took 25 traces parallel to the $xy$-plane and 12 traces in vertical planes.

If you have access to a three-dimensional graphing utility, use it to graph several surfaces.

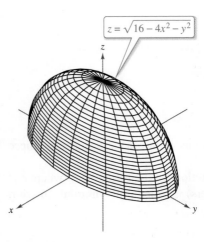

$z = \sqrt{16 - 4x^2 - y^2}$

**Figure 13.4**

# Level Curves

A second way to visualize a function of two variables is to use a **scalar field** in which the scalar

$$z = f(x, y)$$

is assigned to the point $(x, y)$. A scalar field can be characterized by **level curves** (or **contour lines**) along which the value of $f(x, y)$ is constant. For instance, the weather map in Figure 13.5 shows level curves of equal pressure called **isobars.** In weather maps for which the level curves represent points of equal temperature, the level curves are called **isotherms,** as shown in Figure 13.6. Another common use of level curves is in representing electric potential fields. In this type of map, the level curves are called **equipotential lines.**

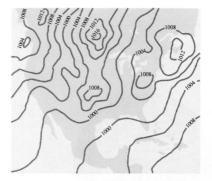

Level curves show the lines of equal pressure (isobars), measured in millibars.

**Figure 13.5**

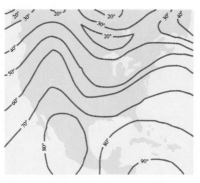

Level curves show the lines of equal temperature (isotherms), measured in degrees Fahrenheit.

**Figure 13.6**

Contour maps are commonly used to show regions on Earth's surface, with the level curves representing the height above sea level. This type of map is called a **topographic map.** For example, the mountain shown in Figure 13.7 is represented by the topographic map in Figure 13.8.

**Figure 13.7**

**Figure 13.8**

A contour map depicts the variation of $z$ with respect to $x$ and $y$ by the spacing between level curves. Much space between level curves indicates that $z$ is changing slowly, whereas little space indicates a rapid change in $z$. Furthermore, to produce a good three-dimensional illusion in a contour map, it is important to choose $c$-values that are *evenly spaced.*

**EXAMPLE 3** Sketching a Contour Map

The hemisphere

$$f(x, y) = \sqrt{64 - x^2 - y^2}$$

is shown in Figure 13.9. Sketch a contour map of this surface using level curves corresponding to $c = 0, 1, 2, \ldots, 8$.

**Solution** For each value of $c$, the equation $f(x, y) = c$ is a circle (or point) in the $xy$-plane. For example, when $c_1 = 0$, the level curve is

$$x^2 + y^2 = 64 \qquad \text{Circle of radius 8}$$

which is a circle of radius 8. Figure 13.10 shows the nine level curves for the hemisphere.

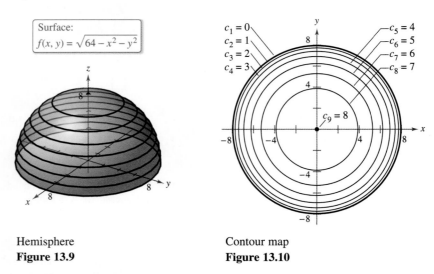

Surface:
$f(x, y) = \sqrt{64 - x^2 - y^2}$

Hemisphere
**Figure 13.9**

Contour map
**Figure 13.10**

**EXAMPLE 4** Sketching a Contour Map

⋅⋅⋅▷ *See LarsonCalculus.com for an interactive version of this type of example.*

The hyperbolic paraboloid

$$z = y^2 - x^2$$

is shown in Figure 13.11. Sketch a contour map of this surface.

**Solution** For each value of $c$, let $f(x, y) = c$ and sketch the resulting level curve in the $xy$-plane. For this function, each of the level curves ($c \neq 0$) is a hyperbola whose asymptotes are the lines $y = \pm x$. When $c < 0$, the transverse axis is horizontal. For instance, the level curve for $c = -4$ is

$$\frac{x^2}{2^2} - \frac{y^2}{2^2} = 1.$$

When $c > 0$, the transverse axis is vertical. For instance, the level curve for $c = 4$ is

$$\frac{y^2}{2^2} - \frac{x^2}{2^2} = 1.$$

When $c = 0$, the level curve is the degenerate conic representing the intersecting asymptotes, as shown in Figure 13.12.

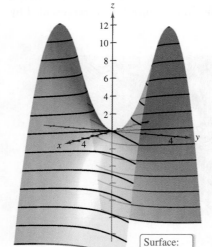

Surface:
$z = y^2 - x^2$

Hyperbolic paraboloid
**Figure 13.11**

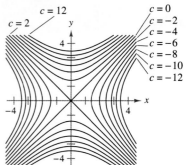

Hyperbolic level curves (at increments of 2)
**Figure 13.12**

One example of a function of two variables used in economics is the **Cobb-Douglas production function.** This function is used as a model to represent the numbers of units produced by varying amounts of labor and capital. If $x$ measures the units of labor and $y$ measures the units of capital, then the number of units produced is

$$f(x, y) = Cx^a y^{1-a}$$

where $C$ and $a$ are constants with $0 < a < 1$.

**EXAMPLE 5**    **The Cobb-Douglas Production Function**

A manufacturer estimates a production function to be

$$f(x, y) = 100x^{0.6}y^{0.4}$$

where $x$ is the number of units of labor and $y$ is the number of units of capital. Compare the production level when $x = 1000$ and $y = 500$ with the production level when $x = 2000$ and $y = 1000$.

**Solution**    When $x = 1000$ and $y = 500$, the production level is

$$f(1000, 500) = 100(1000^{0.6})(500^{0.4})$$
$$\approx 75{,}786.$$

When $x = 2000$ and $y = 1000$, the production level is

$$f(2000, 1000) = 100(2000^{0.6})(1000^{0.4})$$
$$\approx 151{,}572.$$

The level curves of $z = f(x, y)$ are shown in Figure 13.13. Note that by doubling both $x$ and $y$, you double the production level (see Exercise 83). ■

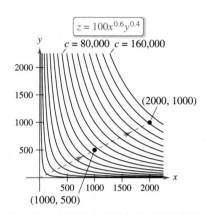

Level curves (at increments of 10,000)
**Figure 13.13**

## Level Surfaces

The concept of a level curve can be extended by one dimension to define a **level surface.** If $f$ is a function of three variables and $c$ is a constant, then the graph of the equation

$$f(x, y, z) = c$$

is a **level surface** of $f$, as shown in Figure 13.14.

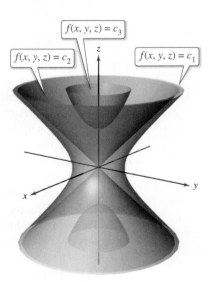

Level surfaces of $f$
**Figure 13.14**

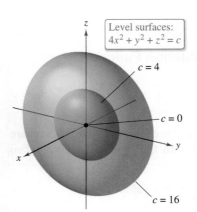

Level surfaces:
$4x^2 + y^2 + z^2 = c$

$c = 4$

$c = 0$

$c = 16$

**Figure 13.15**

**EXAMPLE 6** **Level Surfaces**

Describe the level surfaces of

$$f(x, y, z) = 4x^2 + y^2 + z^2.$$

**Solution** Each level surface has an equation of the form

$$4x^2 + y^2 + z^2 = c. \qquad \text{Equation of level surface}$$

So, the level surfaces are ellipsoids (whose cross sections parallel to the $yz$-plane are circles). As $c$ increases, the radii of the circular cross sections increase according to the square root of $c$. For example, the level surfaces corresponding to the values $c = 0$, $c = 4$, and $c = 16$ are as follows.

$$4x^2 + y^2 + z^2 = 0 \qquad \text{Level surface for } c = 0 \text{ (single point)}$$

$$\frac{x^2}{1} + \frac{y^2}{4} + \frac{z^2}{4} = 1 \qquad \text{Level surface for } c = 4 \text{ (ellipsoid)}$$

$$\frac{x^2}{4} + \frac{y^2}{16} + \frac{z^2}{16} = 1 \qquad \text{Level surface for } c = 16 \text{ (ellipsoid)}$$

These level surfaces are shown in Figure 13.15.

If the function in Example 6 represented the *temperature* at the point $(x, y, z)$, then the level surfaces shown in Figure 13.15 would be called **isothermal surfaces.**

## Computer Graphics

The problem of sketching the graph of a surface in space can be simplified by using a computer. Although there are several types of three-dimensional graphing utilities, most use some form of trace analysis to give the illusion of three dimensions. To use such a graphing utility, you usually need to enter the equation of the surface and the region in the $xy$-plane over which the surface is to be plotted. (You might also need to enter the number of traces to be taken.) For instance, to graph the surface

$$f(x, y) = (x^2 + y^2)e^{1 - x^2 - y^2}$$

you might choose the following bounds for $x$, $y$, and $z$.

$$-3 \le x \le 3 \qquad \text{Bounds for } x$$
$$-3 \le y \le 3 \qquad \text{Bounds for } y$$
$$0 \le z \le 3 \qquad \text{Bounds for } z$$

Figure 13.16 shows a computer-generated graph of this surface using 26 traces taken parallel to the $yz$-plane. To heighten the three-dimensional effect, the program uses a "hidden line" routine. That is, it begins by plotting the traces in the foreground (those corresponding to the largest $x$-values), and then, as each new trace is plotted, the program determines whether all or only part of the next trace should be shown.

The graphs on the next page show a variety of surfaces that were plotted by computer. If you have access to a computer drawing program, use it to reproduce these surfaces. Remember also that the three-dimensional graphics in this text can be viewed and rotated. These rotatable graphs are available at *LarsonCalculus.com.*

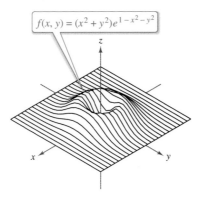

$f(x, y) = (x^2 + y^2)e^{1 - x^2 - y^2}$

**Figure 13.16**

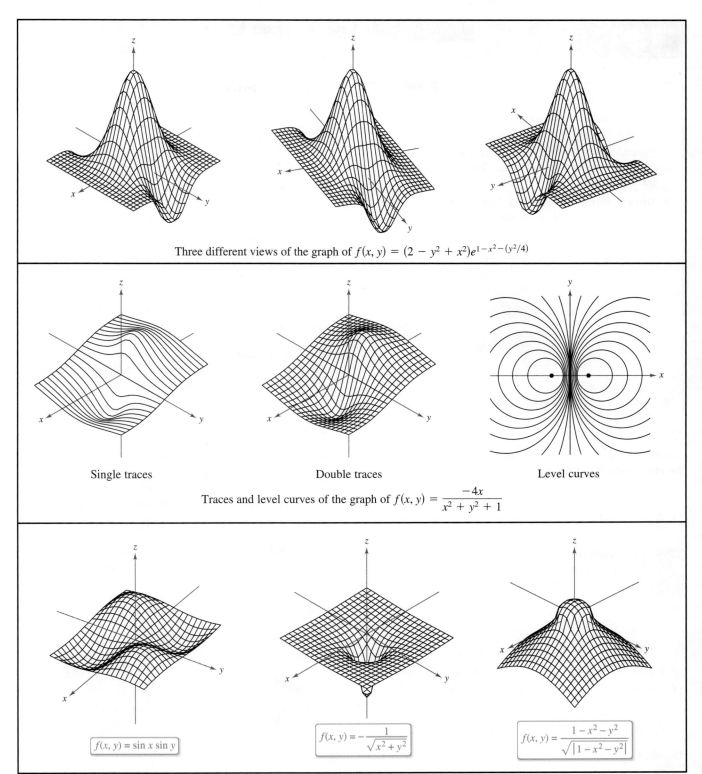

Three different views of the graph of $f(x, y) = (2 - y^2 + x^2)e^{1-x^2-(y^2/4)}$

Single traces

Double traces

Level curves

Traces and level curves of the graph of $f(x, y) = \dfrac{-4x}{x^2 + y^2 + 1}$

$f(x, y) = \sin x \sin y$

$f(x, y) = -\dfrac{1}{\sqrt{x^2 + y^2}}$

$f(x, y) = \dfrac{1 - x^2 - y^2}{\sqrt{|1 - x^2 - y^2|}}$

## 13.1 Exercises

See **CalcChat.com** for tutorial help and worked-out solutions to odd-numbered exercises.

### CONCEPT CHECK

**1. Think About It** Explain why $z^2 = x + 3y$ is not a function of $x$ and $y$.

**2. Function of Two Variables** What is a graph of a function of two variables? How is it interpreted geometrically?

**3. Determining Whether a Graph Is a Function** Use the graph to determine whether $z$ is a function of $x$ and $y$. Explain.

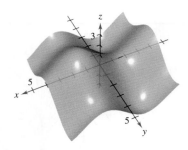

**4. Contour Map** Explain how to sketch a contour map of a function of $x$ and $y$.

**Determining Whether an Equation Is a Function** In Exercises 5–8, determine whether $z$ is a function of $x$ and $y$.

**5.** $x^2z + 3y^2 - xy = 10$   **6.** $xz^2 + 2xy - y^2 = 4$

**7.** $\dfrac{x^2}{4} + \dfrac{y^2}{9} + z^2 = 1$   **8.** $z + x \ln y - 8yz = 0$

**Evaluating a Function** In Exercises 9–20, evaluate the function at the given values of the independent variables. Simplify the results.

**9.** $f(x, y) = 2x - y + 3$   **10.** $f(x, y) = 4 - x^2 - 4y^2$
(a) $f(0, 2)$ (b) $f(-1, 0)$   (a) $f(0, 0)$ (b) $f(0, 1)$
(c) $f(5, 30)$ (d) $f(3, y)$   (c) $f(2, 3)$ (d) $f(1, y)$
(e) $f(x, 4)$ (f) $f(5, t)$   (e) $f(x, 0)$ (f) $f(t, 1)$

**11.** $f(x, y) = xe^y$   **12.** $g(x, y) = \ln|x + y|$
(a) $f(-1, 0)$ (b) $f(0, 2)$   (a) $g(1, 0)$ (b) $g(0, -t^2)$
(c) $f(x, 3)$ (d) $f(t, -y)$   (c) $g(e, 0)$ (d) $g(e, e)$

**13.** $h(x, y, z) = \dfrac{xy}{z}$   **14.** $f(x, y, z) = \sqrt{x + y + z}$
(a) $h(-1, 3, -1)$   (a) $f(2, 2, 5)$
(b) $h(2, 2, 2)$   (b) $f(0, 6, -2)$
(c) $h(4, 4t, t^2)$   (c) $f(8, -7, 2)$
(d) $h(-3, 2, 5)$   (d) $f(0, 1, -1)$

**15.** $f(x, y) = x \sin y$
(a) $f(2, \pi/4)$ (b) $f(3, 1)$ (c) $f(-3, 0)$ (d) $f(4, \pi/2)$

**16.** $V(r, h) = \pi r^2 h$
(a) $V(3, 10)$ (b) $V(5, 2)$ (c) $V(4, 8)$ (d) $V(6, \pi)$

**17.** $g(x, y) = \displaystyle\int_x^y (2t - 3)\, dt$
(a) $g(4, 0)$ (b) $g(4, 1)$ (c) $g\left(4, \tfrac{3}{2}\right)$ (d) $g\left(\tfrac{3}{2}, 0\right)$

**18.** $g(x, y) = \displaystyle\int_x^y \dfrac{1}{t}\, dt$
(a) $g(4, 1)$ (b) $g(6, 3)$ (c) $g(2, 5)$ (d) $g\left(\tfrac{1}{2}, 7\right)$

**19.** $f(x, y) = 2x + y^2$
(a) $\dfrac{f(x + \Delta x, y) - f(x, y)}{\Delta x}$ (b) $\dfrac{f(x, y + \Delta y) - f(x, y)}{\Delta y}$

**20.** $f(x, y) = 3x^2 - 2y$
(a) $\dfrac{f(x + \Delta x, y) - f(x, y)}{\Delta x}$ (b) $\dfrac{f(x, y + \Delta y) - f(x, y)}{\Delta y}$

**Finding the Domain and Range of a Function** In Exercises 21–32, find the domain and range of the function.

**21.** $f(x, y) = 3x^2 - y$   **22.** $f(x, y) = e^{xy}$

**23.** $g(x, y) = x\sqrt{y}$   **24.** $g(x, y) = \dfrac{y}{\sqrt{x}}$

**25.** $z = \dfrac{x + y}{xy}$   **26.** $z = \dfrac{xy}{x + y}$

**27.** $f(x, y) = \sqrt{4 - x^2 - y^2}$   **28.** $f(x, y) = \sqrt{9 - 6x^2 + y^2}$

**29.** $f(x, y) = \arccos(x + y)$   **30.** $f(x, y) = \arcsin(y/x)$

**31.** $f(x, y) = \ln(5 - x - y)$   **32.** $f(x, y) = \ln(xy - 6)$

**33. Think About It** The graphs labeled (a), (b), (c), and (d) are graphs of the function $f(x, y) = -4x/(x^2 + y^2 + 1)$. Match each of the four graphs with the point in space from which the surface is viewed. The four points are $(20, 15, 25)$, $(-15, 10, 20)$, $(20, 20, 0)$, and $(20, 0, 0)$.

(a)
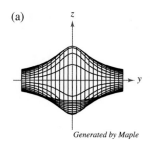
*Generated by Maple*

(b)
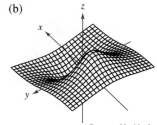
*Generated by Maple*

(c)
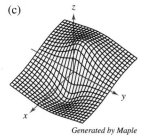
*Generated by Maple*

(d)
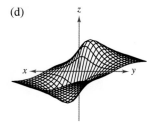
*Generated by Maple*

**34. Think About It** Use the function given in Exercise 33.

(a) Find the domain and range of the function.

(b) Identify the points in the $xy$-plane at which the function value is 0.

(c) Does the surface pass through all the octants of the rectangular coordinate system? Give reasons for your answer.

 **Sketching a Surface** In Exercises 35–42, describe and sketch the surface given by the function.

**35.** $f(x, y) = 4$

**36.** $f(x, y) = 6 - 2x - 3y$

**37.** $f(x, y) = y^2$

**38.** $g(x, y) = \frac{1}{2}y$

**39.** $z = -x^2 - y^2$

**40.** $z = \frac{1}{2}\sqrt{x^2 + y^2}$

**41.** $f(x, y) = e^{-x}$

**42.** $f(x, y) = \begin{cases} xy, & x \geq 0, y \geq 0 \\ 0, & x < 0 \text{ or } y < 0 \end{cases}$

**Graphing a Function Using Technology** In Exercises 43–46, use a computer algebra system to graph the function.

**43.** $z = y^2 - x^2 + 1$

**44.** $z = \frac{1}{12}\sqrt{144 - 16x^2 - 9y^2}$

**45.** $f(x, y) = x^2 e^{(-xy/2)}$

**46.** $f(x, y) = x \sin y$

**Matching** In Exercises 47–50, match the graph of the surface with one of the contour maps. [The contour maps are labeled (a), (b), (c), and (d).]

(a)

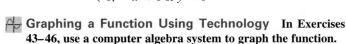

(b)

(c)

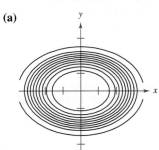

(d)

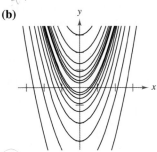

**47.** $f(x, y) = e^{1-x^2-y^2}$

**48.** $f(x, y) = e^{1-x^2+y^2}$

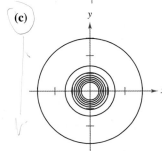

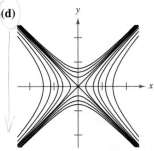

**49.** $f(x, y) = \ln|y - x^2|$

**50.** $f(x, y) = \cos\left(\dfrac{x^2 + 2y^2}{4}\right)$

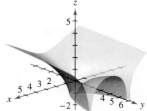

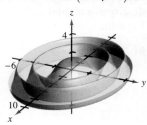

 **Sketching a Contour Map** In Exercises 51–58, describe the level curves of the function. Sketch a contour map of the surface using level curves for the given $c$-values.

**51.** $z = x + y$, $c = -1, 0, 2, 4$

**52.** $z = 6 - 2x - 3y$, $c = 0, 2, 4, 6, 8, 10$

**53.** $z = x^2 + 4y^2$, $c = 0, 1, 2, 3, 4$

**54.** $f(x, y) = \sqrt{9 - x^2 - y^2}$, $c = 0, 1, 2, 3$

**55.** $f(x, y) = xy$, $c = \pm 1, \pm 2, \ldots, \pm 6$

**56.** $f(x, y) = e^{xy/2}$, $c = 2, 3, 4, \frac{1}{2}, \frac{1}{3}, \frac{1}{4}$

**57.** $f(x, y) = x/(x^2 + y^2)$, $c = \pm\frac{1}{2}, \pm 1, \pm\frac{3}{2}, \pm 2$

**58.** $f(x, y) = \ln(x - y)$, $c = 0, \pm\frac{1}{2}, \pm 1, \pm\frac{3}{2}, \pm 2$

**Graphing Level Curves Using Technology** In Exercises 59–62, use a graphing utility to graph six level curves of the function.

**59.** $f(x, y) = x^2 - y^2 + 2$

**60.** $f(x, y) = |xy|$

**61.** $g(x, y) = \dfrac{8}{1 + x^2 + y^2}$

**62.** $h(x, y) = 3 \sin(|x| + |y|)$

**EXPLORING CONCEPTS**

**63. Vertical Line Test** Does the Vertical Line Test apply to functions of two variables? Explain your reasoning.

**64. Using Level Curves** All of the level curves of the surface given by $z = f(x, y)$ are concentric circles. Does this imply that the graph of $f$ is a hemisphere? Illustrate your answer with an example.

**65. Creating a Function** Construct a function whose level curves are lines passing through the origin.

**66. Conjecture** Consider the function $f(x, y) = xy$, for $x \geq 0$ and $y \geq 0$.

(a) Sketch the graph of the surface given by $f$.

(b) Make a conjecture about the relationship between the graphs of $f$ and $g(x, y) = f(x, y) - 3$. Explain your reasoning.

(c) Repeat part (b) for $g(x, y) = -f(x, y)$.

(d) Repeat part (b) for $g(x, y) = \frac{1}{2}f(x, y)$.

(e) On the surface in part (a), sketch the graph of $z = f(x, x)$.

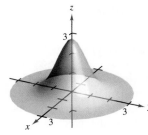

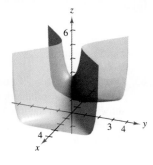

**Writing** In Exercises 67 and 68, use the graphs of the level curves ($c$-values evenly spaced) of the function $f$ to write a description of a possible graph of $f$. Is the graph of $f$ unique? Explain.

67.

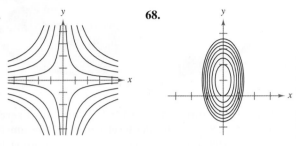

68.

**69. Investment** In 2016, an investment of $1000 was made in a bond earning 6% compounded annually. Assume that the buyer pays tax at rate $R$ and the annual rate of inflation is $I$. In the year 2026, the value $V$ of the investment in constant 2016 dollars is

$$V(I, R) = 1000 \left[ \frac{1 + 0.06(1 - R)}{1 + I} \right]^{10}.$$

Use this function of two variables to complete the table.

|  | Inflation Rate | | |
| --- | --- | --- | --- |
| Tax Rate | 0 | 0.03 | 0.05 |
| 0 | | | |
| 0.28 | | | |
| 0.35 | | | |

**70. Investment** A principal of $5000 is deposited in a savings account that earns interest at a rate of $r$ (written as a decimal), compounded continuously. The amount $A(r, t)$ after $t$ years is

$$A(r, t) = 5000e^{rt}.$$

Use this function of two variables to complete the table.

|  | Number of Years | | | |
| --- | --- | --- | --- | --- |
| Rate | 5 | 10 | 15 | 20 |
| 0.02 | | | | |
| 0.03 | | | | |
| 0.04 | | | | |
| 0.05 | | | | |

 **Sketching a Level Surface** In Exercises 71–76, describe and sketch the graph of the level surface $f(x, y, z) = c$ at the given value of $c$.

**71.** $f(x, y, z) = x - y + z$, $c = 1$

**72.** $f(x, y, z) = 4x + y + 2z$, $c = 4$

**73.** $f(x, y, z) = x^2 + y^2 + z^2$, $c = 9$

**74.** $f(x, y, z) = x^2 + \frac{1}{4}y^2 - z$, $c = 1$

**75.** $f(x, y, z) = 4x^2 + 4y^2 - z^2$, $c = 0$

**76.** $f(x, y, z) = \sin x - z$, $c = 0$

**77. Forestry**

The *Doyle Log Rule* is one of several methods used to determine the lumber yield of a log (in board-feet) in terms of its diameter $d$ (in inches) and its length $L$ (in feet). The number of board-feet is

$$N(d, L) = \left( \frac{d - 4}{4} \right)^2 L.$$

(a) Find the number of board-feet of lumber in a log 22 inches in diameter and 12 feet in length.

(b) Find $N(30, 12)$.

**78. Queuing Model** The average length of time that a customer waits in line for service is

$$W(x, y) = \frac{1}{x - y}, \quad x > y$$

where $y$ is the average arrival rate, written as the number of customers per unit of time, and $x$ is the average service rate, written in the same units. Evaluate each of the following.

(a) $W(15, 9)$  (b) $W(15, 13)$

(c) $W(12, 7)$  (d) $W(5, 2)$

**79. Temperature Distribution** The temperature $T$ (in degrees Celsius) at any point $(x, y)$ on a circular steel plate of radius 10 meters is

$$T = 600 - 0.75x^2 - 0.75y^2$$

where $x$ and $y$ are measured in meters. Sketch the isothermal curves for $T = 0, 100, 200, \ldots, 600$.

**80. Electric Potential** The electric potential $V$ at any point $(x, y)$ is

$$V(x, y) = \frac{5}{\sqrt{25 + x^2 + y^2}}.$$

Sketch the equipotential curves for $V = \frac{1}{2}$, $V = \frac{1}{3}$, and $V = \frac{1}{4}$.

 **Cobb-Douglas Production Function** In Exercises 81 and 82, use the Cobb-Douglas production function to find the production level when $x = 600$ units of labor and $y = 350$ units of capital.

**81.** $f(x, y) = 80x^{0.5}y^{0.5}$       **82.** $f(x, y) = 100x^{0.65}y^{0.35}$

**83. Cobb-Douglas Production Function** Use the Cobb-Douglas production function, $f(x, y) = Cx^a y^{1-a}$, to show that when the number of units of labor and the number of units of capital are doubled, the production level is also doubled.

**84. Cobb-Douglas Production Function** Show that the Cobb-Douglas production function $z = Cx^a y^{1-a}$ can be rewritten as

$$\ln \frac{z}{y} = \ln C + a \ln \frac{x}{y}.$$

**85. Ideal Gas Law** According to the Ideal Gas Law, $PV = kT$, where $P$ is pressure, $V$ is volume, $T$ is temperature (in kelvins), and $k$ is a constant of proportionality. A tank contains 2000 cubic inches of nitrogen at a pressure of 26 pounds per square inch and a temperature of 300 K.

(a) Determine $k$.

(b) Write $P$ as a function of $V$ and $T$ and describe the level curves.

**86. Modeling Data** The table shows the net sales $x$ (in billions of dollars), the total assets $y$ (in billions of dollars), and the shareholder's equity $z$ (in billions of dollars) for Walmart for the years 2010 through 2015. *(Source: Wal-Mart Stores, Inc.)*

| Year | 2010 | 2011 | 2012 | 2013 | 2014 | 2015 |
|------|------|------|------|------|------|------|
| $x$ | 405.0 | 418.5 | 443.4 | 465.6 | 473.1 | 482.2 |
| $y$ | 170.7 | 180.8 | 193.4 | 203.1 | 204.8 | 203.7 |
| $z$ | 70.7 | 68.5 | 71.3 | 76.3 | 76.3 | 81.4 |

A model for the data is $z = f(x, y) = 0.428x - 0.653y + 8.172$.

(a) Complete a fourth row in the table using the model to approximate $z$ for the given values of $x$ and $y$. Compare the approximations with the actual values of $z$.

(b) Which of the two variables in this model has more influence on shareholder's equity? Explain.

(c) Simplify the expression for $f(x, 150)$ and interpret its meaning in the context of the problem.

**87. Meteorology** Meteorologists measure the atmospheric pressure in millibars. From these observations, they create weather maps on which the curves of equal atmospheric pressure (isobars) are drawn (see figure). On the map, the closer the isobars, the higher the wind speed. Match points $A$, $B$, and $C$ with (a) highest pressure, (b) lowest pressure, and (c) highest wind velocity.

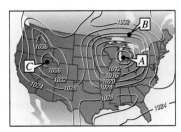

Figure for 87

Figure for 88

**88. Acid Rain** The acidity of rainwater is measured in units called pH. A pH of 7 is neutral, smaller values are increasingly acidic, and larger values are increasingly alkaline. The map shows curves of equal pH and gives evidence that downwind of heavily industrialized areas, the acidity has been increasing. Using the level curves on the map, determine the direction of the prevailing winds in the northeastern United States.

**89. Construction Cost** A rectangular storage box with an open top has a length of $x$ feet, a width of $y$ feet, and a height of $z$ feet. It costs \$4.50 per square foot to build the base and \$2.50 per square foot to build the sides. Write the cost $C$ of constructing the box as a function of $x$, $y$, and $z$.

**90. HOW DO YOU SEE IT?** The contour map of the Southern Hemisphere shown in the figure was computer generated using data collected by satellite instrumentation. Color is used to show the "ozone hole" in Earth's atmosphere. The purple and blue areas represent the lowest levels of ozone, and the green areas represent the highest levels. *(Source: NASA)*

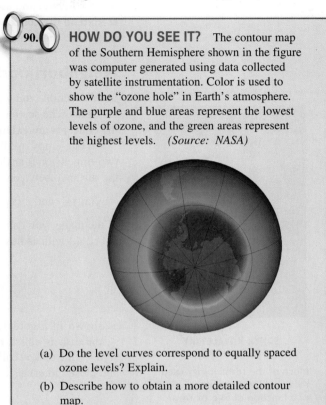

(a) Do the level curves correspond to equally spaced ozone levels? Explain.

(b) Describe how to obtain a more detailed contour map.

**True or False?** In Exercises 91–94, determine whether the statement is true or false. If it is false, explain why or give an example that shows it is false.

**91.** If $f(x_0, y_0) = f(x_1, y_1)$, then $x_0 = x_1$ and $y_0 = y_1$.

**92.** If $f$ is a function, then $f(ax, ay) = a^2 f(x, y)$.

**93.** The equation for a sphere is a function of three variables.

**94.** Two different level curves of the graph of $z = f(x, y)$ can intersect.

**PUTNAM EXAM CHALLENGE**

**95.** Let $f: \mathbb{R}^2 \to \mathbb{R}$ be a function such that

$$f(x, y) + f(y, z) + f(z, x) = 0$$

for all real numbers $x$, $y$, and $z$. Prove that there exists a function $g: \mathbb{R} \to \mathbb{R}$ such that

$$f(x, y) = g(x) - g(y)$$

for all real numbers $x$ and $y$.

This problem was composed by the Committee on the Putnam Prize Competition. © The Mathematical Association of America. All rights reserved.

# 13.2 Limits and Continuity

■ Understand the definition of a neighborhood in the plane.
■ Understand and use the definition of the limit of a function of two variables.
■ Extend the concept of continuity to a function of two variables.
■ Extend the concept of continuity to a function of three variables.

## Neighborhoods in the Plane

In this section, you will study limits and continuity involving functions of two or three variables. The section begins with functions of two variables. At the end of the section, the concepts are extended to functions of three variables.

Your study of the limit of a function of two variables begins by defining a two-dimensional analog to an interval on the real number line. Using the formula for the distance between two points

$$(x, y) \quad \text{and} \quad (x_0, y_0)$$

in the plane, you can define the **δ-neighborhood** about $(x_0, y_0)$ to be the **disk** centered at $(x_0, y_0)$ with radius $\delta > 0$

$$\{(x, y): \sqrt{(x - x_0)^2 + (y - y_0)^2} < \delta\} \qquad \text{Open disk}$$

as shown in Figure 13.17. When this formula contains the *less than* inequality sign, $<$, the disk is called **open,** and when it contains the *less than or equal to* inequality sign, $\le$, the disk is called **closed.** This corresponds to the use of $<$ and $\le$ to define open and closed intervals.

**SONYA KOVALEVSKY
(1850–1891)**

Much of the terminology used to define limits and continuity of a function of two or three variables was introduced by the German mathematician Karl Weierstrass (1815–1897). Weierstrass's rigorous approach to limits and other topics in calculus gained him the reputation as the "father of modern analysis." Weierstrass was a gifted teacher. One of his best-known students was the Russian mathematician Sonya Kovalevsky, who applied many of Weierstrass's techniques to problems in mathematical physics and became one of the first women to gain acceptance as a research mathematician.

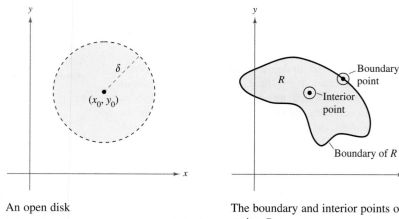

An open disk

**Figure 13.17**

The boundary and interior points of a region $R$

**Figure 13.18**

Let the region $R$ be a set of points in the plane. A point $(x_0, y_0)$ in $R$ is an **interior point** of $R$ if there exists a δ-neighborhood about $(x_0, y_0)$ that lies entirely in $R$, as shown in Figure 13.18. If every point in $R$ is an interior point, then $R$ is an **open region.** A point $(x_0, y_0)$ is a **boundary point** of $R$ if every open disk centered at $(x_0, y_0)$ contains points inside $R$ *and* points outside $R$. If $R$ contains all its boundary points, then $R$ is a **closed region.**

■ **FOR FURTHER INFORMATION** For more information on Sonya Kovalevsky, see the article "S. Kovalevsky: A Mathematical Lesson" by Karen D. Rappaport in *The American Mathematical Monthly.* To view this article, go to *MathArticles.com.*

## Limit of a Function of Two Variables

> ### Definition of the Limit of a Function of Two Variables
>
> Let $f$ be a function of two variables defined, except possibly at $(x_0, y_0)$, on an open disk centered at $(x_0, y_0)$, and let $L$ be a real number. Then
>
> $$\lim_{(x, y) \to (x_0, y_0)} f(x, y) = L$$
>
> if for each $\varepsilon > 0$ there corresponds a $\delta > 0$ such that
>
> $$|f(x, y) - L| < \varepsilon \quad \text{whenever} \quad 0 < \sqrt{(x - x_0)^2 + (y - y_0)^2} < \delta.$$

Graphically, the definition of the limit of a function of two variables implies that for any point $(x, y) \neq (x_0, y_0)$ in the disk of radius $\delta$, the value $f(x, y)$ lies between $L + \varepsilon$ and $L - \varepsilon$, as shown in Figure 13.19.

The definition of the limit of a function of two variables is similar to the definition of the limit of a function of a single variable, yet there is a critical difference. To determine whether a function of a single variable has a limit, you need only test the approach from two directions—from the right and from the left. When the function approaches the same limit from the right and from the left, you can conclude that the limit exists. For a function of two variables, however, the statement

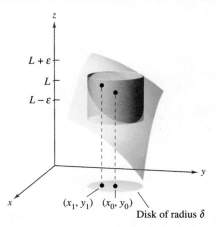

For any $(x, y)$ in the disk of radius $\delta$, the value $f(x, y)$ lies between $L + \varepsilon$ and $L - \varepsilon$.

**Figure 13.19**

$$(x, y) \to (x_0, y_0)$$

means that the point $(x, y)$ is allowed to approach $(x_0, y_0)$ from any direction. If the value of

$$\lim_{(x, y) \to (x_0, y_0)} f(x, y)$$

is not the same for all possible approaches, or **paths**, to $(x_0, y_0)$, then the limit does not exist.

### EXAMPLE 1    Verifying a Limit by the Definition

Show that $\displaystyle\lim_{(x, y) \to (a, b)} x = a$.

**Solution**    Let $f(x, y) = x$ and $L = a$. You need to show that for each $\varepsilon > 0$, there exists a $\delta$-neighborhood about $(a, b)$ such that

$$|f(x, y) - L| = |x - a| < \varepsilon$$

whenever $(x, y) \neq (a, b)$ lies in the neighborhood. You can first observe that from

$$0 < \sqrt{(x - a)^2 + (y - b)^2} < \delta$$

it follows that

$$\begin{aligned}
|f(x, y) - L| &= |x - a| \\
&= \sqrt{(x - a)^2} \\
&\leq \sqrt{(x - a)^2 + (y - b)^2} \\
&< \delta.
\end{aligned}$$

So, you can choose $\delta = \varepsilon$, and the limit is verified.

Limits of functions of several variables have the same properties regarding sums, differences, products, and quotients as do limits of functions of single variables. (See Theorem 1.2 in Section 1.3.) Some of these properties are used in the next example.

**EXAMPLE 2    Finding a Limit**

Find the limit.

$$\lim_{(x,\,y)\to(1,\,2)} \frac{5x^2y}{x^2 + y^2}$$

**Solution**    By using the properties of limits of products and sums, you obtain

$$\lim_{(x,\,y)\to(1,\,2)} 5x^2y = 5(1^2)(2) = 10$$

and

$$\lim_{(x,\,y)\to(1,\,2)} (x^2 + y^2) = (1^2 + 2^2) = 5.$$

Because the limit of a quotient is equal to the quotient of the limits (and the denominator is not 0), you have

$$\lim_{(x,\,y)\to(1,\,2)} \frac{5x^2y}{x^2 + y^2} = \frac{10}{5} = 2.$$

**EXAMPLE 3    Finding a Limit**

Find the limit:    $\displaystyle\lim_{(x,\,y)\to(0,\,0)} \frac{5x^2y}{x^2 + y^2}$.

**Solution**    In this case, the limits of the numerator and of the denominator are both 0, so you cannot determine the existence (or nonexistence) of a limit by taking the limits of the numerator and denominator separately and then dividing. From the graph of $f$ in Figure 13.20, however, it seems reasonable that the limit might be 0. So, you can try applying the definition to $L = 0$. First, note that

$$|y| \le \sqrt{x^2 + y^2}$$

and

$$\frac{x^2}{x^2 + y^2} \le 1.$$

Then, in a $\delta$-neighborhood about $(0, 0)$, you have

$$0 < \sqrt{x^2 + y^2} < \delta$$

and it follows that, for $(x, y) \ne (0, 0)$,

$$|f(x, y) - 0| = \left|\frac{5x^2y}{x^2 + y^2}\right|$$

$$= 5|y|\left(\frac{x^2}{x^2 + y^2}\right)$$

$$\le 5|y|$$

$$\le 5\sqrt{x^2 + y^2}$$

$$< 5\delta.$$

So, you can choose $\delta = \varepsilon/5$ and conclude that

$$\lim_{(x,\,y)\to(0,\,0)} \frac{5x^2y}{x^2 + y^2} = 0.$$

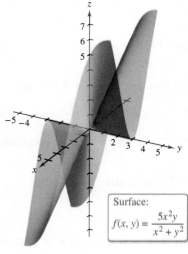

Surface:
$$f(x, y) = \frac{5x^2y}{x^2 + y^2}$$

**Figure 13.20**

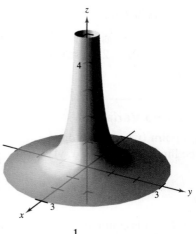

$\displaystyle \lim_{(x,\,y)\to(0,\,0)} \frac{1}{x^2+y^2}$ does not exist.

**Figure 13.21**

For some functions, it is easy to recognize that a limit does not exist. For instance, it is clear that the limit

$$\lim_{(x,\,y)\to(0,\,0)} \frac{1}{x^2+y^2}$$

does not exist because the values of $f(x,y)$ increase without bound as $(x,y)$ approaches $(0,0)$ along *any* path (see Figure 13.21).

For other functions, it is not so easy to recognize that a limit does not exist. For instance, the next example describes a limit that does not exist because the function approaches different values along different paths.

### EXAMPLE 4   A Limit That Does Not Exist

⋮••▷ *See LarsonCalculus.com for an interactive version of this type of example.*

Show that the limit does not exist.

$$\lim_{(x,\,y)\to(0,\,0)} \left(\frac{x^2-y^2}{x^2+y^2}\right)^2$$

**Solution**   The domain of the function

$$f(x,y) = \left(\frac{x^2-y^2}{x^2+y^2}\right)^2$$

consists of all points in the $xy$-plane except for the point $(0,0)$. To show that the limit as $(x,y)$ approaches $(0,0)$ does not exist, consider approaching $(0,0)$ along two different "paths," as shown in Figure 13.22. Along the $x$-axis, every point is of the form

$$(x,0)$$

and the limit along this approach is

$$\lim_{(x,\,0)\to(0,\,0)} \left(\frac{x^2-0^2}{x^2+0^2}\right)^2 = \lim_{(x,\,0)\to(0,\,0)} 1^2 = 1. \qquad \text{Limit along } x\text{-axis}$$

However, when $(x,y)$ approaches $(0,0)$ along the line $y=x$, you obtain

$$\lim_{(x,\,x)\to(0,\,0)} \left(\frac{x^2-x^2}{x^2+x^2}\right)^2 = \lim_{(x,\,x)\to(0,\,0)} \left(\frac{0}{2x^2}\right)^2 = 0. \qquad \text{Limit along line } y=x$$

This means that in any open disk centered at $(0,0)$, there are points $(x,y)$ at which $f$ takes on the value 1 and other points at which $f$ takes on the value 0. For instance,

$$f(x,y) = 1$$

at $(1,0)$, $(0.1,0)$, $(0.01,0)$, and $(0.001,0)$, and

$$f(x,y) = 0$$

at $(1,1)$, $(0.1,0.1)$, $(0.01,0.01)$, and $(0.001,0.001)$. So, $f$ does not have a limit as $(x,y)$ approaches $(0,0)$.

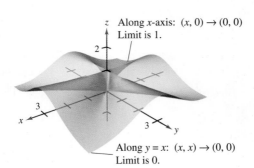

Along $x$-axis: $(x,0) \to (0,0)$
Limit is 1.

Along $y=x$: $(x,x) \to (0,0)$
Limit is 0.

$\displaystyle \lim_{(x,\,y)\to(0,\,0)} \left(\frac{x^2-y^2}{x^2+y^2}\right)^2$ does not exist.

**Figure 13.22**

In Example 4, you could conclude that the limit does not exist because you found two approaches that produced different limits. Be sure you understand that when two approaches produce the same limit, you *cannot* conclude that the limit exists. To form such a conclusion, you must show that the limit is the same along *all* possible approaches.

## Continuity of a Function of Two Variables

Notice in Example 2 that the limit of $f(x, y) = 5x^2y/(x^2 + y^2)$ as $(x, y) \to (1, 2)$ can be evaluated by direct substitution. That is, the limit is $f(1, 2) = 2$. In such cases, the function $f$ is said to be **continuous** at the point $(1, 2)$.

**· · REMARK**  This definition of continuity can be extended to *boundary points* of the open region $R$ by considering a special type of limit in which $(x, y)$ is allowed to approach $(x_0, y_0)$ along paths lying in the region $R$. This notion is similar to that of one-sided limits, as discussed in Chapter 1.

---

**Definition of Continuity of a Function of Two Variables**

A function $f$ of two variables is **continuous at a point** $(x_0, y_0)$ in an open region $R$ if $f(x_0, y_0)$ is defined and is equal to the limit of $f(x, y)$ as $(x, y)$ approaches $(x_0, y_0)$. That is,

$$\lim_{(x, y) \to (x_0, y_0)} f(x, y) = f(x_0, y_0).$$

The function $f$ is **continuous in the open region** $R$ if it is continuous at every point in $R$.

---

In Example 3, it was shown that the function

$$f(x, y) = \frac{5x^2y}{x^2 + y^2}$$

is not continuous at $(0, 0)$. Because the limit at this point exists, however, you can remove the discontinuity by defining $f$ at $(0, 0)$ as being equal to its limit there. Such a discontinuity is called **removable.** In Example 4, the function

$$f(x, y) = \left(\frac{x^2 - y^2}{x^2 + y^2}\right)^2$$

was also shown not to be continuous at $(0, 0)$, but this discontinuity is **nonremovable.**

---

**THEOREM 13.1    Continuous Functions of Two Variables**

If $k$ is a real number and $f(x, y)$ and $g(x, y)$ are continuous at $(x_0, y_0)$, then the following functions are also continuous at $(x_0, y_0)$.

1. Scalar multiple: $kf$                    2. Sum or difference: $f \pm g$
3. Product: $fg$                            4. Quotient: $f/g, \ g(x_0, y_0) \neq 0$

---

Theorem 13.1 establishes the continuity of *polynomial* and *rational* functions at every point in their domains. Furthermore, the continuity of other types of functions can be extended naturally from one to two variables. For instance, the functions whose graphs are shown in Figures 13.23 and 13.24 are continuous at every point in the plane.

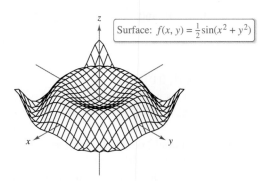

Surface: $f(x, y) = \frac{1}{2}\sin(x^2 + y^2)$

The function $f$ is continuous at every point in the plane.
**Figure 13.23**

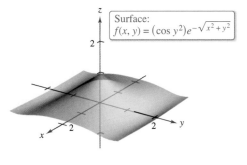

Surface:
$f(x, y) = (\cos y^2)e^{-\sqrt{x^2 + y^2}}$

The function $f$ is continuous at every point in the plane.
**Figure 13.24**

The next theorem states conditions under which a composite function is continuous.

> **THEOREM 13.2 Continuity of a Composite Function**
>
> If $h$ is continuous at $(x_0, y_0)$ and $g$ is continuous at $h(x_0, y_0)$, then the composite function given by $(g \circ h)(x, y) = g(h(x, y))$ is continuous at $(x_0, y_0)$. That is,
>
> $$\lim_{(x, y) \to (x_0, y_0)} g(h(x, y)) = g(h(x_0, y_0)).$$

Note in Theorem 13.2 that $h$ is a function of two variables and $g$ is a function of one variable.

**EXAMPLE 5**   **Testing for Continuity**

Discuss the continuity of each function.

**a.** $f(x, y) = \dfrac{x - 2y}{x^2 + y^2}$      **b.** $g(x, y) = \dfrac{2}{y - x^2}$

**Solution**

**a.** Because a rational function is continuous at every point in its domain, you can conclude that $f$ is continuous at each point in the $xy$-plane except at $(0, 0)$, as shown in Figure 13.25.

**b.** The function

$$g(x, y) = \frac{2}{y - x^2}$$

is continuous except at the points at which the denominator is 0. These points are given by the equation

$$y - x^2 = 0.$$

So, you can conclude that the function is continuous at all points except those lying on the parabola $y = x^2$. Inside this parabola, you have $y > x^2$, and the surface represented by the function lies above the $xy$-plane, as shown in Figure 13.26. Outside the parabola, $y < x^2$, and the surface lies below the $xy$-plane.

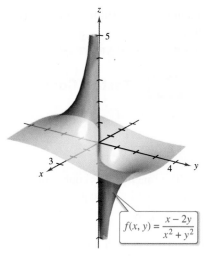

The function $f$ is not continuous at $(0, 0)$.

**Figure 13.25**

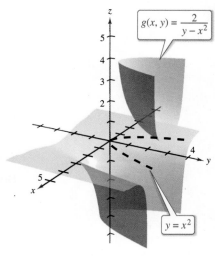

The function $g$ is not continuous on the parabola $y = x^2$.

**Figure 13.26**

## Continuity of a Function of Three Variables

The preceding definitions of limits and continuity can be extended to functions of three variables by considering points $(x, y, z)$ within the *open sphere*

$$(x - x_0)^2 + (y - y_0)^2 + (z - z_0)^2 < \delta^2.$$ Open sphere

The radius of this sphere is $\delta$, and the sphere is centered at $(x_0, y_0, z_0)$, as shown in Figure 13.27.

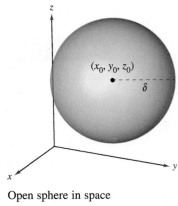

Open sphere in space
**Figure 13.27**

A point $(x_0, y_0, z_0)$ in a region $R$ in space is an **interior point** of $R$ if there exists a $\delta$-sphere about $(x_0, y_0, z_0)$ that lies entirely in $R$. If every point in $R$ is an interior point, then $R$ is called **open.**

---

### Definition of Continuity of a Function of Three Variables

A function $f$ of three variables is **continuous at a point** $(x_0, y_0, z_0)$ in an open region $R$ if $f(x_0, y_0, z_0)$ is defined and is equal to the limit of $f(x, y, z)$ as $(x, y, z)$ approaches $(x_0, y_0, z_0)$. That is,

$$\lim_{(x, y, z) \to (x_0, y_0, z_0)} f(x, y, z) = f(x_0, y_0, z_0).$$

The function $f$ is **continuous in the open region** $R$ if it is continuous at every point in $R$.

---

### EXAMPLE 6  Testing Continuity of a Function of Three Variables

Discuss the continuity of

$$f(x, y, z) = \frac{1}{x^2 + y^2 - z}.$$

**Solution**  The function $f$ is continuous except at the points at which the denominator is 0, which are given by the equation

$$x^2 + y^2 - z = 0.$$

So, $f$ is continuous at each point in space except at the points on the paraboloid

$$z = x^2 + y^2.$$

## 13.2 Exercises

See **CalcChat.com** for tutorial help and worked-out solutions to odd-numbered exercises.

### CONCEPT CHECK

**1. Describing Notation** Write a brief description of the meaning of the notation $\lim\limits_{(x,\,y)\to(-1,\,3)} f(x, y) = 1$.

**2. Limits** Explain how examining limits along different paths might show that a limit does not exist. Does this type of examination show that a limit does exist? Explain.

 **Verifying a Limit by the Definition** In Exercises 3–6, use the definition of the limit of a function of two variables to verify the limit.

**3.** $\lim\limits_{(x,\,y)\to(1,\,0)} x = 1$

**4.** $\lim\limits_{(x,\,y)\to(4,\,-1)} x = 4$

**5.** $\lim\limits_{(x,\,y)\to(1,\,-3)} y = -3$

**6.** $\lim\limits_{(x,\,y)\to(a,\,b)} y = b$

**Using Properties of Limits** In Exercises 7–10, find the indicated limit by using the limits

$$\lim_{(x,\,y)\to(a,\,b)} f(x, y) = 4 \quad \text{and} \quad \lim_{(x,\,y)\to(a,\,b)} g(x, y) = -5.$$

**7.** $\lim\limits_{(x,\,y)\to(a,\,b)} [f(x, y) - g(x, y)]$

**8.** $\lim\limits_{(x,\,y)\to(a,\,b)} \left[ \dfrac{3f(x, y)}{g(x, y)} \right]$

**9.** $\lim\limits_{(x,\,y)\to(a,\,b)} [f(x, y)g(x, y)]$

**10.** $\lim\limits_{(x,\,y)\to(a,\,b)} \left[ \dfrac{f(x, y) + g(x, y)}{f(x, y)} \right]$

 **Limit and Continuity** In Exercises 11–24, find the limit and discuss the continuity of the function.

**11.** $\lim\limits_{(x,\,y)\to(3,\,1)} (x^2 - 2y)$

**12.** $\lim\limits_{(x,\,y)\to(-1,\,1)} (x + 4y^2 + 5)$

**13.** $\lim\limits_{(x,\,y)\to(1,\,2)} e^{xy}$

**14.** $\lim\limits_{(x,\,y)\to(2,\,4)} \dfrac{x + y}{x^2 + 1}$

**15.** $\lim\limits_{(x,\,y)\to(0,\,2)} \dfrac{x}{y}$

**16.** $\lim\limits_{(x,\,y)\to(-1,\,2)} \dfrac{x + y}{x - y}$

**17.** $\lim\limits_{(x,\,y)\to(1,\,1)} \dfrac{xy}{x^2 + y^2}$

**18.** $\lim\limits_{(x,\,y)\to(1,\,1)} \dfrac{x}{\sqrt{x + y}}$

**19.** $\lim\limits_{(x,\,y)\to(\pi/3,\,2)} y \cos xy$

**20.** $\lim\limits_{(x,\,y)\to(\pi,\,-4)} \sin \dfrac{x}{y}$

**21.** $\lim\limits_{(x,\,y)\to(0,\,1)} \dfrac{\arcsin xy}{1 - xy}$

**22.** $\lim\limits_{(x,\,y)\to(0,\,1)} \dfrac{\arccos(x/y)}{1 + xy}$

**23.** $\lim\limits_{(x,\,y,\,z)\to(1,\,3,\,4)} \sqrt{x + y + z}$

**24.** $\lim\limits_{(x,\,y,\,z)\to(-2,\,1,\,0)} xe^{yz}$

 **Finding a Limit** In Exercises 25–36, find the limit (if it exists). If the limit does not exist, explain why.

**25.** $\lim\limits_{(x,\,y)\to(1,\,1)} \dfrac{xy - 1}{1 + xy}$

**26.** $\lim\limits_{(x,\,y)\to(1,\,-1)} \dfrac{x^2y}{1 + xy^2}$

**27.** $\lim\limits_{(x,\,y)\to(0,\,0)} \dfrac{1}{x + y}$

**28.** $\lim\limits_{(x,\,y)\to(0,\,0)} \dfrac{1}{x^2y^2}$

**29.** $\lim\limits_{(x,\,y)\to(0,\,0)} \dfrac{x - y}{\sqrt{x} - \sqrt{y}}$

**30.** $\lim\limits_{(x,\,y)\to(2,\,1)} \dfrac{x - y - 1}{\sqrt{x - y} - 1}$

**31.** $\lim\limits_{(x,\,y)\to(0,\,0)} \dfrac{x + y}{x^2 + y}$

**32.** $\lim\limits_{(x,\,y)\to(0,\,0)} \dfrac{x}{x^2 - y^2}$

**33.** $\lim\limits_{(x,\,y)\to(0,\,0)} \dfrac{x^2}{(x^2 + 1)(y^2 + 1)}$

**34.** $\lim\limits_{(x,\,y)\to(0,\,0)} \ln(x^2 + y^2)$

**35.** $\lim\limits_{(x,\,y,\,z)\to(0,\,0,\,0)} \dfrac{xy + yz + xz}{x^2 + y^2 + z^2}$

**36.** $\lim\limits_{(x,\,y,\,z)\to(0,\,0,\,0)} \dfrac{xy + yz^2 + xz^2}{x^2 + y^2 + z^2}$

### EXPLORING CONCEPTS

**37. Limits** If $f(2, 3) = 4$, can you conclude anything about $\lim\limits_{(x,\,y)\to(2,\,3)} f(x, y)$? Explain.

**38. Limits** If $\lim\limits_{(x,\,y)\to(2,\,3)} f(x, y) = 4$, can you conclude anything about $f(2, 3)$? Explain.

**39. Think About It** Given that $\lim\limits_{(x,\,y)\to(0,\,0)} f(x, y) = 0$, does $\lim\limits_{(x,\,0)\to(0,\,0)} f(x, 0) = 0$? Explain.

**40. HOW DO YOU SEE IT?** The figure shows the graph of $f(x, y) = \ln(x^2 + y^2)$. From the graph, does it appear that the limit at each point exists?

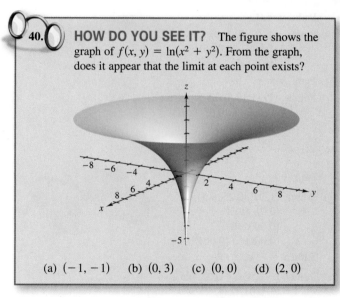

(a) $(-1, -1)$    (b) $(0, 3)$    (c) $(0, 0)$    (d) $(2, 0)$

**Continuity** In Exercises 41 and 42, discuss the continuity of the function and evaluate the limit of $f(x, y)$ (if it exists) as $(x, y) \to (0, 0)$.

**41.** $f(x, y) = e^{xy}$

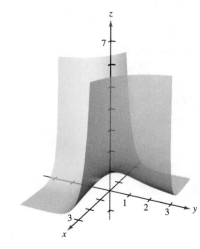

**42.** $f(x, y) = 1 - \dfrac{\cos(x^2 + y^2)}{x^2 + y^2}$

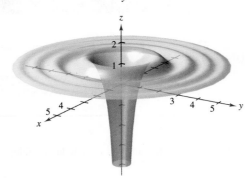

**Limit and Continuity** In Exercises 43–46, use a graphing utility to make a table showing the values of $f(x, y)$ at the given points for each path. Use the result to make a conjecture about the limit of $f(x, y)$ as $(x, y) \to (0, 0)$. Determine analytically whether the limit exists and discuss the continuity of the function.

**43.** $f(x, y) = \dfrac{xy}{x^2 + y^2}$

Path: $y = 0$

Points: $(1, 0)$, $(0.5, 0)$, $(0.1, 0)$, $(0.01, 0)$, $(0.001, 0)$

Path: $y = x$

Points: $(1, 1)$, $(0.5, 0.5)$, $(0.1, 0.1)$, $(0.01, 0.01)$, $(0.001, 0.001)$

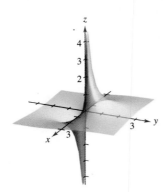

**44.** $f(x, y) = -\dfrac{xy^2}{x^2 + y^4}$

Path: $x = y^2$

Points: $(1, 1)$, $(0.25, 0.5)$, $(0.01, 0.1)$, $(0.0001, 0.01)$, $(0.000001, 0.001)$

Path: $x = -y^2$

Points: $(-1, 1)$, $(-0.25, 0.5)$, $(-0.01, 0.1)$, $(-0.0001, 0.01)$, $(-0.000001, 0.001)$

**45.** $f(x, y) = \dfrac{y}{x^2 + y^2}$

Path: $y = 0$

Points: $(1, 0)$, $(0.5, 0)$, $(0.1, 0)$, $(0.01, 0)$, $(0.001, 0)$

Path: $y = x$

Points: $(1, 1)$, $(0.5, 0.5)$, $(0.1, 0.1)$, $(0.01, 0.01)$, $(0.001, 0.001)$

**46.** $f(x, y) = \dfrac{2x - y^2}{2x^2 + y}$

Path: $y = 0$

Points: $(1, 0)$, $(0.25, 0)$, $(0.01, 0)$, $(0.001, 0)$, $(0.000001, 0)$

Path: $y = x$

Points: $(1, 1)$, $(0.25, 0.25)$, $(0.01, 0.01)$, $(0.001, 0.001)$, $(0.0001, 0.0001)$

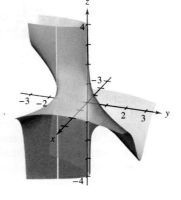

**47. Limit** Consider $\displaystyle\lim_{(x, y)\to(0, 0)} \dfrac{x^2 + y^2}{xy}$ (see figure).

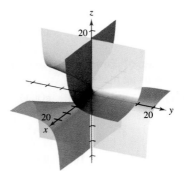

(a) Determine (if possible) the limit along any line of the form $y = ax$.

(b) Determine (if possible) the limit along the parabola $y = x^2$.

(c) Does the limit exist? Explain.

**48. Limit** Consider $\displaystyle\lim_{(x, y)\to(0, 0)} \dfrac{x^2 y}{x^4 + y^2}$ (see figure).

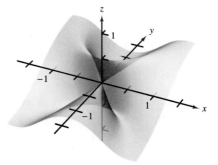

**Comparing Continuity** In Exercises 49 and 50, discuss the continuity of the functions $f$ and $g$. Explain any differences.

**49.** $f(x, y) = \begin{cases} \dfrac{x^4 - y^4}{x^2 + y^2}, & (x, y) \neq (0, 0) \\ 0, & (x, y) = (0, 0) \end{cases}$

$g(x, y) = \begin{cases} \dfrac{x^4 - y^4}{x^2 + y^2}, & (x, y) \neq (0, 0) \\ 1, & (x, y) = (0, 0) \end{cases}$

**50.** $f(x, y) = \begin{cases} \dfrac{x^2 + 2xy^2 + y^2}{x^2 + y^2}, & (x, y) \neq (0, 0) \\ 0, & (x, y) = (0, 0) \end{cases}$

$g(x, y) = \begin{cases} \dfrac{x^2 + 2xy^2 + y^2}{x^2 + y^2}, & (x, y) \neq (0, 0) \\ 1, & (x, y) = (0, 0) \end{cases}$

**Finding a Limit Using Polar Coordinates**  In Exercises 51–56, use polar coordinates to find the limit. [*Hint:* Let $x = r \cos \theta$ and $y = r \sin \theta$, and note that $(x, y) \rightarrow (0, 0)$ implies $r \rightarrow 0$.]

**51.** $\displaystyle\lim_{(x, y) \to (0, 0)} \frac{xy^2}{x^2 + y^2}$

**52.** $\displaystyle\lim_{(x, y) \to (0, 0)} \frac{x^3 + y^3}{x^2 + y^2}$

**53.** $\displaystyle\lim_{(x, y) \to (0, 0)} \frac{x^2 y^2}{x^2 + y^2}$

**54.** $\displaystyle\lim_{(x, y) \to (0, 0)} \frac{x^2 - y^2}{\sqrt{x^2 + y^2}}$

**55.** $\displaystyle\lim_{(x, y) \to (0, 0)} \cos(x^2 + y^2)$

**56.** $\displaystyle\lim_{(x, y) \to (0, 0)} \sin\sqrt{x^2 + y^2}$

**Finding a Limit Using Polar Coordinates**  In Exercises 57–60, use polar coordinates and L'Hôpital's Rule to find the limit.

**57.** $\displaystyle\lim_{(x, y) \to (0, 0)} \frac{\sin\sqrt{x^2 + y^2}}{\sqrt{x^2 + y^2}}$

**58.** $\displaystyle\lim_{(x, y) \to (0, 0)} \frac{\sin(x^2 + y^2)}{x^2 + y^2}$

**59.** $\displaystyle\lim_{(x, y) \to (0, 0)} \frac{1 - \cos(x^2 + y^2)}{x^2 + y^2}$

**60.** $\displaystyle\lim_{(x, y) \to (0, 0)} (x^2 + y^2) \ln(x^2 + y^2)$

 **Continuity**  In Exercises 61–66, discuss the continuity of the function.

**61.** $f(x, y, z) = \dfrac{1}{\sqrt{x^2 + y^2 + z^2}}$

**62.** $f(x, y, z) = \dfrac{z}{x^2 + y^2 - 4}$

**63.** $f(x, y, z) = \dfrac{\sin z}{e^x + e^y}$

**64.** $f(x, y, z) = xy \sin z$

**65.** $f(x, y) = \begin{cases} \dfrac{\sin xy}{xy}, & xy \neq 0 \\ 1, & xy = 0 \end{cases}$

**66.** $f(x, y) = \begin{cases} \dfrac{\sin(x^2 - y^2)}{x^2 - y^2}, & x^2 \neq y^2 \\ 1, & x^2 = y^2 \end{cases}$

**Continuity of a Composite Function**  In Exercises 67–70, discuss the continuity of the composite function $f \circ g$.

**67.** $f(t) = t^2$

$g(x, y) = 2x - 3y$

**68.** $f(t) = \dfrac{1}{t}$

$g(x, y) = x^2 + y^2$

**69.** $f(t) = \dfrac{1}{t}$

$g(x, y) = 2x - 3y$

**70.** $f(t) = \dfrac{1}{1 - t}$

$g(x, y) = x^2 + y^2$

**Finding a Limit**  In Exercises 71–76, find each limit.

(a) $\displaystyle\lim_{\Delta x \to 0} \frac{f(x + \Delta x, y) - f(x, y)}{\Delta x}$

(b) $\displaystyle\lim_{\Delta y \to 0} \frac{f(x, y + \Delta y) - f(x, y)}{\Delta y}$

**71.** $f(x, y) = x^2 - 4y$

**72.** $f(x, y) = 3x^2 + y^2$

**73.** $f(x, y) = \dfrac{x}{y}$

**74.** $f(x, y) = \dfrac{1}{x + y}$

**75.** $f(x, y) = 3x + xy - 2y$

**76.** $f(x, y) = \sqrt{y}(y + 1)$

**Finding a Limit Using Spherical Coordinates**  In Exercises 77 and 78, use spherical coordinates to find the limit. [*Hint:* Let $x = \rho \sin \phi \cos \theta$, $y = \rho \sin \phi \sin \theta$, and $z = \rho \cos \phi$, and note that $(x, y, z) \rightarrow (0, 0, 0)$ implies $\rho \rightarrow 0^+$.]

**77.** $\displaystyle\lim_{(x, y) \to (0, 0, 0)} \frac{xyz}{x^2 + y^2 + z^2}$

**78.** $\displaystyle\lim_{(x, y) \to (0, 0, 0)} \tan^{-1}\left(\frac{1}{x^2 + y^2 + z^2}\right)$

**True or False?**  In Exercises 79–82, determine whether the statement is true or false. If it is false, explain why or give an example that shows it is false.

**79.** A closed region contains all of its boundary points.

**80.** Every point in an open region is an interior point.

**81.** If $f$ is continuous for all nonzero $x$ and $y$, and $f(0, 0) = 0$, then $\displaystyle\lim_{(x, y) \to (0, 0)} f(x, y) = 0$.

**82.** If $g$ is a continuous function of $x$, $h$ is a continuous function of $y$, and $f(x, y) = g(x) + h(y)$, then $f$ is continuous.

**83. Finding a Limit**  Find the following limit.

$$\lim_{(x, y) \to (0, 1)} \tan^{-1}\left[\frac{x^2 + 1}{x^2 + (y - 1)^2}\right]$$

**84. Continuity**  For the function

$$f(x, y) = xy\left(\frac{x^2 - y^2}{x^2 + y^2}\right)$$

define $f(0, 0)$ such that $f$ is continuous at the origin.

**85. Proof**  Prove that

$$\lim_{(x, y) \to (a, b)} [f(x, y) + g(x, y)] = L_1 + L_2$$

where $f(x, y)$ approaches $L_1$ and $g(x, y)$ approaches $L_2$ as $(x, y) \rightarrow (a, b)$.

**86. Proof**  Prove that if $f$ is continuous and $f(a, b) < 0$, then there exists a $\delta$-neighborhood about $(a, b)$ such that $f(x, y) < 0$ for every point $(x, y)$ in the neighborhood.

# 13.3 Partial Derivatives

■ Find and use partial derivatives of a function of two variables.
■ Find and use partial derivatives of a function of three or more variables.
■ Find higher-order partial derivatives of a function of two or three variables.

## Partial Derivatives of a Function of Two Variables

In applications of functions of several variables, the question often arises, "How will the value of a function be affected by a change in one of its independent variables?" You can answer this by considering the independent variables one at a time. For example, to determine the effect of a catalyst in an experiment, a chemist could conduct the experiment several times using varying amounts of the catalyst while keeping constant other variables such as temperature and pressure. You can use a similar procedure to determine the rate of change of a function $f$ with respect to one of its several independent variables. This process is called **partial differentiation,** and the result is referred to as the **partial derivative** of $f$ with respect to the chosen independent variable.

**JEAN LE ROND D'ALEMBERT
(1717–1783)**

The introduction of partial derivatives followed Newton's and Leibniz's work in calculus by several years. Between 1730 and 1760, Leonhard Euler and Jean Le Rond d'Alembert separately published several papers on dynamics, in which they established much of the theory of partial derivatives. These papers used functions of two or more variables to study problems involving equilibrium, fluid motion, and vibrating strings.
*See LarsonCalculus.com to read more of this biography.*

---

**Definition of Partial Derivatives of a Function of Two Variables**

If $z = f(x, y)$, then the **first partial derivatives** of $f$ with respect to $x$ and $y$ are the functions $f_x$ and $f_y$ defined by

$$f_x(x, y) = \lim_{\Delta x \to 0} \frac{f(x + \Delta x, y) - f(x, y)}{\Delta x} \qquad \text{Partial derivative with respect to } x$$

and

$$f_y(x, y) = \lim_{\Delta y \to 0} \frac{f(x, y + \Delta y) - f(x, y)}{\Delta y} \qquad \text{Partial derivative with respect to } y$$

provided the limits exist.

---

This definition indicates that if $z = f(x, y)$, then to find $f_x$, you *consider y constant* and differentiate with respect to $x$. Similarly, to find $f_y$, you *consider x constant* and differentiate with respect to $y$.

**EXAMPLE 1** **Finding Partial Derivatives**

**a.** To find $f_x$ for $f(x, y) = 3x - x^2y^2 + 2x^3y$, consider $y$ to be constant and differentiate with respect to $x$.

$$f_x(x, y) = 3 - 2xy^2 + 6x^2y \qquad \text{Partial derivative with respect to } x$$

To find $f_y$, consider $x$ to be constant and differentiate with respect to $y$.

$$f_y(x, y) = -2x^2y + 2x^3 \qquad \text{Partial derivative with respect to } y$$

**b.** To find $f_x$ for $f(x, y) = (\ln x)(\sin x^2y)$, consider $y$ to be constant and differentiate with respect to $x$.

$$f_x(x, y) = (\ln x)(\cos x^2y)(2xy) + \frac{\sin x^2y}{x} \qquad \text{Partial derivative with respect to } x$$

To find $f_y$, consider $x$ to be constant and differentiate with respect to $y$.

$$f_y(x, y) = (\ln x)(\cos x^2y)(x^2) \qquad \text{Partial derivative with respect to } y$$

**• • REMARK** The notation $\partial z/\partial x$ is read as "the partial derivative of $z$ with respect to $x$," and $\partial z/\partial y$ is read as "the partial derivative of $z$ with respect to $y$."

## Notation for First Partial Derivatives

For $z = f(x, y)$, the partial derivatives $f_x$ and $f_y$ are denoted by

$$\frac{\partial}{\partial x} f(x, y) = f_x(x, y) = z_x = \frac{\partial z}{\partial x} \qquad \text{Partial derivative with respect to } x$$

and

$$\frac{\partial}{\partial y} f(x, y) = f_y(x, y) = z_y = \frac{\partial z}{\partial y}. \qquad \text{Partial derivative with respect to } y$$

The first partials evaluated at the point $(a, b)$ are denoted by

$$\frac{\partial z}{\partial x}\bigg|_{(a, b)} = f_x(a, b)$$

and

$$\frac{\partial z}{\partial y}\bigg|_{(a, b)} = f_y(a, b).$$

**EXAMPLE 2** Finding and Evaluating Partial Derivatives

For $f(x, y) = xe^{x^2 y}$, find $f_x$ and $f_y$, and evaluate each at the point $(1, \ln 2)$.

**Solution** Because

$$f_x(x, y) = xe^{x^2 y}(2xy) + e^{x^2 y} \qquad \text{Partial derivative with respect to } x$$

the partial derivative of $f$ with respect to $x$ at $(1, \ln 2)$ is

$$f_x(1, \ln 2) = e^{\ln 2}(2 \ln 2) + e^{\ln 2}$$
$$= 4 \ln 2 + 2.$$

Because

$$f_y(x, y) = xe^{x^2 y}(x^2)$$
$$= x^3 e^{x^2 y} \qquad \text{Partial derivative with respect to } y$$

the partial derivative of $f$ with respect to $y$ at $(1, \ln 2)$ is

$$f_y(1, \ln 2) = e^{\ln 2}$$
$$= 2.$$

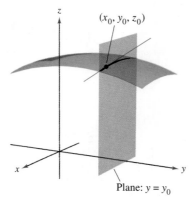

$\dfrac{\partial f}{\partial x}$ = slope in $x$-direction

**Figure 13.28**

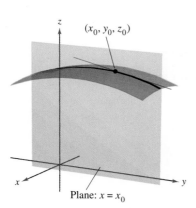

$\dfrac{\partial f}{\partial y}$ = slope in $y$-direction

**Figure 13.29**

The partial derivatives of a function of two variables, $z = f(x, y)$, have a useful geometric interpretation. If $y = y_0$, then $z = f(x, y_0)$ represents the curve formed by intersecting the surface $z = f(x, y)$ with the plane $y = y_0$, as shown in Figure 13.28. Therefore,

$$f_x(x_0, y_0) = \lim_{\Delta x \to 0} \frac{f(x_0 + \Delta x, y_0) - f(x_0, y_0)}{\Delta x}$$

represents the slope of this curve at the point $(x_0, y_0, f(x_0, y_0))$. Note that both the curve and the tangent line lie in the plane $y = y_0$. Similarly,

$$f_y(x_0, y_0) = \lim_{\Delta y \to 0} \frac{f(x_0, y_0 + \Delta y) - f(x_0, y_0)}{\Delta y}$$

represents the slope of the curve given by the intersection of $z = f(x, y)$ and the plane $x = x_0$ at $(x_0, y_0, f(x_0, y_0))$, as shown in Figure 13.29.

Informally, the values of $\partial f/\partial x$ and $\partial f/\partial y$ at the point $(x_0, y_0, z_0)$ denote the **slopes of the surface in the $x$- and $y$-directions,** respectively.

**EXAMPLE 3** **Finding the Slopes of a Surface**

⋯▷ *See LarsonCalculus.com for an interactive version of this type of example.*

Find the slopes in the $x$-direction and in the $y$-direction of the surface

$$f(x, y) = -\frac{x^2}{2} - y^2 + \frac{25}{8}$$

at the point $\left(\frac{1}{2}, 1, 2\right)$.

**Solution** The partial derivatives of $f$ with respect to $x$ and $y$ are

$$f_x(x, y) = -x \quad \text{and} \quad f_y(x, y) = -2y. \qquad \text{Partial derivatives}$$

So, in the $x$-direction, the slope is

$$f_x\left(\frac{1}{2}, 1\right) = -\frac{1}{2} \qquad \text{Figure 13.30}$$

and in the $y$-direction, the slope is

$$f_y\left(\frac{1}{2}, 1\right) = -2. \qquad \text{Figure 13.31}$$

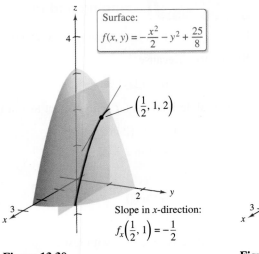

Surface:
$$f(x, y) = -\frac{x^2}{2} - y^2 + \frac{25}{8}$$

$\left(\frac{1}{2}, 1, 2\right)$

Slope in $x$-direction:
$$f_x\left(\frac{1}{2}, 1\right) = -\frac{1}{2}$$

**Figure 13.30**

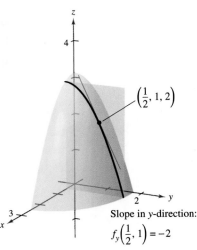

$\left(\frac{1}{2}, 1, 2\right)$

Slope in $y$-direction:
$$f_y\left(\frac{1}{2}, 1\right) = -2$$

**Figure 13.31**

**EXAMPLE 4** **Finding the Slopes of a Surface**

Find the slopes of the surface

$$f(x, y) = 1 - (x - 1)^2 - (y - 2)^2$$

at the point $(1, 2, 1)$ in the $x$-direction and in the $y$-direction.

**Solution** The partial derivatives of $f$ with respect to $x$ and $y$ are

$$f_x(x, y) = -2(x - 1) \quad \text{and} \quad f_y(x, y) = -2(y - 2). \qquad \text{Partial derivatives}$$

So, at the point $(1, 2, 1)$, the slope in the $x$-direction is

$$f_x(1, 2) = -2(1 - 1) = 0$$

and the slope in the $y$-direction is

$$f_y(1, 2) = -2(2 - 2) = 0$$

as shown in Figure 13.32.

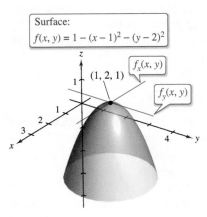

Surface:
$$f(x, y) = 1 - (x - 1)^2 - (y - 2)^2$$

$f_x(x, y)$

$(1, 2, 1)$

$f_y(x, y)$

**Figure 13.32**

No matter how many variables are involved, partial derivatives can be interpreted as *rates of change*.

---

**EXAMPLE 5**    Using Partial Derivatives to Find Rates of Change

The area of a parallelogram with adjacent sides $a$ and $b$ and included angle $\theta$ is given by $A = ab \sin \theta$, as shown in Figure 13.33.

**a.** Find the rate of change of $A$ with respect to $a$ for $a = 10$, $b = 20$, and $\theta = \pi/6$.

**b.** Find the rate of change of $A$ with respect to $\theta$ for $a = 10$, $b = 20$, and $\theta = \pi/6$.

**Solution**

**a.** To find the rate of change of the area with respect to $a$, hold $b$ and $\theta$ constant and differentiate with respect to $a$ to obtain

$$\frac{\partial A}{\partial a} = b \sin \theta. \qquad \text{Find partial derivative with respect to } a.$$

For $a = 10$, $b = 20$, and $\theta = \pi/6$, the rate of change of the area with respect to $a$ is

$$\frac{\partial A}{\partial a} = 20 \sin \frac{\pi}{6} = 10. \qquad \text{Substitute for } b \text{ and } \theta.$$

**b.** To find the rate of change of the area with respect to $\theta$, hold $a$ and $b$ constant and differentiate with respect to $\theta$ to obtain

$$\frac{\partial A}{\partial \theta} = ab \cos \theta. \qquad \text{Find partial derivative with respect to } \theta.$$

For $a = 10$, $b = 20$, and $\theta = \pi/6$, the rate of change of the area with respect to $\theta$ is

$$\frac{\partial A}{\partial \theta} = 200 \cos \frac{\pi}{6} = 100\sqrt{3}. \qquad \text{Substitute for } a, b, \text{ and } \theta. \qquad \blacksquare$$

## Partial Derivatives of a Function of Three or More Variables

The concept of a partial derivative can be extended naturally to functions of three or more variables. For instance, if $w = f(x, y, z)$, then there are three partial derivatives, each of which is formed by holding two of the variables constant. That is, to define the partial derivative of $w$ with respect to $x$, consider $y$ and $z$ to be constant and differentiate with respect to $x$. A similar process is used to find the derivatives of $w$ with respect to $y$ and with respect to $z$.

$$\frac{\partial w}{\partial x} = f_x(x, y, z) = \lim_{\Delta x \to 0} \frac{f(x + \Delta x, y, z) - f(x, y, z)}{\Delta x}$$

$$\frac{\partial w}{\partial y} = f_y(x, y, z) = \lim_{\Delta y \to 0} \frac{f(x, y + \Delta y, z) - f(x, y, z)}{\Delta y}$$

$$\frac{\partial w}{\partial z} = f_z(x, y, z) = \lim_{\Delta z \to 0} \frac{f(x, y, z + \Delta z) - f(x, y, z)}{\Delta z}$$

In general, if $w = f(x_1, x_2, \ldots, x_n)$, then there are $n$ partial derivatives denoted by

$$\frac{\partial w}{\partial x_k} = f_{x_k}(x_1, x_2, \ldots, x_n), \quad k = 1, 2, \ldots, n.$$

To find the partial derivative with respect to one of the variables, hold the other variables constant and differentiate with respect to the given variable.

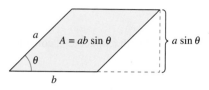

$a$    $A = ab \sin \theta$    $a \sin \theta$

$\theta$

$b$

The area of the parallelogram is $ab \sin \theta$.

**Figure 13.33**

**EXAMPLE 6** **Finding Partial Derivatives**

**a.** To find the partial derivative of $f(x, y, z) = xy + yz^2 + xz$ with respect to $z$, consider $x$ and $y$ to be constant and obtain

$$\frac{\partial}{\partial z}[xy + yz^2 + xz] = 2yz + x.$$

**b.** To find the partial derivative of $f(x, y, z) = z \sin(xy^2 + 2z)$ with respect to $z$, consider $x$ and $y$ to be constant. Then, using the Product Rule, you obtain

$$\frac{\partial}{\partial z}[z \sin(xy^2 + 2z)] = (z)\frac{\partial}{\partial z}[\sin(xy^2 + 2z)] + \sin(xy^2 + 2z)\frac{\partial}{\partial z}[z]$$

$$= (z)[\cos(xy^2 + 2z)](2) + \sin(xy^2 + 2z)$$

$$= 2z \cos(xy^2 + 2z) + \sin(xy^2 + 2z).$$

**c.** To find the partial derivative of

$$f(x, y, z, w) = \frac{x + y + z}{w}$$

with respect to $w$, consider $x$, $y$, and $z$ to be constant and obtain

$$\frac{\partial}{\partial w}\left[\frac{x + y + z}{w}\right] = -\frac{x + y + z}{w^2}.$$ ∎

## Higher-Order Partial Derivatives

As is true for ordinary derivatives, it is possible to take second, third, and higher-order partial derivatives of a function of several variables, provided such derivatives exist. Higher-order derivatives are denoted by the order in which the differentiation occurs. For instance, the function $z = f(x, y)$ has the following second partial derivatives.

**1.** Differentiate twice with respect to $x$:

$$\frac{\partial}{\partial x}\left(\frac{\partial f}{\partial x}\right) = \frac{\partial^2 f}{\partial x^2} = f_{xx}.$$

**2.** Differentiate twice with respect to $y$:

$$\frac{\partial}{\partial y}\left(\frac{\partial f}{\partial y}\right) = \frac{\partial^2 f}{\partial y^2} = f_{yy}.$$

**3.** Differentiate first with respect to $x$ and then with respect to $y$:

$$\frac{\partial}{\partial y}\left(\frac{\partial f}{\partial x}\right) = \frac{\partial^2 f}{\partial y \partial x} = f_{xy}.$$

**4.** Differentiate first with respect to $y$ and then with respect to $x$:

$$\frac{\partial}{\partial x}\left(\frac{\partial f}{\partial y}\right) = \frac{\partial^2 f}{\partial x \partial y} = f_{yx}.$$

**••REMARK** Note that the two types of notation for mixed partials have different conventions for indicating the order of differentiation.

$$\frac{\partial}{\partial y}\left(\frac{\partial f}{\partial x}\right) = \frac{\partial^2 f}{\partial y \partial x} \quad \text{Right-to-left order}$$

$$(f_x)_y = f_{xy} \quad \text{Left-to-right order}$$

You can remember the order by observing that in both notations you differentiate first with respect to the variable "nearest" $f$.

▷ The third and fourth cases are called **mixed partial derivatives.**

**EXAMPLE 7** **Finding Second Partial Derivatives**

Find the second partial derivatives of

$$f(x, y) = 3xy^2 - 2y + 5x^2y^2$$

and determine the value of $f_{xy}(-1, 2)$.

**Solution** Begin by finding the first partial derivatives with respect to $x$ and $y$.

$$f_x(x, y) = 3y^2 + 10xy^2 \quad \text{and} \quad f_y(x, y) = 6xy - 2 + 10x^2y$$

Then, differentiate each of these with respect to $x$ and $y$.

$$f_{xx}(x, y) = 10y^2 \qquad \text{and} \quad f_{yy}(x, y) = 6x + 10x^2$$
$$f_{xy}(x, y) = 6y + 20xy \quad \text{and} \quad f_{yx}(x, y) = 6y + 20xy$$

At $(-1, 2)$, the value of $f_{xy}$ is

$$f_{xy}(-1, 2) = 12 - 40 = -28. \qquad \blacksquare$$

Notice in Example 7 that the two mixed partials are equal. Sufficient conditions for this occurrence are given in Theorem 13.3.

---

**THEOREM 13.3** **Equality of Mixed Partial Derivatives**

If $f$ is a function of $x$ and $y$ such that $f_{xy}$ and $f_{yx}$ are continuous on an open disk $R$, then, for every $(x, y)$ in $R$,

$$f_{xy}(x, y) = f_{yx}(x, y).$$

---

Theorem 13.3 also applies to a function $f$ of *three or more variables* as long as all second partial derivatives are continuous. For example, if

$$w = f(x, y, z) \qquad \text{Function of three variables}$$

and all the second partial derivatives are continuous in an open region $R$, then at each point in $R$, the order of differentiation in the mixed second partial derivatives is irrelevant. If the third partial derivatives of $f$ are also continuous, then the order of differentiation of the mixed third partial derivatives is irrelevant.

**EXAMPLE 8** **Finding Higher-Order Partial Derivatives**

Show that $f_{xz} = f_{zx}$ and $f_{xzz} = f_{zxz} = f_{zzx}$ for the function

$$f(x, y, z) = ye^x + x \ln z.$$

**Solution**

First partials:

$$f_x(x, y, z) = ye^x + \ln z, \quad f_z(x, y, z) = \frac{x}{z}$$

Second partials (note that the first two are equal):

$$f_{xz}(x, y, z) = \frac{1}{z}, \quad f_{zx}(x, y, z) = \frac{1}{z}, \quad f_{zz}(x, y, z) = -\frac{x}{z^2}$$

Third partials (note that all three are equal):

$$f_{xzz}(x, y, z) = -\frac{1}{z^2}, \quad f_{zxz}(x, y, z) = -\frac{1}{z^2}, \quad f_{zzx}(x, y, z) = -\frac{1}{z^2} \qquad \blacksquare$$

## 13.3 Exercises

See **CalcChat.com** for tutorial help and worked-out solutions to odd-numbered exercises.

---

### CONCEPT CHECK

1. **First Partial Derivatives** List three ways of writing the first partial derivative with respect to $x$ of $z = f(x, y)$.

2. **First Partial Derivatives** Sketch a surface representing a function $f$ of two variables $x$ and $y$. Use the sketch to give geometric interpretations of $\partial f/\partial x$ and $\partial f/\partial y$.

3. **Higher-Order Partial Derivatives** Describe the order in which the differentiation of $f(x, y, z)$ occurs for (a) $f_{yxz}$ and (b) $\partial^2 f/\partial x \partial z$.

4. **Mixed Partial Derivatives** If $f$ is a function of $x$ and $y$ such that $f_{xy}$ and $f_{yx}$ are continuous, what is the relationship between the mixed partial derivatives?

---

**Examining a Partial Derivative** In Exercises 5–10, explain whether the Quotient Rule should be used to find the partial derivative. Do not differentiate.

5. $\dfrac{\partial}{\partial x}\left(\dfrac{x^2 y}{y^2 - 3}\right)$

6. $\dfrac{\partial}{\partial y}\left(\dfrac{x^2 y}{y^2 - 3}\right)$

7. $\dfrac{\partial}{\partial y}\left(\dfrac{x - y}{x^2 + 1}\right)$

8. $\dfrac{\partial}{\partial x}\left(\dfrac{x - y}{x^2 + 1}\right)$

9. $\dfrac{\partial}{\partial x}\left(\dfrac{xy}{x^2 + 1}\right)$

10. $\dfrac{\partial}{\partial y}\left(\dfrac{xy}{x^2 + 1}\right)$

 **Finding Partial Derivatives** In Exercises 11–40, find both first partial derivatives.

11. $f(x, y) = 2x - 5y + 3$

12. $f(x, y) = x^2 - 2y^2 + 4$

13. $z = 6x - x^2 y + 8y^2$

14. $f(x, y) = 4x^3 y^{-2}$

15. $z = x\sqrt{y}$

16. $z = 2y^2\sqrt{x}$

17. $z = e^{xy}$

18. $z = e^{x/y}$

19. $z = x^2 e^{2y}$

20. $z = 7ye^{y/x}$

21. $z = \ln\dfrac{x}{y}$

22. $z = \ln\sqrt{xy}$

23. $z = \ln(x^2 + y^2)$

24. $z = \ln\dfrac{x + y}{x - y}$

25. $z = \dfrac{x^2}{2y} + \dfrac{3y^2}{x}$

26. $z = \dfrac{xy}{x^2 + y^2}$

27. $h(x, y) = e^{-(x^2 + y^2)}$

28. $g(x, y) = \ln\sqrt{x^2 + y^2}$

29. $f(x, y) = \sqrt{x^2 + y^2}$

30. $f(x, y) = \sqrt{2x + y^3}$

31. $z = \cos xy$

32. $z = \sin(x + 2y)$

33. $z = \tan(2x - y)$

34. $z = \sin 5x \cos 5y$

35. $z = e^y \sin 8xy$

36. $z = \cos(x^2 + y^2)$

37. $z = \sinh(2x + 3y)$

38. $z = \cosh xy^2$

39. $f(x, y) = \displaystyle\int_x^y (t^2 - 1)\, dt$

40. $f(x, y) = \displaystyle\int_x^y (2t + 1)\, dt + \int_y^x (2t - 1)\, dt$

---

 **Finding Partial Derivatives** In Exercises 41–44, use the limit definition of partial derivatives to find $f_x(x, y)$ and $f_y(x, y)$.

41. $f(x, y) = 3x + 2y$

42. $f(x, y) = x^2 - 2xy + y^2$

43. $f(x, y) = \sqrt{x + y}$

44. $f(x, y) = \dfrac{1}{x + y}$

 **Finding and Evaluating Partial Derivatives** In Exercises 45–52, find $f_x$ and $f_y$, and evaluate each at the given point.

45. $f(x, y) = e^x y^2$, $(\ln 3, 2)$

46. $f(x, y) = x^3 \ln 5y$, $(1, 1)$

47. $f(x, y) = \cos(2x - y)$, $\left(\dfrac{\pi}{4}, \dfrac{\pi}{3}\right)$

48. $f(x, y) = \sin xy$, $\left(2, \dfrac{\pi}{4}\right)$

49. $f(x, y) = \arctan\dfrac{y}{x}$, $(2, -2)$

50. $f(x, y) = \arccos xy$, $(1, 1)$

51. $f(x, y) = \dfrac{xy}{x - y}$, $(2, -2)$

52. $f(x, y) = \dfrac{2xy}{\sqrt{4x^2 + 5y^2}}$, $(1, 1)$

 **Finding the Slopes of a Surface** In Exercises 53–56, find the slopes of the surface in the $x$- and $y$-directions at the given point.

53. $z = xy$

    $(1, 2, 2)$

54. $z = \sqrt{25 - x^2 - y^2}$

    $(3, 0, 4)$

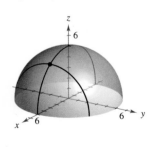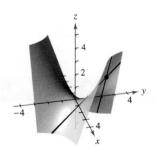

55. $g(x, y) = 4 - x^2 - y^2$

    $(1, 1, 2)$

56. $h(x, y) = x^2 - y^2$

    $(-2, 1, 3)$

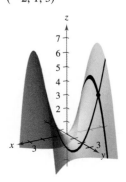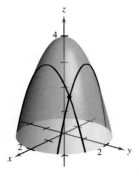

**Finding Partial Derivatives** In Exercises 57–62, find the first partial derivatives with respect to $x$, $y$, and $z$.

**57.** $H(x, y, z) = \sin(x + 2y + 3z)$

**58.** $f(x, y, z) = 3x^2y - 5xyz + 10yz^2$

**59.** $w = \sqrt{x^2 + y^2 + z^2}$

**60.** $w = \dfrac{7xz}{x + y}$

**61.** $F(x, y, z) = \ln\sqrt{x^2 + y^2 + z^2}$

**62.** $G(x, y, z) = \dfrac{1}{\sqrt{1 - x^2 - y^2 - z^2}}$

**Finding and Evaluating Partial Derivatives** In Exercises 63–68, find $f_x$, $f_y$, and $f_z$, and evaluate each at the given point.

**63.** $f(x, y, z) = x^3yz^2$, $(1, 1, 1)$

**64.** $f(x, y, z) = x^2y^3 + 2xyz - 3yz$, $(-2, 1, 2)$

**65.** $f(x, y, z) = \dfrac{\ln x}{yz}$, $(1, -1, -1)$

**66.** $f(x, y, z) = \dfrac{xy}{x + y + z}$, $(3, 1, -1)$

**67.** $f(x, y, z) = z \sin(x + 6y)$, $\left(0, \dfrac{\pi}{2}, -4\right)$

**68.** $f(x, y, z) = \sqrt{3x^2 + y^2 - 2z^2}$, $(1, -2, 1)$

**Using First Partial Derivatives** In Exercises 69–76, find all values of $x$ and $y$ such that $f_x(x, y) = 0$ and $f_y(x, y) = 0$ simultaneously.

**69.** $f(x, y) = x^2 + xy + y^2 - 2x + 2y$

**70.** $f(x, y) = x^2 - xy + y^2 - 5x + y$

**71.** $f(x, y) = x^2 + 4xy + y^2 - 4x + 16y + 3$

**72.** $f(x, y) = x^2 - xy + y^2$

**73.** $f(x, y) = \dfrac{1}{x} + \dfrac{1}{y} + xy$

**74.** $f(x, y) = 3x^3 - 12xy + y^3$

**75.** $f(x, y) = e^{x^2 + xy + y^2}$

**76.** $f(x, y) = \ln(x^2 + y^2 + 1)$

 **Finding Second Partial Derivatives** In Exercises 77–86, find the four second partial derivatives. Observe that the second mixed partials are equal.

**77.** $z = 3xy^2$

**78.** $z = x^2 + 3y^2$

**79.** $z = x^4 - 2xy + 3y^3$

**80.** $z = x^4 - 3x^2y^2 + y^4$

**81.** $z = \sqrt{x^2 + y^2}$

**82.** $z = \ln(x - y)$

**83.** $z = e^x \tan y$

**84.** $z = 2xe^y - 3ye^{-x}$

**85.** $z = \cos xy$

**86.** $z = \arctan \dfrac{y}{x}$

 **Finding Partial Derivatives Using Technology** In Exercises 87–90, use a computer algebra system to find the first and second partial derivatives of the function. Determine whether there exist values of $x$ and $y$ such that $f_x(x, y) = 0$ and $f_y(x, y) = 0$ simultaneously.

**87.** $f(x, y) = x \sec y$

**88.** $f(x, y) = \sqrt{25 - x^2 - y^2}$

**89.** $f(x, y) = \ln \dfrac{x}{x^2 + y^2}$

**90.** $f(x, y) = \dfrac{xy}{x - y}$

**Finding Higher-Order Partial Derivatives** In Exercises 91–94, show that the mixed partial derivatives $f_{xyy}$, $f_{yxy}$, and $f_{yyx}$ are equal.

**91.** $f(x, y, z) = xyz$

**92.** $f(x, y, z) = x^2 - 3xy + 4yz + z^3$

**93.** $f(x, y, z) = e^{-x} \sin yz$

**94.** $f(x, y, z) = \dfrac{2z}{x + y}$

**Laplace's Equation** In Exercises 95–98, show that the function satisfies Laplace's equation $\partial^2z/\partial x^2 + \partial^2z/\partial y^2 = 0$.

**95.** $z = 5xy$

**96.** $z = \tfrac{1}{2}(e^y - e^{-y}) \sin x$

**97.** $z = e^x \sin y$

**98.** $z = \arctan \dfrac{y}{x}$

**Wave Equation** In Exercises 99–102, show that the function satisfies the wave equation $\partial^2z/\partial t^2 = c^2(\partial^2z/\partial x^2)$.

**99.** $z = \sin(x - ct)$

**100.** $z = \cos(4x + 4ct)$

**101.** $z = \ln(x + ct)$

**102.** $z = \sin \omega ct \sin \omega x$

**Heat Equation** In Exercises 103 and 104, show that the function satisfies the heat equation $\partial z/\partial t = c^2(\partial^2z/\partial x^2)$.

**103.** $z = e^{-t} \cos \dfrac{x}{c}$

**104.** $z = e^{-t} \sin \dfrac{x}{c}$

**Cauchy-Riemann Equations** In Exercises 105 and 106, show that the functions $u$ and $v$ satisfy the Cauchy-Riemann equations

$$\dfrac{\partial u}{\partial x} = \dfrac{\partial v}{\partial y} \quad \text{and} \quad \dfrac{\partial u}{\partial y} = -\dfrac{\partial v}{\partial x}.$$

**105.** $u = x^2 - y^2$, $v = 2xy$

**106.** $u = e^x \cos y$, $v = e^x \sin y$

**Using First Partial Derivatives** In Exercises 107 and 108, determine whether there exists a function $f(x, y)$ with the given partial derivatives. Explain your reasoning. If such a function exists, give an example.

**107.** $f_x(x, y) = -3 \sin(3x - 2y)$, $f_y(x, y) = 2 \sin(3x - 2y)$

**108.** $f_x(x, y) = 2x + y$, $f_y(x, y) = x - 4y$

## EXPLORING CONCEPTS

**109. Think About It** Consider $z = f(x, y)$ such that $z_x = z_y$. Does $z = c(x + y)$? Explain.

**110. First Partial Derivatives** Given $z = f(x)g(y)$, find $z_x + z_y$.

**111. Sketching a Graph** Sketch the graph of a function $z = f(x, y)$ whose derivative $f_x$ is always negative and whose derivative $f_y$ is always positive.

**112. Sketching a Graph** Sketch the graph of a function $z = f(x, y)$ whose derivatives $f_x$ and $f_y$ are always positive.

**113. Think About It** The price $P$ (in dollars) of a used car is a function of its initial cost $C$ (in dollars) and its age $A$ (in years). What are the units of $\partial P/\partial A$? Is $\partial P/\partial A$ positive or negative? Explain.

---

**114.** **HOW DO YOU SEE IT?** Use the graph of the surface to determine the sign of each partial derivative. Explain your reasoning.

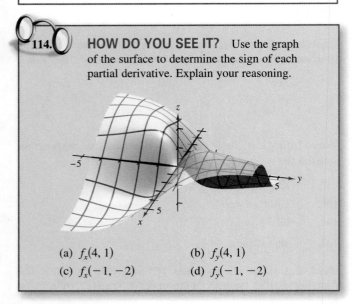

(a) $f_x(4, 1)$         (b) $f_y(4, 1)$
(c) $f_x(-1, -2)$     (d) $f_y(-1, -2)$

---

**115. Area** The area of a triangle is represented by $A = \frac{1}{2}ab \sin \theta$, where $a$ and $b$ are two of the side lengths and $\theta$ is the angle between $a$ and $b$.

(a) Find the rate of change of $A$ with respect to $b$ for $a = 4$, $b = 1$, and $\theta = \pi/4$.

(b) Find the rate of change of $A$ with respect to $\theta$ for $a = 2$, $b = 5$, and $\theta = \pi/3$.

**116. Volume** The volume of a right-circular cone of radius $r$ and height $h$ is represented by $V = \frac{1}{3}\pi r^2 h$.

(a) Find the rate of change of $V$ with respect to $r$ for $r = 2$ and $h = 2$.

(b) Find the rate of change of $V$ with respect to $h$ for $r = 2$ and $h = 2$.

**117. Marginal Revenue** A pharmaceutical corporation has two plants that produce the same over-the-counter medicine. If $x_1$ and $x_2$ are the numbers of units produced at plant 1 and plant 2, respectively, then the total revenue for the product is given by $R = 200x_1 + 200x_2 - 4x_1^2 - 8x_1x_2 - 4x_2^2$. When $x_1 = 4$ and $x_2 = 12$, find (a) the marginal revenue for plant 1, $\partial R/\partial x_1$, and (b) the marginal revenue for plant 2, $\partial R/\partial x_2$.

**118. Marginal Costs**

A company manufactures two types of wood-burning stoves: a freestanding model and a fireplace-insert model. The cost function for producing $x$ freestanding and $y$ fireplace-insert stoves is

$$C = 32\sqrt{xy} + 175x + 205y + 1050.$$

(a) Find the marginal costs ($\partial C/\partial x$ and $\partial C/\partial y$) when $x = 80$ and $y = 20$.

(b) When additional production is required, which model of stove results in the cost increasing at a higher rate? How can this be determined from the cost model?

**119. Psychology** Early in the twentieth century, an intelligence test called the *Stanford-Binet Test* (more commonly known as the IQ test) was developed. In this test, an individual's mental age $M$ is divided by the individual's chronological age $C$ and then the quotient is multiplied by 100. The result is the individual's *IQ*.

$$IQ(M, C) = \frac{M}{C} \times 100$$

Find the partial derivatives of *IQ* with respect to $M$ and with respect to $C$. Evaluate the partial derivatives at the point $(12, 10)$ and interpret the result. *(Source: Adapted from Bernstein/Clark-Stewart/Roy/Wickens, Psychology, Fourth Edition)*

**120. Marginal Productivity** Consider the Cobb-Douglas production function $f(x, y) = 200x^{0.7}y^{0.3}$. When $x = 1000$ and $y = 500$, find (a) the marginal productivity of labor, $\partial f/\partial x$, and (b) the marginal productivity of capital, $\partial f/\partial y$.

**121. Think About It** Let $N$ be the number of applicants to a university, $p$ the charge for food and housing at the university, and $t$ the tuition. Suppose that $N$ is a function of $p$ and $t$ such that $\partial N/\partial p < 0$ and $\partial N/\partial t < 0$. What information is gained by noticing that both partials are negative?

**122. Investment** The value of an investment of $1000 earning 6% compounded annually is

$$V(I, R) = 1000\left[\frac{1 + 0.06(1 - R)}{1 + I}\right]^{10}$$

where $I$ is the annual rate of inflation and $R$ is the tax rate for the person making the investment. Calculate $V_I(0.03, 0.28)$ and $V_R(0.03, 0.28)$. Determine whether the tax rate or the rate of inflation is the greater "negative" factor in the growth of the investment.

**123. Temperature Distribution** The temperature at any point $(x, y)$ on a steel plate is $T = 500 - 0.6x^2 - 1.5y^2$, where $x$ and $y$ are measured in meters. At the point $(2, 3)$, find the rates of change of the temperature with respect to the distances moved along the plate in the directions of the $x$- and $y$-axes.

**124. Apparent Temperature**  A measure of how hot weather feels to an average person is the Apparent Temperature Index. A model for this index is

$$A = 0.885t - 22.4h + 1.20th - 0.544$$

where $A$ is the apparent temperature in degrees Celsius, $t$ is the air temperature, and $h$ is the relative humidity in decimal form.  *(Source: The UMAP Journal)*

(a) Find $\dfrac{\partial A}{\partial t}$ and $\dfrac{\partial A}{\partial h}$ when $t = 30°$ and $h = 0.80$.

(b) Which has a greater effect on $A$, air temperature or humidity? Explain.

**125. Ideal Gas Law**  The Ideal Gas Law states that

$$PV = nRT$$

where $P$ is pressure, $V$ is volume, $n$ is the number of moles of gas, $R$ is a fixed constant (the gas constant), and $T$ is absolute temperature. Show that

$$\frac{\partial T}{\partial P} \cdot \frac{\partial P}{\partial V} \cdot \frac{\partial V}{\partial T} = -1.$$

**126. Marginal Utility**  The utility function $U = f(x, y)$ is a measure of the utility (or satisfaction) derived by a person from the consumption of two products $x$ and $y$. The utility function for two products is

$$U = -5x^2 + xy - 3y^2.$$

(a) Determine the marginal utility of product $x$.

(b) Determine the marginal utility of product $y$.

(c) When $x = 2$ and $y = 3$, should a person consume one more unit of product $x$ or one more unit of product $y$? Explain your reasoning.

(d) Use a computer algebra system to graph the function. Interpret the marginal utilities of products $x$ and $y$ graphically.

**127. Modeling Data**  Personal consumption expenditures (in billions of dollars) for several types of recreation from 2009 through 2014 are shown in the table, where $x$ is the expenditures on amusement parks and campgrounds, $y$ is the expenditures on live entertainment (excluding sports), and $z$ is the expenditures on spectator sports.  *(Source: U.S. Bureau of Economic Analysis)*

| Year | 2009 | 2010 | 2011 | 2012 | 2013 | 2014 |
|------|------|------|------|------|------|------|
| $x$ | 37.2 | 38.8 | 41.3 | 44.6 | 47.0 | 50.3 |
| $y$ | 25.2 | 26.3 | 28.3 | 28.5 | 28.0 | 30.0 |
| $z$ | 18.8 | 19.2 | 20.4 | 20.6 | 21.6 | 22.4 |

A model for the data is given by

$$z = 0.23x + 0.14y + 6.85.$$

(a) Find $\dfrac{\partial z}{\partial x}$ and $\dfrac{\partial z}{\partial y}$.

(b) Interpret the partial derivatives in the context of the problem.

**128. Modeling Data**  The table shows the national health expenditures (in billions of dollars) for the Department of Veterans Affairs $x$, workers' compensation $y$, and Medicaid $z$ from 2009 through 2014.  *(Source: Centers for Medicare and Medicaid Services)*

| Year | 2009 | 2010 | 2011 | 2012 | 2013 | 2014 |
|------|------|------|------|------|------|------|
| $x$ | 42.5 | 45.7 | 48.2 | 49.8 | 52.8 | 57.2 |
| $y$ | 36.0 | 36.1 | 39.1 | 41.7 | 44.1 | 47.3 |
| $z$ | 374.5 | 397.2 | 406.4 | 422.0 | 446.7 | 495.8 |

A model for the data is given by

$$z = -0.120x^2 + 0.657y^2 + 17.70x - 51.53y + 842.5.$$

(a) Find $\dfrac{\partial^2 z}{\partial x^2}$ and $\dfrac{\partial^2 z}{\partial y^2}$.

(b) Determine the concavity of traces parallel to the $xz$-plane. Interpret the result in the context of the problem.

(c) Determine the concavity of traces parallel to the $yz$-plane. Interpret the result in the context of the problem.

**129. Using a Function**  Consider the function defined by

$$f(x, y) = \begin{cases} \dfrac{xy(x^2 - y^2)}{x^2 + y^2}, & (x, y) \neq (0, 0) \\ 0, & (x, y) = (0, 0) \end{cases}.$$

(a) Find $f_x(x, y)$ and $f_y(x, y)$ for $(x, y) \neq (0, 0)$.

(b) Use the definition of partial derivatives to find $f_x(0, 0)$ and $f_y(0, 0)$.

$$\left[ Hint: f_x(0, 0) = \lim_{\Delta x \to 0} \frac{f(\Delta x, 0) - f(0, 0)}{\Delta x} \right]$$

(c) Use the definition of partial derivatives to find $f_{xy}(0, 0)$ and $f_{yx}(0, 0)$.

(d) Using Theorem 13.3 and the result of part (c), what can be said about $f_{xy}$ or $f_{yx}$?

**130. Using a Function**  Consider the function

$$f(x, y) = (x^3 + y^3)^{1/3}.$$

(a) Find $f_x(0, 0)$ and $f_y(0, 0)$.

(b) Determine the points (if any) at which $f_x(x, y)$ or $f_y(x, y)$ fails to exist.

**131. Using a Function**  Consider the function

$$f(x, y) = (x^2 + y^2)^{2/3}.$$

Show that

$$f_x(x, y) = \begin{cases} \dfrac{4x}{3(x^2 + y^2)^{1/3}}, & (x, y) \neq (0, 0) \\ 0, & (x, y) = (0, 0) \end{cases}.$$

**■ FOR FURTHER INFORMATION**  For more information about this problem, see the article "A Classroom Note on a Naturally Occurring Piecewise Defined Function" by Don Cohen in *Mathematics and Computer Education*.

# 13.4   Differentials

■ Understand the concepts of increments and differentials.
■ Extend the concept of differentiability to a function of two variables.
■ Use a differential as an approximation.

## Increments and Differentials

In this section, the concepts of increments and differentials are generalized to functions of two or more variables. Recall from Section 3.9 that for $y = f(x)$, the differential of $y$ was defined as

$$dy = f'(x)\, dx.$$

Similar terminology is used for a function of two variables, $z = f(x, y)$. That is, $\Delta x$ and $\Delta y$ are the **increments of $x$ and $y$,** and the **increment of $z$** is

$$\Delta z = f(x + \Delta x, y + \Delta y) - f(x, y). \qquad \text{Increment of } z$$

---

### Definition of Total Differential

If $z = f(x, y)$ and $\Delta x$ and $\Delta y$ are increments of $x$ and $y$, then the **differentials** of the independent variables $x$ and $y$ are

$$dx = \Delta x \quad \text{and} \quad dy = \Delta y$$

and the **total differential** of the dependent variable $z$ is

$$dz = \frac{\partial z}{\partial x}\, dx + \frac{\partial z}{\partial y}\, dy = f_x(x, y)\, dx + f_y(x, y)\, dy.$$

---

This definition can be extended to a function of three or more variables. For instance, if $w = f(x, y, z, u)$, then $dx = \Delta x$, $dy = \Delta y$, $dz = \Delta z$, $du = \Delta u$, and the total differential of $w$ is

$$dw = \frac{\partial w}{\partial x}\, dx + \frac{\partial w}{\partial y}\, dy + \frac{\partial w}{\partial z}\, dz + \frac{\partial w}{\partial u}\, du.$$

### EXAMPLE 1   Finding the Total Differential

Find the total differential for each function.

**a.** $z = 2x \sin y - 3x^2 y^2$     **b.** $w = x^2 + y^2 + z^2$

**Solution**

**a.** The total differential $dz$ for $z = 2x \sin y - 3x^2 y^2$ is

$$dz = \frac{\partial z}{\partial x}\, dx + \frac{\partial z}{\partial y}\, dy \qquad \text{Total differential } dz$$

$$= (2 \sin y - 6xy^2)\, dx + (2x \cos y - 6x^2 y)\, dy.$$

**b.** The total differential $dw$ for $w = x^2 + y^2 + z^2$ is

$$dw = \frac{\partial w}{\partial x}\, dx + \frac{\partial w}{\partial y}\, dy + \frac{\partial w}{\partial z}\, dz \qquad \text{Total differential } dw$$

$$= 2x\, dx + 2y\, dy + 2z\, dz.$$

## Differentiability

In Section 3.9, you learned that for a *differentiable* function given by $y = f(x)$, you can use the differential $dy = f'(x) \, dx$ as an approximation (for small $\Delta x$) of the value $\Delta y = f(x + \Delta x) - f(x)$. When a similar approximation is possible for a function of two variables, the function is said to be **differentiable.** This is stated explicitly in the next definition.

---

**Definition of Differentiability**

A function $f$ given by $z = f(x, y)$ is **differentiable** at $(x_0, y_0)$ if $\Delta z$ can be written in the form

$$\Delta z = f_x(x_0, y_0) \, \Delta x + f_y(x_0, y_0) \, \Delta y + \varepsilon_1 \Delta x + \varepsilon_2 \Delta y$$

where both $\varepsilon_1$ and $\varepsilon_2 \rightarrow 0$ as

$$(\Delta x, \Delta y) \rightarrow (0, 0).$$

The function $f$ is **differentiable in a region $R$** if it is differentiable at each point in $R$.

---

### EXAMPLE 2  Showing that a Function Is Differentiable

Show that the function

$$f(x, y) = x^2 + 3y$$

is differentiable at every point in the plane.

**Solution**  Letting $z = f(x, y)$, the increment of $z$ at an arbitrary point $(x, y)$ in the plane is

$$
\begin{aligned}
\Delta z &= f(x + \Delta x, y + \Delta y) - f(x, y) && \text{Increment of } z \\
&= (x + \Delta x)^2 + 3(y + \Delta y) - (x^2 + 3y) \\
&= x^2 + 2x\Delta x + (\Delta x)^2 + 3y + 3\Delta y - x^2 - 3y \\
&= 2x\Delta x + (\Delta x)^2 + 3\Delta y \\
&= 2x(\Delta x) + 3(\Delta y) + \Delta x(\Delta x) + 0(\Delta y) \\
&= f_x(x, y)\Delta x + f_y(x, y)\Delta y + \varepsilon_1 \Delta x + \varepsilon_2 \Delta y
\end{aligned}
$$

where $\varepsilon_1 = \Delta x$ and $\varepsilon_2 = 0$. Because $\varepsilon_1 \rightarrow 0$ and $\varepsilon_2 \rightarrow 0$ as $(\Delta x, \Delta y) \rightarrow (0, 0)$, it follows that $f$ is differentiable at every point in the plane. The graph of $f$ is shown in Figure 13.34. ∎

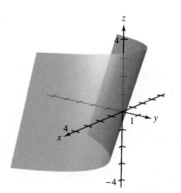

**Figure 13.34**

Be sure you see that the term "differentiable" is used differently for functions of two variables than for functions of one variable. A function of one variable is differentiable at a point when its derivative exists at the point. For a function of two variables, however, the existence of the partial derivatives $f_x$ and $f_y$ does not guarantee that the function is differentiable (see Example 5). The next theorem gives a *sufficient* condition for differentiability of a function of two variables.

---

### THEOREM 13.4  Sufficient Condition for Differentiability

If $f$ is a function of $x$ and $y$, where $f_x$ and $f_y$ are continuous in an open region $R$, then $f$ is differentiable on $R$. A proof of this theorem is given in Appendix A.

---

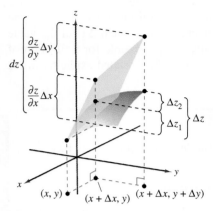

The exact change in $z$ is $\Delta z$. This change can be approximated by the differential $dz$.

**Figure 13.35**

## Approximation by Differentials

Theorem 13.4 tells you that you can choose $(x + \Delta x, y + \Delta y)$ close enough to $(x, y)$ to make $\varepsilon_1 \Delta x$ and $\varepsilon_2 \Delta y$ insignificant. In other words, for small $\Delta x$ and $\Delta y$, you can use the approximation

$$\Delta z \approx dz. \qquad \text{Approximate change in } z$$

This approximation is illustrated graphically in Figure 13.35. Recall that the partial derivatives $\partial z / \partial x$ and $\partial z / \partial y$ can be interpreted as the slopes of the surface in the $x$- and $y$-directions. This means that

$$dz = \frac{\partial z}{\partial x} \Delta x + \frac{\partial z}{\partial y} \Delta y$$

represents the change in height of a plane that is tangent to the surface at the point $(x, y, f(x, y))$. Because a plane in space is represented by a linear equation in the variables $x$, $y$, and $z$, the approximation of $\Delta z$ by $dz$ is called a **linear approximation.** You will learn more about this geometric interpretation in Section 13.7.

---

**EXAMPLE 3**   **Using a Differential as an Approximation**

⋮ ⋯▷ *See LarsonCalculus.com for an interactive version of this type of example.*

Use the differential $dz$ to approximate the change in

$$z = \sqrt{4 - x^2 - y^2}$$

as $(x, y)$ moves from the point $(1, 1)$ to the point $(1.01, 0.97)$. Compare this approximation with the exact change in $z$.

**Solution**   Letting $(x, y) = (1, 1)$ and $(x + \Delta x, y + \Delta y) = (1.01, 0.97)$ produces

$$dx = \Delta x = 0.01 \quad \text{and} \quad dy = \Delta y = -0.03.$$

So, the change in $z$ can be approximated by

$$\Delta z \approx dz = \frac{\partial z}{\partial x} dx + \frac{\partial z}{\partial y} dy = \frac{-x}{\sqrt{4 - x^2 - y^2}} \Delta x + \frac{-y}{\sqrt{4 - x^2 - y^2}} \Delta y.$$

When $x = 1$ and $y = 1$, you have

$$\Delta z \approx -\frac{1}{\sqrt{2}}(0.01) - \frac{1}{\sqrt{2}}(-0.03) = \frac{0.02}{\sqrt{2}} = \sqrt{2}(0.01) \approx 0.0141.$$

In Figure 13.36, you can see that the exact change corresponds to the difference in the heights of two points on the surface of a hemisphere. This difference is given by

$$\Delta z = f(1.01, 0.97) - f(1, 1)$$
$$= \sqrt{4 - (1.01)^2 - (0.97)^2} - \sqrt{4 - 1^2 - 1^2}$$
$$\approx 0.0137.$$

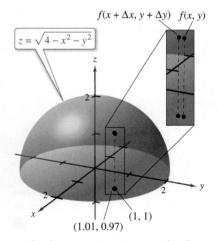

As $(x, y)$ moves from the point $(1, 1)$ to the point $(1.01, 0.97)$, the value of $f(x, y)$ changes by about 0.0137.

**Figure 13.36**

---

A function of three variables $w = f(x, y, z)$ is **differentiable** at $(x, y, z)$ provided that $\Delta w = f(x + \Delta x, y + \Delta y, z + \Delta z) - f(x, y, z)$ can be written in the form

$$\Delta w = f_x \Delta x + f_y \Delta y + f_z \Delta z + \varepsilon_1 \Delta x + \varepsilon_2 \Delta y + \varepsilon_3 \Delta z$$

where $\varepsilon_1$, $\varepsilon_2$, and $\varepsilon_3 \to 0$ as $(\Delta x, \Delta y, \Delta z) \to (0, 0, 0)$. With this definition of differentiability, Theorem 13.4 has the following extension for functions of three variables: If $f$ is a function of $x$, $y$, and $z$, where $f_x$, $f_y$, and $f_z$ are continuous in an open region $R$, then $f$ is differentiable on $R$.

In Section 3.9, you used differentials to approximate the propagated error introduced by an error in measurement. This application of differentials is further illustrated in Example 4.

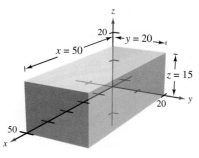

Volume = $xyz$
**Figure 13.37**

EXAMPLE 4    **Error Analysis**

The possible error involved in measuring each dimension of a rectangular box is $\pm0.1$ millimeter. The dimensions of the box are $x = 50$ centimeters, $y = 20$ centimeters, and $z = 15$ centimeters, as shown in Figure 13.37. Use $dV$ to estimate the propagated error and the relative error in the calculated volume of the box.

**Solution**    The volume of the box is $V = xyz$, so

$$dV = \frac{\partial V}{\partial x}\,dx + \frac{\partial V}{\partial y}\,dy + \frac{\partial V}{\partial z}\,dz$$

$$= yz\,dx + xz\,dy + xy\,dz.$$

Using $0.1$ millimeter $= 0.01$ centimeter, you have

$$dx = dy = dz = \pm0.01$$

and the propagated error is approximately

$$dV = (20)(15)(\pm0.01) + (50)(15)(\pm0.01) + (50)(20)(\pm0.01)$$

$$= 300(\pm0.01) + 750(\pm0.01) + 1000(\pm0.01)$$

$$= 2050(\pm0.01)$$

$$= \pm20.5 \text{ cubic centimeters.}$$

Because the measured volume is

$$V = (50)(20)(15) = 15{,}000 \text{ cubic centimeters}$$

the relative error, $\Delta V/V$, is approximately

$$\frac{\Delta V}{V} \approx \frac{dV}{V} = \frac{\pm20.5}{15{,}000} \approx \pm0.0014$$

which is a percent error of about $0.14\%$.

As is true for a function of a single variable, when a function in two or more variables is differentiable at a point, it is also continuous there.

---

**THEOREM 13.5   Differentiability Implies Continuity**

If a function of $x$ and $y$ is differentiable at $(x_0, y_0)$, then it is continuous at $(x_0, y_0)$.

---

**Proof**    Let $f$ be differentiable at $(x_0, y_0)$, where $z = f(x, y)$. Then

$$\Delta z = [f_x(x_0, y_0) + \varepsilon_1]\,\Delta x + [f_y(x_0, y_0) + \varepsilon_2]\,\Delta y$$

where both $\varepsilon_1$ and $\varepsilon_2 \to 0$ as $(\Delta x, \Delta y) \to (0, 0)$. However, by definition, you know that $\Delta z$ is

$$\Delta z = f(x_0 + \Delta x, y_0 + \Delta y) - f(x_0, y_0).$$

Letting $x = x_0 + \Delta x$ and $y = y_0 + \Delta y$ produces

$$f(x, y) - f(x_0, y_0) = [f_x(x_0, y_0) + \varepsilon_1]\,\Delta x + [f_y(x_0, y_0) + \varepsilon_2]\,\Delta y$$

$$= [f_x(x_0, y_0) + \varepsilon_1](x - x_0) + [f_y(x_0, y_0) + \varepsilon_2](y - y_0).$$

Taking the limit as $(x, y) \to (x_0, y_0)$, you have

$$\lim_{(x, y) \to (x_0, y_0)} f(x, y) = f(x_0, y_0)$$

which means that $f$ is continuous at $(x_0, y_0)$.

Remember that the existence of $f_x$ and $f_y$ is not sufficient to guarantee differentiability, as illustrated in the next example.

### EXAMPLE 5    A Function That Is Not Differentiable

For the function

$$f(x, y) = \begin{cases} \dfrac{-3xy}{x^2 + y^2}, & (x, y) \neq (0, 0) \\ 0, & (x, y) = (0, 0) \end{cases}$$

show that $f_x(0, 0)$ and $f_y(0, 0)$ both exist but that $f$ is not differentiable at $(0, 0)$.

**Solution**    You can show that $f$ is not differentiable at $(0, 0)$ by showing that it is not continuous at this point. To see that $f$ is not continuous at $(0, 0)$, look at the values of $f(x, y)$ along two different approaches to $(0, 0)$, as shown in Figure 13.38. Along the line $y = x$, the limit is

$$\lim_{(x, x) \to (0, 0)} f(x, y) = \lim_{(x, x) \to (0, 0)} \frac{-3x^2}{2x^2} = -\frac{3}{2}$$

whereas along $y = -x$, you have

$$\lim_{(x, -x) \to (0, 0)} f(x, y) = \lim_{(x, -x) \to (0, 0)} \frac{3x^2}{2x^2} = \frac{3}{2}.$$

So, the limit of $f(x, y)$ as $(x, y) \to (0, 0)$ does not exist, and you can conclude that $f$ is not continuous at $(0, 0)$. Therefore, by Theorem 13.5, you know that $f$ is not differentiable at $(0, 0)$. On the other hand, by the definition of the partial derivatives $f_x$ and $f_y$, you have

$$f_x(0, 0) = \lim_{\Delta x \to 0} \frac{f(\Delta x, 0) - f(0, 0)}{\Delta x} = \lim_{\Delta x \to 0} \frac{0 - 0}{\Delta x} = 0$$

and

$$f_y(0, 0) = \lim_{\Delta y \to 0} \frac{f(0, \Delta y) - f(0, 0)}{\Delta y} = \lim_{\Delta y \to 0} \frac{0 - 0}{\Delta y} = 0.$$

So, the partial derivatives at $(0, 0)$ exist.

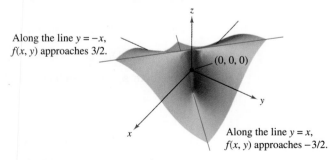

$$f(x, y) = \begin{cases} \dfrac{-3xy}{x^2 + y^2}, & (x, y) \neq (0, 0) \\ 0, & (x, y) = (0, 0) \end{cases}$$

Along the line $y = -x$, $f(x, y)$ approaches $3/2$.

$(0, 0, 0)$

Along the line $y = x$, $f(x, y)$ approaches $-3/2$.

**Figure 13.38**

▷ **TECHNOLOGY**    A graphing utility can be used to graph piecewise-defined functions like the one given in Example 5. For instance, the graph shown at the left was generated by *Mathematica*.

*Generated by Mathematica*

# 13.4 Exercises

See **CalcChat.com** for tutorial help and worked-out solutions to odd-numbered exercises.

**CONCEPT CHECK**

**1. Approximation** Describe the change in accuracy of $dz$ as an approximation of $\Delta z$ as $\Delta x$ and $\Delta y$ increase.

**2. Linear Approximation** What is meant by a linear approximation of $z = f(x, y)$ at the point $P(x_0, y_0)$?

 **Finding a Total Differential** In Exercises 3–8, find the total differential.

3. $z = 5x^3y^2$

4. $z = 2x^3y - 8xy^4$

5. $z = \frac{1}{2}\left(e^{x^2+y^2} - e^{-x^2-y^2}\right)$

6. $z = e^{-x}\tan y$

7. $w = x^2yz^2 + \sin yz$

8. $w = (x + y)/(z - 3y)$

 **Using a Differential as an Approximation** In Exercises 9–14, (a) find $f(2, 1)$ and $f(2.1, 1.05)$ and calculate $\Delta z$, and (b) use the total differential $dz$ to approximate $\Delta z$.

9. $f(x, y) = 2x - 3y$

10. $f(x, y) = x^2 + y^2$

11. $f(x, y) = 16 - x^2 - y^2$

12. $f(x, y) = y/x$

13. $f(x, y) = ye^x$

14. $f(x, y) = x\cos y$

**Approximating an Expression** In Exercises 15–18, find $z = f(x, y)$ and use the total differential to approximate the quantity.

15. $(2.01)^2(9.02) - 2^2 \cdot 9$

16. $\dfrac{1 - (3.05)^2}{(5.95)^2} - \dfrac{1 - 3^2}{6^2}$

17. $\sin[(1.05)^2 + (0.95)^2] - \sin(1^2 + 1^2)$

18. $\sqrt{(4.03)^2 + (3.1)^2} - \sqrt{4^2 + 3^2}$

**EXPLORING CONCEPTS**

**19. Continuity** If $f_x$ and $f_y$ are each continuous in an open region $R$, is $f(x, y)$ continuous in $R$? Explain.

**20. HOW DO YOU SEE IT?** Which point has a greater differential, $(2, 2)$ or $\left(\frac{1}{2}, \frac{1}{2}\right)$? Explain. (Assume that $dx$ and $dy$ are the same for both points.)

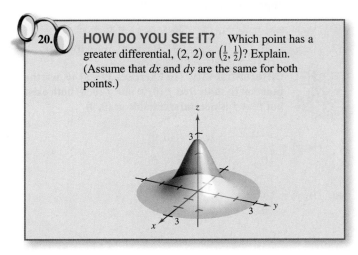

**21. Area** The area of the shaded rectangle in the figure is $A = lh$. The possible errors in the length and height are $\Delta l$ and $\Delta h$, respectively. Find $dA$ and identify the regions in the figure whose areas are given by the terms of $dA$. What region represents the difference between $\Delta A$ and $dA$?

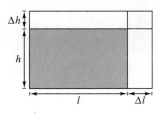

Figure for 21

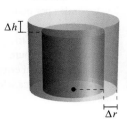

Figure for 22

**22. Volume** The volume of the red right circular cylinder in the figure is $V = \pi r^2 h$. The possible errors in the radius and the height are $\Delta r$ and $\Delta h$, respectively. Find $dV$ and identify the solids in the figure whose volumes are given by the terms of $dV$. What solid represents the difference between $\Delta V$ and $dV$?

**23. Volume** The possible error involved in measuring each dimension of a rectangular box is $\pm 0.02$ inch. The dimensions of the box are 8 inches by 5 inches by 12 inches. Approximate the propagated error and the relative error in the calculated volume of the box.

**24. Volume** The possible error involved in measuring each dimension of a right circular cylinder is $\pm 0.05$ centimeter. The radius is 3 centimeters and the height is 10 centimeters. Approximate the propagated error and the relative error in the calculated volume of the cylinder.

**25. Numerical Analysis** A right circular cone of height $h = 8$ meters and radius $r = 4$ meters is constructed, and in the process, errors $\Delta r$ and $\Delta h$ are made in the radius and height, respectively. Let $V$ be the volume of the cone. Complete the table to show the relationship between $\Delta V$ and $dV$ for the indicated errors.

| $\Delta r$ | $\Delta h$ | $dV$ or $dS$ | $\Delta V$ or $\Delta S$ | $\Delta V - dV$ or $\Delta S - dS$ |
|---|---|---|---|---|
| 0.1 | 0.1 | | | |
| 0.1 | −0.1 | | | |
| 0.001 | 0.002 | | | |
| −0.0001 | 0.0002 | | | |

Table for Exercises 25 and 26

**26. Numerical Analysis** A right circular cone of height $h = 16$ meters and radius $r = 6$ meters is constructed, and in the process, errors of $\Delta r$ and $\Delta h$ are made in the radius and height, respectively. Let $S$ be the lateral surface area of the cone. Complete the table above to show the relationship between $\Delta S$ and $dS$ for the indicated errors.

**27. Wind Chill**

The formula for wind chill $C$ (in degrees Fahrenheit) is given by

$$C = 35.74 + 0.6215T - 35.75v^{0.16} + 0.4275Tv^{0.16}$$

where $v$ is the wind speed in miles per hour and $T$ is the temperature in degrees Fahrenheit. The wind speed is $23 \pm 3$ miles per hour and the temperature is $8° \pm 1°$. Use $dC$ to estimate the maximum possible propagated error and relative error in calculating the wind chill. *(Source: National Oceanic and Atmospheric Administration)*

**28. Resistance** The total resistance $R$ (in ohms) of two resistors connected in parallel is given by

$$\frac{1}{R} = \frac{1}{R_1} + \frac{1}{R_2}.$$

Approximate the change in $R$ as $R_1$ is increased from 10 ohms to 10.5 ohms and $R_2$ is decreased from 15 ohms to 13 ohms.

**29. Power** Electrical power $P$ is given by

$$P = \frac{E^2}{R}$$

where $E$ is voltage and $R$ is resistance. Approximate the maximum percent error in calculating power when 120 volts is applied to a 2000-ohm resistor and the possible percent errors in measuring $E$ and $R$ are 3% and 4%, respectively.

**30. Acceleration** The centripetal acceleration of a particle moving in a circle is $a = v^2/r$, where $v$ is the velocity and $r$ is the radius of the circle. Approximate the maximum percent error in measuring the acceleration due to errors of 3% in $v$ and 2% in $r$.

**31. Volume** A trough is 16 feet long (see figure). Its cross sections are isosceles triangles with each of the two equal sides measuring 18 inches. The angle between the two equal sides is $\theta$.

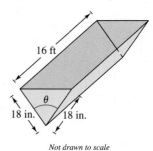

*Not drawn to scale*

(a) Write the volume of the trough as a function of $\theta$ and determine the value of $\theta$ such that the volume is a maximum.

(b) The maximum error in the linear measurements is one-half inch and the maximum error in the angle measure is 2°. Approximate the change in the maximum volume.

**32. Sports** A baseball player in center field is playing approximately 330 feet from a television camera that is behind home plate. A batter hits a fly ball that goes to the wall 420 feet from the camera (see figure).

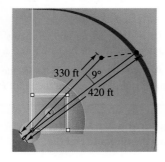

(a) The camera turns 9° to follow the play. Approximate the number of feet that the center fielder has to run to make the catch.

(b) The position of the center fielder could be in error by as much as 6 feet and the maximum error in measuring the rotation of the camera is 1°. Approximate the maximum possible error in the result of part (a).

**33. Inductance** The inductance $L$ (in microhenrys) of a straight nonmagnetic wire in free space is

$$L = 0.00021\left(\ln \frac{2h}{r} - 0.75\right)$$

where $h$ is the length of the wire in millimeters and $r$ is the radius of a circular cross section. Approximate $L$ when $r = 2 \pm \frac{1}{16}$ millimeters and $h = 100 \pm \frac{1}{100}$ millimeters.

**34. Pendulum** The period $T$ of a pendulum of length $L$ is $T = (2\pi\sqrt{L})/\sqrt{g}$, where $g$ is the acceleration due to gravity. A pendulum is moved from the Canal Zone, where $g = 32.09$ feet per second per second, to Greenland, where $g = 32.23$ feet per second per second. Because of the change in temperature, the length of the pendulum changes from 2.5 feet to 2.48 feet. Approximate the change in the period of the pendulum.

 **Differentiability** In Exercises 35–38, show that the function is differentiable by finding values of $\varepsilon_1$ and $\varepsilon_2$ as designated in the definition of differentiability, and verify that both $\varepsilon_1$ and $\varepsilon_2$ approach 0 as $(\Delta x, \Delta y) \to (0, 0)$.

**35.** $f(x, y) = x^2 - 2x + y$  **36.** $f(x, y) = x^2 + y^2$

**37.** $f(x, y) = x^2y$  **38.** $f(x, y) = 5x - 10y + y^3$

 **Differentiability** In Exercises 39 and 40, use the function to show that $f_x(0, 0)$ and $f_y(0, 0)$ both exist but that $f$ is not differentiable at $(0, 0)$.

**39.** $f(x, y) = \begin{cases} \dfrac{3x^2y}{x^4 + y^2}, & (x, y) \neq (0, 0) \\ 0, & (x, y) = (0, 0) \end{cases}$

**40.** $f(x, y) = \begin{cases} \dfrac{5x^2y}{x^3 + y^3}, & (x, y) \neq (0, 0) \\ 0, & (x, y) = (0, 0) \end{cases}$

# 13.5    Chain Rules for Functions of Several Variables

■ Use the Chain Rules for functions of several variables.
■ Find partial derivatives implicitly.

## Chain Rules for Functions of Several Variables

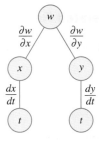

Chain Rule: one independent variable
$w$ is a function of $x$ and $y$, which are
each functions of $t$. This diagram
represents the derivative of $w$ with
respect to $t$.
**Figure 13.39**

Your work with differentials in the preceding section provides the basis for the extension of the Chain Rule to functions of two variables. There are two cases. The first case involves $w$ as a function of $x$ and $y$, where $x$ and $y$ are functions of a single independent variable $t$, as shown in Theorem 13.6.

---

**THEOREM 13.6    Chain Rule: One Independent Variable**

Let $w = f(x, y)$, where $f$ is a differentiable function of $x$ and $y$. If $x = g(t)$ and $y = h(t)$, where $g$ and $h$ are differentiable functions of $t$, then $w$ is a differentiable function of $t$, and

$$\frac{dw}{dt} = \frac{\partial w}{\partial x}\frac{dx}{dt} + \frac{\partial w}{\partial y}\frac{dy}{dt}.$$

The Chain Rule is shown schematically in Figure 13.39. A proof of this theorem is given in Appendix A.

---

**EXAMPLE 1**    Chain Rule: One Independent Variable

Let $w = x^2y - y^2$, where $x = \sin t$ and $y = e^t$. Find $dw/dt$ when $t = 0$.

**Solution**    By the Chain Rule for one independent variable, you have

$$\frac{dw}{dt} = \frac{\partial w}{\partial x}\frac{dx}{dt} + \frac{\partial w}{\partial y}\frac{dy}{dt}$$
$$= 2xy(\cos t) + (x^2 - 2y)e^t$$
$$= 2(\sin t)(e^t)(\cos t) + (\sin^2 t - 2e^t)e^t$$
$$= 2e^t \sin t \cos t + e^t \sin^2 t - 2e^{2t}.$$

When $t = 0$, it follows that

$$\frac{dw}{dt} = -2.$$    ■

The Chain Rules presented in this section provide alternative techniques for solving many problems in single-variable calculus. For instance, in Example 1, you could have used single-variable techniques to find $dw/dt$ by first writing $w$ as a function of $t$,

$$w = x^2y - y^2$$
$$= (\sin t)^2(e^t) - (e^t)^2$$
$$= e^t \sin^2 t - e^{2t}$$

and then differentiating as usual.

$$\frac{dw}{dt} = 2e^t \sin t \cos t + e^t \sin^2 t - 2e^{2t}$$

The Chain Rule in Theorem 13.6 can be extended to any number of variables. For example, if each $x_i$ is a differentiable function of a single variable $t$, then for

$$w = f(x_1, x_2, \ldots, x_n)$$

you have

$$\frac{dw}{dt} = \frac{\partial w}{\partial x_1}\frac{dx_1}{dt} + \frac{\partial w}{\partial x_2}\frac{dx_2}{dt} + \cdots + \frac{\partial w}{\partial x_n}\frac{dx_n}{dt}.$$

**EXAMPLE 2**   **An Application of a Chain Rule to Related Rates**

Two objects are traveling in elliptical paths given by the following parametric equations.

$$x_1 = 4\cos t \quad \text{and} \quad y_1 = 2\sin t \qquad \text{First object}$$
$$x_2 = 2\sin 2t \quad \text{and} \quad y_2 = 3\cos 2t \qquad \text{Second object}$$

At what rate is the distance between the two objects changing when $t = \pi$?

**Solution**   From Figure 13.40, you can see that the distance $s$ between the two objects is given by

$$s = \sqrt{(x_2 - x_1)^2 + (y_2 - y_1)^2}$$

and that when $t = \pi$, you have $x_1 = -4$, $y_1 = 0$, $x_2 = 0$, $y_2 = 3$, and

$$s = \sqrt{(0 + 4)^2 + (3 + 0)^2} = 5.$$

When $t = \pi$, the partial derivatives of $s$ are as follows.

$$\frac{\partial s}{\partial x_1} = \frac{-(x_2 - x_1)}{\sqrt{(x_2 - x_1)^2 + (y_2 - y_1)^2}} = -\frac{1}{5}(0 + 4) = -\frac{4}{5}$$

$$\frac{\partial s}{\partial y_1} = \frac{-(y_2 - y_1)}{\sqrt{(x_2 - x_1)^2 + (y_2 - y_1)^2}} = -\frac{1}{5}(3 - 0) = -\frac{3}{5}$$

$$\frac{\partial s}{\partial x_2} = \frac{(x_2 - x_1)}{\sqrt{(x_2 - x_1)^2 + (y_2 - y_1)^2}} = \frac{1}{5}(0 + 4) = \frac{4}{5}$$

$$\frac{\partial s}{\partial y_2} = \frac{(y_2 - y_1)}{\sqrt{(x_2 - x_1)^2 + (y_2 - y_1)^2}} = \frac{1}{5}(3 - 0) = \frac{3}{5}$$

When $t = \pi$, the derivatives of $x_1$, $y_1$, $x_2$, and $y_2$ are

$$\frac{dx_1}{dt} = -4\sin t = 0$$

$$\frac{dy_1}{dt} = 2\cos t = -2$$

$$\frac{dx_2}{dt} = 4\cos 2t = 4$$

$$\frac{dy_2}{dt} = -6\sin 2t = 0.$$

So, using the appropriate Chain Rule, you know that the distance is changing at a rate of

$$\frac{ds}{dt} = \frac{\partial s}{\partial x_1}\frac{dx_1}{dt} + \frac{\partial s}{\partial y_1}\frac{dy_1}{dt} + \frac{\partial s}{\partial x_2}\frac{dx_2}{dt} + \frac{\partial s}{\partial y_2}\frac{dy_2}{dt}$$

$$= \left(-\frac{4}{5}\right)(0) + \left(-\frac{3}{5}\right)(-2) + \left(\frac{4}{5}\right)(4) + \left(\frac{3}{5}\right)(0)$$

$$= \frac{22}{5}.$$

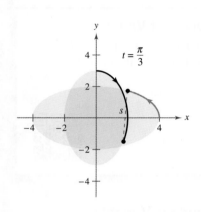

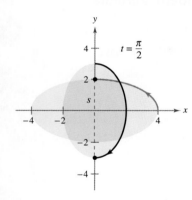

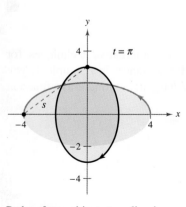

Paths of two objects traveling in elliptical orbits
**Figure 13.40**

In Example 2, note that $s$ is the function of four *intermediate* variables, $x_1$, $y_1$, $x_2$, and $y_2$, each of which is a function of a single variable $t$. Another type of composite function is one in which the intermediate variables are themselves functions of more than one variable. For instance, for $w = f(x, y)$, where $x = g(s, t)$ and $y = h(s, t)$, it follows that $w$ is a function of $s$ and $t$, and you can consider the partial derivatives of $w$ with respect to $s$ and $t$. One way to find these partial derivatives is to write $w$ as a function of $s$ and $t$ explicitly by substituting the equations $x = g(s, t)$ and $y = h(s, t)$ into the equation $w = f(x, y)$. Then you can find the partial derivatives in the usual way, as demonstrated in the next example.

**EXAMPLE 3** **Finding Partial Derivatives by Substitution**

Find $\partial w/\partial s$ and $\partial w/\partial t$ for $w = 2xy$, where $x = s^2 + t^2$ and $y = s/t$.

**Solution** Begin by substituting $x = s^2 + t^2$ and $y = s/t$ into the equation $w = 2xy$ to obtain

$$w = 2xy = 2(s^2 + t^2)\left(\frac{s}{t}\right) = 2\left(\frac{s^3}{t} + st\right).$$

Then, to find $\partial w/\partial s$, hold $t$ constant and differentiate with respect to $s$.

$$\frac{\partial w}{\partial s} = 2\left(\frac{3s^2}{t} + t\right)$$

$$= \frac{6s^2 + 2t^2}{t}$$

Similarly, to find $\partial w/\partial t$, hold $s$ constant and differentiate with respect to $t$ to obtain

$$\frac{\partial w}{\partial t} = 2\left(-\frac{s^3}{t^2} + s\right)$$

$$= 2\left(\frac{-s^3 + st^2}{t^2}\right)$$

$$= \frac{2st^2 - 2s^3}{t^2}.$$ ∎

Theorem 13.7 gives an alternative method for finding the partial derivatives in Example 3 without explicitly writing $w$ as a function of $s$ and $t$.

Chain Rule: two independent variables
**Figure 13.41**

**THEOREM 13.7** **Chain Rule: Two Independent Variables**

Let $w = f(x, y)$, where $f$ is a differentiable function of $x$ and $y$. If $x = g(s, t)$ and $y = h(s, t)$ such that the first partials $\partial x/\partial s$, $\partial x/\partial t$, $\partial y/\partial s$, and $\partial y/\partial t$ all exist, then $\partial w/\partial s$ and $\partial w/\partial t$ exist and are given by

$$\frac{\partial w}{\partial s} = \frac{\partial w}{\partial x}\frac{\partial x}{\partial s} + \frac{\partial w}{\partial y}\frac{\partial y}{\partial s}$$

and

$$\frac{\partial w}{\partial t} = \frac{\partial w}{\partial x}\frac{\partial x}{\partial t} + \frac{\partial w}{\partial y}\frac{\partial y}{\partial t}.$$

The Chain Rule is shown schematically in Figure 13.41.

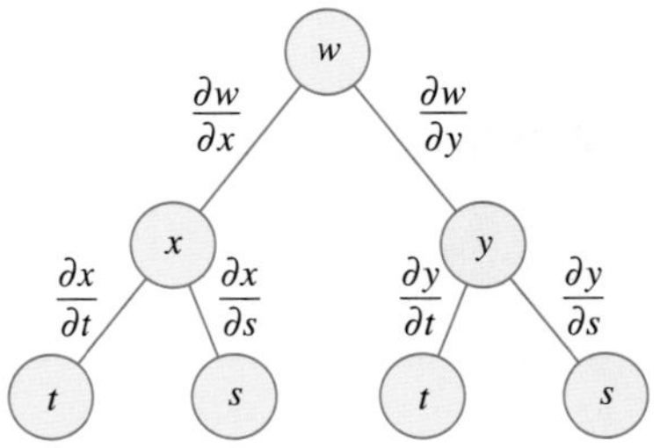

**Proof** To obtain $\partial w/\partial s$, hold $t$ constant and apply Theorem 13.6 to obtain the desired result. Similarly, for $\partial w/\partial t$, hold $s$ constant and apply Theorem 13.6. ∎

**EXAMPLE 4** **The Chain Rule with Two Independent Variables**

⋅⋅⋅▷ *See LarsonCalculus.com for an interactive version of this type of example.*

Use the Chain Rule to find $\partial w/\partial s$ and $\partial w/\partial t$ for

$$w = 2xy$$

where $x = s^2 + t^2$ and $y = s/t$.

**Solution** Note that these same partials were found in Example 3. This time, using Theorem 13.7, you can hold $t$ constant and differentiate with respect to $s$ to obtain

$$\frac{\partial w}{\partial s} = \frac{\partial w}{\partial x}\frac{\partial x}{\partial s} + \frac{\partial w}{\partial y}\frac{\partial y}{\partial s}$$

$$= 2y(2s) + 2x\left(\frac{1}{t}\right)$$

$$= 2\left(\frac{s}{t}\right)(2s) + 2(s^2 + t^2)\left(\frac{1}{t}\right) \qquad \text{Substitute } \frac{s}{t} \text{ for } y \text{ and } s^2 + t^2 \text{ for } x.$$

$$= \frac{4s^2}{t} + \frac{2s^2 + 2t^2}{t}$$

$$= \frac{6s^2 + 2t^2}{t}.$$

Similarly, holding $s$ constant gives

$$\frac{\partial w}{\partial t} = \frac{\partial w}{\partial x}\frac{\partial x}{\partial t} + \frac{\partial w}{\partial y}\frac{\partial y}{\partial t}$$

$$= 2y(2t) + 2x\left(\frac{-s}{t^2}\right)$$

$$= 2\left(\frac{s}{t}\right)(2t) + 2(s^2 + t^2)\left(\frac{-s}{t^2}\right) \qquad \text{Substitute } \frac{s}{t} \text{ for } y \text{ and } s^2 + t^2 \text{ for } x.$$

$$= 4s - \frac{2s^3 + 2st^2}{t^2}$$

$$= \frac{4st^2 - 2s^3 - 2st^2}{t^2}$$

$$= \frac{2st^2 - 2s^3}{t^2}.$$

The Chain Rule in Theorem 13.7 can also be extended to any number of variables. For example, if $w$ is a differentiable function of the $n$ variables

$$x_1, x_2, \ldots, x_n$$

where each $x_i$ is a differentiable function of the $m$ variables $t_1, t_2, \ldots, t_m$, then for

$$w = f(x_1, x_2, \ldots, x_n)$$

you obtain the following.

$$\frac{\partial w}{\partial t_1} = \frac{\partial w}{\partial x_1}\frac{\partial x_1}{\partial t_1} + \frac{\partial w}{\partial x_2}\frac{\partial x_2}{\partial t_1} + \cdots + \frac{\partial w}{\partial x_n}\frac{\partial x_n}{\partial t_1}$$

$$\frac{\partial w}{\partial t_2} = \frac{\partial w}{\partial x_1}\frac{\partial x_1}{\partial t_2} + \frac{\partial w}{\partial x_2}\frac{\partial x_2}{\partial t_2} + \cdots + \frac{\partial w}{\partial x_n}\frac{\partial x_n}{\partial t_2}$$

$$\vdots$$

$$\frac{\partial w}{\partial t_m} = \frac{\partial w}{\partial x_1}\frac{\partial x_1}{\partial t_m} + \frac{\partial w}{\partial x_2}\frac{\partial x_2}{\partial t_m} + \cdots + \frac{\partial w}{\partial x_n}\frac{\partial x_n}{\partial t_m}$$

---

**EXAMPLE 5**  **The Chain Rule for a Function of Three Variables**

Find $\partial w/\partial s$ and $\partial w/\partial t$ when $s = 1$ and $t = 2\pi$ for

$$w = xy + yz + xz$$

where $x = s \cos t$, $y = s \sin t$, and $z = t$.

**Solution**  By extending the result of Theorem 13.7, you have

$$\frac{\partial w}{\partial s} = \frac{\partial w}{\partial x}\frac{\partial x}{\partial s} + \frac{\partial w}{\partial y}\frac{\partial y}{\partial s} + \frac{\partial w}{\partial z}\frac{\partial z}{\partial s}$$

$$= (y + z)(\cos t) + (x + z)(\sin t) + (y + x)(0)$$

$$= (y + z)(\cos t) + (x + z)(\sin t).$$

When $s = 1$ and $t = 2\pi$, you have $x = 1$, $y = 0$, and $z = 2\pi$. So,

$$\frac{\partial w}{\partial s} = (0 + 2\pi)(1) + (1 + 2\pi)(0) = 2\pi.$$

Furthermore,

$$\frac{\partial w}{\partial t} = \frac{\partial w}{\partial x}\frac{\partial x}{\partial t} + \frac{\partial w}{\partial y}\frac{\partial y}{\partial t} + \frac{\partial w}{\partial z}\frac{\partial z}{\partial t}$$

$$= (y + z)(-s \sin t) + (x + z)(s \cos t) + (y + x)(1)$$

and for $s = 1$ and $t = 2\pi$, it follows that

$$\frac{\partial w}{\partial t} = (0 + 2\pi)(0) + (1 + 2\pi)(1) + (0 + 1)(1)$$

$$= 2 + 2\pi.$$

## Implicit Partial Differentiation

This section concludes with an application of the Chain Rule to determine the derivative of a function defined *implicitly*. Let $x$ and $y$ be related by the equation $F(x, y) = 0$, where $y = f(x)$ is a differentiable function of $x$. To find $dy/dx$, you could use the techniques discussed in Section 2.5. You will see, however, that the Chain Rule provides a convenient alternative. Consider the function

$$w = F(x, y) = F(x, f(x)).$$

You can apply Theorem 13.6 to obtain

$$\frac{dw}{dx} = F_x(x, y)\frac{dx}{dx} + F_y(x, y)\frac{dy}{dx}.$$

Because $w = F(x, y) = 0$ for all $x$ in the domain of $f$, you know that

$$\frac{dw}{dx} = 0$$

and you have

$$F_x(x, y)\frac{dx}{dx} + F_y(x, y)\frac{dy}{dx} = 0.$$

Now, if $F_y(x, y) \neq 0$, you can use the fact that $dx/dx = 1$ to conclude that

$$\frac{dy}{dx} = -\frac{F_x(x, y)}{F_y(x, y)}.$$

A similar procedure can be used to find the partial derivatives of functions of several variables that are defined implicitly.

---

**THEOREM 13.8 Chain Rule: Implicit Differentiation**

If the equation $F(x, y) = 0$ defines $y$ implicitly as a differentiable function of $x$, then

$$\frac{dy}{dx} = -\frac{F_x(x, y)}{F_y(x, y)}, \quad F_y(x, y) \neq 0.$$

If the equation $F(x, y, z) = 0$ defines $z$ implicitly as a differentiable function of $x$ and $y$, then

$$\frac{\partial z}{\partial x} = -\frac{F_x(x, y, z)}{F_z(x, y, z)} \quad \text{and} \quad \frac{\partial z}{\partial y} = -\frac{F_y(x, y, z)}{F_z(x, y, z)}, \quad F_z(x, y, z) \neq 0.$$

---

This theorem can be extended to differentiable functions defined implicitly with any number of variables.

**EXAMPLE 6** **Finding a Derivative Implicitly**

Find $dy/dx$ for

$$y^3 + y^2 - 5y - x^2 + 4 = 0.$$

**Solution** Begin by letting

$$F(x, y) = y^3 + y^2 - 5y - x^2 + 4.$$

Then

$$F_x(x, y) = -2x \quad \text{and} \quad F_y(x, y) = 3y^2 + 2y - 5.$$

Using Theorem 13.8, you have

$$\frac{dy}{dx} = -\frac{F_x(x, y)}{F_y(x, y)} = \frac{-(-2x)}{3y^2 + 2y - 5} = \frac{2x}{3y^2 + 2y - 5}.$$

•• **REMARK** Compare the solution to Example 6 with the solution to Example 2 in Section 2.5.

**EXAMPLE 7** **Finding Partial Derivatives Implicitly**

Find $\partial z/\partial x$ and $\partial z/\partial y$ for

$$3x^2z - x^2y^2 + 2z^3 + 3yz - 5 = 0.$$

**Solution** Begin by letting

$$F(x, y, z) = 3x^2z - x^2y^2 + 2z^3 + 3yz - 5.$$

Then

$$F_x(x, y, z) = 6xz - 2xy^2$$
$$F_y(x, y, z) = -2x^2y + 3z$$

and

$$F_z(x, y, z) = 3x^2 + 6z^2 + 3y.$$

Using Theorem 13.8, you have

$$\frac{\partial z}{\partial x} = -\frac{F_x(x, y, z)}{F_z(x, y, z)} = \frac{2xy^2 - 6xz}{3x^2 + 6z^2 + 3y}$$

and

$$\frac{\partial z}{\partial y} = -\frac{F_y(x, y, z)}{F_z(x, y, z)} = \frac{2x^2y - 3z}{3x^2 + 6z^2 + 3y}.$$

# 13.5 Exercises

## CONCEPT CHECK

1. **Chain Rule** Consider $w = f(x, y)$, where $x = g(s, t)$ and $y = h(s, t)$. Describe two ways of finding the partial derivatives $\partial w/\partial s$ and $\partial w/\partial t$.

2. **Implicit Differentiation** Why is using the Chain Rule to determine the derivative of the equation $F(x, y) = 0$ implicitly easier than using the method you learned in Section 2.5?

 **Using the Chain Rule** In Exercises 3–6, find $dw/dt$ using the appropriate Chain Rule. Evaluate $dw/dt$ at the given value of $t$.

| Function | Value |
|---|---|
| 3. $w = x^2 + 5y$ | $t = 2$ |
| $x = 2t, \ y = t$ | |
| 4. $w = \sqrt{x^2 + y^2}$ | $t = 0$ |
| $x = \cos t, \ y = e^t$ | |
| 5. $w = x \sin y$ | $t = 0$ |
| $x = e^t, \ y = \pi - t$ | |
| 6. $w = \ln \dfrac{y}{x}$ | $t = \dfrac{\pi}{4}$ |
| $x = \cos t, \ y = \sin t$ | |

 **Using Different Methods** In Exercises 7–12, find $dw/dt$ (a) by using the appropriate Chain Rule and (b) by converting $w$ to a function of $t$ before differentiating.

7. $w = x - \dfrac{1}{y}, \quad x = e^{2t}, \ y = t^3$

8. $w = \cos(x - y), \quad x = t^2, \ y = 1$

9. $w = x^2 + y^2 + z^2, \quad x = \cos t, \ y = \sin t, \ z = e^t$

10. $w = xy \cos z, \quad x = t, \ y = t^2, \ z = \arccos t$

11. $w = xy + xz + yz, \quad x = t - 1, \ y = t^2 - 1, \ z = t$

12. $w = xy^2 + x^2z + yz^2, \quad x = t^2, \ y = 2t, \ z = 2$

**Projectile Motion** In Exercises 13 and 14, the parametric equations for the paths of two objects are given. At what rate is the distance between the two objects changing at the given value of $t$?

13. $x_1 = 10 \cos 2t, \quad y_1 = 6 \sin 2t$  First object
 $x_2 = 7 \cos t, \quad y_2 = 4 \sin t$  Second object
 $t = \pi/2$

14. $x_1 = 48\sqrt{2}t, \quad y_1 = 48\sqrt{2}t - 16t^2$  First object
 $x_2 = 48\sqrt{3}t, \quad y_2 = 48t - 16t^2$  Second object
 $t = 1$

 **Finding Partial Derivatives** In Exercises 15–18, find $\partial w/\partial s$ and $\partial w/\partial t$ using the appropriate Chain Rule. Evaluate each partial derivative at the given values of $s$ and $t$.

| Function | Values |
|---|---|
| 15. $w = x^2 + y^2$ | $s = 1, \ t = 3$ |
| $x = s + t, \ y = s - t$ | |
| 16. $w = y^3 - 3x^2y$ | $s = -1, \ t = 2$ |
| $x = e^s, \ y = e^t$ | |
| 17. $w = \sin(2x + 3y)$ | $s = 0, \ t = \dfrac{\pi}{2}$ |
| $x = s + t, \ y = s - t$ | |
| 18. $w = x^2 - y^2$ | $s = 3, \ t = \dfrac{\pi}{4}$ |
| $x = s \cos t, \ y = s \sin t$ | |

 **Using Different Methods** In Exercises 19–22, find $\partial w/\partial s$ and $\partial w/\partial t$ (a) by using the appropriate Chain Rule and (b) by converting $w$ to a function of $s$ and $t$ before differentiating.

19. $w = xyz, \quad x = s + t, \ y = s - t, \ z = st^2$

20. $w = x^2 + y^2 + z^2, \quad x = t \sin s, \ y = t \cos s, \ z = st^2$

21. $w = ze^{xy}, \quad x = s - t, \ y = s + t, \ z = st$

22. $w = x \cos yz, \quad x = s^2, \ y = t^2, \ z = s - 2t$

 **Finding a Derivative Implicitly** In Exercises 23–26, differentiate implicitly to find $dy/dx$.

23. $x^2 - xy + y^2 - x + y = 0$

24. $\sec xy + \tan xy + 5 = 0$

25. $\ln\sqrt{x^2 + y^2} + x + y = 4$

26. $\dfrac{x}{x^2 + y^2} - y^2 = 6$

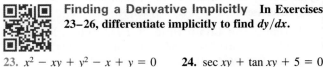 **Finding Partial Derivatives Implicitly** In Exercises 27–34, differentiate implicitly to find the first partial derivatives of $z$.

27. $x^2 + y^2 + z^2 = 1$   28. $xz + yz + xy = 0$

29. $x^2 + 2yz + z^2 = 1$   30. $x + \sin(y + z) = 0$

31. $\tan(x + y) + \cos z = 2$   32. $z = e^x \sin(y + z)$

33. $e^{xz} + xy = 0$

34. $x \ln y + y^2z + z^2 = 8$

**Finding Partial Derivatives Implicitly** In Exercises 35–38, differentiate implicitly to find the first partial derivatives of $w$.

35. $7xy + yz^2 - 4wz + w^2z + w^2x - 6 = 0$

36. $x^2 + y^2 + z^2 - 5yw + 10w^2 = 2$

37. $\cos xy + \sin yz + wz = 20$

38. $w - \sqrt{x - y} - \sqrt{y - z} = 0$

**Homogeneous Functions** A function $f$ is *homogeneous of degree n* when $f(tx, ty) = t^n f(x, y)$. In Exercises 39–42, **(a) show that the function is homogeneous and determine $n$, and (b) show that $xf_x(x, y) + yf_y(x, y) = nf(x, y)$.**

**39.** $f(x, y) = 2x^2 - 5xy$     **40.** $f(x, y) = x^3 - 3xy^2 + y^3$

**41.** $f(x, y) = e^{x/y}$     **42.** $f(x, y) = x \cos \dfrac{x + y}{y}$

**43. Using a Table of Values** Let $w = f(x, y)$, $x = g(t)$, and $y = h(t)$, where $f$, $g$, and $h$ are differentiable. Use the appropriate Chain Rule and the table of values to find $dw/dt$ when $t = 2$.

| $g(2)$ | $h(2)$ | $g'(2)$ | $h'(2)$ | $f_x(4, 3)$ | $f_y(4, 3)$ |
|--------|--------|---------|---------|-------------|-------------|
| 4 | 3 | $-1$ | 6 | $-5$ | 7 |

**44. Using a Table of Values** Let $w = f(x, y)$, $x = g(s, t)$, and $y = h(s, t)$, where $f$, $g$, and $h$ are differentiable. Use the appropriate Chain Rule and the table of values to find $w_s(1, 2)$.

| $g(1, 2)$ | $h(1, 2)$ | $g_s(1, 2)$ | $h_s(1, 2)$ | $f_x(4, 3)$ | $f_y(4, 3)$ |
|-----------|-----------|-------------|-------------|-------------|-------------|
| 4 | 3 | $-3$ | 5 | $-5$ | 7 |

**EXPLORING CONCEPTS**

**45. Using the Chain Rule** Show that $\dfrac{\partial w}{\partial u} + \dfrac{\partial w}{\partial v} = 0$ for $w = f(x, y)$, $x = u - v$, and $y = v - u$.

**46. Using the Chain Rule** Demonstrate the result of Exercise 45 for $w = (x - y) \sin(y - x)$.

**47. Using the Chain Rule** Let $F(u, v)$ be a function of two variables. Find a formula for $f'(x)$ when (a) $f(x) = F(4x, 4)$ and (b) $f(x) = F(-2x, x^2)$.

**48. HOW DO YOU SEE IT?** The path of an object represented by $w = f(x, y)$ is shown, where $x$ and $y$ are functions of $t$. The point on the graph represents the position of the object.

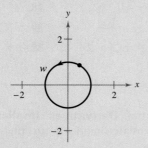

Determine whether each of the following is positive, negative, or zero.

(a) $\dfrac{dx}{dt}$     (b) $\dfrac{dy}{dt}$

**49. Volume and Surface Area** The radius of a right circular cylinder is increasing at a rate of 6 inches per minute, and the height is decreasing at a rate of 4 inches per minute. What are the rates of change of the volume and surface area when the radius is 12 inches and the height is 36 inches?

**50. Ideal Gas Law** The Ideal Gas Law is

$$PV = mRT$$

where $P$ is the pressure, $V$ is the volume, $m$ is the constant mass, $R$ is a constant, $T$ is the temperature, and $P$ and $V$ are functions of time. Find $dT/dt$, the rate at which the temperature changes with respect to time.

**51. Moment of Inertia** An annular cylinder has an inside radius of $r_1$ and an outside radius of $r_2$ (see figure). Its moment of inertia is

$$I = \tfrac{1}{2}m(r_1^2 + r_2^2)$$

where $m$ is the mass. The two radii are increasing at a rate of 2 centimeters per second. Find the rate at which $I$ is changing at the instant the radii are 6 centimeters and 8 centimeters. (Assume mass is a constant.)

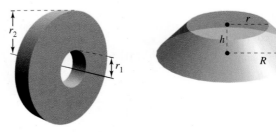

Figure for 51          Figure for 52

**52. Volume and Surface Area** The two radii of the frustum of a right circular cone are increasing at a rate of 4 centimeters per minute, and the height is increasing at a rate of 12 centimeters per minute (see figure). Find the rates at which the volume and surface area are changing when the two radii are 15 centimeters and 25 centimeters and the height is 10 centimeters.

**53. Cauchy-Riemann Equations** Given the functions $u(x, y)$ and $v(x, y)$, verify that the **Cauchy-Riemann equations**

$$\frac{\partial u}{\partial x} = \frac{\partial v}{\partial y} \quad \text{and} \quad \frac{\partial u}{\partial y} = -\frac{\partial v}{\partial x}$$

can be written in polar coordinate form as

$$\frac{\partial u}{\partial r} = \frac{1}{r} \cdot \frac{\partial v}{\partial \theta} \quad \text{and} \quad \frac{\partial v}{\partial r} = -\frac{1}{r} \cdot \frac{\partial u}{\partial \theta}.$$

**54. Cauchy-Riemann Equations** Demonstrate the result of Exercise 53 for the functions

$$u = \ln \sqrt{x^2 + y^2} \quad \text{and} \quad v = \arctan \frac{y}{x}.$$

**55. Homogeneous Function** Show that if $f(x, y)$ is homogeneous of degree $n$, then

$$xf_x(x, y) + yf_y(x, y) = nf(x, y).$$

[*Hint:* Let $g(t) = f(tx, ty) = t^n f(x, y)$. Find $g'(t)$ and then let $t = 1$.]

# 13.6 Directional Derivatives and Gradients

- Find and use directional derivatives of a function of two variables.
- Find the gradient of a function of two variables.
- Use the gradient of a function of two variables in applications.
- Find directional derivatives and gradients of functions of three variables.

## Directional Derivative

You are standing on the hillside represented by $z = f(x, y)$ in Figure 13.42 and want to determine the hill's incline toward the $z$-axis. You already know how to determine the slopes in two different directions—the slope in the $y$-direction is given by the partial derivative $f_y(x, y)$, and the slope in the $x$-direction is given by the partial derivative $f_x(x, y)$. In this section, you will see that these two partial derivatives can be used to find the slope in *any* direction.

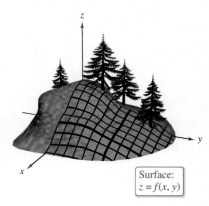

Surface:
$z = f(x, y)$

**Figure 13.42**

To determine the slope at a point on a surface, you will define a new type of derivative called a **directional derivative**. Begin by letting $z = f(x, y)$ be a *surface* and $P(x_0, y_0)$ be a *point* in the domain of $f$, as shown in Figure 13.43. The "direction" of the directional derivative is given by a unit vector

$$\mathbf{u} = \cos\theta\mathbf{i} + \sin\theta\mathbf{j}$$

where $\theta$ is the angle the vector makes with the positive $x$-axis. To find the desired slope, reduce the problem to two dimensions by intersecting the surface with a vertical plane passing through the point $P$ and parallel to $\mathbf{u}$, as shown in Figure 13.44. This vertical plane intersects the surface to form a curve $C$. The slope of the surface at $(x_0, y_0, f(x_0, y_0))$ in the direction of $\mathbf{u}$ is defined as the slope of the curve $C$ at that point.

Informally, you can write the slope of the curve $C$ as a limit that looks much like those used in single-variable calculus. The vertical plane used to form $C$ intersects the $xy$-plane in a line $L$, represented by the parametric equations

$$x = x_0 + t\cos\theta$$

and

$$y = y_0 + t\sin\theta$$

so that for any value of $t$, the point $Q(x, y)$ lies on the line $L$. For each of the points $P$ and $Q$, there is a corresponding point on the surface.

$(x_0, y_0, f(x_0, y_0))$     Point above $P$
$(x, y, f(x, y))$     Point above $Q$

Moreover, because the distance between $P$ and $Q$ is

$$\sqrt{(x - x_0)^2 + (y - y_0)^2} = \sqrt{(t\cos\theta)^2 + (t\sin\theta)^2}$$
$$= |t|$$

you can write the slope of the secant line through $(x_0, y_0, f(x_0, y_0))$ and $(x, y, f(x, y))$ as

$$\frac{f(x, y) - f(x_0, y_0)}{t} = \frac{f(x_0 + t\cos\theta, y_0 + t\sin\theta) - f(x_0, y_0)}{t}.$$

Finally, by letting $t$ approach 0, you arrive at the definition on the next page.

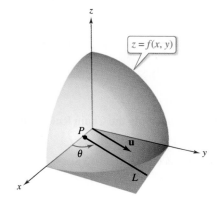

$z = f(x, y)$

**Figure 13.43**

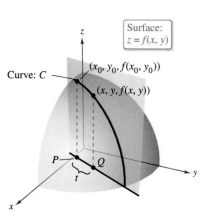

Surface:
$z = f(x, y)$

Curve: $C$

$(x_0, y_0, f(x_0, y_0))$

$(x, y, f(x, y))$

**Figure 13.44**

•••••••••••••••▷

**•• REMARK** Be sure you understand that the directional derivative represents the *rate of change of a function* in the direction of the unit vector $\mathbf{u} = \cos\theta\mathbf{i} + \sin\theta\mathbf{j}$. Geometrically, you can interpret the directional derivative as giving the *slope of a surface* in the direction of $\mathbf{u}$ at a point on the surface. (See Figure 13.46.)

---

**Definition of Directional Derivative**

Let $f$ be a function of two variables $x$ and $y$ and let $\mathbf{u} = \cos\theta\mathbf{i} + \sin\theta\mathbf{j}$ be a unit vector. Then the **directional derivative of $f$ in the direction of u,** denoted by $D_{\mathbf{u}}f$, is

$$D_{\mathbf{u}}f(x, y) = \lim_{t \to 0} \frac{f(x + t\cos\theta, y + t\sin\theta) - f(x, y)}{t}$$

provided this limit exists.

---

Calculating directional derivatives by this definition is similar to finding the derivative of a function of one variable by the limit process (see Section 2.1). A simpler formula for finding directional derivatives involves the partial derivatives $f_x$ and $f_y$.

---

**THEOREM 13.9 Directional Derivative**

If $f$ is a differentiable function of $x$ and $y$, then the directional derivative of $f$ in the direction of the unit vector $\mathbf{u} = \cos\theta\mathbf{i} + \sin\theta\mathbf{j}$ is

$$D_{\mathbf{u}}f(x, y) = f_x(x, y)\cos\theta + f_y(x, y)\sin\theta.$$

---

**Proof** For a fixed point $(x_0, y_0)$, let

$$x = x_0 + t\cos\theta \quad \text{and} \quad y = y_0 + t\sin\theta.$$

Then, let $g(t) = f(x, y)$. Because $f$ is differentiable, you can apply the Chain Rule given in Theorem 13.6 to obtain

$$g'(t) = f_x(x, y)x'(t) + f_y(x, y)y'(t) \qquad \text{Apply Chain Rule (Theorem 13.6).}$$
$$= f_x(x, y)\cos\theta + f_y(x, y)\sin\theta.$$

If $t = 0$, then $x = x_0$ and $y = y_0$, so

$$g'(0) = f_x(x_0, y_0)\cos\theta + f_y(x_0, y_0)\sin\theta.$$

By the definition of $g'(t)$, it is also true that

$$g'(0) = \lim_{t \to 0} \frac{g(t) - g(0)}{t}$$
$$= \lim_{t \to 0} \frac{f(x_0 + t\cos\theta, y_0 + t\sin\theta) - f(x_0, y_0)}{t}.$$

Consequently, $D_{\mathbf{u}}f(x_0, y_0) = f_x(x_0, y_0)\cos\theta + f_y(x_0, y_0)\sin\theta.$ ∎

There are infinitely many directional derivatives of a surface at a given point—one for each direction specified by $\mathbf{u}$, as shown in Figure 13.45. Two of these are the partial derivatives $f_x$ and $f_y$.

**1.** Direction of positive $x$-axis $(\theta = 0)$: $\mathbf{u} = \cos 0\mathbf{i} + \sin 0\mathbf{j} = \mathbf{i}$

$$D_{\mathbf{i}}f(x, y) = f_x(x, y)\cos 0 + f_y(x, y)\sin 0 = f_x(x, y)$$

**2.** Direction of positive $y$-axis $\left(\theta = \dfrac{\pi}{2}\right)$: $\mathbf{u} = \cos\dfrac{\pi}{2}\mathbf{i} + \sin\dfrac{\pi}{2}\mathbf{j} = \mathbf{j}$

$$D_{\mathbf{j}}f(x, y) = f_x(x, y)\cos\frac{\pi}{2} + f_y(x, y)\sin\frac{\pi}{2} = f_y(x, y)$$

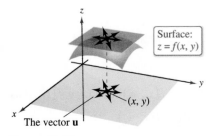

Surface:
$z = f(x, y)$

$(x, y)$

The vector $\mathbf{u}$

**Figure 13.45**

EXAMPLE 1    **Finding a Directional Derivative**

Find the directional derivative of

$$f(x, y) = 4 - x^2 - \frac{1}{4}y^2 \qquad \text{Surface}$$

at $(1, 2)$ in the direction of

$$\mathbf{u} = \left(\cos\frac{\pi}{3}\right)\mathbf{i} + \left(\sin\frac{\pi}{3}\right)\mathbf{j}. \qquad \text{Direction}$$

**Solution**  Because $f_x(x, y) = -2x$ and $f_y(x, y) = -y/2$ are continuous, $f$ is differentiable, and you can apply Theorem 13.9.

$$D_{\mathbf{u}}f(x, y) = f_x(x, y)\cos\theta + f_y(x, y)\sin\theta = (-2x)\cos\theta + \left(-\frac{y}{2}\right)\sin\theta$$

Evaluating at $\theta = \pi/3$, $x = 1$, and $y = 2$ produces

$$D_{\mathbf{u}}f(1, 2) = (-2)\left(\frac{1}{2}\right) + (-1)\left(\frac{\sqrt{3}}{2}\right)$$

$$= -1 - \frac{\sqrt{3}}{2}$$

$$\approx -1.866. \qquad \text{See Figure 13.46.}$$

Note in Figure 13.46 that you can interpret the directional derivative as giving the slope of the surface at the point $(1, 2, 2)$ in the direction of the unit vector $\mathbf{u}$. ∎

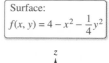

Surface:
$f(x, y) = 4 - x^2 - \frac{1}{4}y^2$

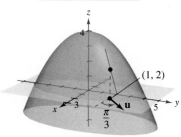

**Figure 13.46**

You have been specifying direction by a unit vector $\mathbf{u}$. When the direction is given by a vector whose length is not 1, you must normalize the vector before applying the formula in Theorem 13.9.

EXAMPLE 2    **Finding a Directional Derivative**

⋯▷ *See LarsonCalculus.com for an interactive version of this type of example.*

Find the directional derivative of

$$f(x, y) = x^2 \sin 2y \qquad \text{Surface}$$

at $(1, \pi/2)$ in the direction of

$$\mathbf{v} = 3\mathbf{i} - 4\mathbf{j}. \qquad \text{Direction}$$

**Solution**  Because $f_x(x, y) = 2x \sin 2y$ and $f_y(x, y) = 2x^2 \cos 2y$ are continuous, $f$ is differentiable, and you can apply Theorem 13.9. Begin by finding a unit vector in the direction of $\mathbf{v}$.

$$\mathbf{u} = \frac{\mathbf{v}}{\|\mathbf{v}\|} = \frac{3}{5}\mathbf{i} - \frac{4}{5}\mathbf{j} = \cos\theta\mathbf{i} + \sin\theta\mathbf{j}$$

Using this unit vector, you have

$$D_{\mathbf{u}}f(x, y) = (2x \sin 2y)(\cos\theta) = (2x^2 \cos 2y)(\sin\theta)$$

$$D_{\mathbf{u}}f\left(1, \frac{\pi}{2}\right) = (2 \sin \pi)\left(\frac{3}{5}\right) + (2 \cos \pi)\left(-\frac{4}{5}\right)$$

$$= (0)\left(\frac{3}{5}\right) + (-2)\left(-\frac{4}{5}\right)$$

$$= \frac{8}{5}. \qquad \text{See Figure 13.47.}$$

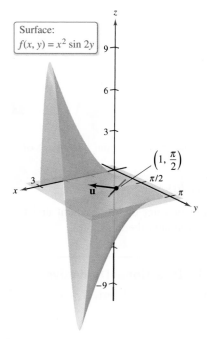

Surface:
$f(x, y) = x^2 \sin 2y$

**Figure 13.47**

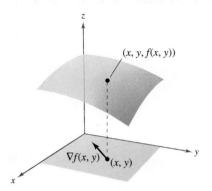

The gradient of $f$ is a vector in the $xy$-plane.

**Figure 13.48**

## The Gradient of a Function of Two Variables

The **gradient** of a function of two variables is a vector-valued function of two variables. This function has many important uses, some of which are described later in this section.

---

**Definition of Gradient of a Function of Two Variables**

Let $z = f(x, y)$ be a function of $x$ and $y$ such that $f_x$ and $f_y$ exist. Then the **gradient of $f$,** denoted by $\nabla f(x, y)$, is the vector

$$\nabla f(x, y) = f_x(x, y)\mathbf{i} + f_y(x, y)\mathbf{j}.$$

(The symbol $\nabla f$ is read as "del $f$.") Another notation for the gradient is given by **grad** $f(x, y)$. In Figure 13.48, note that for each $(x, y)$, the gradient $\nabla f(x, y)$ is a vector in the plane (not a vector in space).

---

Notice that no value is assigned to the symbol $\nabla$ by itself. It is an operator in the same sense that $d/dx$ is an operator. When $\nabla$ operates on $f(x, y)$, it produces the vector $\nabla f(x, y)$.

---

**EXAMPLE 3** **Finding the Gradient of a Function**

Find the gradient of

$$f(x, y) = y \ln x + xy^2$$

at the point $(1, 2)$.

**Solution** Using

$$f_x(x, y) = \frac{y}{x} + y^2 \quad \text{and} \quad f_y(x, y) = \ln x + 2xy$$

you have

$$\nabla f(x, y) = f_x(x, y)\mathbf{i} + f_y(x, y)\mathbf{j}$$
$$= \left(\frac{y}{x} + y^2\right)\mathbf{i} + (\ln x + 2xy)\mathbf{j}.$$

At the point $(1, 2)$, the gradient is

$$\nabla f(1, 2) = \left(\frac{2}{1} + 2^2\right)\mathbf{i} + [\ln 1 + 2(1)(2)]\mathbf{j}$$
$$= 6\mathbf{i} + 4\mathbf{j}.$$

Because the gradient of $f$ is a vector, you can write the directional derivative of $f$ in the direction of $\mathbf{u}$ as

$$D_{\mathbf{u}}f(x, y) = [f_x(x, y)\mathbf{i} + f_y(x, y)\mathbf{j}] \cdot (\cos \theta \mathbf{i} + \sin \theta \mathbf{j}).$$

In other words, the directional derivative is the dot product of the gradient and the direction vector. This useful result is summarized in the next theorem.

---

**THEOREM 13.10** **Alternative Form of the Directional Derivative**

If $f$ is a differentiable function of $x$ and $y$, then the directional derivative of $f$ in the direction of the unit vector $\mathbf{u}$ is

$$D_{\mathbf{u}} f(x, y) = \nabla f(x, y) \cdot \mathbf{u}.$$

**EXAMPLE 4**   **Using $\nabla f(x, y)$ to Find a Directional Derivative**

Find the directional derivative of $f(x, y) = 3x^2 - 2y^2$ at $\left(-\frac{3}{4}, 0\right)$ in the direction from $P\left(-\frac{3}{4}, 0\right)$ to $Q(0, 1)$.

**Solution**   Because the partials of $f$ are continuous, $f$ is differentiable and you can apply Theorem 13.10. A vector in the specified direction is

$$\overrightarrow{PQ} = \left(0 + \frac{3}{4}\right)\mathbf{i} + (1 - 0)\mathbf{j} = \frac{3}{4}\mathbf{i} + \mathbf{j}$$

and a unit vector in this direction is

$$\mathbf{u} = \frac{\overrightarrow{PQ}}{\|\overrightarrow{PQ}\|} = \frac{3}{5}\mathbf{i} + \frac{4}{5}\mathbf{j}. \qquad \text{Unit vector in direction of } \overrightarrow{PQ}$$

Because $\nabla f(x, y) = f_x(x, y)\mathbf{i} + f_y(x, y)\mathbf{j} = 6x\mathbf{i} - 4y\mathbf{j}$, the gradient at $\left(-\frac{3}{4}, 0\right)$ is

$$\nabla f\left(-\frac{3}{4}, 0\right) = -\frac{9}{2}\mathbf{i} + 0\mathbf{j}. \qquad \text{Gradient at } \left(-\frac{3}{4}, 0\right)$$

Consequently, at $\left(-\frac{3}{4}, 0\right)$, the directional derivative is

$$\begin{aligned} D_{\mathbf{u}}f\left(-\frac{3}{4}, 0\right) &= \nabla f\left(-\frac{3}{4}, 0\right) \cdot \mathbf{u} \\ &= \left(-\frac{9}{2}\mathbf{i} + 0\mathbf{j}\right) \cdot \left(\frac{3}{5}\mathbf{i} + \frac{4}{5}\mathbf{j}\right) \\ &= -\frac{27}{10}. \qquad \text{Directional derivative at } \left(-\frac{3}{4}, 0\right) \end{aligned}$$

See Figure 13.49.

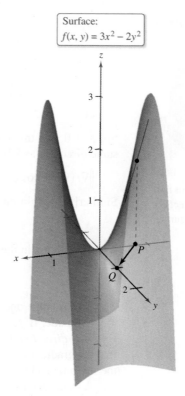

Surface:
$f(x, y) = 3x^2 - 2y^2$

**Figure 13.49**

## Applications of the Gradient

You have already seen that there are many directional derivatives at the point $(x, y)$ on a surface. In many applications, you may want to know in which direction to move so that $f(x, y)$ increases most rapidly. This direction is called the direction of *steepest ascent*, and it is given by the gradient, as stated in the next theorem.

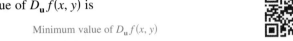

**REMARK** Property 2 of Theorem 13.11 says that at the point $(x, y)$, $f$ increases most rapidly in the direction of the gradient, $\nabla f(x, y)$.

---

**THEOREM 13.11 Properties of the Gradient**

Let $f$ be differentiable at the point $(x, y)$.

1. If $\nabla f(x, y) = \mathbf{0}$, then $D_{\mathbf{u}} f(x, y) = 0$ for all $\mathbf{u}$.
2. The direction of *maximum* increase of $f$ is given by $\nabla f(x, y)$. The maximum value of $D_{\mathbf{u}} f(x, y)$ is

   $\quad \|\nabla f(x, y)\|.$  $\qquad$ Maximum value of $D_{\mathbf{u}} f(x, y)$

3. The direction of *minimum* increase of $f$ is given by $-\nabla f(x, y)$. The minimum value of $D_{\mathbf{u}} f(x, y)$ is

   $\quad -\|\nabla f(x, y)\|.$  $\qquad$ Minimum value of $D_{\mathbf{u}} f(x, y)$

---

**Proof** If $\nabla f(x, y) = \mathbf{0}$, then for any direction (any $\mathbf{u}$), you have

$$D_{\mathbf{u}} f(x, y) = \nabla f(x, y) \cdot \mathbf{u}$$
$$= (0\mathbf{i} + 0\mathbf{j}) \cdot (\cos \theta \mathbf{i} + \sin \theta \mathbf{j})$$
$$= 0.$$

If $\nabla f(x, y) \neq \mathbf{0}$, then let $\phi$ be the angle between $\nabla f(x, y)$ and a unit vector $\mathbf{u}$. Using the dot product, you can apply Theorem 11.5 to conclude that

$$D_{\mathbf{u}} f(x, y) = \nabla f(x, y) \cdot \mathbf{u}$$
$$= \|\nabla f(x, y)\| \, \|\mathbf{u}\| \cos \phi$$
$$= \|\nabla f(x, y)\| \cos \phi$$

and it follows that the maximum value of $D_{\mathbf{u}} f(x, y)$ will occur when

$$\cos \phi = 1.$$

So, $\phi = 0$, and the maximum value of the directional derivative occurs when $\mathbf{u}$ has the same direction as $\nabla f(x, y)$. Moreover, this largest value of $D_{\mathbf{u}} f(x, y)$ is precisely

$$\|\nabla f(x, y)\| \cos \phi = \|\nabla f(x, y)\|.$$

Similarly, the minimum value of $D_{\mathbf{u}} f(x, y)$ can be obtained by letting

$$\phi = \pi$$

so that $\mathbf{u}$ points in the direction opposite that of $\nabla f(x, y)$, as shown in Figure 13.50.

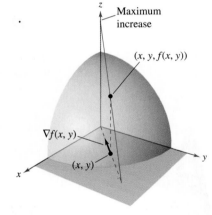

The gradient of $f$ is a vector in the $xy$-plane that points in the direction of maximum increase on the surface given by $z = f(x, y)$.
**Figure 13.50**

To visualize one of the properties of the gradient, imagine a skier coming down a mountainside. If $f(x, y)$ denotes the altitude of the skier, then $-\nabla f(x, y)$ indicates the *compass direction* the skier should take to ski the path of steepest descent. (Remember that the gradient indicates direction in the $xy$-plane and does not itself point up or down the mountainside.)

As another illustration of the gradient, consider the temperature $T(x, y)$ at any point $(x, y)$ on a flat metal plate. In this case, $\nabla T(x, y)$ gives the direction of greatest temperature increase at the point $(x, y)$, as illustrated in the next example.

Level curves:
$T(x, y) = 20 - 4x^2 - y^2$

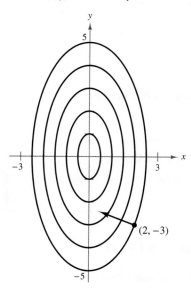

The direction of most rapid increase in temperature at $(2, -3)$ is given by $-16\mathbf{i} + 6\mathbf{j}$.

**Figure 13.51**

**EXAMPLE 5**    **Finding the Direction of Maximum Increase**

The temperature in degrees Celsius on the surface of a metal plate is

$$T(x, y) = 20 - 4x^2 - y^2$$

where $x$ and $y$ are measured in centimeters. In what direction from $(2, -3)$ does the temperature increase most rapidly? What is this rate of increase?

**Solution**    The gradient is

$$\nabla T(x, y) = T_x(x, y)\mathbf{i} + T_y(x, y)\mathbf{j} = -8x\mathbf{i} - 2y\mathbf{j}.$$

It follows that the direction of maximum increase is given by

$$\nabla T(2, -3) = -16\mathbf{i} + 6\mathbf{j}$$

as shown in Figure 13.51, and the rate of increase is

$$\|\nabla T(2, -3)\| = \sqrt{256 + 36} = \sqrt{292} \approx 17.09° \text{ per centimeter.} \qquad \blacksquare$$

The solution presented in Example 5 can be misleading. Although the gradient points in the direction of maximum temperature increase, it does not necessarily point toward the hottest spot on the plate. In other words, the gradient provides a local solution to finding an increase relative to the temperature at the point $(2, -3)$. *Once you leave that position, the direction of maximum increase may change.*

**EXAMPLE 6**    **Finding the Path of a Heat-Seeking Particle**

A heat-seeking particle is located at the point $(2, -3)$ on a metal plate whose temperature at $(x, y)$ is

$$T(x, y) = 20 - 4x^2 - y^2.$$

Find the path of the particle as it continuously moves in the direction of maximum temperature increase.

**Solution**    Let the path be represented by the position vector

$$\mathbf{r}(t) = x(t)\mathbf{i} + y(t)\mathbf{j}.$$

A tangent vector at each point $(x(t), y(t))$ is given by

$$\mathbf{r}'(t) = \frac{dx}{dt}\mathbf{i} + \frac{dy}{dt}\mathbf{j}.$$

Because the particle seeks maximum temperature increase, the directions of $\mathbf{r}'(t)$ and $\nabla T(x, y) = -8x\mathbf{i} - 2y\mathbf{j}$ are the same at each point on the path. So,

$$-8x = k\frac{dx}{dt} \quad \text{and} \quad -2y = k\frac{dy}{dt}$$

where $k$ depends on $t$. By solving each equation for $dt/k$ and equating the results, you obtain

$$\frac{dx}{-8x} = \frac{dy}{-2y}.$$

The solution of this differential equation is $x = Cy^4$. Because the particle starts at the point $(2, -3)$, you can determine that $C = 2/81$. So, the path of the heat-seeking particle is

$$x = \frac{2}{81}y^4.$$

The path is shown in Figure 13.52. $\qquad \blacksquare$

Level curves:
$T(x, y) = 20 - 4x^2 - y^2$

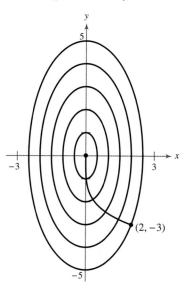

Path followed by a heat-seeking particle

**Figure 13.52**

In Figure 13.52, the path of the particle (determined by the gradient at each point) appears to be orthogonal to each of the level curves. This becomes clear when you consider that the temperature $T(x, y)$ is constant along a given level curve. So, at any point $(x, y)$ on the curve, the rate of change of $T$ in the direction of a unit tangent vector $\mathbf{u}$ is 0, and you can write

$$\nabla f(x, y) \cdot \mathbf{u} = D_{\mathbf{u}} T(x, y) = 0. \qquad \text{\small u is a unit tangent vector.}$$

Because the dot product of $\nabla f(x, y)$ and $\mathbf{u}$ is 0, you can conclude that they must be orthogonal. This result is stated in the next theorem.

---

**THEOREM 13.12    Gradient Is Normal to Level Curves**

If $f$ is differentiable at $(x_0, y_0)$ and $\nabla f(x_0, y_0) \neq \mathbf{0}$, then $\nabla f(x_0, y_0)$ is normal to the level curve through $(x_0, y_0)$.

---

**EXAMPLE 7    Finding a Normal Vector to a Level Curve**

Sketch the level curve corresponding to $c = 0$ for the function given by

$$f(x, y) = y - \sin x$$

and find a normal vector at several points on the curve.

**Solution**    The level curve for $c = 0$ is given by

$$0 = y - \sin x \quad \Longrightarrow \quad y = \sin x$$

as shown in Figure 13.53(a). Because the gradient of $f$ at $(x, y)$ is

$$\nabla f(x, y) = f_x(x, y)\mathbf{i} + f_y(x, y)\mathbf{j}$$
$$= -\cos x\,\mathbf{i} + \mathbf{j}$$

you can use Theorem 13.12 to conclude that $\nabla f(x, y)$ is normal to the level curve at the point $(x, y)$. Some gradients are

$$\nabla f(-\pi, 0) = \mathbf{i} + \mathbf{j}$$

$$\nabla f\left(-\frac{2\pi}{3}, -\frac{\sqrt{3}}{2}\right) = \frac{1}{2}\mathbf{i} + \mathbf{j}$$

$$\nabla f\left(-\frac{\pi}{2}, -1\right) = \mathbf{j}$$

$$\nabla f\left(-\frac{\pi}{3}, -\frac{\sqrt{3}}{2}\right) = -\frac{1}{2}\mathbf{i} + \mathbf{j}$$

$$\nabla f(0, 0) = -\mathbf{i} + \mathbf{j}$$

$$\nabla f\left(\frac{\pi}{3}, \frac{\sqrt{3}}{2}\right) = -\frac{1}{2}\mathbf{i} + \mathbf{j}$$

$$\nabla f\left(\frac{\pi}{2}, 1\right) = \mathbf{j}$$

$$\nabla f\left(\frac{2\pi}{3}, \frac{\sqrt{3}}{2}\right) = \frac{1}{2}\mathbf{i} + \mathbf{j}$$

and

$$\nabla f(\pi, 0) = \mathbf{i} + \mathbf{j}.$$

These are shown in Figure 13.53(b).

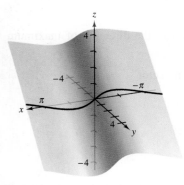

**(a)** The surface is given by $f(x, y) = y - \sin x$.

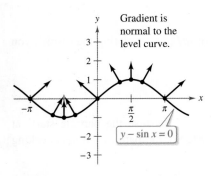

Gradient is normal to the level curve.

$y - \sin x = 0$

**(b)** The level curve is given by $f(x, y) = 0$.
**Figure 13.53**

## Functions of Three Variables

The definitions of the directional derivative and the gradient can be extended naturally to functions of three or more variables. As often happens, some of the geometric interpretation is lost in the generalization from functions of two variables to those of three variables. For example, you cannot interpret the directional derivative of a function of three variables as representing slope.

The definitions and properties of the directional derivative and the gradient of a function of three variables are listed below.

---

### Directional Derivative and Gradient for Three Variables

Let $f$ be a function of $x$, $y$, and $z$ with continuous first partial derivatives. The **directional derivative of** $f$ in the direction of a unit vector

$$\mathbf{u} = a\mathbf{i} + b\mathbf{j} + c\mathbf{k}$$

is given by

$$D_{\mathbf{u}}f(x, y, z) = af_x(x, y, z) + bf_y(x, y, z) + cf_z(x, y, z).$$

The **gradient of** $f$ is defined as

$$\nabla f(x, y, z) = f_x(x, y, z)\mathbf{i} + f_y(x, y, z)\mathbf{j} + f_z(x, y, z)\mathbf{k}.$$

Properties of the gradient are as follows.

1. $D_{\mathbf{u}}f(x, y, z) = \nabla f(x, y, z) \cdot \mathbf{u}$
2. If $\nabla f(x, y, z) = \mathbf{0}$, then $D_{\mathbf{u}}f(x, y, z) = 0$ for all $\mathbf{u}$.
3. The direction of *maximum* increase of $f$ is given by $\nabla f(x, y, z)$. The maximum value of $D_{\mathbf{u}}f(x, y, z)$ is

$$\|\nabla f(x, y, z)\|. \qquad \text{Maximum value of } D_{\mathbf{u}}f(x, y, z)$$

4. The direction of *minimum* increase of $f$ is given by $-\nabla f(x, y, z)$. The minimum value of $D_{\mathbf{u}}f(x, y, z)$ is

$$-\|\nabla f(x, y, z)\|. \qquad \text{Minimum value of } D_{\mathbf{u}}f(x, y, z)$$

---

You can generalize Theorem 13.12 to functions of three variables. Under suitable hypotheses,

$$\nabla f(x_0, y_0, z_0)$$

is normal to the level surface through $(x_0, y_0, z_0)$.

### EXAMPLE 8    Finding the Gradient of a Function

Find $\nabla f(x, y, z)$ for the function

$$f(x, y, z) = x^2 + y^2 - 4z$$

and find the direction of maximum increase of $f$ at the point $(2, -1, 1)$.

**Solution**    The gradient is

$$\nabla f(x, y, z) = f_x(x, y, z)\mathbf{i} + f_y(x, y, z)\mathbf{j} + f_z(x, y, z)\mathbf{k}$$
$$= 2x\mathbf{i} + 2y\mathbf{j} - 4\mathbf{k}.$$

So, it follows that the direction of maximum increase at $(2, -1, 1)$ is

$$\nabla f(2, -1, 1) = 4\mathbf{i} - 2\mathbf{j} - 4\mathbf{k}. \qquad \text{See Figure 13.54.}$$

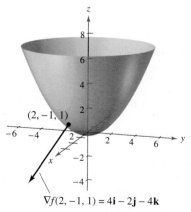

$\nabla f(2, -1, 1) = 4\mathbf{i} - 2\mathbf{j} - 4\mathbf{k}$

Level surface and gradient at $(2, -1, 1)$ for $f(x, y, z) = x^2 + y^2 - 4z$

**Figure 13.54**

# 13.6 Exercises

See **CalcChat.com** for tutorial help and worked-out solutions to odd-numbered exercises.

**CONCEPT CHECK**

1. **Directional Derivative** For a function $f(x, y)$, when does the directional derivative at the point $(x_0, y_0)$ equal the partial derivative with respect to $x$ at the point $(x_0, y_0)$? What does this mean graphically?

2. **Gradient** What is the meaning of the gradient of a function $f$ at a point $(x, y)$?

 **Finding a Directional Derivative** In Exercises 3–6, use Theorem 13.9 to find the directional derivative of the function at $P$ in the direction of the unit vector $\mathbf{u} = \cos\theta\mathbf{i} + \sin\theta\mathbf{j}$.

3. $f(x, y) = x^2 + y^2$, $P(1, -2)$, $\theta = \dfrac{\pi}{4}$

4. $f(x, y) = \dfrac{y}{x + y}$, $P(3, 0)$, $\theta = -\dfrac{\pi}{6}$

5. $f(x, y) = \sin(2x + y)$, $P(0, \pi)$, $\theta = -\dfrac{5\pi}{6}$

6. $g(x, y) = xe^y$, $P(0, 2)$, $\theta = \dfrac{2\pi}{3}$

 **Finding a Directional Derivative** In Exercises 7–10, use Theorem 13.9 to find the directional derivative of the function at $P$ in the direction of $\mathbf{v}$.

7. $f(x, y) = 3x - 4xy + 9y$, $P(1, 2)$, $\mathbf{v} = \frac{3}{5}\mathbf{i} + \frac{4}{5}\mathbf{j}$

8. $f(x, y) = x^3 - y^3$, $P(4, 3)$, $\mathbf{v} = \dfrac{\sqrt{2}}{2}(\mathbf{i} + \mathbf{j})$

9. $g(x, y) = \sqrt{x^2 + y^2}$, $P(3, 4)$, $\mathbf{v} = 3\mathbf{i} - 4\mathbf{j}$

10. $h(x, y) = e^{-(x^2+y^2)}$, $P(0, 0)$, $\mathbf{v} = \mathbf{i} + \mathbf{j}$

**Finding a Directional Derivative** In Exercises 11–14, use Theorem 13.9 to find the directional derivative of the function at $P$ in the direction of $\overrightarrow{PQ}$.

11. $f(x, y) = x^2 + 3y^2$, $P(1, 1)$, $Q(4, 5)$

12. $f(x, y) = \cos(x + y)$, $P(0, \pi)$, $Q\left(\dfrac{\pi}{2}, 0\right)$

13. $f(x, y) = e^y \sin x$, $P(0, 0)$, $Q(2, 1)$

14. $f(x, y) = \sin 2x \cos y$, $P(\pi, 0)$, $Q\left(\dfrac{\pi}{2}, \pi\right)$

 **Finding the Gradient of a Function** In Exercises 15–20, find the gradient of the function at the given point.

15. $f(x, y) = 3x + 5y^2 + 1$, $(2, 1)$

16. $g(x, y) = 2xe^{y/x}$, $(2, 0)$

17. $z = \dfrac{\ln(x^2 - y)}{x} - 4$, $(2, 3)$

18. $z = \cos(x^2 + y^2)$, $(3, -4)$

19. $w = 6xy - y^2 + 2xyz^3$, $(-1, 5, -1)$

20. $w = x\tan(y + z)$, $(4, 3, -1)$

**Finding a Directional Derivative** In Exercises 21–24, use the gradient to find the directional derivative of the function at $P$ in the direction of $\mathbf{v}$.

21. $f(x, y) = xy$, $P(0, -2)$, $\mathbf{v} = \frac{1}{2}(\mathbf{i} + \sqrt{3}\mathbf{j})$

22. $h(x, y) = e^{-3x} \sin y$, $P\left(1, \dfrac{\pi}{2}\right)$, $\mathbf{v} = -\mathbf{i}$

23. $f(x, y, z) = x^2 + y^2 + z^2$, $P(1, 1, 1)$, $\mathbf{v} = \dfrac{\sqrt{3}}{3}(\mathbf{i} - \mathbf{j} + \mathbf{k})$

24. $f(x, y, z) = xy + yz + xz$, $P(1, 2, -1)$, $\mathbf{v} = 2\mathbf{i} + \mathbf{j} - \mathbf{k}$

 **Finding a Directional Derivative** In Exercises 25–28, use the gradient to find the directional derivative of the function at $P$ in the direction of $\overrightarrow{PQ}$.

25. $g(x, y) = x^2 + y^2 + 1$, $P(1, 2)$, $Q(2, 3)$

26. $f(x, y) = 3x^2 - y^2 + 4$, $P(-1, 4)$, $Q(3, 6)$

27. $g(x, y, z) = xye^z$, $P(2, 4, 0)$, $Q(0, 0, 0)$

28. $h(x, y, z) = \ln(x + y + z)$, $P(1, 0, 0)$, $Q(4, 3, 1)$

 **Using Properties of the Gradient** In Exercises 29–38, find the gradient of the function and the maximum value of the directional derivative at the given point.

29. $f(x, y) = y^2 - x\sqrt{y}$, $(0, 3)$

30. $f(x, y) = \dfrac{x + y}{y + 1}$, $(0, 1)$

31. $h(x, y) = x\tan y$, $\left(2, \dfrac{\pi}{4}\right)$

32. $h(x, y) = y\cos(x - y)$, $\left(0, \dfrac{\pi}{3}\right)$

33. $f(x, y) = \sin x^2 y^3$, $\left(\dfrac{1}{\pi}, \pi\right)$

34. $g(x, y) = \ln\sqrt[3]{x^2 + y^2}$, $(1, 2)$

35. $f(x, y, z) = \sqrt{x^2 + y^2 + z^2}$, $(1, 4, 2)$

36. $w = \dfrac{1}{\sqrt{1 - x^2 - y^2 - z^2}}$, $(0, 0, 0)$

37. $w = xy^2z^2$, $(2, 1, 1)$

38. $f(x, y, z) = xe^{yz}$, $(2, 0, -4)$

 **Finding a Normal Vector to a Level Curve** In Exercises 39–42, find a normal vector to the level curve $f(x, y) = c$ at $P$.

39. $f(x, y) = 6 - 2x - 3y$
$c = 6$, $P(0, 0)$

40. $f(x, y) = x^2 + y^2$
$c = 25$, $P(3, 4)$

**41.** $f(x, y) = xy$

$c = -3$,   $P(-1, 3)$

**42.** $f(x, y) = \dfrac{x}{x^2 + y^2}$

$c = \frac{1}{2}$,   $P(1, 1)$

**Using a Function**   In Exercises 43–46, (a) **find the gradient of the function at $P$, (b) find a unit normal vector to the level curve $f(x, y) = c$ at $P$, (c) find the tangent line to the level curve $f(x, y) = c$ at $P$, and (d) sketch the level curve, the unit normal vector, and the tangent line in the $xy$-plane.**

**43.** $f(x, y) = 4x^2 - y$

$c = 6$,   $P(2, 10)$

**44.** $f(x, y) = x - y^2$

$c = 3$,   $P(4, -1)$

**45.** $f(x, y) = 3x^2 - 2y^2$

$c = 1$,   $P(1, 1)$

**46.** $f(x, y) = 9x^2 + 4y^2$

$c = 40$,   $P(2, -1)$

**47. Using a Function**   Consider the function

$$f(x, y) = 3 - \frac{x}{3} - \frac{y}{2}.$$

(a) Sketch the graph of $f$ in the first octant and plot the point $(3, 2, 1)$ on the surface.

(b) Find $D_{\mathbf{u}} f(3, 2)$, where $\mathbf{u} = \cos \theta \mathbf{i} + \sin\theta \mathbf{j}$, using each given value of $\theta$.

(i) $\theta = \dfrac{\pi}{4}$   (ii) $\theta = \dfrac{2\pi}{3}$   (iii) $\theta = \dfrac{4\pi}{3}$   (iv) $\theta = -\dfrac{\pi}{6}$

(c) Find $D_{\mathbf{u}} f(3, 2)$, where $\mathbf{u} = \dfrac{\mathbf{v}}{\|\mathbf{v}\|}$, using each given vector $\mathbf{v}$.

(i) $\mathbf{v} = \mathbf{i} + \mathbf{j}$   (ii) $\mathbf{v} = -3\mathbf{i} - 4\mathbf{j}$

(iii) $\mathbf{v}$ is the vector from $(1, 2)$ to $(-2, 6)$.

(iv) $\mathbf{v}$ is the vector from $(3, 2)$ to $(4, 5)$.

(d) Find $\nabla f(x, y)$.

(e) Find the maximum value of the directional derivative at $(3, 2)$.

(f) Find a unit vector $\mathbf{u}$ orthogonal to $\nabla f(3, 2)$ and calculate $D_{\mathbf{u}} f(3, 2)$. Discuss the geometric meaning of the result.

**48. Using a Function**   Consider the function

$$f(x, y) = 9 - x^2 - y^2.$$

(a) Sketch the graph of $f$ in the first octant and plot the point $(1, 2, 4)$ on the surface.

(b) Find $D_{\mathbf{u}} f(1, 2)$, where $\mathbf{u} = \cos \theta \mathbf{i} + \sin \theta \mathbf{j}$, using each given value of $\theta$.

(i) $\theta = -\dfrac{\pi}{4}$   (ii) $\theta = \dfrac{\pi}{3}$   (iii) $\theta = \dfrac{3\pi}{4}$   (iv) $\theta = -\dfrac{\pi}{2}$

(c) Find $D_{\mathbf{u}} f(1, 2)$, where $\mathbf{u} = \dfrac{\mathbf{v}}{\|\mathbf{v}\|}$, using each given vector $\mathbf{v}$.

(i) $\mathbf{v} = 3\mathbf{i} + \mathbf{j}$   (ii) $\mathbf{v} = -8\mathbf{i} - 6\mathbf{j}$

(iii) $\mathbf{v}$ is the vector from $(-1, -1)$ to $(3, 5)$.

(iv) $\mathbf{v}$ is the vector from $(-2, 0)$ to $(1, 3)$.

(d) Find $\nabla f(1, 2)$.

(e) Find the maximum value of the directional derivative at $(1, 2)$.

(f) Find a unit vector $\mathbf{u}$ orthogonal to $\nabla f(1, 2)$ and calculate $D_{\mathbf{u}} f(1, 2)$. Discuss the geometric meaning of the result.

**49. Investigation**   Consider the function

$$f(x, y) = x^2 - y^2$$

at the point $(4, -3, 7)$.

(a) Use a computer algebra system to graph the surface represented by the function.

(b) Determine the directional derivative $D_{\mathbf{u}} f(4, -3)$ as a function of $\theta$, where $\mathbf{u} = \cos \theta \mathbf{i} + \sin \theta \mathbf{j}$. Use a computer algebra system to graph the function on the interval $[0, 2\pi)$.

(c) Approximate the zeros of the function in part (b) and interpret each in the context of the problem.

(d) Approximate the critical numbers of the function in part (b) and interpret each in the context of the problem.

(e) Find $\|\nabla f(4, -3)\|$ and explain its relationship to your answers in part (d).

(f) Use a computer algebra system to graph the level curve of the function $f$ at the level $c = 7$. On this curve, graph the vector in the direction of $\nabla f(4, -3)$ and state its relationship to the level curve.

**50. Investigation**   Consider the function

$$f(x, y) = \frac{8y}{1 + x^2 + y^2}.$$

(a) Analytically verify that the level curve of $f(x, y)$ at the level $c = 2$ is a circle.

(b) At the point $\left(\sqrt{3}, 2\right)$ on the level curve for which $c = 2$, sketch the vector showing the direction of the greatest rate of increase of the function. To print a graph of the level curve, go to *MathGraphs.com*.

(c) At the point $\left(\sqrt{3}, 2\right)$ on the level curve for which $c = 2$, sketch a vector such that the directional derivative is 0.

(d) Use a computer algebra system to graph the surface to verify your answers in parts (a)–(c).

**EXPLORING CONCEPTS**

**51. Think About It**   Consider $\mathbf{v} = 3\mathbf{u}$. Is the directional derivative of a differentiable function $f(x, y)$ in the direction of $\mathbf{v}$ at the point $(x_0, y_0)$ three times the directional derivative of $f$ in the direction of $\mathbf{u}$ at the point $(x_0, y_0)$? Explain.

**52. Sketching a Graph and a Vector**   Sketch the graph of a surface and select a point $P$ on the surface. Sketch a vector in the $xy$-plane giving the direction of steepest ascent on the surface at $P$.

**53. Topography**   The surface of a mountain is modeled by the equation

$$h(x, y) = 5000 - 0.001x^2 - 0.004y^2.$$

A mountain climber is at the point $(500, 300, 4390)$. In what direction should the climber move in order to ascend at the greatest rate?

**54.** **HOW DO YOU SEE IT?** The figure shows a topographic map carried by a group of hikers. Sketch the paths of steepest descent when the hikers start at point $A$ and when they start at point $B$. (To print an enlarged copy of the graph, go to *MathGraphs.com*.)

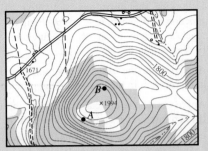

**55. Temperature** The temperature at the point $(x, y)$ on a metal plate is $T(x, y) = x/(x^2 + y^2)$. Find the direction of greatest increase in heat from the point $(3, 4)$.

**56. Temperature** The temperature at the point $(x, y)$ on a metal plate is $T(x, y) = 400e^{-(x^2+y)/2}$, $x \geq 0$, $y \geq 0$.

 (a) Use a computer algebra system to graph the temperature distribution function.

(b) Find the directions of no change in heat on the plate from the point $(3, 5)$.

(c) Find the direction of greatest increase in heat from the point $(3, 5)$.

**Finding the Direction of Maximum Increase** In Exercises 57 and 58, the temperature in degrees Celsius on the surface of a metal plate is given by $T(x, y)$, where $x$ and $y$ are measured in centimeters. Find the direction from point $P$ where the temperature increases most rapidly and this rate of increase.

**57.** $T(x, y) = 80 - 3x^2 - y^2$, $P(-1, 5)$

**58.** $T(x, y) = 50 - x^2 - 4y^2$, $P(2, -1)$

 **Finding the Path of a Heat-Seeking Particle** In Exercises 59 and 60, find the path of a heat-seeking particle placed at point $P$ on a metal plate whose temperature at $(x, y)$ is $T(x, y)$.

**59.** $T(x, y) = 400 - 2x^2 - y^2$, $P(10, 10)$

**60.** $T(x, y) = 100 - x^2 - 2y^2$, $P(4, 3)$

**True or False?** In Exercises 61–64, determine whether the statement is true or false. If it is false, explain why or give an example that shows it is false.

**61.** If $f(x, y) = \sqrt{1 - x^2 - y^2}$, then $D_{\mathbf{u}} f(0, 0) = 0$ for any unit vector $\mathbf{u}$.

**62.** If $f(x, y) = x + y$, then $-1 \leq D_{\mathbf{u}} f(x, y) \leq 1$.

**63.** If $D_{\mathbf{u}} f(x, y)$ exists, then $D_{\mathbf{u}} f(x, y) = -D_{\mathbf{u}} f(x, y)$.

**64.** If $D_{\mathbf{u}} f(x_0, y_0) = c$ for any unit vector $\mathbf{u}$, then $c = 0$.

**65. Finding a Function** Find a function $f$ such that

$$\nabla f = e^x \cos y \mathbf{i} - e^x \sin y \mathbf{j} + z \mathbf{k}.$$

**66. Ocean Floor**

A team of oceanographers is mapping the ocean floor to assist in the recovery of a sunken ship. Using sonar, they develop the model

$$D = 250 + 30x^2 + 50 \sin \frac{\pi y}{2}, \quad 0 \leq x \leq 2, \quad 0 \leq y \leq 2$$

where $D$ is the depth in meters, and $x$ and $y$ are the distances in kilometers.

(a) Use a computer algebra system to graph $D$.

(b) Because the graph in part (a) is showing depth, it is not a map of the ocean floor. How could the model be changed so that the graph of the ocean floor could be obtained?

(c) What is the depth of the ship if it is located at the coordinates $x = 1$ and $y = 0.5$?

(d) Determine the steepness of the ocean floor in the positive $x$-direction from the position of the ship.

(e) Determine the steepness of the ocean floor in the positive $y$-direction from the position of the ship.

(f) Determine the direction of the greatest rate of change of depth from the position of the ship.

**67. Using a Function** Consider the function

$$f(x, y) = \sqrt[3]{xy}.$$

(a) Show that $f$ is continuous at the origin.

(b) Show that $f_x$ and $f_y$ exist at the origin but that the directional derivatives at the origin in all other directions do not exist.

 (c) Use a computer algebra system to graph $f$ near the origin to verify your answers in parts (a) and (b). Explain.

**68. Directional Derivative** Consider the function

$$f(x, y) = \begin{cases} \dfrac{4xy}{x^2 + y^2}, & (x, y) \neq (0, 0) \\ 0, & (x, y) \neq (0, 0) \end{cases}$$

and the unit vector

$$\mathbf{u} = \frac{1}{\sqrt{2}}(\mathbf{i} + \mathbf{j}).$$

Does the directional derivative of $f$ at $P(0, 0)$ in the direction of $\mathbf{u}$ exist? If $f(0, 0)$ were defined as 2 instead of 0, would the directional derivative exist? Explain.

# 13.7  Tangent Planes and Normal Lines

- Find equations of tangent planes and normal lines to surfaces.
- Find the angle of inclination of a plane in space.
- Compare the gradients $\nabla f(x, y)$ and $\nabla F(x, y, z)$.

## Exploration

***Billiard Balls and Normal Lines***  In each of the three figures below, the cue ball is about to strike a stationary ball at point $P$. Explain how you can use the normal line to the stationary ball at point $P$ to describe the resulting motion of each of the two balls. Assuming that each cue ball has the same speed, which stationary ball will acquire the greatest speed? Which will acquire the least? Explain your reasoning.

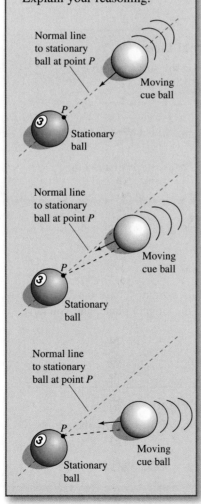

## Tangent Plane and Normal Line to a Surface

So far, you have represented surfaces in space primarily by equations of the form

$$z = f(x, y). \qquad \text{Equation of a surface } S$$

In the development to follow, however, it is convenient to use the more general representation $F(x, y, z) = 0$. For a surface $S$ given by $z = f(x, y)$, you can convert to the general form by defining $F$ as

$$F(x, y, z) = f(x, y) - z.$$

Because $f(x, y) - z = 0$, you can consider $S$ to be the level surface of $F$ given by

$$F(x, y, z) = 0. \qquad \text{Alternative equation of surface } S$$

### EXAMPLE 1  Writing an Equation of a Surface

For the function

$$F(x, y\ z) = x^2 + y^2 + z^2 - 4$$

describe the level surface given by

$$F(x, y, z) = 0.$$

**Solution**  The level surface given by $F(x, y, z) = 0$ can be written as

$$x^2 + y^2 + z^2 = 4$$

which is a sphere of radius 2 whose center is at the origin.  ∎

You have seen many examples of the usefulness of normal lines in applications involving curves. Normal lines are equally important in analyzing surfaces and solids. For example, consider the collision of two billiard balls. When a stationary ball is struck at a point $P$ on its surface, it moves along the **line of impact** determined by $P$ and the center of the ball. The impact can occur in *two* ways. When the cue ball is moving along the line of impact, it stops dead and imparts all of its momentum to the stationary ball, as shown in Figure 13.55.

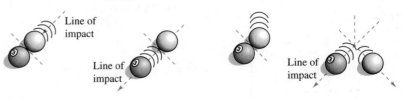

**Figure 13.55**          **Figure 13.56**

When the cue ball is not moving along the line of impact, it is deflected to one side or the other and retains part of its momentum. The part of the momentum that is transferred to the stationary ball occurs along the line of impact, *regardless* of the direction of the cue ball, as shown in Figure 13.56. This line of impact is called the **normal line** to the surface of the ball at the point $P$.

In the process of finding a normal line to a surface, you are also able to solve the problem of finding a **tangent plane** to the surface. Let $S$ be a surface given by

$$F(x, y, z) = 0$$

and let $P(x_0, y_0, z_0)$ be a point on $S$. Let $C$ be a curve on $S$ through $P$ that is defined by the vector-valued function

$$\mathbf{r}(t) = x(t)\mathbf{i} + y(t)\mathbf{j} + z(t)\mathbf{k}.$$

Then, for all $t$,

$$F(x(t), y(t), z(t)) = 0.$$

If $F$ is differentiable and $x'(t)$, $y'(t)$, and $z'(t)$ all exist, then it follows from the Chain Rule that

$$0 = F'(t)$$
$$= F_x(x, y, z)x'(t) + F_y(x, y, z)y'(t) + F_z(x, y, z)z'(t).$$

At $(x_0, y_0, z_0)$, the equivalent vector form is

$$0 = \underbrace{\nabla F(x_0, y_0, z_0)}_{\text{Gradient}} \cdot \underbrace{\mathbf{r}'(t_0)}_{\text{Tangent vector}}.$$

This result means that the gradient at $P$ is orthogonal to the tangent vector of every curve on $S$ through $P$. So, all tangent lines on $S$ lie in a plane that is normal to $\nabla F(x_0, y_0, z_0)$ and contains $P$, as shown in Figure 13.57.

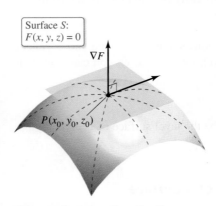

Surface $S$:
$F(x, y, z) = 0$

$\nabla F$

$P(x_0, y_0, z_0)$

Tangent plane to surface $S$ at $P$
**Figure 13.57**

**REMARK** In the remainder of this section, assume $\nabla F(x_0, y_0, z_0)$ to be nonzero unless stated otherwise.

---

**Definitions of Tangent Plane and Normal Line**

Let $F$ be differentiable at the point $P(x_0, y_0, z_0)$ on the surface $S$ given by $F(x, y, z) = 0$ such that

$$\nabla F(x_0, y_0, z_0) \neq \mathbf{0}.$$

1. The plane through $P$ that is normal to $\nabla F(x_0, y_0, z_0)$ is called the **tangent plane to $S$ at $P$.**
2. The line through $P$ having the direction of $\nabla F(x_0, y_0, z_0)$ is called the **normal line to $S$ at $P$.**

---

To find an equation for the tangent plane to $S$ at $(x_0, y_0, z_0)$, let $(x, y, z)$ be an arbitrary point in the tangent plane. Then the vector

$$\mathbf{v} = (x - x_0)\mathbf{i} + (y - y_0)\mathbf{j} + (z - z_0)\mathbf{k}$$

lies in the tangent plane. Because $\nabla F(x_0, y_0, z_0)$, is normal to the tangent plane at $(x_0, y_0, z_0)$, it must be orthogonal to every vector in the tangent plane, and you have

$$\nabla F(x_0, y_0, z_0) \cdot \mathbf{v} = 0$$

which leads to the next theorem.

---

**THEOREM 13.13 Equation of Tangent Plane**

If $F$ is differentiable at $(x_0, y_0, z_0)$, then an equation of the tangent plane to the surface given by $F(x, y, z) = 0$ at $(x_0, y_0, z_0)$ is

$$F_x(x_0, y_0, z_0)(x - x_0) + F_y(x_0, y_0, z_0)(y - y_0) + F_z(x_0, y_0, z_0)(z - z_0) = 0.$$

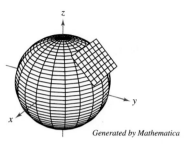
**EXAMPLE 2**    **Finding an Equation of a Tangent Plane**

Find an equation of the tangent plane to the hyperboloid

$$z^2 - 2x^2 - 2y^2 = 12$$

at the point $(1, -1, 4)$.

**Solution**   Begin by writing the equation of the surface as

$$z^2 - 2x^2 - 2y^2 - 12 = 0.$$

Then, considering

$$F(x, y, z) = z^2 - 2x^2 - 2y^2 - 12$$

you have

$$F_x(x, y, z) = -4x, \quad F_y(x, y, z) = -4y, \quad \text{and} \quad F_z(x, y, z) = 2z.$$

At the point $(1, -1, 4)$, the partial derivatives are

$$F_x(1, -1, 4) = -4, \quad F_y(1, -1, 4) = 4, \quad \text{and} \quad F_z(1, -1, 4) = 8.$$

So, an equation of the tangent plane at $(1, -1, 4)$ is

$$-4(x - 1) + 4(y + 1) + 8(z - 4) = 0$$
$$-4x + 4 + 4y + 4 + 8z - 32 = 0$$
$$-4x + 4y + 8z - 24 = 0$$
$$x - y - 2z + 6 = 0.$$

Figure 13.58 shows a portion of the hyperboloid and the tangent plane.

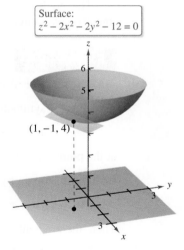

Surface:
$z^2 - 2x^2 - 2y^2 - 12 = 0$

$(1, -1, 4)$

Tangent plane to surface
**Figure 13.58**

To find an equation of the tangent plane at a point on a surface given by
$z = f(x, y)$, you can define the function $F$ by

$$F(x, y, z) = f(x, y) - z.$$

Then $S$ is given by the level surface $F(x, y, z) = 0$, and by Theorem 13.13, an equation
of the tangent plane to $S$ at the point $(x_0, y_0, z_0)$ is

$$f_x(x_0, y_0)(x - x_0) + f_y(x_0, y_0)(y - y_0) - (z - z_0) = 0.$$

**Finding an Equation of the Tangent Plane**

Find an equation of the tangent plane to the paraboloid

$$z = 1 - \frac{1}{10}(x^2 + 4y^2)$$

at the point $\left(1, 1, \frac{1}{2}\right)$.

**Solution**  From $z = f(x, y) = 1 - \frac{1}{10}(x^2 + 4y^2)$, you obtain

$$f_x(x, y) = -\frac{x}{5} \quad \Longrightarrow \quad f_x(1, 1) = -\frac{1}{5}$$

and

$$f_y(x, y) = -\frac{4y}{5} \quad \Longrightarrow \quad f_y(1, 1) = -\frac{4}{5}.$$

So, an equation of the tangent plane at $\left(1, 1, \frac{1}{2}\right)$ is

$$f_x(1, 1)(x - 1) + f_y(1, 1)(y - 1) - \left(z - \frac{1}{2}\right) = 0$$

$$-\frac{1}{5}(x - 1) - \frac{4}{5}(y - 1) - \left(z - \frac{1}{2}\right) = 0$$

$$-\frac{1}{5}x - \frac{4}{5}y - z + \frac{3}{2} = 0.$$

This tangent plane is shown in Figure 13.59.

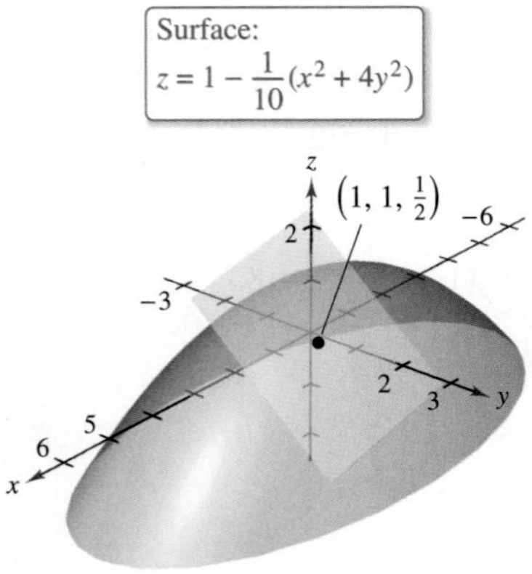

Surface:
$$z = 1 - \frac{1}{10}(x^2 + 4y^2)$$

$\left(1, 1, \frac{1}{2}\right)$

**Figure 13.59**

The gradient $\nabla F(x, y, z)$ provides a convenient way to find equations of normal lines, as shown in Example 4.

**Finding an Equation of a Normal Line to a Surface**

⋅ ⋅ ⋅▷ *See LarsonCalculus.com for an interactive version of this type of example.*

Find a set of symmetric equations for the normal line to the surface

$$xyz = 12$$

at the point $(2, -2, -3)$.

**Solution**  Begin by letting

$$F(x, y, z) = xyz - 12.$$

Then, the gradient is given by

$$\nabla F(x, y, z) = F_x(x, y, z)\mathbf{i} + F_y(x, y, z)\mathbf{j} + F_z(x, y, z)\mathbf{k}$$
$$= yz\mathbf{i} + xz\mathbf{j} + xy\mathbf{k}$$

and at the point $(2, -2, -3)$, you have

$$\nabla F(2, -2, -3) = (-2)(-3)\mathbf{i} + (2)(-3)\mathbf{j} + (2)(-2)\mathbf{k}$$
$$= 6\mathbf{i} - 6\mathbf{j} - 4\mathbf{k}.$$

The normal line at $(2, -2, -3)$ has direction numbers $6$, $-6$, and $-4$, and the corresponding set of symmetric equations is

$$\frac{x - 2}{6} = \frac{y + 2}{-6} = \frac{z + 3}{-4}.$$

See Figure 13.60.

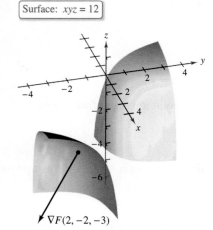

Surface: $xyz = 12$

$\nabla F(2, -2, -3)$

**Figure 13.60**

Knowing that the gradient $\nabla F(x, y, z)$ is normal to the surface given by $F(x, y, z) = 0$ allows you to solve a variety of problems dealing with surfaces and curves in space.

**EXAMPLE 5**  **Finding the Equation of a Tangent Line to a Curve**

Find a set of parametric equations for the tangent line to the curve of intersection of the ellipsoid

$$x^2 + 2y^2 + 2z^2 = 20 \qquad\qquad \text{Ellipsoid}$$

and the paraboloid

$$x^2 + y^2 + z = 4 \qquad\qquad \text{Paraboloid}$$

at the point $(0, 1, 3)$, as shown in Figure 13.61.

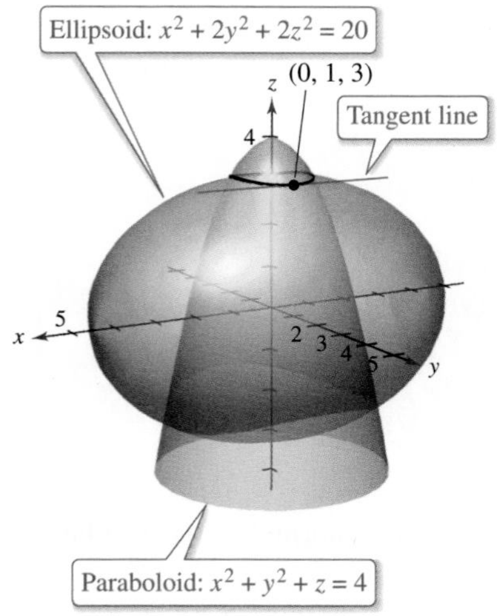

**Solution**  Begin by finding the gradients to both surfaces at the point $(0, 1, 3)$.

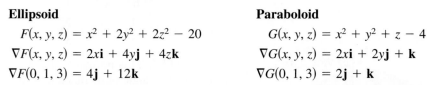

Ellipsoid: $x^2 + 2y^2 + 2z^2 = 20$
$z$ $(0, 1, 3)$
Tangent line
Paraboloid: $x^2 + y^2 + z = 4$

**Figure 13.61**

| **Ellipsoid** | **Paraboloid** |
|---|---|
| $F(x, y, z) = x^2 + 2y^2 + 2z^2 - 20$ | $G(x, y, z) = x^2 + y^2 + z - 4$ |
| $\nabla F(x, y, z) = 2x\mathbf{i} + 4y\mathbf{j} + 4z\mathbf{k}$ | $\nabla G(x, y, z) = 2x\mathbf{i} + 2y\mathbf{j} + \mathbf{k}$ |
| $\nabla F(0, 1, 3) = 4\mathbf{j} + 12\mathbf{k}$ | $\nabla G(0, 1, 3) = 2\mathbf{j} + \mathbf{k}$ |

The cross product of these two gradients is a vector that is tangent to both surfaces at the point $(0, 1, 3)$.

$$\nabla F(0, 1, 3) \times \nabla G(0, 1, 3) = \begin{vmatrix} \mathbf{i} & \mathbf{j} & \mathbf{k} \\ 0 & 4 & 12 \\ 0 & 2 & 1 \end{vmatrix} = -20\mathbf{i}$$

So, the tangent line to the curve of intersection of the two surfaces at the point $(0, 1, 3)$ is a line that is parallel to the $x$-axis and passes through the point $(0, 1, 3)$. Because $-20\mathbf{i} = -20(\mathbf{i} + 0\mathbf{j} + 0\mathbf{k})$, the direction numbers are 1, 0, and 0. So a set of parametric equations for the tangent line passing through the point $(0, 1, 3)$ is $x = t$, $y = 1$, and $z = 3$. ∎

## The Angle of Inclination of a Plane

Another use of the gradient $\nabla F(x, y, z)$ is to determine the angle of inclination of the tangent plane to a surface. The **angle of inclination** of a plane is defined as the angle $\theta$ ($0 \leq \theta \leq \pi/2$) between the given plane and the $xy$-plane, as shown in Figure 13.62. (The angle of inclination of a horizontal plane is defined as zero.) Because the vector $\mathbf{k}$ is normal to the $xy$-plane, you can use the formula for the cosine of the angle between two planes (given in Section 11.5) to conclude that the angle of inclination of a plane with normal vector $\mathbf{n}$ is

$$\cos \theta = \frac{|\mathbf{n} \cdot \mathbf{k}|}{\|\mathbf{n}\| \|\mathbf{k}\|} = \frac{|\mathbf{n} \cdot \mathbf{k}|}{\|\mathbf{n}\|}. \qquad\qquad \text{Angle of inclination of a plane}$$

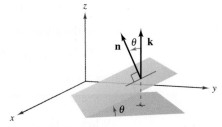

The angle of inclination
**Figure 13.62**

**EXAMPLE 6** **Finding the Angle of Inclination of a Tangent Plane**

Find the angle of inclination of the tangent plane to the ellipsoid

$$\frac{x^2}{12} + \frac{y^2}{12} + \frac{z^2}{3} = 1$$

at the point $(2, 2, 1)$.

**Solution** Begin by letting

$$F(x, y, z) = \frac{x^2}{12} + \frac{y^2}{12} + \frac{z^2}{3} - 1.$$

Then, the gradient of $F$ at the point $(2, 2, 1)$ is

$$\nabla F(x, y, z) = \frac{x}{6}\mathbf{i} + \frac{y}{6}\mathbf{j} + \frac{2z}{3}\mathbf{k}$$

$$\nabla F(2, 2, 1) = \frac{1}{3}\mathbf{i} + \frac{1}{3}\mathbf{j} + \frac{2}{3}\mathbf{k}.$$

Because $\nabla F(2, 2, 1)$ is normal to the tangent plane and $\mathbf{k}$ is normal to the $xy$-plane, it follows that the angle of inclination of the tangent plane is

$$\cos\theta = \frac{|\nabla F(2, 2, 1) \cdot \mathbf{k}|}{\|\nabla F(2, 2, 1)\|} = \frac{2/3}{\sqrt{(1/3)^2 + (1/3)^2 + (2/3)^2}} = \sqrt{\frac{2}{3}}$$

which implies that

$$\theta = \arccos\sqrt{\frac{2}{3}} \approx 35.3°$$

as shown in Figure 13.63.

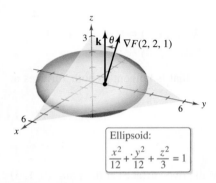

Ellipsoid:
$$\frac{x^2}{12} + \frac{y^2}{12} + \frac{z^2}{3} = 1$$

**Figure 13.63**

A special case of the procedure shown in Example 6 is worth noting. The angle of inclination $\theta$ of the tangent plane to the surface $z = f(x, y)$ at $(x_0, y_0, z_0)$ is

$$\cos\theta = \frac{1}{\sqrt{[f_x(x_0, y_0)]^2 + [f_y(x_0, y_0)]^2 + 1}}.$$    Alternative formula for angle of inclination (See Exercise 63.)

## A Comparison of the Gradients $\nabla f(x, y)$ and $\nabla F(x, y, z)$

This section concludes with a comparison of the gradients $\nabla f(x, y)$ and $\nabla F(x, y, z)$. In the preceding section, you saw that the gradient of a function $f$ of two variables is normal to the level curves of $f$. Specifically, Theorem 13.12 states that if $f$ is differentiable at $(x_0, y_0)$ and $\nabla f(x_0, y_0) \neq \mathbf{0}$, then $\nabla f(x_0, y_0)$ is normal to the level curve through $(x_0, y_0)$. Having developed normal lines to surfaces, you can now extend this result to a function of three variables. The proof of Theorem 13.14 is left as an exercise (see Exercise 64).

---

**THEOREM 13.14** **Gradient Is Normal to Level Surfaces**

If $F$ is differentiable at $(x_0, y_0, z_0)$ and

$$\nabla F(x_0, y_0, z_0) \neq \mathbf{0}$$

then $\nabla F(x_0, y_0, z_0)$ is normal to the level surface through $(x_0, y_0, z_0)$.

---

When working with the gradients $\nabla f(x, y)$ and $\nabla F(x, y, z)$, be sure you remember that $\nabla f(x, y)$ is a vector in the $xy$-plane and $\nabla F(x, y, z)$ is a vector in space.

# 13.7 Exercises

See CalcChat.com for tutorial help and worked-out solutions to odd-numbered exercises.

## CONCEPT CHECK

**1. Tangent Vector** Consider a point $(x_0, y_0, z_0)$ on a surface given by $F(x, y, z) = 0$. What is the relationship between $\nabla F(x_0, y_0, z_0)$ and any tangent vector $\mathbf{v}$ at $(x_0, y_0, z_0)$? How do you represent this relationship mathematically?

**2. Normal Line** Consider a point $(x_0, y_0, z_0)$ on a surface given by $F(x, y, z) = 0$. What is the relationship between $\nabla F(x_0, y_0, z_0)$ and the normal line through $(x_0, y_0, z_0)$?

  **Describing a Surface** In Exercises 3–6, describe the level surface $F(x, y, z) = 0$.

**3.** $F(x, y, z) = 3x - 5y + 3z - 15$

**4.** $F(x, y, z) = 36 - x^2 - y^2 - z^2$

**5.** $F(x, y, z) = 4x^2 + 9y^2 - 4z^2$

**6.** $F(x, y, z) = 16x^2 - 9y^2 + 36z$

**Finding an Equation of a Tangent Plane** In Exercises 7–16, find an equation of the tangent plane to the surface at the given point.

**7.** $z = x^2 + y^2 + 3$

$(2, 1, 8)$

**8.** $f(x, y) = \dfrac{y}{x}$

$(1, 2, 2)$

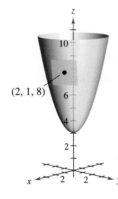

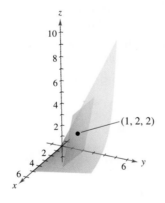

**9.** $z = \sqrt{x^2 + y^2}, \quad (3, 4, 5)$

**10.** $g(x, y) = \arctan \dfrac{y}{x}, \quad (1, 0, 0)$

**11.** $g(x, y) = x^2 + y^2, \quad (1, -1, 2)$

**12.** $f(x, y) = x^2 - 2xy + y^2, \quad (1, 2, 1)$

**13.** $h(x, y) = \ln \sqrt{x^2 + y^2}, \quad (3, 4, \ln 5)$

**14.** $f(x, y) = \sin x \cos y, \quad \left(\dfrac{\pi}{3}, \dfrac{\pi}{6}, \dfrac{3}{4}\right)$

**15.** $x^2 + y^2 - 5z^2 = 15, \quad (-4, -2, 1)$

**16.** $x^2 + 2z^2 = y^2, \quad (1, 3, -2)$

 **Finding an Equation of a Tangent Plane and a Normal Line** In Exercises 17–26, **(a)** find an equation of the tangent plane to the surface at the given point and **(b)** find a set of symmetric equations for the normal line to the surface at the given point.

**17.** $x + y + z = 9, \quad (3, 3, 3)$

**18.** $x^2 + y^2 + z^2 = 9, \quad (1, 2, 2)$

**19.** $x^2 + 2y^2 + z^2 = 7, \quad (1, -1, 2)$

**20.** $z = 16 - x^2 - y^2, \quad (2, 2, 8)$

**21.** $z = x^2 - y^2, \quad (3, 2, 5)$

**22.** $xy - z = 0, \quad (-2, -3, 6)$

**23.** $xyz = 10, \quad (1, 2, 5)$

**24.** $6xy = z, \quad (-1, 1, -6)$

**25.** $z = ye^{2xy}, \quad (0, 2, 2)$

**26.** $y \ln xz^2 = 2, \quad (e, 2, 1)$

 **Finding the Equation of a Tangent Line to a Curve** In Exercises 27–32, find a set of parametric equations for the tangent line to the curve of intersection of the surfaces at the given point.

**27.** $x^2 + y^2 = 2, \quad z = x, \quad (1, 1, 1)$

**28.** $z = x^2 + y^2, \quad z = 4 - y, \quad (2, -1, 5)$

**29.** $x^2 + z^2 = 25, \quad y^2 + z^2 = 25, \quad (3, 3, 4)$

**30.** $z = \sqrt{x^2 + y^2}, \quad 5x - 2y + 3z = 22, \quad (3, 4, 5)$

**31.** $x^2 + y^2 + z^2 = 14, \quad x - y - z = 0, \quad (3, 1, 2)$

**32.** $z = x^2 + y^2, \quad x + y + 6z = 33, \quad (1, 2, 5)$

 **Finding the Angle of Inclination of a Tangent Plane** In Exercises 33–36, find the angle of inclination of the tangent plane to the surface at the given point.

**33.** $3x^2 + 2y^2 - z = 15, \quad (2, 2, 5)$

**34.** $2xy - z^3 = 0, \quad (2, 2, 2)$

**35.** $x^2 - y^2 + z = 0, \quad (1, 2, 3)$

**36.** $x^2 + y^2 = 5, \quad (2, 1, 3)$

**Horizontal Tangent Plane** In Exercises 37–42, find the point(s) on the surface at which the tangent plane is horizontal.

**37.** $z = 3 - x^2 - y^2 + 6y$

**38.** $z = 3x^2 + 2y^2 - 3x + 4y - 5$

**39.** $z = x^2 - xy + y^2 - 2x - 2y$

**40.** $z = 4x^2 + 4xy - 2y^2 + 8x - 5y - 4$

**41.** $z = 5xy$

**42.** $z = xy + \dfrac{1}{x} + \dfrac{1}{y}$

**Tangent Surfaces** In Exercises 43 and 44, show that the surfaces are tangent to each other at the given point by showing that the surfaces have the same tangent plane at this point.

**43.** $x^2 + 2y^2 + 3z^2 = 3$, $x^2 + y^2 + z^2 + 6x - 10y + 14 = 0$, $(-1, 1, 0)$

**44.** $x^2 + y^2 + z^2 - 8x - 12y + 4z + 42 = 0$, $x^2 + y^2 + 2z = 7$, $(2, 3, -3)$

**Perpendicular Tangent Planes** In Exercises 45 and 46, (a) show that the surfaces intersect at the given point and (b) show that the surfaces have perpendicular tangent planes at this point.

**45.** $z = 2xy^2$, $8x^2 - 5y^2 - 8z = -13$, $(1, 1, 2)$

**46.** $x^2 + y^2 + z^2 + 2x - 4y - 4z - 12 = 0$, $4x^2 + y^2 + 16z^2 = 24$, $(1, -2, 1)$

---

**EXPLORING CONCEPTS**

**47. Tangent Plane** The tangent plane to the surface represented by $F(x, y, z) = 0$ at a point $P$ is also tangent to the surface represented by $G(x, y, z) = 0$ at $P$. Is $\nabla F(x, y, z) = \nabla G(x, y, z)$ at $P$? Explain.

**48. Normal Lines** For some surfaces, the normal lines at any point pass through the same geometric object. What is the common geometric object for a sphere? What is the common geometric object for a right circular cylinder? Explain.

---

**49. Using an Ellipsoid** Find a point on the ellipsoid $3x^2 + y^2 + 3z^2 = 1$ where the tangent line is parallel to the plane $-12x + 2y + 6z = 0$.

**50. Using a Hyperboloid** Find a point on the hyperboloid $x^2 + 4y^2 - z^2 = 1$ where the tangent plane is parallel to the plane $x + 4y - z = 0$.

**51. Using an Ellipsoid** Find a point on the ellipsoid $x^2 + 4y^2 + z^2 = 9$ where the tangent plane is perpendicular to the line with parametric equations

$x = 2 - 4t$, $y = 1 + 8t$, and $z = 3 - 2t$.

**52. HOW DO YOU SEE IT?** The graph shows the ellipsoid $x^2 + 4y^2 + z^2 = 16$. Use the graph to determine the equation of the tangent plane at each of the given points.

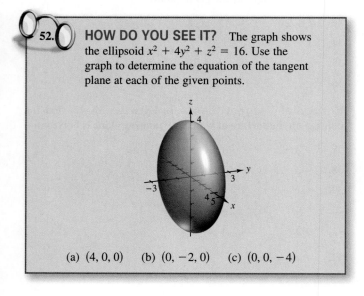

(a) $(4, 0, 0)$    (b) $(0, -2, 0)$    (c) $(0, 0, -4)$

**53. Investigation** Consider the function

$$f(x, y) = \frac{4xy}{(x^2 + 1)(y^2 + 1)}$$

on the intervals $-2 \le x \le 2$ and $0 \le y \le 3$.

(a) Find a set of parametric equations of the normal line and an equation of the tangent plane to the surface at the point $(1, 1, 1)$.

(b) Repeat part (a) for the point $\left(-1, 2, -\frac{4}{5}\right)$.

(c) Use a computer algebra system to graph the surface, the normal lines, and the tangent planes found in parts (a) and (b).

**54. Investigation** Consider the function

$$f(x, y) = \frac{\sin y}{x}$$

on the intervals $-3 \le x \le 3$ and $0 \le y \le 2\pi$.

(a) Find a set of parametric equations of the normal line and an equation of the tangent plane to the surface at the point $\left(2, \frac{\pi}{2}, \frac{1}{2}\right)$.

(b) Repeat part (a) for the point $\left(-\frac{2}{3}, \frac{3\pi}{2}, \frac{3}{2}\right)$.

(c) Use a computer algebra system to graph the surface, the normal lines, and the tangent planes found in parts (a) and (b).

**55. Using Functions** Consider the functions

$$f(x, y) = 6 - x^2 - \frac{y^2}{4} \quad \text{and} \quad g(x, y) = 2x + y.$$

(a) Find a set of parametric equations of the tangent line to the curve of intersection of the surfaces at the point $(1, 2, 4)$ and find the angle between the gradients of $f$ and $g$.

(b) Use a computer algebra system to graph the surfaces and the tangent line found in part (a).

**56. Using Functions** Consider the functions

$$f(x, y) = \sqrt{16 - x^2 - y^2 + 2x - 4y}$$

and

$$g(x, y) = \frac{\sqrt{2}}{2}\sqrt{1 - 3x^2 + y^2 + 6x + 4y}.$$

(a) Use a computer algebra system to graph the first-octant portion of the surfaces represented by $f$ and $g$.

(b) Find one first-octant point on the curve of intersection and show that the surfaces are orthogonal at this point.

(c) These surfaces are orthogonal along the curve of intersection. Does part (b) prove this fact? Explain.

**Writing a Tangent Plane** In Exercises 57 and 58, show that the tangent plane to the quadric surface at the point $(x_0, y_0, z_0)$ can be written in the given form.

**57.** Ellipsoid: $\dfrac{x^2}{a^2} + \dfrac{y^2}{b^2} + \dfrac{z^2}{c^2} = 1$

Tangent plane: $\dfrac{x_0 x}{a^2} + \dfrac{y_0 y}{b^2} + \dfrac{z_0 z}{c^2} = 1$

**58.** Hyperboloid: $\dfrac{x^2}{a^2} + \dfrac{y^2}{b^2} - \dfrac{z^2}{c^2} = 1$

Tangent plane: $\dfrac{x_0 x}{a^2} + \dfrac{y_0 y}{b^2} - \dfrac{z_0 z}{c^2} = 1$

**59. Tangent Planes of a Cone** Show that any tangent plane to the cone

$$z^2 = a^2 x^2 + b^2 y^2$$

passes through the origin.

**60. Tangent Planes** Let $f$ be a differentiable function and consider the surface

$$z = xf\left(\frac{y}{x}\right).$$

Show that the tangent plane at any point $P(x_0, y_0, z_0)$ on the surface passes through the origin.

**61. Approximation** Consider the following approximations for a function $f(x, y)$ centered at $(0, 0)$.

**Linear Approximation:**

$$P_1(x, y) = f(0, 0) + f_x(0, 0)x + f_y(0, 0)y$$

**Quadratic Approximation:**

$$P_2(x, y) = f(0, 0) + f_x(0, 0)x + f_y(0, 0)y +$$
$$\tfrac{1}{2}f_{xx}(0, 0)x^2 + f_{xy}(0, 0)xy + \tfrac{1}{2}f_{yy}(0, 0)y^2$$

[Note that the linear approximation is the tangent plane to the surface at $(0, 0, f(0, 0))$.]

(a) Find the linear approximation of $f(x, y) = e^{x-y}$ centered at $(0, 0)$.

(b) Find the quadratic approximation of $f(x, y) = e^{x-y}$ centered at $(0, 0)$.

(c) When $x = 0$ in the quadratic approximation, you obtain the second-degree Taylor polynomial for what function? Answer the same question for $y = 0$.

(d) Complete the table.

| $x$ | $y$ | $f(x, y)$ | $P_1(x, y)$ | $P_2(x, y)$ |
|---|---|---|---|---|
| 0 | 0 | | | |
| 0 | 0.1 | | | |
| 0.2 | 0.1 | | | |
| 0.2 | 0.5 | | | |
| 1 | 0.5 | | | |

(e) Use a computer algebra system to graph the surfaces $z = f(x, y)$, $z = P_1(x, y)$, and $z = P_2(x, y)$.

**62. Approximation** Repeat Exercise 61 for the function $f(x, y) = \cos(x + y)$.

**63. Proof** Prove that the angle of inclination $\theta$ of the tangent plane to the surface $z = f(x, y)$ at the point $(x_0, y_0, z_0)$ is given by

$$\cos\theta = \frac{1}{\sqrt{[f_x(x_0, y_0)]^2 + [f_y(x_0, y_0)]^2 + 1}}.$$

**64. Proof** Prove Theorem 13.14.

---

**SECTION PROJECT**

## Wildflowers

The diversity of wildflowers in a meadow can be measured by counting the numbers of daisies, buttercups, shooting stars, and so on. When there are $n$ types of wildflowers, each with a proportion $p_i$ of the total population, it follows that

$$p_1 + p_2 + \cdots + p_n = 1.$$

The measure of diversity of the population is defined as

$$H = -\sum_{i=1}^{n} p_i \log_2 p_i.$$

In this definition, it is understood that $p_i \log_2 p_i = 0$ when $p_i = 0$. The tables show proportions of wildflowers in a meadow in May, June, August, and September.

**May**

| Flower type | 1 | 2 | 3 | 4 |
|---|---|---|---|---|
| Proportion | $\frac{5}{16}$ | $\frac{5}{16}$ | $\frac{5}{16}$ | $\frac{1}{16}$ |

**June**

| Flower type | 1 | 2 | 3 | 4 |
|---|---|---|---|---|
| Proportion | $\frac{1}{4}$ | $\frac{1}{4}$ | $\frac{1}{4}$ | $\frac{1}{4}$ |

**August**

| Flower type | 1 | 2 | 3 | 4 |
|---|---|---|---|---|
| Proportion | $\frac{1}{4}$ | 0 | $\frac{1}{4}$ | $\frac{1}{2}$ |

**September**

| Flower type | 1 | 2 | 3 | 4 |
|---|---|---|---|---|
| Proportion | 0 | 0 | 0 | 1 |

(a) Determine the wildflower diversity for each month. How would you interpret September's diversity? Which month had the greatest diversity?

(b) When the meadow contains 10 types of wildflowers in roughly equal proportions, is the diversity of the population greater than or less than the diversity of a similar distribution of 4 types of flowers? What type of distribution (of 10 types of wildflowers) would produce maximum diversity?

(c) Let $H_n$ represent the maximum diversity of $n$ types of wildflowers. Does $H_n$ approach a limit as $n$ approaches $\infty$?

■ **FOR FURTHER INFORMATION** Biologists use the concept of diversity to measure the proportions of different types of organisms within an environment. For more information on this technique, see the article "Information Theory and Biological Diversity" by Steven Kolmes and Kevin Mitchell in the *UMAP Modules*.

# 13.8 Extrema of Functions of Two Variables

■ Find absolute and relative extrema of a function of two variables.
■ Use the Second Partials Test to find relative extrema of a function of two variables.

## Absolute Extrema and Relative Extrema

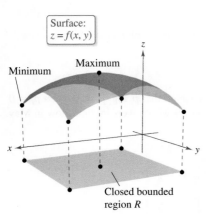

Surface:
$z = f(x, y)$

Minimum — Maximum

Closed bounded
region $R$

$R$ contains point(s) at which $f(x, y)$ is a minimum and point(s) at which $f(x, y)$ is a maximum.
**Figure 13.64**

In Chapter 3, you studied techniques for finding the extreme values of a function of a single variable. In this section, you will extend these techniques to functions of two variables. For example, in Theorem 13.15 below, the Extreme Value Theorem for a function of a single variable is extended to a function of two variables.

Consider the continuous function $f$ of two variables, defined on a closed bounded region $R$ in the $xy$-plane. The values $f(a, b)$ and $f(c, d)$ such that

$$f(a, b) \leq f(x, y) \leq f(c, d) \qquad \text{(a, b) and (c, d) are in R.}$$

for all $(x, y)$ in $R$ are called the **minimum** and **maximum** of $f$ in the region $R$, as shown in Figure 13.64. Recall from Section 13.2 that a region in the plane is *closed* when it contains all of its boundary points. The Extreme Value Theorem deals with a region in the plane that is both closed and *bounded*. A region in the plane is **bounded** when it is a subregion of a closed disk in the plane.

---

**THEOREM 13.15 Extreme Value Theorem**

Let $f$ be a continuous function of two variables $x$ and $y$ defined on a closed bounded region $R$ in the $xy$-plane.

1. There is at least one point in $R$ at which $f$ takes on a minimum value.
2. There is at least one point in $R$ at which $f$ takes on a maximum value.

---

A minimum is also called an **absolute minimum** and a maximum is also called an **absolute maximum.** As in single-variable calculus, there is a distinction made between absolute extrema and **relative extrema.**

---

**Definition of Relative Extrema**

Let $f$ be a function defined on a region $R$ containing $(x_0, y_0)$.

1. The function $f$ has a **relative minimum** at $(x_0, y_0)$ if

$$f(x, y) \geq f(x_0, y_0)$$

for all $(x, y)$ in an *open* disk containing $(x_0, y_0)$.

2. The function $f$ has a **relative maximum** at $(x_0, y_0)$ if

$$f(x, y) \leq f(x_0, y_0)$$

for all $(x, y)$ in an *open* disk containing $(x_0, y_0)$.

---

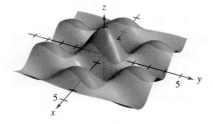

Relative extrema
**Figure 13.65**

To say that $f$ has a relative maximum at $(x_0, y_0)$ means that the point $(x_0, y_0, z_0)$ is at least as high as all nearby points on the graph of

$$z = f(x, y).$$

Similarly, $f$ has a relative minimum at $(x_0, y_0)$ when $(x_0, y_0, z_0)$ is at least as low as all nearby points on the graph. (See Figure 13.65.)

To locate relative extrema of $f$, you can investigate the points at which the gradient of $f$ is **0** or the points at which one of the partial derivatives does not exist. Such points are called **critical points** of $f$.

---

**Definition of Critical Point**

Let $f$ be defined on an open region $R$ containing $(x_0, y_0)$. The point $(x_0, y_0)$ is a **critical point** of $f$ if one of the following is true.

**1.** $f_x(x_0, y_0) = 0$ and $f_y(x_0, y_0) = 0$

**2.** $f_x(x_0, y_0)$ or $f_y(x_0, y_0)$ does not exist.

---

**KARL WEIERSTRASS**
**(1815–1897)**

Although the Extreme Value Theorem had been used by earlier mathematicians, the first to provide a rigorous proof was the German mathematician Karl Weierstrass. Weierstrass also provided rigorous justifications for many other mathematical results already in common use. We are indebted to him for much of the logical foundation on which modern calculus is built.
*See LarsonCalculus.com to read more of this biography.*

Recall from Theorem 13.11 that if $f$ is differentiable and

$$\nabla f(x_0, y_0) = f_x(x_0, y_0)\mathbf{i} + f_y(x_0, y_0)\mathbf{j} = 0\mathbf{i} + 0\mathbf{j}$$

then every directional derivative at $(x_0, y_0)$ must be 0. This implies that the function has a horizontal tangent plane at the point $(x_0, y_0)$, as shown in Figure 13.66. It appears that such a point is a likely location of a relative extremum. This is confirmed by Theorem 13.16.

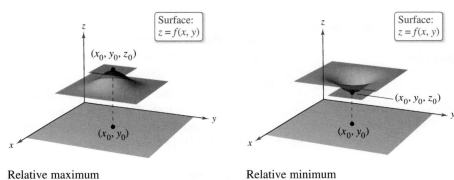

Relative maximum          Relative minimum
**Figure 13.66**

---

**THEOREM 13.16   Relative Extrema Occur Only at Critical Points**

If $f$ has a relative extremum at $(x_0, y_0)$ on an open region $R$, then $(x_0, y_0)$ is a critical point of $f$.

---

**Exploration**

Use a graphing utility to graph $z = x^3 - 3xy + y^3$ using the bounds $0 \leq x \leq 3$, $0 \leq y \leq 3$, and $-3 \leq z \leq 3$. This view makes it appear as though the surface has an absolute minimum. Does the surface have an absolute minimum? Why or why not?

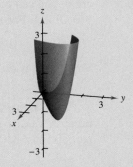

**Finding a Relative Extremum**

$\vdots \cdots \triangleright$ *See LarsonCalculus.com for an interactive version of this type of example.*

Determine the relative extrema of

$$f(x, y) = 2x^2 + y^2 + 8x - 6y + 20.$$

**Solution** Begin by finding the critical points of $f$. Because

$$f_x(x, y) = 4x + 8 \qquad \text{Partial with respect to } x$$

and

$$f_y(x, y) = 2y - 6 \qquad \text{Partial with respect to } y$$

are defined for all $x$ and $y$, the only critical points are those for which both first partial derivatives are 0. To locate these points, set $f_x(x, y)$ and $f_y(x, y)$ equal to 0, and solve the equations

$$4x + 8 = 0 \quad \text{and} \quad 2y - 6 = 0$$

to obtain the critical point $(-2, 3)$. By completing the square for $f$, you can see that for all $(x, y) \neq (-2, 3)$

$$f(x, y) = 2(x + 2)^2 + (y - 3)^2 + 3 > 3.$$

So, a relative *minimum* of $f$ occurs at $(-2, 3)$. The value of the relative minimum is $f(-2, 3) = 3$, as shown in Figure 13.67.

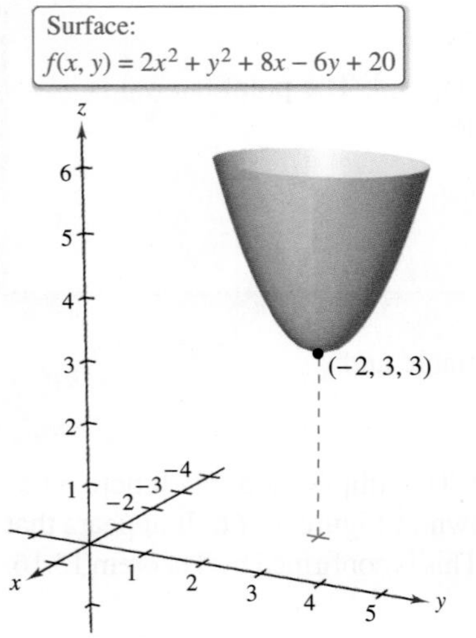

Surface:
$f(x, y) = 2x^2 + y^2 + 8x - 6y + 20$

The function $z = f(x, y)$ has a relative minimum at $(-2, 3)$.
**Figure 13.67**

Example 1 shows a relative minimum occurring at one type of critical point—the type for which both $f_x(x, y)$ and $f_y(x, y)$ are 0. The next example concerns a relative maximum that occurs at the other type of critical point—the type for which either $f_x(x, y)$ or $f_y(x, y)$ does not exist.

**Finding a Relative Extremum**

Determine the relative extrema of

$$f(x, y) = 1 - (x^2 + y^2)^{1/3}.$$

**Solution** Because

$$f_x(x, y) = -\frac{2x}{3(x^2 + y^2)^{2/3}} \qquad \text{Partial with respect to } x$$

and

$$f_y(x, y) = -\frac{2y}{3(x^2 + y^2)^{2/3}} \qquad \text{Partial with respect to } y$$

it follows that both partial derivatives exist for all points in the $xy$-plane except for $(0, 0)$. Moreover, because the partial derivatives cannot both be 0 unless both $x$ and $y$ are 0, you can conclude that $(0, 0)$ is the only critical point. In Figure 13.68, note that $f(0, 0)$ is 1. For all other $(x, y)$, it is clear that

$$f(x, y) = 1 - (x^2 + y^2)^{1/3} < 1.$$

So, $f$ has a relative *maximum* at $(0, 0)$.

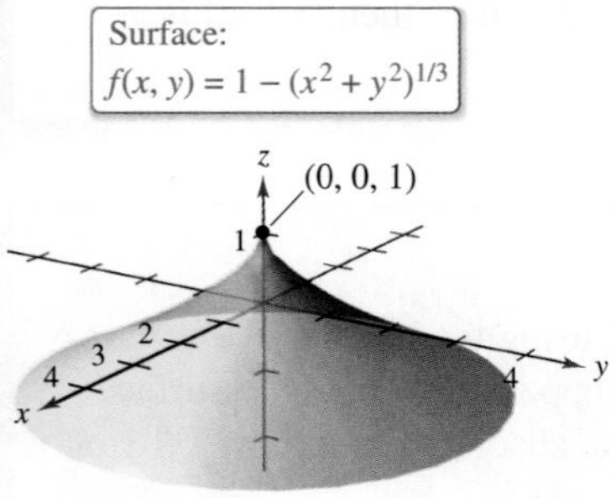

Surface:
$f(x, y) = 1 - (x^2 + y^2)^{1/3}$

$f_x(x, y)$ and $f_y(x, y)$ are undefined at $(0, 0)$.
**Figure 13.68**

In Example 2, $f_x(x, y) = 0$ for every point on the $y$-axis other than $(0, 0)$. However, because $f_y(x, y)$ is nonzero, these are not critical points. Remember that *one* of the partials must not exist or *both* must be 0 in order to yield a critical point.

## The Second Partials Test

Theorem 13.16 tells you that to find relative extrema, you need only examine values of $f(x, y)$ at critical points. However, as is true for a function of one variable, the critical points of a function of two variables do not always yield relative maxima or minima. Some critical points yield **saddle points,** which are neither relative maxima nor relative minima.

As an example of a critical point that does not yield a relative extremum, consider the hyperbolic paraboloid

$$f(x, y) = y^2 - x^2$$

as shown in Figure 13.69. At the point $(0, 0)$, both partial derivatives

$$f_x(x, y) = -2x \quad \text{and} \quad f_y(x, y) = 2y$$

are 0. The function $f$ does not, however, have a relative extremum at this point because in any open disk centered at $(0, 0)$, the function takes on both negative values (along the $x$-axis) *and* positive values (along the $y$-axis), So, the point $(0, 0, 0)$ is a saddle point of the surface. (The term "saddle point" comes from the fact that surfaces such as the one shown in Figure 13.69 resemble saddles.)

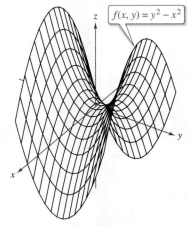

Saddle point at $(0, 0, 0)$:
$f_x(0, 0) = f_y(0, 0) = 0$
**Figure 13.69**

For the functions in Examples 1 and 2, it was relatively easy to determine the relative extrema, because each function was either given, or able to be written, in completed square form. For more complicated functions, algebraic arguments are less convenient and it is better to rely on the analytic means presented in the following Second Partials Test. This is the two-variable counterpart of the Second Derivative Test for functions of one variable. The proof of this theorem is best left to a course in advanced calculus.

---

**THEOREM 13.17    Second Partials Test**

Let $f$ have continuous second partial derivatives on an open region containing a point $(a, b)$ for which

$$f_x(a, b) = 0 \quad \text{and} \quad f_y(a, b) = 0.$$

To test for relative extrema of $f$, consider the quantity

$$d = f_{xx}(a, b)f_{yy}(a, b) - [f_{xy}(a, b)]^2.$$

1. If $d > 0$ and $f_{xx}(a, b) > 0$, then $f$ has a **relative minimum** at $(a, b)$.
2. If $d > 0$ and $f_{xx}(a, b) < 0$, then $f$ has a **relative maximum** at $(a, b)$.
3. If $d < 0$, then $(a, b, f(a, b))$ is a **saddle point.**
4. The test is inconclusive if $d = 0$.

---

**• • REMARK** If $d > 0$, then $f_{xx}(a, b)$ and $f_{yy}(a, b)$ must have the same sign. This means that $f_{xx}(a, b)$ can be replaced by $f_{yy}(a, b)$ in the first two parts of the test.

A convenient device for remembering the formula for $d$ in the Second Partials Test is given by the $2 \times 2$ determinant

$$d = \begin{vmatrix} f_{xx}(a, b) & f_{xy}(a, b) \\ f_{yx}(a, b) & f_{yy}(a, b) \end{vmatrix}$$

where $f_{xy}(a, b) = f_{yx}(a, b)$ by Theorem 13.3.

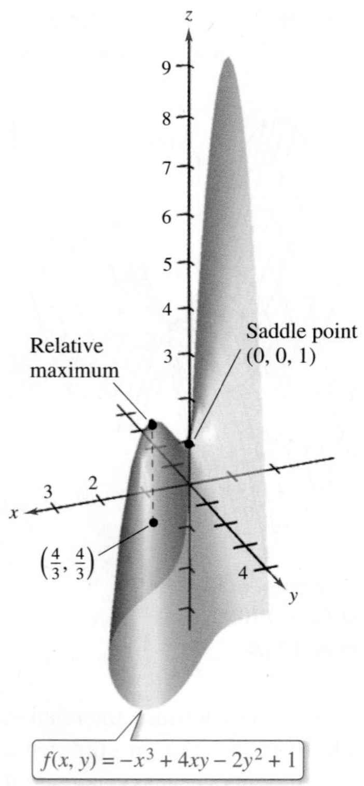

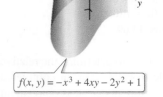

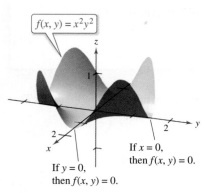

**Relative maximum**

Saddle point
$(0, 0, 1)$

$\left(\frac{4}{3}, \frac{4}{3}\right)$

$f(x, y) = -x^3 + 4xy - 2y^2 + 1$

**Figure 13.70**

---

EXAMPLE 3 **Using the Second Partials Test**

Find the relative extrema of $f(x, y) = -x^3 + 4xy - 2y^2 + 1$.

**Solution** Begin by finding the critical points of $f$. Because

$$f_x(x, y) = -3x^2 + 4y \quad \text{and} \quad f_y(x, y) = 4x - 4y$$

exist for all $x$ and $y$, the only critical points are those for which both first partial derivatives are 0. To locate these points, set $f_x(x, y)$ and $f_y(x, y)$ equal to 0 to obtain

$$-3x^2 + 4y = 0 \quad \text{and} \quad 4x - 4y = 0.$$

From the second equation, you know that $x = y$, and, by substitution into the first equation, you obtain two solutions: $y = x = 0$ and $y = x = \frac{4}{3}$. Because

$$f_{xx}(x, y) = -6x, \quad f_{yy}(x, y) = -4, \quad \text{and} \quad f_{xy}(x, y) = 4$$

it follows that, for the critical point $(0, 0)$,

$$d = f_{xx}(0, 0)f_{yy}(0, 0) - [f_{xy}(0, 0)]^2 = 0 - 16 < 0$$

and, by the Second Partials Test, you can conclude that $(0, 0, 1)$ is a saddle point of $f$. Furthermore, for the critical point $\left(\frac{4}{3}, \frac{4}{3}\right)$,

$$d = f_{xx}\left(\frac{4}{3}, \frac{4}{3}\right)f_{yy}\left(\frac{4}{3}, \frac{4}{3}\right) - \left[f_{xy}\left(\frac{4}{3}, \frac{4}{3}\right)\right]^2$$
$$= -8(-4) - 16$$
$$= 16$$
$$> 0$$

and because $f_{xx}\left(\frac{4}{3}, \frac{4}{3}\right) = -8 < 0$, you can conclude that $f$ has a relative maximum at $\left(\frac{4}{3}, \frac{4}{3}\right)$, as shown in Figure 13.70. ∎

The Second Partials Test can fail to find relative extrema in two ways. If either of the first partial derivatives does not exist, you cannot use the test. Also, if

$$d = f_{xx}(a, b)f_{yy}(a, b) - [f_{xy}(a, b)]^2 = 0$$

the test fails. In such cases, you can try a sketch or some other approach, as demonstrated in the next example.

EXAMPLE 4 **Failure of the Second Partials Test**

Find the relative extrema of $f(x, y) = x^2y^2$.

**Solution** Because $f_x(x, y) = 2xy^2$ and $f_y(x, y) = 2x^2y$, you know that both partial derivatives are 0 when $x = 0$ or $y = 0$. That is, every point along the $x$- or $y$-axis is a critical point. Moreover, because

$$f_{xx}(x, y) = 2y^2, \quad f_{yy}(x, y) = 2x^2, \quad \text{and} \quad f_{xy}(x, y) = 4xy$$

you know that

$$d = f_{xx}(x, y)f_{yy}(x, y) - [f_{xy}(x, y)]^2$$
$$= 4x^2y^2 - 16x^2y^2$$
$$= -12x^2y^2$$

which is 0 when either $x = 0$ or $y = 0$. So, the Second Partials Test fails. However, because $f(x, y) = 0$ for every point along the $x$- or $y$-axis and $f(x, y) = x^2y^2 > 0$ for all other points, you can conclude that each of these critical points yields an absolute minimum, as shown in Figure 13.71. ∎

$f(x, y) = x^2y^2$

If $x = 0$,
then $f(x, y) = 0$.

If $y = 0$,
then $f(x, y) = 0$.

**Figure 13.71**

Absolute extrema of a function can occur in two ways. First, some relative extrema also happen to be absolute extrema. For instance, in Example 1, $f(-2, 3)$ is an absolute minimum of the function. (On the other hand, the relative maximum found in Example 3 is not an absolute maximum of the function.) Second, absolute extrema can occur at a boundary point of the domain. This is illustrated in Example 5.

### EXAMPLE 5    Finding Absolute Extrema

Find the absolute extrema of the function

$$f(x, y) = \sin xy$$

on the closed region given by

$$0 \le x \le \pi \quad \text{and} \quad 0 \le y \le 1.$$

**Solution**    From the partial derivatives

$$f_x(x, y) = y \cos xy \quad \text{and} \quad f_y(x, y) = x \cos xy$$

you can see that each point lying on the hyperbola $xy = \pi/2$ is a critical point. These points each yield the value

$$f(x, y) = \sin \frac{\pi}{2} = 1$$

which you know is the absolute maximum, as shown in Figure 13.72. The only other critical point of $f$ *lying in the given region* is $(0, 0)$. It yields an absolute minimum of 0, because

$$0 \le xy \le \pi$$

implies that

$$0 \le \sin xy \le 1.$$

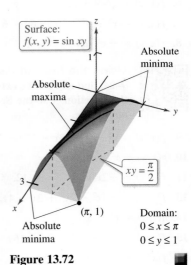

To locate other absolute extrema, you should consider the four boundaries of the region formed by taking traces with the vertical planes $x = 0$, $x = \pi$, $y = 0$, and $y = 1$. In doing this, you will find that $\sin xy = 0$ at all points on the $x$-axis, at all points on the $y$-axis, and at the point $(\pi, 1)$. Each of these points yields an absolute minimum for the surface, as shown in Figure 13.72.

**Figure 13.72**

The concepts of relative extrema and critical points can be extended to functions of three or more variables. When all first partial derivatives of

$$w = f(x_1, x_2, x_3, \ldots, x_n)$$

exist, it can be shown that a relative maximum or minimum can occur at $(x_1, x_2, x_3, \ldots, x_n)$ only when every first partial derivative is 0 at that point. This means that the critical points are obtained by solving the following system of equations.

$$f_{x_1}(x_1, x_2, x_3, \ldots, x_n) = 0$$
$$f_{x_2}(x_1, x_2, x_3, \ldots, x_n) = 0$$
$$\vdots$$
$$f_{x_n}(x_1, x_2, x_3, \ldots, x_n) = 0$$

The extension of Theorem 13.17 to three or more variables is also possible, although you will not study such an extension in this text.

# 13.8 Exercises

See **CalcChat.com** for tutorial help and worked-out solutions to odd-numbered exercises.

**CONCEPT CHECK**

**1. Function of Two Variables** For a function of two variables, describe (a) relative minimum, (b) relative maximum, (c) critical point, and (d) saddle point.

**2. Second Partials Test** Under what condition does the Second Partials Test fail?

**Finding Relative Extrema** In Exercises 3–8, identify any extrema of the function by recognizing its given form or its form after completing the square. Verify your results by using the partial derivatives to locate any critical points and test for relative extrema.

**3.** $g(x, y) = (x - 1)^2 + (y - 3)^2$

**4.** $g(x, y) = 5 - (x - 6)^2 - (y + 2)^2$

**5.** $f(x, y) = \sqrt{x^2 + y^2 + 1}$

**6.** $f(x, y) = \sqrt{49 - (x - 2)^2 - y^2}$

**7.** $f(x, y) = x^2 + y^2 + 2x - 6y + 6$

**8.** $f(x, y) = -x^2 - y^2 + 10x + 12y - 64$

**Using the Second Partials Test** In Exercises 9–24, find all relative extrema and saddle points of the function. Use the Second Partials Test where applicable.

**9.** $f(x, y) = x^2 + y^2 + 8x - 12y - 3$

**10.** $g(x, y) = x^2 - y^2 - x - y$

**11.** $f(x, y) = -2x^4y^4$     **12.** $f(x, y) = \frac{1}{2}xy$

**13.** $f(x, y) = -3x^2 - 2y^2 + 3x - 4y + 5$

**14.** $h(x, y) = x^2 - 3xy - y^2$

**15.** $f(x, y) = 7x^2 + 2y^2 - 7x + 16y - 13$

**16.** $f(x, y) = x^5 + y^5$

**17.** $z = x^2 + xy + \frac{1}{2}y^2 - 2x + y$

**18.** $z = -5x^2 + 4xy - y^2 + 16x + 10$

**19.** $f(x, y) = -4(x^2 + y^2 + 81)^{1/4}$

**20.** $h(x, y) = (x^2 + y^2)^{1/3} + 2$

**21.** $f(x, y) = x^2 - xy - y^2 - 3x - y$

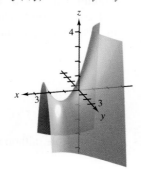

**22.** $f(x, y) = 2xy - \frac{1}{2}(x^4 + y^4) + 1$

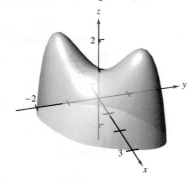

**23.** $z = e^{-x} \sin y$

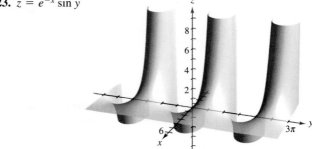

**24.** $z = \left(\frac{1}{2} - x^2 + y^2\right)e^{1 - x^2 - y^2}$

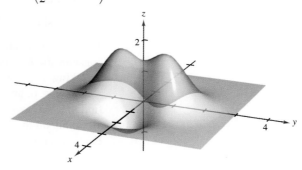

**Finding Relative Extrema and Saddle Points Using Technology** In Exercises 25–28, use a computer algebra system to graph the surface and locate any relative extrema and saddle points.

**25.** $z = \dfrac{-4x}{x^2 + y^2 + 1}$

**26.** $z = \cos x + \sin y, \quad -\pi/2 < x < \pi/2, \ -\pi < y < \pi$

**27.** $z = (x^2 + 4y^2)e^{1 - x^2 - y^2}$     **28.** $z = e^{xy}$

**Finding Relative Extrema** In Exercises 29 and 30, examine the function for extrema without using the derivative tests and use a computer algebra system to graph the surface and verify your answers. (*Hint:* By observation, determine whether it is possible for $z$ to be negative. When is $z$ equal to 0?)

**29.** $z = \dfrac{(x - y)^4}{x^2 + y^2}$         **30.** $z = \dfrac{(x^2 - y^2)^2}{x^2 + y^2}$

**Think About It**  In Exercises 31–34, determine whether there is a relative maximum, a relative minimum, a saddle point, or insufficient information to determine the nature of the function $f(x, y)$ at the critical point $(x_0, y_0)$.

**31.** $f_{xx}(x_0, y_0) = 9,\quad f_{yy}(x_0, y_0) = 4,\quad f_{xy}(x_0, y_0) = 6$

**32.** $f_{xx}(x_0, y_0) = -3,\quad f_{yy}(x_0, y_0) = -8,\quad f_{xy}(x_0, y_0) = 2$

**33.** $f_{xx}(x_0, y_0) = -9,\quad f_{yy}(x_0, y_0) = 6,\quad f_{xy}(x_0, y_0) = 10$

**34.** $f_{xx}(x_0, y_0) = 25,\quad f_{yy}(x_0, y_0) = 8,\quad f_{xy}(x_0, y_0) = 10$

 **Finding Relative Extrema and Saddle Points**  In Exercises 35–38, (a) find the critical points, (b) test for relative extrema, (c) list the critical points for which the Second Partials Test fails, and (d) use a computer algebra system to graph the function, labeling any extrema and saddle points.

**35.** $f(x, y) = x^3 + y^3$

**36.** $f(x, y) = x^3 + y^3 - 6x^2 + 9x^2 + 12x + 27y + 19$

**37.** $f(x, y) = (x - 1)^2(y + 4)^2$

**38.** $f(x, y) = x^{2/3} + y^{2/3}$

**Finding Absolute Extrema**  In Exercises 39–46, find the absolute extrema of the function over the region $R$. (In each case, $R$ contains the boundaries.) Use a computer algebra system to confirm your results.

**39.** $f(x, y) = x^2 - 4xy + 5$

$R = \{(x, y) : 1 \le x \le 4, 0 \le y \le 2\}$

**40.** $f(x, y) = x^2 + xy, \quad R = \{(x, y) : |x| \le 2, |y| \le 1\}$

**41.** $f(x, y) = 12 - 3x - 2y$

$R$: The triangular region in the $xy$-plane with vertices $(2, 0)$, $(0, 1)$, and $(1, 2)$

**42.** $f(x, y) = (2x - y)^2$

$R$: The triangular region in the $xy$-plane with vertices $(2, 0)$, $(0, 1)$, and $(1, 2)$

**43.** $f(x, y) = 3x^2 + 2y^2 - 4y$

$R$: The region in the $xy$-plane bounded by the graphs of $y = x^2$ and $y = 4$

**44.** $f(x, y) = 2x - 2xy + y^2$

$R$: The region in the $xy$-plane bounded by the graphs of $y = x^2$ and $y = 1$

**45.** $f(x, y) = x^2 + 2xy + y^2, \quad R = \{(x, y) : |x| \le 2, |y| \le 1\}$

**46.** $f(x, y) = \dfrac{4xy}{(x^2 + 1)(y^2 + 1)}$

$R = \{(x, y) : 0 \le x \le 1, 0 \le y \le 1\}$

**Examining a Function**  In Exercises 47 and 48, find the critical points of the function and, from the form of the function, determine whether a relative maximum or a relative minimum occurs at each point.

**47.** $f(x, y, z) = x^2 + (y - 3)^2 + (z + 1)^2$

**48.** $f(x, y, z) = 9 - [x(y - 1)(z + 2)]^2$

**EXPLORING CONCEPTS**

**49. Using the Second Partials Test**  A function $f$ has continuous second partial derivatives on an open region containing the critical point $(3, 7)$. The function has a minimum at $(3, 7)$, and $d > 0$ for the Second Partials Test. Determine the interval for $f_{xy}(3, 7)$ when $f_{xx}(3, 7) = 2$ and $f_{yy}(3, 7) = 8$.

**50. Using the Second Partials Test**  A function $f$ has continuous second partial derivatives on an open region containing the critical point $(a, b)$. If $f_{xx}(a, b)$ and $f_{yy}(a, b)$ have opposite signs, what is implied? Explain.

**Sketching a Graph**  In Exercises 51 and 52, sketch the graph of an arbitrary function $f$ satisfying the given conditions. State whether the function has any extrema or saddle points. (There are many correct answers.)

**51.** All of the first and second partial derivatives of $f$ are 0.

**52.** $f_x(x, y) > 0$ and $f_y(x, y) < 0$ for all $(x, y)$.

**53. Comparing Functions**  Consider the functions

$$f(x, y) = x^2 - y^2 \quad \text{and} \quad g(x, y) = x^2 + y^2.$$

(a) Show that both functions have a critical point at $(0, 0)$.

(b) Explain how $f$ and $g$ behave differently at this critical point.

**54.** **HOW DO YOU SEE IT?**  Determine whether each labeled point is an absolute maximum, an absolute minimum, or neither.

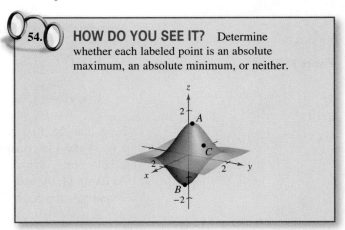

**True or False?**  In Exercises 55–58, determine whether the statement is true or false. If it is false, explain why or give an example that shows it is false.

**55.** If $f$ has a relative maximum at $(x_0, y_0, z_0)$, then

$$f_x(x_0, y_0) = f_y(x_0, y_0) = 0.$$

**56.** If $f_x(x_0, y_0) = f_y(x_0, y_0) = 0$, then $f$ has a relative extremum at $(x_0, y_0, z_0)$.

**57.** Between any two relative minima of $f$, there must be at least one relative maximum of $f$.

**58.** If $f$ is continuous for all $x$ and $y$ and has two relative minima, then $f$ must have at least one relative maximum.

# 13.9 Applications of Extrema

- Solve optimization problems involving functions of several variables.
- Use the method of least squares.

## Applied Optimization Problems

In this section, you will study a few of the many applications of extrema of functions of two (or more) variables.

### EXAMPLE 1    Finding Maximum Volume

⋅⋅⋅▷ *See LarsonCalculus.com for an interactive version of this type of example.*

A rectangular box is resting on the $xy$-plane with one vertex at the origin. The opposite vertex lies in the plane

$$6x + 4y + 3z = 24$$

as shown in Figure 13.73. Find the maximum volume of the box.

**Solution**  Let $x$, $y$, and $z$ represent the length, width, and height of the box. Because one vertex of the box lies in the plane $6x + 4y + 3z = 24$, you know that $z = \frac{1}{3}(24 - 6x - 4y)$. So, you can write the volume $xyz$ of the box as a function of two variables.

$$V(x, y) = (x)(y)\left[\tfrac{1}{3}(24 - 6x - 4y)\right]$$
$$= \tfrac{1}{3}(24xy - 6x^2y - 4xy^2)$$

Next, find the first partial derivatives of $V$.

$$V_x(x, y) = \frac{1}{3}(24y - 12xy - 4y^2) = \frac{y}{3}(24 - 12x - 4y)$$

$$V_y(x, y) = \frac{1}{3}(24x - 6x^2 - 8xy) = \frac{x}{3}(24 - 6x - 8y)$$

Note that the first partial derivatives are defined for all $x$ and $y$. So, by setting $V_x(x, y)$ and $V_y(x, y)$ equal to 0 and solving the equations $\frac{1}{3}y(24 - 12x - 4y) = 0$ and $\frac{1}{3}x(24 - 6x - 8y) = 0$, you obtain the critical points $(0, 0)$, $(4, 0)$, $(0, 6)$, and $\left(\frac{4}{3}, 2\right)$. At $(0, 0)$, $(4, 0)$, and $(0, 6)$, the volume is 0, so these points do not yield a maximum volume. At the point $\left(\frac{4}{3}, 2\right)$, you can apply the Second Partials Test.

$$V_{xx}(x, y) = -4y$$

$$V_{yy}(x, y) = \frac{-8x}{3}$$

$$V_{xy}(x, y) = \frac{1}{3}(24 - 12x - 8y)$$

Because

$$V_{xx}\left(\tfrac{4}{3}, 2\right)V_{yy}\left(\tfrac{4}{3}, 2\right) - \left[V_{xy}\left(\tfrac{4}{3}, 2\right)\right]^2 = (-8)\left(-\tfrac{32}{9}\right) - \left(-\tfrac{8}{3}\right)^2 = \tfrac{64}{3} > 0$$

and

$$V_{xx}\left(\tfrac{4}{3}, 2\right) = -8 < 0$$

you can conclude from the Second Partials Test that the maximum volume is

$$V\left(\tfrac{4}{3}, 2\right) = \tfrac{1}{3}\left[24\left(\tfrac{4}{3}\right)(2) - 6\left(\tfrac{4}{3}\right)^2(2) - 4\left(\tfrac{4}{3}\right)(2^2)\right] = \tfrac{64}{9} \text{ cubic units.}$$

Note that the volume is 0 at the boundary points of the triangular domain of $V$. ■

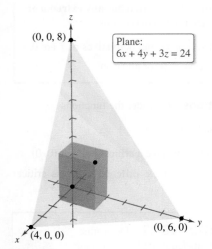

$(0, 0, 8)$

Plane:
$6x + 4y + 3z = 24$

$(0, 6, 0)$

$(4, 0, 0)$

**Figure 13.73**

⋅⋅**REMARK**  In many applied problems, the domain of the function to be optimized is a closed bounded region. To find minimum or maximum points, you must not only test critical points, but also consider the values of the function at points on the boundary.

Applications of extrema in economics and business often involve more than one independent variable. For instance, a company may produce several models of one type of product. The price per unit and profit per unit are usually different for each model. Moreover, the demand for each model is often a function of the prices of the other models (as well as its own price). The next example illustrates an application involving two products.

### EXAMPLE 2    Finding the Maximum Profit

A manufacturer determines that the profit $P$ (in dollars) obtained by producing and selling $x$ units of Product 1 and $y$ units of Product 2 is approximated by the model

$$P(x, y) = 8x + 10y - (0.001)(x^2 + xy + y^2) - 10{,}000.$$

Find the production level that produces a maximum profit. What is the maximum profit?

**Solution**    The partial derivatives of the profit function are

$$P_x(x, y) = 8 - (0.001)(2x + y)$$

and

$$P_y(x, y) = 10 - (0.001)(x + 2y).$$

By setting these partial derivatives equal to 0, you obtain the following system of equations.

$$8 - (0.001)(2x + y) = 0$$
$$10 - (0.001)(x + 2y) = 0$$

After simplifying, this system of linear equations can be written as

$$2x + \ y = \ \ 8000$$
$$x + 2y = 10{,}000.$$

Solving this system produces $x = 2000$ and $y = 4000$. The second partial derivatives of $P$ are

$$P_{xx}(2000, 4000) = -0.002$$
$$P_{yy}(2000, 4000) = -0.002$$
$$P_{xy}(2000, 4000) = -0.001.$$

Because $P_{xx} < 0$ and

$$P_{xx}(2000, 4000)P_{yy}(2000, 4000) - [P_{xy}(2000, 4000)]^2 = (-0.002)^2 - (-0.001)^2$$

is greater than 0, you can conclude that the production level of $x = 2000$ units and $y = 4000$ units yields a *maximum* profit. The maximum profit is

$$P(2000, 4000)$$
$$= 8(2000) + 10(4000) - (0.001)[2000^2 + 2000(4000) + 4000^2] - 10{,}000$$
$$= \$18{,}000.$$

In Example 2, it was assumed that the manufacturing plant is able to produce the required number of units to yield a maximum profit. In actual practice, the production would be bounded by physical constraints. You will study such constrained optimization problems in the next section.

■ **FOR FURTHER INFORMATION**    For more information on the use of mathematics in economics, see the article "Mathematical Methods of Economics" by Joel Franklin in *The American Mathematical Monthly*. To view this article, go to *MathArticles.com*.

## The Method of Least Squares

Many of the examples in this text have involved **mathematical models.** For instance, Example 2 involves a quadratic model for profit. There are several ways to develop such models; one is called the **method of least squares.**

In constructing a model to represent a particular phenomenon, the goals are simplicity and accuracy. Of course, these goals often conflict. For instance, a simple linear model for the points in Figure 13.74 is

$$y = 1.9x - 5.$$

However, Figure 13.75 shows that by choosing the slightly more complicated quadratic model

$$y = 0.20x^2 - 0.7x + 1$$

you can achieve greater accuracy.

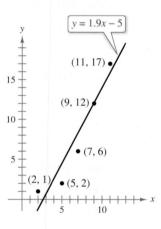

**Figure 13.74**          **Figure 13.75**

As a measure of how well the model $y = f(x)$ fits the collection of points

$$\{(x_1, y_1), (x_2, y_2), (x_3, y_3), \ldots, (x_n, y_n)\}$$

you can add the squares of the differences between the actual $y$-values and the values given by the model to obtain the **sum of the squared errors**

$$S = \sum_{i=1}^{n} [f(x_i) - y_i]^2. \qquad \text{Sum of the squared errors}$$

Graphically, $S$ can be interpreted as the sum of the squares of the vertical distances between the graph of $f$ and the given points in the plane, as shown in Figure 13.76. If the model is perfect, then $S = 0$. However, when perfection is not feasible, you can settle for a model that minimizes $S$. For instance, the sum of the squared errors for the linear model in Figure 13.74 is

$$S = 17.6.$$

**• • REMARK** A method for finding the least squares regression quadratic for a collection of data is described in Exercise 31.

Statisticians call the *linear model* that minimizes $S$ the **least squares regression line.** The proof that this line actually minimizes $S$ involves the minimizing of a function of two variables.

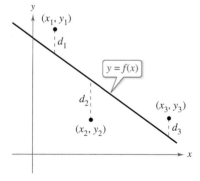

Sum of the squared errors:
$$S = d_1^2 + d_2^2 + d_3^2$$
**Figure 13.76**

**THEOREM 13.18   Least Squares Regression Line**

The **least squares regression line** for $\{(x_1, y_1), (x_2, y_2), \ldots, (x_n, y_n)\}$ is given by $f(x) = ax + b$, where

$$a = \frac{n\sum_{i=1}^{n} x_i y_i - \sum_{i=1}^{n} x_i \sum_{i=1}^{n} y_i}{n\sum_{i=1}^{n} x_i^2 - \left(\sum_{i=1}^{n} x_i\right)^2} \quad \text{and} \quad b = \frac{1}{n}\left(\sum_{i=1}^{n} y_i - a\sum_{i=1}^{n} x_i\right).$$

**Proof**   Let $S(a, b)$ represent the sum of the squared errors for the model

$$f(x) = ax + b$$

and the given set of points. That is,

$$S(a, b) = \sum_{i=1}^{n} [f(x_i) - y_i]^2$$

$$= \sum_{i=1}^{n} (ax_i + b - y_i)^2$$

where the points $(x_i, y_i)$ represent constants. Because $S$ is a function of $a$ and $b$, you can use the methods discussed in the preceding section to find the minimum value of $S$. Specifically, the first partial derivatives of $S$ are

$$S_a(a, b) = \sum_{i=1}^{n} 2x_i(ax_i + b - y_i)$$

$$= 2a\sum_{i=1}^{n} x_i^2 + 2b\sum_{i=1}^{n} x_i - 2\sum_{i=1}^{n} x_i y_i$$

and

$$S_b(a, b) = \sum_{i=1}^{n} 2(ax_i + b - y_i)$$

$$= 2a\sum_{i=1}^{n} x_i + 2nb - 2\sum_{i=1}^{n} y_i.$$

By setting these two partial derivatives equal to 0, you obtain the values of $a$ and $b$ that are listed in the theorem. It is left to you to apply the Second Partials Test (see Exercise 41) to verify that these values of $a$ and $b$ yield a minimum. ∎

If the $x$-values are symmetrically spaced about the $y$-axis, then $\Sigma x_i = 0$ and the formulas for $a$ and $b$ simplify to

$$a = \frac{\sum_{i=1}^{n} x_i y_i}{\sum_{i=1}^{n} x_i^2}$$

and

$$b = \frac{1}{n}\sum_{i=1}^{n} y_i.$$

This simplification is often possible with a translation of the $x$-values. For instance, given that the $x$-values in a data collection consist of the values 9, 10, 11, 12, and 13, you could let 11 be represented by 0.

EXAMPLE 3    **Finding the Least Squares Regression Line**

Find the least squares regression line for the points

$$(-3, 0), \quad (-1, 1), \quad (0, 2), \quad \text{and} \quad (2, 3).$$

**Solution**    The table shows the calculations involved in finding the least squares regression line using $n = 4$.

| $x$ | $y$ | $xy$ | $x^2$ |
|---|---|---|---|
| $-3$ | $0$ | $0$ | $9$ |
| $-1$ | $1$ | $-1$ | $1$ |
| $0$ | $2$ | $0$ | $0$ |
| $2$ | $3$ | $6$ | $4$ |
| $\displaystyle\sum_{i=1}^{n} x_i = -2$ | $\displaystyle\sum_{i=1}^{n} y_i = 6$ | $\displaystyle\sum_{i=1}^{n} x_i y_i = 5$ | $\displaystyle\sum_{i=1}^{n} x_i^2 = 14$ |

Applying Theorem 13.18 produces

$$a = \frac{n \displaystyle\sum_{i=1}^{n} x_i y_i - \displaystyle\sum_{i=1}^{n} x_i \displaystyle\sum_{i=1}^{n} y_i}{n \displaystyle\sum_{i=1}^{n} x_i^2 - \left(\displaystyle\sum_{i=1}^{n} x_i\right)^2}$$

$$= \frac{4(5) - (-2)(6)}{4(14) - (-2)^2}$$

$$= \frac{8}{13}$$

and

$$b = \frac{1}{n}\left(\sum_{i=1}^{n} y_i - a \sum_{i=1}^{n} x_i\right)$$

$$= \frac{1}{4}\left[6 - \frac{8}{13}(-2)\right]$$

$$= \frac{47}{26}.$$

> **TECHNOLOGY**    Many calculators have "built-in" least squares regression programs. If your calculator has such a program, use it to duplicate the results of Example 3.

The least squares regression line is

$$f(x) = \frac{8}{13}x + \frac{47}{26}$$

as shown in Figure 13.77.

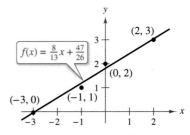

Least squares regression line
**Figure 13.77**

# 13.9 Exercises

See CalcChat.com for tutorial help and worked-out solutions to odd-numbered exercises.

## CONCEPT CHECK

**1. Applied Optimization Problems** In your own words, state the problem-solving strategy for applied minimum and maximum problems.

**2. Method of Least Squares** In your own words, describe the method of least squares for finding mathematical models.

 **Finding Minimum Distance** In Exercises 3 and 4, find the minimum distance from the point to the plane $x - y + z = 3$. (*Hint:* To simplify the computations, minimize the square of the distance.)

**3.** $(1, -3, 2)$          **4.** $(4, 0, 6)$

**Finding Minimum Distance** In Exercises 5 and 6, find the minimum distance from the point to the surface $z = \sqrt{1 - 2x - 2y}$. (*Hint:* To simplify the computations, minimize the square of the distance.)

**5.** $(-2, -2, 0)$          **6.** $(-4, 1, 0)$

 **Finding Positive Numbers** In Exercises 7–10, find three positive integers $x$, $y$, and $z$ that satisfy the given conditions.

**7.** The product is 27, and the sum is a minimum.

**8.** The sum is 32, and $P = xy^2z$ is a maximum.

**9.** The sum is 30, and the sum of the squares is a minimum.

**10.** The product is 1, and the sum of the squares is a minimum.

**11. Cost** A home improvement contractor is painting the walls and ceiling of a rectangular room. The volume of the room is 668.25 cubic feet. The cost of wall paint is $0.06 per square foot and the cost of ceiling paint is $0.11 per square foot. Find the room dimensions that result in a minimum cost for the paint. What is the minimum cost for the paint?

**12. Maximum Volume** The material for constructing the base of an open box costs 1.5 times as much per unit area as the material for constructing the sides. For a fixed amount of money $C$, find the dimensions of the box of largest volume that can be made.

**13. Volume and Surface Area** Show that a rectangular box of given volume and minimum surface area is a cube.

**14. Maximum Volume** Show that the rectangular box of maximum volume inscribed in a sphere of radius $r$ is a cube.

**15. Maximum Revenue** A company manufactures running shoes and basketball shoes. The total revenue (in thousands of dollars) from $x_1$ units of running shoes and $x_2$ units of basketball shoes is

$$R = -5x_1^2 - 8x_2^2 - 2x_1x_2 + 42x_1 + 102x_2$$

where $x_1$ and $x_2$ are in thousands of units. Find $x_1$ and $x_2$ so as to maximize the revenue.

**16. Maximum Profit** A corporation manufactures candles at two locations. The cost of producing $x_1$ units at location 1 is $C_1 = 0.02x_1^2 + 4x_1 + 500$ and the cost of producing $x_2$ units at location 2 is $C_2 = 0.05x_2^2 + 4x_2 + 275$. The candles sell for $15 per unit. Find the quantity that should be produced at each location to maximize the profit $P = 15(x_1 + x_2) - C_1 - C_2$.

**17. Hardy-Weinberg Law** Common blood types are determined genetically by three alleles A, B, and O. (An allele is any of a group of possible mutational forms of a gene.) A person whose blood type is AA, BB, or OO is homozygous. A person whose blood type is AB, AO, or BO is heterozygous. The Hardy-Weinberg Law states that the proportion $P$ of heterozygous individuals in any given population is

$$P(p, q, r) = 2pq + 2pr + 2qr$$

where $p$ represents the percent of allele A in the population, $q$ represents the percent of allele B in the population, and $r$ represents the percent of allele O in the population. Use the fact that

$$p + q + r = 1$$

to show that the maximum proportion of heterozygous individuals in any population is $\frac{2}{3}$.

**18. Shannon Diversity Index** One way to measure species diversity is to use the Shannon diversity index $H$. If a habitat consists of three species, A, B, and C, then its Shannon diversity index is

$$H = -x \ln x - y \ln y - z \ln z$$

where $x$ is the percent of species A in the habitat, $y$ is the percent of species B in the habitat, and $z$ is the percent of species C in the habitat. Use the fact that

$$x + y + z = 1$$

to show that the maximum value of $H$ occurs when $x = y = z = \frac{1}{3}$. What is the maximum value of $H$?

**19. Minimum Cost** A water line is to be built from point $P$ to point $S$ and must pass through regions where construction costs differ (see figure). The cost per kilometer (in dollars) is $3k$ from $P$ to $Q$, $2k$ from $Q$ to $R$, and $k$ from $R$ to $S$. Find $x$ and $y$ such that the total cost $C$ will be minimized.

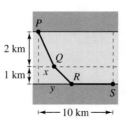     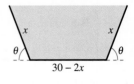

Figure for 19            Figure for 20

**20. Area** A trough with trapezoidal cross sections is formed by turning up the edges of a 30-inch-wide sheet of aluminum (see figure). Find the cross section of maximum area.

**Finding the Least Squares Regression Line** In Exercises 21–24, (a) find the least squares regression line and (b) calculate $S$, the sum of the squared errors. Use the regression capabilities of a graphing utility to verify your results.

21.
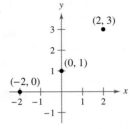

22.
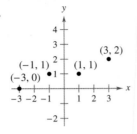

23.
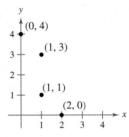

24.
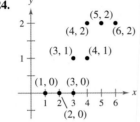

**Finding the Least Squares Regression Line** In Exercises 25–28, find the least squares regression line for the points. Use the regression capabilities of a graphing utility to verify your results. Use the graphing utility to plot the points and graph the regression line.

25. $(0, 0)$, $(1, 1)$, $(3, 6)$, $(4, 8)$, $(5, 9)$

26. $(0, 4)$, $(4, 1)$, $(7, -3)$

27. $(0, 6)$, $(4, 3)$, $(5, 0)$, $(8, -4)$, $(10, -5)$

28. $(6, 4)$, $(1, 2)$, $(3, 3)$, $(8, 6)$, $(11, 8)$, $(13, 8)$

29. **Modeling Data** The table shows the gross income tax collections (in billions of dollars) by the Internal Revenue Service for individuals $x$ and businesses $y$ for selected years. *(Source: U.S. Internal Revenue Service)*

| Year | 1980 | 1985 | 1990 | 1995 |
|---|---|---|---|---|
| Individual, $x$ | 288 | 397 | 540 | 676 |
| Business, $y$ | 72 | 77 | 110 | 174 |

| Year | 2000 | 2005 | 2010 | 2015 |
|---|---|---|---|---|
| Individual, $x$ | 1137 | 1108 | 1164 | 1760 |
| Business, $y$ | 236 | 307 | 278 | 390 |

(a) Use the regression capabilities of a graphing utility to find the least squares regression line for the data.

(b) Use the model to estimate the business income taxes collected when the individual income taxes collected is $1300 billion.

(c) In 1975, the individual income taxes collected was $156 billion and the business income taxes collected was $46 billion. Describe how including this information would affect the model.

30. **Modeling Data** The ages $x$ (in years) and systolic blood pressures $y$ (in mmHg) of seven men are shown in the table.

| Age, $x$ | 16 | 25 | 39 | 45 | 49 | 64 | 70 |
|---|---|---|---|---|---|---|---|
| Systolic Blood Pressure, $y$ | 109 | 122 | 150 | 165 | 159 | 183 | 199 |

(a) Use the regression capabilities of a graphing utility to find the least squares regression line for the data.

(b) Use a graphing utility to plot the data and graph the model.

(c) Use the model to approximate the change in systolic blood pressure for each one-year increase in age.

(d) A 30-year-old man has a systolic blood pressure of 180 mmHg. Describe how including this information would affect the model.

**EXPLORING CONCEPTS**

31. **Method of Least Squares** Find a system of equations whose solution yields the coefficients $a$, $b$, and $c$ for the least squares regression quadratic

$$y = ax^2 + bx + c$$

for the points $(x_1, y_1), (x_2, y_2), \ldots, (x_n, y_n)$ by minimizing the sum

$$S(a, b, c) = \sum_{i=1}^{n} (y_i - ax_i^2 - bx_i - c)^2.$$

32. **HOW DO YOU SEE IT?** Match the regression equation with the appropriate graph. Explain your reasoning. (Note that the $x$- and $y$-axes are broken.)

(a) $y = 0.22x - 7.5$

(b) $y = -0.35x + 11.5$

(c) $y = 0.09x + 19.8$

(d) $y = -1.29x + 89.8$

(i)
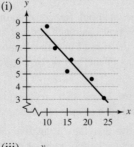

(ii)
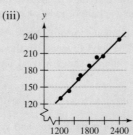

(iii)

(iv)
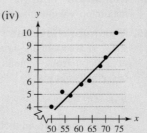

**Finding the Least Squares Regression Quadratic  In Exercises 33–36, use the result of Exercise 31 to find the least squares regression quadratic for the points. Use the regression capabilities of a graphing utility to verify your results. Use the graphing utility to plot the points and graph the least squares regression quadratic.**

**33.** $(-2, 0)$, $(-1, 0)$, $(0, 1)$, $(1, 2)$, $(2, 5)$

**34.** $(-4, 5)$, $(-2, 6)$, $(2, 6)$, $(4, 2)$

**35.** $(0, 0)$, $(2, 2)$, $(3, 6)$, $(4, 12)$

**36.** $(0, 10)$, $(1, 9)$, $(2, 6)$, $(3, 0)$

**37. Modeling Data**  After a new turbocharger for an automobile engine was developed, the following experimental data were obtained for speed $y$ in miles per hour at two-second time intervals $x$.

| Time, $x$ | 0 | 2 | 4 | 6 | 8 | 10 |
|---|---|---|---|---|---|---|
| Speed, $y$ | 0 | 15 | 30 | 50 | 65 | 70 |

(a) Use the result of Exercise 31 to find the least squares regression quadratic for the data.

(b) Use a graphing utility to plot the points and graph the model.

**38. Modeling Data**  The table shows the total numbers of enrollees $y$ (in millions) for the Veterans Health Administration for 2010 through 2014. Let $x = 0$ represent the year 2010.  *(Source: U.S. Department of Veterans Affairs)*

| Year, $x$ | 2010 | 2011 | 2012 | 2013 | 2014 |
|---|---|---|---|---|---|
| Total Enrollees, $y$ | 8.3 | 8.6 | 8.8 | 8.9 | 9.1 |

(a) Use the regression capabilities of a graphing utility to find the least squares regression line for the data.

(b) Use the regression capabilities of a graphing utility to find the least squares regression quadratic for the data.

(c) Use a graphing utility to plot the data and graph the models.

(d) Use both models to forecast the total number of enrollees for the year 2025. How do the two models differ as you extrapolate into the future?

**39. Modeling Data**  A meteorologist measures the atmospheric pressure $P$ (in kilograms per square meter) at altitude $h$ (in kilometers). The data are shown below.

| Altitude, $h$ | 0 | 5 | 10 | 15 | 20 |
|---|---|---|---|---|---|
| Pressure, $P$ | 10,332 | 5583 | 2376 | 1240 | 517 |

(a) Use the regression capabilities of a graphing utility to find the least squares regression line for the points $(h, \ln P)$.

(b) The result in part (a) is an equation of the form $\ln P = ah + b$. Write this logarithmic form in exponential form.

(c) Use a graphing utility to plot the original data and graph the exponential model in part (b).

**40. Modeling Data**  The endpoints of the interval over which distinct vision is possible are called the near point and far point of the eye. With increasing age, these points normally change. The table shows the approximate near points $y$ (in inches) for various ages $x$ (in years).  *(Source: Ophthalmology & Physiological Optics)*

| Age, $x$ | 16 | 32 | 44 | 50 | 60 |
|---|---|---|---|---|---|
| Near Point, $y$ | 3.0 | 4.7 | 9.8 | 19.7 | 39.4 |

(a) Find a rational model for the data by taking the reciprocals of the near points to generate the points $(x, 1/y)$. Use the regression capabilities of a graphing utility to find the least squares regression line for the revised data. The resulting line has the form $1/y = ax + b$. Solve for $y$.

(b) Use a graphing utility to plot the data and graph the model.

(c) Do you think the model can be used to predict the near point for a person who is 70 years old? Explain.

**41. Using the Second Partials Test**  Use the Second Partials Test to verify that the formulas for $a$ and $b$ given in Theorem 13.18 yield a minimum.

$$\left[ \text{Hint: Use the fact that } n\sum_{i=1}^{n} x_i^2 \geq \left( \sum_{i=1}^{n} x_i \right)^2. \right]$$

## SECTION PROJECT

## Building a Pipeline

An oil company wishes to construct a pipeline from its offshore facility $A$ to its refinery $B$. The offshore facility is 2 miles from shore, and the refinery is 1 mile inland. Furthermore, $A$ and $B$ are 5 miles apart, as shown in the figure.

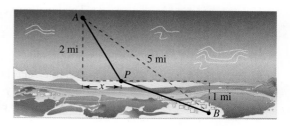

The cost of building the pipeline is $3 million per mile in the water and $4 million per mile on land. So, the cost of the pipeline depends on the location of point $P$, where it meets the shore. What would be the most economical route of the pipeline?

Imagine that you are to write a report to the oil company about this problem. Let $x$ be the distance shown in the figure. Determine the cost of building the pipeline from $A$ to $P$ and the cost of building it from $P$ to $B$. Analyze some sample pipeline routes and their corresponding costs. For instance, what is the cost of the most direct route? Then use calculus to determine the route of the pipeline that minimizes the cost. Explain all steps of your development and include any relevant graphs.

# 13.10  Lagrange Multipliers

- Understand the Method of Lagrange Multipliers.
- Use Lagrange multipliers to solve constrained optimization problems.
- Use the Method of Lagrange Multipliers with two constraints.

## Lagrange Multipliers

> **LAGRANGE MULTIPLIERS**
>
> The Method of Lagrange Multipliers is named after the French mathematician Joseph-Louis Lagrange. Lagrange first introduced the method in his famous paper on mechanics, written when he was just 19 years old.

Many optimization problems have restrictions, or **constraints,** on the values that can be used to produce the optimal solution. Such constraints tend to complicate optimization problems because the optimal solution can occur at a boundary point of the domain. In this section, you will study an ingenious technique for solving such problems. It is called the **Method of Lagrange Multipliers.**

To see how this technique works, consider the problem of finding the rectangle of maximum area that can be inscribed in the ellipse

$$\frac{x^2}{3^2} + \frac{y^2}{4^2} = 1.$$

Let $(x, y)$ be the vertex of the rectangle in the first quadrant, as shown in Figure 13.78. Because the rectangle has sides of lengths $2x$ and $2y$, its area is given by

$$f(x, y) = 4xy. \qquad \text{Objective function}$$

You want to find $x$ and $y$ such that $f(x, y)$ is a maximum. Your choice of $(x, y)$ is restricted to first-quadrant points that lie on the ellipse

$$\frac{x^2}{3^2} + \frac{y^2}{4^2} = 1. \qquad \text{Constraint}$$

Now, consider the constraint equation to be a fixed level curve of

$$g(x, y) = \frac{x^2}{3^2} + \frac{y^2}{4^2}.$$

The level curves of $f$ represent a family of hyperbolas

$$f(x, y) = 4xy = k.$$

In this family, the level curves that meet the constraint correspond to the hyperbolas that intersect the ellipse. Moreover, to maximize $f(x, y)$, you want to find the hyperbola that just barely satisfies the constraint. The level curve that does this is the one that is *tangent* to the ellipse, as shown in Figure 13.79.

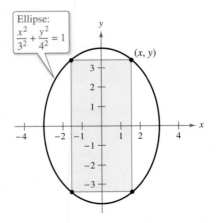

Objective function: $f(x, y) = 4xy$

**Figure 13.78**

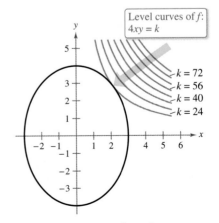

Constraint: $g(x, y) = \dfrac{x^2}{3^2} + \dfrac{y^2}{4^2} = 1$

**Figure 13.79**

To find the appropriate hyperbola, use the fact that two curves are tangent at a point if and only if their gradients are parallel. This means that $\nabla f(x, y)$ must be a scalar multiple of $\nabla g(x, y)$ at the point of tangency. In the context of constrained optimization problems, this scalar is denoted by $\lambda$ (the lowercase Greek letter lambda).

$$\nabla f(x, y) = \lambda \nabla g(x, y)$$

The scalar $\lambda$ is called a **Lagrange multiplier.** Theorem 13.19 gives the necessary conditions for the existence of such multipliers.

• • • • • • • • • • • • • • • ▷
• •
**REMARK** Lagrange's Theorem can be shown to be true for functions of three variables, using a similar argument with level surfaces and Theorem 13.14.

---

**THEOREM 13.19  Lagrange's Theorem**

Let $f$ and $g$ have continuous first partial derivatives such that $f$ has an extremum at a point $(x_0, y_0)$ on the smooth constraint curve $g(x, y) = c$. If $\nabla g(x_0, y_0) \neq \mathbf{0}$, then there is a real number $\lambda$ such that

$$\nabla f(x_0, y_0) = \lambda \nabla g(x_0, y_0).$$

---

**Proof**  To begin, represent the smooth curve given by $g(x, y) = c$ by the vector-valued function

$$\mathbf{r}(t) = x(t)\mathbf{i} + y(t)\mathbf{j}, \quad \mathbf{r}'(t) \neq \mathbf{0}$$

where $x'$ and $y'$ are continuous on an open interval $I$. Define the function $h$ as $h(t) = f(x(t), y(t))$. Then, because $f(x_0, y_0)$ is an extreme value of $f$, you know that

$$h(t_0) = f(x(t_0), y(t_0)) = f(x_0, y_0)$$

is an extreme value of $h$. This implies that $h'(t_0) = 0$, and, by the Chain Rule,

$$h'(t_0) = f_x(x_0, y_0)x'(t_0) + f_y(x_0, y_0)y'(t_0) = \nabla f(x_0, y_0) \cdot \mathbf{r}'(t_0) = 0.$$

So, $\nabla f(x_0, y_0)$ is orthogonal to $\mathbf{r}'(t_0)$. Moreover, by Theorem 13.12, $\nabla g(x_0, y_0)$ is also orthogonal to $\mathbf{r}'(t_0)$. Consequently, the gradients $\nabla f(x_0, y_0)$ and $\nabla g(x_0, y_0)$ are parallel, and there must exist a scalar $\lambda$ such that

$$\nabla f(x_0, y_0) = \lambda \nabla g(x_0, y_0). \qquad \blacksquare$$

The Method of Lagrange Multipliers uses Theorem 13.19 to find the extreme values of a function $f$ subject to a constraint.

• • • • • • • • • • • • • • ▷
• •
**REMARK**  As you will see in Examples 1 and 2, the Method of Lagrange Multipliers requires solving systems of nonlinear equations. This often can require some tricky algebraic manipulation.

---

**Method of Lagrange Multipliers**

Let $f$ and $g$ satisfy the hypothesis of Lagrange's Theorem, and let $f$ have a minimum or maximum subject to the constraint $g(x, y) = c$. To find the minimum or maximum of $f$, use these steps.

1. Simultaneously solve the equations $\nabla f(x, y) = \lambda \nabla g(x, y)$ and $g(x, y) = c$ by solving the following system of equations.

$$f_x(x, y) = \lambda g_x(x, y)$$
$$f_y(x, y) = \lambda g_y(x, y)$$
$$g(x, y) = c$$

2. Evaluate $f$ at each solution point obtained in the first step. The greatest value yields the maximum of $f$ subject to the constraint $g(x, y) = c$, and the least value yields the minimum of $f$ subject to the constraint $g(x, y) = c$.

---

## Constrained Optimization Problems

In the problem at the beginning of this section, you wanted to maximize the area of a rectangle that is inscribed in an ellipse. Example 1 shows how to use Lagrange multipliers to solve this problem.

### EXAMPLE 1 Using a Lagrange Multiplier with One Constraint

Find the maximum value of $f(x, y) = 4xy$, where $x > 0$ and $y > 0$, subject to the constraint $(x^2/3^2) + (y^2/4^2) = 1$.

**Solution** To begin, let

$$g(x, y) = \frac{x^2}{3^2} + \frac{y^2}{4^2} = 1.$$

By equating $\nabla f(x, y) = 4y\mathbf{i} + 4x\mathbf{j}$ and $\lambda \nabla g(x, y) = (2\lambda x/9)\mathbf{i} + (\lambda y/8)\mathbf{j}$, you obtain the following system of equations.

$$4y = \frac{2}{9}\lambda x \qquad f_x(x, y) = \lambda g_x(x, y)$$

$$4x = \frac{1}{8}\lambda y \qquad f_y(x, y) = \lambda g_y(x, y)$$

$$\frac{x^2}{3^2} + \frac{y^2}{4^2} = 1 \qquad \text{Constraint}$$

•••••••••••••••▷

•• **REMARK** Note in Example 1 that writing the constraint as

$$\frac{x^2}{3^2} + \frac{y^2}{4^2} = 1$$

or

$$\frac{x^2}{3^2} + \frac{y^2}{4^2} - 1 = 0$$

does not affect the solution—the constant is eliminated when you form $\nabla g$.

From the first equation, you obtain $\lambda = 18y/x$, and substitution into the second equation produces

$$4x = \frac{1}{8}\left(\frac{18y}{x}\right)y \implies x^2 = \frac{9}{16}y^2.$$

Substituting this value for $x^2$ into the third equation produces

$$\frac{1}{9}\left(\frac{9}{16}y^2\right) + \frac{1}{16}y^2 = 1 \implies y^2 = 8 \implies y = \pm 2\sqrt{2}.$$

Because $y > 0$, choose the positive value and find that

$$x^2 = \frac{9}{16}y^2$$

$$= \frac{9}{16}(8)$$

$$= \frac{9}{2}$$

$$x = \pm\frac{3}{\sqrt{2}}.$$

Because $x > 0$, choose the positive value. So, the maximum value of $f$ is

$$f\left(\frac{3}{\sqrt{2}}, 2\sqrt{2}\right) = 4\left(\frac{3}{\sqrt{2}}\right)(2\sqrt{2}) = 24.$$ ∎

Example 1 can also be solved using the techniques you learned in Chapter 3. To see how, try to find the maximum value of $A = 4xy$ given that $(x^2/3^2) + (y^2/4^2) = 1$. To begin, solve the second equation for $y$ to obtain $y = \frac{4}{3}\sqrt{9 - x^2}$. Then substitute into the first equation to obtain $A = 4x\left(\frac{4}{3}\sqrt{9 - x^2}\right)$. Finally, use the techniques of Chapter 3 to maximize $A$.

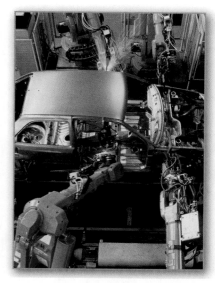

For some industrial applications, a robot can cost more than the annual wages and benefits for one employee. So, manufacturers must carefully balance the amount of money spent on labor and capital.

**EXAMPLE 2**    **A Business Application**

The Cobb-Douglas production function (see Section 13.1) for a manufacturer is given by

$$f(x, y) = 100x^{3/4}y^{1/4} \qquad \text{Objective function}$$

where $x$ represents the units of labor (at \$150 per unit) and $y$ represents the units of capital (at \$250 per unit). The total cost of labor and capital is limited to \$50,000. Find the maximum production level for this manufacturer.

**Solution**    The gradient of $f$ is

$$\nabla f(x, y) = 75x^{-1/4}y^{1/4}\mathbf{i} + 25x^{3/4}y^{-3/4}\mathbf{j}.$$

The limit on the cost of labor and capital produces the constraint

$$g(x, y) = 150x + 250y = 50,000. \qquad \text{Constraint}$$

So, $\lambda\nabla g(x, y) = 150\lambda\mathbf{i} + 250\lambda\mathbf{j}$. This gives rise to the following system of equations.

$$75x^{-1/4}y^{1/4} = 150\lambda \qquad f_x(x, y) = \lambda g_x(x, y)$$
$$25x^{3/4}y^{-3/4} = 250\lambda \qquad f_y(x, y) = \lambda g_y(x, y)$$
$$150x + 250y = 50,000 \qquad \text{Constraint}$$

By solving for $\lambda$ in the first equation

$$\lambda = \frac{75x^{-1/4}y^{1/4}}{150} = \frac{x^{-1/4}y^{1/4}}{2}$$

and substituting into the second equation, you obtain

$$25x^{3/4}y^{-3/4} = 250\left(\frac{x^{-1/4}y^{1/4}}{2}\right)$$
$$25x = 125y \qquad \text{Multiply by } x^{1/4}y^{3/4}.$$
$$x = 5y.$$

By substituting this value for $x$ in the third equation, you have

$$150(5y) + 250y = 50,000$$
$$1000y = 50,000$$
$$y = 50 \text{ units of capital.}$$

This means that the value of $x$ is

$$x = 5(50)$$
$$= 250 \text{ units of labor.}$$

So, the maximum production level is

$$f(250, 50) = 100(250)^{3/4}(50)^{1/4}$$
$$\approx 16,719 \text{ units of product.}$$

**■ FOR FURTHER INFORMATION**
For more information on the use of Lagrange multipliers in economics, see the article "Lagrange Multiplier Problems in Economics" by John V. Baxley and John C. Moorhouse in *The American Mathematical Monthly.* To view this article, go to *MathArticles.com.*

Economists call the Lagrange multiplier obtained in a production function the **marginal productivity of money.** For instance, in Example 2, the marginal productivity of money at $x = 250$ and $y = 50$ is

$$\lambda = \frac{x^{-1/4}y^{1/4}}{2} = \frac{(250)^{-1/4}(50)^{1/4}}{2} \approx 0.334$$

which means that for each additional dollar spent on production, an additional 0.334 unit of the product can be produced.

| EXAMPLE 3 | **Lagrange Multipliers and Three Variables** |

$\vdots\cdots\triangleright$ *See LarsonCalculus.com for an interactive version of this type of example.*

Find the minimum value of

$$f(x, y, z) = 2x^2 + y^2 + 3z^2 \qquad \text{Objective function}$$

subject to the constraint $2x - 3y - 4z = 49$.

**Solution**    Let $g(x, y, z) = 2x - 3y - 4z = 49$. Then, because

$$\nabla f(x, y, z) = 4x\mathbf{i} + 2y\mathbf{j} + 6z\mathbf{k}$$

and

$$\lambda\nabla g(x, y, z) = 2\lambda\mathbf{i} - 3\lambda\mathbf{j} - 4\lambda\mathbf{k}$$

you obtain the following system of equations.

$$
\begin{aligned}
4x &= 2\lambda & f_x(x, y, z) &= \lambda g_x(x, y, z) \\
2y &= -3\lambda & f_y(x, y, z) &= \lambda g_y(x, y, z) \\
6z &= -4\lambda & f_z(x, y, z) &= \lambda g_z(x, y, z) \\
2x - 3y - 4z &= 49 & &\text{Constraint}
\end{aligned}
$$

The solution of this system is $x = 3$, $y = -9$, and $z = -4$. So, the optimum value of $f$ is

$$
\begin{aligned}
f(3, -9, -4) &= 2(3)^2 + (-9)^2 + 3(-4)^2 \\
&= 147.
\end{aligned}
$$

From the original function and constraint, it is clear that $f(x, y, z)$ has no maximum. So, the optimum value of $f$ determined above is a minimum. ■

A graphical interpretation of constrained optimization problems in two variables was given at the beginning of this section. In three variables, the interpretation is similar, except that level surfaces are used instead of level curves. For instance, in Example 3, the level surfaces of $f$ are ellipsoids centered at the origin, and the constraint

$$2x - 3y - 4z = 49$$

is a plane. The minimum value of $f$ is represented by the ellipsoid that is tangent to the constraint plane, as shown in Figure 13.80.

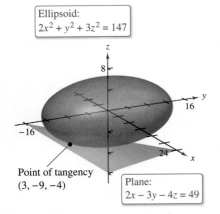

Ellipsoid:
$2x^2 + y^2 + 3z^2 = 147$

Point of tangency
$(3, -9, -4)$

Plane:
$2x - 3y - 4z = 49$

**Figure 13.80**

| EXAMPLE 4 | **Optimization Inside a Region** |

Find the extreme values of

$$f(x, y) = x^2 + 2y^2 - 2x + 3 \qquad \text{Objective function}$$

subject to the constraint $x^2 + y^2 \leq 10$.

**Solution**    To solve this problem, you can break the constraint into two cases.

**a.** For points *on the circle* $x^2 + y^2 = 10$, you can use Lagrange multipliers to find that the maximum value of $f(x, y)$ is 24—this value occurs at $(-1, 3)$ and at $(-1, -3)$. In a similar way, you can determine that the minimum value of $f(x, y)$ is approximately 6.675—this value occurs at $(\sqrt{10}, 0)$.

**b.** For points *inside the circle,* you can use the techniques discussed in Section 13.8 to conclude that the function has a relative minimum of 2 at the point $(1, 0)$.

By combining these two results, you can conclude that $f$ has a maximum of 24 at $(-1, \pm3)$ and a minimum of 2 at $(1, 0)$, as shown in Figure 13.81. ■

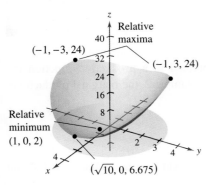

Relative maxima
$(-1, -3, 24)$
$(-1, 3, 24)$
Relative minimum
$(1, 0, 2)$
$(\sqrt{10}, 0, 6.675)$

**Figure 13.81**

## The Method of Lagrange Multipliers with Two Constraints

For optimization problems involving *two* constraint functions $g$ and $h$, you can introduce a second Lagrange multiplier, $\mu$ (the lowercase Greek letter mu), and then solve the equation

$$\nabla f = \lambda \nabla g + \mu \nabla h$$

where the gradients are not parallel, as illustrated in Example 5.

**EXAMPLE 5**  **Optimization with Two Constraints**

Let $T(x, y, z) = 20 + 2x + 2y + z^2$ represent the temperature at each point on the sphere

$$x^2 + y^2 + z^2 = 11.$$

Find the extreme temperatures on the curve formed by the intersection of the plane $x + y + z = 3$ and the sphere.

**Solution**  The two constraints are

$$g(x, y, z) = x^2 + y^2 + z^2 = 11 \quad \text{and} \quad h(x, y, z) = x + y + z = 3.$$

Using

$$\nabla T(x, y, z) = 2\mathbf{i} + 2\mathbf{j} + 2z\mathbf{k}$$
$$\lambda \nabla g(x, y, z) = 2\lambda x\mathbf{i} + 2\lambda y\mathbf{j} + 2\lambda z\mathbf{k}$$

and

$$\mu \nabla h(x, y, z) = \mu\mathbf{i} + \mu\mathbf{j} + \mu\mathbf{k}$$

you can write the following system of equations.

| | |
|---|---|
| $2 = 2\lambda x + \mu$ | $T_x(x, y, z) = \lambda g_x(x, y, z) + \mu h_x(x, y, z)$ |
| $2 = 2\lambda y + \mu$ | $T_y(x, y, z) = \lambda g_y(x, y, z) + \mu h_y(x, y, z)$ |
| $2z = 2\lambda z + \mu$ | $T_z(x, y, z) = \lambda g_z(x, y, z) + \mu h_z(x, y, z)$ |
| $x^2 + y^2 + z^2 = 11$ | Constraint 1 |
| $x + y + z = 3$ | Constraint 2 |

By subtracting the second equation from the first, you obtain the following system.

$$\lambda(x - y) = 0$$
$$2z(1 - \lambda) - \mu = 0$$
$$x^2 + y^2 + z^2 = 11$$
$$x + y + z = 3$$

> **• • REMARK**  The systems of equations that arise when the Method of Lagrange Multipliers is used are not, in general, linear systems, and finding the solutions often requires ingenuity.

From the first equation, you can conclude that $\lambda = 0$ or $x = y$. For $\lambda = 0$, you can show that the critical points are $(3, -1, 1)$ and $(-1, 3, 1)$. (Try doing this—it takes a little work.) For $\lambda \neq 0$, then $x = y$ and you can show that the critical points occur when $x = y = (3 \pm 2\sqrt{3})/3$ and $z = (3 \mp 4\sqrt{3})/3$. Finally, to find the optimal solutions, compare the temperatures at the four critical points.

$$T(3, -1, 1) = T(-1, 3, 1) = 25$$
$$T\left(\frac{3 - 2\sqrt{3}}{3}, \frac{3 - 2\sqrt{3}}{3}, \frac{3 + 4\sqrt{3}}{3}\right) = \frac{91}{3} \approx 30.33$$
$$T\left(\frac{3 + 2\sqrt{3}}{3}, \frac{3 + 2\sqrt{3}}{3}, \frac{3 - 4\sqrt{3}}{3}\right) = \frac{91}{3} \approx 30.33$$

So, $T = 25$ is the minimum temperature and $T = \frac{91}{3}$ is the maximum temperature on the curve.

# 13.10 Exercises

See **CalcChat.com** for tutorial help and worked-out solutions to odd-numbered exercises.

### CONCEPT CHECK

1. **Constrained Optimization Problems** Explain what is meant by constrained optimization problems.

2. **Method of Lagrange Multipliers** In your own words, describe the Method of Lagrange Multipliers for solving constrained optimization problems.

 **Using Lagrange Multipliers** In Exercises 3–10, use Lagrange multipliers to find the indicated extrema, assuming that $x$ and $y$ are positive.

3. Maximize $f(x, y) = xy$

Constraint: $x + y = 10$

4. Minimize $f(x, y) = 2x + y$

Constraint: $xy = 32$

5. Minimize $f(x, y) = x^2 + y^2$

Constraint: $x + 2y - 5 = 0$

6. Maximize $f(x, y) = x^2 - y^2$

Constraint: $2y - x^2 = 0$

7. Maximize $f(x, y) = 2x + 2xy + y$

Constraint: $2x + y = 100$

8. Minimize $f(x, y) = 3x + y + 10$

Constraint: $x^2 y = 6$

9. Maximize $f(x, y) = \sqrt{6 - x^2 - y^2}$

Constraint: $x + y - 2 = 0$

10. Minimize $f(x, y) = \sqrt{x^2 + y^2}$

Constraint: $2x + 4y - 15 = 0$

 **Using Lagrange Multipliers** In Exercises 11–14, use Lagrange multipliers to find the indicated extrema, assuming that $x$, $y$, and $z$ are positive.

11. Minimize $f(x, y, z) = x^2 + y^2 + z^2$

Constraint: $x + y + z - 9 = 0$

12. Maximize $f(x, y, z) = xyz$

Constraint: $x + y + z - 3 = 0$

13. Minimize $f(x, y, z) = x^2 + y^2 + z^2$

Constraint: $x + y + z = 1$

14. Maximize $f(x, y, z) = x + y + z$

Constraint: $x^2 + y^2 + z^2 = 1$

 **Using Lagrange Multipliers** In Exercises 15 and 16, use Lagrange multipliers to find any extrema of the function subject to the constraint $x^2 + y^2 \leq 1$.

15. $f(x, y) = x^2 + 3xy + y^2$    16. $f(x, y) = e^{-xy/4}$

 **Using Lagrange Multipliers** In Exercises 17 and 18, use Lagrange multipliers to find the indicated extrema of $f$ subject to two constraints, assuming that $x$, $y$, and $z$ are nonnegative.

17. Maximize $f(x, y, z) = xyz$

Constraints: $x + y + z = 32$, $x - y + z = 0$

18. Minimize $f(x, y, z) = x^2 + y^2 + z^2$

Constraints: $x + 2z = 6$, $x + y = 12$

 **Finding Minimum Distance** In Exercises 19–28, use Lagrange multipliers to find the minimum distance from the curve or surface to the indicated point. (*Hint:* To simplify the computations, minimize the square of the distance.)

| Curve | Point |
|---|---|
| 19. Line: $x + y = 1$ | $(0, 0)$ |
| 20. Line: $2x + 3y = -1$ | $(0, 0)$ |
| 21. Line: $x - y = 4$ | $(0, 2)$ |
| 22. Line: $x + 4y = 3$ | $(1, 0)$ |
| 23. Parabola: $y = x^2$ | $(0, 3)$ |
| 24. Parabola: $y = x^2$ | $(-3, 0)$ |
| 25. Circle: $x^2 + (y - 1)^2 = 9$ | $(4, 4)$ |
| 26. Circle: $(x - 4)^2 + y^2 = 4$ | $(0, 10)$ |

| Surface | Point |
|---|---|
| 27. Plane: $x + y + z = 1$ | $(2, 1, 1)$ |
| 28. Cone: $z = \sqrt{x^2 + y^2}$ | $(4, 0, 0)$ |

**Intersection of Surfaces** In Exercises 29 and 30, use Lagrange multipliers to find the highest point on the curve of intersection of the surfaces.

29. Cone: $x^2 + y^2 - z^2 = 0$    30. Sphere: $x^2 + y^2 + z^2 = 36$

Plane: $x + 2z = 4$    Plane: $2x + y - z = 2$

**Using Lagrange Multipliers** In Exercises 31–38, use Lagrange multipliers to solve the indicated exercise in Section 13.9.

31. Exercise 3    32. Exercise 4

33. Exercise 7    34. Exercise 8

35. Exercise 11    36. Exercise 12

37. Exercise 17    38. Exercise 18

39. **Maximum Volume** Use Lagrange multipliers to find the dimensions of a rectangular box of maximum volume that can be inscribed (with edges parallel to the coordinate axes) in the ellipsoid

$$\frac{x^2}{a^2} + \frac{y^2}{b^2} + \frac{z^2}{c^2} = 1.$$

**40.** **HOW DO YOU SEE IT?** The graphs show the constraint and several level curves of the objective function. Use the graph to approximate the indicated extrema.

(a) Maximize $z = xy$    (b) Minimize $z = x^2 + y^2$
Constraint: $2x + y = 4$    Constraint: $x + y - 4 = 0$

## EXPLORING CONCEPTS

**41. Method of Lagrange Multipliers** Explain why you cannot use Lagrange multipliers to find the minimum of the function $f(x, y) = x$ subject to the constraint $y^2 + x^4 - x^3 = 0$.

**42. Method of Lagrange Multipliers** Draw the level curves for $f(x, y) = x^2 + y^2 = c$ for $c = 1, 2, 3$, and $4$, and sketch the constraint $x + y = 2$. Explain analytically how you know that the extremum of $f(x, y) = x^2 + y^2$ at $(1, 1)$ is a minimum instead of a maximum.

**43. Minimum Cost** A cargo container (in the shape of a rectangular solid) must have a volume of 480 cubic feet. The bottom will cost \$5 per square foot to construct, and the sides and the top will cost \$3 per square foot to construct. Use Lagrange multipliers to find the dimensions of the container of this size that has minimum cost.

**44. Geometric and Arithmetic Means**

(a) Use Lagrange multipliers to prove that the product of three positive numbers $x$, $y$, and $z$, whose sum has the constant value $S$, is a maximum when the three numbers are equal. Use this result to prove that

$$\sqrt[3]{xyz} \le \frac{x + y + z}{3}.$$

(b) Generalize the result of part (a) to prove that the product $x_1 x_2 x_3 \cdots x_n$ is a maximum when

$$x_1 = x_2 = x_3 = \cdots = x_n, \quad \sum_{i=1}^{n} x_i = S, \text{ and all } x_i \ge 0.$$

Then prove that

$$\sqrt[n]{x_1 x_2 x_3 \cdots x_n} \le \frac{x_1 + x_2 + x_3 + \cdots + x_n}{n}.$$

This shows that the geometric mean is never greater than the arithmetic mean.

**45. Minimum Surface Area** Use Lagrange multipliers to find the dimensions of a right circular cylinder with volume $V_0$ and minimum surface area.

**46. Temperature** Let $T(x, y, z) = 100 + x^2 + y^2$ represent the temperature at each point on the sphere

$$x^2 + y^2 + z^2 = 50.$$

Use Lagrange multipliers to find the maximum temperature on the curve formed by the intersection of the sphere and the plane $x - z = 0$.

**47. Refraction of Light** When light waves traveling in a transparent medium strike the surface of a second transparent medium, they tend to "bend" in order to follow the path of minimum time. This tendency is called *refraction* and is described by **Snell's Law of Refraction,**

$$\frac{\sin \theta_1}{v_1} = \frac{\sin \theta_2}{v_2}$$

where $\theta_1$ and $\theta_2$ are the magnitudes of the angles shown in the figure, and $v_1$ and $v_2$ are the velocities of light in the two media. Use Lagrange multipliers to derive this law using $x + y = a$.

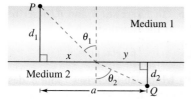

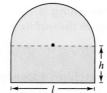

Figure for 47        Figure for 48

**48. Area and Perimeter** A semicircle is on top of a rectangle (see figure). When the area is fixed and the perimeter is a minimum, or when the perimeter is fixed and the area is a maximum, use Lagrange multipliers to verify that the length of the rectangle is twice its height.

 **Production Level** In Exercises 49 and 50, use Lagrange multipliers to find the maximum production level when the total cost of labor (at \$112 per unit) and capital (at \$60 per unit) is limited to \$250,000, where $P$ is the production function, $x$ is the number of units of labor, and $y$ is the number of units of capital.

**49.** $P(x, y) = 100x^{0.25}y^{0.75}$    **50.** $P(x, y) = 100x^{0.4}y^{0.6}$

**Cost** In Exercises 51 and 52, use Lagrange multipliers to find the minimum cost of producing 50,000 units of a product, where $P$ is the production function, $x$ is the number of units of labor (at \$72 per unit), and $y$ is the number of units of capital (at \$80 per unit).

**51.** $P(x, y) = 100x^{0.25}y^{0.75}$    **52.** $P(x, y) = 100x^{0.6}y^{0.4}$

## PUTNAM EXAM CHALLENGE

**53.** A can buoy is to be made of three pieces, namely, a cylinder and two equal cones, the altitude of each cone being equal to the altitude of the cylinder. For a given area of surface, what shape will have the greatest volume?

This problem was composed by the Committee on the Putnam Prize Competition.
© The Mathematical Association of America. All rights reserved.

**Evaluating a Function** In Exercises 1 and 2, evaluate the function at the given values of the independent variables. Simplify the results.

1. $f(x, y) = x^2y - 3$

   (a) $f(0, 4)$    (b) $f(2, -1)$    (c) $f(-3, 2)$    (d) $f(x, 7)$

2. $f(x, y) = 6 - 4x - 2y^2$

   (a) $f(0, 2)$    (b) $f(5, 0)$    (c) $f(-1, -2)$    (d) $f(-3, y)$

**Finding the Domain and Range of a Function** In Exercises 3 and 4, find the domain and range of the function.

3. $f(x, y) = \dfrac{\sqrt{x}}{y}$

4. $f(x, y) = \sqrt{36 - x^2 - y^2}$

**Sketching a Surface** In Exercises 5 and 6, describe and sketch the surface given by the function.

5. $f(x, y) = -2$          6. $g(x, y) = x$

**Sketching a Contour Map** In Exercises 7 and 8, describe the level curves of the function. Sketch a contour map of the surface using level curves for the given $c$-values.

7. $z = 3 - 2x + y$,   $c = 0, 2, 4, 6, 8$

8. $z = 2x^2 + y^2$,   $c = 1, 2, 3, 4, 5$

9. **Conjecture** Consider the function $f(x, y) = x^2 + y^2$.

   (a) Sketch the graph of the surface given by $f$.

   (b) Make a conjecture about the relationship between the graphs of $f$ and $g(x, y) = f(x, y) + 2$. Explain your reasoning.

   (c) Make a conjecture about the relationship between the graphs of $f$ and $g(x, y) = f(x, y - 2)$. Explain your reasoning.

   (d) On the surface in part (a), sketch the graphs of $z = f(1, y)$ and $z = f(x, 1)$.

10. **Cobb-Douglas Production Function** A manufacturer estimates that its production can be modeled by

$$f(x, y) = 100x^{0.8}y^{0.2}$$

   where $x$ is the number of units of labor and $y$ is the number of units of capital.

   (a) Find the production level when $x = 100$ and $y = 200$.

   (b) Find the production level when $x = 500$ and $y = 1500$.

**Sketching a Level Surface** In Exercises 11 and 12, describe and sketch the graph of the level surface $f(x, y, z) = c$ at the given value of $c$.

11. $f(x, y, z) = x^2 - y + z^2$,   $c = 2$

12. $f(x, y, z) = 4x^2 - y^2 + 4z^2$,   $c = 0$

**Limit and Continuity** In Exercises 13–18, find the limit (if it exists) and discuss the continuity of the function.

13. $\displaystyle\lim_{(x, y)\to(1, 1)} \dfrac{xy}{x^2 + y^2}$     14. $\displaystyle\lim_{(x, y)\to(1, 1)} \dfrac{xy}{x^2 - y^2}$

15. $\displaystyle\lim_{(x, y)\to(0, 0)} \dfrac{y + xe^{-y^2}}{1 + x^2}$     16. $\displaystyle\lim_{(x, y)\to(0, 0)} \dfrac{x^2y}{x^4 + y^2}$

17. $\displaystyle\lim_{(x, y, z)\to(-3, 1, 2)} \dfrac{\ln z}{xy - z}$     18. $\displaystyle\lim_{(x, y, z)\to(1, 3, \pi)} \sin\dfrac{xz}{2y}$

**Finding Partial Derivatives** In Exercises 19–26, find all first partial derivatives.

19. $f(x, y) = 5x^3 + 7y - 3$     20. $f(x, y) = 4x^2 - 2xy + y^2$

21. $f(x, y) = e^x \cos y$     22. $f(x, y) = \dfrac{xy}{x + y}$

23. $f(x, y) = y^3 e^{y/x}$     24. $z = \ln(x^2 + y^2 + 1)$

25. $f(x, y, z) = 2xz^2 + 6xyz$     26. $w = \sqrt{x^2 - y^2 - z^2}$

**Finding and Evaluating Partial Derivatives** In Exercises 27–30, find all first partial derivatives, and evaluate each at the given point.

27. $f(x, y) = x^2 - y$,   $(0, 2)$     28. $f(x, y) = xe^{2y}$,   $(-1, 1)$

29. $f(x, y, z) = xy \cos xz$,   $(2, 3, -\pi/3)$

30. $f(x, y, z) = \sqrt{x^2 + y - z^2}$,   $(-3, -3, 1)$

**Finding Second Partial Derivatives** In Exercises 31–34, find the four second partial derivatives. Observe that the second mixed partials are equal.

31. $f(x, y) = 3x^2 - xy + 2y^3$     32. $h(x, y) = \dfrac{x}{x + y}$

33. $h(x, y) = x \sin y + y \cos x$     34. $g(x, y) = \cos(x - 2y)$

35. **Finding the Slopes of a Surface** Find the slopes of the surface $z = x^2 \ln(y + 1)$ in the $x$- and $y$-directions at the point $(2, 0, 0)$.

36. **Marginal Revenue** A company has two plants that produce the same lawn mower. If $x_1$ and $x_2$ are the numbers of units produced at plant 1 and plant 2, respectively, then the total revenue for the product is given by

$$R = 300x_1 + 300x_2 - 5x_1^2 - 10x_1x_2 - 5x_2^2.$$

   When $x_1 = 5$ and $x_2 = 8$, find (a) the marginal revenue for plant 1, $\partial R/\partial x_1$, and (b) the marginal revenue for plant 2, $\partial R/\partial x_2$.

**Finding a Total Differential** In Exercises 37–40, find the total differential.

37. $z = x \sin xy$             38. $z = 5x^4y^3$

39. $w = 3xy^2 - 2x^3yz^2$     40. $w = \dfrac{3x + 4y}{y + 3z}$

**Using a Differential as an Approximation** In Exercises 41 and 42, (a) find $f(2, 1)$ and $f(2.1, 1.05)$ and calculate $\Delta z$, and (b) use the total differential $dz$ to approximate $\Delta z$.

**41.** $f(x, y) = 4x + 2y$
**42.** $f(x, y) = 36 - x^2 - y^2$

**43. Volume** The possible error involved in measuring each dimension of a right circular cone is $\pm\frac{1}{8}$ inch. The radius is 2 inches and the height is 5 inches. Approximate the propagated error and the relative error in the calculated volume of the cone.

**44. Lateral Surface Area** Approximate the propagated error and the relative error in the calculated lateral surface area of the cone in Exercise 43. $\left(\text{The lateral surface area is given by } A = \pi r\sqrt{r^2 + h^2}.\right)$

**Differentiability** In Exercises 45 and 46, show that the function is differentiable by finding values of $\varepsilon_1$ and $\varepsilon_2$ as designated in the definition of differentiability, and verify that both $\varepsilon_1$ and $\varepsilon_2$ approach 0 as $(\Delta x, \Delta y) \to (0, 0)$.

**45.** $f(x, y) = 6x - y^2$

**46.** $f(x, y) = xy^2$

**Using Different Methods** In Exercises 47–50, find $dw/dt$ (a) by using the appropriate Chain Rule and (b) by converting $w$ to a function of $t$ before differentiating.

**47.** $w = \ln(x^2 + y)$,   $x = 2t$,   $y = 4 - t$
**48.** $w = y^2 - x$,   $x = \cos t$,   $y = \sin t$
**49.** $w = x^2z + y + z$,   $x = e^t$,   $y = t$,   $z = t^2$
**50.** $w = \sin x + y^2z + 2z$,   $x = \arcsin(t - 1)$,   $y = t^3$,   $z = 3$

**Using Different Methods** In Exercises 51 and 52, find $\partial w/\partial r$ and $\partial w/\partial t$ (a) by using the appropriate Chain Rule and (b) by converting $w$ to a function of $r$ and $t$ before differentiating.

**51.** $w = \dfrac{xy}{z}$,   $x = 2r + t$,   $y = rt$,   $z = 2r - t$
**52.** $w = x^2 + y^2 + z^2$,   $x = r \cos t$,   $y = r \sin t$,   $z = t$

**Finding a Derivative Implicitly** In Exercises 53 and 54, differentiate implicitly to find $dy/dx$.

**53.** $x^3 - xy + 5y = 0$

**54.** $\dfrac{xy^2}{x + y} = 3$

**Finding Partial Derivatives Implicitly** In Exercises 55 and 56, differentiate implicitly to find the first partial derivatives of $z$.

**55.** $x^2 + xy + y^2 + yz + z^2 = 0$

**56.** $xz^2 - y \sin z = 0$

**Finding a Directional Derivative** In Exercises 57 and 58, use Theorem 13.9 to find the directional derivative of the function at $P$ in the direction of $\mathbf{v}$.

**57.** $f(x, y) = x^2y$,   $P(-5, 5)$,   $\mathbf{v} = 3\mathbf{i} - 4\mathbf{j}$
**58.** $f(x, y) = \frac{1}{4}y^2 - x^2$,   $P(1, 4)$,   $\mathbf{v} = 2\mathbf{i} + \mathbf{j}$

**Finding a Directional Derivative** In Exercises 59 and 60, use the gradient to find the directional derivative of the function at $P$ in the direction of $\mathbf{v}$.

**59.** $w = y^2 + xz$,   $P(1, 2, 2)$,   $\mathbf{v} = 2\mathbf{i} - \mathbf{j} + 2\mathbf{k}$
**60.** $w = 5x^2 + 2xy - 3y^2z$,   $P(1, 0, 1)$,   $\mathbf{v} = \mathbf{i} + \mathbf{j} - \mathbf{k}$

**Using Properties of the Gradient** In Exercises 61–66, find the gradient of the function and the maximum value of the directional derivative at the given point.

**61.** $z = x^2y$,   $(2, 1)$
**62.** $z = e^{-x} \cos y$,   $\left(0, \dfrac{\pi}{4}\right)$
**63.** $z = \dfrac{y}{x^2 + y^2}$,   $(1, 1)$
**64.** $z = \dfrac{x^2}{x - y}$,   $(2, 1)$
**65.** $w = x^4y - y^2z^2$,   $\left(-1, \frac{1}{2}, 2\right)$
**66.** $w = e^{\sqrt{x+y+z^2}}$,   $(5, 0, 2)$

**Using a Function** In Exercises 67 and 68, (a) find the gradient of the function at $P$, (b) find a unit normal vector to the level curve $f(x, y) = c$ at $P$, (c) find the tangent line to the level curve $f(x, y) = c$ at $P$, and (d) sketch the level curve, the unit normal vector, and the tangent line in the $xy$-plane.

**67.** $f(x, y) = 9x^2 - 4y^2$
$c = 65$,   $P(3, 2)$

**68.** $f(x, y) = 4y \sin x - y$
$c = 3$,   $P\left(\dfrac{\pi}{2}, 1\right)$

**Finding an Equation of a Tangent Plane** In Exercises 69–72, find an equation of the tangent plane to the surface at the given point.

**69.** $z = x^2 + y^2 + 2$,   $(1, 3, 12)$
**70.** $9x^2 + y^2 + 4z^2 = 25$,   $(0, -3, 2)$
**71.** $z = -9 + 4x - 6y - x^2 - y^2$,   $(2, -3, 4)$
**72.** $f(x, y) = \sqrt{25 - y^2}$,   $(2, 3, 4)$

**Finding an Equation of a Tangent Plane and a Normal Line** In Exercises 73 and 74, (a) find an equation of the tangent plane to the surface at the given point and (b) find a set of symmetric equations for the normal line to the surface at the given point.

**73.** $f(x, y) = x^2y$,   $(2, 1, 4)$
**74.** $z = \sqrt{9 - x^2 - y^2}$,   $(1, 2, 2)$

**Finding the Angle of Inclination of a Tangent Plane** In Exercises 75 and 76, find the angle of inclination of the tangent plane to the surface at the given point.

**75.** $x^2 + y^2 + z^2 = 14$,   $(2, 1, 3)$
**76.** $xy + yz^2 = 32$,   $(-4, 1, 6)$

**Horizontal Tangent Plane** In Exercises 77 and 78, find the point(s) on the surface at which the tangent plane is horizontal.

**77.** $z = 9 - 2x^2 + y^3$
**78.** $z = 2xy + 3x + 5y$

**Using the Second Partials Test** In Exercises 79–84, find all relative extrema and saddle points of the function. Use the Second Partials Test where applicable.

**79.** $f(x, y) = -x^2 - 4y^2 + 8x - 8y - 11$

**80.** $f(x, y) = x^2 - y^2 - 16x - 16y$

**81.** $f(x, y) = 2x^2 + 6xy + 9y^2 + 8x + 14$

**82.** $f(x, y) = x^6 y^6$

**83.** $f(x, y) = xy + \dfrac{1}{x} + \dfrac{1}{y}$

**84.** $f(x, y) = -8x^2 + 4xy - y^2 + 12x + 7$

**85. Finding Minimum Distance** Find the minimum distance from the point $(2, 1, 4)$ to the surface $x + y + z = 4$. (*Hint:* To simplify the computations, minimize the square of the distance.)

**86. Finding Positive Numbers** Find three positive integers, $x$, $y$, and $z$, such that the product is 64 and the sum is a minimum.

**87. Maximum Revenue** A company manufactures two types of bicycles, a racing bicycle and a mountain bicycle. The total revenue (in thousands of dollars) from $x_1$ units of racing bicycles and $x_2$ units of mountain bicycles is

$$R = -6x_1^2 - 10x_2^2 - 2x_1x_2 + 32x_1 + 84x_2$$

where $x_1$ and $x_2$ are in thousands of units. Find $x_1$ and $x_2$ so as to maximize the revenue.

**88. Maximum Profit** A corporation manufactures digital cameras at two locations. The cost of producing $x_1$ units at location 1 is $C_1 = 0.05x_1^2 + 15x_1 + 5400$ and the cost of producing $x_2$ units at location 2 is $C_2 = 0.03x_2^2 + 15x_2 + 6100$. The digital cameras sell for $180 per unit. Find the quantity that should be produced at each location to maximize the profit $P = 180(x_1 + x_2) - C_1 - C_2$.

**Finding the Least Squares Regression Line** In Exercises 89 and 90, find the least squares regression line for the points. Use the regression capabilities of a graphing utility to verify your results. Use the graphing utility to plot the points and graph the regression line.

**89.** $(0, 4), (1, 5), (3, 6), (6, 8), (8, 10)$

**90.** $(0, 10), (2, 8), (4, 7), (7, 5), (9, 3), (12, 0)$

**91. Modeling Data** An agronomist used four test plots to determine the relationship between the wheat yield $y$ (in bushels per acre) and the amount of fertilizer $x$ (in pounds per acre). The results are shown in the table.

| Fertilizer, $x$ | 100 | 150 | 200 | 250 |
|---|---|---|---|---|
| Yield, $y$ | 35 | 44 | 50 | 56 |

(a) Use the regression capabilities of a graphing utility to find the least squares regression line for the data.

(b) Use the model to approximate the wheat yield for a fertilizer application of 175 pounds per acre.

**92. Modeling Data** The table shows the yield $y$ (in milligrams) of a chemical reaction after $t$ minutes.

| Minutes, $t$ | 1 | 2 | 3 | 4 |
|---|---|---|---|---|
| Yield, $y$ | 1.2 | 7.1 | 9.9 | 13.1 |

| Minutes, $t$ | 5 | 6 | 7 | 8 |
|---|---|---|---|---|
| Yield, $y$ | 15.5 | 16.0 | 17.9 | 18.0 |

(a) Use the regression capabilities of a graphing utility to find the least squares regression line for the data. Then use the graphing utility to plot the data and graph the model.

(b) Use a graphing utility to plot the points $(\ln t, y)$. Do these points appear to follow a linear pattern more closely than the plot of the given data in part (a)?

(c) Use the regression capabilities of a graphing utility to find the least squares regression line for the points $(\ln t, y)$ and obtain the logarithmic model $y = a + b \ln t$.

(d) Use a graphing utility to plot the original data and graph the linear and logarithmic models. Which is a better model? Explain.

**Using Lagrange Multipliers** In Exercises 93–98, use Lagrange multipliers to find the indicated extrema, assuming that $x$ and $y$ are positive.

**93.** Minimize $f(x, y) = x^2 + y^2$

Constraint: $x + y - 8 = 0$

**94.** Maximize $f(x, y) = xy$

Constraint: $x + 3y - 6 = 0$

**95.** Maximize $f(x, y) = 2x + 3xy + y$

Constraint: $x + 2y = 29$

**96.** Minimize $f(x, y) = x^2 - y^2$

Constraint: $x - 2y + 6 = 0$

**97.** Maximize $f(x, y) = 2xy$

Constraint: $2x + y = 12$

**98.** Minimize $f(x, y) = 3x^2 - y^2$

Constraint: $2x - 2y + 5 = 0$

**99. Minimum Cost** A water line is to be built from point $P$ to point $S$ and must pass through regions where construction costs differ (see figure). The cost per kilometer (in dollars) is $3k$ from $P$ to $Q$, $2k$ from $Q$ to $R$, and $k$ from $R$ to $S$. For simplicity, let $k = 1$. Use Lagrange multipliers to find $x$, $y$, and $z$ such that the total cost $C$ will be minimized.

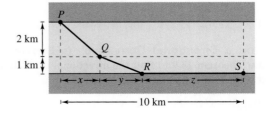

# P.S. Problem Solving

See **CalcChat.com** for tutorial help and worked-out solutions to odd-numbered exercises.

**1. Area**   **Heron's Formula** states that the area of a triangle with sides of lengths $a$, $b$, and $c$ is given by

$$A = \sqrt{s(s-a)(s-b)(s-c)}$$

where $s = \dfrac{a+b+c}{2}$, as shown in the figure.

(a) Use Heron's Formula to find the area of the triangle with vertices $(0, 0)$, $(3, 4)$, and $(6, 0)$.

(b) Show that among all triangles having a fixed perimeter, the triangle with the largest area is an equilateral triangle.

(c) Show that among all triangles having a fixed area, the triangle with the smallest perimeter is an equilateral triangle.

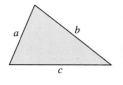   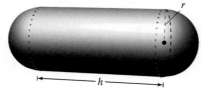

Figure for 1                    Figure for 2

**2. Minimizing Material**   An industrial container is in the shape of a cylinder with hemispherical ends, as shown in the figure. The container must hold 1000 liters of fluid. Determine the radius $r$ and length $h$ that minimize the amount of material used in the construction of the tank.

**3. Tangent Plane**   Let $P(x_0, y_0, z_0)$ be a point in the first octant on the surface $xyz = 1$, as shown in the figure.

(a) Find the equation of the tangent plane to the surface at the point $P$.

(b) Show that the volume of the tetrahedron formed by the three coordinate planes and the tangent plane is constant, independent of the point of tangency (see figure).

**4. Using Functions**   Use a graphing utility to graph the functions

$$f(x) = \sqrt[3]{x^3 - 1} \quad \text{and} \quad g(x) = x$$

in the same viewing window.

(a) Show that

$$\lim_{x \to \infty} [f(x) - g(x)] = 0 \quad \text{and} \quad \lim_{x \to -\infty} [f(x) - g(x)] = 0.$$

(b) Find the point on the graph of $f$ that is farthest from the graph of $g$.

**5. Finding Maximum and Minimum Values**

(a) Let $f(x, y) = x - y$ and $g(x, y) = x^2 + y^2 = 4$. Graph various level curves of $f$ and the constraint $g$ in the $xy$-plane. Use the graph to determine the maximum value of $f$ subject to the constraint $g = 4$. Then verify your answer using Lagrange multipliers.

(b) Let $f(x, y) = x - y$ and $g(x, y) = x^2 + y^2 = 0$. Find the maximum and minimum values of $f$ subject to the constraint $g = 0$. Does the Method of Lagrange Multipliers work in this case? Explain.

**6. Minimizing Costs**   A heated storage room has the shape of a rectangular prism and has a volume of 1000 cubic feet, as shown in the figure. Because warm air rises, the heat loss per unit of area through the ceiling is five times as great as the heat loss through the floor. The heat loss through the four walls is three times as great as the heat loss through the floor. Determine the room dimensions that will minimize heat loss and therefore minimize heating costs.

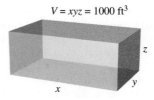

$$V = xyz = 1000 \text{ ft}^3$$

**7. Minimizing Costs**   Repeat Exercise 6 assuming that the heat loss through the walls and ceiling remain the same, but the floor is insulated so that there is no heat loss through the floor.

**8. Temperature**   Consider a circular plate of radius 1 given by $x^2 + y^2 \le 1$, as shown in the figure. The temperature at any point $P(x, y)$ on the plate is $T(x, y) = 2x^2 + y^2 - y + 10$.

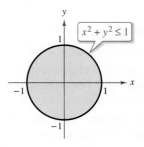

$x^2 + y^2 \le 1$

(a) Sketch the isotherm $T(x, y) = 10$. To print an enlarged copy of the graph, go to *MathGraphs.com*.

(b) Find the hottest and coldest points on the plate.

**9. Cobb-Douglas Production Function**   Consider the Cobb-Douglas production function

$$f(x, y) = Cx^a y^{1-a}, \quad 0 < a < 1.$$

(a) Show that $f$ satisfies the equation $x\dfrac{\partial f}{\partial x} + y\dfrac{\partial f}{\partial y} = f$.

(b) Show that $f(tx, ty) = tf(x, y)$.

**10. Minimizing Area** Consider the ellipse

$$\frac{x^2}{a^2} + \frac{y^2}{b^2} = 1$$

that encloses the circle $x^2 + y^2 = 2x$. Find values of $a$ and $b$ that minimize the area of the ellipse.

**11. Projectile Motion** A projectile is launched at an angle of $45°$ with the horizontal and with an initial velocity of 64 feet per second. A television camera is located in the plane of the path of the projectile 50 feet behind the launch site (see figure).

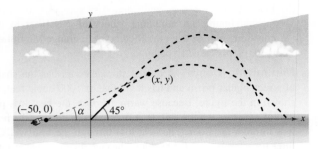

(a) Find parametric equations for the path of the projectile in terms of the parameter $t$ representing time.

(b) Write the angle $\alpha$ that the camera makes with the horizontal in terms of $x$ and $y$ and in terms of $t$.

(c) Use the results of part (b) to find $\dfrac{d\alpha}{dt}$.

(d) Use a graphing utility to graph $\alpha$ in terms of $t$. Is the graph symmetric to the axis of the parabolic arch of the projectile? At what time is the rate of change of $\alpha$ greatest?

(e) At what time is the angle $\alpha$ maximum? Does this occur when the projectile is at its greatest height?

**12. Distance** Consider the distance $d$ between the launch site and the projectile in Exercise 11.

(a) Write the distance $d$ in terms of $x$ and $y$ and in terms of the parameter $t$.

(b) Use the results of part (a) to find the rate of change of $d$.

(c) Find the rate of change of the distance when $t = 2$.

(d) When is the rate of change of $d$ minimum during the flight of the projectile? Does this occur at the time when the projectile reaches its maximum height?

**13. Finding Extrema and Saddle Points Using Technology** Consider the function

$$f(x, y) = (\alpha x^2 + \beta y^2)e^{-(x^2+y^2)}, \quad 0 < |\alpha| < \beta.$$

(a) Use a computer algebra system to graph the function for $\alpha = 1$ and $\beta = 2$, and identify any extrema or saddle points.

(b) Use a computer algebra system to graph the function for $\alpha = -1$ and $\beta = 2$, and identify any extrema or saddle points.

(c) Generalize the results in parts (a) and (b) for the function $f$.

**14. Proof** Prove that if $f$ is a differentiable function such that $\nabla f(x_0, y_0) = \mathbf{0}$, then the tangent plane at $(x_0, y_0)$ is horizontal.

**15. Area** The figure shows a rectangle that is approximately $l = 6$ centimeters long and $h = 1$ centimeter high.

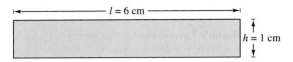

(a) Draw a rectangular strip along the rectangular region showing a small increase in length.

(b) Draw a rectangular strip along the rectangular region showing a small increase in height.

(c) Use the results in parts (a) and (b) to identify the measurement that has more effect on the area $A$ of the rectangle.

(d) Verify your answer in part (c) analytically by comparing the value of $dA$ when $dl = 0.01$ and when $dh = 0.01$.

**16. Tangent Planes** Let $f$ be a differentiable function of one variable. Show that all tangent planes to the surface $z = yf(x/y)$ intersect in a common point.

**17. Wave Equation** Show that

$$u(x, t) = \frac{1}{2}[\sin(x - t) + \sin(x + t)]$$

is a solution to the one-dimensional wave equation

$$\frac{\partial^2 u}{\partial t^2} = \frac{\partial^2 u}{\partial x^2}.$$

**18. Wave Equation** Show that

$$u(x, t) = \frac{1}{2}[f(x - ct) + f(x + ct)]$$

is a solution to the one-dimensional wave equation

$$\frac{\partial^2 u}{\partial t^2} = c^2\frac{\partial^2 u}{\partial x^2}.$$

(This equation describes the small transverse vibration of an elastic string such as those on certain musical instruments.)

**19. Verifying Equations** Consider the function $w = f(x, y)$, where $x = r \cos \theta$ and $y = r \sin \theta$. Verify each of the following.

(a) $\dfrac{\partial w}{\partial x} = \dfrac{\partial w}{\partial r} \cos \theta - \dfrac{\partial w}{\partial \theta} \dfrac{\sin \theta}{r}$

$\dfrac{\partial w}{\partial y} = \dfrac{\partial w}{\partial r} \sin \theta + \dfrac{\partial w}{\partial \theta} \dfrac{\cos \theta}{r}$

(b) $\left(\dfrac{\partial w}{\partial x}\right)^2 + \left(\dfrac{\partial w}{\partial y}\right)^2 = \left(\dfrac{\partial w}{\partial r}\right)^2 + \left(\dfrac{1}{r^2}\right)\left(\dfrac{\partial w}{\partial \theta}\right)^2$

**20. Using a Function** Demonstrate the result of Exercise 19(b) for

$$w = \arctan \frac{y}{x}.$$

**21. Laplace's Equation** Rewrite Laplace's equation

$$\frac{\partial^2 u}{\partial x^2} + \frac{\partial^2 u}{\partial y^2} + \frac{\partial^2 u}{\partial z^2} = 0$$

in cylindrical coordinates.

# 14 Multiple Integration

Modeling Data *(Exercise 36, p. 1012)*

Center of Pressure on a Sail
*(Section Project, p. 1005)*

Glacier *(Exercise 58, p. 997)*

Population
*(Exercise 55, p. 996)*

Average Production *(Exercise 57, p. 988)*

# 14.1 Iterated Integrals and Area in the Plane

■ Evaluate an iterated integral.
■ Use an iterated integral to find the area of a plane region.

## Iterated Integrals

In Chapters 14 and 15, you will study several applications of integration involving functions of several variables. Chapter 14 is like Chapter 7 in that it surveys the use of integration to find plane areas, volumes, surface areas, moments, and centers of mass.

In Chapter 13, you saw that it is meaningful to differentiate functions of several variables with respect to one variable while holding the other variables constant. You can *integrate* functions of several variables by a similar procedure. For example, consider the partial derivative $f_x(x, y) = 2xy$. By considering $y$ constant, you can integrate with respect to $x$ to obtain

$$f(x, y) = \int f_x(x, y)\, dx \qquad \text{Integrate with respect to } x.$$

$$= \int 2xy\, dx \qquad \text{Hold } y \text{ constant.}$$

$$= y \int 2x\, dx \qquad \text{Factor out constant } y.$$

$$= y(x^2) + C(y) \qquad \text{Antiderivative of } 2x \text{ is } x^2.$$

$$= x^2y + C(y). \qquad C(y) \text{ is a function of } y.$$

The "constant" of integration, $C(y)$, is a function of $y$. In other words, by integrating with respect to $x$, you are able to recover $f(x, y)$ only partially. The total recovery of a function of $x$ and $y$ from its partial derivatives is a topic you will study in Chapter 15. For now, you will focus on extending definite integrals to functions of several variables. For instance, by considering $y$ constant, you can apply the Fundamental Theorem of Calculus to evaluate

$$\int_1^{2y} 2xy\, dx = x^2y \Big]_1^{2y} = (2y)^2y - (1)^2y = 4y^3 - y.$$

$x$ is the variable of integration and $y$ is fixed.    Replace $x$ by the limits of integration.    The result is a function of $y$.

Similarly, you can integrate with respect to $y$ by holding $x$ fixed. Both procedures are summarized as follows.

$$\int_{h_1(y)}^{h_2(y)} f_x(x, y)\, dx = f(x, y) \Big]_{h_1(y)}^{h_2(y)} = f(h_2(y), y) - f(h_1(y), y) \qquad \text{With respect to } x$$

$$\int_{g_1(x)}^{g_2(x)} f_y(x, y)\, dy = f(x, y) \Big]_{g_1(x)}^{g_2(x)} = f(x, g_2(x)) - f(x, g_1(x)) \qquad \text{With respect to } y$$

Note that the variable of integration cannot appear in either limit of integration. For instance, it makes no sense to write

$$\int_0^x y\, dx.$$

**EXAMPLE 1**  **Integrating with Respect to y**

Evaluate $\displaystyle\int_{1}^{x} (2xy + 3y^2)\, dy$.

**Solution**  Considering $x$ to be constant and integrating with respect to $y$, you have

$$\int_{1}^{x} (2xy + 3y^2)\, dy = \left[ xy^2 + y^3 \right]_{1}^{x} \qquad \text{Integrate with respect to } y.$$

$$= (2x^3) - (x + 1)$$

$$= 2x^3 - x - 1.$$

. . . . . . . . . . . . . . . ▷

**· · REMARK**  Remember that you can check an antiderivative using differentiation. For instance, in Example 1, you can verify that

$$xy^2 + y^3$$

is the correct anitderivative by finding

$$\frac{\partial}{\partial y}[xy^2 + y^3].$$

Notice in Example 1 that the integral defines a function of $x$ and can *itself* be integrated, as shown in the next example.

**EXAMPLE 2**  **The Integral of an Integral**

Evaluate $\displaystyle\int_{1}^{2} \left[ \int_{1}^{x} (2xy + 3y^2)\, dy \right] dx$.

**Solution**  Using the result of Example 1, you have

$$\int_{1}^{2} \left[ \int_{1}^{x} (2xy + 3y^2)\, dy \right] dx = \int_{1}^{2} (2x^3 - x - 1)\, dx$$

$$= \left[ \frac{x^4}{2} - \frac{x^2}{2} - x \right]_{1}^{2} \qquad \text{Integrate with respect to } x.$$

$$= 4 - (-1)$$

$$= 5.$$

The integral in Example 2 is an **iterated integral.** The brackets used in Example 2 are normally not written. Instead, iterated integrals are usually written simply as

$$\int_{a}^{b} \int_{g_1(x)}^{g_2(x)} f(x, y)\, dy\, dx \quad \text{and} \quad \int_{c}^{d} \int_{h_1(y)}^{h_2(y)} f(x, y)\, dx\, dy.$$

The **inside limits of integration** can be variable with respect to the outer variable of integration. However, the **outside limits of integration** *must be* constant with respect to both variables of integration. After performing the inside integration, you obtain a "standard" definite integral, and the second integration produces a real number. The limits of integration for an iterated integral identify two sets of boundary intervals for the variables. For instance, in Example 2, the outside limits indicate that $x$ lies in the interval $1 \le x \le 2$ and the inside limits indicate that $y$ lies in the interval $1 \le y \le x$. Together, these two intervals determine the **region of integration $R$** of the iterated integral, as shown in Figure 14.1.

Because an iterated integral is just a special type of definite integral—one in which the integrand is also an integral—you can use the properties of definite integrals to evaluate iterated integrals.

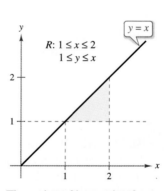

The region of integration for

$$\int_{1}^{2} \int_{1}^{x} f(x, y)\, dy\, dx$$

**Figure 14.1**

## Area of a Plane Region

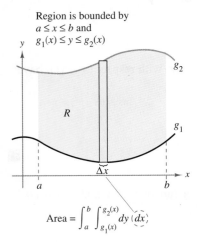

Region is bounded by
$a \le x \le b$ and
$g_1(x) \le y \le g_2(x)$

Area $= \displaystyle\int_a^b \int_{g_1(x)}^{g_2(x)} dy \, (dx)$

**Vertically simple region**
**Figure 14.2**

In the remainder of this section, you will take another look at the problem of finding the area of a plane region. Consider the plane region $R$ bounded by $a \le x \le b$ and $g_1(x) \le y \le g_2(x)$, as shown in Figure 14.2. The area of $R$ is

$$\int_a^b \left[ g_2(x) - g_1(x) \right] dx. \qquad \text{Area of } R$$

Using the Fundamental Theorem of Calculus, you can rewrite the integrand $g_2(x) - g_1(x)$ as a definite integral. Specifically, consider $x$ to be fixed and let $y$ vary from $g_1(x)$ to $g_2(x)$, and you can write

$$\int_{g_1(x)}^{g_2(x)} dy = y \Big]_{g_1(x)}^{g_2(x)} = g_2(x) - g_1(x).$$

Combining these two integrals, you can write the area of the region $R$ as an iterated integral

$$\int_a^b \int_{g_1(x)}^{g_2(x)} dy \, dx = \int_a^b y \Big]_{g_1(x)}^{g_2(x)} dx$$

$$= \int_a^b \left[ g_2(x) - g_1(x) \right] dx. \qquad \text{Area of } R$$

Placing a representative rectangle in the region $R$ helps determine both the order and the limits of integration. A vertical rectangle implies the order $dy \, dx$, with the inside limits of integration corresponding to the upper and lower bounds of the rectangle, as shown in Figure 14.2. This type of region is **vertically simple,** because the outside limits of integration represent the vertical lines

$$x = a \quad \text{and} \quad x = b.$$

Similarly, a horizontal rectangle implies the order $dx \, dy$, with the inside limits of integration determined by the left and right bounds of the rectangle, as shown in Figure 14.3. This type of region is **horizontally simple,** because the outside limits of integration represent the horizontal lines

$$y = c \quad \text{and} \quad y = d.$$

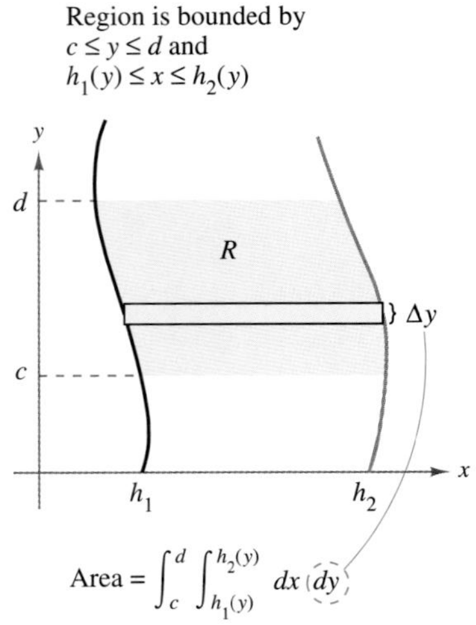

Region is bounded by
$c \le y \le d$ and
$h_1(y) \le x \le h_2(y)$

Area $= \displaystyle\int_c^d \int_{h_1(y)}^{h_2(y)} dx \, (dy)$

**Horizontally simple region**
**Figure 14.3**

The iterated integrals used for these two types of simple regions are summarized as follows.

· · · · · · · · · · · · ▷

**· · REMARK** Be sure you see that the orders of integration of these two integrals are different—the order $dy \, dx$ corresponds to a vertically simple region, and the order $dx \, dy$ corresponds to a horizontally simple region.

### Area of a Region in the Plane

**1.** If $R$ is defined by $a \le x \le b$ and $g_1(x) \le y \le g_2(x)$, where $g_1$ and $g_2$ are continuous on $[a, b]$, then the area of $R$ is

$$A = \int_a^b \int_{g_1(x)}^{g_2(x)} dy \, dx. \qquad \text{Figure 14.2 (vertically simple)}$$

**2.** If $R$ is defined by $c \le y \le d$ and $h_1(y) \le x \le h_2(y)$, where $h_1$ and $h_2$ are continuous on $[c, d]$, then the area of $R$ is

$$A = \int_c^d \int_{h_1(y)}^{h_2(y)} dx \, dy. \qquad \text{Figure 14.3 (horizontally simple)}$$

If all four limits of integration happen to be constants, then the region of integration is rectangular, as shown in Example 3.

**EXAMPLE 3**    **The Area of a Rectangular Region**

Use an iterated integral to represent the area of the rectangle shown in Figure 14.4.

**Solution**    The region shown in Figure 14.4 is both vertically simple and horizontally simple, so you can use either order of integration. By choosing the order $dy\, dx$, you obtain the following.

$$\int_a^b \int_c^d dy\, dx = \int_a^b y\Big]_c^d dx \qquad \text{Integrate with respect to } y.$$

$$= \int_a^b (d - c)\, dx$$

$$= \Big[(d - c)x\Big]_a^b \qquad \text{Integrate with respect to } x.$$

$$= (d - c)(b - a)$$

Notice that this answer is consistent with what you know from geometry.

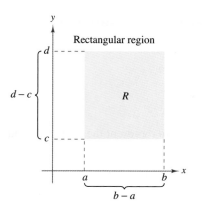

**Figure 14.4**

**EXAMPLE 4**    **Finding Area by an Iterated Integral**

Use an iterated integral to find the area of the region bounded by the graphs of

$$f(x) = \sin x \qquad\qquad\qquad \text{Sine curve forms upper boundary.}$$

and

$$g(x) = \cos x \qquad\qquad\qquad \text{Cosine curve forms lower boundary.}$$

between $x = \pi/4$ and $x = 5\pi/4$.

**Solution**    Because $f$ and $g$ are given as functions of $x$, a vertical representative rectangle is convenient, and you can choose $dy\, dx$ as the order of integration, as shown in Figure 14.5. The outside limits of integration are

$$\frac{\pi}{4} \le x \le \frac{5\pi}{4}.$$

Moreover, because the rectangle is bounded above by $f(x) = \sin x$ and below by $g(x) = \cos x$, you have

$$\text{Area of } R = \int_{\pi/4}^{5\pi/4} \int_{\cos x}^{\sin x} dy\, dx$$

$$= \int_{\pi/4}^{5\pi/4} y\Big]_{\cos x}^{\sin x} dx \qquad \text{Integrate with respect to } y.$$

$$= \int_{\pi/4}^{5\pi/4} (\sin x - \cos x)\, dx$$

$$= \Big[-\cos x - \sin x\Big]_{\pi/4}^{5\pi/4} \qquad \text{Integrate with respect to } x.$$

$$= 2\sqrt{2}.$$

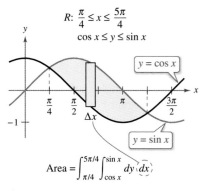

**Figure 14.5**

The region of integration of an iterated integral need not have any straight lines as boundaries. For instance, the region of integration shown in Figure 14.5 is *vertically simple* even though it has no vertical lines as left and right boundaries. The quality that makes the region vertically simple is that it is bounded above and below by the graphs of *functions of x*.

One order of integration will often produce a simpler integration problem than the other order. For instance, try reworking Example 4 with the order $dx\,dy$. You may be surprised to see that the task is formidable. However, if you succeed, you will see that the answer is the same. In other words, the order of integration affects the ease of integration but not the value of the integral.

### EXAMPLE 5    Comparing Different Orders of Integration

⋮⋅⋅▷  *See LarsonCalculus.com for an interactive version of this type of example.*

Sketch the region whose area is represented by the integral

$$\int_0^2 \int_{y^2}^4 dx\,dy.$$

Then find another iterated integral using the order $dy\,dx$ to represent the same area and show that both integrals yield the same value.

**Solution**    From the given limits of integration, you know that

$$y^2 \le x \le 4 \qquad\qquad \text{Inner limits of integration}$$

which means that the region $R$ is bounded on the left by the parabola $x = y^2$ and on the right by the line $x = 4$. Furthermore, because

$$0 \le y \le 2 \qquad\qquad \text{Outer limits of integration}$$

you know that $R$ is bounded below by the $x$-axis, as shown in Figure 14.6(a). The value of this integral is

$$\int_0^2 \int_{y^2}^4 dx\,dy = \int_0^2 x\Big]_{y^2}^4 dy \qquad \text{Integrate with respect to } x.$$

$$= \int_0^2 (4 - y^2)\,dy$$

$$= \left[4y - \frac{y^3}{3}\right]_0^2 \qquad \text{Integrate with respect to } y.$$

$$= \frac{16}{3}.$$

To change the order of integration to $dy\,dx$, place a vertical rectangle in the region, as shown in Figure 14.6(b). From this, you can see that the constant bounds $0 \le x \le 4$ serve as the outer limits of integration. By solving for $y$ in the equation $x = y^2$, you can conclude that the inner bounds are $0 \le y \le \sqrt{x}$. So, the area of the region can also be represented by

$$\int_0^4 \int_0^{\sqrt{x}} dy\,dx.$$

By evaluating this integral, you can see that it has the same value as the original integral.

$$\int_0^4 \int_0^{\sqrt{x}} dy\,dx = \int_0^4 y\Big]_0^{\sqrt{x}} dx \qquad \text{Integrate with respect to } y.$$

$$= \int_0^4 \sqrt{x}\,dx$$

$$= \frac{2}{3}x^{3/2}\Big]_0^4 \qquad \text{Integrate with respect to } x.$$

$$= \frac{16}{3}$$

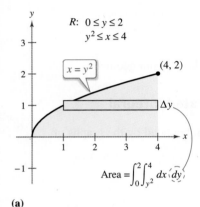

$R:\ 0 \le y \le 2$
$\quad\ y^2 \le x \le 4$

$x = y^2$

$(4, 2)$

$\Delta y$

$\text{Area} = \int_0^2 \int_{y^2}^4 dx\,dy$

**(a)**

$R:\ 0 \le x \le 4$
$\quad\ 0 \le y \le \sqrt{x}$

$y = \sqrt{x}$

$(4, 2)$

$\Delta x$

$\text{Area} = \int_0^4 \int_0^{\sqrt{x}} dy\,dx$

**(b)**

**Figure 14.6**

Sometimes it is not possible to calculate the area of a region with a single iterated integral. In these cases, you can divide the region into subregions such that the area of each subregion can be calculated by an iterated integral. The total area is then the sum of the iterated integrals.

▷ **TECHNOLOGY** Some computer software can perform symbolic integration for integrals such as those in Example 6. If you have access to such software, use it to evaluate the integrals in the exercises and examples given in this section.

| EXAMPLE 6 | **An Area Represented by Two Iterated Integrals**

Find the area of the region $R$ that lies below the parabola

$$y = 4x - x^2 \qquad \text{Parabola forms upper boundary.}$$

above the $x$-axis, and above the line

$$y = -3x + 6. \qquad \text{Line and } x\text{-axis form lower boundary.}$$

**Solution** Begin by dividing $R$ into the two subregions $R_1$ and $R_2$ shown in Figure 14.7.

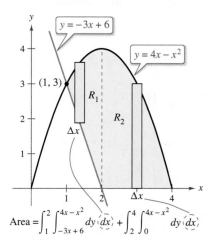

$$\text{Area} = \int_1^2 \int_{-3x+6}^{4x-x^2} dy\, dx + \int_2^4 \int_0^{4x-x^2} dy\, dx$$

**Figure 14.7**

• • **REMARK** In Examples 3 through 6, be sure you see the benefit of sketching the region of integration. You should develop the habit of making sketches to help you determine the limits of integration for all iterated integrals in this chapter.
• • • • • • • • • • • • • • • • ▷

In both regions, it is convenient to use vertical rectangles, and you have

$$\begin{aligned}
\text{Area} &= \int_1^2 \int_{-3x+6}^{4x-x^2} dy\, dx + \int_2^4 \int_0^{4x-x^2} dy\, dx \\[2mm]
&= \int_1^2 (4x - x^2 + 3x - 6)\, dx + \int_2^4 (4x - x^2)\, dx \\[2mm]
&= \left[ \frac{7x^2}{2} - \frac{x^3}{3} - 6x \right]_1^2 + \left[ 2x^2 - \frac{x^3}{3} \right]_2^4 \\[2mm]
&= \left( 14 - \frac{8}{3} - 12 - \frac{7}{2} + \frac{1}{3} + 6 \right) + \left( 32 - \frac{64}{3} - 8 + \frac{8}{3} \right) \\[2mm]
&= \frac{15}{2}.
\end{aligned}$$

The area of the region is $15/2$ square units. Try checking this using the procedure for finding the area between two curves, as presented in Section 7.1. ∎

At this point, you may be wondering why you would need iterated integrals. After all, you already know how to use conventional integration to find the area of a region in the plane. (For instance, compare the solution to Example 4 in this section with that given in Example 3 in Section 7.1.) The need for iterated integrals will become clear in the next section. In this section, primary attention is given to procedures for finding the limits of integration of the region of an iterated integral, and the following exercise set is designed to develop skill in this important procedure.

## 14.1 Exercises

See **CalcChat.com** for tutorial help and worked-out solutions to odd-numbered exercises.

**CONCEPT CHECK**

**1. Iterated Integral** Explain what is meant by an iterated integral. How is it evaluated?

**2. Region of Integration** Sketch the region of integration for the iterated integral.

$$\int_1^2 \int_0^{2-x} f(x, y) \, dy \, dx.$$

 **Evaluating an Integral** In Exercises 3–10, evaluate the integral.

**3.** $\int_0^x (2x - y) \, dy$

**4.** $\int_x^{x^2} \frac{y}{x} \, dy$

**5.** $\int_0^{\sqrt{4-x^2}} x^2 y \, dy$

**6.** $\int_{x^3}^{\sqrt{x}} (x^2 + 3y^2) \, dy$

**7.** $\int_{e^y}^y \frac{y \ln x}{x} \, dx, \quad y > 0$

**8.** $\int_{-\sqrt{1-y^2}}^{\sqrt{1-y^2}} (x^2 + y^2) \, dx$

**9.** $\int_0^{x^3} ye^{-y/x} \, dy$

**10.** $\int_y^{\pi/2} \sin^3 x \cos y \, dx$

 **Evaluating an Iterated Integral** In Exercises 11–28, evaluate the iterated integral.

**11.** $\int_0^1 \int_0^2 (x + y) \, dy \, dx$

**12.** $\int_{-1}^1 \int_{-2}^2 (x^2 - y^2) \, dy \, dx$

**13.** $\int_0^{\pi/4} \int_0^1 y \cos x \, dy \, dx$

**14.** $\int_0^{\ln 4} \int_0^{\ln 3} e^{x+y} \, dy \, dx$

**15.** $\int_0^2 \int_0^{6x^2} x^3 \, dy \, dx$

**16.** $\int_0^1 \int_0^y (6x + 5y^3) \, dx \, dy$

**17.** $\int_0^{\pi/2} \int_0^{\cos x} (1 + \sin x) \, dy \, dx$

**18.** $\int_1^4 \int_1^{\sqrt{x}} 2ye^{-x} \, dy \, dx$

**19.** $\int_0^1 \int_0^x \sqrt{1 - x^2} \, dy \, dx$

**20.** $\int_{-4}^4 \int_0^{x^2} \sqrt{64 - x^3} \, dy \, dx$

**21.** $\int_0^1 \int_0^{\sqrt{1-y^2}} (x + y) \, dx \, dy$

**22.** $\int_0^2 \int_{3y^2-6y}^{2y-y^2} 3y \, dx \, dy$

**23.** $\int_0^2 \int_0^{\sqrt{4-y^2}} \frac{2}{\sqrt{4-y^2}} \, dx \, dy$

**24.** $\int_1^3 \int_0^y \frac{4}{x^2 + y^2} \, dx \, dy$

**25.** $\int_0^{\pi/2} \int_0^{2\cos\theta} r \, dr \, d\theta$

**26.** $\int_0^{\pi/4} \int_{\sqrt{3}}^{\sqrt{3}\cos\theta} r \, dr \, d\theta$

**27.** $\int_0^{\pi/2} \int_0^{\sin\theta} \theta r \, dr \, d\theta$

**28.** $\int_0^{\pi/4} \int_0^{\cos\theta} 3r^2 \sin\theta \, dr \, d\theta$

**Evaluating an Improper Iterated Integral** In Exercises 29–32, evaluate the improper iterated integral.

**29.** $\int_1^\infty \int_0^{1/x} y \, dy \, dx$

**30.** $\int_0^3 \int_0^\infty \frac{x^2}{1 + y^2} \, dy \, dx$

**31.** $\int_1^\infty \int_1^\infty \frac{1}{xy} \, dx \, dy$

**32.** $\int_0^\infty \int_0^\infty xye^{-(x^2+y^2)} \, dx \, dy$

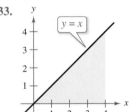 **Finding the Area of a Region** In Exercises 33–36, use an iterated integral to find the area of the region.

**33.**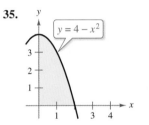

**34.**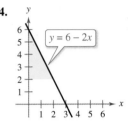

**35.**

**36.**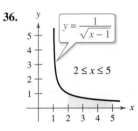

 **Finding the Area of a Region** In Exercises 37–42, use an iterated integral to find the area of the region bounded by the graphs of the equations.

**37.** $y = 9 - x^2, \quad y = 0$

**38.** $2x - 3y = 0, \quad x + y = 5, \quad y = 0$

**39.** $\sqrt{x} + \sqrt{y} = 2, \quad x = 0, \quad y = 0$

**40.** $y = x^{3/2}, \quad y = 2x$

**41.** $y = 4 - x^2, \quad y = x + 2$

**42.** $y = x, \quad y = 2x, \quad x = 2$

**Changing the Order of Integration** In Exercises 43–50, sketch the region $R$ of integration and change the order of integration.

**43.** $\int_0^4 \int_0^y f(x, y) \, dx \, dy$

**44.** $\int_0^4 \int_{\sqrt{y}}^2 f(x, y) \, dx \, dy$

**45.** $\int_{-2}^2 \int_0^{\sqrt{4-x^2}} f(x, y) \, dy \, dx$

**46.** $\int_0^2 \int_0^{4-x^2} f(x, y) \, dy \, dx$

**47.** $\int_1^{10} \int_0^{\ln y} f(x, y) \, dx \, dy$

**48.** $\int_{-1}^2 \int_0^{e^{-x}} f(x, y) \, dy \, dx$

**49.** $\int_{-1}^1 \int_{x^2}^1 f(x, y) \, dy \, dx$

**50.** $\int_{-\pi/2}^{\pi/2} \int_0^{\cos x} f(x, y) \, dy \, dx$

**Changing the Order of Integration** In Exercises 51–60, sketch the region $R$ whose area is given by the iterated integral. Then change the order of integration and show that both orders yield the same area.

**51.** $\displaystyle\int_0^1 \int_0^2 dy\, dx$

**52.** $\displaystyle\int_1^2 \int_2^4 dx\, dy$

**53.** $\displaystyle\int_0^1 \int_{2y}^2 dx\, dy$

**54.** $\displaystyle\int_0^9 \int_{\sqrt{x}}^3 dy\, dx$

**55.** $\displaystyle\int_0^1 \int_{-\sqrt{1-y^2}}^{\sqrt{1-y^2}} dx\, dy$

**56.** $\displaystyle\int_{-2}^2 \int_{-\sqrt{4-x^2}}^{\sqrt{4-x^2}} dy\, dx$

**57.** $\displaystyle\int_0^2 \int_0^x dy\, dx + \int_2^4 \int_0^{4-x} dy\, dx$

**58.** $\displaystyle\int_0^4 \int_0^{x/2} dy\, dx + \int_4^6 \int_0^{6-x} dy\, dx$

**59.** $\displaystyle\int_0^1 \int_{y^2}^{\sqrt[3]{y}} dx\, dy$

**60.** $\displaystyle\int_{-2}^2 \int_0^{4-y^2} dx\, dy$

**Changing the Order of Integration** In Exercises 61–66, sketch the region of integration. Then evaluate the iterated integral. (*Hint:* Note that it is necessary to change the order of integration.)

**61.** $\displaystyle\int_0^2 \int_x^2 x\sqrt{1+y^3}\, dy\, dx$

**62.** $\displaystyle\int_0^4 \int_{\sqrt{x}}^2 \frac{3}{2+y^3}\, dy\, dx$

**63.** $\displaystyle\int_0^1 \int_{2x}^2 4e^{y^2}\, dy\, dx$

**64.** $\displaystyle\int_0^2 \int_x^2 e^{-y^2}\, dy\, dx$

**65.** $\displaystyle\int_0^1 \int_y^1 \sin x^2\, dx\, dy$

**66.** $\displaystyle\int_0^2 \int_{y^2}^4 \sqrt{x}\sin x\, dx\, dy$

---

**EXPLORING CONCEPTS**

**67. Area of a Circle** Write an iterated integral that represents the area of a circle of radius 5 centered at the origin. Verify that your integral produces the correct area.

**68. Using Different Methods** Express the area of the region bounded by $x = \sqrt{4 - 4y^2}$, $y = 1$, and $x = 2$ in at least two different ways, one of which is an iterated integral. Do not find the area of the region.

**69. Think About It** Determine whether each expression represents the area of the shaded region (see figure).

(a) $\displaystyle\int_0^5 \int_y^{\sqrt{50-y^2}} dy\, dx$

(b) $\displaystyle\int_0^5 \int_x^{\sqrt{50-x^2}} dy\, dx$

(c) $\displaystyle\int_0^5 \int_0^y dx\, dy + \int_5^{5\sqrt{2}} \int_0^{\sqrt{50-y^2}} dx\, dy$

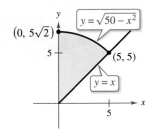

**70.** **HOW DO YOU SEE IT?** Use each order of integration to write an iterated integral that represents the area of the region $R$ (see figure).

(a) Area $= \displaystyle\iint dx\, dy$

(b) Area $= \displaystyle\iint dy\, dx$

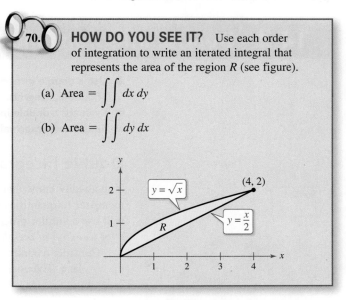

**Evaluating an Iterated Integral Using Technology** In Exercises 71–76, use a computer algebra system to evaluate the iterated integral.

**71.** $\displaystyle\int_0^1 \int_y^{2y} \sin(x+y)\, dx\, dy$

**72.** $\displaystyle\int_0^2 \int_0^{4-x^2} e^{xy}\, dy\, dx$

**73.** $\displaystyle\int_0^4 \int_0^y \frac{2}{(x+1)(y+1)}\, dx\, dy$

**74.** $\displaystyle\int_0^2 \int_x^2 \sqrt{16 - x^3 - y^3}\, dy\, dx$

**75.** $\displaystyle\int_0^{2\pi} \int_0^{1+\cos\theta} 6r^2 \cos\theta\, dr\, d\theta$

**76.** $\displaystyle\int_0^{\pi/2} \int_0^{1+\sin\theta} 15\theta r\, dr\, d\theta$

**Comparing Different Orders of Integration Using Technology** In Exercises 77 and 78, (a) sketch the region of integration, (b) change the order of integration, and (c) use a computer algebra system to show that both orders yield the same value.

**77.** $\displaystyle\int_0^2 \int_{y^3}^{4\sqrt{2y}} (x^2 y - xy^2)\, dx\, dy$

**78.** $\displaystyle\int_0^2 \int_{\sqrt{4-x^2}}^{4-(x^2/4)} \frac{xy}{x^2+y^2+1}\, dy\, dx$

**True or False?** In Exercises 79 and 80, determine whether the statement is true or false. If it is false, explain why or give an example that shows it is false.

**79.** $\displaystyle\int_a^b \int_c^d f(x,y)\, dy\, dx = \int_c^d \int_a^b f(x,y)\, dx\, dy$

**80.** $\displaystyle\int_0^1 \int_0^x f(x,y)\, dy\, dx = \int_0^1 \int_0^y f(x,y)\, dx\, dy$

# 14.2 Double Integrals and Volume

■ Use a double integral to represent the volume of a solid region and use properties of double integrals.
■ Evaluate a double integral as an iterated integral.
■ Find the average value of a function over a region.

## Double Integrals and Volume of a Solid Region

You already know that a definite integral over an *interval* uses a limit process to assign measures to quantities such as area, volume, arc length, and mass. In this section, you will use a similar process to define the **double integral** of a function of two variables over a *region in the plane*.

Consider a continuous function $f$ such that $f(x, y) \geq 0$ for all $(x, y)$ in a region $R$ in the $xy$-plane. The goal is to find the volume of the solid region lying between the surface given by

$$z = f(x, y) \qquad \text{Surface lying above the } xy\text{-plane}$$

and the $xy$-plane, as shown in Figure 14.8. You can begin by superimposing a rectangular grid over the region, as shown in Figure 14.9. The rectangles lying entirely within $R$ form an **inner partition** $\Delta$, whose **norm** $\|\Delta\|$ is defined as the length of the longest diagonal of the $n$ rectangles. Next, choose a point $(x_i, y_i)$ in each rectangle and form the rectangular prism whose height is

$$f(x_i, y_i) \qquad \text{Height of } i\text{th prism}$$

as shown in Figure 14.10. Because the area of the $i$th rectangle is

$$\Delta A_i \qquad \text{Area of } i\text{th rectangle}$$

it follows that the volume of the $i$th prism is

$$f(x_i, y_i)\, \Delta A_i \qquad \text{Volume of } i\text{th prism}$$

and you can approximate the volume of the solid region by the Riemann sum of the volumes of all $n$ prisms,

$$\sum_{i=1}^{n} f(x_i, y_i)\, \Delta A_i \qquad \text{Riemann sum}$$

as shown in Figure 14.11. This approximation can be improved by tightening the mesh of the grid to form smaller and smaller rectangles, as shown in Example 1.

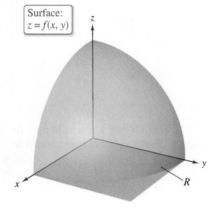

Surface:
$z = f(x, y)$

**Figure 14.8**

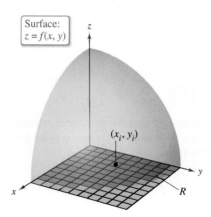

The rectangles lying within $R$ form an inner partition of $R$.
**Figure 14.9**

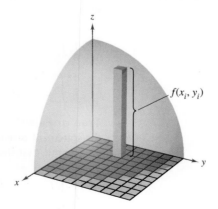

Rectangular prism whose base has an area of $\Delta A_i$ and whose height is $f(x_i, y_i)$
**Figure 14.10**

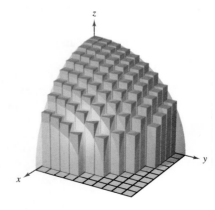

Volume approximated by rectangular prisms
**Figure 14.11**

EXAMPLE 1    **Approximating the Volume of a Solid**

Approximate the volume of the solid lying between the paraboloid

$$f(x, y) = 1 - \frac{1}{2}x^2 - \frac{1}{2}y^2$$

and the square region $R$ given by $0 \le x \le 1, 0 \le y \le 1$. Use a partition made up of squares whose sides have a length of $\frac{1}{4}$.

**Solution**    Begin by forming the specified partition of $R$. For this partition, it is convenient to choose the centers of the subregions as the points at which to evaluate $f(x, y)$.

| | | | |
|---|---|---|---|
| $\left(\frac{1}{8}, \frac{1}{8}\right)$ | $\left(\frac{1}{8}, \frac{3}{8}\right)$ | $\left(\frac{1}{8}, \frac{5}{8}\right)$ | $\left(\frac{1}{8}, \frac{7}{8}\right)$ |
| $\left(\frac{3}{8}, \frac{1}{8}\right)$ | $\left(\frac{3}{8}, \frac{3}{8}\right)$ | $\left(\frac{3}{8}, \frac{5}{8}\right)$ | $\left(\frac{3}{8}, \frac{7}{8}\right)$ |
| $\left(\frac{5}{8}, \frac{1}{8}\right)$ | $\left(\frac{5}{8}, \frac{3}{8}\right)$ | $\left(\frac{5}{8}, \frac{5}{8}\right)$ | $\left(\frac{5}{8}, \frac{7}{8}\right)$ |
| $\left(\frac{7}{8}, \frac{1}{8}\right)$ | $\left(\frac{7}{8}, \frac{3}{8}\right)$ | $\left(\frac{7}{8}, \frac{5}{8}\right)$ | $\left(\frac{7}{8}, \frac{7}{8}\right)$ |

Because the area of each square is $\Delta A_i = \frac{1}{16}$, you can approximate the volume by the sum

$$\sum_{i=1}^{16} f(x_i, y_i) \, \Delta A_i = \sum_{i=1}^{16} \left(1 - \frac{1}{2}x_i^2 - \frac{1}{2}y_i^2\right)\left(\frac{1}{16}\right) \approx 0.672.$$

This approximation is shown graphically in Figure 14.12. The exact volume of the solid is $\frac{2}{3}$ (see Example 2). You can obtain a better approximation by using a finer partition. For example, with a partition of squares with sides of length $\frac{1}{10}$, the approximation is 0.668. ■

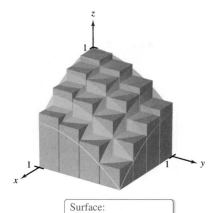

Surface:
$f(x, y) = 1 - \frac{1}{2}x^2 - \frac{1}{2}y^2$

**Figure 14.12**

▷ **TECHNOLOGY**    Some three-dimensional graphing utilities are capable of graphing figures such as that shown in Figure 14.12. For instance, the graph shown at the right was drawn with a computer program. In this graph, note that each of the rectangular prisms lies within the solid region.

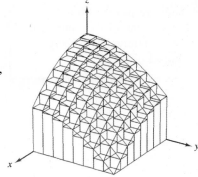

In Example 1, note that by using finer partitions, you obtain better approximations of the volume. This observation suggests that you could obtain the exact volume by taking a limit. That is,

$$\text{Volume} = \lim_{\|\Delta\| \to 0} \sum_{i=1}^{n} f(x_i, y_i) \, \Delta A_i.$$

The precise meaning of this limit is that the limit is equal to $L$ if for every $\varepsilon > 0$, there exists a $\delta > 0$ such that

$$\left| L - \sum_{i=1}^{n} f(x_i, y_i) \, \Delta A_i \right| < \varepsilon$$

for all partitions $\Delta$ of the plane region $R$ (that satisfy $\|\Delta\| < \delta$) and for all possible choices of $x_i$ and $y_i$ in the $i$th region.

Using the limit of a Riemann sum to define volume is a special case of using the limit to define a **double integral.** The general case, however, does not require that the function be positive or continuous.

## Exploration

The entries in the table represent the depths (in yards) of earth at the centers of the squares in the figure below.

| x \ y | 10 | 20 | 30 |
|---|---|---|---|
| 10 | 100 | 90 | 70 |
| 20 | 70 | 70 | 40 |
| 30 | 50 | 50 | 40 |
| 40 | 40 | 50 | 30 |

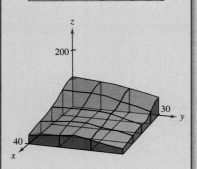

Approximate the number of cubic yards of earth in the first octant. (This exploration was submitted by Robert Vojack.)

### Definition of Double Integral

If $f$ is defined on a closed, bounded region $R$ in the $xy$-plane, then the **double integral of $f$ over $R$** is

$$\int_R \int f(x, y)\, dA = \lim_{\|\Delta\| \to 0} \sum_{i=1}^{n} f(x_i, y_i)\, \Delta A_i$$

provided the limit exists. If the limit exists, then $f$ is **integrable** over $R$.

Having defined a double integral, you will see that a definite integral is occasionally referred to as a **single integral.**

Sufficient conditions for the double integral of $f$ on the region $R$ to exist are that $R$ can be written as a union of a finite number of nonoverlapping subregions (see Figure 14.13) that are vertically or horizontally simple *and* that $f$ is continuous on the region $R$. This means that the intersection of two nonoverlapping regions is a set that has an area of 0. In Figure 14.13, the area of the line segment common to $R_1$ and $R_2$ is 0.

A double integral can be used to find the volume of a solid region that lies between the $xy$-plane and the surface given by $z = f(x, y)$.

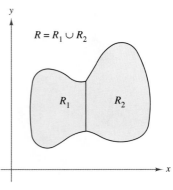

The two regions $R_1$ and $R_2$ are nonoverlapping.

**Figure 14.13**

### Volume of a Solid Region

If $f$ is integrable over a plane region $R$ and $f(x, y) \geq 0$ for all $(x, y)$ in $R$, then the volume of the solid region that lies above $R$ and below the graph of $f$ is

$$V = \int_R \int f(x, y)\, dA.$$

Double integrals share many properties of single integrals.

### THEOREM 14.1   Properties of Double Integrals

Let $f$ and $g$ be continuous over a closed, bounded plane region $R$, and let $c$ be a constant.

1. $\displaystyle \int_R \int cf(x, y)\, dA = c \int_R \int f(x, y)\, dA$

2. $\displaystyle \int_R \int [f(x, y) \pm g(x, y)]\, dA = \int_R \int f(x, y)\, dA \pm \int_R \int g(x, y)\, dA$

3. $\displaystyle \int_R \int f(x, y)\, dA \geq 0, \quad \text{if } f(x, y) \geq 0$

4. $\displaystyle \int_R \int f(x, y)\, dA \geq \int_R \int g(x, y)\, dA, \quad \text{if } f(x, y) \geq g(x, y)$

5. $\displaystyle \int_R \int f(x, y)\, dA = \int_{R_1} \int f(x, y)\, dA + \int_{R_2} \int f(x, y)\, dA$, where $R$ is the union

   of two nonoverlapping subregions $R_1$ and $R_2$.

## Evaluation of Double Integrals

Normally, the first step in evaluating a double integral is to rewrite it as an iterated integral. To show how this is done, a geometric model of a double integral is used as the volume of a solid.

Consider the solid region bounded by the plane $z = f(x, y) = 2 - x - 2y$ and the three coordinate planes, as shown in Figure 14.14. Each vertical cross section taken parallel to the $yz$-plane is a triangular region whose base has a length of $y = (2 - x)/2$ and whose height is $z = 2 - x$. This implies that for a fixed value of $x$, the area of the triangular cross section is

$$A(x) = \frac{1}{2}(\text{base})(\text{height}) = \frac{1}{2}\left(\frac{2 - x}{2}\right)(2 - x) = \frac{(2 - x)^2}{4}.$$

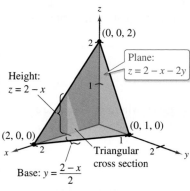

Height:
$z = 2 - x$

$(0, 0, 2)$

Plane:
$z = 2 - x - 2y$

$(2, 0, 0)$

$(0, 1, 0)$

Triangular cross section

Base: $y = \dfrac{2 - x}{2}$

Volume: $\displaystyle\int_0^2 A(x)\, dx$

**Figure 14.14**

By the formula for the volume of a solid with known cross sections (see Section 7.2), the volume of the solid is

$$\text{Volume} = \int_a^b A(x)\, dx \qquad \text{Formula for volume}$$

$$= \int_0^2 \frac{(2 - x)^2}{4}\, dx \qquad \text{Substitute.}$$

$$= -\frac{(2 - x)^3}{12}\bigg]_0^2 \qquad \text{Integrate with respect to } x.$$

$$= \frac{2}{3}. \qquad \text{Volume of solid region (See Figure 14.14.)}$$

This procedure works no matter how $A(x)$ is obtained. In particular, you can find $A(x)$ by integration, as shown in Figure 14.15. That is, you consider $x$ to be constant and integrate $z = 2 - x - 2y$ from 0 to $(2 - x)/2$ to obtain

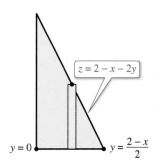

$z = 2 - x - 2y$

$y = 0$

$y = \dfrac{2 - x}{2}$

Triangular cross section
**Figure 14.15**

$$A(x) = \int_0^{(2-x)/2} (2 - x - 2y)\, dy \qquad \text{Apply formula for area.}$$

$$= \left[(2 - x)y - y^2\right]_0^{(2-x)/2} \qquad \text{Integrate with respect to } y.$$

$$= \frac{(2 - x)^2}{4}. \qquad \text{Area of triangular cross section (See Figure 14.15.)}$$

Combining these results, you have the *iterated integral*

$$\text{Volume} = \int_R \int f(x, y)\, dA = \int_0^2 \int_0^{(2-x)/2} (2 - x - 2y)\, dy\, dx.$$

To understand this procedure better, it helps to imagine the integration as two sweeping motions. For the inner integration, a vertical line sweeps out the area of a cross section. For the outer integration, the triangular cross section sweeps out the volume, as shown in Figure 14.16.

 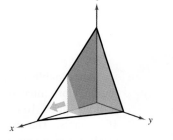 

Integrate with respect to $y$ to obtain the area of the cross section.

Integrate with respect to $x$ to obtain the volume of the solid.

**Figure 14.16**

The next theorem was proved by the Italian mathematician Guido Fubini (1879–1943). The theorem states that if $R$ is a vertically or horizontally simple region and $f$ is continuous on $R$, then the double integral of $f$ on $R$ is equal to an iterated integral.

---

**THEOREM 14.2   Fubini's Theorem**

Let $f$ be continuous on a plane region $R$.

1. If $R$ is defined by $a \leq x \leq b$ and $g_1(x) \leq y \leq g_2(x)$, where $g_1$ and $g_2$ are continuous on $[a, b]$, then

$$\int\int_R f(x, y)\, dA = \int_a^b \int_{g_1(x)}^{g_2(x)} f(x, y)\, dy\, dx.$$

2. If $R$ is defined by $c \leq y \leq d$ and $h_1(y) \leq x \leq h_2(y)$, where $h_1$ and $h_2$ are continuous on $[c, d]$, then

$$\int\int_R f(x, y)\, dA = \int_c^d \int_{h_1(y)}^{h_2(y)} f(x, y)\, dx\, dy.$$

---

**EXAMPLE 2**   **Evaluating a Double Integral as an Iterated Integral**

Evaluate

$$\int\int_R \left(1 - \frac{1}{2}x^2 - \frac{1}{2}y^2\right) dA$$

where $R$ is the region given by

$$0 \leq x \leq 1, \quad 0 \leq y \leq 1.$$

**Solution**   Because the region $R$ is a square, it is both vertically and horizontally simple, and you can use either order of integration. Choose $dy\, dx$ by placing a vertical representative rectangle in the region (see the figure at the right). This produces the following.

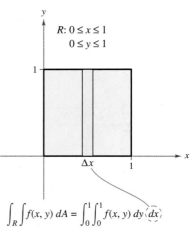

$$\int\int_R f(x, y)\, dA = \int_0^1 \int_0^1 f(x, y)\, dy\, dx$$

$$\int\int_R \left(1 - \frac{1}{2}x^2 - \frac{1}{2}y^2\right) dA = \int_0^1 \int_0^1 \left(1 - \frac{1}{2}x^2 - \frac{1}{2}y^2\right) dy\, dx$$

$$= \int_0^1 \left[\left(1 - \frac{1}{2}x^2\right)y - \frac{y^3}{6}\right]_0^1 dx$$

$$= \int_0^1 \left(\frac{5}{6} - \frac{1}{2}x^2\right) dx$$

$$= \left[\frac{5}{6}x - \frac{x^3}{6}\right]_0^1$$

$$= \frac{2}{3}$$

The double integral evaluated in Example 2 represents the volume of the solid region approximated in Example 1. Note that the approximation obtained in Example 1 is quite good $\left(0.672 \text{ vs. } \frac{2}{3}\right)$, even though you used a partition consisting of only 16 squares. The error resulted because the centers of the square subregions were used as the points in the approximation. This is comparable to the Midpoint Rule approximation of a single integral.

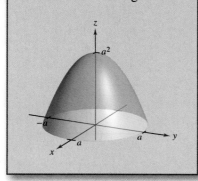
• • • • • • • • • • • • • ▷

**• • REMARK**   In Example 3, note the usefulness of Wallis's Formula to evaluate $\int_0^{\pi/2} \cos^n \theta \, d\theta$. You may want to review this formula in Section 8.3.

The difficulty of evaluating a single integral $\int_a^b f(x) \, dx$ usually depends on the function $f$, not on the interval $[a, b]$. This is a major difference between single and double integrals. In the next example, you will integrate a function similar to the one in Examples 1 and 2. Notice that a change in the region $R$ produces a much more difficult integration problem.

**EXAMPLE 3**   **Finding Volume by a Double Integral**

Find the volume of the solid region bounded by the paraboloid $z = 4 - x^2 - 2y^2$ and the *xy*-plane, as shown in Figure 14.17(a).

**Solution**   By letting $z = 0$, you can see that the base of the region in the *xy*-plane is the ellipse $x^2 + 2y^2 = 4$, as shown in Figure 14.17(b). This plane region is both vertically and horizontally simple, so the order $dy \, dx$ is appropriate.

***Variable bounds for y:***   $-\sqrt{\dfrac{4 - x^2}{2}} \le y \le \sqrt{\dfrac{4 - x^2}{2}}$

***Constant bounds for x:***   $-2 \le x \le 2$

The volume is

$$
\begin{aligned}
V &= \int_{-2}^{2} \int_{-\sqrt{(4-x^2)/2}}^{\sqrt{(4-x^2)/2}} (4 - x^2 - 2y^2) \, dy \, dx && \text{See Figure 14.17(b).}\\[2mm]
&= \int_{-2}^{2} \left[ (4 - x^2)y - \frac{2y^3}{3} \right]_{-\sqrt{(4-x^2)/2}}^{\sqrt{(4-x^2)/2}} dx \\[2mm]
&= \frac{4}{3\sqrt{2}} \int_{-2}^{2} (4 - x^2)^{3/2} \, dx \\[2mm]
&= \frac{4}{3\sqrt{2}} \int_{-\pi/2}^{\pi/2} 16 \cos^4 \theta \, d\theta && x = 2\sin\theta \\[2mm]
&= \frac{64}{3\sqrt{2}} (2) \int_{0}^{\pi/2} \cos^4 \theta \, d\theta \\[2mm]
&= \frac{128}{3\sqrt{2}} \left( \frac{3\pi}{16} \right) && \text{Wallis's Formula} \\[2mm]
&= 4\sqrt{2}\,\pi.
\end{aligned}
$$

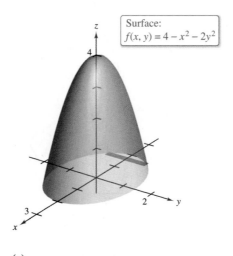

Surface:
$f(x, y) = 4 - x^2 - 2y^2$

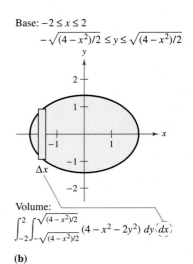

Base: $-2 \le x \le 2$
$-\sqrt{(4-x^2)/2} \le y \le \sqrt{(4-x^2)/2}$

Volume:
$\int_{-2}^{2} \int_{-\sqrt{(4-x^2)/2}}^{\sqrt{(4-x^2)/2}} (4 - x^2 - 2y^2) \, dy \, dx$

**(a)**                                             **(b)**

**Figure 14.17**

In Examples 2 and 3, the problems could be solved with either order of integration because the regions were both vertically and horizontally simple. Moreover, had you used the order $dx\,dy$, you would have obtained integrals of comparable difficulty. There are, however, some occasions when one order of integration is much more convenient than the other. Example 4 shows such a case.

### EXAMPLE 4   Comparing Different Orders of Integration

⋮ ⋯ ▷ *See LarsonCalculus.com for an interactive version of this type of example.*

Find the volume of the solid region bounded by the surface

$$f(x, y) = e^{-x^2} \qquad \text{Surface}$$

and the planes $z = 0$, $y = 0$, $y = x$, and $x = 1$, as shown in Figure 14.18.

**Solution**   The base of the solid region in the $xy$-plane is bounded by the lines $y = 0$, $x = 1$, and $y = x$. The two possible orders of integration are shown in Figure 14.19.

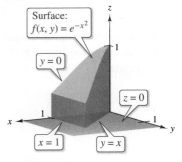

Surface:
$f(x, y) = e^{-x^2}$

$y = 0$

$z = 0$

$x = 1$     $y = x$

Base is bounded by $y = 0$, $y = x$, and $x = 1$.

**Figure 14.18**

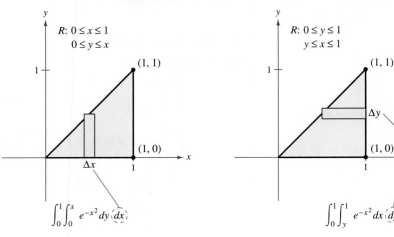

**Figure 14.19**

By setting up the corresponding iterated integrals, you can see that the order $dx\,dy$ requires the antiderivative

$$\int e^{-x^2}\,dx$$

which is not an elementary function. On the other hand, the order $dy\,dx$ produces

$$\int_0^1 \int_0^x e^{-x^2}\,dy\,dx = \int_0^1 e^{-x^2}\,y\,\Big]_0^x\,dx \qquad \text{Integrate with respect to } y.$$

$$= \int_0^1 xe^{-x^2}\,dx$$

$$= -\frac{1}{2}e^{-x^2}\,\Big]_0^1 \qquad \text{Integrate with respect to } x.$$

$$= -\frac{1}{2}\left(\frac{1}{e} - 1\right)$$

$$= \frac{e - 1}{2e} \qquad \text{Volume of solid region (See Figure 14.18.)}$$

$$\approx 0.316.$$

▷ **TECHNOLOGY**   Try using a symbolic integration utility to evaluate the iterated
⋮ integral in Example 4.

EXAMPLE 5 **Volume of a Region Bounded by Two Surfaces**

Find the volume of the solid region bounded above by the paraboloid

$$z = 1 - x^2 - y^2 \qquad \text{Paraboloid}$$

and below by the plane

$$z = 1 - y \qquad \text{Plane}$$

as shown in Figure 14.20.

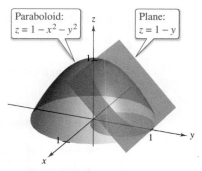

Paraboloid:
$z = 1 - x^2 - y^2$

Plane:
$z = 1 - y$

**Figure 14.20**

**Solution**  Equating $z$-values, you can determine that the intersection of the two surfaces occurs on the right circular cylinder given by

$$1 - y = 1 - x^2 - y^2 \quad \Longrightarrow \quad x^2 = y - y^2.$$

So, the region $R$ in the $xy$-plane is a circle, as shown in Figure 14.21. Because the volume of the solid region is the difference between the volume under the paraboloid and the volume under the plane, you have

$$\text{Volume} = (\text{volume under paraboloid}) - (\text{volume under plane})$$

$$= \int_0^1 \int_{-\sqrt{y-y^2}}^{\sqrt{y-y^2}} (1 - x^2 - y^2)\, dx\, dy - \int_0^1 \int_{-\sqrt{y-y^2}}^{\sqrt{y-y^2}} (1 - y)\, dx\, dy$$

$$= \int_0^1 \int_{-\sqrt{y-y^2}}^{\sqrt{y-y^2}} (y - y^2 - x^2)\, dx\, dy$$

$$= \int_0^1 \left[ (y - y^2)x - \frac{x^3}{3} \right]_{-\sqrt{y-y^2}}^{\sqrt{y-y^2}} dy$$

$$= \frac{4}{3} \int_0^1 (y - y^2)^{3/2}\, dy$$

$$= \left(\frac{4}{3}\right)\left(\frac{1}{8}\right) \int_0^1 [1 - (2y - 1)^2]^{3/2}\, dy$$

$$= \frac{1}{6} \int_{-\pi/2}^{\pi/2} \frac{\cos^4 \theta}{2}\, d\theta \qquad 2y - 1 = \sin \theta$$

$$= \frac{1}{6} \int_0^{\pi/2} \cos^4 \theta\, d\theta$$

$$= \left(\frac{1}{6}\right)\left(\frac{3\pi}{16}\right) \qquad \text{Wallis's Formula}$$

$$= \frac{\pi}{32}.$$

$R:\ 0 \le y \le 1$
$-\sqrt{y - y^2} \le x \le \sqrt{y - y^2}$

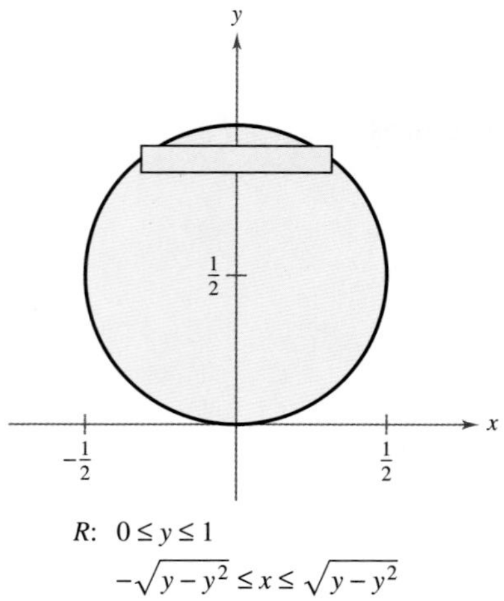

**Figure 14.21**

## Average Value of a Function

Recall from Section 4.4 that for a function $f$ in one variable, the average value of $f$ on the interval $[a, b]$ is

$$\frac{1}{b - a}\int_{a}^{b} f(x)\, dx.$$

Given a function $f$ in two variables, you can find the average value of $f$ over the plane region $R$ as shown in the following definition.

---

**Definition of the Average Value of a Function Over a Region**

If $f$ is integrable over the plane region $R$, then the **average value** of $f$ over $R$ is

$$\text{Average value} = \frac{1}{A}\int_{R}\int f(x, y)\, dA$$

where $A$ is the area of $R$.

---

**EXAMPLE 6**   **Finding the Average Value of a Function**

Find the average value of

$$f(x, y) = \frac{1}{2}xy$$

over the plane region $R$, where $R$ is a rectangle with vertices

$$(0, 0),\ (4, 0),\ (4, 3),\quad \text{and}\quad (0, 3).$$

**Solution**   The area of the rectangular region $R$ is

$$A = (4)(3) = 12$$

as shown in Figure 14.22. The bounds for $x$ are

$$0 \le x \le 4$$

and the bounds for $y$ are

$$0 \le y \le 3.$$

So, the average value is

$$\text{Average value} = \frac{1}{A}\int_{R}\int f(x, y)\, dA$$

$$= \frac{1}{12}\int_{0}^{4}\int_{0}^{3} \frac{1}{2}xy\, dy\, dx$$

$$= \frac{1}{12}\int_{0}^{4} \frac{1}{4}xy^{2}\Big]_{0}^{3}\, dx$$

$$= \left(\frac{1}{12}\right)\left(\frac{9}{4}\right)\int_{0}^{4} x\, dx$$

$$= \frac{3}{16}\left[\frac{1}{2}x^{2}\right]_{0}^{4}$$

$$= \left(\frac{3}{16}\right)(8)$$

$$= \frac{3}{2}.$$

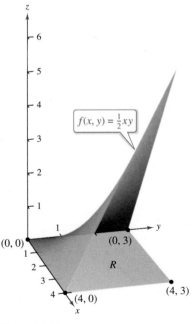

Figure 14.22

# 14.2 Exercises

**CONCEPT CHECK**

**1. Approximating the Volume of a Solid** In your own words, describe the process of using an inner partition to approximate the volume of a solid region lying above the $xy$-plane. How can the approximation be improved?

**2. Fubini's Theorem** What is the benefit of Fubini's Theorem when evaluating a double integral?

**Approximation** In Exercises 3–6, approximate the integral $\int_R\int f(x, y)\, dA$ by dividing the rectangle $R$ with vertices $(0, 0)$, $(4, 0)$, $(4, 2)$, and $(0, 2)$ into eight equal squares and finding the sum
$$\sum_{i=1}^{8} f(x_i, y_i)\, \Delta A_i,$$
where $(x_i, y_i)$ is the center of the $i$th square. Evaluate the iterated integral and compare it with the approximation.

**3.** $\displaystyle\int_0^4\int_0^2 (x + y)\, dy\, dx$    **4.** $\displaystyle\frac{1}{2}\int_0^4\int_0^2 x^2 y\, dy\, dx$

**5.** $\displaystyle\int_0^4\int_0^2 (x^2 + y^2)\, dy\, dx$    **6.** $\displaystyle\int_0^4\int_0^2 \frac{1}{(x + 1)(y + 1)}\, dy\, dx$

**Evaluating a Double Integral** In Exercises 7–12, sketch the region $R$ and evaluate the iterated integral $\int_R\int f(x, y)\, dA$.

**7.** $\displaystyle\int_0^2\int_0^1 (1 - 4x + 8y)\, dy\, dx$    **8.** $\displaystyle\int_0^\pi\int_0^{\pi/2} \sin^2 x \cos^2 y\, dy\, dx$

**9.** $\displaystyle\int_0^6\int_{y/2}^3 (x + y)\, dx\, dy$

**10.** $\displaystyle\int_0^4\int_{y/2}^{\sqrt{y}} x^2 y^2\, dx\, dy$

**11.** $\displaystyle\int_{-3}^3\int_{-\sqrt{9-x^2}}^{\sqrt{9-x^2}} (x + y)\, dy\, dx$

**12.** $\displaystyle\int_0^1\int_{y-1}^0 e^{x+y}\, dx\, dy + \int_0^1\int_0^{1-y} e^{x+y}\, dx\, dy$

**Evaluating a Double Integral** In Exercises 13–20, set up integrals for both orders of integration. Use the more convenient order to evaluate the integral over the plane region $R$.

**13.** $\displaystyle\int_R\int xy\, dA$

$R$: rectangle with vertices $(0, 0)$, $(0, 5)$, $(3, 5)$, $(3, 0)$

**14.** $\displaystyle\int_R\int \sin x \sin y\, dA$

$R$: rectangle with vertices $(-\pi, 0)$, $(\pi, 0)$, $(\pi, \pi/2)$, $(-\pi, \pi/2)$

**15.** $\displaystyle\int_R\int \frac{y}{x^2 + y^2}\, dA$

$R$: trapezoid bounded by $y = x$, $y = 2x$, $x = 1$, $x = 2$

**16.** $\displaystyle\int_R\int xe^y\, dA$

$R$: triangle bounded by $y = 4 - x$, $y = 0$, $x = 0$

**17.** $\displaystyle\int_R\int -2y\, dA$

$R$: region bounded by $y = 4 - x^2$, $y = 4 - x$

**18.** $\displaystyle\int_R\int \frac{y}{1 + x^2}\, dA$

$R$: region bounded by $y = 0$, $y = \sqrt{x}$, $x = 4$

**19.** $\displaystyle\int_R\int x\, dA$

$R$: sector of a circle in the first quadrant bounded by $y = \sqrt{25 - x^2}$, $3x - 4y = 0$, $y = 0$

**20.** $\displaystyle\int_R\int (x^2 + y^2)\, dA$

$R$: semicircle bounded by $y = \sqrt{4 - x^2}$, $y = 0$

**Finding Volume** In Exercises 21–26, use a double integral to find the volume of the indicated solid.

**21.**

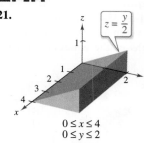

$0 \le x \le 4$
$0 \le y \le 2$

**22.**

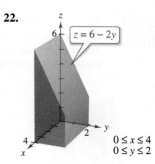

$0 \le x \le 4$
$0 \le y \le 2$

**23.**

**24.**

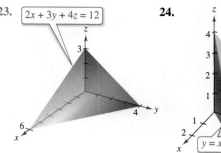

**25.**

**26.**

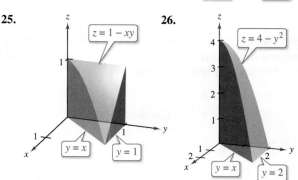

**Finding Volume** In Exercises 27 and 28, use an improper double integral to find the volume of the indicated solid.

**27.**

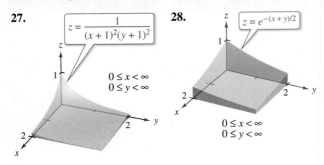

$$z = \frac{1}{(x+1)^2(y+1)^2}$$

$0 \le x < \infty$
$0 \le y < \infty$

**28.**

$z = e^{-(x+y)/2}$

$0 \le x < \infty$
$0 \le y < \infty$

**Finding Volume** In Exercises 29–34, set up and evaluate a double integral to find the volume of the solid bounded by the graphs of the equations.

**29.** $z = xy$, $z = 0$, $y = x^3$, $x = 1$, first octant

**30.** $z = 0$, $z = x^2$, $x = 0$, $x = 2$, $y = 0$, $y = 4$

**31.** $z = x + y$, $x^2 + y^2 = 4$, first octant

**32.** $z = \dfrac{1}{1 + y^2}$, $x = 0$, $x = 2$, $y \ge 0$

**33.** $y = 4 - x^2$, $z = 4 - x^2$, first octant

**34.** $x^2 + z^2 = 1$, $y^2 + z^2 = 1$, first octant

 **Volume of a Region Bounded by Two Surfaces** In Exercises 35–40, set up a double integral to find the volume of the solid region bounded by the graphs of the equations. Do not evaluate the integral.

**35.**

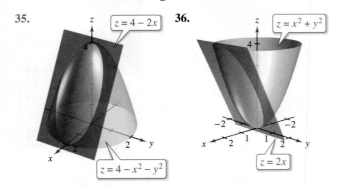

$z = 4 - 2x$

$z = 4 - x^2 - y^2$

**36.**

$z = x^2 + y^2$

$z = 2x$

**37.** $z = x^2 + y^2$, $x^2 + y^2 = 4$, $z = 0$

**38.** $z = \sin^2 x$, $z = 0$, $0 \le x \le \pi$, $0 \le y \le 5$

**39.** $z = x^2 + 2y^2$, $z = 4y$

**40.** $z = x^2 + y^2$, $z = 18 - x^2 - y^2$

**Finding Volume Using Technology** In Exercises 41–44, use a computer algebra system to find the volume of the solid bounded by the graphs of the equations.

**41.** $z = 9 - x^2 - y^2$, $z = 0$

**42.** $x^2 = 9 - y$, $z^2 = 9 - y$, first octant

**43.** $z = \dfrac{2}{1 + x^2 + y^2}$, $z = 0$, $y = 0$, $x = 0$, $y = -0.5x + 1$

**44.** $z = \ln(1 + x + y)$, $z = 0$, $y = 0$, $x = 0$, $x = 4 - \sqrt{y}$

**Evaluating an Iterated Integral** In Exercises 45–50, sketch the region of integration. Then evaluate the iterated integral, changing the order of integration if necessary.

**45.** $\displaystyle \int_0^1 \int_{y/2}^{1/2} e^{-x^2} \, dx \, dy$

**46.** $\displaystyle \int_0^{\ln 10} \int_{e^x}^{10} \frac{1}{\ln y} \, dy \, dx$

**47.** $\displaystyle \int_{-2}^2 \int_{-\sqrt{4-x^2}}^{\sqrt{4-x^2}} \sqrt{4 - y^2} \, dy \, dx$

**48.** $\displaystyle \int_0^3 \int_{y/3}^1 \frac{1}{1 + x^4} \, dx \, dy$

**49.** $\displaystyle \int_0^2 \int_{2x}^4 \sin y^2 \, dy \, dx$

**50.** $\displaystyle \int_0^2 \int_{x^2/2}^2 \sqrt{y} \cos y \, dy \, dx$

 **Average Value** In Exercises 51–56, find the average value of $f(x, y)$ over the plane region $R$.

**51.** $f(x, y) = x$

   $R$: rectangle with vertices $(0, 0)$, $(4, 0)$, $(4, 2)$, $(0, 2)$

**52.** $f(x, y) = 2xy$

   $R$: rectangle with vertices $(0, 1)$, $(1, 1)$, $(1, 6)$, $(0, 6)$

**53.** $f(x, y) = x^2 + y^2$

   $R$: square with vertices $(0, 0)$, $(2, 0)$, $(2, 2)$, $(0, 2)$

**54.** $f(x, y) = \dfrac{1}{x + y}$, $R$: triangle with vertices $(0, 0)$, $(1, 0)$, $(1, 1)$

**55.** $f(x, y) = e^{x+y}$, $R$: triangle with vertices $(0, 0)$, $(0, 1)$, $(1, 1)$

**56.** $f(x, y) = \sin(x + y)$

   $R$: rectangle with vertices $(0, 0)$, $(\pi, 0)$, $(\pi, \pi)$, $(0, \pi)$

**57. Average Production**

The Cobb-Douglas production function for an automobile manufacturer is $f(x, y) = 100x^{0.6}y^{0.4}$, where $x$ is the number of units of labor and $y$ is the number of units of capital. Estimate the average production level when the number of units of labor $x$ varies between 200 and 250 and the number of units of capital $y$ varies between 300 and 325.

**58. Average Temperature** The temperature in degrees Celsius on the surface of a metal plate is $T(x, y) = 20 - 4x^2 - y^2$, where $x$ and $y$ are measured in centimeters. Estimate the average temperature when $x$ varies between 0 and 2 centimeters and $y$ varies between 0 and 4 centimeters.

**EXPLORING CONCEPTS**

**59. Volume** Let $R$ be a region in the $xy$-plane whose area is $B$. When $f(x, y) = k$ for every point $(x, y)$ in $R$, what is the value of $\int_R \int f(x, y) \, dA$? Explain.

**60. Volume** Let the plane region $R$ be a unit circle and let the maximum value of $f$ on $R$ be 6. Is the greatest possible value of $\int_R \int f(x, y) \, dy \, dx$ equal to 6? Why or why not? If not, what is the greatest possible value?

**Probability** A *joint density function* of the continuous random variables $x$ and $y$ is a function $f(x, y)$ satisfying the following properties.

(a) $f(x, y) \geq 0$ for all $(x, y)$

(b) $\displaystyle\int_{-\infty}^{\infty} \int_{-\infty}^{\infty} f(x, y) \, dA = 1$

(c) $P[(x, y) \in R] = \displaystyle\int_{R}\int f(x, y) \, dA$

In Exercises 61–64, show that the function is a joint density function and find the required probability.

**61.** $f(x, y) = \begin{cases} \frac{1}{3}, & 0 \leq x \leq 1, 1 \leq y \leq 4 \\ 0, & \text{elsewhere} \end{cases}$

$P(0 \leq x \leq 1, 1 \leq y \leq 3)$

**62.** $f(x, y) = \begin{cases} \frac{1}{5}xy, & 0 \leq x \leq 2, 0 \leq y \leq \sqrt{5} \\ 0, & \text{elsewhere} \end{cases}$

$P(0 \leq x \leq 1, 0 \leq y \leq 2)$

**63.** $f(x, y) = \begin{cases} \frac{1}{27}(9 - x - y), & 0 \leq x \leq 3, 3 \leq y \leq 6 \\ 0, & \text{elsewhere} \end{cases}$

$P(0 \leq x \leq 1, 3 \leq y \leq 6)$

**64.** $f(x, y) = \begin{cases} e^{-x-y}, & x \geq 0, y \geq 0 \\ 0, & \text{elsewhere} \end{cases}$

$P(0 \leq x \leq 1, x \leq y \leq 1)$

**65. Proof** Let $f$ be a continuous function such that $0 \leq f(x, y) \leq 1$ over a region $R$ of area 1. Prove that $0 \leq \int_{R}\int f(x, y) \, dA \leq 1$.

**66. Finding Volume** Find the volume of the solid in the first octant bounded by the coordinate planes and the plane

$$\frac{x}{a} + \frac{y}{b} + \frac{z}{c} = 1$$

where $a > 0$, $b > 0$, and $c > 0$.

**67. Approximation** The table shows values of a function $f$ over a square region $R$. Divide the region into 16 equal squares and select $(x_i, y_i)$ to be the point in the $i$th square closest to the origin. Approximate the value of the integral below. Compare this approximation with that obtained by using the point in the $i$th square farthest from the origin.

$$\int_0^4 \int_0^4 f(x, y) \, dy \, dx$$

| x \ y | 0 | 1 | 2 | 3 | 4 |
|---|---|---|---|---|---|
| 0 | 32 | 31 | 28 | 23 | 16 |
| 1 | 31 | 30 | 27 | 22 | 15 |
| 2 | 28 | 27 | 24 | 19 | 12 |
| 3 | 23 | 22 | 19 | 14 | 7 |
| 4 | 16 | 15 | 12 | 7 | 0 |

**68.** 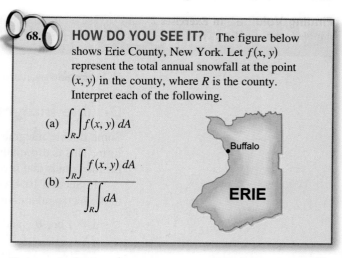 **HOW DO YOU SEE IT?** The figure below shows Erie County, New York. Let $f(x, y)$ represent the total annual snowfall at the point $(x, y)$ in the county, where $R$ is the county. Interpret each of the following.

(a) $\displaystyle\int_{R}\int f(x, y) \, dA$

(b) $\dfrac{\displaystyle\int_{R}\int f(x, y) \, dA}{\displaystyle\int_{R}\int dA}$

**True or False?** In Exercises 69 and 70, determine whether the statement is true or false. If it is false, explain why or give an example that shows it is false.

**69.** The volume of the sphere $x^2 + y^2 + z^2 = 1$ is given by the integral

$$V = 8\int_0^1 \int_0^1 \sqrt{1 - x^2 - y^2} \, dx \, dy.$$

**70.** If $f(x, y) \leq g(x, y)$ for all $(x, y)$ in $R$, and both $f$ and $g$ are continuous over $R$, then $\int_{R}\int f(x, y) \, dA \leq \int_{R}\int g(x, y) \, dA$.

**71. Maximizing a Double Integral** Determine the region $R$ in the $xy$-plane that maximizes the value of

$$\int_{R}\int (9 - x^2 - y^2) \, dA.$$

**72. Minimizing a Double Integral** Determine the region $R$ in the $xy$-plane that minimizes the value of

$$\int_{R}\int (x^2 + y^2 - 4) \, dA.$$

**73. Average Value** Let

$$f(x) = \int_1^x e^{t^2} \, dt.$$

Find the average value of $f$ on the interval $[0, 1]$.

**74. Using Geometry** Use a geometric argument to show that

$$\int_0^3 \int_0^{\sqrt{9-y^2}} \sqrt{9 - x^2 - y^2} \, dx \, dy = \frac{9\pi}{2}.$$

---

**PUTNAM EXAM CHALLENGE**

**75.** Evaluate $\int_0^a \int_0^b e^{\max\{b^2x^2, \, a^2y^2\}} \, dy \, dx$, where $a$ and $b$ are positive.

**76.** Show that if $\lambda > \frac{1}{2}$ there does not exist a real-valued function $u$ such that for all $x$ in the closed interval $0 \leq x \leq 1$, $u(x) = 1 + \lambda\int_x^1 u(y)u(y - x) \, dy$.

# 14.3 Change of Variables: Polar Coordinates

■ Write and evaluate double integrals in polar coordinates.

## Double Integrals in Polar Coordinates

Some double integrals are *much* easier to evaluate in polar form than in rectangular form. This is especially true for regions such as circles, cardioids, and rose curves, and for integrands that involve $x^2 + y^2$.

In Section 10.4, you learned that the polar coordinates $(r, \theta)$ of a point are related to the rectangular coordinates $(x, y)$ of the point as follows.

$$x = r \cos \theta \quad \text{and} \quad y = r \sin \theta$$

$$r^2 = x^2 + y^2 \quad \text{and} \quad \tan \theta = \frac{y}{x}$$

### EXAMPLE 1   Using Polar Coordinates to Describe a Region

Use polar coordinates to describe each region shown in Figure 14.23.

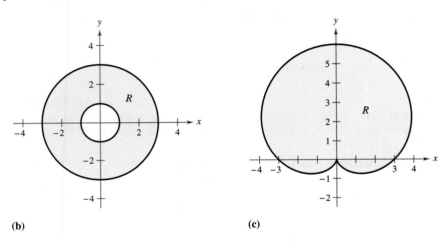

(a)

(b)

(c)

**Figure 14.23**

### Solution

**a.** The region $R$ is a quarter circle of radius 2. It can be described in polar coordinates as

$$R = \{(r, \theta): 0 \leq r \leq 2, \quad 0 \leq \theta \leq \pi/2\}.$$

**b.** The region $R$ consists of all points between concentric circles of radii 1 and 3. It can be described in polar coordinates as

$$R = \{(r, \theta): 1 \leq r \leq 3, \quad 0 \leq \theta \leq 2\pi\}.$$

**c.** The region $R$ is a cardioid with $a = b = 3$. It can be described in polar coordinates as

$$R = \{(r, \theta): 0 \leq r \leq 3 + 3 \sin \theta, \quad 0 \leq \theta \leq 2\pi\}. \qquad ■$$

The regions in Example 1 are special cases of **polar sectors**

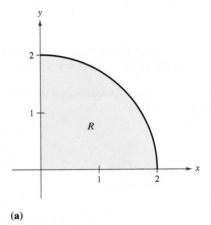

Polar sector

**Figure 14.24**

$$R = \{(r, \theta): r_1 \leq r \leq r_2, \ \theta_1 \leq \theta \leq \theta_2\} \qquad \text{Polar sector}$$

as shown in Figure 14.24.

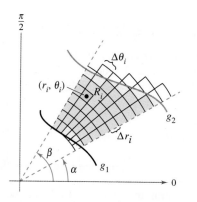

Polar grid superimposed over region $R$
**Figure 14.25**

To define a double integral of a continuous function $z = f(x, y)$ in polar coordinates, consider a region $R$ bounded by the graphs of

$$r = g_1(\theta) \quad \text{and} \quad r = g_2(\theta)$$

and the lines $\theta = \alpha$ and $\theta = \beta$. Instead of partitioning $R$ into small rectangles, use a partition of small polar sectors. On $R$, superimpose a polar grid made of rays and circular arcs, as shown in Figure 14.25. The polar sectors $R_i$ lying entirely within $R$ form an **inner polar partition** $\Delta$, whose **norm** $\|\Delta\|$ is the length of the longest diagonal of the $n$ polar sectors.

Consider a specific polar sector $R_i$, as shown in Figure 14.26. It can be shown (see Exercise 68) that the area of $R_i$ is

$$\Delta A_i = r_i \, \Delta r_i \, \Delta \theta_i \qquad \text{Area of } R_i$$

where $\Delta r_i = r_2 - r_1$ and $\Delta \theta_i = \theta_2 - \theta_1$. This implies that the volume of the solid of height $f(r_i \cos \theta_i, r_i \sin \theta_i)$ above $R_i$ is approximately

$$f(r_i \cos \theta_i, r_i \sin \theta_i) r_i \, \Delta r_i \, \Delta \theta_i$$

and you have

$$\int_R \int f(x, y) \, dA \approx \sum_{i=1}^{n} f(r_i \cos \theta_i, r_i \sin \theta_i) r_i \, \Delta r_i \, \Delta \theta_i.$$

The sum on the right can be interpreted as a Riemann sum for

$$f(r \cos \theta, r \sin \theta) r.$$

The region $R$ corresponds to a *horizontally simple* region $S$ in the $r\theta$-plane, as shown in Figure 14.27. The polar sectors $R_i$ correspond to rectangles $S_i$, and the area $\Delta A_i$ of $S_i$ is $\Delta r_i \, \Delta \theta_i$. So, the right-hand side of the equation corresponds to the double integral

$$\int_S \int f(r \cos \theta, r \sin \theta) r \, dA.$$

From this, you can apply Theorem 14.2 to write

$$\int_R \int f(x, y) \, dA = \int_S \int f(r \cos \theta, r \sin \theta) r \, dA$$

$$= \int_\alpha^\beta \int_{g_1(\theta)}^{g_2(\theta)} f(r \cos \theta, r \sin \theta) r \, dr \, d\theta.$$

This suggests the theorem on the next page, the proof of which is discussed in Section 14.8.

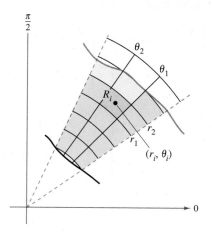

The polar sector $R_i$ is the set of all points $(r, \theta)$ such that $r_1 \le r \le r_2$ and $\theta_1 \le \theta \le \theta_2$.
**Figure 14.26**

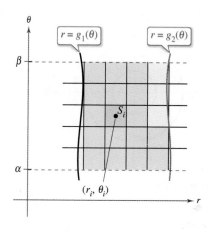

Horizontally simple region $S$
**Figure 14.27**

---

### THEOREM 14.3 Change of Variables to Polar Form

Let $R$ be a plane region consisting of all points $(x, y) = (r \cos \theta, r \sin \theta)$ satisfying the conditions $0 \le g_1(\theta) \le r \le g_2(\theta)$, $\alpha \le \theta \le \beta$, where $0 \le (\beta - \alpha) \le 2\pi$. If $g_1$ and $g_2$ are continuous on $[\alpha, \beta]$ and $f$ is continuous on $R$, then

$$\int_R \int f(x, y)\, dA = \int_\alpha^\beta \int_{g_1(\theta)}^{g_2(\theta)} f(r \cos \theta, r \sin \theta) r\, dr\, d\theta.$$

---

**Exploration**

***Volume of a Paraboloid Sector*** In the Exploration on page 983, you were asked to summarize the different ways you know of finding the volume of the solid bounded by the paraboloid

$$z = a^2 - x^2 - y^2, a > 0$$

and the $xy$-plane. You now know another way. Use it to find the volume of the solid.

If $z = f(x, y)$ is nonnegative on $R$, then the integral in Theorem 14.3 can be interpreted as the volume of the solid region between the graph of $f$ and the region $R$. When using the integral in Theorem 14.3, be certain not to omit the extra factor of $r$ in the integrand.

The region $R$ is restricted to two basic types, ***r*-simple** regions and ***θ*-simple** regions, as shown in Figure 14.28.

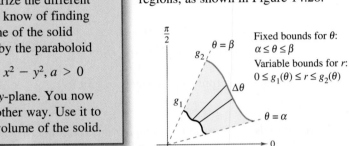

*r*-Simple region

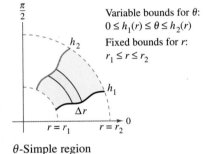

*θ*-Simple region

**Figure 14.28**

---

### EXAMPLE 2 Evaluating a Double Polar Integral

Let $R$ be the annular region lying between the two circles $x^2 + y^2 = 1$ and $x^2 + y^2 = 5$. Evaluate the integral

$$\int_R \int (x^2 + y)\, dA.$$

**Solution** The polar boundaries are $1 \le r \le \sqrt{5}$ and $0 \le \theta \le 2\pi$, as shown in Figure 14.29. Furthermore, $x^2 = (r \cos \theta)^2$ and $y = r \sin \theta$. So, you have

$$\int_R \int (x^2 + y)\, dA = \int_0^{2\pi} \int_1^{\sqrt{5}} (r^2 \cos^2 \theta + r \sin \theta) r\, dr\, d\theta$$

$$= \int_0^{2\pi} \int_1^{\sqrt{5}} (r^3 \cos^2 \theta + r^2 \sin \theta)\, dr\, d\theta$$

$$= \int_0^{2\pi} \left( \frac{r^4}{4} \cos^2 \theta + \frac{r^3}{3} \sin \theta \right) \Big]_1^{\sqrt{5}} d\theta$$

$$= \int_0^{2\pi} \left( 6 \cos^2 \theta + \frac{5\sqrt{5} - 1}{3} \sin \theta \right) d\theta$$

$$= \int_0^{2\pi} \left( 3 + 3 \cos 2\theta + \frac{5\sqrt{5} - 1}{3} \sin \theta \right) d\theta$$

$$= \left( 3\theta + \frac{3 \sin 2\theta}{2} - \frac{5\sqrt{5} - 1}{3} \cos \theta \right) \Big]_0^{2\pi}$$

$$= 6\pi.$$

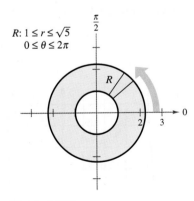

$R: 1 \le r \le \sqrt{5}$
$0 \le \theta \le 2\pi$

**Figure 14.29**

In Example 2, be sure to notice the factor of $r$ with $dr\ d\theta$ in the integrand. This comes from the formula for the area of a polar sector. In differential notation, you can write

$$dA = r\ dr\ d\theta$$

which indicates that the area of a polar sector increases as you move away from the origin.

**EXAMPLE 3**    **Change of Variables to Polar Coordinates**

Use polar coordinates to find the volume of the solid region bounded above by the hemisphere

$$z = \sqrt{16 - x^2 - y^2} \qquad \text{Hemisphere forms upper surface.}$$

and below by the circular region $R$ given by

$$x^2 + y^2 \le 4 \qquad \text{Circular region forms lower surface.}$$

as shown in Figure 14.30.

**Solution**   In Figure 14.30, you can see that $R$ has the bounds

$$-\sqrt{4 - y^2} \le x \le \sqrt{4 - y^2}, \quad -2 \le y \le 2$$

and that $0 \le z \le \sqrt{16 - x^2 - y^2}$. In polar coordinates, the bounds are

$$0 \le r \le 2 \quad \text{and} \quad 0 \le \theta \le 2\pi$$

with height $z = \sqrt{16 - x^2 - y^2} = \sqrt{16 - r^2}$. Consequently, the volume $V$ is

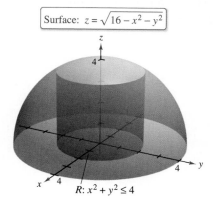

Surface: $z = \sqrt{16 - x^2 - y^2}$

$R: x^2 + y^2 \le 4$

**Figure 14.30**

$$
\begin{aligned}
V &= \int_R \int f(x, y)\ dA & \text{Formula for volume} \\[2mm]
&= \int_0^{2\pi} \int_0^2 \sqrt{16 - r^2}\ r\ dr\ d\theta & \text{Polar coordinates} \\[2mm]
&= -\frac{1}{3} \int_0^{2\pi} (16 - r^2)^{3/2} \Big]_0^2\ d\theta & \text{Integrate with respect to } r. \\[2mm]
&= -\frac{1}{3} \int_0^{2\pi} \left(24\sqrt{3} - 64\right) d\theta \\[2mm]
&= -\frac{8}{3}\left(3\sqrt{3} - 8\right)\theta\Big]_0^{2\pi} & \text{Integrate with respect to } \theta. \\[2mm]
&= \frac{16\pi}{3}\left(8 - 3\sqrt{3}\right) \\[2mm]
&\approx 46.979.
\end{aligned}
$$

**REMARK**   To see the benefit of polar coordinates in Example 3, you should try to evaluate the corresponding rectangular iterated integral

$$\int_{-2}^2 \int_{-\sqrt{4 - y^2}}^{\sqrt{4 - y^2}} \sqrt{16 - x^2 - y^2}\ dx\ dy.$$

▷ **TECHNOLOGY**   Any computer algebra system that can evaluate double integrals in rectangular coordinates can also evaluate double integrals in polar coordinates. The reason this is true is that once you have formed the iterated integral, its value is not changed by using different variables. In other words, if you use a computer algebra system to evaluate

$$\int_0^{2\pi} \int_0^2 \sqrt{16 - x^2}\ x\ dx\ dy$$

you should obtain the same value as that obtained in Example 3.

Just as with rectangular coordinates, the double integral

$$\int_R \int dA$$

can be used to find the area of a region in the plane.

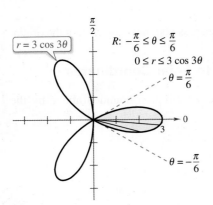

**Figure 14.31**

See LarsonCalculus.com for an interactive version of this type of example.

| | **EXAMPLE 4** | **Finding Areas of Polar Regions** |

To use a double integral to find the area enclosed by the graph of $r = 3 \cos 3\theta$, let $R$ be one petal of the curve shown in Figure 14.31. This region is $r$-simple, and the boundaries are $-\pi/6 \le \theta \le \pi/6$ and $0 \le r \le 3 \cos 3\theta$. So, the area of one petal is

$$\frac{1}{3}A = \iint_R dA = \int_{-\pi/6}^{\pi/6} \int_0^{3\cos 3\theta} r \, dr \, d\theta$$

$$= \int_{-\pi/6}^{\pi/6} \frac{r^2}{2} \Big]_0^{3\cos 3\theta} d\theta \qquad \text{Integrate with respect to } r.$$

$$= \frac{9}{2} \int_{-\pi/6}^{\pi/6} \cos^2 3\theta \, d\theta$$

$$= \frac{9}{4} \int_{-\pi/6}^{\pi/6} (1 + \cos 6\theta) \, d\theta$$

$$= \frac{9}{4} \left[ \theta + \frac{1}{6} \sin 6\theta \right]_{-\pi/6}^{\pi/6} \qquad \text{Integrate with respect to } \theta.$$

$$= \frac{3\pi}{4}.$$

So, the total area is $A = 9\pi/4$. ∎

As illustrated in Example 4, the area of a region in the plane can be represented by

$$A = \int_\alpha^\beta \int_{g_1(\theta)}^{g_2(\theta)} r \, dr \, d\theta.$$

For $g_1(\theta) = 0$, you obtain

$$A = \int_\alpha^\beta \int_0^{g_2(\theta)} r \, dr \, d\theta = \int_\alpha^\beta \frac{r^2}{2} \Big]_0^{g_2(\theta)} d\theta = \frac{1}{2} \int_\alpha^\beta [g_2(\theta)]^2 \, d\theta$$

which agrees with Theorem 10.13.

So far in this section, all of the examples of iterated integrals in polar form have been of the form

$$\int_\alpha^\beta \int_{g_1(\theta)}^{g_2(\theta)} f(r \cos \theta, r \sin \theta) r \, dr \, d\theta$$

in which the order of integration is with respect to $r$ first. Sometimes you can obtain a simpler integration problem by integrating with respect to $\theta$ first.

| | **EXAMPLE 5** | **Integrating with Respect to $\theta$ First** |

Find the area of the region bounded above by the spiral $r = \pi/(3\theta)$ and below by the polar axis, between $r = 1$ and $r = 2$.

**Solution** The region is shown in Figure 14.32. The polar boundaries for the region are

$$1 \le r \le 2 \quad \text{and} \quad 0 \le \theta \le \frac{\pi}{3r}.$$

So, the area of the region can be evaluated as follows.

$$A = \int_1^2 \int_0^{\pi/(3r)} r \, d\theta \, dr = \int_1^2 r\theta \Big]_0^{\pi/(3r)} dr = \int_1^2 \frac{\pi}{3} \, dr = \frac{\pi r}{3} \Big]_1^2 = \frac{\pi}{3} \quad ∎$$

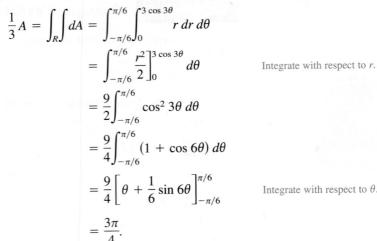

$\theta$-Simple region
**Figure 14.32**

## 14.3 Exercises

See CalcChat.com for tutorial help and worked-out solutions to odd-numbered exercises.

**CONCEPT CHECK**

**Choosing a Coordinate System** In Exercises 1 and 2, the region $R$ for the integral $\int_R \int f(x, y)\, dA$ is shown. State whether you would use rectangular or polar coordinates to evaluate the integral.

1.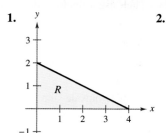

2.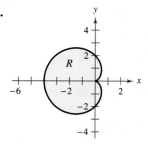

3. **Describing Regions** In your own words, describe $r$-simple regions and $\theta$-simple regions.

4. **Using Polar Coordinates** Sketch the region of integration represented by the double integral

$$\int_0^{2\pi} \int_3^6 f(r, \theta) r \, dr \, d\theta.$$

 **Describing a Region** In Exercises 5–8, use polar coordinates to describe the region shown.

5.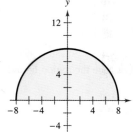

6.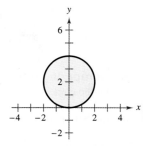

7.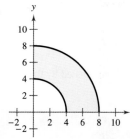

8.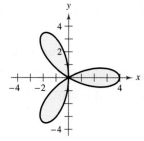

 **Evaluating a Double Integral** In Exercises 9–16, evaluate the double integral $\int_R \int f(r, \theta)\, dA$ and sketch the region $R$.

9. $\displaystyle\int_0^\pi \int_0^{2\cos\theta} r \, dr \, d\theta$

10. $\displaystyle\int_0^{\pi/2} \int_0^{\sin\theta} r^2 \, dr \, d\theta$

11. $\displaystyle\int_0^{2\pi} \int_0^1 6r^2 \sin\theta \, dr \, d\theta$

12. $\displaystyle\int_0^{\pi/4} \int_0^4 r^2 \sin\theta \cos\theta \, dr \, d\theta$

13. $\displaystyle\int_0^{\pi/2} \int_1^3 \sqrt{9 - r^2}\, r \, dr \, d\theta$

14. $\displaystyle\int_0^{\pi/2} \int_0^3 re^{-r^2} \, dr \, d\theta$

15. $\displaystyle\int_0^{\pi/2} \int_0^{1+\sin\theta} \theta r \, dr \, d\theta$

16. $\displaystyle\int_0^{\pi/2} \int_0^{1-\cos\theta} (\sin\theta) r \, dr \, d\theta$

 **Converting to Polar Coordinates** In Exercises 17–26, evaluate the iterated integral by converting to polar coordinates.

17. $\displaystyle\int_0^3 \int_0^{\sqrt{9-y^2}} y \, dx \, dy$

18. $\displaystyle\int_0^2 \int_0^{\sqrt{4-x^2}} x \, dy \, dx$

19. $\displaystyle\int_{-2}^2 \int_0^{\sqrt{4-x^2}} (x^2 + y^2) \, dy \, dx$

20. $\displaystyle\int_0^1 \int_{-\sqrt{x-x^2}}^{\sqrt{x-x^2}} (x^2 + y^2) \, dy \, dx$

21. $\displaystyle\int_0^1 \int_0^{\sqrt{1-x^2}} (x^2 + y^2)^{3/2} \, dy \, dx$

22. $\displaystyle\int_0^2 \int_y^{\sqrt{8-y^2}} \sqrt{x^2 + y^2} \, dx \, dy$

23. $\displaystyle\int_0^2 \int_0^{\sqrt{2x-x^2}} xy \, dy \, dx$

24. $\displaystyle\int_0^4 \int_0^{\sqrt{4y-y^2}} x^2 \, dx \, dy$

25. $\displaystyle\int_{-1}^1 \int_0^{\sqrt{1-x^2}} \cos(x^2 + y^2) \, dy \, dx$

26. $\displaystyle\int_0^{\sqrt{6}} \int_0^{\sqrt{6-x^2}} \sin\sqrt{x^2 + y^2} \, dy \, dx$

**Converting to Polar Coordinates** In Exercises 27 and 28, write the sum of the two iterated integrals as a single iterated integral by converting to polar coordinates. Evaluate the resulting iterated integral.

27. $\displaystyle\int_0^2 \int_0^x \sqrt{x^2 + y^2} \, dy \, dx + \int_2^{2\sqrt{2}} \int_0^{\sqrt{8-x^2}} \sqrt{x^2 + y^2} \, dy \, dx$

28. $\displaystyle\int_0^{(5\sqrt{2})/2} \int_0^x xy \, dy \, dx + \int_{(5\sqrt{2})/2}^5 \int_0^{\sqrt{25-x^2}} xy \, dy \, dx$

**Converting to Polar Coordinates** In Exercises 29–32, use polar coordinates to set up and evaluate the double integral $\int_R \int f(x, y)\, dA$.

**29.** $f(x, y) = x + y$

R: $x^2 + y^2 \le 36$, $x \ge 0$, $y \ge 0$

**30.** $f(x, y) = e^{-(x^2 + y^2)/2}$

R: $x^2 + y^2 \le 25$, $x \ge 0$

**31.** $f(x, y) = \arctan \dfrac{y}{x}$

R: $x^2 + y^2 \ge 1$, $x^2 + y^2 \le 4$, $0 \le y \le x$

**32.** $f(x, y) = 9 - x^2 - y^2$

R: $x^2 + y^2 \le 9$, $x \ge 0$, $y \ge 0$

 **Volume** In Exercises 33–38, use a double integral in polar coordinates to find the volume of the solid bounded by the graphs of the equations.

**33.** $z = xy$, $x^2 + y^2 = 1$, first octant

**34.** $z = x^2 + y^2 + 3$, $z = 0$, $x^2 + y^2 = 1$

**35.** $z = \sqrt{x^2 + y^2}$, $z = 0$, $x^2 + y^2 = 25$

**36.** $z = \ln(x^2 + y^2)$, $z = 0$, $x^2 + y^2 \ge 1$, $x^2 + y^2 \le 4$

**37.** Inside the hemisphere $z = \sqrt{16 - x^2 - y^2}$ and inside the cylinder $x^2 + y^2 - 4x = 0$

**38.** Inside the hemisphere $z = \sqrt{16 - x^2 - y^2}$ and outside the cylinder $x^2 + y^2 = 1$

**39. Volume** Use a double integral in polar coordinates to find $a$ such that the volume inside the hemisphere $z = \sqrt{16 - x^2 - y^2}$ and outside the cylinder $x^2 + y^2 = a^2$ is one-half the volume of the hemisphere.

**40. Volume** Use a double integral in polar coordinates to find the volume of a sphere of radius $a$.

 **Area** In Exercises 41–46, use a double integral to find the area of the shaded region.

**41.**

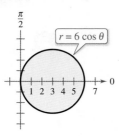

$r = 6 \cos \theta$

**42.**

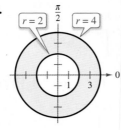

$r = 2$, $r = 4$

**43.**

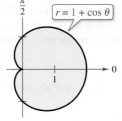

$r = 1 + \cos \theta$

**44.**

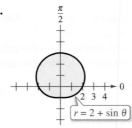

$r = 2 + \sin \theta$

**45.**

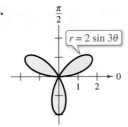

$r = 2 \sin 3\theta$

**46.**

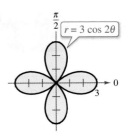

$r = 3 \cos 2\theta$

**Area** In Exercises 47–52, sketch a graph of the region bounded by the graphs of the equations. Then use a double integral to find the area of the region.

**47.** Inside the circle $r = 2 \cos \theta$ and outside the circle $r = 1$

**48.** Inside the cardioid $r = 2 + 2 \cos \theta$ and outside the circle $r = 1$

**49.** Inside the circle $r = 3 \cos \theta$ and outside the cardioid $r = 1 + \cos \theta$

**50.** Inside the cardioid $r = 1 + \cos \theta$ and outside the circle $r = 3 \cos \theta$

**51.** Inside the rose curve $r = 4 \sin 3\theta$ and outside the circle $r = 2$

**52.** Inside the circle $r = 2$ and outside the cardioid $r = 2 - 2 \cos \theta$

**EXPLORING CONCEPTS**

**53. Area** Express the area of the region in the figure using the sum of two double polar integrals. Then find the area of the region without using integrals.

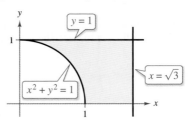

**54. Comparing Integrals** Let $R$ be the region bounded by the circle $x^2 + y^2 = 9$.

(a) Set up the integral $\displaystyle\int_R \int f(x, y)\, dA$.

(b) Convert the integral in part (a) to polar coordinates.

(c) Which integral would you choose to evaluate? Why?

**55. Population**

The population density of a city is approximated by the model

$f(x, y) = 4000e^{-0.01(x^2 + y^2)}$

for the region $x^2 + y^2 \le 49$, where $x$ and $y$ are measured in miles. Integrate the density function over the indicated circular region to approximate the population of the city.

**56. HOW DO YOU SEE IT?**  Each figure shows a region of integration for the double integral $\int_R \int f(x, y)\, dA$. For each region, state whether horizontal representative elements, vertical representative elements, or polar sectors would yield the easiest method for obtaining the limits of integration. Explain your reasoning.

(a)

(b)

(c)

**57. Volume**  Determine the diameter of a hole that is drilled vertically through the center of the solid bounded by the graphs of the equations $z = 25e^{-(x^2 + y^2)/4}$, $z = 0$, and $x^2 + y^2 = 16$ when one-tenth of the volume of the solid is removed.

**58. Glacier**

Horizontal cross sections of a piece of ice that broke from a glacier are in the shape of a quarter of a circle with a radius of approximately 50 feet. The base is divided into 20 subregions, as shown in the figure. At the center of each subregion, the height of the ice is measured, yielding the following points in cylindrical coordinates.

$(5, \frac{\pi}{16}, 7)$, $(15, \frac{\pi}{16}, 8)$, $(25, \frac{\pi}{16}, 10)$, $(35, \frac{\pi}{16}, 12)$, $(45, \frac{\pi}{16}, 9)$,

$(5, \frac{3\pi}{16}, 9)$, $(15, \frac{3\pi}{16}, 10)$, $(25, \frac{3\pi}{16}, 14)$, $(35, \frac{3\pi}{16}, 15)$, $(45, \frac{3\pi}{16}, 10)$,

$(5, \frac{5\pi}{16}, 9)$, $(15, \frac{5\pi}{16}, 11)$, $(25, \frac{5\pi}{16}, 15)$, $(35, \frac{5\pi}{16}, 18)$, $(45, \frac{5\pi}{16}, 14)$,

$(5, \frac{7\pi}{16}, 5)$, $(15, \frac{7\pi}{16}, 8)$, $(25, \frac{7\pi}{16}, 11)$, $(35, \frac{7\pi}{16}, 16)$, $(45, \frac{7\pi}{16}, 12)$

(a) Approximate the volume of the piece of ice.

(b) Ice weighs approximately 57 pounds per cubic foot. Approximate the weight of the piece of ice.

(c) There are 7.48 gallons of water per cubic foot. Approximate the number of gallons of water in the piece of ice.

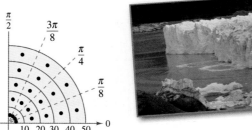

**Approximation**  In Exercises 59 and 60, use a computer algebra system to approximate the iterated integral.

**59.** $\displaystyle\int_{\pi/4}^{\pi/2} \int_0^5 r\sqrt{1 + r^3} \sin\sqrt{\theta}\, dr\, d\theta$

**60.** $\displaystyle\int_0^{\pi/4} \int_0^4 5re^{\sqrt{r\theta}}\, dr\, d\theta$

**True or False?**  In Exercises 61 and 62, determine whether the statement is true or false. If it is false, explain why or give an example that shows it is false.

**61.** If $\int_R \int f(r, \theta)\, dA > 0$, then $f(r, \theta) > 0$ for all $(r, \theta)$ in $R$.

**62.** If $f(r, \theta)$ is a constant function and the area of the region $S$ is twice that of the region $R$, then

$$2\int_R \int f(r, \theta)\, dA = \int_S \int f(r, \theta)\, dA.$$

**63. Probability**  The value of the integral

$$I = \int_{-\infty}^{\infty} e^{-x^2/2}\, dx$$

is required in the development of the normal probability density function.

(a) Use polar coordinates to evaluate the improper integral.

$$I^2 = \left(\int_{-\infty}^{\infty} e^{-x^2/2}\, dx\right)\left(\int_{-\infty}^{\infty} e^{-y^2/2}\, dy\right)$$

$$= \int_{-\infty}^{\infty} \int_{-\infty}^{\infty} e^{-(x^2 + y^2)/2}\, dA$$

(b) Use the result of part (a) to determine $I$.

**FOR FURTHER INFORMATION**  For more information on this problem, see the article "Integrating $e^{-x^2}$ Without Polar Coordinates" by William Dunham in *Mathematics Teacher*. To view this article, go to *MathArticles.com*.

**64. Evaluating Integrals**  Use the result of Exercise 63 and a change of variables to evaluate each integral. No integration is required.

(a) $\displaystyle\int_{-\infty}^{\infty} e^{-x^2}\, dx$   (b) $\displaystyle\int_{-\infty}^{\infty} e^{-4x^2}\, dx$

**65. Think About It**  Consider the region $R$ bounded by the graphs of $y = 2$, $y = 4$, $y = x$, and $y = \sqrt{3}x$ and the double integral $\int_R \int f(x, y)\, dA$. Determine the limits of integration when the region $R$ is divided into (a) horizontal representative elements, (b) vertical representative elements, and (c) polar sectors.

**66. Think About It**  Repeat Exercise 65 for a region $R$ bounded by the graph of the equation $(x - 2)^2 + y^2 = 4$.

**67. Probability**  Find $k$ such that the function

$$f(x, y) = \begin{cases} ke^{-(x^2 + y^2)}, & x \geq 0, y \geq 0 \\ 0, & \text{elsewhere} \end{cases}$$

is a probability density function. (*Hint:* Show that $\int_R \int f(x, y)\, dA = 1$.)

**68. Area**  Show that the area $A$ of the polar sector $R$ (see figure) is $A = r\Delta r\Delta\theta$, where $r = (r_1 + r_2)/2$ is the average radius of $R$.

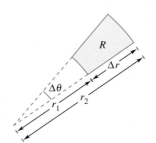

# 14.4 Center of Mass and Moments of Inertia

■ Find the mass of a planar lamina using a double integral.
■ Find the center of mass of a planar lamina using double integrals.
■ Find moments of inertia using double integrals.

## Mass

Section 7.6 discussed several applications of integration involving a lamina of *constant* density $\rho$. For example, if the lamina corresponding to the region $R$, as shown in Figure 14.33, has a constant density $\rho$, then the mass of the lamina is given by

$$\text{Mass} = \rho A = \rho \int_R\!\!\int dA = \int_R\!\!\int \rho \, dA. \qquad \text{Constant density}$$

If not otherwise stated, a lamina is assumed to have a constant density. In this section, however, you will extend the definition of the term *lamina* to include thin plates of *variable* density. Double integrals can be used to find the mass of a lamina of variable density, where the density at $(x, y)$ is given by the **density function** $\rho$.

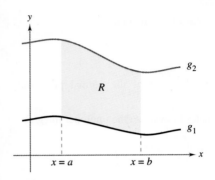

Lamina of constant density $\rho$
**Figure 14.33**

---

**Definition of Mass of a Planar Lamina of Variable Density**

If $\rho$ is a continuous density function on the lamina corresponding to a plane region $R$, then the mass $m$ of the lamina is given by

$$m = \int_R\!\!\int \rho(x, y) \, dA. \qquad \text{Variable density}$$

---

Density is normally expressed as mass per unit volume. For a planar lamina, however, density is mass per unit surface area.

### EXAMPLE 1    Finding the Mass of a Planar Lamina

Find the mass of the triangular lamina with vertices $(0, 0)$, $(0, 3)$, and $(2, 3)$, given that the density at $(x, y)$ is $\rho(x, y) = 2x + y$.

**Solution**    As shown in Figure 14.34, region $R$ has the boundaries $x = 0$, $y = 3$, and $y = 3x/2$ (or $x = 2y/3$). Therefore, the mass of the lamina is

$$
\begin{aligned}
m &= \int_R\!\!\int (2x + y) \, dA \\
&= \int_0^3 \int_0^{2y/3} (2x + y) \, dx \, dy \\
&= \int_0^3 \left[ x^2 + xy \right]_0^{2y/3} dy \qquad \text{Integrate with respect to } x. \\
&= \frac{10}{9} \int_0^3 y^2 \, dy \\
&= \frac{10}{9} \left[ \frac{y^3}{3} \right]_0^3 \qquad \text{Integrate with respect to } y. \\
&= 10.
\end{aligned}
$$

In Figure 14.34, note that the planar lamina is shaded so that the darkest shading corresponds to the densest part.

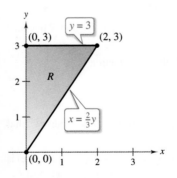

Lamina of variable density
$\rho(x, y) = 2x + y$
**Figure 14.34**

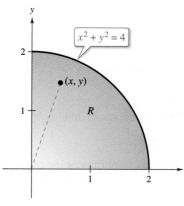

Density at $(x, y)$: $\rho(x, y) = k\sqrt{x^2 + y^2}$
**Figure 14.35**

| EXAMPLE 2 | **Finding Mass by Polar Coordinates** |

Find the mass of the lamina corresponding to the first-quadrant portion of the circle

$$x^2 + y^2 = 4$$

where the density at the point $(x, y)$ is proportional to the distance between the point and the origin, as shown in Figure 14.35.

**Solution**    At any point $(x, y)$, the density of the lamina is

$$\rho(x, y) = k\sqrt{(x - 0)^2 + (y - 0)^2}$$
$$= k\sqrt{x^2 + y^2}$$

where $k$ is the constant of proportionality. Because $0 \le x \le 2$ and $0 \le y \le \sqrt{4 - x^2}$, the mass is given by

$$m = \int_R \int k\sqrt{x^2 + y^2}\, dA$$
$$= \int_0^2 \int_0^{\sqrt{4 - x^2}} k\sqrt{x^2 + y^2}\, dy\, dx.$$

To simplify the integration, you can convert to polar coordinates, using the bounds

$$0 \le \theta \le \pi/2 \quad \text{and} \quad 0 \le r \le 2.$$

So, the mass is

$$m = \int_R \int k\sqrt{x^2 + y^2}\, dA$$

$$= \int_0^{\pi/2} \int_0^2 k\sqrt{r^2}\, r\, dr\, d\theta \qquad \text{Polar coordinates}$$

$$= \int_0^{\pi/2} \int_0^2 kr^2\, dr\, d\theta \qquad \text{Simplify integrand.}$$

$$= \int_0^{\pi/2} \frac{kr^3}{3}\Big]_0^2 d\theta \qquad \text{Integrate with respect to } r.$$

$$= \frac{8k}{3} \int_0^{\pi/2} d\theta$$

$$= \frac{8k}{3}\Big[\theta\Big]_0^{\pi/2} \qquad \text{Integrate with respect } \theta.$$

$$= \frac{4\pi k}{3}.$$

▷ **TECHNOLOGY**   On many occasions, this text has mentioned the benefits of computer programs that perform symbolic integration. Even if you use such a program regularly, you should remember that its greatest benefit comes only in the hands of a knowledgeable user. For instance, notice how much simpler the integral in Example 2 becomes when it is converted to polar form.

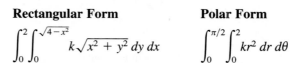

    **Rectangular Form**          **Polar Form**

$$\int_0^2 \int_0^{\sqrt{4 - x^2}} k\sqrt{x^2 + y^2}\, dy\, dx \qquad \int_0^{\pi/2} \int_0^2 kr^2\, dr\, d\theta$$

If you have access to software that performs symbolic integration, use it to evaluate both integrals. Some software programs cannot handle the first integral, but any program that can handle double integrals can evaluate the second integral.

## Moments and Center of Mass

For a lamina of variable density, moments of mass are defined in a manner similar to that used for the uniform density case. For a partition $\Delta$ of a lamina corresponding to a plane region $R$, consider the $i$th rectangle $R_i$ of one area $\Delta A_i$, as shown in Figure 14.36. Assume that the mass of $R_i$ is concentrated at one of its interior points $(x_i, y_i)$. The moment of mass of $R_i$ with respect to the $x$-axis can be approximated by

$$(\text{Mass})(y_i) \approx [\rho(x_i, y_i)\, \Delta A_i](y_i).$$

Similarly, the moment of mass with respect to the $y$-axis can be approximated by

$$(\text{Mass})(x_i) \approx [\rho(x_i, y_i)\Delta A_i](x_i).$$

By forming the Riemann sum of all such products and taking the limits as the norm of $\Delta$ approaches 0, you obtain the following definitions of moments of mass with respect to the $x$- and $y$-axes.

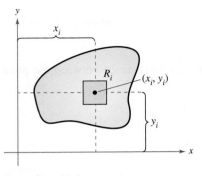

$M_x = (\text{mass})(y_i)$
$M_y = (\text{mass})(x_i)$
**Figure 14.36**

---

### Moments and Center of Mass of a Variable Density Planar Lamina

Let $\rho$ be a continuous density function on the planar lamina $R$. The **moments of mass** with respect to the $x$- and $y$-axes are

$$M_x = \int_R\!\!\int (y)\rho(x, y)\, dA$$

and

$$M_y = \int_R\!\!\int (x)\rho(x, y)\, dA.$$

If $m$ is the mass of the lamina, then the **center of mass** is

$$(\bar{x}, \bar{y}) = \left( \frac{M_y}{m}, \frac{M_x}{m} \right).$$

If $R$ represents a simple plane region rather than a lamina, then the point $(\bar{x}, \bar{y})$ is called the **centroid** of the region.

---

For some planar laminas with a constant density $\rho$, you can determine the center of mass (or one of its coordinates) using symmetry rather than using integration. For instance, consider the laminas of constant density shown in Figure 14.37. Using symmetry, you can see that $\bar{y} = 0$ for the first lamina and $\bar{x} = 0$ for the second lamina.

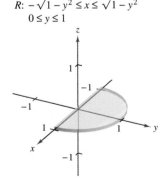

$R: 0 \leq x \leq 1$
$-\sqrt{1 - x^2} \leq y \leq \sqrt{1 - x^2}$

$R: -\sqrt{1 - y^2} \leq x \leq \sqrt{1 - y^2}$
$0 \leq y \leq 1$

Lamina of constant density that is symmetric with respect to the $x$-axis

Lamina of constant density that is symmetric with respect to the $y$-axis

**Figure 14.37**

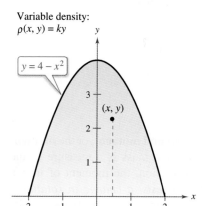

Variable density:
$\rho(x, y) = ky$

$y = 4 - x^2$

$(x, y)$

Parabolic region of variable density
**Figure 14.38**

**EXAMPLE 3**   **Finding the Center of Mass**

⋮ ⬝ ⬝ ⬝ ▷  *See LarsonCalculus.com for an interactive version of this type of example.*

Find the center of mass of the lamina corresponding to the parabolic region

$$0 \le y \le 4 - x^2 \qquad \text{Parabolic region}$$

where the density at the point $(x, y)$ is proportional to the distance between $(x, y)$ and the $x$-axis, as shown in Figure 14.38.

**Solution**   The lamina is symmetric with respect to the $y$-axis and $\rho(x, y) = ky$, where $k$ is the constant of proportionality. So, the center of mass lies on the $y$-axis and $\bar{x} = 0$. To find $\bar{y}$, first find the mass of the lamina.

$$m = \int_{-2}^{2} \int_{0}^{4-x^2} ky \, dy \, dx$$

$$= \frac{k}{2} \int_{-2}^{2} y^2 \Big]_{0}^{4-x^2} dx \qquad \text{Integrate with respect to } y.$$

$$= \frac{k}{2} \int_{-2}^{2} (16 - 8x^2 + x^4) \, dx$$

$$= \frac{k}{2} \left[ 16x - \frac{8x^3}{3} + \frac{x^5}{5} \right]_{-2}^{2} \qquad \text{Integrate with respect to } x.$$

$$= k\left( 32 - \frac{64}{3} + \frac{32}{5} \right)$$

$$= \frac{256k}{15} \qquad \text{Mass of the limina}$$

Next, find the moment of mass about the $x$-axis.

$$M_x = \int_{-2}^{2} \int_{0}^{4-x^2} (y)(ky) \, dy \, dx$$

$$= \frac{k}{3} \int_{-2}^{2} y^3 \Big]_{0}^{4-x^2} dx \qquad \text{Integrate with respect to } y.$$

$$= \frac{k}{3} \int_{-2}^{2} (64 - 48x^2 + 12x^4 - x^6) \, dx$$

$$= \frac{k}{3} \left[ 64x - 16x^3 + \frac{12x^5}{5} - \frac{x^7}{7} \right]_{-2}^{2} \qquad \text{Integrate with respect to } x.$$

$$= \frac{4096k}{105} \qquad \text{Moment of mass about } x\text{-axis}$$

So,

$$\bar{y} = \frac{M_x}{m} = \frac{4096k/105}{256k/15} = \frac{16}{7}$$

and the center of mass is $\left(0, \frac{16}{7}\right)$.   ∎

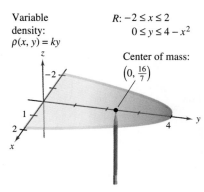

Variable
density:
$\rho(x, y) = ky$

$R: -2 \le x \le 2$
$0 \le y \le 4 - x^2$

Center of mass:
$\left(0, \frac{16}{7}\right)$

**Figure 14.39**

Although you can think of the moments $M_x$ and $M_y$ as measuring the tendency to rotate about the $x$- or $y$-axis, the calculation of moments is usually an intermediate step toward a more tangible goal. The use of the moments $M_x$ and $M_y$ is typical—to find the center of mass. Determination of the center of mass is useful in a variety of applications that allow you to treat a lamina as if its mass were concentrated at just one point. Intuitively, you can think of the center of mass as the balancing point of the lamina. For instance, the lamina in Example 3 should balance on the point of a pencil placed at $\left(0, \frac{16}{7}\right)$, as shown in Figure 14.39.

## Moments of Inertia

The moments of $M_x$ and $M_y$ used in determining the center of mass of a lamina are sometimes called the **first moments** about the $x$- and $y$-axes. In each case, the moment is the product of a mass times a distance.

$$M_x = \int_R\!\!\int (y)\rho(x, y)\, dA \qquad\qquad M_y = \int_R\!\!\int (x)\rho(x, y)\, dA$$

Distance   Mass        Distance   Mass
to $x$-axis            to $y$-axis

You will now look at another type of moment—the **second moment**, or the **moment of inertia** of a lamina about a line. In the same way that mass is a measure of the tendency of matter to resist a change in straight-line motion, the moment of inertia about a line is a *measure of the tendency of matter to resist a change in rotational motion*. For example, when a particle of mass $m$ is a distance $d$ from a fixed line, its moment of inertia about the line is defined as

$$I = md^2 = (\text{mass})(\text{distance})^2.$$

As with moments of mass, you can generalize this concept to obtain the moments of inertia about the $x$- and $y$-axes of a lamina of variable density. These second moments are denoted by $I_x$ and $I_y$, and in each case the moment is the product of a mass times the square of a distance.

$$I_x = \int_R\!\!\int (y^2)\rho(x, y)\, dA \qquad\qquad I_y = \int_R\!\!\int (x^2)\rho(x, y)\, dA$$

Square of distance   Mass        Square of distance   Mass
to $x$-axis                      to $y$-axis

The sum of the moments $I_x$ and $I_y$ is called the **polar moment of inertia** and is denoted by $I_0$. For a lamina in the $xy$-plane, $I_0$ represents the moment of inertia of the lamina about the $z$-axis. The term "polar moment of inertia" stems from the fact that the square of the polar distance $r$ is used in the calculation.

$$I_0 = \int_R\!\!\int (x^2 + y^2)\rho(x, y)\, dA = \int_R\!\!\int (r^2)\rho(x, y)\, dA$$

### EXAMPLE 4  Finding the Moment of Inertia

Find the moment of inertia about the $x$-axis of the lamina in Example 3.

**Solution**   From the definition of moment of inertia, you have

$$
\begin{aligned}
I_x &= \int_{-2}^{2}\int_{0}^{4-x^2} (y^2)(ky)\, dy\, dx \\[2mm]
&= \frac{k}{4}\int_{-2}^{2} y^4 \Big]_{0}^{4-x^2} dx && \text{Integrate with respect to } y. \\[2mm]
&= \frac{k}{4}\int_{-2}^{2} (256 - 256x^2 + 96x^4 - 16x^6 + x^8)\, dx \\[2mm]
&= \frac{k}{4}\left[ 256x - \frac{256x^3}{3} + \frac{96x^5}{5} - \frac{16x^7}{7} + \frac{x^9}{9} \right]_{-2}^{2} && \text{Integrate with respect to } x. \\[2mm]
&= \frac{32{,}768k}{315}. && \text{Moment of inertia about } x\text{-axis}
\end{aligned}
$$

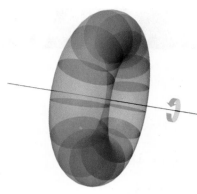

Planar lamina revolving at $\omega$ radians per second
**Figure 14.40**

The moment of inertia $I$ of a revolving lamina can be used to measure its kinetic energy. For example, suppose a planar lamina is revolving about a line with an **angular speed** of $\omega$ radians per second, as shown in Figure 14.40. The kinetic energy $E$ of the revolving lamina is

$$E = \frac{1}{2}I\omega^2. \qquad \text{Kinetic energy for rotational motion}$$

On the other hand, the kinetic energy $E$ of a mass $m$ moving in a straight line at a velocity $v$ is

$$E = \frac{1}{2}mv^2. \qquad \text{Kinetic energy for linear motion}$$

So, the kinetic energy of a mass moving in a straight line is proportional to its mass, but the kinetic energy of a mass revolving about an axis is proportional to its moment of inertia.

The **radius of gyration** $\bar{\bar{r}}$ of a revolving mass $m$ with moment of inertia $I$ is defined as

$$\bar{\bar{r}} = \sqrt{\frac{I}{m}}. \qquad \text{Radius of gyration}$$

If the entire mass were located at a distance $\bar{\bar{r}}$ from its axis of revolution, it would have the same moment of inertia and, consequently, the same kinetic energy. For instance, the radius of gyration of the lamina in Example 4 about the $x$-axis is

$$\bar{\bar{y}} = \sqrt{\frac{I_x}{m}} = \sqrt{\frac{32{,}768k/315}{256k/15}} = \sqrt{\frac{128}{21}} \approx 2.469.$$

### EXAMPLE 5  Finding the Radius of Gyration

Find the radius of gyration about the $y$-axis for the lamina corresponding to the region $R: 0 \le y \le \sin x, 0 \le x \le \pi$, where the density at $(x, y)$ is given by $\rho(x, y) = x$.

**Solution**  The region $R$ is shown in Figure 14.41. By integrating $\rho(x, y) = x$ over the region $R$, you can determine that the mass of the region is $\pi$. The moment of inertia about the $y$-axis is

$$\begin{aligned}
I_y &= \int_0^\pi \int_0^{\sin x} x^3 \, dy \, dx \\
&= \int_0^\pi x^3 y \Big]_0^{\sin x} dx && \text{Integrate with respect to } y. \\
&= \int_0^\pi x^3 \sin x \, dx \\
&= \left[ (3x^2 - 6)(\sin x) - (x^3 - 6x)(\cos x) \right]_0^\pi && \text{Integrate with respect to } x. \\
&= \pi^3 - 6\pi. && \text{Moment of inertia about } y\text{-axis}
\end{aligned}$$

So, the radius of gyration about the $y$-axis is

$$\begin{aligned}
\bar{\bar{x}} &= \sqrt{\frac{I_y}{m}} \\
&= \sqrt{\frac{\pi^3 - 6\pi}{\pi}} \\
&= \sqrt{\pi^2 - 6} \\
&\approx 1.967. && \text{Radius of gyration about } y\text{-axis}
\end{aligned}$$

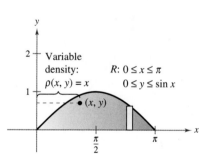

**Figure 14.41**

# 14.4 Exercises

See CalcChat.com for tutorial help and worked-out solutions to odd-numbered exercises.

**CONCEPT CHECK**

1. **Mass of a Planar Lamina** Explain when you should use a double integral to find the mass of a planar lamina.

2. **Moment of Inertia** Describe what the moment of inertia measures.

**Finding the Mass of a Lamina** In Exercises 3–6, find the mass of the lamina described by the inequalities, given that its density is $\rho(x, y) = xy$.

3. $0 \le x \le 2,\ 0 \le y \le 2$

4. $0 \le x \le 2,\ 0 \le y \le 4 - x^2$

5. $0 \le x \le 1,\ 0 \le y \le \sqrt{1 - x^2}$

6. $x \ge 0,\ 3 \le y \le 3 + \sqrt{9 - x^2}$

**Finding the Center of Mass** In Exercises 7–10, find the mass and center of mass of the lamina corresponding to the region $R$ for each density.

7. $R$: square with vertices $(0, 0),\ (a, 0),\ (0, a),\ (a, a)$

   (a) $\rho = k$  (b) $\rho = ky$  (c) $\rho = kx$

8. $R$: rectangle with vertices $(0, 0),\ (a, 0),\ (0, b),\ (a, b)$

   (a) $\rho = kxy$  (b) $\rho = k(x^2 + y^2)$

9. $R$: triangle with vertices $(0, 0),\ (0, a),\ (a, a)$

   (a) $\rho = k$  (b) $\rho = ky$  (c) $\rho = kx$

10. $R$: triangle with vertices $(0, 0),\ (a/2, a),\ (a, 0)$

   (a) $\rho = k$  (b) $\rho = kxy$

11. **Translations in the Plane** Translate the lamina in Exercise 7 to the right five units and determine the resulting center of mass.

12. **Conjecture** Use the result of Exercise 11 to make a conjecture about the change in the center of mass when a lamina of constant density is translated $c$ units horizontally or $d$ units vertically. Is the conjecture true when the density is not constant? Explain.

**Finding the Center of Mass** In Exercises 13–24, find the mass and center of mass of the lamina bounded by the graphs of the equations for the given density.

13. $y = \sqrt{x},\ y = 0,\ x = 1,\ \rho = ky$

14. $y = x^2,\ y = 0,\ x = 2,\ \rho = kxy$

15. $y = 4/x,\ y = 0,\ x = 1,\ x = 4,\ \rho = kx^2$

16. $y = \dfrac{1}{1 + x^2},\ y = 0,\ x = -1,\ x = 1,\ \rho = k$

17. $y = e^x,\ y = 0,\ x = 0,\ x = 1,\ \rho = k$

18. $y = e^{-x},\ y = 0,\ x = 0,\ x = 1,\ \rho = ky^2$

19. $y = 4 - x^2,\ y = 0,\ \rho = ky$

20. $x = 9 - y^2,\ x = 0,\ \rho = kx$

21. $y = \sin\dfrac{\pi x}{3},\ y = 0,\ x = 0,\ x = 3,\ \rho = k$

22. $y = \cos\dfrac{\pi x}{8},\ y = 0,\ x = 0,\ x = 4,\ \rho = ky$

23. $y = \sqrt{36 - x^2},\ 0 \le y \le x,\ \rho = k$

24. $x^2 + y^2 = 16,\ x \ge 0,\ y \ge 0,\ \rho = k(x^2 + y^2)$

**Finding the Center of Mass Using Technology** In Exercises 25–28, use a computer algebra system to find the mass and center of mass of the lamina bounded by the graphs of the equations for the given density.

25. $y = e^{-x},\ y = 0,\ x = 0,\ x = 2,\ \rho = kxy$

26. $y = \ln x,\ y = 0,\ x = 1,\ x = e,\ \rho = k/x$

27. $r = 2\cos 3\theta,\ -\dfrac{\pi}{6} \le \theta \le \dfrac{\pi}{6},\ \rho = k$

28. $r = 1 + \cos\theta,\ \rho = k$

**Finding the Radius of Gyration About Each Axis** In Exercises 29–34, verify the given moment(s) of inertia and find $\bar{x}$ and $\bar{y}$. Assume that each lamina has a density of $\rho = 1$ gram per square centimeter. (These regions are common shapes used in engineering.)

29. Rectangle

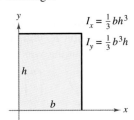

$I_x = \frac{1}{3}bh^3$

$I_y = \frac{1}{3}b^3h$

30. Right triangle

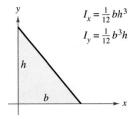

$I_x = \frac{1}{12}bh^3$

$I_y = \frac{1}{12}b^3h$

31. Circle

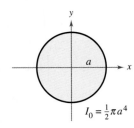

$I_0 = \frac{1}{2}\pi a^4$

32. Semicircle

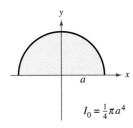

$I_0 = \frac{1}{4}\pi a^4$

33. Quarter circle

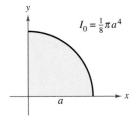

$I_0 = \frac{1}{8}\pi a^4$

34. Ellipse

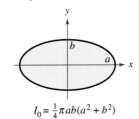

$I_0 = \frac{1}{4}\pi ab(a^2 + b^2)$

**Finding Moments of Inertia and Radii of Gyration** In Exercises 35–38, find $I_x$, $I_y$, $I_0$, $\bar{\bar{x}}$, and $\bar{\bar{y}}$ for the lamina bounded by the graphs of the equations.

**35.** $y = 4 - x^2$, $y = 0$, $x > 0$, $\rho = kx$

**36.** $y = x$, $y = x^2$, $\rho = kxy$

**37.** $y = \sqrt{x}$, $y = 0$, $x = 4$, $\rho = kxy$

**38.** $y = x^2$, $y^2 = x$, $\rho = kx$

**Finding a Moment of Inertia Using Technology** In Exercises 39–42, set up the double integral required to find the moment of inertia about the given line of the lamina bounded by the graphs of the equations for the given density. Use a computer algebra system to evaluate the double integral.

**39.** $x^2 + y^2 = b^2$, $\rho = k$, line: $x = a$ $(a > b)$

**40.** $y = \sqrt{x}$, $y = 0$, $x = 4$, $\rho = kx$, line: $x = 6$

**41.** $y = \sqrt{a^2 - x^2}$, $y = 0$, $\rho = ky$, line: $y = a$

**42.** $y = 4 - x^2$, $y = 0$, $\rho = k$, line: $y = 2$

**Hydraulics** In Exercises 43–46, determine the location of the horizontal axis $y_a$ at which a vertical gate in a dam is to be hinged so that there is no moment causing rotation under the indicated loading (see figure). The model for $y_a$ is

$$y_a = \bar{y} - \frac{I_{\bar{y}}}{hA}$$

where $\bar{y}$ is the $y$-coordinate of the centroid of the gate, $I_{\bar{y}}$ is the moment of inertia of the gate about the line $y = \bar{y}$, $h$ is the depth of the centroid below the surface, and $A$ is the area of the gate.

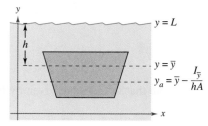

**43.**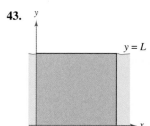

**44.**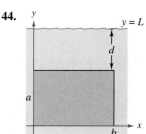

**45.**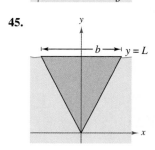

**46.**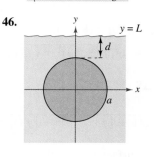

## EXPLORING CONCEPTS

**47. Polar Moment of Inertia** What does it mean for an object to have a greater polar moment of inertia than another object?

**48. HOW DO YOU SEE IT?** The center of mass of the lamina of constant density shown in the figure is $\left(2, \frac{8}{5}\right)$. Make a conjecture about how the center of mass $(\bar{x}, \bar{y})$ changes for each given nonconstant density $\rho(x, y)$. Explain. (Make your conjecture *without* performing any calculations.)

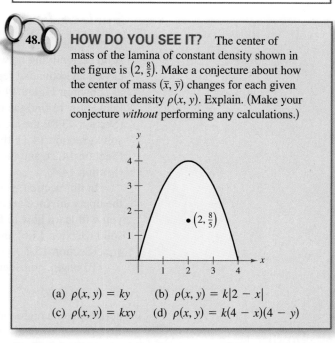

(a) $\rho(x, y) = ky$     (b) $\rho(x, y) = k|2 - x|$

(c) $\rho(x, y) = kxy$    (d) $\rho(x, y) = k(4 - x)(4 - y)$

**49. Proof** Prove the following Theorem of Pappus: Let $R$ be a region in a plane and let $L$ be a line in the same plane such that $L$ does not intersect the interior of $R$. If $r$ is the distance between the centroid of $R$ and the line, then the volume $V$ of the solid of revolution formed by revolving $R$ about the line is $V = 2\pi rA$, where $A$ is the area of $R$.

## SECTION PROJECT

## Center of Pressure on a Sail

The center of pressure on a sail is the point $(x_p, y_p)$ at which the total aerodynamic force may be assumed to act. If the sail is represented by a plane region $R$, then the center of pressure is

$$x_p = \frac{\int_R\int xy\, dA}{\int_R\int y\, dA} \quad \text{and} \quad y_p = \frac{\int_R\int y^2\, dA}{\int_R\int y\, dA}.$$

Consider a triangular sail with vertices at $(0, 0)$, $(2, 1)$, and $(0, 5)$. Verify the value of each integral.

(a) $\displaystyle\int_R\int y\, dA = 10$

(b) $\displaystyle\int_R\int xy\, dA = \frac{35}{6}$

(c) $\displaystyle\int_R\int y^2\, dA = \frac{155}{6}$

Calculate the coordinates $(x_p, y_p)$ of the center of pressure. Sketch a graph of the sail and indicate the location of the center of pressure.

# 14.5 Surface Area

■ Use a double integral to find the area of a surface.

## Surface Area

At this point, you know a great deal about the solid region lying between a surface and a closed and bounded region $R$ in the $xy$-plane, as shown in Figure 14.42. For example, you know how to find the extrema of $f$ on $R$ (Section 13.8), the area of the base $R$ of the solid (Section 14.1), the volume of the solid (Section 14.2), and the centroid of the base $R$ (Section 14.4).

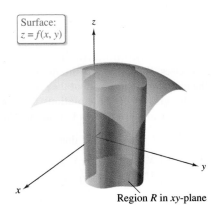

Surface:
$z = f(x, y)$

Region $R$ in $xy$-plane

**Figure 14.42**

In this section, you will learn how to find the upper **surface area** of the solid. Later, you will learn how to find the centroid of the solid (Section 14.6) and the lateral surface area (Section 15.2).

To begin, consider a surface $S$ given by

$$z = f(x, y) \qquad \text{Surface defined over a region } R$$

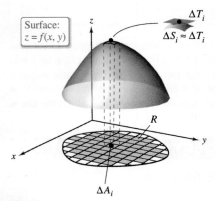

Surface:
$z = f(x, y)$

$\Delta T_i$

$\Delta S_i \approx \Delta T_i$

$R$

$\Delta A_i$

**Figure 14.43**

defined over a region $R$. Assume that $R$ is closed and bounded and that $f$ has continuous first partial derivatives. To find the surface area, construct an inner partition of $R$ consisting of $n$ rectangles, where the area of the $i$th rectangle $R_i$ is $\Delta A_i = \Delta x_i \Delta y_i$, as shown in Figure 14.43. In each $R_i$, let $(x_i, y_i)$ be the point that is closest to the origin. At the point $(x_i, y_i, z_i) = (x_i, y_i, f(x_i, y_i))$ on the surface $S$, construct a tangent plane $T_i$. The area of the portion of the tangent plane that lies directly above $R_i$ is approximately equal to the area of the surface lying directly above $R_i$. That is, $\Delta T_i \approx \Delta S_i$. So, the surface area of $S$ is approximated by

$$\sum_{i=1}^{n} \Delta S_i \approx \sum_{i=1}^{n} \Delta T_i.$$

To find the area of the parallelogram $\Delta T_i$, note that its sides are given by the vectors

$$\mathbf{u} = \Delta x_i \mathbf{i} + f_x(x_i, y_i)\,\Delta x_i \mathbf{k}$$

and

$$\mathbf{v} = \Delta y_i \mathbf{j} + f_y(x_i, y_i)\,\Delta y_i \mathbf{k}.$$

From Theorem 11.8, the area of $\Delta T_i$ is given by $\|\mathbf{u} \times \mathbf{v}\|$, where

$$\mathbf{u} \times \mathbf{v} = \begin{vmatrix} \mathbf{i} & \mathbf{j} & \mathbf{k} \\ \Delta x_i & 0 & f_x(x_i, y_i)\,\Delta x_i \\ 0 & \Delta y_i & f_y(x_i, y_i)\,\Delta y_i \end{vmatrix}$$

$$= -f_x(x_i, y_i)\,\Delta x_i \Delta y_i \mathbf{i} - f_y(x_i, y_i)\,\Delta x_i \Delta y_i \mathbf{j} + \Delta x_i \Delta y_i \mathbf{k}$$

$$= [-f_x(x_i, y_i)\mathbf{i} - f_y(x_i, y_i)\mathbf{j} + \mathbf{k}]\,\Delta A_i.$$

So, the area of $\Delta T_i$ is $\|\mathbf{u} \times \mathbf{v}\| = \sqrt{[f_x(x_i, y_i)]^2 + [f_y(x_i, y_i)]^2 + 1}\,\Delta A_i$, and

$$\text{Surface area of } S \approx \sum_{i=1}^{n} \Delta S_i$$

$$\approx \sum_{i=1}^{n} \sqrt{1 + [f_x(x_i, y_i)]^2 + [f_y(x_i, y_i)]^2}\,\Delta A_i.$$

This suggests the definition of surface area on the next page.

---

**Definition of Surface Area**

If $f$ and its first partial derivatives are continuous on the closed region $R$ in the $xy$-plane, then the **area of the surface $S$** given by $z = f(x, y)$ over $R$ is defined as

$$\text{Surface area} = \iint_R dS$$

$$= \iint_R \sqrt{1 + [f_x(x, y)]^2 + [f_y(x, y)]^2}\, dA.$$

---

As an aid to remembering the double integral for surface area, it is helpful to note its similarity to the integral for arc length.

> •• **REMARK** Note that the differential $ds$ of arc length in the $xy$-plane is
>
> $$\sqrt{1 + [f'(x)]^2}\, dx$$
>
> and the differential $dS$ of surface area in space is
>
> $$\sqrt{1 + [f_x(x, y)]^2 + [f_y(x, y)]^2}\, dA.$$
> •••••••••••••••••••▷

**Length on $x$-axis:** $\qquad \displaystyle\int_a^b dx$

**Arc length in $xy$-plane:** $\qquad \displaystyle\int_a^b ds = \int_a^b \sqrt{1 + [f'(x)]^2}\, dx$

**Area in $xy$-plane:** $\qquad \displaystyle\iint_R dA$

**Surface area in space:** $\qquad \displaystyle\iint_R dS = \iint_R \sqrt{1 + [f_x(x, y)]^2 + [f_y(x, y)]^2}\, dA$

Like integrals for arc length, integrals for surface area are often very difficult to evaluate. However, one type that is easily evaluated is demonstrated in the next example.

---

**EXAMPLE 1**    **The Surface Area of a Plane Region**

Find the surface area of the portion of the plane

$$z = 2 - x - y$$

that lies above the circle $x^2 + y^2 \leq 1$ in the first quadrant, as shown in Figure 14.44.

**Solution**    Note that $f(x, y) = 2 - x - y$, $f_x(x, y) = -1$, and $f_y(x, y) = -1$ are continuous on the region $R$. So, the surface area is given by

$$S = \iint_R \sqrt{1 + [f_x(x, y)]^2 + [f_y(x, y)]^2}\, dA \qquad \text{Formula for surface area}$$

$$= \iint_R \sqrt{1 + (-1)^2 + (-1)^2}\, dA \qquad \text{Substitute.}$$

$$= \iint_R \sqrt{3}\, dA$$

$$= \sqrt{3} \iint_R dA.$$

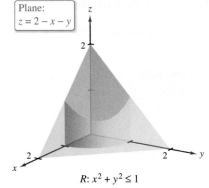

Plane:
$z = 2 - x - y$

$R: x^2 + y^2 \leq 1$

**Figure 14.44**

Note that the last integral is $\sqrt{3}$ times the area of the region $R$. Because $R$ is a quarter circle of radius 1, the area of $R$ is $\frac{1}{4}\pi(1^2)$ or $\pi/4$. So, the area of $S$ is

$$S = \sqrt{3}\,(\text{area of } R)$$

$$= \sqrt{3}\left(\frac{\pi}{4}\right)$$

$$= \frac{\sqrt{3}\,\pi}{4}.$$

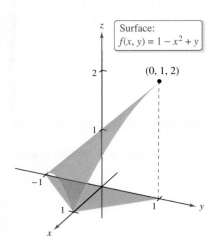

Surface:
$f(x, y) = 1 - x^2 + y$

**Figure 14.45**

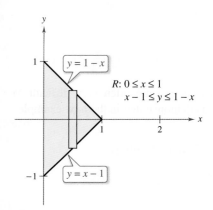

$R: 0 \le x \le 1$
$x - 1 \le y \le 1 - x$

$y = 1 - x$

$y = x - 1$

**Figure 14.46**

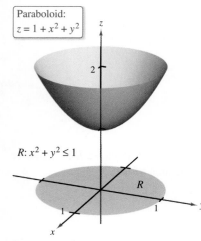

Paraboloid:
$z = 1 + x^2 + y^2$

$R: x^2 + y^2 \le 1$

**Figure 14.47**

**EXAMPLE 2**   **Finding Surface Area**

⋯▷ *See LarsonCalculus.com for an interactive version of this type of example.*

Find the area of the portion of the surface $f(x, y) = 1 - x^2 + y$ that lies above the triangular region with vertices $(1, 0, 0)$, $(0, -1, 0)$, and $(0, 1, 0)$, as shown in Figure 14.45.

**Solution**   Because $f_x(x, y) = -2x$ and $f_y(x, y) = 1$, you have

$$S = \int_R\int \sqrt{1 + [f_x(x, y)]^2 + [f_y(x, y)]^2} \, dA = \int_R\int \sqrt{1 + 4x^2 + 1} \, dA.$$

In Figure 14.46, you can see that the bounds for $R$ are $0 \le x \le 1$ and $x - 1 \le y \le 1 - x$. So, the integral becomes

$$S = \int_0^1 \int_{x-1}^{1-x} \sqrt{2 + 4x^2} \, dy \, dx \qquad \text{Apply formula for surface area.}$$

$$= \int_0^1 y\sqrt{2 + 4x^2} \, \Big]_{x-1}^{1-x} dx$$

$$= \int_0^1 \left[ (1 - x)\sqrt{2 + 4x^2} - (x - 1)\sqrt{2 + 4x^2} \right] dx$$

$$= \int_0^1 \left( 2\sqrt{2 + 4x^2} - 2x\sqrt{2 + 4x^2} \right) dx \qquad \begin{array}{l}\text{Integration tables (Appendix B),} \\ \text{Formula 26 and Power Rule}\end{array}$$

$$= \left[ x\sqrt{2 + 4x^2} + \ln\left(2x + \sqrt{2 + 4x^2}\right) - \frac{(2 + 4x^2)^{3/2}}{6} \right]_0^1$$

$$= \sqrt{6} + \ln\left(2 + \sqrt{6}\right) - \sqrt{6} - \ln\sqrt{2} + \frac{1}{3}\sqrt{2}$$

$$\approx 1.618.$$

**EXAMPLE 3**   **Change of Variables to Polar Coordinates**

Find the surface area of the paraboloid $z = 1 + x^2 + y^2$ that lies above the unit circle, as shown in Figure 14.47.

**Solution**   Because $f_x(x, y) = 2x$ and $f_y(x, y) = 2y$, you have

$$S = \int_R\int \sqrt{1 + [f_x(x, y)]^2 + [f_y(x, y)]^2} \, dA = \int_R\int \sqrt{1 + 4x^2 + 4y^2} \, dA.$$

You can convert to polar coordinates by letting $x = r\cos\theta$ and $y = r\sin\theta$. Then, because the region $R$ is bounded by $0 \le r \le 1$ and $0 \le \theta \le 2\pi$, you have

$$S = \int_0^{2\pi} \int_0^1 \sqrt{1 + 4r^2} \, r \, dr \, d\theta \qquad \text{Polar coordinates}$$

$$= \int_0^{2\pi} \frac{1}{12}(1 + 4r^2)^{3/2} \, \Big]_0^1 d\theta \qquad \text{Integrate with respect to } r.$$

$$= \int_0^{2\pi} \frac{5\sqrt{5} - 1}{12} \, d\theta$$

$$= \frac{5\sqrt{5} - 1}{12} \theta \, \Big]_0^{2\pi} \qquad \text{Integrate with respect to } \theta.$$

$$= \frac{\pi(5\sqrt{5} - 1)}{6}$$

$$\approx 5.33.$$

Hemisphere:
$f(x, y) = \sqrt{25 - x^2 - y^2}$

$R: x^2 + y^2 \le 9$

**Figure 14.48**

**EXAMPLE 4    Finding Surface Area**

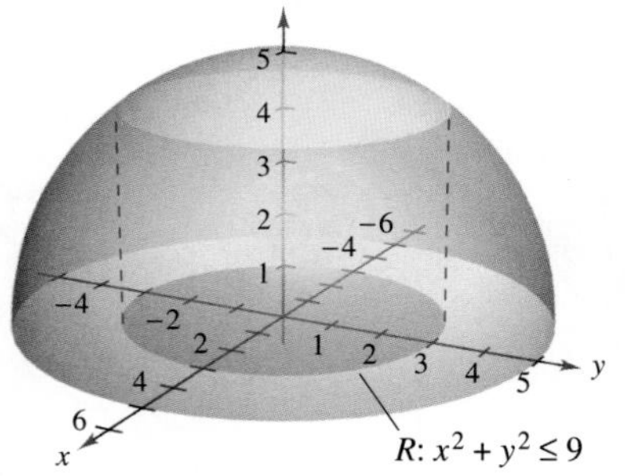

Find the surface area $S$ of the portion of the hemisphere

$$f(x, y) = \sqrt{25 - x^2 - y^2} \qquad \text{Hemisphere}$$

that lies above the region $R$ bounded by the circle $x^2 + y^2 \le 9$, as shown in Figure 14.48.

**Solution**    The first partial derivatives of $f$ are

$$f_x(x, y) = \frac{-x}{\sqrt{25 - x^2 - y^2}}$$

and

$$f_y(x, y) = \frac{-y}{\sqrt{25 - x^2 - y^2}}$$

and, from the formula for surface area, you have

$$dS = \sqrt{1 + [f_x(x, y)]^2 + [f_y(x, y)]^2}\, dA$$

$$= \sqrt{1 + \left(\frac{-x}{\sqrt{25 - x^2 - y^2}}\right)^2 + \left(\frac{-y}{\sqrt{25 - x^2 - y^2}}\right)^2}\, dA$$

$$= \frac{5}{\sqrt{25 - x^2 - y^2}}\, dA.$$

So, the surface area is

$$S = \iint_R \frac{5}{\sqrt{25 - x^2 - y^2}}\, dA.$$

You can convert to polar coordinates by letting $x = r \cos\theta$ and $y = r \sin\theta$. Then, because the region $R$ is bounded by $0 \le r \le 3$ and $0 \le \theta \le 2\pi$, you obtain

$$S = \int_0^{2\pi} \int_0^3 \frac{5}{\sqrt{25 - r^2}}\, r\, dr\, d\theta \qquad \text{Polar coordinates}$$

$$= 5 \int_0^{2\pi} \left. -\sqrt{25 - r^2}\, \right]_0^3 d\theta \qquad \text{Integrate with respect to } r.$$

$$= 5 \int_0^{2\pi} d\theta$$

$$= 10\pi. \qquad \text{Integrate with respect to } \theta.$$

The procedure used in Example 4 can be extended to find the surface area of a sphere by using the region $R$ bounded by the circle $x^2 + y^2 \le a^2$, where $0 < a < 5$, as shown in Figure 14.49. The surface area of the portion of the hemisphere

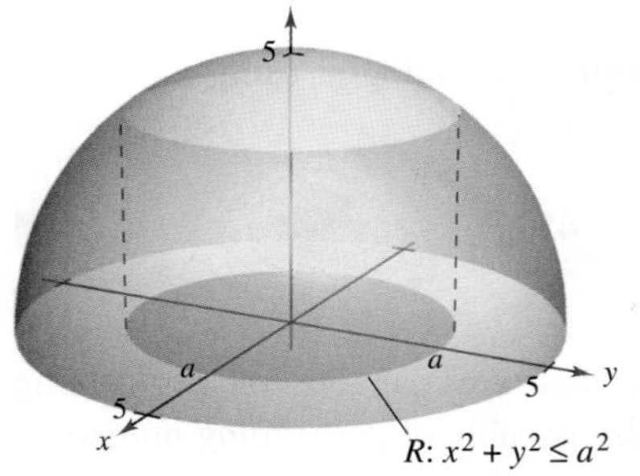

$$f(x, y) = \sqrt{25 - x^2 - y^2}$$

lying above the circular region can be shown to be

$$S = \iint_R \frac{5}{\sqrt{25 - x^2 - y^2}}\, dA$$

$$= \int_0^{2\pi} \int_0^a \frac{5}{\sqrt{25 - r^2}}\, r\, dr\, d\theta$$

$$= 10\pi\left(5 - \sqrt{25 - a^2}\right).$$

By taking the limit as $a$ approaches 5 and doubling the result, you obtain a total area of $100\pi$. (The surface area of a sphere of radius $r$ is $S = 4\pi r^2$.)

Hemisphere:
$f(x, y) = \sqrt{25 - x^2 - y^2}$

$R: x^2 + y^2 \le a^2$

**Figure 14.49**

You can use Simpson's Rule or the Trapezoidal Rule to approximate the value of a double integral, *provided* you can get through the first integration. This is demonstrated in the next example.

**EXAMPLE 5**  **Approximating Surface Area by Simpson's Rule**

Find the area of the surface of the paraboloid

$$f(x, y) = 2 - x^2 - y^2$$  Paraboloid

that lies above the square region bounded by

$$-1 \le x \le 1 \quad \text{and} \quad -1 \le y \le 1$$

as shown in Figure 14.50.

**Solution**  Using the partial derivatives

$$f_x(x, y) = -2x \quad \text{and} \quad f_y(x, y) = -2y$$

you have a surface area of

$$S = \int_R \int \sqrt{1 + [f_x(x, y)]^2 + [f_y(x, y)]^2} \, dA$$  Formula for surface area

$$= \int_R \int \sqrt{1 + (-2x)^2 + (-2y)^2} \, dA$$  Substitute.

$$= \int_R \int \sqrt{1 + 4x^2 + 4y^2} \, dA.$$  Simplify.

In polar coordinates, the line $x = 1$ is given by

$$r \cos \theta = 1 \quad \text{or} \quad r = \sec \theta$$

and you can determine from Figure 14.51 that one-fourth of the region $R$ is bounded by

$$0 \le r \le \sec \theta \quad \text{and} \quad -\frac{\pi}{4} \le \theta \le \frac{\pi}{4}.$$

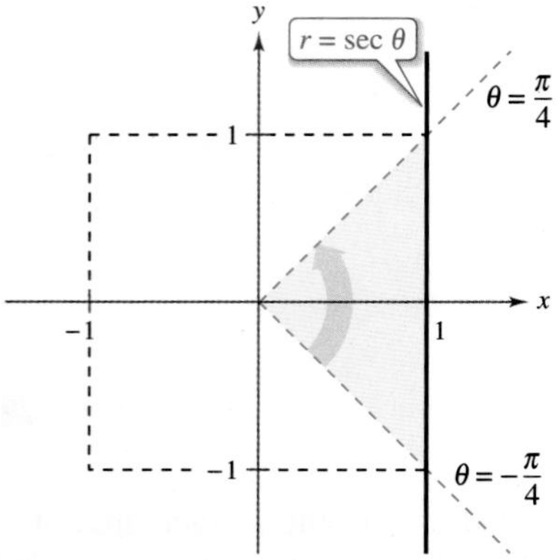

Letting $x = r \cos \theta$ and $y = r \sin \theta$ produces

$$\frac{1}{4} S = \frac{1}{4} \int_R \int \sqrt{1 + 4x^2 + 4y^2} \, dA$$  One-fourth of surface area

$$= \int_{-\pi/4}^{\pi/4} \int_0^{\sec \theta} \sqrt{1 + 4r^2} \, r \, dr \, d\theta$$  Polar coordinates

$$= \int_{-\pi/4}^{\pi/4} \frac{1}{12}(1 + 4r^2)^{3/2} \Big]_0^{\sec \theta} d\theta$$  Integrate with respect to $r$.

$$= \frac{1}{12} \int_{-\pi/4}^{\pi/4} [(1 + 4 \sec^2 \theta)^{3/2} - 1] \, d\theta.$$

After multiplying each side by 4, you can approximate the integral using Simpson's Rule with $n = 10$ to find that the area of the surface is

$$S = 4\left(\frac{1}{12}\right) \int_{-\pi/4}^{\pi/4} [(1 + 4 \sec^2 \theta)^{3/2} - 1] \, d\theta \approx 7.450. \quad \blacksquare$$

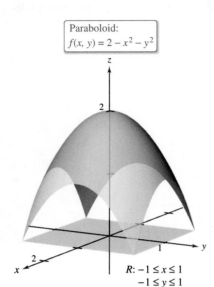

Paraboloid:
$$f(x, y) = 2 - x^2 - y^2$$

$R: -1 \le x \le 1$
$-1 \le y \le 1$

**Figure 14.50**

$r = \sec \theta$

$\theta = \dfrac{\pi}{4}$

$\theta = -\dfrac{\pi}{4}$

One-fourth of the region $R$ is bounded by $0 \le r \le \sec \theta$ and $-\dfrac{\pi}{4} \le \theta \le \dfrac{\pi}{4}$.

**Figure 14.51**

▷ **TECHNOLOGY**  Most computer programs that are capable of performing symbolic integration for multiple integrals are also capable of performing numerical approximation techniques. If you have access to such software, use it to approximate the value of the integral in Example 5.

## 14.5 Exercises

See **CalcChat.com** for tutorial help and worked-out solutions to odd-numbered exercises.

### CONCEPT CHECK

1. **Surface Area**  What is the differential of surface area, $dS$, in space?

2. **Numerical Integration**  Write a double integral that represents the surface area of the portion of the plane $z = 3$ that lies above the rectangular region with vertices $(0, 0)$, $(4, 0)$, $(0, 5)$, and $(4, 5)$. Then find the surface area without integrating.

 **Finding Surface Area**  In Exercises 3–16, find the area of the surface given by $z = f(x, y)$ that lies above the region $R$.

3. $f(x, y) = 2x + 2y$

   $R$: triangle with vertices $(0, 0)$, $(4, 0)$, $(0, 4)$

4. $f(x, y) = 15 + 2x - 3y$

   $R$: square with vertices $(0, 0)$, $(3, 0)$, $(0, 3)$, $(3, 3)$

5. $f(x, y) = 4 + 5x + 6y$, $R = \{(x, y): x^2 + y^2 \le 4\}$

6. $f(x, y) = 12 + 2x - 3y$, $R = \{(x, y): x^2 + y^2 \le 9\}$

7. $f(x, y) = 9 - x^2$

   $R$: square with vertices $(0, 0)$, $(2, 0)$, $(0, 2)$, $(2, 2)$

8. $f(x, y) = y^2$

   $R$: square with vertices $(0, 0)$, $(3, 0)$, $(0, 3)$, $(3, 3)$

9. $f(x, y) = 3 + 2x^{3/2}$

   $R$: rectangle with vertices $(0, 0)$, $(0, 4)$, $(1, 4)$, $(1, 0)$

10. $f(x, y) = 2 + \frac{2}{3}y^{3/2}$

    $R = \{(x, y): 0 \le x \le 2, 0 \le y \le 2 - x\}$

11. $f(x, y) = \ln|\sec x|$

    $R = \left\{(x, y): 0 \le x \le \dfrac{\pi}{4}, 0 \le y \le \tan x\right\}$

12. $f(x, y) = 13 + x^2 - y^2$, $R = \{(x, y): x^2 + y^2 \le 4\}$

13. $f(x, y) = \sqrt{x^2 + y^2}$, $R = \{(x, y): 0 \le f(x, y) \le 1\}$

14. $f(x, y) = xy$, $R = \{(x, y): x^2 + y^2 \le 16\}$

15. $f(x, y) = \sqrt{a^2 - x^2 - y^2}$

    $R = \{(x, y): x^2 + y^2 \le b^2, \; 0 < b < a\}$

16. $f(x, y) = \sqrt{a^2 - x^2 - y^2}$

    $R = \{(x, y): x^2 + y^2 \le a^2\}$

 **Finding Surface Area**  In Exercises 17–20, find the area of the surface.

17. The portion of the plane $z = 12 - 3x - 2y$ in the first octant

18. The portion of the paraboloid $z = 16 - x^2 - y^2$ in the first octant

19. The portion of the sphere $x^2 + y^2 + z^2 = 25$ inside the cylinder $x^2 + y^2 = 9$

20. The portion of the cone $z = 2\sqrt{x^2 + y^2}$ inside the cylinder $x^2 + y^2 = 4$

 **Finding Surface Area Using Technology**  In Exercises 21–26, write a double integral that represents the surface area of $z = f(x, y)$ that lies above the region $R$. Use a computer algebra system to evaluate the double integral.

21. $f(x, y) = 2y + x^2$, $R$: triangle with vertices $(0, 0)$, $(1, 0)$, $(1, 1)$

22. $f(x, y) = 2x + y^2$, $R$: triangle with vertices $(0, 0)$, $(2, 0)$, $(2, 2)$

23. $f(x, y) = 9 - x^2 - y^2$, $R = \{(x, y): 0 \le f(x, y)\}$

24. $f(x, y) = x^2 + y^2$, $R = \{(x, y): 0 \le f(x, y) \le 16\}$

25. $f(x, y) = 4 - x^2 - y^2$

    $R = \{(x, y): 0 \le x \le 1, 0 \le y \le 1\}$

26. $f(x, y) = \frac{2}{3}x^{3/2} + \cos x$

    $R = \{(x, y): 0 \le x \le 1, 0 \le y \le 1\}$

**Setting Up a Double Integral**  In Exercises 27–30, set up a double integral that represents the area of the surface given by $z = f(x, y)$ that lies above the region $R$.

27. $f(x, y) = e^{xy}$, $R = \{(x, y): 0 \le x \le 4, 0 \le y \le 10\}$

28. $f(x, y) = x^2 - 3xy - y^2$

    $R = \{(x, y): 0 \le x \le 4, 0 \le y \le x\}$

29. $f(x, y) = e^{-x} \sin y$, $R = \{(x, y): x^2 + y^2 \le 4\}$

30. $f(x, y) = \cos(x^2 + y^2)$, $R = \left\{(x, y): x^2 + y^2 \le \dfrac{\pi}{2}\right\}$

### EXPLORING CONCEPTS

31. **Surface Area**  Will the surface area of the graph of a function $z = f(x, y)$ that lies above a region $R$ increase when the graph is shifted $k$ units vertically? Explain using the partial derivatives of $z$.

32. **HOW DO YOU SEE IT?**  Consider the surface $f(x, y) = x^2 + y^2$ (see figure) and the surface area of $f$ that lies above each region $R$. Without integrating, order the surface areas from least to greatest. Explain.

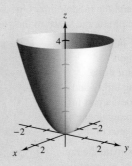

(a) $R$: rectangle with vertices $(0, 0)$, $(2, 0)$, $(2, 2)$, $(0, 2)$

(b) $R$: triangle with vertices $(0, 0)$, $(2, 0)$, $(0, 2)$

(c) $R = \{(x, y): x^2 + y^2 \le 4, \text{ first quadrant only}\}$

**33. Surface Area** Answer each question about the surface area $S$ on a surface given by a positive function $z = f(x, y)$ that lies above a region $R$ in the $xy$-plane. Explain each answer.

(a) Is it possible for $S$ to equal the area of $R$?

(b) Can $S$ be greater than the area of $R$?

(c) Can $S$ be less than the area of $R$?

**34. Surface Area** Consider the surface $f(x, y) = x + y$. What is the relationship between the area of the surface that lies above the region

$$R_1 = \{(x, y): x^2 + y^2 \le 1\}$$

and the area of the surface that lies above the region

$$R_2 = \{(x, y): x^2 + y^2 \le 4\}?$$

**35. Product Design** A company produces a spherical object of radius 25 centimeters. A hole of radius 4 centimeters is drilled through the center of the object.

(a) Find the volume of the object.

(b) Find the outer surface area of the object.

• • **36. Modeling Data** • • • • • • • • • • • • • • • •

A company builds a warehouse with dimensions 30 feet by 50 feet. The symmetrical shape and selected heights of the roof are shown in the figure.

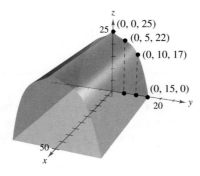

(a) Use the regression capabilities of a graphing utility to find a model of the form

$$z = ay^3 + by^2 + cy + d$$

for the roof line.

(b) Use the numerical integration capabilities of a graphing utility and the model in part (a) to approximate the volume of storage space in the warehouse.

(c) Use the numerical integration capabilities of a graphing utility and the model in part (a) to approximate the surface area of the roof.

(d) Approximate the arc length of the roof line and find the surface area of the roof by multiplying the arc length by the length of the warehouse. Compare the results and the integrations with those found in part (c).

**37. Surface Area** Find the surface area of the solid of intersection of the cylinders $x^2 + z^2 = 1$ and $y^2 + z^2 = 1$ (see figure).

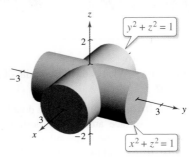

**38. Surface Area** Show that the surface area of the cone $z = k\sqrt{x^2 + y^2}$, $k > 0$, that lies above the circular region $x^2 + y^2 \le r^2$ in the $xy$-plane is $\pi r^2 \sqrt{k^2 + 1}$ (see figure).

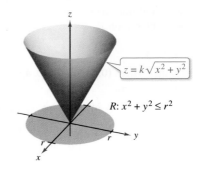

**SECTION PROJECT**

## Surface Area in Polar Coordinates

(a) Use the formula for surface area in rectangular coordinates to derive the following formula for surface area in polar coordinates, where $z = f(x, y) = f(r\cos\theta, r\sin\theta)$. (*Hint:* You will need to use the Chain Rule for functions of two variables.)

$$S = \int\int_R \sqrt{1 + f_r^2 + \frac{1}{r^2}f_\theta^2}\, r\, dr\, d\theta$$

(b) Use the formula from part (a) to find the surface area of the paraboloid $z = x^2 + y^2$ that lies above the circular region $x^2 + y^2 \le 4$ in the $xy$-plane (see figure).

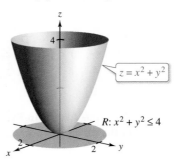

(c) Use the formula from part (a) to find the surface area of $z = xy$ that lies above the circular region $x^2 + y^2 \le 16$ in the $xy$-plane. Compare your answer with your answer to Exercise 14.

# 14.6   Triple Integrals and Applications

- Use a triple integral to find the volume of a solid region.
- Find the center of mass and moments of inertia of a solid region.

## Triple Integrals

The procedure used to define a **triple integral** follows that used for double integrals. Consider a function $f$ of three variables that is continuous over a bounded solid region $Q$. Then encompass $Q$ with a network of boxes and form the **inner partition** consisting of all boxes lying entirely within $Q$, as shown in Figure 14.52. The volume of the $i$th box is

$$\Delta V_i = \Delta x_i \Delta y_i \Delta z_i. \qquad \text{Volume of } i\text{th box}$$

The **norm** $\|\Delta\|$ of the partition is the length of the longest diagonal of the $n$ boxes in the partition. Choose a point $(x_i, y_i, z_i)$ in each box and form the Riemann sum

$$\sum_{i=1}^{n} f(x_i, y_i, z_i) \, \Delta V_i.$$

Taking the limit as $\|\Delta\| \to 0$ leads to the following definition.

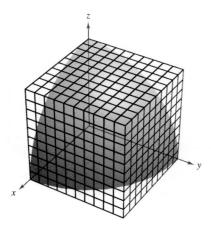

Solid region $Q$

> ### Definition of Triple Integral
>
> If $f$ is continuous over a bounded solid region $Q$, then the **triple integral of $f$ over $Q$** is defined as
>
> $$\iiint\limits_{Q} f(x, y, z) \, dV = \lim_{\|\Delta\| \to 0} \sum_{i=1}^{n} f(x_i, y_i, z_i) \, \Delta V_i$$
>
> provided the limit exists. The **volume** of the solid region $Q$ is given by
>
> $$\text{Volume of } Q = \iiint\limits_{Q} dV.$$

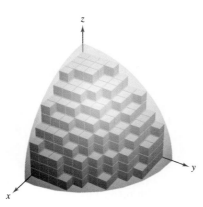

Volume of $Q \approx \displaystyle\sum_{i=1}^{n} \Delta V_i$

**Figure 14.52**

Some of the properties of double integrals in Theorem 14.1 can be restated in terms of triple integrals.

**1.** $\displaystyle\iiint\limits_{Q} cf(x, y, z) \, dV = c \iiint\limits_{Q} f(x, y, z) \, dV$

**2.** $\displaystyle\iiint\limits_{Q} [f(x, y, z) \pm g(x, y, z)] \, dV = \iiint\limits_{Q} f(x, y, z) \, dV \pm \iiint\limits_{Q} g(x, y, z) \, dV$

**3.** $\displaystyle\iiint\limits_{Q} f(x, y, z) \, dV = \iiint\limits_{Q_1} f(x, y, z) \, dV + \iiint\limits_{Q_2} f(x, y, z) \, dV$

In the properties above, $Q$ is the union of two nonoverlapping solid subregions $Q_1$ and $Q_2$. If the solid region $Q$ is simple, then the triple integral $\iiint f(x, y, z) \, dV$ can be evaluated with an iterated integral using one of the six possible orders of integration listed below.

$$dx \, dy \, dz \quad dy \, dx \, dz \quad dz \, dx \, dy$$
$$dx \, dz \, dy \quad dy \, dz \, dx \quad dz \, dy \, dx$$

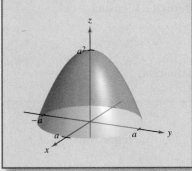
The following version of Fubini's Theorem describes a region that is considered simple with respect to the order *dz dy dx*. Similar versions of this theorem can be given for the other five orders.

---

**THEOREM 14.4  Evaluation by Iterated Integrals**

Let *f* be continuous on a solid region *Q* defined by

$$a \le x \le b,$$
$$h_1(x) \le y \le h_2(x),$$
$$g_1(x, y) \le z \le g_2(x, y)$$

where $h_1$, $h_2$, $g_1$, and $g_2$ are continuous functions. Then,

$$\iiint\limits_{Q} f(x, y, z)\, dV = \int_a^b \int_{h_1(x)}^{h_2(x)} \int_{g_1(x, y)}^{g_2(x, y)} f(x, y, z)\, dz\, dy\, dx.$$

---

To evaluate a triple iterated integral in the order *dz dy dx*, hold *both x* and *y* constant for the innermost integration. Then hold *x* constant for the second integration.

**EXAMPLE 1  Evaluating a Triple Iterated Integral**

Evaluate the triple iterated integral

$$\int_0^2 \int_0^x \int_0^{x+y} e^x(y + 2z)\, dz\, dy\, dx.$$

**Solution**  For the first integration, hold *x* and *y* constant and integrate with respect to *z*.

$$\int_0^2 \int_0^x \int_0^{x+y} e^x(y + 2z)\, dz\, dy\, dx = \int_0^2 \int_0^x e^x(yz + z^2)\Big]_0^{x+y} dy\, dx$$

$$= \int_0^2 \int_0^x e^x(x^2 + 3xy + 2y^2)\, dy\, dx$$

For the second integration, hold *x* constant and integrate with respect to *y*.

$$\int_0^2 \int_0^x e^x(x^2 + 3xy + 2y^2)\, dy\, dx = \int_0^2 \left[ e^x\left( x^2 y + \frac{3xy^2}{2} + \frac{2y^3}{3} \right) \right]_0^x dx$$

$$= \frac{19}{6} \int_0^2 x^3 e^x\, dx$$

Finally, integrate with respect to *x*.

$$\frac{19}{6} \int_0^2 x^3 e^x\, dx = \frac{19}{6} \left[ e^x(x^3 - 3x^2 + 6x - 6) \right]_0^2$$

$$= 19\left( \frac{e^2}{3} + 1 \right)$$

$$\approx 65.797$$

• • **REMARK**  To do the last integration in Example 1, use integration by parts three times.

Example 1 demonstrates the integration order *dz dy dx*. For other orders, you can follow a similar procedure. For instance, to evaluate a triple iterated integral in the order *dx dy dz*, hold both *y* and *z* constant for the innermost integration and integrate with respect to *x*. Then, for the second integration, hold *z* constant and integrate with respect to *y*. Finally, for the third integration, integrate with respect to *z*.

To find the limits for a particular order of integration, it is generally advisable first to determine the innermost limits, which may be functions of the outer two variables. Then, by projecting the solid $Q$ onto the coordinate plane of the outer two variables, you can determine their limits of integration by the methods used for double integrals. For instance, to evaluate

$$\iiint_Q f(x, y, z)\, dz\, dy\, dx$$

first determine the limits for $z$; the integral then has the form

$$\iint \left[ \int_{g_1(x, y)}^{g_2(x, y)} f(x, y, z)\, dz \right] dy\, dx.$$

By projecting the solid $Q$ onto the $xy$-plane, you can determine the limits for $x$ and $y$ as you did for double integrals, as shown in Figure 14.53.

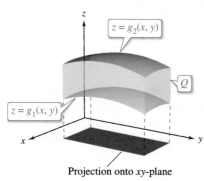

Solid region $Q$ lies between two surfaces.
**Figure 14.53**

### EXAMPLE 2  Using a Triple Integral to Find Volume

Find the volume of the ellipsoid given by $4x^2 + 4y^2 + z^2 = 16$.

**Solution**  Because $x$, $y$, and $z$ play similar roles in the equation, the order of integration is probably immaterial, and you can arbitrarily choose $dz\, dy\, dx$. Moreover, you can simplify the calculation by considering only the portion of the ellipsoid lying in the first octant, as shown in Figure 14.54. From the order $dz\, dy\, dx$, you first determine the bounds for $z$.

$$0 \le z \le 2\sqrt{4 - x^2 - y^2} \qquad \text{Bounds for } z$$

In Figure 14.55, you can see that the bounds for $x$ and $y$ are

$$0 \le x \le 2 \quad \text{and} \quad 0 \le y \le \sqrt{4 - x^2}. \qquad \text{Bounds for } x \text{ and } y$$

So, the volume of the ellipsoid is

$$V = \iiint_Q dV \qquad \text{Formula for volume}$$

$$= 8 \int_0^2 \int_0^{\sqrt{4-x^2}} \int_0^{2\sqrt{4-x^2-y^2}} dz\, dy\, dx \qquad \text{Convert to iterated integral.}$$

$$= 8 \int_0^2 \int_0^{\sqrt{4-x^2}} z \Big]_0^{2\sqrt{4-x^2-y^2}} dy\, dx$$

$$= 16 \int_0^2 \int_0^{\sqrt{4-x^2}} \sqrt{(4 - x^2) - y^2}\, dy\, dx \qquad \begin{array}{l}\text{Integration tables (Appendix B),} \\ \text{Formula 37}\end{array}$$

$$= 8 \int_0^2 \left[ y\sqrt{4 - x^2 - y^2} + (4 - x^2) \arcsin\left( \frac{y}{\sqrt{4 - x^2}} \right) \right]_0^{\sqrt{4-x^2}} dx$$

$$= 8 \int_0^2 \left[ 0 + (4 - x^2) \arcsin(1) - 0 - 0 \right] dx$$

$$= 8 \int_0^2 (4 - x^2)\left( \frac{\pi}{2} \right) dx$$

$$= 4\pi \left[ 4x - \frac{x^3}{3} \right]_0^2$$

$$= \frac{64\pi}{3}.$$

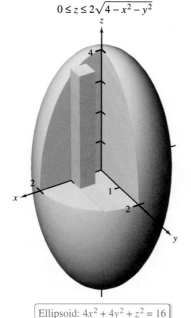

$$0 \le z \le 2\sqrt{4 - x^2 - y^2}$$

Ellipsoid: $4x^2 + 4y^2 + z^2 = 16$

**Figure 14.54**

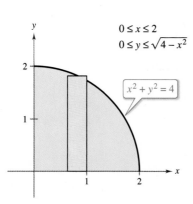

$$0 \le x \le 2$$
$$0 \le y \le \sqrt{4 - x^2}$$

$x^2 + y^2 = 4$

**Figure 14.55**

Example 2 is unusual in that all six possible orders of integration produce integrals of comparable difficulty. Try setting up some other possible orders of integration to find the volume of the elipsoid. For instance, the order $dx\,dy\,dz$ yields the integral

$$V = 8 \int_0^4 \int_0^{\sqrt{16-z^2}/2} \int_0^{\sqrt{16-4y^2-z^2}/2} dx\,dy\,dz.$$

The evaluation of this integral yields the same volume obtained in Example 2. This is always the case—the order of integration does not affect the value of the integral. However, the order of integration often does affect the complexity of the integral. In Example 3, the given order of integration is not convenient, so you can change the order to simplify the problem.

## EXAMPLE 3    Changing the Order of Integration

Evaluate $\displaystyle\int_0^{\sqrt{\pi/2}} \int_x^{\sqrt{\pi/2}} \int_1^3 \sin(y^2)\,dz\,dy\,dx.$

**Solution**   Note that after one integration in the given order, you would encounter the integral $2\int \sin(y^2)\,dy$, which is not an elementary function. To avoid this problem, change the order of integration to $dz\,dx\,dy$ so that $y$ is the outer variable. From Figure 14.56, you can see that the solid region $Q$ is

$$0 \le x \le \sqrt{\frac{\pi}{2}}$$

$$x \le y \le \sqrt{\frac{\pi}{2}}$$

$$1 \le z \le 3$$

and the projection of $Q$ in the $xy$-plane yields the bounds

$$0 \le y \le \sqrt{\frac{\pi}{2}}$$

and

$$0 \le x \le y.$$

So, evaluating the triple integral using the order $dz\,dx\,dy$ produces

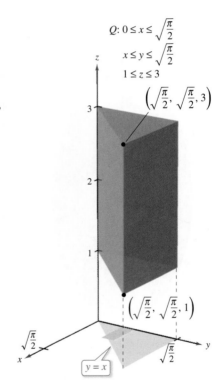

**Figure 14.56**

$$\int_0^{\sqrt{\pi/2}} \int_0^y \int_1^3 \sin(y^2)\,dz\,dx\,dy = \int_0^{\sqrt{\pi/2}} \int_0^y z \sin(y^2)\Big]_1^3 dx\,dy$$

$$= 2\int_0^{\sqrt{\pi/2}} \int_0^y \sin(y^2)\,dx\,dy$$

$$= 2\int_0^{\sqrt{\pi/2}} x \sin(y^2)\Big]_0^y dy$$

$$= 2\int_0^{\sqrt{\pi/2}} y \sin(y^2)\,dy$$

$$= -\cos(y^2)\Big]_0^{\sqrt{\pi/2}}$$

$$= 1.$$

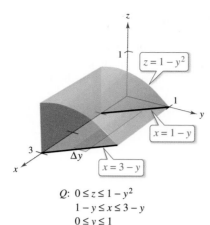

$$Q: 0 \le z \le 1 - y^2$$
$$1 - y \le x \le 3 - y$$
$$0 \le y \le 1$$

**Figure 14.57**

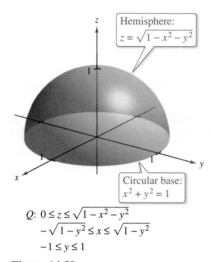

$$Q: 0 \le z \le \sqrt{1 - x^2 - y^2}$$
$$-\sqrt{1 - y^2} \le x \le \sqrt{1 - y^2}$$
$$-1 \le y \le 1$$

**Figure 14.58**

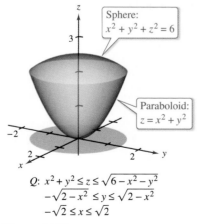

$$Q: x^2 + y^2 \le z \le \sqrt{6 - x^2 - y^2}$$
$$-\sqrt{2 - x^2} \le y \le \sqrt{2 - x^2}$$
$$-\sqrt{2} \le x \le \sqrt{2}$$

**Figure 14.59**

**EXAMPLE 4**　**Determining the Limits of Integration**

Set up a triple integral for the volume of each solid region.

**a.** The region in the first octant bounded above by the cylinder $z = 1 - y^2$ and lying between the vertical planes $x + y = 1$ and $x + y = 3$

**b.** The upper hemisphere $z = \sqrt{1 - x^2 - y^2}$

**c.** The region bounded below by the paraboloid $z = x^2 + y^2$ and above by the sphere $x^2 + y^2 + z^2 = 6$

**Solution**

**a.** In Figure 14.57, note that the solid is bounded below by the $xy$-plane $(z = 0)$ and above by the cylinder $z = 1 - y^2$. So,

$$0 \le z \le 1 - y^2. \qquad \text{Bounds for } z$$

Projecting the region onto the $xy$-plane produces a parallelogram. Because two sides of the parallelogram are parallel to the $x$-axis, you have the following bounds:

$$1 - y \le x \le 3 - y \quad \text{and} \quad 0 \le y \le 1. \qquad \text{Bounds for } x \text{ and } y$$

So, the volume of the region is given by

$$V = \iiint_Q dV = \int_0^1 \int_{1-y}^{3-y} \int_0^{1-y^2} dz \, dx \, dy.$$

**b.** For the upper hemisphere $z = \sqrt{1 - x^2 - y^2}$, you have

$$0 \le z \le \sqrt{1 - x^2 - y^2}. \qquad \text{Bounds for } z$$

In Figure 14.58, note that the projection of the hemisphere onto the $xy$-plane is the circle

$$x^2 + y^2 = 1$$

and you can use either order $dx \, dy$ or $dy \, dx$. Choosing the first produces

$$-\sqrt{1 - y^2} \le x \le \sqrt{1 - y^2} \quad \text{and} \quad -1 \le y \le 1 \qquad \text{Bounds for } x \text{ and } y$$

which implies that the volume of the region is given by

$$V = \iiint_Q dV = \int_{-1}^1 \int_{-\sqrt{1-y^2}}^{\sqrt{1-y^2}} \int_0^{\sqrt{1-x^2-y^2}} dz \, dx \, dy.$$

**c.** For the region bounded below by the paraboloid $z = x^2 + y^2$ and above by the sphere $x^2 + y^2 + z^2 = 6$, you have

$$x^2 + y^2 \le z \le \sqrt{6 - x^2 - y^2}. \qquad \text{Bounds for } z$$

The sphere and the paraboloid intersect at $z = 2$. Moreover, you can see in Figure 14.59 that the projection of the solid region onto the $xy$-plane is the circle

$$x^2 + y^2 = 2.$$

Using the order $dy \, dx$ produces

$$-\sqrt{2 - x^2} \le y \le \sqrt{2 - x^2} \quad \text{and} \quad -\sqrt{2} \le x \le \sqrt{2} \qquad \text{Bounds for } x \text{ and } y$$

which implies that the volume of the region is given by

$$V = \iiint_Q dV = \int_{-\sqrt{2}}^{\sqrt{2}} \int_{-\sqrt{2-x^2}}^{\sqrt{2-x^2}} \int_{x^2+y^2}^{\sqrt{6-x^2-y^2}} dz \, dy \, dx. \qquad \blacksquare$$

## Center of Mass and Moments of Inertia

In the remainder of this section, two important engineering applications of triple integrals are discussed. Consider a solid region $Q$ whose density is given by the **density function** $\rho$. The **center of mass** of a solid region $Q$ of mass $m$ is given by $(\bar{x}, \bar{y}, \bar{z})$, where

$$m = \iiint_Q \rho(x, y, z)\, dV \qquad \text{Mass of the solid}$$

$$M_{yz} = \iiint_Q x\rho(x, y, z)\, dV \qquad \text{First moment about } yz\text{-plane}$$

$$M_{xz} = \iiint_Q y\rho(x, y, z)\, dV \qquad \text{First moment about } xz\text{-plane}$$

$$M_{xy} = \iiint_Q z\rho(x, y, z)\, dV \qquad \text{First moment about } xy\text{-plane}$$

and

$$\bar{x} = \frac{M_{yz}}{m}, \quad \bar{y} = \frac{M_{xz}}{m}, \quad \bar{z} = \frac{M_{xy}}{m}.$$

The quantities $M_{yz}$, $M_{xz}$, and $M_{xy}$ are called the **first moments** of the region $Q$ about the $yz$-, $xz$-, and $xy$-planes, respectively.

The first moments for solid regions are taken about a plane, whereas the second moments for solids are taken about a line. The **second moments** (or **moments of inertia**) about the $x$-, $y$-, and $z$-axes are

$$I_x = \iiint_Q (y^2 + z^2)\rho(x, y, z)\, dV \qquad \text{Moment of inertia about } x\text{-axis}$$

$$I_y = \iiint_Q (x^2 + z^2)\rho(x, y, z)\, dV \qquad \text{Moment of inertia about } y\text{-axis}$$

and

$$I_z = \iiint_Q (x^2 + y^2)\rho(x, y, z)\, dV. \qquad \text{Moment of inertia about } z\text{-axis}$$

For problems requiring the calculation of all three moments, considerable effort can be saved by applying the additive property of triple integrals and writing

$$I_x = I_{xz} + I_{xy}, \quad I_y = I_{yz} + I_{xy}, \quad \text{and} \quad I_z = I_{yz} + I_{xz},$$

where $I_{xy}$, $I_{xz}$, and $I_{yz}$ are

$$I_{xy} = \iiint_Q z^2\rho(x, y, z)\, dV$$

$$I_{xz} = \iiint_Q y^2\rho(x, y, z)\, dV$$

and

$$I_{yz} = \iiint_Q x^2\rho(x, y, z)\, dV.$$

· · · · · · · · · · · · · · · ▷

**REMARK** In engineering and physics, the moment of inertia of a mass is used to find the time required for the mass to reach a given speed of rotation about an axis, as shown in Figure 14.60. The greater the moment of inertia, the longer a force must be applied for the mass to reach the given speed.

**Figure 14.60**

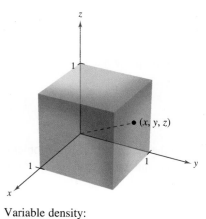

Variable density:
$\rho(x, y, z) = k(x^2 + y^2 + z^2)$
**Figure 14.61**

EXAMPLE 5    **Finding the Center of Mass of a Solid Region**

• • • • ▷ *See LarsonCalculus.com for an interactive version of this type of example.*

Find the center of mass of the unit cube shown in Figure 14.61, given that the density at the point $(x, y, z)$ is proportional to the square of its distance from the origin.

**Solution**    Because the density at $(x, y, z)$ is proportional to the square of the distance between $(0, 0, 0)$ and $(x, y, z)$, you have

$$\rho(x, y, z) = k(x^2 + y^2 + z^2)$$

where $k$ is the constant of proportionality. You can use this density function to find the mass of the cube. Because of the symmetry of the region, any order of integration will produce an integral of comparable difficulty.

$$m = \int_0^1 \int_0^1 \int_0^1 k(x^2 + y^2 + z^2)\, dz\, dy\, dx \qquad \text{Apply formula for mass of a solid.}$$

$$= k \int_0^1 \int_0^1 \left[ (x^2 + y^2)z + \frac{z^3}{3} \right]_0^1 dy\, dx \qquad \text{Integrate with respect to } z.$$

$$= k \int_0^1 \int_0^1 \left( x^2 + y^2 + \frac{1}{3} \right) dy\, dx$$

$$= k \int_0^1 \left[ \left( x^2 + \frac{1}{3} \right)y + \frac{y^3}{3} \right]_0^1 dx \qquad \text{Integrate with respect to } y.$$

$$= k \int_0^1 \left( x^2 + \frac{2}{3} \right) dx$$

$$= k \left[ \frac{x^3}{3} + \frac{2x}{3} \right]_0^1 \qquad \text{Integrate with respect to } x.$$

$$= k$$

The first moment about the $yz$-plane is

$$M_{yz} = k \int_0^1 \int_0^1 \int_0^1 x(x^2 + y^2 + z^2)\, dz\, dy\, dx \qquad \begin{array}{l}\text{Apply formula for first moment}\\\text{about } yz\text{-plane.}\end{array}$$

$$= k \int_0^1 x \left[ \int_0^1 \int_0^1 (x^2 + y^2 + z^2)\, dz\, dy \right] dx. \qquad \text{Factor.}$$

Note that $x$ can be factored out of the two inner integrals, because it is constant with respect to $y$ and $z$. After factoring, the two inner integrals are the same as for the mass $m$. Therefore, you have

$$M_{yz} = k \int_0^1 x \left( x^2 + \frac{2}{3} \right) dx$$

$$= k \left[ \frac{x^4}{4} + \frac{x^2}{3} \right]_0^1 \qquad \text{Integrate with respect to } x.$$

$$= \frac{7k}{12}. \qquad \text{First moment about } yz\text{-plane}$$

So,

$$\bar{x} = \frac{M_{yz}}{m} = \frac{7k/12}{k} = \frac{7}{12}.$$

Finally, from the nature of $\rho$ and the symmetry of $x$, $y$, and $z$ in this solid region, you have $\bar{x} = \bar{y} = \bar{z}$, and the center of mass is $\left( \frac{7}{12}, \frac{7}{12}, \frac{7}{12} \right)$.

**Moments of Inertia for a Solid Region**

Find the moments of inertia about the $x$- and $y$-axes for the solid region lying between the hemisphere

$$z = \sqrt{4 - x^2 - y^2}$$

and the $xy$-plane, given that the density at $(x, y, z)$ is proportional to the distance between $(x, y, z)$ and the $xy$-plane.

**Solution**    The density of the region is given by

$$\rho(x, y, z) = kz$$

where $k$ is the constant of proportionality. Considering the symmetry of this problem, you know that $I_x = I_y$, and you need to find only one moment, such as $I_x$. From Figure 14.62, choose the order $dz\, dy\, dx$ and write

$$I_x = \iiint_Q (y^2 + z^2)\rho(x, y, z)\, dV \qquad \text{Moment of inertia about } x\text{-axis}$$

$$= \int_{-2}^{2} \int_{-\sqrt{4-x^2}}^{\sqrt{4-x^2}} \int_{0}^{\sqrt{4-x^2-y^2}} (y^2 + z^2)(kz)\, dz\, dy\, dx$$

$$= k \int_{-2}^{2} \int_{-\sqrt{4-x^2}}^{\sqrt{4-x^2}} \left[ \frac{y^2 z^2}{2} + \frac{z^4}{4} \right]_{0}^{\sqrt{4-x^2-y^2}} dy\, dx$$

$$= k \int_{-2}^{2} \int_{-\sqrt{4-x^2}}^{\sqrt{4-x^2}} \left[ \frac{y^2(4 - x^2 - y^2)}{2} + \frac{(4 - x^2 - y^2)^2}{4} \right] dy\, dx$$

$$= \frac{k}{4} \int_{-2}^{2} \int_{-\sqrt{4-x^2}}^{\sqrt{4-x^2}} \left[ (4 - x^2)^2 - y^4 \right] dy\, dx$$

$$= \frac{k}{4} \int_{-2}^{2} \left[ (4 - x^2)^2 y - \frac{y^5}{5} \right]_{-\sqrt{4-x^2}}^{\sqrt{4-x^2}} dx$$

$$= \frac{k}{4} \int_{-2}^{2} \frac{8}{5}(4 - x^2)^{5/2}\, dx$$

$$= \frac{4k}{5} \int_{0}^{2} (4 - x^2)^{5/2}\, dx$$

$$= \frac{4k}{5} \int_{0}^{\pi/2} 64 \cos^6 \theta\, d\theta \qquad \text{Trigonometric substitution: } x = 2 \sin \theta$$

$$= \left( \frac{256k}{5} \right)\left( \frac{5\pi}{32} \right) \qquad \text{Wallis's Formula}$$

$$= 8k\pi.$$

So, $I_x = 8k\pi = I_y$.    ■

In Example 6, notice that the moments of inertia about the $x$- and $y$-axes are equal to each other. The moment about the $z$-axis, however, is different. Does it seem that the moment of inertia about the $z$-axis should be less than or greater than the moments calculated in Example 6? By performing the calculations, you can determine that

$$I_z = \frac{16}{3} k\pi.$$

This tells you that the solid shown in Figure 14.62 has a greater resistance to rotation about the $x$- or $y$-axis than about the $z$-axis.

$0 \le z \le \sqrt{4 - x^2 - y^2}$
$-\sqrt{4 - x^2} \le y \le \sqrt{4 - x^2}$
$-2 \le x \le 2$

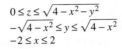

Hemisphere:
$z = \sqrt{4 - x^2 - y^2}$

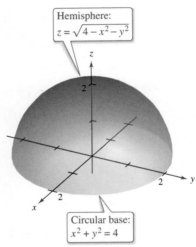

Circular base:
$x^2 + y^2 = 4$

Variable density: $\rho(x, y, z) = kz$
**Figure 14.62**

# 14.6 Exercises

See **CalcChat.com** for tutorial help and worked-out solutions to odd-numbered exercises.

---

**CONCEPT CHECK**

**1. Triple Integrals**   What does $Q = \iiint\limits_{Q} dV$ represent?

**2. Changing the Order of Integration**   Why is it beneficial to be able to change the order of integration for a triple integral? Explain.

---

 **Evaluating a Triple Iterated Integral**   In Exercises 3–10, evaluate the triple iterated integral.

**3.** $\displaystyle\int_0^3 \int_0^2 \int_0^1 (x + y + z)\, dx\, dz\, dy$

**4.** $\displaystyle\int_0^2 \int_0^1 \int_{-1}^2 xyz^3\, dx\, dy\, dz$

**5.** $\displaystyle\int_0^1 \int_0^x \int_0^{\sqrt{xy}} x\, dz\, dy\, dx$

**6.** $\displaystyle\int_0^9 \int_0^{y/3} \int_0^{\sqrt{y^2 - 9x^2}} z\, dz\, dx\, dy$

**7.** $\displaystyle\int_1^4 \int_0^1 \int_0^x 2ze^{-x^2}\, dy\, dx\, dz$

**8.** $\displaystyle\int_1^4 \int_1^{e^2} \int_0^{1/xz} \ln z\, dy\, dz\, dx$

**9.** $\displaystyle\int_{-3}^4 \int_0^{\pi/2} \int_0^{1+3x} x \cos y\, dz\, dy\, dx$

**10.** $\displaystyle\int_0^{\pi/2} \int_0^{y/2} \int_0^{1/y} \sin y\, dz\, dx\, dy$

**Evaluating a Triple Iterated Integral Using Technology**   In Exercises 11 and 12, use a computer algebra system to evaluate the triple iterated integral.

**11.** $\displaystyle\int_0^3 \int_{-\sqrt{9-y^2}}^{\sqrt{9-y^2}} \int_0^{y^2} y\, dz\, dx\, dy$

**12.** $\displaystyle\int_0^3 \int_0^{2-(2y/3)} \int_0^{6-2y-3z} ze^{-x^2y^2}\, dx\, dz\, dy$

**Setting Up a Triple Integral**   In Exercises 13–18, set up a triple integral for the volume of the solid. Do not evaluate the integral.

**13.** The solid in the first octant bounded by the coordinate planes and the plane $z = 7 - x - 2y$

**14.** The solid bounded by $z = 9 - x^2$, $z = 0$, $y = 0$, and $y = 2x$

**15.** The solid bounded by $z = 6 - x^2 - y^2$ and $z = 0$

**16.** The solid bounded by $z = \sqrt{1 - x^2 - y^2}$ and $z = 0$

**17.** The solid that is the common interior below the sphere $x^2 + y^2 + z^2 = 80$ and above the paraboloid $z = \frac{1}{2}(x^2 + y^2)$

**18.** The solid bounded above by the cylinder $z = 4 - x^2$ and below by the paraboloid $z = x^2 + 3y^2$

---

 **Volume**   In Exercises 19–24, use a triple integral to find the volume of the solid bounded by the graphs of the equations.

**19.**

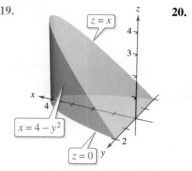

**20.**

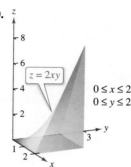

$0 \le x \le 2$
$0 \le y \le 2$

**21.** $z = 6x^2$, $y = 3 - 3x$, first octant

**22.** $z = 9 - x^3$, $y = -x^2 + 2$, $y = 0$, $z = 0$, $x \ge 0$

**23.** $z = 2 - y$, $z = 4 - y^2$, $x = 0$, $x = 3$, $y = 0$

**24.** $z = \sqrt{x}$, $y = x + 2$, $y = x^2$, first octant

 **Changing the Order of Integration**   In Exercises 25–30, sketch the solid whose volume is given by the iterated integral. Then rewrite the integral using the indicated order of integration.

**25.** $\displaystyle\int_0^1 \int_{-1}^0 \int_0^{y^2} dz\, dy\, dx$

Rewrite using $dy\, dz\, dx$.

**26.** $\displaystyle\int_{-1}^1 \int_{y^2}^1 \int_0^{1-x} dz\, dx\, dy$

Rewrite using $dx\, dz\, dy$.

**27.** $\displaystyle\int_0^4 \int_0^{(4-x)/2} \int_0^{(12-3x-6y)/4} dz\, dy\, dx$

Rewrite using $dy\, dx\, dz$.

**28.** $\displaystyle\int_0^3 \int_0^{\sqrt{9-x^2}} \int_0^{6-x-y} dz\, dy\, dx$

Rewrite using $dz\, dx\, dy$.

**29.** $\displaystyle\int_0^1 \int_y^1 \int_0^{\sqrt{1-y^2}} dz\, dx\, dy$

Rewrite using $dz\, dy\, dx$.

**30.** $\displaystyle\int_0^2 \int_{2x}^4 \int_0^{\sqrt{y^2 - 4x^2}} dz\, dy\, dx$

Rewrite using $dx\, dy\, dz$.

**Orders of Integration**   In Exercises 31–34, write a triple integral for $f(x, y, z) = xyz$ over the solid region $Q$ for each of the six possible orders of integration. Then evaluate one of the triple integrals.

**31.** $Q = \{(x, y, z)\colon 0 \le x \le 1, 0 \le y \le 5x, 0 \le z \le 3\}$

**32.** $Q = \{(x, y, z)\colon 0 \le x \le 2, x^2 \le y \le 4, 0 \le z \le 2 - x\}$

**33.** $Q = \{(x, y, z)\colon x^2 + y^2 \le 9, 0 \le z \le 4\}$

**34.** $Q = \{(x, y, z)\colon 0 \le x \le 1, 0 \le y \le 1 - x^2, 0 \le z \le 6\}$

**Orders of Integration** In Exercises 35 and 36, the figure shows the region of integration for the given integral. Rewrite the integral as an equivalent iterated integral in the five other orders.

**35.** $\int_0^1 \int_0^{1-y^2} \int_0^{1-y} dz\, dx\, dy$

**36.** $\int_0^3 \int_0^x \int_0^{9-x^2} dz\, dy\, dx$

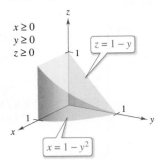

$x \geq 0$
$y \geq 0$
$z \geq 0$

$z = 1 - y$

$x = 1 - y^2$

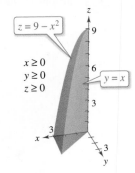

$z = 9 - x^2$

$x \geq 0$
$y \geq 0$
$z \geq 0$

$y = x$

**Center of Mass** In Exercises 37–40, find the mass and the indicated coordinate of the center of mass of the solid region $Q$ of density $\rho$ bounded by the graphs of the equations.

**37.** Find $\bar{x}$ using $\rho(x, y, z) = k$.

$Q$: $2x + 3y + 6z = 12$, $x = 0$, $y = 0$, $z = 0$

**38.** Find $\bar{y}$ using $\rho(x, y, z) = ky$.

$Q$: $3x + 3y + 5z = 15$, $x = 0$, $y = 0$, $z = 0$

**39.** Find $\bar{z}$ using $\rho(x, y, z) = kx$.

$Q$: $z = 4 - x$, $z = 0$, $y = 0$, $y = 4$, $x = 0$

**40.** Find $\bar{y}$ using $\rho(x, y, z) = k$.

$Q$: $\dfrac{x}{a} + \dfrac{y}{b} + \dfrac{z}{c} = 1 \,(a, b, c > 0)$, $x = 0$, $y = 0$, $z = 0$

**Center of Mass** In Exercises 41 and 42, set up the triple integrals for finding the mass and the center of mass of the solid of density $\rho$ bounded by the graphs of the equations. Do not evaluate the integrals.

**41.** $x = 0$, $x = b$, $y = 0$, $y = b$, $z = 0$, $z = b$, $\rho(x, y, z) = kxy$

**42.** $x = 0$, $x = a$, $y = 0$, $y = b$, $z = 0$, $z = c$, $\rho(x, y, z) = kz$

**Think About It** The center of mass of a solid of constant density is shown in the figure. In Exercises 43–46, make a conjecture about how the center of mass $(\bar{x}, \bar{y}, \bar{z})$ will change for the nonconstant density $\rho(x, y, z)$. Explain. (Make your conjecture *without* performing any calculations.)

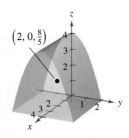

$\left(2, 0, \frac{8}{5}\right)$

**43.** $\rho(x, y, z) = kx$

**44.** $\rho(x, y, z) = kz$

**45.** $\rho(x, y, z) = k(y + 2)$

**46.** $\rho(x, y, z) = kxz^2(y + 2)^2$

**Centroid** In Exercises 47–52, find the centroid of the solid region bounded by the graphs of the equations or described by the figure. Use a computer algebra system to evaluate the triple integrals. (Assume uniform density and find the center of mass.)

**47.** $z = \dfrac{h}{r}\sqrt{x^2 + y^2}$, $z = h$

**48.** $y = \sqrt{9 - x^2}$, $z = y$, $z = 0$

**49.** $z = \sqrt{16 - x^2 - y^2}$, $z = 0$

**50.** $z = \dfrac{1}{y^2 + 1}$, $z = 0$, $x = -2$, $x = 2$, $y = 0$, $y = 1$

**51.**

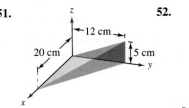

12 cm

20 cm

5 cm

**52.**

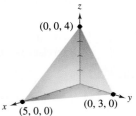

$(0, 0, 4)$

$(5, 0, 0)$

$(0, 3, 0)$

**Moments of Inertia** In Exercises 53–56, find $I_x$, $I_y$, and $I_z$ for the solid of given density. Use a computer algebra system to evaluate the triple integrals.

**53.** (a) $\rho = k$

(b) $\rho = kxyz$

**54.** (a) $\rho(x, y, z) = k$

(b) $\rho(x, y, z) = k(x^2 + y^2)$

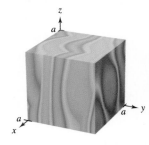

$a$

$a$

$a$

$\frac{a}{2}$

$\frac{a}{2}$

$\frac{a}{2}$

**55.** (a) $\rho(x, y, z) = k$

(b) $\rho = ky$

**56.** (a) $\rho = kz$

(b) $\rho = k(4 - z)$

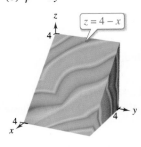

$z = 4 - x$

$z = 4 - y^2$

**Moments of Inertia** In Exercises 57 and 58, verify the moments of inertia for the solid of uniform density. Use a computer algebra system to evaluate the triple integrals.

**57.** $I_x = \frac{1}{12}m(3a^2 + L^2)$

$I_y = \frac{1}{2}ma^2$

$I_z = \frac{1}{12}m(3a^2 + L^2)$

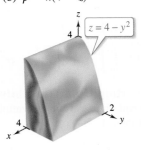

$L$

$a$

$\frac{L}{2}$

**58.** $I_x = \frac{1}{12}m(a^2 + b^2)$

   $I_y = \frac{1}{12}m(b^2 + c^2)$

   $I_z = \frac{1}{12}m(a^2 + c^2)$

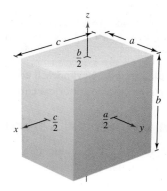

**Moments of Inertia**   In Exercises 59 and 60, set up a triple integral for the moment of inertia about the $z$-axis of the solid region $Q$ of density $\rho$. Do not evaluate the integral.

**59.** $Q = \{(x, y, z): -1 \le x \le 1, -1 \le y \le 1, 0 \le z \le 1 - x\}$

   $\rho = \sqrt{x^2 + y^2 + z^2}$

**60.** $Q = \{(x, y, z): x^2 + y^2 \le 1, 0 \le z \le 4 - x^2 - y^2\}$

   $\rho = kx^2$

**Setting Up Triple Integrals**   In Exercises 61 and 62, use the description of the solid region to set up the triple integral for (a) the mass, (b) the center of mass, and (c) the moment of inertia about the $z$-axis. Do not evaluate the integrals.

**61.** The solid bounded by $z = 4 - x^2 - y^2$ and $z = 0$ with density $\rho(x, y, z) = kz$

**62.** The solid in the first octant bounded by the coordinate planes and $x^2 + y^2 + z^2 = 25$ with density $\rho(x, y, z) = kxy$

**Average Value**   In Exercises 63–66, find the average value of the function over the given solid region. The average value of a continuous function $f(x, y, z)$ over a solid region $Q$ is

$$\text{Average value} = \frac{1}{V} \iiint\limits_{Q} f(x, y, z)\, dV$$

where $V$ is the volume of the solid region $Q$.

**63.** $f(x, y, z) = z^2 + 4$ over the cube in the first octant bounded by the coordinate planes and the planes $x = 1$, $y = 1$, and $z = 1$

**64.** $f(x, y, z) = xyz$ over the cube in the first octant bounded by the coordinate planes and the planes $x = 4$, $y = 4$, and $z = 4$

**65.** $f(x, y, z) = x + y + z$ over the tetrahedron in the first octant with vertices $(0, 0, 0)$, $(2, 0, 0)$, $(0, 2, 0)$, and $(0, 0, 2)$

**66.** $f(x, y, z) = x + y$ over the solid bounded by the sphere $x^2 + y^2 + z^2 = 3$

**EXPLORING CONCEPTS**

**67. Moment of Inertia**   Determine whether the moment of inertia about the $y$-axis of the cylinder in Exercise 57 will increase or decrease for the nonconstant density $\rho(x, y, z) = \sqrt{x^2 + z^2}$.

**68. Using Different Methods**   Find the volume of the sphere $x^2 + y^2 + z^2 = 9$ using the shell method and using a triple integral. Compare your answers.

**EXPLORING CONCEPTS   (continued)**

**69. Think About It**   Which of the integrals below is equal to $\int_1^3 \int_0^2 \int_{-1}^1 f(x, y, z)\, dz\, dy\, dx$? Explain.

(a) $\int_1^3 \int_0^2 \int_{-1}^1 f(x, y, z)\, dz\, dx\, dy$

(b) $\int_{-1}^1 \int_0^2 \int_1^3 f(x, y, z)\, dx\, dy\, dz$

(c) $\int_0^2 \int_1^3 \int_{-1}^1 f(x, y, z)\, dy\, dx\, dz$

**70. HOW DO YOU SEE IT?**   Consider two solids of equal weight, as shown below.

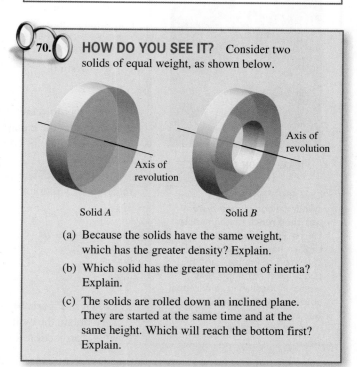

Solid $A$          Solid $B$

(a) Because the solids have the same weight, which has the greater density? Explain.

(b) Which solid has the greater moment of inertia? Explain.

(c) The solids are rolled down an inclined plane. They are started at the same time and at the same height. Which will reach the bottom first? Explain.

**71. Maximizing a Triple Integral**   Find the solid region $Q$ where the triple integral

$$\iiint\limits_{Q} (1 - 2x^2 - y^2 - 3z^2)\, dV$$

is a maximum. Use a computer algebra system to approximate the maximum value. What is the exact maximum value?

**72. Finding a Value**   Solve for $a$ in the triple integral.

$$\int_0^1 \int_0^{3-a-y^2} \int_a^{4-x-y^2} dz\, dx\, dy = \frac{14}{15}$$

**PUTNAM EXAM CHALLENGE**

**73.** Evaluate

$$\lim_{n \to \infty} \int_0^1 \int_0^1 \cdots \int_0^1 \cos^2\left\{\frac{\pi}{2n}(x_1 + x_2 + \cdots + x_n)\right\} dx_1\, dx_2 \cdots dx_n.$$

# 14.7 Triple Integrals in Other Coordinates

■ Write and evaluate a triple integral in cylindrical coordinates.
■ Write and evaluate a triple integral in spherical coordinates.

## Triple Integrals in Cylindrical Coordinates

Many common solid regions, such as spheres, ellipsoids, cones, and paraboloids, can yield difficult triple integrals in rectangular coordinates. In fact, it is precisely this difficulty that led to the introduction of nonrectangular coordinate systems. In this section, you will learn how to use *cylindrical* and *spherical* coordinates to evaluate triple integrals.

Recall from Section 11.7 that the rectangular conversion equations for cylindrical coordinates are

$$x = r \cos \theta$$
$$y = r \sin \theta$$
$$z = z.$$

An easy way to remember these conversions is to note that the equations for $x$ and $y$ are the same as in polar coordinates and $z$ is unchanged.

In this coordinate system, the simplest solid region is a cylindrical block determined by

$$r_1 \leq r \leq r_2$$
$$\theta_1 \leq \theta \leq \theta_2$$

and

$$z_1 \leq z \leq z_2$$

as shown in Figure 14.63.

**PIERRE SIMON DE LAPLACE (1749–1827)**

One of the first to use a cylindrical coordinate system was the French mathematician Pierre Simon de Laplace. Laplace has been called the "Newton of France," and he published many important works in mechanics, differential equations, and probability.
*See LarsonCalculus.com to read more of this biography.*

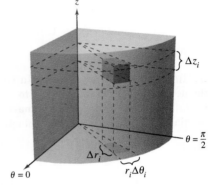

Volume of cylindrical block:
$\Delta V_i = r_i \Delta r_i \Delta \theta_i \Delta z_i$
**Figure 14.63**

To obtain the cylindrical coordinate form of a triple integral, consider a solid region $Q$ whose projection $R$ onto the $xy$-plane can be described in polar coordinates. That is,

$$Q = \{(x, y, z): (x, y) \text{ is in } R, \quad h_1(x, y) \leq z \leq h_2(x, y)\}$$

and

$$R = \{(r, \theta): \theta_1 \leq \theta \leq \theta_2, \quad g_1(\theta) \leq r \leq g_2(\theta)\}.$$

If $f$ is a continuous function on the solid $Q$, then you can write the triple integral of $f$ over $Q$ as

$$\iiint_Q f(x, y, z) \, dV = \iint_R \left[ \int_{h_1(x, y)}^{h_2(x, y)} f(x, y, z) \, dz \right] dA$$

where the double integral over $R$ is evaluated in polar coordinates. That is, $R$ is a plane region that is either $r$-simple or $\theta$-simple (see Section 14.3). If $R$ is $r$-simple, then the iterated form of the triple integral in cylindrical form is

$$\iiint_Q f(x, y, z) \, dV = \int_{\theta_1}^{\theta_2} \int_{g_1(\theta)}^{g_2(\theta)} \int_{h_1(r \cos \theta, r \sin \theta)}^{h_2(r \cos \theta, r \sin \theta)} f(r \cos \theta, r \sin \theta, z) r \, dz \, dr \, d\theta.$$

This is only one of six possible orders of integration. The other five are $dz \, d\theta \, dr$, $dr \, dz \, d\theta$, $dr \, d\theta \, dz$, $d\theta \, dz \, dr$, and $d\theta \, dr \, dz$.

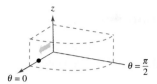

Integrate with respect to $r$.

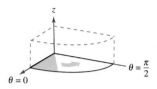

Integrate with respect to $\theta$.

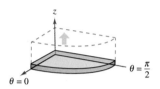

Integrate with respect to $z$.
**Figure 14.64**

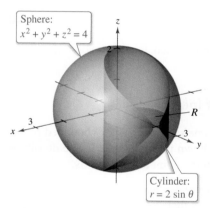

**Figure 14.65**

To visualize a particular order of integration, it helps to view the iterated integral in terms of three sweeping motions, each adding another dimension to the solid. For instance, in the order $dr\, d\theta\, dz$, the first integration occurs in the $r$-direction as a point sweeps out a ray. Then, as $\theta$ increases, the line sweeps out a sector. Finally, as $z$ increases, the sector sweeps out a solid wedge, as shown in Figure 14.64.

**Exploration**

***Volume of a Paraboloid Sector*** In the Explorations on pages 983, 992, and 1014, you were asked to summarize the different ways you know of finding the volume of the solid bounded by the paraboloid

$$z = a^2 - x^2 - y^2, \quad a > 0$$

and the $xy$-plane. You now know one more way. Use it to find the volume of the solid. Compare the different methods. What are the advantages and disadvantages of each?

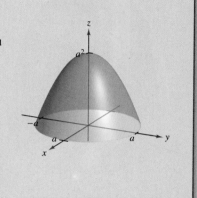

**EXAMPLE 1** **Finding Volume in Cylindrical Coordinates**

Find the volume of the solid region $Q$ cut from the sphere $x^2 + y^2 + z^2 = 4$ by the cylinder $r = 2 \sin \theta$, as shown in Figure 14.65.

**Solution** Because $x^2 + y^2 + z^2 = r^2 + z^2 = 4$, the bounds on $z$ are

$$-\sqrt{4 - r^2} \le z \le \sqrt{4 - r^2}. \qquad \text{Bounds for } z$$

Let $R$ be the circular projection of the solid onto the $r\theta$-plane. Then the bounds on $R$ are

$$0 \le r \le 2 \sin \theta \quad \text{and} \quad 0 \le \theta \le \pi. \qquad \text{Bounds for } R$$

So, the volume of $Q$ is

$$V = \int_0^\pi \int_0^{2 \sin \theta} \int_{-\sqrt{4-r^2}}^{\sqrt{4-r^2}} r\, dz\, dr\, d\theta \qquad \text{Apply formula for volume.}$$

$$= 2 \int_0^{\pi/2} \int_0^{2 \sin \theta} \int_{-\sqrt{4-r^2}}^{\sqrt{4-r^2}} r\, dz\, dr\, d\theta \qquad \text{Use symmetry to rewrite bounds for } \theta.$$

$$= 2 \int_0^{\pi/2} \int_0^{2 \sin \theta} 2r\sqrt{4 - r^2}\, dr\, d\theta \qquad \text{Integrate with respect to } z.$$

$$= 2 \int_0^{\pi/2} -\frac{2}{3} (4 - r^2)^{3/2} \Big]_0^{2 \sin \theta} d\theta \qquad \text{Integrate with respect to } r.$$

$$= \frac{4}{3} \int_0^{\pi/2} (8 - 8 \cos^3 \theta)\, d\theta$$

$$= \frac{32}{3} \int_0^{\pi/2} [1 - (\cos \theta)(1 - \sin^2 \theta)]\, d\theta \qquad \begin{array}{l}\text{Factor and use trigonometric}\\ \text{identity } \cos^2 \theta = 1 - \sin^2 \theta.\end{array}$$

$$= \frac{32}{3} \left[ \theta - \sin \theta + \frac{\sin^3 \theta}{3} \right]_0^{\pi/2} \qquad \text{Integrate with respect to } \theta.$$

$$= \frac{16}{9}(3\pi - 4)$$

$$\approx 9.644.$$

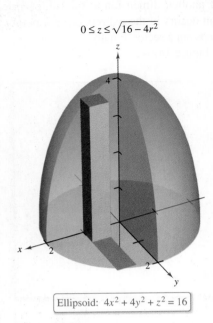

$0 \le z \le \sqrt{16 - 4r^2}$

Ellipsoid: $4x^2 + 4y^2 + z^2 = 16$

**Figure 14.66**

**EXAMPLE 2**    **Finding Mass in Cylindrical Coordinates**

Find the mass of the ellipsoid $Q$ given by $4x^2 + 4y^2 + z^2 = 16$, lying above the $xy$-plane. The density at a point in the solid is proportional to the distance between the point and the $xy$-plane.

**Solution**    The density function is $\rho(r, \theta, z) = kz$, where $k$ is the constant of proportionality. The bounds on $z$ are

$$0 \le z \le \sqrt{16 - 4x^2 - 4y^2} = \sqrt{16 - 4r^2}$$

where $0 \le r \le 2$ and $0 \le \theta \le 2\pi$, as shown in Figure 14.66. The mass of the solid is

$$m = \int_0^{2\pi} \int_0^2 \int_0^{\sqrt{16-4r^2}} kzr \, dz \, dr \, d\theta \qquad \text{Apply formula for mass of a solid.}$$

$$= \frac{k}{2} \int_0^{2\pi} \int_0^2 z^2 r \Big]_0^{\sqrt{16-4r^2}} dr \, d\theta \qquad \text{Integrate with respect to } z.$$

$$= \frac{k}{2} \int_0^{2\pi} \int_0^2 (16r - 4r^3) \, dr \, d\theta$$

$$= \frac{k}{2} \int_0^{2\pi} \left[ 8r^2 - r^4 \right]_0^2 d\theta \qquad \text{Integrate with respect to } r.$$

$$= 8k \int_0^{2\pi} d\theta$$

$$= 16\pi k. \qquad \text{Integrate with respect to } \theta.$$

Integration in cylindrical coordinates is useful when factors involving $x^2 + y^2$ appear in the integrand, as illustrated in Example 3.

**EXAMPLE 3**    **Finding a Moment of Inertia**

Find the moment of inertia about the axis of symmetry of the solid $Q$ bounded by the paraboloid $z = x^2 + y^2$ and the plane $z = 4$, as shown in Figure 14.67. The density at each point is proportional to the distance between the point and the $z$-axis.

**Solution**    Because the $z$-axis is the axis of symmetry and $\rho(x, y, z) = k\sqrt{x^2 + y^2}$, where $k$ is the constant of proportionality, it follows that

$$I_z = \iiint_Q k(x^2 + y^2)\sqrt{x^2 + y^2} \, dV. \qquad \text{Moment of inertia about } z\text{-axis}$$

In cylindrical coordinates, $0 \le r \le \sqrt{x^2 + y^2} = \sqrt{z}$ and $0 \le \theta \le 2\pi$. So, you have

$$I_z = k \int_0^4 \int_0^{2\pi} \int_0^{\sqrt{z}} r^2(r)r \, dr \, d\theta \, dz \qquad \text{Cylindrical coordinates}$$

$$= k \int_0^4 \int_0^{2\pi} \frac{r^5}{5} \Big]_0^{\sqrt{z}} d\theta \, dz \qquad \text{Integrate with respect to } r.$$

$$= k \int_0^4 \int_0^{2\pi} \frac{z^{5/2}}{5} d\theta \, dz$$

$$= \frac{k}{5} \int_0^4 z^{5/2} (2\pi) \, dz \qquad \text{Integrate with respect to } \theta.$$

$$= \frac{2\pi k}{5} \left[ \frac{2}{7} z^{7/2} \right]_0^4 \qquad \text{Integrate with respect to } z.$$

$$= \frac{512k\pi}{35}.$$

$Q$: Bounded by
$z = x^2 + y^2$
and
$z = 4$

**Figure 14.67**

## Triple Integrals in Spherical Coordinates

Triple integrals involving spheres or cones are often easier to evaluate by converting to spherical coordinates. Recall from Section 11.7 that the rectangular conversion equations for spherical coordinates are

$$x = \rho \sin \phi \cos \theta$$
$$y = \rho \sin \phi \sin \theta$$
$$z = \rho \cos \phi.$$

.  .  .  .  .  .  .  .  .  .  .  .  .  .  .  .  ▷

**·· REMARK** The Greek letter $\rho$ used in spherical coordinates is not related to density. Rather, it is the three-dimensional analog of the $r$ used in polar coordinates. For problems involving spherical coordinates and a density function, this text uses a different symbol to denote density.

In this coordinate system, the simplest region is a spherical block determined by

$$\{(\rho, \theta, \phi): \rho_1 \le \rho \le \rho_2, \quad \theta_1 \le \theta \le \theta_2, \quad \phi_1 \le \phi \le \phi_2\}$$

where $\rho_1 \ge 0$, $\theta_2 - \theta_1 \le 2\pi$, and $0 \le \phi_1 \le \phi_2 \le \pi$, as shown in Figure 14.68. If $(\rho, \theta, \phi)$ is a point in the interior of such a block, then the volume of the block can be approximated by $\Delta V \approx \rho^2 \sin \phi \Delta\rho \Delta\phi \Delta\theta$. (See Exercise 8 in the Problem Solving exercises at the end of this chapter.)

Using the usual process involving an inner partition, summation, and a limit, you can develop a triple integral in spherical coordinates for a continuous function $f$ defined on the solid region $Q$. This formula, shown below, can be modified for different orders of integration and generalized to include regions with variable boundaries.

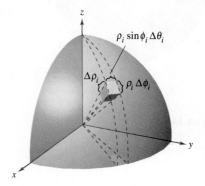

Spherical block: $\Delta V_i \approx \rho_i^2 \sin \phi_i \Delta\rho_i \Delta\phi_i \Delta\theta_i$
**Figure 14.68**

$$\iiint\limits_{Q} f(x, y, z)\, dV = \int_{\theta_1}^{\theta_2} \int_{\phi_1}^{\phi_2} \int_{\rho_1}^{\rho_2} f(\rho \sin \phi \cos \theta, \rho \sin \phi \sin \theta, \rho \cos \phi)\rho^2 \sin \phi\, d\rho\, d\phi\, d\theta$$

Like triple integrals in cylindrical coordinates, triple integrals in spherical coordinates are evaluated with iterated integrals. As with cylindrical coordinates, you can visualize a particular order of integration by viewing the iterated integral in terms of three sweeping motions, each adding another dimension to the solid. For instance, the iterated integral

$$\int_0^{2\pi} \int_0^{\pi/4} \int_0^3 \rho^2 \sin \phi\, d\rho\, d\phi\, d\theta$$

(which is used in Example 4) is illustrated in Figure 14.69.

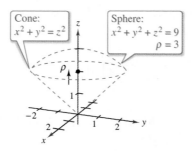

$\rho$ varies from 0 to 3 with $\phi$ and $\theta$ held constant.
**Figure 14.69**

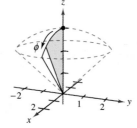

$\phi$ varies from 0 to $\pi/4$ with $\theta$ held constant.

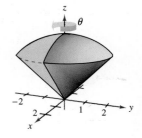

$\theta$ varies from 0 to $2\pi$.

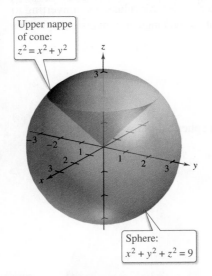

Upper nappe of cone:
$z^2 = x^2 + y^2$

Sphere:
$x^2 + y^2 + z^2 = 9$

**Figure 14.70**

**EXAMPLE 4**   **Finding Volume in Spherical Coordinates**

Find the volume of the solid region $Q$ bounded below by the upper nappe of the cone $z^2 = x^2 + y^2$ and above by the sphere $x^2 + y^2 + z^2 = 9$, as shown in Figure 14.70.

**Solution**   In spherical coordinates, the equation of the sphere is

$$\rho^2 = x^2 + y^2 + z^2 = 9 \quad \Longrightarrow \quad \rho = 3.$$

Furthermore, the sphere and cone intersect when

$$(x^2 + y^2) + z^2 = (z^2) + z^2 = 9 \quad \Longrightarrow \quad z = \frac{3}{\sqrt{2}}$$

and, because $z = \rho \cos \phi$, it follows that

$$\left(\frac{3}{\sqrt{2}}\right)\left(\frac{1}{3}\right) = \cos \phi \quad \Longrightarrow \quad \phi = \frac{\pi}{4}.$$

Consequently, you can use the integration order $d\rho \, d\phi \, d\theta$, where $0 \le \rho \le 3$, $0 \le \phi \le \pi/4$, and $0 \le \theta \le 2\pi$. The volume is

$$\begin{aligned}
V &= \int_0^{2\pi}\int_0^{\pi/4}\int_0^3 \rho^2 \sin \phi \, d\rho \, d\phi \, d\theta & &\text{Apply formula for volume.}\\
&= \int_0^{2\pi}\int_0^{\pi/4} 9 \sin \phi \, d\phi \, d\theta & &\text{Integrate with respect to } \rho.\\
&= 9\int_0^{2\pi} -\cos \phi \Big]_0^{\pi/4} d\theta & &\text{Integrate with respect to } \phi.\\
&= 9\int_0^{2\pi}\left(1 - \frac{\sqrt{2}}{2}\right) d\theta & &\\
&= 9\pi(2 - \sqrt{2}) & &\text{Integrate with respect to } \theta.\\
&\approx 16.563.
\end{aligned}$$

**EXAMPLE 5**   **Finding the Center of Mass of a Solid Region**

$\cdots \triangleright$ *See LarsonCalculus.com for an interactive version of this type of example.*

Find the center of mass of the solid region $Q$ of uniform density from Example 4.

**Solution**   Because the density is uniform, you can consider the density at the point $(x, y, z)$ to be $k$. By symmetry, the center of mass lies on the $z$-axis, and you need only calculate $\bar{z} = M_{xy}/m$, where $m = kV = 9k\pi(2 - \sqrt{2})$. Because $z = \rho \cos \phi$, it follows that

$$\begin{aligned}
M_{xy} &= \iiint_Q kz \, dV = k\int_0^3\int_0^{2\pi}\int_0^{\pi/4} (\rho \cos \phi)\rho^2 \sin \phi \, d\phi \, d\theta \, d\rho\\
&= k\int_0^3\int_0^{2\pi} \rho^3 \frac{\sin^2 \phi}{2}\Big]_0^{\pi/4} d\theta \, d\rho\\
&= \frac{k}{4}\int_0^3\int_0^{2\pi} \rho^3 \, d\theta \, d\rho = \frac{k\pi}{2}\int_0^3 \rho^3 \, d\rho = \frac{81k\pi}{8}.
\end{aligned}$$

So,

$$\bar{z} = \frac{M_{xy}}{m} = \frac{81k\pi/8}{9k\pi(2 - \sqrt{2})} = \frac{9(2 + \sqrt{2})}{16} \approx 1.920$$

and the center of mass is approximately $(0, 0, 1.920)$.

# 14.7 Exercises

**CONCEPT CHECK**

**1. Volume** Explain why triple integrals that represent the volumes of solids are sometimes easier to evaluate in cylindrical or spherical coordinates instead of rectangular coordinates.

**2. Differential of Volume** What is the differential of volume, $dV$, for (a) cylindrical coordinates and (b) spherical coordinates? Choose one order of integration for each system.

**Evaluating a Triple Iterated Integral** In Exercises 3–8, evaluate the triple iterated integral.

3. $\displaystyle\int_{-1}^{5}\int_{0}^{\pi/2}\int_{0}^{3} r\cos\theta\, dr\, d\theta\, dz$

4. $\displaystyle\int_{0}^{\pi/4}\int_{0}^{6}\int_{0}^{6-r} rz\, dz\, dr\, d\theta$

5. $\displaystyle\int_{0}^{\pi/2}\int_{0}^{\cos\theta}\int_{0}^{3+r^2} 2r\sin\theta\, dz\, dr\, d\theta$

6. $\displaystyle\int_{0}^{\pi/2}\int_{0}^{\pi}\int_{0}^{2} e^{-\rho^3}\rho^2\, d\rho\, d\theta\, d\phi$

7. $\displaystyle\int_{0}^{2\pi}\int_{0}^{\pi/2}\int_{0}^{\sin\phi} \rho\cos\phi\, d\rho\, d\phi\, d\theta$

8. $\displaystyle\int_{0}^{\pi/4}\int_{0}^{\pi/4}\int_{0}^{\cos\theta} \rho^2\sin\phi\cos\phi\, d\rho\, d\theta\, d\phi$

 **Evaluating a Triple Iterated Integral Using Technology** In Exercises 9 and 10, use a computer algebra system to evaluate the triple iterated integral.

9. $\displaystyle\int_{0}^{4}\int_{0}^{z}\int_{0}^{\pi/2} re^r\, d\theta\, dr\, dz$

10. $\displaystyle\int_{0}^{\pi/2}\int_{0}^{\pi}\int_{0}^{\sin\theta} 2\rho^2\cos\phi\, d\rho\, d\theta\, d\phi$

**Volume** In Exercises 11–14, sketch the solid region whose volume is given by the iterated integral and evaluate the iterated integral.

11. $\displaystyle\int_{0}^{\pi/2}\int_{0}^{3}\int_{0}^{e^{-r^2}} r\, dz\, dr\, d\theta$

12. $\displaystyle\int_{0}^{2\pi}\int_{0}^{2\sqrt{2}}\int_{r^2-2}^{6} r\, dz\, dr\, d\theta$

13. $\displaystyle\int_{0}^{2\pi}\int_{\pi/6}^{\pi/2}\int_{0}^{4} \rho^2\sin\phi\, d\rho\, d\phi\, d\theta$

14. $\displaystyle\int_{0}^{2\pi}\int_{0}^{\pi}\int_{2}^{5} \rho^2\sin\phi\, d\rho\, d\phi\, d\theta$

 **Volume** In Exercises 15–20, use cylindrical coordinates to find the volume of the solid.

15. Solid inside both $x^2 + y^2 + z^2 = 36$ and $(x-3)^2 + y^2 = 9$

16. Solid inside $x^2 + y^2 + z^2 = 16$ and outside $z = \sqrt{x^2 + y^2}$

17. Solid bounded above by $z = 2x$ and below by $z = 2x^2 + 2y^2$

18. Solid bounded above by $z = 2 - x^2 - y^2$ and below by $z = x^2 + y^2$

19. Solid bounded by the graphs of the sphere $r^2 + z^2 = 25$ and the cylinder $r = 5\cos\theta$

20. Solid inside the sphere $x^2 + y^2 + z^2 = 4$ and above the upper nappe of the cone $z^2 = x^2 + y^2$

 **Mass** In Exercises 21 and 22, use cylindrical coordinates to find the mass of the solid $Q$ of density $\rho$.

21. $Q = \{(x, y, z): 0 \le z \le 9 - x - 2y,\ x^2 + y^2 \le 4\}$
$\rho(x, y, z) = k\sqrt{x^2 + y^2}$

22. $Q = \{(x, y, z): 0 \le z \le 12e^{-(x^2+y^2)},\ x^2 + y^2 \le 4,\ x \ge 0,\ y \ge 0\}$
$\rho(x, y, z) = k$

**Using Cylindrical Coordinates** In Exercises 23–28, use cylindrical coordinates to find the indicated characteristic of the cone shown in the figure.

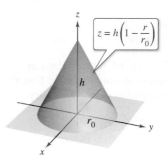

23. Find the volume of the cone.

24. Find the centroid of the cone.

25. Find the center of mass of the cone, assuming that its density at any point is proportional to the distance between the point and the axis of the cone. Use a computer algebra system to evaluate the triple integral.

26. Find the center of mass of the cone, assuming that its density at any point is proportional to the distance between the point and the base. Use a computer algebra system to evaluate the triple integral.

27. Assume that the cone has uniform density and show that the moment of inertia about the $z$-axis is

$$I_z = \tfrac{3}{10}mr_0^2.$$

28. Assume that the density of the cone is $\rho(x, y, z) = k\sqrt{x^2 + y^2}$ and find the moment of inertia about the $z$-axis.

**Moment of Inertia** In Exercises 29 and 30, use cylindrical coordinates to verify the given moment of inertia of the solid of uniform density.

29. Cylindrical shell: $I_z = \tfrac{1}{2}m(a^2 + b^2)$

$0 < a \le r \le b,\quad 0 \le z \le h$

**30.** Right circular cylinder: $I_z = \frac{3}{2}ma^2$

$r = 2a \sin \theta, \quad 0 \le z \le h$

(Use a computer algebra system to evaluate the triple integral.)

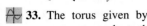 **Volume** In Exercises 31–34, use spherical coordinates to find the volume of the solid.

**31.** Solid inside $x^2 + y^2 + z^2 = 9$, outside $z = \sqrt{x^2 + y^2}$, and above the $xy$-plane

**32.** Solid bounded above by $x^2 + y^2 + z^2 = z$ and below by $z = \sqrt{x^2 + y^2}$

**33.** The torus given by $\rho = 4 \sin \phi$ (Use a computer algebra system to evaluate the triple integral.)

**34.** The solid between the spheres

$$x^2 + y^2 + z^2 = a^2 \quad \text{and} \quad x^2 + y^2 + z^2 = b^2, b > a,$$

and inside the cone $z^2 = x^2 + y^2$

**Mass** In Exercises 35 and 36, use spherical coordinates to find the mass of the sphere $x^2 + y^2 + z^2 = a^2$ with the given density.

**35.** The density at any point is proportional to the distance between the point and the origin.

**36.** The density at any point is proportional to the distance between the point and the $z$-axis.

 **Center of Mass** In Exercises 37 and 38, use spherical coordinates to find the center of mass of the solid of uniform density.

**37.** Hemispherical solid of radius $r$

**38.** Solid lying between two concentric hemispheres of radii $r$ and $R$, where $r < R$

**Moment of Inertia** In Exercises 39 and 40, use spherical coordinates to find the moment of inertia about the $z$-axis of the solid of uniform density.

**39.** Solid bounded by the hemisphere $\rho = \cos \phi$, $\frac{\pi}{4} \le \phi \le \frac{\pi}{2}$, and the cone $\phi = \frac{\pi}{4}$

**40.** Solid lying between two concentric hemispheres of radii $r$ and $R$, where $r < R$

**Converting Coordinates** In Exercises 41–44, convert the integral from rectangular coordinates to both cylindrical and spherical coordinates, and evaluate the simplest iterated integral.

**41.** $\displaystyle\int_{-2}^{2}\int_{-\sqrt{4-x^2}}^{\sqrt{4-x^2}}\int_{x^2+y^2}^{4} x\, dz\, dy\, dx$

**42.** $\displaystyle\int_{0}^{2}\int_{0}^{\sqrt{4-x^2}}\int_{0}^{\sqrt{16-x^2-y^2}} \sqrt{x^2+y^2}\, dz\, dy\, dx$

**43.** $\displaystyle\int_{-1}^{1}\int_{-\sqrt{1-x^2}}^{\sqrt{1-x^2}}\int_{1}^{1+\sqrt{1-x^2-y^2}} x\, dz\, dy\, dx$

**44.** $\displaystyle\int_{0}^{3}\int_{0}^{\sqrt{9-x^2}}\int_{0}^{\sqrt{9-x^2-y^2}} \sqrt{x^2+y^2+z^2}\, dz\, dy\, dx$

**EXPLORING CONCEPTS**

**45. Using Coordinates** Describe the surface whose equation is a coordinate equal to a constant for each of the coordinates in (a) the cylindrical coordinate system and (b) the spherical coordinate system.

**46.** **HOW DO YOU SEE IT?** The solid is bounded below by the upper nappe of a cone and above by a sphere (see figure). Would it be easier to use cylindrical coordinates or spherical coordinates to find the volume of the solid? Explain.

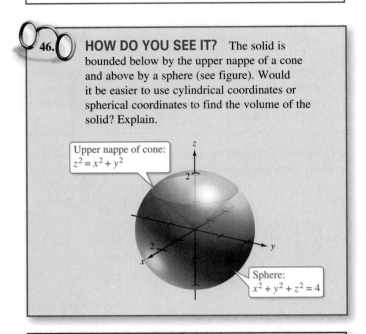

Upper nappe of cone: $z^2 = x^2 + y^2$

Sphere: $x^2 + y^2 + z^2 = 4$

**PUTNAM EXAM CHALLENGE**

**47.** Find the volume of the region of points $(x, y, z)$ such that $(x^2 + y^2 + z^2 + 8)^2 \le 36(x^2 + y^2)$.

This problem was composed by the Committee on the Putnam Prize Competition.
© The Mathematical Association of America. All rights reserved.

**SECTION PROJECT**

## Wrinkled and Bumpy Spheres

In parts (a) and (b), find the volume of the wrinkled sphere or bumpy sphere. These solids are used as models for tumors.

(a) Wrinkled sphere

$\rho = 1 + 0.2 \sin 8\theta \sin \phi$

$0 \le \theta \le 2\pi, 0 \le \phi \le \pi$

(b) Bumpy sphere

$\rho = 1 + 0.2 \sin 8\theta \sin 4\phi$

$0 \le \theta \le 2\pi, 0 \le \phi \le \pi$

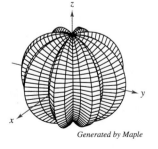

*Generated by Maple*

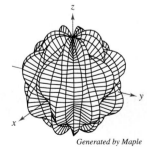

*Generated by Maple*

**■ FOR FURTHER INFORMATION** For more information on these types of spheres, see the article "Heat Therapy for Tumors" by Leah Edelstein-Keshet in *The UMAP Journal*.

# 14.8 Change of Variables: Jacobians

■ Understand the concept of a Jacobian.
■ Use a Jacobian to change variables in a double integral.

## Jacobians

**CARL GUSTAV JACOBI**
**(1804–1851)**

The Jacobian is named after the German mathematician Carl Gustav Jacobi. Jacobi is known for his work in many areas of mathematics, but his interest in integration stemmed from the problem of finding the circumference of an ellipse. *See LarsonCalculus.com to read more of this biography.*

For the single integral

$$\int_a^b f(x)\, dx$$

you can change variables by letting $x = g(u)$, so that $dx = g'(u)\, du$, and obtain

$$\int_a^b f(x)\, dx = \int_c^d f(g(u))g'(u)\, du$$

where $a = g(c)$ and $b = g(d)$. Note that the change of variables process introduces an additional factor $g'(u)$ into the integrand. This also occurs in the case of double integrals

$$\int_R \int f(x, y)\, dA = \int_S \int f(g(u, v), h(u, v)) \underbrace{\left| \frac{\partial x}{\partial u} \frac{\partial y}{\partial v} - \frac{\partial y}{\partial u} \frac{\partial x}{\partial v} \right|}_{\text{Jacobian}}\, du\, dv$$

where the change of variables $x = g(u, v)$ and $y = h(u, v)$ introduces a factor called the Jacobian of $x$ and $y$ with respect to $u$ and $v$. In defining the Jacobian, it is convenient to use the determinant notation shown below.

---

### Definition of the Jacobian

If $x = g(u, v)$ and $y = h(u, v)$, then the **Jacobian** of $x$ and $y$ with respect to $u$ and $v$, denoted by $\partial(x, y)/\partial(u, v)$, is

$$\frac{\partial(x, y)}{\partial(u, v)} = \begin{vmatrix} \dfrac{\partial x}{\partial u} & \dfrac{\partial x}{\partial v} \\ \dfrac{\partial y}{\partial u} & \dfrac{\partial y}{\partial v} \end{vmatrix} = \frac{\partial x}{\partial u} \frac{\partial y}{\partial v} - \frac{\partial y}{\partial u} \frac{\partial x}{\partial v}.$$

---

**EXAMPLE 1** **The Jacobian for Rectangular-to-Polar Conversion**

Find the Jacobian for the change of variables defined by

$$x = r \cos \theta \quad \text{and} \quad y = r \sin \theta.$$

**Solution** From the definition of the Jacobian, you obtain

$$\frac{\partial(x, y)}{\partial(r, \theta)} = \begin{vmatrix} \dfrac{\partial x}{\partial r} & \dfrac{\partial x}{\partial \theta} \\ \dfrac{\partial y}{\partial r} & \dfrac{\partial y}{\partial \theta} \end{vmatrix} \qquad \text{Definition of Jacobian}$$

$$= \begin{vmatrix} \cos \theta & -r \sin \theta \\ \sin \theta & r \cos \theta \end{vmatrix} \qquad \text{Substitute.}$$

$$= r \cos^2 \theta + r \sin^2 \theta \qquad \text{Find determinant.}$$

$$= r(\cos^2 \theta + \sin^2 \theta) \qquad \text{Factor.}$$

$$= r \qquad \text{Trigonometric identity}$$

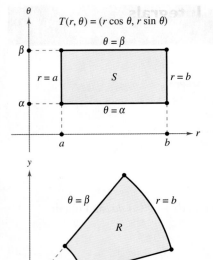

$T(r, \theta) = (r \cos \theta, r \sin \theta)$

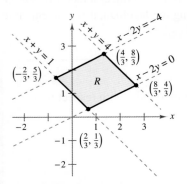

S in the region in the $r\theta$-plane that corresponds to R in the xy-plane.

**Figure 14.71**

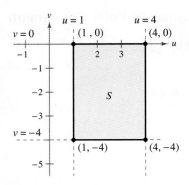

Region R in the xy-plane

**Figure 14.72**

Example 1 points out that the change of variables from rectangular to polar coordinates for a double integral can be written as

$$\iint_R f(x, y)\, dA = \iint_S f(r \cos \theta, r \sin \theta) r\, dr\, d\theta, \; r > 0$$

$$= \iint_S f(r \cos \theta, r \sin \theta) \left| \frac{\partial(x, y)}{\partial(r, \theta)} \right| dr\, d\theta$$

where S is the region in the $r\theta$-plane that corresponds to the region R in the xy-plane, as shown in Figure 14.71. This formula is similar to that found in Theorem 14.3 on page 992.

In general, a change of variables using a one-to-one **transformation** T from a region S in the uv-plane to a region R in the xy-plane is given by

$$T(u, v) = (x, y) = (g(u, v), h(u, v))$$

where g and h have continuous first partial derivatives in the region S. Note that the point $(u, v)$ lies in S and the point $(x, y)$ lies in R. In most cases, you are hunting for a transformation in which the region S is simpler than the region R.

---

**EXAMPLE 2** **Finding a Change of Variables to Simplify a Region**

Let R be the region bounded by the lines

$$x - 2y = 0, \quad x - 2y = -4, \quad x + y = 4, \quad \text{and} \quad x + y = 1$$

as shown in Figure 14.72. Find a transformation T from a region S to R such that S is a rectangular region (with sides parallel to the u- or v-axis).

**Solution** To begin, let $u = x + y$ and $v = x - 2y$. Solving this system of equations for x and y produces $T(u, v) = (x, y)$, where

$$x = \frac{1}{3}(2u + v) \quad \text{and} \quad y = \frac{1}{3}(u - v).$$

The four boundaries for R in the xy-plane give rise to the following bounds for S in the uv-plane.

| Bounds in the xy-Plane | | Bounds in the uv-Plane |
|---|---|---|
| $x + y = 1$ | ⇨ | $u = 1$ |
| $x + y = 4$ | ⇨ | $u = 4$ |
| $x - 2y = 0$ | ⇨ | $v = 0$ |
| $x - 2y = -4$ | ⇨ | $v = -4$ |

The region S is shown in Figure 14.73. Note that the transformation

$$T(u, v) = (x, y) = \left( \frac{1}{3}[2u + v], \frac{1}{3}[u - v] \right)$$

maps the vertices of the region S onto the vertices of the region R, as shown below.

$$T(1, 0) = \left( \frac{1}{3}[2(1) + 0], \frac{1}{3}[1 - 0] \right) = \left( \frac{2}{3}, \frac{1}{3} \right)$$

$$T(4, 0) = \left( \frac{1}{3}[2(4) + 0], \frac{1}{3}[4 - 0] \right) = \left( \frac{8}{3}, \frac{4}{3} \right)$$

$$T(4, -4) = \left( \frac{1}{3}[2(4) - 4], \frac{1}{3}[4 - (-4)] \right) = \left( \frac{4}{3}, \frac{8}{3} \right)$$

$$T(1, -4) = \left( \frac{1}{3}[2(1) - 4], \frac{1}{3}[1 - (-4)] \right) = \left( -\frac{2}{3}, \frac{5}{3} \right)$$

Region S in the uv-plane

**Figure 14.73**

## Change of Variables for Double Integrals

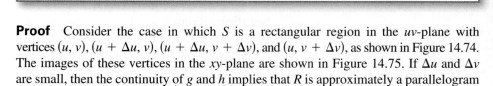

> **THEOREM 14.5   Change of Variables for Double Integrals**
>
> Let $R$ be a vertically or horizontally simple region in the $xy$-plane, and let $S$ be a vertically or horizontally simple region in the $uv$-plane. Let $T$ from $S$ to $R$ be given by $T(u, v) = (x, y) = (g(u, v), h(u, v))$, where $g$ and $h$ have continuous first partial derivatives. Assume that $T$ is one-to-one except possibly on the boundary of $S$. If $f$ is continuous on $R$, and $\partial(x, y)/\partial(u, v)$ is nonzero on $S$, then
>
> $$\int_R \int f(x, y) \, dx \, dy = \int_S \int f(g(u, v), h(u, v)) \left| \frac{\partial(x, y)}{\partial(u, v)} \right| \, du \, dv.$$

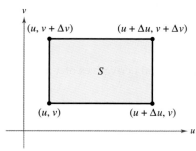

Area of $S = \Delta u \, \Delta v$
$\Delta u > 0, \Delta v > 0$
**Figure 14.74**

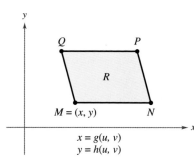

The vertices in the $xy$-plane are
$M(g(u, v), h(u, v))$,
$N(g(u + \Delta u, v), h(u + \Delta u, v))$,
$P(g(u + \Delta u, v + \Delta v),$
$h(u + \Delta u, v + \Delta v))$, and
$Q(g(u, v + \Delta v), h(u, v + \Delta v))$.
**Figure 14.75**

**Proof**   Consider the case in which $S$ is a rectangular region in the $uv$-plane with vertices $(u, v), (u + \Delta u, v), (u + \Delta u, v + \Delta v)$, and $(u, v + \Delta v)$, as shown in Figure 14.74. The images of these vertices in the $xy$-plane are shown in Figure 14.75. If $\Delta u$ and $\Delta v$ are small, then the continuity of $g$ and $h$ implies that $R$ is approximately a parallelogram determined by the vectors $\overrightarrow{MN}$ and $\overrightarrow{MQ}$. So, the area of $R$ is

$$\Delta A \approx \|\overrightarrow{MN} \times \overrightarrow{MQ}\|.$$

Moreover, for small $\Delta u$ and $\Delta v$, the partial derivatives of $g$ and $h$ with respect to $u$ can be approximated by

$$g_u(u, v) \approx \frac{g(u + \Delta u, v) - g(u, v)}{\Delta u} \quad \text{and} \quad h_u(u, v) \approx \frac{h(u + \Delta u, v) - h(u, v)}{\Delta u}.$$

Consequently,

$$\begin{aligned}
\overrightarrow{MN} &= [g(u + \Delta u, v) - g(u, v)]\mathbf{i} + [h(u + \Delta u, v) - h(u, v)]\mathbf{j} \\
&\approx [g_u(u, v) \, \Delta u]\mathbf{i} + [h_u(u, v) \, \Delta u]\mathbf{j} \\
&= \frac{\partial x}{\partial u} \Delta u \mathbf{i} + \frac{\partial y}{\partial u} \Delta u \mathbf{j}.
\end{aligned}$$

Similarly, you can approximate $\overrightarrow{MQ}$ by $\dfrac{\partial x}{\partial v} \Delta v \mathbf{i} + \dfrac{\partial y}{\partial v} \Delta v \mathbf{j}$, which implies that

$$\overrightarrow{MN} \times \overrightarrow{MQ} \approx
\begin{vmatrix}
\mathbf{i} & \mathbf{j} & \mathbf{k} \\
\dfrac{\partial x}{\partial u} \Delta u & \dfrac{\partial y}{\partial u} \Delta u & 0 \\
\dfrac{\partial x}{\partial v} \Delta v & \dfrac{\partial y}{\partial v} \Delta v & 0
\end{vmatrix}
=
\begin{vmatrix}
\dfrac{\partial x}{\partial u} & \dfrac{\partial y}{\partial u} \\
\dfrac{\partial x}{\partial v} & \dfrac{\partial y}{\partial v}
\end{vmatrix}
\Delta u \, \Delta v \mathbf{k}.$$

It follows that, in Jacobian notation,

$$\Delta A \approx \|\overrightarrow{MN} \times \overrightarrow{MQ}\| \approx \left| \frac{\partial(x, y)}{\partial(u, v)} \right| \Delta u \, \Delta v.$$

Because this approximation improves as $\Delta u$ and $\Delta v$ approach 0, the limiting case can be written as

$$dA \approx \|\overrightarrow{MN} \times \overrightarrow{MQ}\| \approx \left| \frac{\partial(x, y)}{\partial(u, v)} \right| \, du \, dv.$$

So,

$$\int_R \int f(x, y) \, dx \, dy = \int_S \int f(g(u, v), h(u, v)) \left| \frac{\partial(x, y)}{\partial(u, v)} \right| \, du \, dv. \qquad \blacksquare$$

The next two examples show how a change of variables can simplify the integration process. The simplification can occur in various ways. You can make a change of variables to simplify either the *region R* or the *integrand f(x, y)*, or both.

### EXAMPLE 3   Using a Change of Variables to Simplify a Region

• • • • ▷ *See LarsonCalculus.com for an interactive version of this type of example.*

Let $R$ be the region bounded by the lines

$$x - 2y = 0, \quad x - 2y = -4, \quad x + y = 4, \quad \text{and} \quad x + y = 1$$

as shown in Figure 14.76. Evaluate the double integral

$$\int_R \int 3xy \, dA.$$

**Solution**   From Example 2, you can use the change of variables

$$x = \frac{1}{3}(2u + v) \quad \text{and} \quad y = \frac{1}{3}(u - v).$$

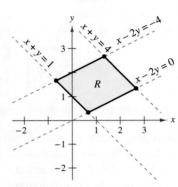

**Figure 14.76**

(Note that the region $S$ is shown in Figure 14.77.) The partial derivatives of $x$ and $y$ are

$$\frac{\partial x}{\partial u} = \frac{2}{3}, \quad \frac{\partial x}{\partial v} = \frac{1}{3}, \quad \frac{\partial y}{\partial u} = \frac{1}{3}, \quad \text{and} \quad \frac{\partial y}{\partial v} = -\frac{1}{3}$$

which implies that the Jacobian is

$$\frac{\partial(x, y)}{\partial(u, v)} = \begin{vmatrix} \dfrac{\partial x}{\partial u} & \dfrac{\partial x}{\partial v} \\ \dfrac{\partial y}{\partial u} & \dfrac{\partial y}{\partial v} \end{vmatrix}$$

$$= \begin{vmatrix} \dfrac{2}{3} & \dfrac{1}{3} \\ \dfrac{1}{3} & -\dfrac{1}{3} \end{vmatrix}$$

$$= -\frac{2}{9} - \frac{1}{9}$$

$$= -\frac{1}{3}.$$

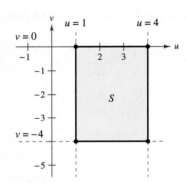

**Figure 14.77**

So, by Theorem 14.5, you obtain

$$\int_R \int 3xy \, dA = \int_S \int 3\left[\frac{1}{3}(2u + v)\frac{1}{3}(u - v)\right]\left|\frac{\partial(x, y)}{\partial(u, v)}\right| dv \, du$$

$$= \int_1^4 \int_{-4}^0 \frac{1}{9}(2u^2 - uv - v^2) \, dv \, du$$

$$= \frac{1}{9}\int_1^4 \left[2u^2v - \frac{uv^2}{2} - \frac{v^3}{3}\right]_{-4}^0 du$$

$$= \frac{1}{9}\int_1^4 \left(8u^2 + 8u - \frac{64}{3}\right) du$$

$$= \frac{1}{9}\left[\frac{8u^3}{3} + 4u^2 - \frac{64}{3}u\right]_1^4$$

$$= \frac{164}{9}.$$

EXAMPLE 4   **Change of Variables: Simplifying an Integrand**

Let $R$ be the region bounded by the square with vertices $(0, 1)$, $(1, 2)$, $(2, 1)$, and $(1, 0)$. Evaluate the integral

$$\iint_R (x + y)^2 \sin^2(x - y) \, dA.$$

**Solution**   Note that the sides of $R$ lie on the lines $x + y = 1$, $x - y = 1$, $x + y = 3$, and $x - y = -1$, as shown in Figure 14.78. Letting $u = x + y$ and $v = x - y$, you can determine the bounds for region $S$ in the $uv$-plane to be

$$1 \le u \le 3 \quad \text{and} \quad -1 \le v \le 1$$

as shown in Figure 14.79. Solving for $x$ and $y$ in terms of $u$ and $v$ produces

$$x = \frac{1}{2}(u + v) \quad \text{and} \quad y = \frac{1}{2}(u - v).$$

The partial derivatives of $x$ and $y$ are

$$\frac{\partial x}{\partial u} = \frac{1}{2}, \quad \frac{\partial x}{\partial v} = \frac{1}{2}, \quad \frac{\partial y}{\partial u} = \frac{1}{2}, \quad \text{and} \quad \frac{\partial y}{\partial v} = -\frac{1}{2}$$

which implies that the Jacobian is

$$\frac{\partial(x, y)}{\partial(u, v)} = \begin{vmatrix} \dfrac{\partial x}{\partial u} & \dfrac{\partial x}{\partial v} \\[2mm] \dfrac{\partial y}{\partial u} & \dfrac{\partial y}{\partial v} \end{vmatrix} = \begin{vmatrix} \dfrac{1}{2} & \dfrac{1}{2} \\[2mm] \dfrac{1}{2} & -\dfrac{1}{2} \end{vmatrix} = -\frac{1}{4} - \frac{1}{4} = -\frac{1}{2}.$$

By Theorem 14.5, it follows that

$$\iint_R (x + y)^2 \sin^2(x - y) \, dA = \int_{-1}^{1} \int_{1}^{3} u^2 (\sin^2 v)\left(\frac{1}{2}\right) du \, dv$$

$$= \frac{1}{2} \int_{-1}^{1} (\sin^2 v) \frac{u^3}{3} \bigg]_1^3 dv$$

$$= \frac{13}{3} \int_{-1}^{1} \sin^2 v \, dv$$

$$= \frac{13}{6} \int_{-1}^{1} (1 - \cos 2v) \, dv$$

$$= \frac{13}{6} \left[ v - \frac{1}{2} \sin 2v \right]_{-1}^{1}$$

$$= \frac{13}{6} \left[ 2 - \frac{1}{2} \sin 2 + \frac{1}{2} \sin(-2) \right]$$

$$= \frac{13}{6}(2 - \sin 2)$$

$$\approx 2.363.$$

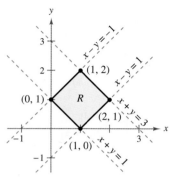

Region $R$ in the $xy$-plane
**Figure 14.78**

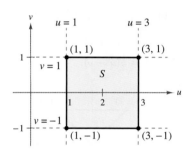

Region $S$ in the $uv$-plane
**Figure 14.79**

In each of the change of variables examples in this section, the region $S$ has been a rectangle with sides parallel to the $u$- or $v$-axis. Occasionally, a change of variables can be used for other types of regions. For instance, letting $T(u, v) = \left(x, \frac{1}{2}y\right)$ changes the circular region $u^2 + v^2 = 1$ to the elliptical region

$$x^2 + \frac{y^2}{4} = 1.$$

# 14.8 Exercises

See **CalcChat.com** for tutorial help and worked-out solutions to odd-numbered exercises.

## CONCEPT CHECK

**1. Jacobian** Describe how to find the Jacobian of $x$ and $y$ with respect to $u$ and $v$ for $x = g(u, v)$ and $y = h(u, v)$.

**2. Change of Variable** When is it beneficial to use the Jacobian to change variables in a double integral?

 **Finding a Jacobian** In Exercises 3–10, find the Jacobian $\partial(x, y)/\partial(u, v)$ for the indicated change of variables.

**3.** $x = -\frac{1}{2}(u - v)$, $y = \frac{1}{2}(u + v)$

**4.** $x = 5u - v$, $y = 3u + 4v$

**5.** $x = u - v^2$, $y = u + v$

**6.** $x = uv - 2u$, $y = uv$

**7.** $x = u \cos \theta - v \sin \theta$, $y = u \sin \theta + v \cos \theta$

**8.** $x = u + 1$, $y = 9v$

**9.** $x = e^u \sin v$, $y = e^u \cos v$

**10.** $x = u/v$, $y = u + v$

 **Using a Transformation** In Exercises 11–14, sketch the image $S$ in the $uv$-plane of the region $R$ in the $xy$-plane using the given transformations.

**11.** $x = 3u + 2v$
$y = 3v$

**12.** $x = \frac{1}{3}(4u - v)$
$y = \frac{1}{3}(u - v)$

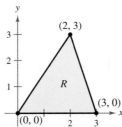

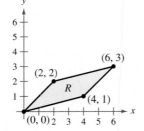

**13.** $x = \frac{1}{2}(u + v)$
$y = \frac{1}{2}(u - v)$

**14.** $x = \frac{1}{3}(v - u)$
$y = \frac{1}{3}(2v + u)$

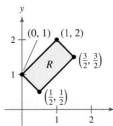

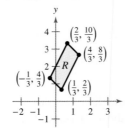

**Verifying a Change of Variables** In Exercises 15 and 16, verify the result of the indicated example by setting up the integral using $dy\, dx$ or $dx\, dy$ for $dA$. Then use a computer algebra system to evaluate the integral.

**15.** Example 3     **16.** Example 4

 **Evaluating a Double Integral Using a Change of Variables** In Exercises 17–22, use the indicated change of variables to evaluate the double integral.

**17.** $\displaystyle\iint_R 4(x^2 + y^2)\, dA$

$x = \frac{1}{2}(u + v)$
$y = \frac{1}{2}(u - v)$

**18.** $\displaystyle\iint_R (2y - x)\, dA$

$x = \frac{1}{2}(v - u)$
$y = \frac{1}{2}(3u - v)$

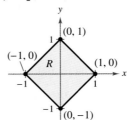

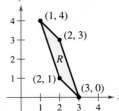

**19.** $\displaystyle\iint_R y(x - y)\, dA$

$x = u + v$
$y = u$

**20.** $\displaystyle\iint_R 4(x + y)e^{x-y}\, dA$

$x = \frac{1}{2}(u + v)$
$y = \frac{1}{2}(u - v)$

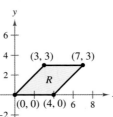

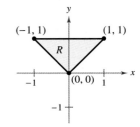

**21.** $\displaystyle\iint_R e^{-xy/2}\, dA$

$x = \sqrt{\dfrac{v}{u}}, \quad y = \sqrt{uv}$

**22.** $\displaystyle\iint_R y \sin xy\, dA$

$x = \dfrac{u}{v}, \quad y = v$

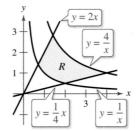

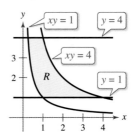

**Finding Volume Using a Change of Variables** In Exercises 23–30, use a change of variables to find the volume of the solid region lying below the surface $z = f(x, y)$ and above the plane region $R$.

**23.** $f(x, y) = 9xy$

$R$: region bounded by the square with vertices $(1, 0)$, $(0, 1)$, $(1, 2)$, $(2, 1)$

**24.** $f(x, y) = (3x + 2y)^2 \sqrt{2y - x}$

    $R$: region bounded by the parallelogram with vertices $(0, 0)$, $(-2, 3)$, $(2, 5)$, $(4, 2)$

**25.** $f(x, y) = (x + y)e^{x-y}$

    $R$: region bounded by the square with vertices $(4, 0)$, $(6, 2)$, $(4, 4)$, $(2, 2)$

**26.** $f(x, y) = (x + y)^2 \sin^2(x - y)$

    $R$: region bounded by the square with vertices $(\pi, 0)$, $(3\pi/2, \pi/2)$, $(\pi, \pi)$, $(\pi/2, \pi/2)$

**27.** $f(x, y) = \sqrt{(x - y)(x + 4y)}$

    $R$: region bounded by the parallelogram with vertices $(0, 0)$, $(1, 1)$, $(5, 0)$, $(4, -1)$

**28.** $f(x, y) = (3x + 2y)(2y - x)^{3/2}$

    $R$: region bounded by the parallelogram with vertices $(0, 0)$, $(-2, 3)$, $(2, 5)$, $(4, 2)$

**29.** $f(x, y) = \sqrt{x + y}$

    $R$: region bounded by the triangle with vertices $(0, 0)$, $(a, 0)$, $(0, a)$, where $a > 0$

**30.** $f(x, y) = \dfrac{xy}{1 + x^2y^2}$

    $R$: region bounded by the graphs of $xy = 1$, $xy = 4$, $x = 1$, $x = 4$ (*Hint:* Let $x = u$, $y = v/u$.)

**EXPLORING CONCEPTS**

**31. Using a Transformation** The substitutions $u = 2x - y$ and $v = x + y$ make the region $R$ (see figure) into a simpler region $S$ in the $uv$-plane. Determine the total number of sides of $S$ that are parallel to either the $u$-axis or the $v$-axis.

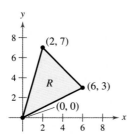

**32.** **HOW DO YOU SEE IT?** The region $R$ is transformed into a simpler region $S$ (see figure). Which substitution can be used to make the transformation?

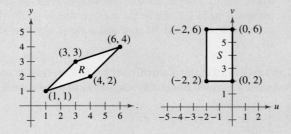

  (a) $u = 3y - x$, $v = y - x$  (b) $u = y - x$, $v = 3y - x$

**33. Using an Ellipse** Consider the region $R$ in the $xy$-plane bounded by the ellipse $(x^2/a^2) + (y^2/b^2) = 1$ and the transformations $x = au$ and $y = bv$.

  (a) Sketch the graph of the region $R$ and its image $S$ under the given transformation.

  (b) Find $\dfrac{\partial(x, y)}{\partial(u, v)}$.

  (c) Find the area of the ellipse using the indicated change of variables.

**34. Volume** Use the result of Exercise 33 to find the volume of each dome-shaped solid lying below the surface $z = f(x, y)$ and above the elliptical region $R$. (*Hint:* After making the change of variables given by the results in Exercise 33, make a second change of variables to polar coordinates.)

  (a) $f(x, y) = 16 - x^2 - y^2$; $R: \dfrac{x^2}{16} + \dfrac{y^2}{9} \le 1$

  (b) $f(x, y) = A \cos\left(\dfrac{\pi}{2}\sqrt{\dfrac{x^2}{a^2} + \dfrac{y^2}{b^2}}\right)$; $R: \dfrac{x^2}{a^2} + \dfrac{y^2}{b^2} \le 1$

**Finding a Jacobian** **In Exercises 35–40, find the Jacobian**

$$\dfrac{\partial(x, y, z)}{\partial(u, v, w)}$$

**for the indicated change of variables. If**

$$x = f(u, v, w), \quad y = g(u, v, w), \quad \text{and} \quad z = h(u, v, w)$$

**then the Jacobian of $x$, $y$, and $z$ with respect to $u$, $v$, and $w$ is**

$$\dfrac{\partial(x, y, z)}{\partial(u, v, w)} = \begin{vmatrix} \dfrac{\partial x}{\partial u} & \dfrac{\partial x}{\partial v} & \dfrac{\partial x}{\partial w} \\ \dfrac{\partial y}{\partial u} & \dfrac{\partial y}{\partial v} & \dfrac{\partial y}{\partial w} \\ \dfrac{\partial z}{\partial u} & \dfrac{\partial z}{\partial v} & \dfrac{\partial z}{\partial w} \end{vmatrix}.$$

**35.** $x = u(1 - v)$, $y = uv(1 - w)$, $z = uvw$

**36.** $x = 4u - v$, $y = 4v - w$, $z = u + w$

**37.** $x = \frac{1}{2}(u + v)$, $y = \frac{1}{2}(u - v)$, $z = 2uvw$

**38.** $x = u - v + w$, $y = 2uv$, $z = u + v + w$

**39. Spherical Coordinates**

    $x = \rho \sin \phi \cos \theta$, $y = \rho \sin \phi \sin \theta$, $z = \rho \cos \phi$

**40. Cylindrical Coordinates**

    $x = r \cos \theta$, $y = r \sin \theta$, $z = z$

**PUTNAM EXAM CHALLENGE**

**41.** Let $A$ be the area of the region in the first quadrant bounded by the line $y = \frac{1}{2}x$, the $x$-axis, and the ellipse $\frac{1}{9}x^2 + y^2 = 1$. Find the positive number $m$ such that $A$ is equal to the area of the region in the first quadrant bounded by the line $y = mx$, the $y$-axis, and the ellipse $\frac{1}{9}x^2 + y^2 = 1$.

**Evaluating an Integral**  In Exercises 1 and 2, evaluate the integral.

1. $\displaystyle\int_0^{3x} \sin(xy)\, dy$

2. $\displaystyle\int_y^{y^2} \frac{x}{y+1}\, dx$

**Evaluating an Iterated Integral**  In Exercises 3–6, evaluate the iterated integral.

3. $\displaystyle\int_0^1 \int_0^{1+x} (3x + 2y)\, dy\, dx$

4. $\displaystyle\int_0^2 \int_{x^2}^{2x} (x^2 + 2y)\, dy\, dx$

5. $\displaystyle\int_0^1 \int_0^{\sqrt{1-x^4}} x^3\, dy\, dx$

6. $\displaystyle\int_0^1 \int_0^{2y} (9 + 3x^2 + 3y^2)\, dx\, dy$

**Finding the Area of a Region**  In Exercises 7–10, use an iterated integral to find the area of the region bounded by the graphs of the equations.

7. $x + 3y = 3,\ x = 0,\ y = 0$

8. $y = 6x - x^2,\ y = x^2 - 2x$

9. $y = x,\ y = 2x + 2,\ x = 0,\ x = 4$

10. $x = y^2 + 1,\ x = 0,\ y = 0,\ y = 2$

**Changing the Order of Integration**  In Exercises 11–14, sketch the region $R$ whose area is given by the iterated integral. Then change the order of integration and show that both orders yield the same area.

11. $\displaystyle\int_1^5 \int_0^4 dy\, dx$

12. $\displaystyle\int_{-3}^3 \int_0^{9-y^2} dx\, dy$

13. $\displaystyle\int_0^2 \int_{y/2}^{3-y} dx\, dy$

14. $\displaystyle\int_0^3 \int_0^x dy\, dx + \int_3^6 \int_0^{6-x} dy\, dx$

**Evaluating a Double Integral**  In Exercises 15 and 16, set up integrals for both orders of integration. Use the more convenient order to evaluate the integral over the plane region $R$.

15. $\displaystyle\iint_R 4xy\, dA$

    $R$: rectangle with vertices $(0, 0),\ (0, 4),\ (2, 4),\ (2, 0)$

16. $\displaystyle\iint_R 6x^2\, dA$

    $R$: region bounded by $y = 0,\ y = \sqrt{x},\ x = 1$

**Finding Volume**  In Exercises 17–20, use a double integral to find the volume of the indicated solid.

17.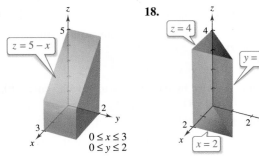
    $z = 5 - x$
    $0 \le x \le 3$
    $0 \le y \le 2$

18.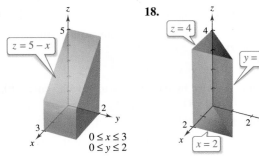
    $z = 4$
    $y = x$
    $x = 2$

19.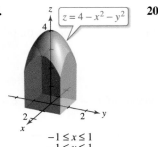
    $z = 4 - x^2 - y^2$
    $-1 \le x \le 1$
    $-1 \le y \le 1$

20.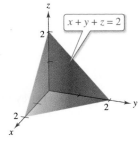
    $x + y + z = 2$

**Average Value**  In Exercises 21 and 22, find the average value of $f(x, y)$ over the plane region $R$.

21. $f(x) = 16 - x^2 - y^2$

    $R$: rectangle with vertices $(2, 2),\ (-2, 2),\ (-2, -2),\ (2, -2)$

22. $f(x) = 2x^2 + y^2$

    $R$: square with vertices $(0, 0),\ (3, 0),\ (3, 3),\ (0, 3)$

23. **Average Temperature**  The temperature in degrees Celsius on the surface of a metal plate is

    $$T(x, y) = 40 - 6x^2 - y^2$$

    where $x$ and $y$ are measured in centimeters. Estimate the average temperature when $x$ varies between 0 and 3 centimeters and $y$ varies between 0 and 5 centimeters.

24. **Average Profit**  A firm's profit $P$ (in dollars) from marketing two television models is

    $$P = 192x + 576y - x^2 - 5y^2 - 2xy - 5000$$

    where $x$ and $y$ represent the numbers of units of the two television models. Estimate the average weekly profit when $x$ varies between 40 and 50 units and $y$ varies between 45 and 60 units.

**Converting to Polar Coordinates**  In Exercises 25 and 26, evaluate the iterated integral by converting to polar coordinates.

25. $\displaystyle\int_0^{\sqrt{5}} \int_0^{\sqrt{5-x^2}} \sqrt{x^2 + y^2}\, dy\, dx$

26. $\displaystyle\int_0^4 \int_0^{\sqrt{16-y^2}} (x^2 + y^2)\, dx\, dy$

**Volume**    In Exercises 27 and 28, use a double integral in polar coordinates to find the volume of the solid bounded by the graphs of the equations.

**27.** $z = xy^2$, $x^2 + y^2 = 9$, first octant

**28.** $z = \sqrt{25 - x^2 - y^2}$, $z = 0$, $x^2 + y^2 = 16$

**Area**    In Exercises 29 and 30, use a double integral to find the area of the shaded region.

**29.**

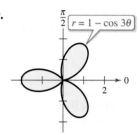

$$\frac{\pi}{2} \quad r = 1 - \cos 3\theta$$

$$0$$

$$2$$

**30.**

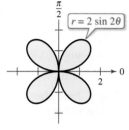

$$\frac{\pi}{2}$$

$$r = 2 \sin 2\theta$$

$$0$$

$$2$$

**Area**    In Exercises 31 and 32, sketch a graph of the region bounded by the graphs of the equations. Then use a double integral to find the area of the region.

**31.** Inside the limaçon $r = 3 + 2 \cos \theta$ and outside the circle $r = 4$

**32.** Inside the circle $r = 3 \sin \theta$ and outside the cardioid $r = 1 + \sin \theta$

**33. Area and Volume**    Consider the region $R$ in the $xy$-plane bounded by $(x^2 + y^2)^2 = 9(x^2 - y^2)$.

    (a) Convert the equation to polar coordinates. Use a graphing utility to graph the equation.

    (b) Use a double integral to find the area of the region $R$.

    (c) Use a computer algebra system to find the volume of the solid region bounded above by the hemisphere $z = \sqrt{9 - x^2 - y^2}$ and below by the region $R$.

**34. Converting to Polar Coordinates**    Write the sum of the two iterated integrals as a single iterated integral by converting to polar coordinates. Evaluate the resulting iterated integral.

$$\int_0^{8/\sqrt{13}} \int_0^{3x/2} xy \, dy \, dx + \int_{8/\sqrt{13}}^4 \int_0^{\sqrt{16-x^2}} xy \, dy \, dx$$

**Finding the Mass of a Lamina**    In Exercises 35 and 36, find the mass of the lamina described by the inequalities, given that its density is $\rho(x, y) = x + 3y$.

**35.** $0 \le x \le 1$, $0 \le y \le 2$

**36.** $x \ge 0$, $0 \le y \le \sqrt{4 - x^2}$

**Finding the Center of Mass**    In Exercises 37–40, find the mass and center of mass of the lamina bounded by the graphs of the equations for the given density.

**37.** $y = x^3$, $y = 0$, $x = 2$, $\rho = kx$

**38.** $y = \dfrac{2}{x}$, $y = 0$, $x = 1$, $x = 2$, $\rho = ky$

**39.** $y = 2x$, $y = 2x^3$, $x \ge 0$, $y \ge 0$, $\rho = kxy$

**40.** $y = 6 - x$, $y = 0$, $x = 0$, $\rho = kx^2$

**Finding Moments of Inertia and Radii of Gyration**    In Exercises 41 and 42, find $I_x, I_y, I_0, \bar{x}$, and $\bar{y}$ for the lamina bounded by the graphs of the equations.

**41.** $y = 0$, $y = 2$, $x = 0$, $x = 3$, $\rho = kx$

**42.** $y = 4 - x^2$, $y = 0$, $x > 0$, $\rho = ky$

**Finding Surface Area**    In Exercises 43–46, find the area of the surface given by $z = f(x, y)$ that lies above the region $R$.

**43.** $f(x, y) = 25 - x^2 - y^2$

    $R = \{(x, y): x^2 + y^2 \le 25\}$

**44.** $f(x, y) = 8 + 4x - 5y$

    $R = \{(x, y): x^2 + y^2 \le 1\}$

**45.** $f(x, y) = 9 - y^2$

    $R$: triangle with vertices $(-3, 3)$, $(0, 0)$, $(3, 3)$

**46.** $f(x, y) = 4 - x^2$

    $R$: triangle with vertices $(-2, 2)$, $(0, 0)$, $(2, 2)$

**47. Building Design**    A new auditorium is built with a foundation in the shape of one-fourth of a circle of radius 50 feet. So, it forms a region $R$ bounded by the graph of $x^2 + y^2 = 50^2$ with $x \ge 0$ and $y \ge 0$. The following equations are models for the floor and ceiling.

    Floor: $z = \dfrac{x + y}{5}$

    Ceiling: $z = 20 + \dfrac{xy}{100}$

    (a) Calculate the volume of the room, which is needed to determine the heating and cooling requirements.

    (b) Find the surface area of the ceiling.

**48. Surface Area**    The roof over the stage of an open air theater at a theme park is modeled by

$$f(x, y) = 25\left[1 + e^{-(x^2+y^2)/1000} \cos^2\left(\frac{x^2 + y^2}{1000}\right)\right]$$

    where the stage is a semicircle bounded by the graphs of $y = \sqrt{50^2 - x^2}$ and $y = 0$.

    (a) Use a computer algebra system to graph the surface.

    (b) Use a computer algebra system to approximate the number of square feet of roofing required to cover the surface.

**Evaluating a Triple Iterated Integral**    In Exercises 49–52, evaluate the triple iterated integral.

**49.** $\displaystyle\int_0^4 \int_0^1 \int_0^2 (2x + y + 4z) \, dy \, dz \, dx$

**50.** $\displaystyle\int_0^1 \int_0^{1+\sqrt{y}} \int_0^{xy} y \, dz \, dx \, dy$

**51.** $\displaystyle\int_0^2 \int_1^2 \int_0^1 (e^x + y^2 + z^2) \, dx \, dy \, dz$

**52.** $\displaystyle\int_0^3 \int_{\pi/2}^{\pi} \int_2^5 z \sin x \, dy \, dx \, dz$

**Evaluating a Triple Iterated Integral Using Technology** In Exercises 53 and 54, use a computer algebra system to evaluate the triple iterated integral.

**53.** $\displaystyle\int_{-1}^{1}\int_{-\sqrt{1-x^2}}^{\sqrt{1-x^2}}\int_{-\sqrt{1-x^2-y^2}}^{\sqrt{1-x^2-y^2}} (x^2 + y^2)\, dz\, dy\, dx$

**54.** $\displaystyle\int_{0}^{2}\int_{0}^{\sqrt{4-x^2}}\int_{0}^{\sqrt{4-x^2-y^2}} xyz\, dz\, dy\, dx$

**Volume** In Exercises 55 and 56, use a triple integral to find the volume of the solid bounded by the graphs of the equations.

**55.** $z = xy$, $z = 0$, $0 \le x \le 3$, $0 \le y \le 4$

**56.** $z = 8 - x - y$, $z = 0$, $y = x$, $y = 3$, $x = 0$

**Changing the Order of Integration** In Exercises 57 and 58, sketch the solid whose volume is given by the iterated integral. Then rewrite the integral using the indicated order of integration.

**57.** $\displaystyle\int_{0}^{1}\int_{0}^{y}\int_{0}^{\sqrt{1-x^2}} dz\, dx\, dy$

Rewrite using the order $dz\, dy\, dx$.

**58.** $\displaystyle\int_{0}^{6}\int_{0}^{6-x}\int_{0}^{6-x-y} dz\, dy\, dx$

Rewrite using the order $dy\, dx\, dz$.

**Center of Mass** In Exercises 59 and 60, find the mass and the indicated coordinate of the center of mass of the solid region $Q$ of density $\rho$ bounded by the graphs of the equations.

**59.** Find $\bar{x}$ using $\rho(x, y, z) = k$.

   $Q$: $x + y + z = 10$, $x = 0$, $y = 0$, $z = 0$

**60.** Find $\bar{y}$ using $\rho(x, y, z) = kx$.

   $Q$: $z = 5 - y$, $z = 0$, $y = 0$, $x = 0$, $x = 5$

**Evaluating a Triple Iterated Integral** In Exercises 61–64, evaluate the triple iterated integral.

**61.** $\displaystyle\int_{0}^{3}\int_{\pi/6}^{\pi/3}\int_{0}^{4} r\cos\theta\, dr\, d\theta\, dz$

**62.** $\displaystyle\int_{0}^{\pi/2}\int_{0}^{3}\int_{0}^{4-z} z\, dr\, dz\, d\theta$

**63.** $\displaystyle\int_{0}^{\pi}\int_{0}^{\pi/2}\int_{0}^{\sin\theta} \rho^2\sin\theta\cos\theta\, d\rho\, d\theta\, d\phi$

**64.** $\displaystyle\int_{0}^{\pi/4}\int_{0}^{\pi/4}\int_{0}^{\cos\phi} \cos\theta\, d\rho\, d\phi\, d\theta$

**Evaluating a Triple Iterated Integral Using Technology** In Exercises 65 and 66, use a computer algebra system to evaluate the triple iterated integral.

**65.** $\displaystyle\int_{0}^{\pi}\int_{0}^{2}\int_{0}^{3} \sqrt{z^2 + 4}\, dz\, dr\, d\theta$

**66.** $\displaystyle\int_{0}^{\pi/2}\int_{0}^{\pi/2}\int_{0}^{\cos\phi} \rho^2\cos\theta\, d\rho\, d\theta\, d\phi$

**Volume** In Exercises 67 and 68, use cylindrical coordinates to find the volume of the solid.

**67.** Solid bounded above by $z = 8 - x^2 - y^2$ and below by $z = x^2 + y^2$

**68.** Solid bounded above by $3x^2 + 3y^2 + z^2 = 45$ and below by the $xy$-plane

**Volume** In Exercises 69 and 70, use spherical coordinates to find the volume of the solid.

**69.** Solid bounded above by $x^2 + y^2 + z^2 = 4$ and below by $z^2 = 3x^2 + 3y^2$

**70.** Solid bounded above by $x^2 + y^2 + z^2 = 36$ and below by $z = \sqrt{x^2 + y^2}$

**Finding a Jacobian** In Exercises 71–74, find the Jacobian $\partial(x, y)/\partial(u, v)$ for the indicated change of variables.

**71.** $x = 3uv$, $y = 2(u - v)$

**72.** $x = u^2 + v^2$, $y = u^2 - v^2$

**73.** $x = u\sin\theta + v\cos\theta$, $y = u\cos\theta + v\sin\theta$

**74.** $x = uv$, $y = \dfrac{v}{u}$

**Evaluating a Double Integral Using a Change of Variables** In Exercises 75–78, use the indicated change of variables to evaluate the double integral.

**75.** $\displaystyle\iint_{R} \ln(x + y)\, dA$

   $x = \dfrac{1}{2}(u + v)$

   $y = \dfrac{1}{2}(u - v)$

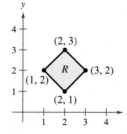

**76.** $\displaystyle\iint_{R} 16xy\, dA$

   $x = \dfrac{1}{4}(u + v)$

   $y = \dfrac{1}{2}(v - u)$

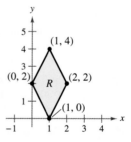

**77.** $\displaystyle\iint_{R} (xy + x^2)\, dA$

   $x = u$

   $y = \dfrac{1}{3}(u - v)$

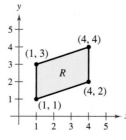

**78.** $\displaystyle\iint_{R} \dfrac{x}{1 + x^2 y^2}\, dA$

   $x = u$

   $y = \dfrac{v}{u}$

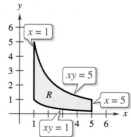

# P.S. Problem Solving

See **CalcChat.com** for tutorial help and worked-out solutions to odd-numbered exercises.

1. **Volume**  Find the volume of the solid of intersection of the three cylinders $x^2 + z^2 = 1$, $y^2 + z^2 = 1$, and $x^2 + y^2 = 1$ (see figure).

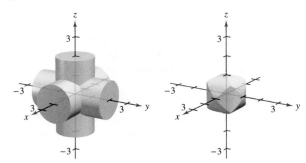

2. **Surface Area**  Let $a$, $b$, $c$, and $d$ be positive real numbers. The portion of the plane $ax + by + cz = d$ in the first octant is shown in the figure. Show that the surface area of this portion of the plane is equal to

$$\frac{A(R)}{c}\sqrt{a^2 + b^2 + c^2}$$

where $A(R)$ is the area of the triangular region $R$ in the $xy$-plane, as shown in the figure.

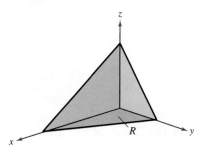

3. **Using a Change of Variables**  The figure shows the region $R$ bounded by the curves

$$y = \sqrt{x}, \ y = \sqrt{2x}, \ y = \frac{x^2}{3}, \text{ and } y = \frac{x^2}{4}.$$

Use the change of variables $x = u^{1/3}v^{2/3}$ and $y = u^{2/3}v^{1/3}$ to find the area of the region $R$.

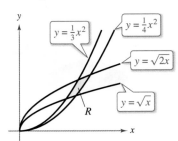

4. **Proof**  Prove that $\displaystyle\lim_{n\to\infty} \int_0^1 \int_0^1 x^n y^n \, dx \, dy = 0$.

5. **Deriving a Sum**  Derive Euler's famous result that was mentioned in Section 9.3,

$$\sum_{n=1}^{\infty} \frac{1}{n^2} = \frac{\pi^2}{6}$$

by completing each step.

(a) Prove that

$$\int \frac{dv}{2 - u^2 + v^2} = \frac{1}{\sqrt{2 - u^2}} \arctan \frac{v}{\sqrt{2 - u^2}} + C.$$

(b) Prove that $\displaystyle I_1 = \int_0^{\sqrt{2}/2} \int_{-u}^{u} \frac{2}{2 - u^2 + v^2} \, dv \, du = \frac{\pi^2}{18}$

by using the substitution $u = \sqrt{2} \sin \theta$.

(c) Prove that

$$I_2 = \int_{\sqrt{2}/2}^{\sqrt{2}} \int_{u-\sqrt{2}}^{-u+\sqrt{2}} \frac{2}{2 - u^2 + v^2} \, dv \, du$$

$$= 4 \int_{\pi/6}^{\pi/2} \arctan \frac{1 - \sin \theta}{\cos \theta} \, d\theta$$

by using the substitution $u = \sqrt{2} \sin \theta$.

(d) Prove the trigonometric identity

$$\frac{1 - \sin \theta}{\cos \theta} = \tan\left[\frac{(\pi/2) - \theta}{2}\right].$$

(e) Prove that $\displaystyle I_2 = \int_{\sqrt{2}/2}^{\sqrt{2}} \int_{u-\sqrt{2}}^{-u+\sqrt{2}} \frac{2}{2 - u^2 + v^2} \, dv \, du = \frac{\pi^2}{9}$.

(f) Use the formula for the sum of an infinite geometric series to verify that

$$\sum_{n=1}^{\infty} \frac{1}{n^2} = \int_0^1 \int_0^1 \frac{1}{1 - xy} \, dx \, dy.$$

(g) Use the change of variables

$$u = \frac{x + y}{\sqrt{2}} \quad \text{and} \quad v = \frac{y - x}{\sqrt{2}}$$

to prove that

$$\sum_{n=1}^{\infty} \frac{1}{n^2} = \int_0^1 \int_0^1 \frac{1}{1 - xy} \, dx \, dy = I_1 + I_2 = \frac{\pi^2}{6}.$$

6. **Evaluating a Double Integral**  Evaluate the integral

$$\int_0^{\infty} \int_0^{\infty} \frac{1}{(1 + x^2 + y^2)^2} \, dx \, dy.$$

7. **Evaluating Double Integrals**  Evaluate the integrals

$$\int_0^1 \int_0^1 \frac{x - y}{(x + y)^3} \, dx \, dy \quad \text{and} \quad \int_0^1 \int_0^1 \frac{x - y}{(x + y)^3} \, dy \, dx.$$

Are the results the same? Why or why not?

8. **Volume**  Show that the volume of a spherical block can be approximated by $\Delta V \approx \rho^2 \sin \phi \, \Delta\rho \, \Delta\phi \, \Delta\theta$. (*Hint:* See Section 14.7, page 1027.)

**Evaluating an Integral** In Exercises 9 and 10, evaluate the integral. (*Hint:* See Exercise 63 in Section 14.3.)

**9.** $\int_0^\infty x^2 e^{-x^2}\, dx$

**10.** $\int_0^1 \sqrt{\ln \frac{1}{x}}\, dx$

**11. Joint Density Function** Consider the function

$$f(x, y) = \begin{cases} ke^{-(x+y)/a}, & x \geq 0, y \geq 0 \\ 0, & \text{elsewhere.} \end{cases}$$

Find the relationship between the positive constants $a$ and $k$ such that $f$ is a joint density function of the continuous random variables $x$ and $y$. (*Hint:* See Exercises 61–64 in Section 14.2)

**12. Volume** Find the volume of the solid generated by revolving the region in the first quadrant bounded by $y = e^{-x^2}$ about the $y$-axis. Use this result to find

$$\int_{-\infty}^\infty e^{-x^2}\, dx.$$

**13. Volume and Surface Area** From 1963 to 1986, the volume of the Great Salt Lake approximately tripled while its top surface area approximately doubled. Read the article "Relations between Surface Area and Volume in Lakes" by Daniel Cass and Gerald Wildenberg in *The College Mathematics Journal*. Then give examples of solids that have "water levels" $a$ and $b$ such that $V(b) = 3V(a)$ and $A(b) = 2A(a)$, where $V$ is volume and $A$ is area (see figure).

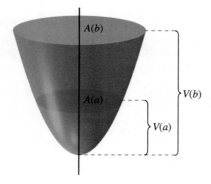

**14. Proof** The angle between a plane $P$ and the $xy$-plane is $\theta$, where $0 \leq \theta < \pi/2$. The projection of a rectangular region in $P$ onto the $xy$-plane is a rectangle whose sides have lengths $\Delta x$ and $\Delta y$, as shown in the figure. Prove that the area of the rectangular region in $P$ is $\sec \theta\, \Delta x\, \Delta y$.

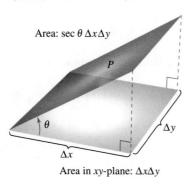

Area: $\sec \theta\, \Delta x \Delta y$

Area in $xy$-plane: $\Delta x \Delta y$

**15. Surface Area** Use the result of Exercise 14 to order the planes in ascending order of their surface areas for a fixed region $R$ in the $xy$-plane. Explain your ordering without doing any calculations.

(a) $z_1 = 2 + x$

(b) $z_2 = 5$

(c) $z_3 = 10 - 5x + 9y$

(d) $z_4 = 3 + x - 2y$

**16. Sprinkler** Consider a circular lawn with a radius of 10 feet, as shown in the figure. Assume that a sprinkler distributes water in a radial fashion according to the formula

$$f(r) = \frac{r}{16} - \frac{r^2}{160}$$

(measured in cubic feet of water per hour per square foot of lawn), where $r$ is the distance in feet from the sprinkler. Find the amount of water that is distributed in 1 hour in the following two annular regions.

$A = \{(r, \theta): 4 \leq r \leq 5, 0 \leq \theta \leq 2\pi\}$

$B = \{(r, \theta): 9 \leq r \leq 10, 0 \leq \theta \leq 2\pi\}$

Is the distribution of water uniform? Determine the amount of water the entire lawn receives in 1 hour.

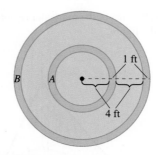

**17. Changing the Order of Integration** Sketch the solid whose volume is given by the sum of the iterated integrals

$$\int_0^6 \int_{z/2}^3 \int_{z/2}^y dx\, dy\, dz + \int_0^6 \int_3^{(12-z)/2} \int_{z/2}^{6-y} dx\, dy\, dz.$$

Then write the volume as a single iterated integral in the order $dy\, dz\, dx$ and find the volume of the solid.

**18. Volume** The figure shows a solid bounded below by the plane $z = 2$ and above by the sphere $x^2 + y^2 + z^2 = 8$.

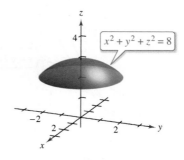

(a) Find the volume of the solid using cylindrical coordinates.

(b) Find the volume of the solid using spherical coordinates.

# 15 Vector Analysis

Finding Surface Area
*(Example 6, p. 1093)*

Work *(Exercise 35, p. 1077)*

Mass of a Spring *(Example 5, p. 1059)*

Building Design
*(Exercise 74, p. 1068)*

Earth's Magnetic Field *(Exercise 78, p. 1054)*

# 15.1 Vector Fields

■ Understand the concept of a vector field.
■ Determine whether a vector field is conservative.
■ Find the curl of a vector field.
■ Find the divergence of a vector field.

## Vector Fields

In Chapter 12, you studied vector-valued functions—functions that assign a vector to a *real number.* There you saw that vector-valued functions of real numbers are useful in representing curves and motion along a curve. In this chapter, you will study two other types of vector-valued functions—functions that assign a vector to a *point in the plane* or a *point in space.* Such functions are called **vector fields,** and they are useful in representing various types of **force fields** and **velocity fields.**

> ### Definition of Vector Field
>
> A **vector field over a plane region $R$** is a function $\mathbf{F}$ that assigns a vector $\mathbf{F}(x, y)$ to each point in $R$.
>
> A **vector field over a solid region $Q$ in space** is a function $\mathbf{F}$ that assigns a vector $\mathbf{F}(x, y, z)$ to each point in $Q$.

Although a vector field consists of infinitely many vectors, you can get a good idea of what the vector field looks like by sketching several representative vectors $\mathbf{F}(x, y)$ whose initial points are $(x, y)$.

The *gradient* is one example of a vector field. For instance, if

$$f(x, y) = x^2y + 3xy^3$$

then the gradient of $f$

$$\nabla f(x, y) = f_x(x, y)\mathbf{i} + f_y(x, y)\mathbf{j}$$
$$= (2xy + 3y^3)\mathbf{i} + (x^2 + 9xy^2)\mathbf{j} \qquad \text{Vector field in the plane}$$

is a vector field in the plane. From Chapter 13, the graphical interpretation of this field is a family of vectors, each of which points in the direction of maximum increase along the surface given by $z = f(x, y)$.

Similarly, if

$$f(x, y, z) = x^2 + y^2 + z^2$$

then the gradient of $f$

$$\nabla f(x, y, z) = f_x(x, y, z)\mathbf{i} + f_y(x, y, z)\mathbf{j} + f_z(x, y, z)\mathbf{k}$$

$$= 2x\mathbf{i} + 2y\mathbf{j} + 2z\mathbf{k} \qquad \text{Vector field in space}$$

is a vector field in space. Note that the component functions for this particular vector field are $2x$, $2y$, and $2z$.

A vector field

$$\mathbf{F}(x, y, z) = M(x, y, z)\mathbf{i} + N(x, y, z)\mathbf{j} + P(x, y, z)\mathbf{k}$$

is **continuous** at a point if and only if each of its component functions $M$, $N$, and $P$ is continuous at that point.

Velocity field

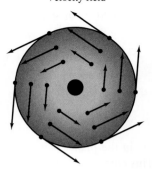

Rotating wheel
**Figure 15.1**

Air flow vector field
**Figure 15.2**

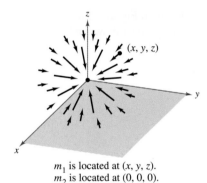

$m_1$ is located at $(x, y, z)$.
$m_2$ is located at $(0, 0, 0)$.

Gravitational force field
**Figure 15.3**

Some common *physical* examples of vector fields are **velocity fields, gravitational fields,** and **electric force fields.**

1. *Velocity fields* describe the motions of systems of particles in the plane or in space. For instance, Figure 15.1 shows the vector field determined by a wheel rotating on an axle. Notice that the velocity vectors are determined by the locations of their initial points—the farther a point is from the axle, the greater its velocity. Velocity fields are also determined by the flow of liquids through a container or by the flow of air currents around a moving object, as shown in Figure 15.2.

2. *Gravitational fields* are defined by **Newton's Law of Gravitation,** which states that the force of attraction exerted on a particle of mass $m_1$ located at $(x, y, z)$ by a particle of mass $m_2$ located at $(0, 0, 0)$ is

$$\mathbf{F}(x, y, z) = \frac{-Gm_1m_2}{x^2 + y^2 + z^2}\mathbf{u}$$

where $G$ is the gravitational constant and $\mathbf{u}$ is the unit vector in the direction from the origin to $(x, y, z)$. In Figure 15.3, you can see that the gravitational field $\mathbf{F}$ has the properties that $\mathbf{F}(x, y, z)$ always points toward the origin, and that the magnitude of $\mathbf{F}(x, y, z)$ is the same at all points equidistant from the origin. A vector field with these two properties is called a **central force field.** Using the position vector

$$\mathbf{r} = x\mathbf{i} + y\mathbf{j} + z\mathbf{k}$$

for the point $(x, y, z)$, you can write the gravitational field $\mathbf{F}$ as

$$\mathbf{F}(x, y, z) = \frac{-Gm_1m_2}{\|\mathbf{r}\|^2}\left(\frac{\mathbf{r}}{\|\mathbf{r}\|}\right) = \frac{-Gm_1m_2}{\|\mathbf{r}\|^2}\mathbf{u}.$$

3. *Electric force fields* are defined by **Coulomb's Law,** which states that the force exerted on a particle with electric charge $q_1$ located at $(x, y, z)$ by a particle with electric charge $q_2$ located at $(0, 0, 0)$ is

$$\mathbf{F}(x, y, z) = \frac{cq_1q_2}{\|\mathbf{r}\|^2}\mathbf{u}$$

where $\mathbf{r} = x\mathbf{i} + y\mathbf{j} + z\mathbf{k}$, $\mathbf{u} = \mathbf{r}/\|\mathbf{r}\|$, and $c$ is a constant that depends on the choice of units for $\|\mathbf{r}\|$, $q_1$, and $q_2$.

Note that an electric force field has the same form as a gravitational field. That is,

$$\mathbf{F}(x, y, z) = \frac{k}{\|\mathbf{r}\|^2}\mathbf{u}.$$

Such a force field is called an **inverse square field.**

---

**Definition of Inverse Square Field**

Let $\mathbf{r}(t) = x(t)\mathbf{i} + y(t)\mathbf{j} + z(t)\mathbf{k}$ be a position vector. The vector field $\mathbf{F}$ is an **inverse square field** if

$$\mathbf{F}(x, y, z) = \frac{k}{\|\mathbf{r}\|^2}\mathbf{u}$$

where $k$ is a real number and

$$\mathbf{u} = \frac{\mathbf{r}}{\|\mathbf{r}\|}$$

is a unit vector in the direction of $\mathbf{r}$.

Because vector fields consist of infinitely many vectors, it is not possible to create a sketch of the entire field. Instead, when you sketch a vector field, your goal is to sketch representative vectors that help you visualize the field.

### EXAMPLE 1   Sketching a Vector Field

Sketch some vectors in the vector field

$$\mathbf{F}(x, y) = -y\mathbf{i} + x\mathbf{j}.$$

**Solution**   You could plot vectors at several random points in the plane. It is more enlightening, however, to plot vectors of equal magnitude. This corresponds to finding level curves in scalar fields. In this case, vectors of equal magnitude lie on circles.

$$\|\mathbf{F}\| = c \qquad \text{Vectors of length } c$$
$$\sqrt{x^2 + y^2} = c$$
$$x^2 + y^2 = c^2 \qquad \text{Equation of circle}$$

To begin making the sketch, choose a value for $c$ and plot several vectors on the resulting circle. For instance, the following vectors occur on the unit circle.

| Point | Vector |
|---|---|
| $(1, 0)$ | $\mathbf{F}(1, 0) = \mathbf{j}$ |
| $(0, 1)$ | $\mathbf{F}(0, 1) = -\mathbf{i}$ |
| $(-1, 0)$ | $\mathbf{F}(-1, 0) = -\mathbf{j}$ |
| $(0, -1)$ | $\mathbf{F}(0, -1) = \mathbf{i}$ |

These and several other vectors in the vector field are shown in Figure 15.4. Note in the figure that this vector field is similar to that given by the rotating wheel shown in Figure 15.1.

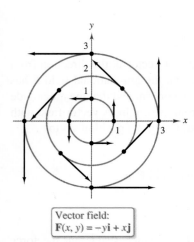

Vector field:
$\mathbf{F}(x, y) = -y\mathbf{i} + x\mathbf{j}$

**Figure 15.4**

### EXAMPLE 2   Sketching a Vector Field

Sketch some vectors in the vector field

$$\mathbf{F}(x, y) = 2x\mathbf{i} + y\mathbf{j}.$$

**Solution**   For this vector field, vectors of equal magnitude lie on ellipses given by

$$\|\mathbf{F}\| = c$$
$$\sqrt{(2x)^2 + (y)^2} = c$$

which implies that

$$4x^2 + y^2 = c^2. \qquad \text{Equation of ellipse}$$

For $c = 1$, sketch several vectors $2x\mathbf{i} + y\mathbf{j}$ of magnitude 1 at points on the ellipse given by

$$4x^2 + y^2 = 1.$$

For $c = 2$, sketch several vectors $2x\mathbf{i} + y\mathbf{j}$ of magnitude 2 at points on the ellipse given by

$$4x^2 + y^2 = 4.$$

These vectors are shown in Figure 15.5.

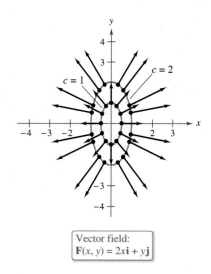

Vector field:
$\mathbf{F}(x, y) = 2x\mathbf{i} + y\mathbf{j}$

**Figure 15.5**

▷ **TECHNOLOGY**   A computer algebra system can be used to graph vectors in a vector field. If you have access to a computer algebra system, use it to graph several representative vectors for the vector field in Example 2.

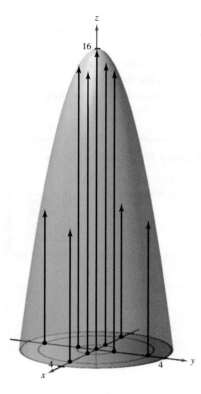

Velocity field:
$$\mathbf{v}(x, y, z) = (16 - x^2 - y^2)\mathbf{k}$$

**Figure 15.6**

---

EXAMPLE 3    **Sketching a Velocity Field**

Sketch some vectors in the velocity field

$$\mathbf{v}(x, y, z) = (16 - x^2 - y^2)\mathbf{k}$$

where $x^2 + y^2 \leq 16$.

**Solution**  You can imagine that $\mathbf{v}$ describes the velocity of a liquid flowing through a tube of radius 4. Vectors near the $z$-axis are longer than those near the edge of the tube. For instance, at the point $(0, 0, 0)$, the velocity vector is $\mathbf{v}(0, 0, 0) = 16\mathbf{k}$, whereas at the point $(0, 3, 0)$, the velocity vector is $\mathbf{v}(0, 3, 0) = 7\mathbf{k}$. Figure 15.6 shows these and several other vectors for the velocity field. From the figure, you can see that the speed of the liquid is greater near the center of the tube than near the edges of the tube.

## Conservative Vector Fields

Notice in Figure 15.5 that all the vectors appear to be normal to the level curve from which they emanate. Because this is a property of gradients, it is natural to ask whether the vector field

$$\mathbf{F}(x, y) = 2x\mathbf{i} + y\mathbf{j}$$

is the *gradient* of some differentiable function $f$. The answer is that some vector fields can be represented as the gradients of differentiable functions and some cannot—those that can are called **conservative** vector fields.

---

**Definition of Conservative Vector Field**

A vector field $\mathbf{F}$ is called **conservative** when there exists a differentiable function $f$ such that $\mathbf{F} = \nabla f$. The function $f$ is called the **potential function** for $\mathbf{F}$.

---

EXAMPLE 4    **Conservative Vector Fields**

**a.** The vector field given by $\mathbf{F}(x, y) = 2x\mathbf{i} + y\mathbf{j}$ is conservative. To see this, consider the potential function $f(x, y) = x^2 + \frac{1}{2}y^2$. Because

$$\nabla f = 2x\mathbf{i} + y\mathbf{j} = \mathbf{F}$$

it follows that $\mathbf{F}$ is conservative.

**b.** Every inverse square field is conservative. To see this, let

$$\mathbf{F}(x, y, z) = \frac{k}{\|\mathbf{r}\|^2}\mathbf{u} \quad \text{and} \quad f(x, y, z) = \frac{-k}{\sqrt{x^2 + y^2 + z^2}}$$

where $\mathbf{u} = \mathbf{r}/\|\mathbf{r}\|$. Because

$$\nabla f = \frac{kx}{(x^2 + y^2 + z^2)^{3/2}}\mathbf{i} + \frac{ky}{(x^2 + y^2 + z^2)^{3/2}}\mathbf{j} + \frac{kz}{(x^2 + y^2 + z^2)^{3/2}}\mathbf{k}$$

$$= \frac{k}{x^2 + y^2 + z^2}\left(\frac{x\mathbf{i} + y\mathbf{j} + z\mathbf{k}}{\sqrt{x^2 + y^2 + z^2}}\right)$$

$$= \frac{k}{\|\mathbf{r}\|^2}\left(\frac{\mathbf{r}}{\|\mathbf{r}\|}\right)$$

$$= \frac{k}{\|\mathbf{r}\|^2}\mathbf{u}$$

it follows that $\mathbf{F}$ is conservative.

As can be seen in Example 4(b), many important vector fields, including gravitational fields and electric force fields, are conservative. Most of the terminology in this chapter comes from physics. For example, the term "conservative" is derived from the classic physical law regarding the conservation of energy. This law states that the sum of the kinetic energy and the potential energy of a particle moving in a conservative force field is constant. (The kinetic energy of a particle is the energy due to its motion, and the potential energy is the energy due to its position in the force field.)

The next theorem gives a necessary and sufficient condition for a vector field *in the plane* to be conservative.

• • • • • • • • • • • • • • • ▷

• **REMARK** Theorem 15.1 is valid on *simply connected* domains. A plane region $R$ is simply connected when every simple closed curve in $R$ encloses only points that are in $R$. (See Figure 15.26 in Section 15.4.)

---

**THEOREM 15.1   Test for Conservative Vector Field in the Plane**

Let $M$ and $N$ have continuous first partial derivatives on an open disk $R$. The vector field $\mathbf{F}(x, y) = M\mathbf{i} + N\mathbf{j}$ is conservative if and only if

$$\frac{\partial N}{\partial x} = \frac{\partial M}{\partial y}.$$

---

**Proof**   To prove that the given condition is necessary for $\mathbf{F}$ to be conservative, suppose there exists a potential function $f$ such that

$$\mathbf{F}(x, y) = \nabla f(x, y) = M\mathbf{i} + N\mathbf{j}.$$

Then you have

$$f_x(x, y) = M \implies f_{xy}(x, y) = \frac{\partial M}{\partial y}$$

$$f_y(x, y) = N \implies f_{yx}(x, y) = \frac{\partial N}{\partial x}$$

and, by the equivalence of the mixed partials $f_{xy}$ and $f_{yx}$, you can conclude that $\partial N/\partial x = \partial M/\partial y$ for all $(x, y)$ in $R$. The sufficiency of this condition is proved in Section 15.4. ∎

---

**EXAMPLE 5**   **Testing for Conservative Vector Fields in the Plane**

Determine whether the vector field given by $\mathbf{F}$ is conservative.

**a.** $\mathbf{F}(x, y) = x^2 y\mathbf{i} + xy\mathbf{j}$

**b.** $\mathbf{F}(x, y) = 2x\mathbf{i} + y\mathbf{j}$

**Solution**

**a.** The vector field

$$\mathbf{F}(x, y) = x^2 y\mathbf{i} + xy\mathbf{j}$$

is not conservative because

$$\frac{\partial M}{\partial y} = \frac{\partial}{\partial y}[x^2 y] = x^2 \quad \text{and} \quad \frac{\partial N}{\partial x} = \frac{\partial}{\partial x}[xy] = y.$$

**b.** The vector field

$$\mathbf{F}(x, y) = 2x\mathbf{i} + y\mathbf{j}$$

is conservative because

$$\frac{\partial M}{\partial y} = \frac{\partial}{\partial y}[2x] = 0 \quad \text{and} \quad \frac{\partial N}{\partial x} = \frac{\partial}{\partial x}[y] = 0.$$

Theorem 15.1 tells you whether a vector field **F** is conservative. It does not tell you how to find a potential function of **F**. The problem is comparable to antidifferentiation. Sometimes you will be able to find a potential function by simple inspection. For instance, in Example 4, you observed that

$$f(x, y) = x^2 + \frac{1}{2}y^2$$

has the property that

$$\nabla f(x, y) = 2x\mathbf{i} + y\mathbf{j}.$$

### EXAMPLE 6    Finding a Potential Function for F(x, y)

Find a potential function for

$$\mathbf{F}(x, y) = 2xy\mathbf{i} + (x^2 - y)\mathbf{j}.$$

**Solution**    From Theorem 15.1, it follows that **F** is conservative because

$$\frac{\partial}{\partial y}[2xy] = 2x \quad \text{and} \quad \frac{\partial}{\partial x}[x^2 - y] = 2x.$$

If $f$ is a function whose gradient is equal to $\mathbf{F}(x, y)$, then

$$\nabla f(x, y) = 2xy\mathbf{i} + (x^2 - y)\mathbf{j}$$

which implies that

$$f_x(x, y) = 2xy$$

and

$$f_y(x, y) = x^2 - y.$$

To reconstruct the function $f$ from these two partial derivatives, integrate $f_x(x, y)$ with respect to $x$

$$f(x, y) = \int f_x(x, y)\, dx = \int 2xy\, dx = x^2y + g(y)$$

and integrate $f_y(x, y)$ with respect to $y$

$$f(x, y) = \int f_y(x, y)\, dy = \int (x^2 - y)\, dy = x^2y - \frac{y^2}{2} + h(x).$$

Notice that $g(y)$ is constant with respect to $x$ and $h(x)$ is constant with respect to $y$. To find a single expression that represents $f(x, y)$, let

$$g(y) = -\frac{y^2}{2} + K_1 \quad \text{and} \quad h(x) = K_2.$$

Then you can write

$$f(x, y) = x^2y - \frac{y^2}{2} + K. \qquad K = K_1 + K_2$$

You can check this result by forming the gradient of $f$. You will see that it is equal to the original function **F**. ◼

Notice that the solution to Example 6 is comparable to that given by an indefinite integral. That is, the solution represents a family of potential functions, any two of which differ by a constant. To find a unique solution, you would have to be given an initial condition that is satisfied by the potential function.

## Curl of a Vector Field

Theorem 15.1 has a counterpart for vector fields in space. Before stating that result, the definition of the **curl of a vector field** in space is given.

---

**Definition of Curl of a Vector Field**

The curl of $\mathbf{F}(x, y, z) = M\mathbf{i} + N\mathbf{j} + P\mathbf{k}$ is

$$\text{curl } \mathbf{F}(x, y, z) = \nabla \times \mathbf{F}(x, y, z)$$
$$= \left( \frac{\partial P}{\partial y} - \frac{\partial N}{\partial z} \right)\mathbf{i} - \left( \frac{\partial P}{\partial x} - \frac{\partial M}{\partial z} \right)\mathbf{j} + \left( \frac{\partial N}{\partial x} - \frac{\partial M}{\partial y} \right)\mathbf{k}.$$

If curl $\mathbf{F} = \mathbf{0}$, then $\mathbf{F}$ is said to be **irrotational.**

---

The cross product notation used for curl comes from viewing the gradient $\nabla f$ as the result of the **differential operator** $\nabla$ acting on the function $f$. In this context, you can use the following determinant form as an aid in remembering the formula for curl.

$$\text{curl } \mathbf{F}(x, y, z) = \nabla \times \mathbf{F}(x, y, z)$$

$$= \begin{vmatrix} \mathbf{i} & \mathbf{j} & \mathbf{k} \\ \dfrac{\partial}{\partial x} & \dfrac{\partial}{\partial y} & \dfrac{\partial}{\partial z} \\ M & N & P \end{vmatrix}$$

$$= \left( \frac{\partial P}{\partial y} - \frac{\partial N}{\partial z} \right)\mathbf{i} - \left( \frac{\partial P}{\partial x} - \frac{\partial M}{\partial z} \right)\mathbf{j} + \left( \frac{\partial N}{\partial x} - \frac{\partial M}{\partial y} \right)\mathbf{k}$$

---

**EXAMPLE 7**     **Finding the Curl of a Vector Field**

⋯▷ *See LarsonCalculus.com for an interactive version of this type of example.*

Find curl $\mathbf{F}$ of the vector field

$$\mathbf{F}(x, y, z) = 2xy\mathbf{i} + (x^2 + z^2)\mathbf{j} + 2yz\mathbf{k}.$$

Is $\mathbf{F}$ irrotational?

**Solution**   The curl of $\mathbf{F}$ is

$$\text{curl } \mathbf{F}(x, y, z) = \nabla \times \mathbf{F}(x, y, z)$$

$$= \begin{vmatrix} \mathbf{i} & \mathbf{j} & \mathbf{k} \\ \dfrac{\partial}{\partial x} & \dfrac{\partial}{\partial y} & \dfrac{\partial}{\partial z} \\ 2xy & x^2 + z^2 & 2yz \end{vmatrix}$$

$$= \begin{vmatrix} \dfrac{\partial}{\partial y} & \dfrac{\partial}{\partial z} \\ x^2 + z^2 & 2yz \end{vmatrix} \mathbf{i} - \begin{vmatrix} \dfrac{\partial}{\partial x} & \dfrac{\partial}{\partial z} \\ 2xy & 2yz \end{vmatrix} \mathbf{j} + \begin{vmatrix} \dfrac{\partial}{\partial x} & \dfrac{\partial}{\partial y} \\ 2xy & x^2 + z^2 \end{vmatrix} \mathbf{k}$$

$$= (2z - 2z)\mathbf{i} - (0 - 0)\mathbf{j} + (2x - 2x)\mathbf{k}$$

$$= \mathbf{0}.$$

Because curl $\mathbf{F} = \mathbf{0}$, $\mathbf{F}$ is irrotational.

---

▷ **TECHNOLOGY**   Some computer algebra systems have a command that can be used to find the curl of a vector field. If you have access to a computer algebra system that has such a command, use it to find the curl of the vector field in Example 7.

Later in this chapter, you will assign a physical interpretation to the curl of a vector field. But for now, the primary use of curl is shown in the following test for conservative vector fields in space. The test states that for a vector field in space, the curl is **0** at every point in its domain if and only if **F** is conservative. The proof is similar to that given for Theorem 15.1.

· · · · · · · · · · · · · · · · · · ▷
**· · REMARK**   Theorem 15.2 is valid for *simply connected* domains in space. A simply connected domain in space is a domain *D* for which every simple closed curve in *D* can be shrunk to a point in *D* without leaving *D*.

---

**THEOREM 15.2    Test for Conservative Vector Field in Space**

Suppose that *M*, *N*, and *P* have continuous first partial derivatives in an open sphere *Q* in space. The vector field

$$\mathbf{F}(x, y, z) = M\mathbf{i} + N\mathbf{j} + P\mathbf{k}$$

is conservative if and only if

$$\text{curl } \mathbf{F}(x, y, z) = \mathbf{0}.$$

That is, **F** is conservative if and only if

$$\frac{\partial P}{\partial y} = \frac{\partial N}{\partial z}, \quad \frac{\partial P}{\partial x} = \frac{\partial M}{\partial z}, \quad \text{and} \quad \frac{\partial N}{\partial x} = \frac{\partial M}{\partial y}.$$

---

From Theorem 15.2, you can see that the vector field given in Example 7 is conservative because curl $\mathbf{F}(x, y, z) = \mathbf{0}$. Try showing that the vector field

$$\mathbf{F}(x, y, z) = x^3 y^2 z \mathbf{i} + x^2 z \mathbf{j} + x^2 y \mathbf{k}$$

is not conservative—you can do this by showing that its curl is

$$\text{curl } \mathbf{F}(x, y, z) = (x^3 y^2 - 2xy)\mathbf{j} + (2xz - 2x^3 yz)\mathbf{k} \neq \mathbf{0}.$$

For vector fields in space that pass the test for being conservative, you can find a potential function by following the same pattern used in the plane (as demonstrated in Example 6).

· · · · · · · · · · · · · · · · · · ▷
**· · REMARK**   Examples 6 and 8 are illustrations of a type of problem called *recovering a function from its gradient*. If you go on to take a course in differential equations, you will study other methods for solving this type of problem. One popular method gives an interplay between successive "partial integrations" and partial differentiations.

**EXAMPLE 8**   **Finding a Potential Function for F(*x, y, z*)**

Find a potential function for

$$\mathbf{F}(x, y, z) = 2xy\mathbf{i} + (x^2 + z^2)\mathbf{j} + 2yz\mathbf{k}.$$

**Solution**   From Example 7, you know that the vector field given by **F** is conservative. If *f* is a function such that $\mathbf{F}(x, y, z) = \nabla f(x, y, z)$, then

$$f_x(x, y, z) = 2xy, \quad f_y(x, y, z) = x^2 + z^2, \quad \text{and} \quad f_z(x, y, z) = 2yz$$

and integrating with respect to *x*, *y*, and *z* separately produces

$$f(x, y, z) = \int M \, dx = \int 2xy \, dx = x^2 y + g(y, z)$$

$$f(x, y, z) = \int N \, dy = \int (x^2 + z^2) \, dy = x^2 y + yz^2 + h(x, z)$$

$$f(x, y, z) = \int P \, dz = \int 2yz \, dz = yz^2 + k(x, y).$$

Comparing these three versions of $f(x, y, z)$, you can conclude that

$$g(y, z) = yz^2 + K_1, \quad h(x, z) = K_2, \quad \text{and} \quad k(x, y) = x^2 y + K_3.$$

So, $f(x, y, z)$ is given by

$$f(x, y, z) = x^2 y + yz^2 + K. \qquad K = K_1 + K_2 + K_3$$

## Divergence of a Vector Field

You have seen that the curl of a vector field $\mathbf{F}$ is itself a vector field. Another important function defined on a vector field is **divergence,** which is a scalar function.

---

**Definition of Divergence of a Vector Field**

The **divergence** of $\mathbf{F}(x, y) = M\mathbf{i} + N\mathbf{j}$ is

$$\text{div } \mathbf{F}(x, y) = \nabla \cdot \mathbf{F}(x, y) = \frac{\partial M}{\partial x} + \frac{\partial N}{\partial y}. \qquad \text{Plane}$$

The **divergence** of $\mathbf{F}(x, y, z) = M\mathbf{i} + N\mathbf{j} + P\mathbf{k}$ is

$$\text{div } \mathbf{F}(x, y, z) = \nabla \cdot \mathbf{F}(x, y, z) = \frac{\partial M}{\partial x} + \frac{\partial N}{\partial y} + \frac{\partial P}{\partial z}. \qquad \text{Space}$$

If div $\mathbf{F} = 0$, then $\mathbf{F}$ is said to be **divergence free.**

---

The dot product notation used for divergence comes from considering $\nabla$ as a **differential operator,** as follows.

$$\nabla \cdot \mathbf{F}(x, y, z) = \left[\left(\frac{\partial}{\partial x}\right)\mathbf{i} + \left(\frac{\partial}{\partial y}\right)\mathbf{j} + \left(\frac{\partial}{\partial z}\right)\mathbf{k}\right] \cdot (M\mathbf{i} + N\mathbf{j} + P\mathbf{k})$$

$$= \frac{\partial M}{\partial x} + \frac{\partial N}{\partial y} + \frac{\partial P}{\partial z}$$

▷ **TECHNOLOGY** Some computer algebra systems have a command that can be used to find the divergence of a vector field. If you have access to a computer algebra system that has such a command, use it to find the divergence of the vector field in Example 9.

### EXAMPLE 9  Finding the Divergence of a Vector Field

Find the divergence at $(2, 1, -1)$ for the vector field

$$\mathbf{F}(x, y, z) = x^3y^2z\mathbf{i} + x^2z\mathbf{j} + x^2y\mathbf{k}.$$

**Solution** The divergence of $\mathbf{F}$ is

$$\text{div } \mathbf{F}(x, y, z) = \frac{\partial}{\partial x}[x^3y^2z] + \frac{\partial}{\partial y}[x^2z] + \frac{\partial}{\partial z}[x^2y] = 3x^2y^2z.$$

At the point $(2, 1, -1)$, the divergence is

$$\text{div } \mathbf{F}(2, 1, -1) = 3(2^2)(1^2)(-1) = -12. \qquad ■$$

Divergence can be viewed as a type of derivative of $\mathbf{F}$ in that, for vector fields representing velocities of moving particles, the divergence measures the rate of particle flow per unit volume at a point. In hydrodynamics (the study of fluid motion), a velocity field that is divergence free is called **incompressible.** In the study of electricity and magnetism, a vector field that is divergence free is called **solenoidal.**

There are many important properties of the divergence and curl of a vector field $\mathbf{F}$ [see Exercise 77(a)–(g)]. One that is used often is described in Theorem 15.3. You are asked to prove this theorem in Exercise 77(h).

---

**THEOREM 15.3  Divergence and Curl**

If $\mathbf{F}(x, y, z) = M\mathbf{i} + N\mathbf{j} + P\mathbf{k}$ is a vector field and $M$, $N$, and $P$ have continuous second partial derivatives, then

$$\text{div(curl } \mathbf{F}) = 0.$$

---

# 15.1 Exercises

See **CalcChat.com** for tutorial help and worked-out solutions to odd-numbered exercises.

**CONCEPT CHECK**

**1. Vector Field**   Define a vector field in the plane and in space. Give some physical examples of vector fields.

**2. Conservative Vector Field**   What is a conservative vector field? How do you test whether a vector field is conservative in the plane and in space?

**3. Potential Function**   Describe how to find a potential function for a vector field that is conservative.

**4. Vector Field**   A vector field in space is conservative. Is the vector field irrotational? Explain.

**Matching**   In Exercises 5–8, match the vector field with its graph. [The graphs are labeled (a), (b), (c), and (d).]

(a)

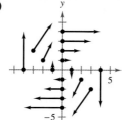

(b)

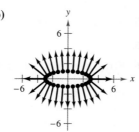

(c)

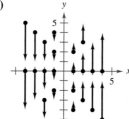

(d)

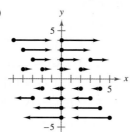

**5.** $\mathbf{F}(x, y) = y\mathbf{i}$
**6.** $\mathbf{F}(x, y) = x\mathbf{j}$
**7.** $\mathbf{F}(x, y) = y\mathbf{i} - x\mathbf{j}$
**8.** $\mathbf{F}(x, y) = x\mathbf{i} + 3y\mathbf{j}$

 **Sketching a Vector Field**   In Exercises 9–14, find $\|\mathbf{F}\|$ and sketch several representative vectors in the vector field.

**9.** $\mathbf{F}(x, y) = \mathbf{i} + \mathbf{j}$
**10.** $\mathbf{F}(x, y) = y\mathbf{i} - 2x\mathbf{j}$
**11.** $\mathbf{F}(x, y) = -\mathbf{i} + 3y\mathbf{j}$
**12.** $\mathbf{F}(x, y) = y\mathbf{i} + x\mathbf{j}$
**13.** $\mathbf{F}(x, y, z) = \mathbf{i} + \mathbf{j} + \mathbf{k}$
**14.** $\mathbf{F}(x, y, z) = x\mathbf{i} + y\mathbf{j} + z\mathbf{k}$

**Graphing a Vector Field Using Technology**   In Exercises 15–18, use a computer algebra system to graph several representative vectors in the vector field.

**15.** $\mathbf{F}(x, y) = \frac{1}{8}(2xy\mathbf{i} + y^2\mathbf{j})$

**16.** $\mathbf{F}(x, y) = \langle 2y - x, 2y + x \rangle$

**17.** $\mathbf{F}(x, y, z) = \dfrac{x\mathbf{i} + y\mathbf{j} + z\mathbf{k}}{\sqrt{x^2 + y^2 + z^2}}$

**18.** $\mathbf{F}(x, y, z) = \langle x, -y, z \rangle$

 **Finding a Conservative Vector Field**   In Exercises 19–28, find the conservative vector field for the potential function by finding its gradient.

**19.** $f(x, y) = x^2 + 2y^2$
**20.** $f(x, y) = x^3 - 2xy$
**21.** $g(x, y) = 5x^2 + 3xy + y^2$
**22.** $g(x, y) = \sin 3x \cos 4y$
**23.** $f(x, y, z) = 6xyz$
**24.** $f(x, y, z) = \sqrt{x^2 y + z^2}$
**25.** $g(x, y, z) = z + ye^{x^2}$
**26.** $g(x, y, z) = \dfrac{y}{z} + \dfrac{z}{x} - \dfrac{xz}{y}$
**27.** $h(x, y, z) = xy \ln(x + y)$
**28.** $h(x, y, z) = x \arcsin yz$

 **Testing for a Conservative Vector Field**   In Exercises 29–36, determine whether the vector field is conservative.

**29.** $\mathbf{F}(x, y) = xy^2\mathbf{i} + x^2y\mathbf{j}$
**30.** $\mathbf{F}(x, y) = \dfrac{1}{x^2}(y\mathbf{i} - x\mathbf{j})$
**31.** $\mathbf{F}(x, y) = \sin y\mathbf{i} + x \sin y\mathbf{j}$
**32.** $\mathbf{F}(x, y) = 5y^2(y\mathbf{i} + 2x\mathbf{j})$
**33.** $\mathbf{F}(x, y) = \dfrac{1}{xy}(y\mathbf{i} - x\mathbf{j})$
**34.** $\mathbf{F}(x, y) = \dfrac{2}{y^2}e^{2x/y}(y\mathbf{i} - x\mathbf{j})$
**35.** $\mathbf{F}(x, y) = \dfrac{\mathbf{i} + \mathbf{j}}{\sqrt{x^2 + y^2}}$
**36.** $\mathbf{F}(x, y) = \dfrac{y\mathbf{i} + x\mathbf{j}}{\sqrt{1 + xy}}$

 **Finding a Potential Function**   In Exercises 37–44, determine whether the vector field is conservative. If it is, find a potential function for the vector field.

**37.** $\mathbf{F}(x, y) = (3y - x^2)\mathbf{i} + (3x + y)\mathbf{j}$
**38.** $\mathbf{F}(x, y) = (x^3 + e^y)\mathbf{i} + (xe^y - 6)\mathbf{j}$
**39.** $\mathbf{F}(x, y) = xe^{x^2y}(2y\mathbf{i} + x\mathbf{j})$
**40.** $\mathbf{F}(x, y) = \dfrac{1}{y^2}(y\mathbf{i} - 2x\mathbf{j})$
**41.** $\mathbf{F}(x, y) = \dfrac{2y}{x}\mathbf{i} - \dfrac{x^2}{y^2}\mathbf{j}$
**42.** $\mathbf{F}(x, y) = \dfrac{x\mathbf{i} + y\mathbf{j}}{x^2 + y^2}$
**43.** $\mathbf{F}(x, y) = \sin y\mathbf{i} + x \cos y\mathbf{j}$
**44.** $\mathbf{F}(x, y) = (\ln y + 2)\mathbf{i} + \dfrac{x}{y}\mathbf{j}$

 **Finding the Curl of a Vector Field**   In Exercises 45–48, find the curl of the vector field at the given point.

**45.** $\mathbf{F}(x, y, z) = xyz\mathbf{i} + xyz\mathbf{j} + xyz\mathbf{k};\ (2, 1, 3)$
**46.** $\mathbf{F}(x, y, z) = x^2z\mathbf{i} - 2xz\mathbf{j} + yz\mathbf{k};\ (2, -1, 3)$
**47.** $\mathbf{F}(x, y, z) = e^x \sin y\mathbf{i} - e^x \cos y\mathbf{j};\ (0, 0, 1)$
**48.** $\mathbf{F}(x, y, z) = e^{-xyz}(\mathbf{i} + \mathbf{j} + \mathbf{k});\ (3, 2, 0)$

**Finding the Curl of a Vector Field Using Technology**   In Exercises 49 and 50, use a computer algebra system to find the curl of the vector field.

**49.** $\mathbf{F}(x, y, z) = \arctan\left(\dfrac{x}{y}\right)\mathbf{i} + \ln\sqrt{x^2 + y^2}\,\mathbf{j} + \mathbf{k}$

**50.** $\mathbf{F}(x, y, z) = \dfrac{yz}{y - z}\mathbf{i} + \dfrac{xz}{x - z}\mathbf{j} + \dfrac{xy}{x - y}\mathbf{k}$

**Finding a Potential Function** In Exercises 51–56, determine whether the vector field is conservative. If it is, find a potential function for the vector field.

**51.** $F(x, y, z) = (3x^2 + yz)i + (3y^2 + xz)j + (3z^2 + xy)k$

**52.** $F(x, y, z) = y^2z^3i + 2xyz^3j + 3xy^2z^2k$

**53.** $F(x, y, z) = \sin z\,i + \sin x\,j + \sin y\,k$

**54.** $F(x, y, z) = ye^zi + ze^xj + xe^yk$

**55.** $F(x, y, z) = \dfrac{z}{y}i - \dfrac{xz}{y^2}j + \left(\dfrac{x}{y} - 1\right)k$

**56.** $F(x, y, z) = \dfrac{x}{x^2 + y^2}i + \dfrac{y}{x^2 + y^2}j + k$

**Finding the Divergence of a Vector Field** In Exercises 57–60, find the divergence of the vector field.

**57.** $F(x, y) = x^2i + 2y^2j$      **58.** $F(x, y) = xe^xi - x^2y^2j$

**59.** $F(x, y, z) = \sin^2 x\,i + z\cos z\,j + z^3k$

**60.** $F(x, y, z) = \ln(x^2 + y^2)i + xyj + \ln(y^2 + z^2)k$

**Finding the Divergence of a Vector Field** In Exercises 61–64, find the divergence of the vector field at the given point.

**61.** $F(x, y, z) = xyzi + xz^2j + 3yz^2k$;  $(2, 4, 1)$

**62.** $F(x, y, z) = x^2zi - 2xzj + yzk$;  $(2, -1, 3)$

**63.** $F(x, y, z) = e^x \sin y\,i - e^x \cos y\,j + z^2k$;  $(3, 0, 0)$

**64.** $F(x, y, z) = \ln(xyz)(i + j + k)$;  $(3, 2, 1)$

---

**EXPLORING CONCEPTS**

**Think About It** In Exercises 65–67, consider a scalar function $f$ and a vector field $F$ in space. Determine whether the expression is a vector field, a scalar function, or neither. Explain.

**65.** $\text{curl}(\nabla f)$          **66.** $\text{div}[\text{curl}(\nabla f)]$

**67.** $\text{curl}(\text{div } F)$

---

**68.** **HOW DO YOU SEE IT?** Several representative vectors in the vector fields

$$F(x, y) = \frac{xi + yj}{\sqrt{x^2 + y^2}} \quad \text{and} \quad G(x, y) = \frac{xi - yj}{\sqrt{x^2 + y^2}}$$

are shown below. Match each vector field with its graph. Explain your reasoning.

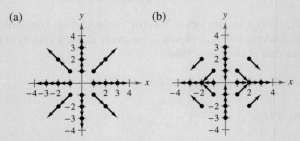

(a)                              (b)

---

**Curl of a Cross Product** In Exercises 69 and 70, find $\text{curl}(F \times G) = \nabla \times (F \times G)$.

**69.** $F(x, y, z) = i + 3xj + 2yk$   **70.** $F(x, y, z) = xi - zk$

   $G(x, y, z) = xi - yj + zk$      $G(x, y, z) = x^2i + yj + z^2k$

**Curl of the Curl of a Vector Field** In Exercises 71 and 72, find $\text{curl}(\text{curl } F) = \nabla \times (\nabla \times F)$.

**71.** $F(x, y, z) = xyzi + yj + zk$

**72.** $F(x, y, z) = x^2zi - 2xzj + yzk$

**Divergence of a Cross Product** In Exercises 73 and 74, find $\text{div}(F \times G) = \nabla \cdot (F \times G)$.

**73.** $F(x, y, z) = i + 3xj + 2yk$

   $G(x, y, z) = xi - yj + zk$

**74.** $F(x, y, z) = xi - zk$

   $G(x, y, z) = x^2i + yj + z^2k$

**Divergence of the Curl of a Vector Field** In Exercises 75 and 76, find $\text{div}(\text{curl } F) = \nabla \cdot (\nabla \times F)$.

**75.** $F(x, y, z) = xyzi + yj + zk$

**76.** $F(x, y, z) = x^2zi - 2xzj + yzk$

**77.** **Proof** In parts (a)–(h), prove the property for vector fields $F$ and $G$ and scalar function $f$. (Assume that the required partial derivatives are continuous.)

(a) $\text{curl}(F + G) = \text{curl } F + \text{curl } G$

(b) $\text{curl}(\nabla f) = \nabla \times (\nabla f) = 0$

(c) $\text{div}(F + G) = \text{div } F + \text{div } G$

(d) $\text{div}(F \times G) = (\text{curl } F) \cdot G - F \cdot (\text{curl } G)$

(e) $\nabla \times [\nabla f + (\nabla \times F)] = \nabla \times (\nabla \times F)$

(f) $\nabla \times (fF) = f(\nabla \times F) + (\nabla f) \times F$

(g) $\text{div}(fF) = f\,\text{div } F + \nabla f \cdot F$

(h) $\text{div}(\text{curl } F) = 0$  (Theorem 15.3)

**78. Earth's Magnetic Field**

A cross section of Earth's magnetic field can be represented as a vector field in which the center of Earth is located at the origin and the positive $y$-axis points in the direction of the magnetic north pole. The equation for this field is

$$F(x, y) = M(x, y)i + N(x, y)j$$

$$= \frac{m}{(x^2 + y^2)^{5/2}}[3xyi + (2y^2 - x^2)j]$$

where $m$ is the magnetic moment of Earth. Show that this vector field is conservative.

# 15.2    Line Integrals

■ Understand and use the concept of a piecewise smooth curve.
■ Write and evaluate a line integral.
■ Write and evaluate a line integral of a vector field.
■ Write and evaluate a line integral in differential form.

## Piecewise Smooth Curves

A classic property of gravitational fields is that, subject to certain physical constraints, the work done by gravity on an object moving between two points in the field is independent of the path taken by the object. One of the constraints is that the **path** must be a piecewise smooth curve. Recall that a plane curve $C$ given by

$$\mathbf{r}(t) = x(t)\mathbf{i} + y(t)\mathbf{j}, \quad a \le t \le b$$

is **smooth** when

$$\frac{dx}{dt} \quad \text{and} \quad \frac{dy}{dt}$$

are continuous on $[a, b]$ and not simultaneously 0 on $(a, b)$. Similarly, a space curve $C$ given by

$$\mathbf{r}(t) = x(t)\mathbf{i} + y(t)\mathbf{j} + z(t)\mathbf{k}, \quad a \le t \le b$$

is **smooth** when

$$\frac{dx}{dt}, \quad \frac{dy}{dt}, \quad \text{and} \quad \frac{dz}{dt}$$

are continuous on $[a, b]$ and not simultaneously 0 on $(a, b)$. A curve $C$ is **piecewise smooth** when the interval $[a, b]$ can be partitioned into a finite number of subintervals, on each of which $C$ is smooth.

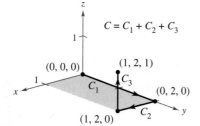

Figure 15.7

**EXAMPLE 1**    Finding a Piecewise Smooth Parametrization

Find a piecewise smooth parametrization of the graph of $C$ shown in Figure 15.7.

**Solution**    Because $C$ consists of three line segments $C_1$, $C_2$, and $C_3$, you can construct a smooth parametrization for each segment and piece them together by making the last $t$-value in $C_i$ correspond to the first $t$-value in $C_{i+1}$.

$C_1$: $x(t) = 0,$        $y(t) = 2t,$      $z(t) = 0,$      $0 \le t \le 1$
$C_2$: $x(t) = t - 1,$    $y(t) = 2,$      $z(t) = 0,$      $1 \le t \le 2$
$C_3$: $x(t) = 1,$        $y(t) = 2,$      $z(t) = t - 2,$   $2 \le t \le 3$

So, $C$ is given by

$$\mathbf{r}(t) = \begin{cases} 2t\mathbf{j}, & 0 \le t \le 1 \\ (t-1)\mathbf{i} + 2\mathbf{j}, & 1 \le t \le 2. \\ \mathbf{i} + 2\mathbf{j} + (t-2)\mathbf{k}, & 2 \le t \le 3 \end{cases}$$

Because $C_1$, $C_2$, and $C_3$ are smooth, it follows that $C$ is piecewise smooth.    ■

Recall that parametrization of a curve induces an **orientation** to the curve. For instance, in Example 1, the curve is oriented such that the positive direction is from $(0, 0, 0)$, following the curve to $(1, 2, 1)$. Try finding a parametrization that induces the opposite orientation.

**JOSIAH WILLARD GIBBS
(1839–1903)**

Many physicists and mathematicians have contributed to the theory and applications described in this chapter—Newton, Gauss, Laplace, Hamilton, and Maxwell, among others. However, the use of vector analysis to describe these results is attributed primarily to the American mathematical physicist Josiah Willard Gibbs. *See LarsonCalculus.com to read more of this biography.*

## Line Integrals

Up to this point in the text, you have studied various types of integrals. For a single integral

$$\int_a^b f(x)\, dx \qquad \text{Integrate over interval } [a, b].$$

you integrated over the interval $[a, b]$. Similarly, for a double integral

$$\iint_R f(x, y)\, dA \qquad \text{Integrate over region } R.$$

you integrated over the region $R$ in the plane. In this section, you will study a new type of integral called a **line integral**

$$\int_C f(x, y)\, ds \qquad \text{Integrate over curve } C.$$

for which you integrate over a piecewise smooth curve $C$. (The terminology is somewhat unfortunate—this type of integral might be better described as a "curve integral.")

To introduce the concept of a line integral, consider the mass of a wire of finite length, given by a curve $C$ in space. The density (mass per unit length) of the wire at the point $(x, y, z)$ is given by $f(x, y, z)$. Partition the curve $C$ by the points

$$P_0, P_1, \ldots, P_n$$

producing $n$ subarcs, as shown in Figure 15.8. The length of the $i$th subarc is given by $\Delta s_i$. Next, choose a point $(x_i, y_i, z_i)$ in each subarc. If the length of each subarc is small, then the total mass of the wire can be approximated by the sum

$$\text{Mass of wire} \approx \sum_{i=1}^n f(x_i, y_i, z_i)\, \Delta s_i.$$

By letting $\|\Delta\|$ denote the length of the longest subarc and letting $\|\Delta\|$ approach 0, it seems reasonable that the limit of this sum approaches the mass of the wire. This leads to the next definition.

Partitioning of curve $C$
**Figure 15.8**

---

### Definition of Line Integral

If $f$ is defined in a region containing a smooth curve $C$ of finite length, then the **line integral of $f$ along $C$** is given by

$$\int_C f(x, y)\, ds = \lim_{\|\Delta\| \to 0} \sum_{i=1}^n f(x_i, y_i)\, \Delta s_i \qquad \text{Plane}$$

or

$$\int_C f(x, y, z)\, ds = \lim_{\|\Delta\| \to 0} \sum_{i=1}^n f(x_i, y_i, z_i)\, \Delta s_i \qquad \text{Space}$$

provided this limit exists.

---

As with the integrals discussed in Chapter 14, evaluation of a line integral is best accomplished by converting it to a definite integral. It can be shown that if $f$ is *continuous*, then the limit given above exists and is the same for all smooth parametrizations of $C$.

To evaluate a line integral over a plane curve $C$ given by $\mathbf{r}(t) = x(t)\mathbf{i} + y(t)\mathbf{j}$, use the fact that

$$ds = \|\mathbf{r}'(t)\| \, dt = \sqrt{[x'(t)]^2 + [y'(t)]^2} \, dt.$$

A similar formula holds for a space curve, as indicated in Theorem 15.4.

---

**THEOREM 15.4    Evaluation of a Line Integral as a Definite Integral**

Let $f$ be continuous in a region containing a smooth curve $C$. If $C$ is given by $\mathbf{r}(t) = x(t)\mathbf{i} + y(t)\mathbf{j}$, where $a \le t \le b$, then

$$\int_C f(x, y) \, ds = \int_a^b f(x(t), y(t)) \sqrt{[x'(t)]^2 + [y'(t)]^2} \, dt.$$

If $C$ is given by $\mathbf{r}(t) = x(t)\mathbf{i} + y(t)\mathbf{j} + z(t)\mathbf{k}$, where $a \le t \le b$, then

$$\int_C f(x, y, z) \, ds = \int_a^b f(x(t), y(t), z(t)) \sqrt{[x'(t)]^2 + [y'(t)]^2 + [z'(t)]^2} \, dt.$$

---

Note that if $f(x, y, z) = 1$, then the line integral gives the arc length of the curve $C$, as defined in Section 12.5. That is,

$$\int_C 1 \, ds = \int_a^b \|\mathbf{r}'(t)\| \, dt = \text{length of curve } C.$$

---

**EXAMPLE 2    Evaluating a Line Integral**

Evaluate

$$\int_C (x^2 - y + 3z) \, ds$$

where $C$ is the line segment shown in Figure 15.9.

**Solution**  Begin by writing a parametric form of the equation of the line segment:

$$x = t, \quad y = 2t, \quad \text{and} \quad z = t, \quad 0 \le t \le 1.$$

Therefore, $x'(t) = 1$, $y'(t) = 2$, and $z'(t) = 1$, which implies that

$$\sqrt{[x'(t)]^2 + [y'(t)]^2 + [z'(t)]^2} = \sqrt{1^2 + 2^2 + 1^2} = \sqrt{6}.$$

So, the line integral takes the following form.

$$\begin{aligned}
\int_C (x^2 - y + 3z) \, ds &= \int_0^1 (t^2 - 2t + 3t) \sqrt{6} \, dt \\
&= \sqrt{6} \int_0^1 (t^2 + t) \, dt \\
&= \sqrt{6} \left[ \frac{t^3}{3} + \frac{t^2}{2} \right]_0^1 \\
&= \frac{5\sqrt{6}}{6}
\end{aligned}$$

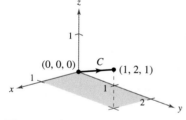

**Figure 15.9**

The value of the line integral in Example 2 does not depend on the parametrization of the line segment $C$; any smooth parametrization will produce the same value. To convince yourself of this, try some other parametrizations, such as $x = 1 + 2t$, $y = 2 + 4t$, and $z = 1 + 2t$, $-\frac{1}{2} \le t \le 0$, or $x = -t$, $y = -2t$, and $z = -t$, $-1 \le t \le 0$.

Let $C$ be a path composed of smooth curves $C_1, C_2, \ldots, C_n$. If $f$ is continuous on $C$, then it can be shown that

$$\int_C f(x, y)\, ds = \int_{C_1} f(x, y)\, ds + \int_{C_2} f(x, y)\, ds + \cdots + \int_{C_n} f(x, y)\, ds.$$

This property is used in Example 3.

**EXAMPLE 3**    **Evaluating a Line Integral Over a Path**

Evaluate

$$\int_C x\, ds$$

where $C$ is the piecewise smooth curve shown in Figure 15.10.

**Solution**    Begin by integrating up the line $y = x$, using the following parametrization.

$$C_1\colon\; x = t, \quad y = t, \quad 0 \le t \le 1$$

For this curve, $\mathbf{r}(t) = t\mathbf{i} + t\mathbf{j}$, which implies that $x'(t) = 1$ and $y'(t) = 1$. So,

$$\sqrt{[x'(t)]^2 + [y'(t)]^2} = \sqrt{2}$$

and you have

$$\int_{C_1} x\, ds = \int_0^1 t\sqrt{2}\, dt = \frac{\sqrt{2}}{2}t^2 \Big]_0^1 = \frac{\sqrt{2}}{2}.$$

Next, integrate down the parabola $y = x^2$, using the parametrization

$$C_2\colon x = 1 - t, \quad y = (1 - t)^2, \quad 0 \le t \le 1.$$

For this curve,

$$\mathbf{r}(t) = (1 - t)\mathbf{i} + (1 - t)^2\mathbf{j}$$

which implies that $x'(t) = -1$ and $y'(t) = -2(1 - t)$. So,

$$\sqrt{[x'(t)]^2 + [y'(t)]^2} = \sqrt{1 + 4(1 - t)^2}$$

and you have

$$\int_{C_2} x\, ds = \int_0^1 (1 - t)\sqrt{1 + 4(1 - t)^2}\, dt$$

$$= -\frac{1}{8}\left[\frac{2}{3}[1 + 4(1 - t)^2]^{3/2}\right]_0^1$$

$$= \frac{1}{12}(5^{3/2} - 1).$$

Consequently,

$$\int_C x\, ds = \int_{C_1} x\, ds + \int_{C_2} x\, ds = \frac{\sqrt{2}}{2} + \frac{1}{12}(5^{3/2} - 1) \approx 1.56.$$

For parametrizations given by $\mathbf{r}(t) = x(t)\mathbf{i} + y(t)\mathbf{j} + z(t)\mathbf{k}$, it is helpful to remember the form of $ds$ as

$$ds = \|\mathbf{r}'(t)\|\, dt = \sqrt{[x'(t)]^2 + [y'(t)]^2 + [z'(t)]^2}\, dt.$$

This is demonstrated in Example 4.

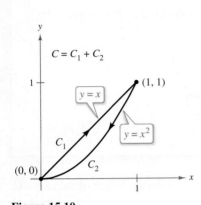

**Figure 15.10**

**EXAMPLE 4** **Evaluating a Line Integral**

Evaluate $\displaystyle\int_C (x + 2)\, ds$, where $C$ is the curve represented by

$$\mathbf{r}(t) = t\mathbf{i} + \frac{4}{3}t^{3/2}\mathbf{j} + \frac{1}{2}t^2\mathbf{k}, \quad 0 \le t \le 2.$$

**Solution** Because $\mathbf{r}'(t) = \mathbf{i} + 2t^{1/2}\mathbf{j} + t\mathbf{k}$ and

$$\|\mathbf{r}'(t)\| = \sqrt{[x'(t)^2] + [y'(t)]^2 + [z'(t)]^2} = \sqrt{1 + 4t + t^2}$$

it follows that

$$\int_C (x + 2)\, ds = \int_0^2 (t + 2)\sqrt{1 + 4t + t^2}\, dt$$

$$= \frac{1}{2}\int_0^2 2(t + 2)(1 + 4t + t^2)^{1/2}\, dt$$

$$= \frac{1}{3}\left[(1 + 4t + t^2)^{3/2}\right]_0^2$$

$$= \frac{1}{3}\left(13\sqrt{13} - 1\right)$$

$$\approx 15.29.$$

The next example shows how a line integral can be used to find the mass of a spring whose density varies. In Figure 15.11, note that the density of this spring increases as the spring spirals up the $z$-axis.

**EXAMPLE 5** **Finding the Mass of a Spring**

Find the mass of a spring in the shape of the circular helix

$$\mathbf{r}(t) = \frac{1}{\sqrt{2}}(\cos t\mathbf{i} + \sin t\mathbf{j} + t\mathbf{k})$$

where $0 \le t \le 6\pi$ and the density of the spring is

$$\rho(x, y, z) = 1 + z$$

as shown in Figure 15.11.

**Solution** Because

$$\|\mathbf{r}'(t)\| = \frac{1}{\sqrt{2}}\sqrt{(-\sin t)^2 + (\cos t)^2 + (1)^2} = 1$$

it follows that the mass of the spring is

$$\text{Mass} = \int_C (1 + z)\, ds$$

$$= \int_0^{6\pi} \left(1 + \frac{t}{\sqrt{2}}\right) dt$$

$$= \left[t + \frac{t^2}{2\sqrt{2}}\right]_0^{6\pi}$$

$$= 6\pi\left(1 + \frac{3\pi}{\sqrt{2}}\right)$$

$$\approx 144.47.$$

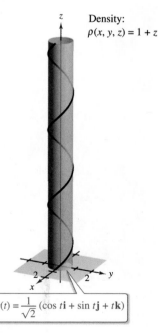

Density:
$\rho(x, y, z) = 1 + z$

$\mathbf{r}(t) = \dfrac{1}{\sqrt{2}}(\cos t\mathbf{i} + \sin t\mathbf{j} + t\mathbf{k})$

**Figure 15.11**

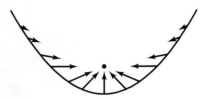

Inverse square force field **F**

Vectors along a parabolic path in the
force field **F**

**Figure 15.12**

## Line Integrals of Vector Fields

One of the most important physical applications of line integrals is that of finding the **work** done on an object moving in a force field. For example, Figure 15.12 shows an inverse square force field similar to the gravitational field of the sun. Note that the magnitude of the force along a circular path about the center is constant, whereas the magnitude of the force along a parabolic path varies from point to point.

To see how a line integral can be used to find work done in a force field **F**, consider an object moving along a path $C$ in the field, as shown in Figure 15.13. To determine the work done by the force, you need consider only that part of the force that is acting in the same direction as that in which the object is moving (or the opposite direction). This means that at each point on $C$, you can consider the projection **F · T** of the force vector **F** onto the unit tangent vector **T**. On a small subarc of length $\Delta s_i$, the increment of work is

$$\Delta W_i = (\text{force})(\text{distance})$$
$$\approx [\mathbf{F}(x_i, y_i, z_i) \cdot \mathbf{T}(x_i, y_i, z_i)] \, \Delta s_i$$

where $(x_i, y_i, z_i)$ is a point in the $i$th subarc. Consequently, the total work done is given by the integral

$$W = \int_C \mathbf{F}(x, y, z) \cdot \mathbf{T}(x, y, z) \, ds.$$

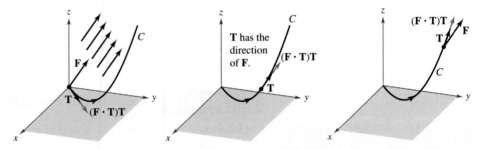

At each point on $C$, the force in the direction of motion is $(\mathbf{F} \cdot \mathbf{T})\mathbf{T}$.
**Figure 15.13**

This line integral appears in other contexts and is the basis of the definition of the **line integral of a vector field** shown below. Note in the definition that

$$\mathbf{F} \cdot \mathbf{T} \, ds = \mathbf{F} \cdot \frac{\mathbf{r}'(t)}{\|\mathbf{r}'(t)\|} \|\mathbf{r}'(t)\| \, dt$$
$$= \mathbf{F} \cdot \mathbf{r}'(t) \, dt$$
$$= \mathbf{F} \cdot d\mathbf{r}.$$

---

### Definition of the Line Integral of a Vector Field

Let **F** be a continuous vector field defined on a smooth curve $C$ given by

$$\mathbf{r}(t), \quad a \leq t \leq b.$$

The **line integral** of **F** on $C$ is given by

$$\int_C \mathbf{F} \cdot d\mathbf{r} = \int_C \mathbf{F} \cdot \mathbf{T} \, ds$$
$$= \int_a^b \mathbf{F}(x(t), y(t), z(t)) \cdot \mathbf{r}'(t) \, dt.$$

**EXAMPLE 6** **Work Done by a Force**

⋅⋅⋅⋅▷ *See LarsonCalculus.com for an interactive version of this type of example.*

Find the work done by the force field

$$\mathbf{F}(x, y, z) = -\frac{1}{2}x\mathbf{i} - \frac{1}{2}y\mathbf{j} + \frac{1}{4}\mathbf{k} \qquad \text{Force field } \mathbf{F}$$

on a particle as it moves along the helix given by

$$\mathbf{r}(t) = \cos t\,\mathbf{i} + \sin t\,\mathbf{j} + t\mathbf{k} \qquad \text{Space curve } C$$

from the point $(1, 0, 0)$ to the point $(-1, 0, 3\pi)$, as shown in Figure 15.14.

**Solution** Because

$$\mathbf{r}(t) = x(t)\mathbf{i} + y(t)\mathbf{j} + z(t)\mathbf{k}$$
$$= \cos t\,\mathbf{i} + \sin t\,\mathbf{j} + t\mathbf{k}$$

it follows that

$$x(t) = \cos t, \quad y(t) = \sin t, \quad \text{and} \quad z(t) = t.$$

So, the force field can be written as

$$\mathbf{F}(x(t), y(t), z(t)) = -\frac{1}{2}\cos t\,\mathbf{i} - \frac{1}{2}\sin t\,\mathbf{j} + \frac{1}{4}\mathbf{k}.$$

To find the work done by the force field in moving a particle along the curve $C$, use the fact that

$$\mathbf{r}'(t) = -\sin t\,\mathbf{i} + \cos t\,\mathbf{j} + \mathbf{k}$$

and write the following.

$$W = \int_C \mathbf{F} \cdot d\mathbf{r}$$

$$= \int_a^b \mathbf{F}(x(t), y(t), z(t)) \cdot \mathbf{r}'(t)\, dt$$

$$= \int_0^{3\pi} \left(-\frac{1}{2}\cos t\,\mathbf{i} - \frac{1}{2}\sin t\,\mathbf{j} + \frac{1}{4}\mathbf{k}\right) \cdot (-\sin t\,\mathbf{i} + \cos t\,\mathbf{j} + \mathbf{k})\, dt$$

$$= \int_0^{3\pi} \left(\frac{1}{2}\sin t\cos t - \frac{1}{2}\sin t\cos t + \frac{1}{4}\right) dt$$

$$= \int_0^{3\pi} \frac{1}{4}\, dt$$

$$= \frac{1}{4}t\,\Big]_0^{3\pi}$$

$$= \frac{3\pi}{4}$$

In Example 6, note that the $x$- and $y$-components of the force field end up contributing nothing to the total work. This occurs because *in this particular example*, the $z$-component of the force field is the only portion of the force that is acting in the same (or opposite) direction in which the particle is moving (see Figure 15.15).

▷ **TECHNOLOGY** Figure 15.15 shows a computer-generated view of the force field in Example 6. The figure indicates that each vector in the force field points toward the $z$-axis.

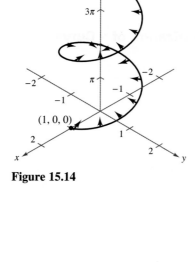

**Figure 15.14**

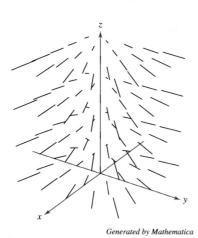

*Generated by Mathematica*

**Figure 15.15**

For line integrals of vector functions, the orientation of the curve $C$ is important. When the orientation of the curve is reversed, the unit tangent vector $\mathbf{T}(t)$ is changed to $-\mathbf{T}(t)$, and you obtain

$$\int_{-C} \mathbf{F} \cdot d\mathbf{r} = -\int_{C} \mathbf{F} \cdot d\mathbf{r}.$$

### EXAMPLE 7    Orientation and Parametrization of a Curve

Let $\mathbf{F}(x, y) = y\mathbf{i} + x^2\mathbf{j}$ and evaluate the line integral

$$\int_{C} \mathbf{F} \cdot d\mathbf{r}$$

for each parabolic curve shown in Figure 15.16.

**a.** $C_1$: $\mathbf{r}_1(t) = (4 - t)\mathbf{i} + (4t - t^2)\mathbf{j}$,  $0 \le t \le 3$
**b.** $C_2$: $\mathbf{r}_2(t) = t\mathbf{i} + (4t - t^2)\mathbf{j}$,  $1 \le t \le 4$

**Solution**

**a.** Because $\mathbf{r}_1'(t) = -\mathbf{i} + (4 - 2t)\mathbf{j}$ and

$$\mathbf{F}(x(t), y(t)) = (4t - t^2)\mathbf{i} + (4 - t)^2\mathbf{j}$$

the line integral is

$$\int_{C_1} \mathbf{F} \cdot d\mathbf{r} = \int_0^3 [(4t - t^2)\mathbf{i} + (4 - t)^2\mathbf{j}] \cdot [-\mathbf{i} + (4 - 2t)\mathbf{j}]\, dt$$

$$= \int_0^3 (-4t + t^2 + 64 - 64t + 20t^2 - 2t^3)\, dt$$

$$= \int_0^3 (-2t^3 + 21t^2 - 68t + 64)\, dt$$

$$= \left[ -\frac{t^4}{2} + 7t^3 - 34t^2 + 64t \right]_0^3$$

$$= \frac{69}{2}.$$

**b.** Because $\mathbf{r}_2'(t) = \mathbf{i} + (4 - 2t)\mathbf{j}$ and

$$\mathbf{F}(x(t), y(t)) = (4t - t^2)\mathbf{i} + t^2\mathbf{j}$$

the line integral is

$$\int_{C_2} \mathbf{F} \cdot d\mathbf{r} = \int_1^4 [(4t - t^2)\mathbf{i} + t^2\mathbf{j}] \cdot [\mathbf{i} + (4 - 2t)\mathbf{j}]\, dt$$

$$= \int_1^4 (4t - t^2 + 4t^2 - 2t^3)\, dt$$

$$= \int_1^4 (-2t^3 + 3t^2 + 4t)\, dt$$

$$= \left[ -\frac{t^4}{2} + t^3 + 2t^2 \right]_1^4$$

$$= -\frac{69}{2}.$$

The answer in part (b) is the negative of that in part (a) because $C_1$ and $C_2$ represent opposite orientations of the same parabolic segment.

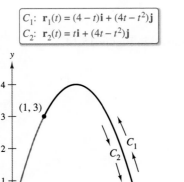

$C_1$: $\mathbf{r}_1(t) = (4 - t)\mathbf{i} + (4t - t^2)\mathbf{j}$
$C_2$: $\mathbf{r}_2(t) = t\mathbf{i} + (4t - t^2)\mathbf{j}$

**Figure 15.16**

•• **REMARK** Although the value of the line integral in Example 7 depends on the orientation of $C$, it does not depend on the parametrization of $C$. To see this, let $C_3$ be represented by

$$\mathbf{r}_3(t) = (t + 2)\mathbf{i} + (4 - t^2)\mathbf{j}$$

where $-1 \le t \le 2$. The graph of this curve is the same parabolic segment shown in Figure 15.16. Does the value of the line integral over $C_3$ agree with the value over $C_1$ or $C_2$? Why or why not?

## Line Integrals in Differential Form

A second commonly used form of line integrals is derived from the vector field notation used in Section 15.1. If $\mathbf{F}$ is a vector field of the form $\mathbf{F}(x, y) = M\mathbf{i} + N\mathbf{j}$ and $C$ is given by $\mathbf{r}(t) = x(t)\mathbf{i} + y(t)\mathbf{j}$, then $\mathbf{F} \cdot d\mathbf{r}$ is often written as $M\,dx + N\,dy$.

$$\int_C \mathbf{F} \cdot d\mathbf{r} = \int_C \mathbf{F} \cdot \frac{d\mathbf{r}}{dt}\,dt$$

$$= \int_a^b (M\mathbf{i} + N\mathbf{j}) \cdot (x'(t)\mathbf{i} + y'(t)\mathbf{j})\,dt$$

$$= \int_a^b \left(M\frac{dx}{dt} + N\frac{dy}{dt}\right) dt$$

$$= \int_C (M\,dx + N\,dy)$$

$\cdots\cdots\cdots\cdots\cdots\cdots\triangleright$

**·· REMARK**   The parentheses are often omitted from this differential form, as shown below.

$$\int_C M\,dx + N\,dy$$

In three variables, the differential form is

$$\int_C M\,dx + N\,dy + P\,dz.$$

This **differential form** can be extended to three variables.

---

**EXAMPLE 8**    **Evaluating a Line Integral in Differential Form**

Let $C$ be the circle of radius 3 given by

$$\mathbf{r}(t) = 3\cos t\,\mathbf{i} + 3\sin t\,\mathbf{j}, \quad 0 \le t \le 2\pi$$

as shown in Figure 15.17. Evaluate the line integral

$$\int_C y^3\,dx + (x^3 + 3xy^2)\,dy.$$

**Solution**   Because $x = 3\cos t$ and $y = 3\sin t$, you have $dx = -3\sin t\,dt$ and $dy = 3\cos t\,dt$. So, the line integral is

$$\int_C M\,dx + N\,dy$$

$$= \int_C y^3\,dx + (x^3 + 3xy^2)\,dy$$

$$= \int_0^{2\pi} [(27\sin^3 t)(-3\sin t) + (27\cos^3 t + 81\cos t\sin^2 t)(3\cos t)]\,dt$$

$$= 81\int_0^{2\pi} (\cos^4 t - \sin^4 t + 3\cos^2 t\sin^2 t)\,dt$$

$$= 81\int_0^{2\pi} \left(\cos^2 t - \sin^2 t + \frac{3}{4}\sin^2 2t\right) dt-$$

$$= 81\int_0^{2\pi} \left[\cos 2t + \frac{3}{4}\left(\frac{1 - \cos 4t}{2}\right)\right] dt$$

$$= 81\left[\frac{\sin 2t}{2} + \frac{3}{8}t - \frac{3\sin 4t}{32}\right]_0^{2\pi}$$

$$= \frac{243\pi}{4}.$$

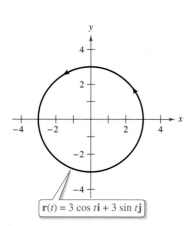

$\boxed{\mathbf{r}(t) = 3\cos t\,\mathbf{i} + 3\sin t\,\mathbf{j}}$

**Figure 15.17**

The orientation of $C$ affects the value of the differential form of a line integral. Specifically, if $-C$ has the orientation opposite to that of $C$, then

$$\int_{-C} M\,dx + N\,dy = -\int_C M\,dx + N\,dy.$$

So, of the three line integral forms presented in this section, the orientation of $C$ does not affect the form $\int_C f(x, y)\,ds$, but it does affect the vector form and the differential form.

For curves represented by $y = g(x)$, $a \leq x \leq b$, you can let $x = t$ and obtain the parametric form

$$x = t \quad \text{and} \quad y = g(t), \quad a \leq t \leq b.$$

Because $dx = dt$ for this form, you have the option of evaluating the line integral in the variable $x$ or the variable $t$. This is demonstrated in Example 9.

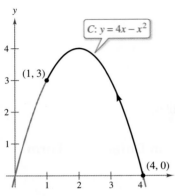

$C: y = 4x - x^2$

(1, 3)

(4, 0)

**Figure 15.18**

### EXAMPLE 9  Evaluating a Line Integral in Differential Form

Evaluate

$$\int_C y \, dx + x^2 \, dy$$

where $C$ is the parabolic arc given by $y = 4x - x^2$ from $(4, 0)$ to $(1, 3)$, as shown in Figure 15.18.

**Solution**  Rather than converting to the parameter $t$, you can simply retain the variable $x$ and write

$$y = 4x - x^2 \quad \implies \quad dy = (4 - 2x) \, dx.$$

Then, in the direction from $(4, 0)$ to $(1, 3)$, the line integral is

$$\int_C y \, dx + x^2 \, dy = \int_4^1 \left[ (4x - x^2) \, dx + x^2(4 - 2x) \, dx \right]$$

$$= \int_4^1 (4x + 3x^2 - 2x^3) \, dx$$

$$= \left[ 2x^2 + x^3 - \frac{x^4}{2} \right]_4^1$$

$$= \frac{69}{2}. \qquad \text{See Example 7.}$$

### Exploration

***Finding Lateral Surface Area***  The figure below shows a piece of tin that has been cut from a circular cylinder. The base of the circular cylinder is modeled by $x^2 + y^2 = 9$. At any point $(x, y)$ on the base, the height of the object is

$$f(x, y) = 1 + \cos \frac{\pi x}{4}.$$

Explain how to use a line integral to find the surface area of the piece of tin.

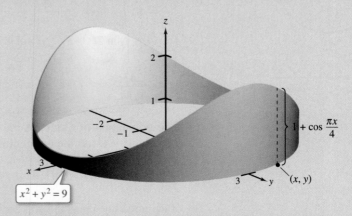

$1 + \cos \dfrac{\pi x}{4}$

$(x, y)$

$x^2 + y^2 = 9$

## 15.2 Exercises

See **CalcChat.com** for tutorial help and worked-out solutions to odd-numbered exercises.

---

**CONCEPT CHECK**

**1. Line Integral** What is the physical interpretation of each line integral?

(a) $\int_C 1 \, ds$

(b) $\int_C f(x, y, z) \, ds$, where $f(x, y, z)$ is the density of a string of finite length

**2. Orientation of a Curve** Describe how reversing the orientation of a curve $C$ affects $\int_C \mathbf{F} \cdot d\mathbf{r}$.

---

 **Finding a Piecewise Smooth Parametrization** In Exercises 3–8, find a piecewise smooth parametrization of the path $C$. (There is more than one correct answer.)

**3.**

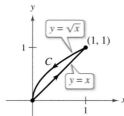

**4.**
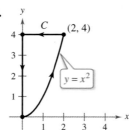

**5.**
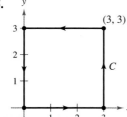

**6.**
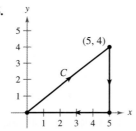

**7.**
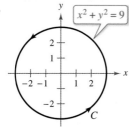

**8.**
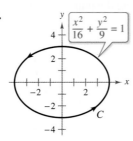

 **Evaluating a Line Integral** In Exercises 9–12, (a) find a parametrization of the path $C$, and (b) evaluate $\int_C (x^2 + y^2) \, ds$.

**9.** $C$: line segment from $(0, 0)$ to $(1, 1)$

**10.** $C$: line segment from $(0, 0)$ to $(2, 4)$

**11.** $C$: counterclockwise around the circle $x^2 + y^2 = 1$ from $(1, 0)$ to $(0, 1)$

**12.** $C$: counterclockwise around the circle $x^2 + y^2 = 4$ from $(2, 0)$ to $(-2, 0)$

---

**Evaluating a Line Integral** In Exercises 13–16, (a) find a piecewise smooth parametrization of the path $C$, and (b) evaluate $\int_C \left(2x + 3\sqrt{y}\right) ds$.

**13.** $C$: line segments from $(0, 0)$ to $(1, 0)$ and $(1, 0)$ to $(2, 4)$

**14.** $C$: line segments from $(0, 1)$ to $(0, 4)$ and $(0, 4)$ to $(3, 3)$

**15.** $C$: counterclockwise around the triangle with vertices $(0, 0)$, $(1, 0)$, and $(0, 1)$

**16.** $C$: counterclockwise around the square with vertices $(0, 0)$, $(2, 0)$, $(2, 2)$, and $(0, 2)$

**Evaluating a Line Integral** In Exercises 17 and 18, (a) find a piecewise smooth parametrization of the path $C$ shown in the figure and (b) evaluate $\int_C (2x + y^2 - z) \, ds$.

**17.**
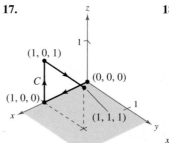

**18.**
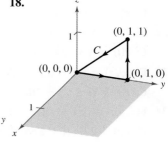

 **Evaluating a Line Integral** In Exercises 19–22, evaluate the line integral along the given path.

**19.** $\int_C xy \, ds$

$C$: $\mathbf{r}(t) = 4t\mathbf{i} + 3t\mathbf{j}$

$0 \le t \le 1$

**20.** $\int_C 3(x - y) \, ds$

$C$: $\mathbf{r}(t) = t\mathbf{i} + (2 - t)\mathbf{j}$

$0 \le t \le 2$

**21.** $\int_C (x^2 + y^2 + z^2) \, ds$

$C$: $\mathbf{r}(t) = \sin t\mathbf{i} + \cos t\mathbf{j} + 2\mathbf{k}$

$0 \le t \le \dfrac{\pi}{2}$

**22.** $\int_C 2xyz \, ds$

$C$: $\mathbf{r}(t) = 12t\mathbf{i} + 5t\mathbf{j} + 84t\mathbf{k}$

$0 \le t \le 1$

**Mass** In Exercises 23 and 24, find the total mass of a spring with density $\rho$ in the shape of the circular helix

$$\mathbf{r}(t) = 2\cos t\mathbf{i} + 2\sin t\mathbf{j} + t\mathbf{k}, \quad 0 \le t \le 4\pi.$$

**23.** $\rho(x, y, z) = \frac{1}{2}(x^2 + y^2 + z^2)$

**24.** $\rho(x, y, z) = z$

**Mass** In Exercises 25–28, find the total mass of the wire with density $\rho$ whose shape is modeled by **r**.

25. $\mathbf{r}(t) = \cos t\mathbf{i} + \sin t\mathbf{j}, \ 0 \le t \le \pi, \ \rho(x, y) = x + y + 2$

26. $\mathbf{r}(t) = t^2\mathbf{i} + 2t\mathbf{j}, \ 0 \le t \le 1, \ \rho(x, y) = \frac{3}{4}y$

27. $\mathbf{r}(t) = t^2\mathbf{i} + 2t\mathbf{j} + t\mathbf{k}, \ 1 \le t \le 3, \ \rho(x, y, z) = kz \ (k > 0)$

28. $\mathbf{r}(t) = 2\cos t\mathbf{i} + 2\sin t\mathbf{j} + 3t\mathbf{k}, \ 0 \le t \le 2\pi,$
    $\rho(x, y, z) = k + z \ (k > 0)$

 **Evaluating a Line Integral of a Vector Field**
In Exercises 29–34, evaluate $\int_C \mathbf{F} \cdot d\mathbf{r}$.

29. $\mathbf{F}(x, y) = x\mathbf{i} + y\mathbf{j}$
    $C: \ \mathbf{r}(t) = (3t + 1)\mathbf{i} + t\mathbf{j}, \ 0 \le t \le 1$

30. $\mathbf{F}(x, y) = xy\mathbf{i} + y\mathbf{j}$
    $C: \ \mathbf{r}(t) = 4\cos t\mathbf{i} + 4\sin t\mathbf{j}, \ 0 \le t \le \dfrac{\pi}{2}$

31. $\mathbf{F}(x, y) = x^2\mathbf{i} + 4y\mathbf{j}$
    $C: \ \mathbf{r}(t) = e^t\mathbf{i} + t^2\mathbf{j}, \ 0 \le t \le 2$

32. $\mathbf{F}(x, y) = 3x\mathbf{i} + 4y\mathbf{j}$
    $C: \ \mathbf{r}(t) = t\mathbf{i} + \sqrt{4 - t^2}\mathbf{j}, \ -2 \le t \le 2$

33. $\mathbf{F}(x, y, z) = xy\mathbf{i} + xz\mathbf{j} + yz\mathbf{k}$
    $C: \ \mathbf{r}(t) = t\mathbf{i} + t^2\mathbf{j} + 2t\mathbf{k}, \ 0 \le t \le 1$

34. $\mathbf{F}(x, y, z) = x^2\mathbf{i} + y^2\mathbf{j} + z^2\mathbf{k}$
    $C: \ \mathbf{r}(t) = 2\sin t\mathbf{i} + 2\cos t\mathbf{j} + \frac{1}{2}t^2\mathbf{k}, \ 0 \le t \le \pi$

**Evaluating a Line Integral of a Vector Field Using Technology** In Exercises 35 and 36, use a computer algebra system to evaluate $\int_C \mathbf{F} \cdot d\mathbf{r}$.

35. $\mathbf{F}(x, y, z) = x^2 z\mathbf{i} + 6y\mathbf{j} + yz^2\mathbf{k}$
    $C: \ \mathbf{r}(t) = t\mathbf{i} + t^2\mathbf{j} + \ln t\mathbf{k}, \ 1 \le t \le 3$

36. $\mathbf{F}(x, y, z) = \dfrac{x\mathbf{i} + y\mathbf{j} + z\mathbf{k}}{\sqrt{x^2 + y^2 + z^2}}$
    $C: \ \mathbf{r}(t) = t\mathbf{i} + t\mathbf{j} + e^t\mathbf{k}, \ 0 \le t \le 2$

 **Work** In Exercises 37–42, find the work done by the force field **F** on a particle moving along the given path.

37. $\mathbf{F}(x, y) = x\mathbf{i} + 2y\mathbf{j}$
    $C: \ x = t, \ y = t^3$ from $(0, 0)$ to $(2, 8)$

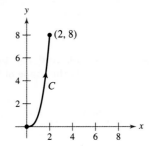

Figure for 37          Figure for 38

38. $\mathbf{F}(x, y) = x^2\mathbf{i} - xy\mathbf{j}$
    $C: \ x = \cos^3 t, \ y = \sin^3 t$ from $(1, 0)$ to $(0, 1)$

39. $\mathbf{F}(x, y) = x\mathbf{i} + y\mathbf{j}$
    $C:$ counterclockwise around the triangle with vertices $(0, 0)$, $(1, 0)$, and $(0, 1)$

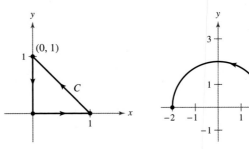

Figure for 39          Figure for 40

40. $\mathbf{F}(x, y) = -y\mathbf{i} - x\mathbf{j}$
    $C:$ counterclockwise around the semicircle $y = \sqrt{4 - x^2}$ from $(2, 0)$ to $(-2, 0)$

41. $\mathbf{F}(x, y, z) = x\mathbf{i} + y\mathbf{j} - 5z\mathbf{k}$
    $C: \ \mathbf{r}(t) = 2\cos t\mathbf{i} + 2\sin t\mathbf{j} + t\mathbf{k}, \ 0 \le t \le 2\pi$

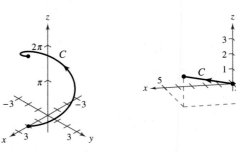

Figure for 41          Figure for 42

42. $\mathbf{F}(x, y, z) = yz\mathbf{i} + xz\mathbf{j} + xy\mathbf{k}$
    $C:$ line from $(0, 0, 0)$ to $(5, 3, 2)$

**Work** In Exercises 43–46, determine whether the work done along the path $C$ is positive, negative, or zero. Explain.

43.

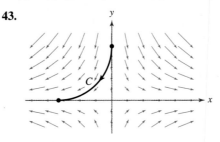

44.

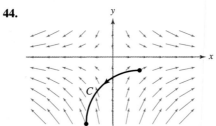

**45.**

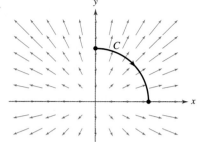

**46.**

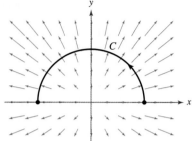

 **Evaluating a Line Integral of a Vector Field** In Exercises 47 and 48, evaluate $\int_C \mathbf{F} \cdot d\mathbf{r}$ for each curve. Discuss the orientation of the curve and its effect on the value of the integral.

**47.** $\mathbf{F}(x, y) = x^2\mathbf{i} + xy\mathbf{j}$

(a) $C_1$: $\mathbf{r}_1(t) = 2t\mathbf{i} + (t - 1)\mathbf{j}$, $1 \le t \le 3$

(b) $C_2$: $\mathbf{r}_2(t) = 2(3 - t)\mathbf{i} + (2 - t)\mathbf{j}$, $0 \le t \le 2$

**48.** $\mathbf{F}(x, y) = x^2y\mathbf{i} + xy^{3/2}\mathbf{j}$

(a) $C_1$: $\mathbf{r}_1(t) = (t + 1)\mathbf{i} + t^2\mathbf{j}$, $0 \le t \le 2$

(b) $C_2$: $\mathbf{r}_2(t) = (1 + 2\cos t)\mathbf{i} + (4\cos^2 t)\mathbf{j}$, $0 \le t \le \pi/2$

**Demonstrating a Property** In Exercises 49–52, demonstrate the property that $\int_C \mathbf{F} \cdot d\mathbf{r} = 0$ regardless of the initial and terminal points of $C$, where the tangent vector $\mathbf{r}'(t)$ is orthogonal to the force field $\mathbf{F}$.

**49.** $\mathbf{F}(x, y) = y\mathbf{i} - x\mathbf{j}$

$C$: $\mathbf{r}(t) = t\mathbf{i} - 2t\mathbf{j}$

**50.** $\mathbf{F}(x, y) = -3y\mathbf{i} + x\mathbf{j}$

$C$: $\mathbf{r}(t) = t\mathbf{i} - t^3\mathbf{j}$

**51.** $\mathbf{F}(x, y) = (x^3 - 2x^2)\mathbf{i} + \left(x - \dfrac{y}{2}\right)\mathbf{j}$

$C$: $\mathbf{r}(t) = t\mathbf{i} + t^2\mathbf{j}$

**52.** $\mathbf{F}(x, y) = x\mathbf{i} + y\mathbf{j}$

$C$: $\mathbf{r}(t) = 3\sin t\mathbf{i} + 3\cos t\mathbf{j}$

**Evaluating a Line Integral in Differential Form** In Exercises 53–56, evaluate the line integral along the path $C$ given by $x = 2t$, $y = 4t$, where $0 \le t \le 1$.

**53.** $\displaystyle\int_C (x + 3y^2)\, dy$

**54.** $\displaystyle\int_C (x^3 + 2y)\, dx$

**55.** $\displaystyle\int_C xy\, dx + y\, dy$

**56.** $\displaystyle\int_C (y - x)\, dx + 5x^2y^2\, dy$

 **Evaluating a Line Integral in Differential Form** In Exercises 57–64, evaluate

$$\int_C (2x - y)\, dx + (x + 3y)\, dy.$$

**57.** $C$: $x$-axis from $x = 0$ to $x = 5$

**58.** $C$: $y$-axis from $y = 0$ to $y = 2$

**59.** $C$: line segments from $(0, 0)$ to $(3, 0)$ and $(3, 0)$ to $(3, 3)$

**60.** $C$: line segments from $(0, 0)$ to $(0, -3)$ and $(0, -3)$ to $(2, -3)$

**61.** $C$: arc on $y = 1 - x^2$ from $(0, 1)$ to $(1, 0)$

**62.** $C$: arc on $y = x^{3/2}$ from $(0, 0)$ to $(4, 8)$

**63.** $C$: parabolic path $x = t$, $y = 2t^2$ from $(0, 0)$ to $(2, 8)$

**64.** $C$: elliptic path $x = 4\sin t$, $y = 3\cos t$ from $(0, 3)$ to $(4, 0)$

**Lateral Surface Area** In Exercises 65–72, find the area of the lateral surface (see figure) over the curve $C$ in the $xy$-plane and under the surface $z = f(x, y)$, where

$$\text{Lateral surface area} = \int_C f(x, y)\, ds.$$

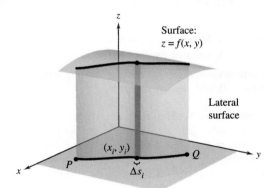

$C$: Curve in $xy$-plane

**65.** $f(x, y) = h$, $C$: line from $(0, 0)$ to $(3, 4)$

**66.** $f(x, y) = y$, $C$: line from $(0, 0)$ to $(4, 4)$

**67.** $f(x, y) = xy$, $C$: $x^2 + y^2 = 1$ from $(1, 0)$ to $(0, 1)$

**68.** $f(x, y) = x + y$, $C$: $x^2 + y^2 = 1$ from $(1, 0)$ to $(0, 1)$

**69.** $f(x, y) = h$, $C$: $y = 1 - x^2$ from $(1, 0)$ to $(0, 1)$

**70.** $f(x, y) = y + 1$, $C$: $y = 1 - x^2$ from $(1, 0)$ to $(0, 1)$

**71.** $f(x, y) = xy$, $C$: $y = 1 - x^2$ from $(1, 0)$ to $(0, 1)$

**72.** $f(x, y) = x^2 - y^2 + 4$, $C$: $x^2 + y^2 = 4$

**73. Engine Design** A tractor engine has a steel component with a circular base modeled by the vector-valued function

$$\mathbf{r}(t) = 2\cos t\mathbf{i} + 2\sin t\mathbf{j}.$$

Its height is given by $z = 1 + y^2$. (All measurements of the component are in centimeters.)

(a) Find the lateral surface area of the component.

(b) The component is in the form of a shell of thickness 0.2 centimeter. Use the result of part (a) to approximate the amount of steel used to manufacture the component.

(c) Draw a sketch of the component.

**74. Building Design**

The ceiling of a building has a height above the floor given by $z = 20 + \frac{1}{4}x$. One of the walls follows a path modeled by $y = x^{3/2}$. Find the surface area of the wall for $0 \le x \le 40$. (All measurements are in feet.)

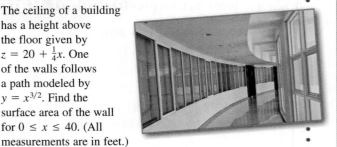

**Moments of Inertia** **Consider a wire of density $\rho(x, y)$ given by the space curve**

$$C: \mathbf{r}(t) = x(t)\mathbf{i} + y(t)\mathbf{j}, \quad 0 \le t \le b.$$

**The moments of inertia about the $x$- and $y$-axes are given by**

$$I_x = \int_C y^2 \rho(x, y)\, ds \quad \text{and} \quad I_y = \int_C x^2 \rho(x, y)\, ds.$$

**In Exercises 75 and 76, find the moments of inertia for the wire of density $\rho$.**

**75.** A wire lies along $\mathbf{r}(t) = a \cos t\mathbf{i} + a \sin t\mathbf{j}$, where $0 \le t \le 2\pi$ and $a > 0$, with density $\rho(x, y) = 1$.

**76.** A wire lies along $\mathbf{r}(t) = a \cos t\mathbf{i} + a \sin t\mathbf{j}$, where $0 \le t \le 2\pi$ and $a > 0$, with density $\rho(x, y) = y$.

**77. Investigation** The top outer edge of a solid with vertical sides that is resting on the $xy$-plane is modeled by $\mathbf{r}(t) = 3 \cos t\mathbf{i} + 3 \sin t\mathbf{j} + (1 + \sin^2 2t)\mathbf{k}$, where all measurements are in centimeters. The intersection of the plane $y = b$, where $-3 < b < 3$, with the top of the solid is a horizontal line.

(a) Use a computer algebra system to graph the solid.

(b) Use a computer algebra system to approximate the lateral surface area of the solid.

(c) Find (if possible) the volume of the solid.

**78. Work** A particle moves along the path $y = x^2$ from the point $(0, 0)$ to the point $(1, 1)$. The force field $\mathbf{F}$ is measured at five points along the path, and the results are shown in the table. Use Simpson's Rule or a graphing utility to approximate the work done by the force field.

| $(x, y)$ | $(0, 0)$ | $\left(\frac{1}{4}, \frac{1}{16}\right)$ | $\left(\frac{1}{2}, \frac{1}{4}\right)$ | $\left(\frac{3}{4}, \frac{9}{16}\right)$ | $(1, 1)$ |
| --- | --- | --- | --- | --- | --- |
| $\mathbf{F}(x, y)$ | $\langle 5, 0 \rangle$ | $\langle 3.5, 1 \rangle$ | $\langle 2, 2 \rangle$ | $\langle 1.5, 3 \rangle$ | $\langle 1, 5 \rangle$ |

**79. Work** Find the work done by a person weighing 175 pounds walking exactly one revolution up a circular helical staircase of radius 3 feet when the person rises 10 feet.

**80. Investigation** Determine the value of $c$ such that the work done by the force field $\mathbf{F}(x, y) = 15[(4 - x^2y)\mathbf{i} - xy\mathbf{j}]$ on an object moving along the parabolic path $y = c(1 - x^2)$ between the points $(-1, 0)$ and $(1, 0)$ is a minimum. Compare the result with the work required to move the object along the straight-line path connecting the points.

**EXPLORING CONCEPTS**

**81. Think About It** A path $C$ is given by $x = t$, $y = 2t$, where $0 \le t \le 1$. Are $\int_C (x + y)\, dx$ and $\int_C (x + y)\, dy$ equivalent? Explain.

**82. Line Integrals** Let $\mathbf{F}(x, y) = 2x\mathbf{i} + xy^2\mathbf{j}$ and consider the curve $y = x^2$ from $(0, 0)$ to $(2, 4)$ in the $xy$-plane. Set up and evaluate line integrals of the forms $\int_C \mathbf{F} \cdot d\mathbf{r}$ and $\int_C M\, dx + N\, dy$. Compare your results. Which method do you prefer? Explain.

**83. Ordering Surfaces** Order the surfaces in ascending order of the lateral surface area under the surface and over the curve $y = \sqrt{x}$ from $(0, 0)$ to $(4, 2)$ in the $xy$-plane. Explain your ordering without doing any calculations.

(a) $z_1 = 2 + x$

(b) $z_2 = 5 + x$

(c) $z_3 = 2$

(d) $z_4 = 10 + x + 2y$

**84. HOW DO YOU SEE IT?** For each of the following, determine whether the work done in moving an object from the first to the second point through the force field shown in the figure is positive, negative, or zero. Explain your answer. (In the figure, the circles have radii 1, 2, 3, 4, 5, and 6.)

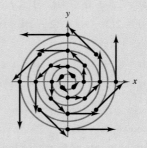

(a) From $(-3, -3)$ to $(3, 3)$

(b) From $(-3, 0)$ to $(0, 3)$

(c) From $(5, 0)$ to $(0, 3)$

**True or False?** **In Exercises 85 and 86, determine whether the statement is true or false. If it is false, explain why or give an example that shows it is false.**

**85.** If $C$ is given by $x = t$, $y = t$, where $0 \le t \le 1$, then

$$\int_C xy\, ds = \int_0^1 t^2\, dt.$$

**86.** If $C_2 = -C_1$, then $\int_{C_1} f(x, y)\, ds + \int_{C_2} f(x, y)\, ds = 0$.

**87. Work** Consider a particle that moves through the force field

$$\mathbf{F}(x, y) = (y - x)\mathbf{i} + xy\mathbf{j}$$

from the point $(0, 0)$ to the point $(0, 1)$ along the curve $x = kt(1 - t)$, $y = t$. Find the value of $k$ such that the work done by the force field is 1.

# 15.3 Conservative Vector Fields and Independence of Path

■ Understand and use the Fundamental Theorem of Line Integrals.
■ Understand the concept of independence of path.
■ Understand the concept of conservation of energy.

## Fundamental Theorem of Line Integrals

The discussion at the beginning of Section 15.2 pointed out that in a gravitational field, the work done by gravity on an object moving between two points in the field is independent of the path taken by the object. In this section, you will study an important generalization of this result—it is called the **Fundamental Theorem of Line Integrals.** To begin, an example is presented in which the line integral of a *conservative vector field* is evaluated over three different paths.

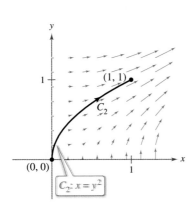

(a)

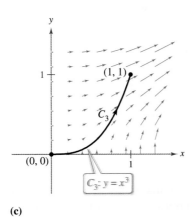

(b)

**EXAMPLE 1** **Line Integral of a Conservative Vector Field**

Find the work done by the force field

$$\mathbf{F}(x, y) = \frac{1}{2}xy\mathbf{i} + \frac{1}{4}x^2\mathbf{j}$$

on a particle that moves from $(0, 0)$ to $(1, 1)$ along each path, as shown in Figure 15.19.

**a.** $C_1$: $y = x$     **b.** $C_2$: $x = y^2$     **c.** $C_3$: $y = x^3$

**Solution** Note that $\mathbf{F}$ is conservative because the first partial derivatives are equal.

$$\frac{\partial}{\partial y}\left[\frac{1}{2}xy\right] = \frac{1}{2}x \quad \text{and} \quad \frac{\partial}{\partial x}\left[\frac{1}{4}x^2\right] = \frac{1}{2}x$$

**a.** Let $\mathbf{r}(t) = t\mathbf{i} + t\mathbf{j}$ for $0 \leq t \leq 1$, so that

$$d\mathbf{r} = (\mathbf{i} + \mathbf{j})\,dt \quad \text{and} \quad \mathbf{F}(x, y) = \frac{1}{2}t^2\mathbf{i} + \frac{1}{4}t^2\mathbf{j}.$$

Then the work done is

$$W = \int_{C_1} \mathbf{F} \cdot d\mathbf{r} = \int_0^1 \frac{3}{4}t^2\,dt = \frac{1}{4}t^3\Big]_0^1 = \frac{1}{4}.$$

**b.** Let $\mathbf{r}(t) = t\mathbf{i} + \sqrt{t}\mathbf{j}$ for $0 \leq t \leq 1$, so that

$$d\mathbf{r} = \left(\mathbf{i} + \frac{1}{2\sqrt{t}}\mathbf{j}\right)dt \quad \text{and} \quad \mathbf{F}(x, y) = \frac{1}{2}t^{3/2}\mathbf{i} + \frac{1}{4}t^2\mathbf{j}.$$

Then the work done is

$$W = \int_{C_2} \mathbf{F} \cdot d\mathbf{r} = \int_0^1 \frac{5}{8}t^{3/2}\,dt = \frac{1}{4}t^{5/2}\Big]_0^1 = \frac{1}{4}.$$

**c.** Let $\mathbf{r}(t) = \frac{1}{2}t\mathbf{i} + \frac{1}{8}t^3\mathbf{j}$ for $0 \leq t \leq 2$, so that

$$d\mathbf{r} = \left(\frac{1}{2}\mathbf{i} + \frac{3}{8}t^2\mathbf{j}\right)dt \quad \text{and} \quad \mathbf{F}(x, y) = \frac{1}{32}t^4\mathbf{i} + \frac{1}{16}t^2\mathbf{j}.$$

Then the work done is

$$W = \int_{C_3} \mathbf{F} \cdot d\mathbf{r} = \int_0^2 \frac{5}{128}t^4\,dt = \frac{1}{128}t^5\Big]_0^2 = \frac{1}{4}.$$

So, the work done by the conservative vector field $\mathbf{F}$ is the same for each path. ■

(c)
**Figure 15.19**

In Example 1, note that the vector field $\mathbf{F}(x, y) = \frac{1}{2}xy\mathbf{i} + \frac{1}{4}x^2\mathbf{j}$ is conservative because $\mathbf{F}(x, y) = \nabla f(x, y)$, where $f(x, y) = \frac{1}{4}x^2y$. In such cases, the next theorem states that the value of $\int_C \mathbf{F} \cdot d\mathbf{r}$ is given by

$$\int_C \mathbf{F} \cdot d\mathbf{r} = f(x(1), y(1)) - f(x(0), y(0))$$

$$= \frac{1}{4} - 0$$

$$= \frac{1}{4}.$$

▷

**REMARK** Notice how the Fundamental Theorem of Line Integrals is similar to the Fundamental Theorem of Calculus (see Section 4.4), which states that

$$\int_a^b f(x)\, dx = F(b) - F(a)$$

where $F'(x) = f(x)$.

---

**THEOREM 15.5  Fundamental Theorem of Line Integrals**

Let $C$ be a piecewise smooth curve lying in an open region $R$ and given by

$$\mathbf{r}(t) = x(t)\mathbf{i} + y(t)\mathbf{j}, \quad a \le t \le b.$$

If $\mathbf{F}(x, y) = M\mathbf{i} + N\mathbf{j}$ is conservative in $R$, and $M$ and $N$ are continuous in $R$, then

$$\int_C \mathbf{F} \cdot d\mathbf{r} = \int_C \nabla f \cdot d\mathbf{r} = f(x(b), y(b)) - f(x(a), y(a))$$

where $f$ is a potential function of $\mathbf{F}$. That is, $\mathbf{F}(x, y) = \nabla f(x, y)$.

---

**Proof**  A proof is provided only for a smooth curve. For piecewise smooth curves, the procedure is carried out separately on each smooth portion. Because

$$\mathbf{F}(x, y) = \nabla f(x, y) = f_x(x, y)\mathbf{i} + f_y(x, y)\mathbf{j}$$

it follows that

$$\int_C \mathbf{F} \cdot d\mathbf{r} = \int_a^b \mathbf{F} \cdot \frac{d\mathbf{r}}{dt}\, dt$$

$$= \int_a^b \left[ f_x(x, y) \frac{dx}{dt} + f_y(x, y) \frac{dy}{dt} \right] dt$$

and, by the Chain Rule (see Theorem 13.6 in Section 13.5), you have

$$\int_C \mathbf{F} \cdot d\mathbf{r} = \int_a^b \frac{d}{dt}[f(x(t), y(t))]\, dt$$

$$= f(x(b), y(b)) - f(x(a), y(a)).$$

The last step is an application of the Fundamental Theorem of Calculus. ∎

In space, the Fundamental Theorem of Line Integrals takes the following form. Let $C$ be a piecewise smooth curve lying in an open region $Q$ and given by

$$\mathbf{r}(t) = x(t)\mathbf{i} + y(t)\mathbf{j} + z(t)\mathbf{k}, \quad a \le t \le b.$$

If $\mathbf{F}(x, y, z) = M\mathbf{i} + N\mathbf{j} + P\mathbf{k}$ is conservative and $M$, $N$, and $P$ are continuous, then

$$\int_C \mathbf{F} \cdot d\mathbf{r} = \int_C \nabla f \cdot d\mathbf{r} = f(x(b), y(b), z(b)) - f(x(a), y(a), z(a))$$

where $\mathbf{F}(x, y, z) = \nabla f(x, y, z)$.

The Fundamental Theorem of Line Integrals states that if the vector field $\mathbf{F}$ is conservative, then the line integral between any two points is simply the difference in the values of the *potential* function $f$ at these points.

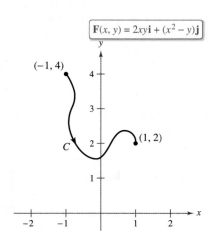

$\boxed{F(x, y) = 2xy\mathbf{i} + (x^2 - y)\mathbf{j}}$

(-1, 4)

(1, 2)

$C$

**Figure 15.20**

**EXAMPLE 2**    **Using the Fundamental Theorem of Line Integrals**

Evaluate $\displaystyle\int_C \mathbf{F} \cdot d\mathbf{r}$, where $C$ is a piecewise smooth curve from $(-1, 4)$ to $(1, 2)$ and

$$\mathbf{F}(x, y) = 2xy\mathbf{i} + (x^2 - y)\mathbf{j}$$

as shown in Figure 15.20.

**Solution**    From Example 6 in Section 15.1, you know that $\mathbf{F}$ is the gradient of $f$, where

$$f(x, y) = x^2 y - \frac{y^2}{2} + K.$$

Consequently, $\mathbf{F}$ is conservative, and by the Fundamental Theorem of Line Integrals, it follows that

$$\int_C \mathbf{F} \cdot d\mathbf{r} = f(1, 2) - f(-1, 4)$$

$$= \left[ 1^2(2) - \frac{2^2}{2} \right] - \left[ (-1)^2(4) - \frac{4^2}{2} \right]$$

$$= 4.$$

Note that it is unnecessary to include a constant $K$ as part of $f$, because it is canceled by subtraction.

**EXAMPLE 3**    **Using the Fundamental Theorem of Line Integrals**

Evaluate $\displaystyle\int_C \mathbf{F} \cdot d\mathbf{r}$, where $C$ is a piecewise smooth curve from $(1, 1, 0)$ to $(0, 2, 3)$ and

$$\mathbf{F}(x, y, z) = 2xy\mathbf{i} + (x^2 + z^2)\mathbf{j} + 2yz\mathbf{k}$$

as shown in Figure 15.21.

**Solution**    From Example 8 in Section 15.1, you know that $\mathbf{F}$ is the gradient of $f$, where

$$f(x, y, z) = x^2 y + yz^2 + K.$$

Consequently, $\mathbf{F}$ is conservative, and by the Fundamental Theorem of Line Integrals, it follows that

$$\int_C \mathbf{F} \cdot d\mathbf{r} = f(0, 2, 3) - f(1, 1, 0)$$

$$= [(0)^2(2) + (2)(3)^2] - [(1)^2(1) + (1)(0)^2]$$

$$= 17.$$

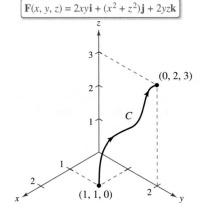

$\boxed{F(x, y, z) = 2xy\mathbf{i} + (x^2 + z^2)\mathbf{j} + 2yz\mathbf{k}}$

(0, 2, 3)

$C$

(1, 1, 0)

**Figure 15.21**

In Examples 2 and 3, be sure you see that the value of the line integral is the same for any smooth curve $C$ that has the given initial and terminal points. For instance, in Example 3, try evaluating the line integral for the curve given by

$$\mathbf{r}(t) = (1 - t)\mathbf{i} + (1 + t)\mathbf{j} + 3t\mathbf{k}.$$

You should obtain

$$\int_C \mathbf{F} \cdot d\mathbf{r} = \int_0^1 (30t^2 + 16t - 1) \, dt$$

$$= 17.$$

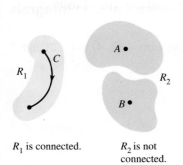

$R_1$ is connected. $R_2$ is not connected.

**Figure 15.22**

## Independence of Path

From the Fundamental Theorem of Line Integrals, it is clear that if **F** is continuous and conservative in an open region $R$, then the value of $\int_C \mathbf{F} \cdot d\mathbf{r}$ is the same for every piecewise smooth curve $C$ from one fixed point in $R$ to another fixed point in $R$. This result is described by saying that the line integral $\int_C \mathbf{F} \cdot d\mathbf{r}$ is **independent of path** in the region $R$.

A region in the plane (or in space) is **connected** when any two points in the region can be joined by a piecewise smooth curve lying entirely within the region, as shown in Figure 15.22. In open regions that are *connected,* the path independence of $\int_C \mathbf{F} \cdot d\mathbf{r}$ is equivalent to the condition that **F** is conservative.

---

**THEOREM 15.6 Independence of Path and Conservative Vector Fields**

If **F** is continuous on an open connected region, then the line integral

$$\int_C \mathbf{F} \cdot d\mathbf{r}$$

is independent of path if and only if **F** is conservative.

---

**Proof** If **F** is conservative, then, by the Fundamental Theorem of Line Integrals, the line integral is independent of path. Now establish the converse for a plane region $R$. Let $\mathbf{F}(x, y) = M\mathbf{i} + N\mathbf{j}$, and let $(x_0, y_0)$ be a fixed point in $R$. For any point $(x, y)$ in $R$, choose a piecewise smooth curve $C$ running from $(x_0, y_0)$ to $(x, y)$, and define $f$ by

$$f(x, y) = \int_C \mathbf{F} \cdot d\mathbf{r}$$

$$= \int_C M\, dx + N\, dy.$$

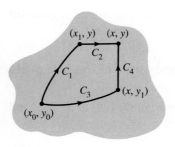

**Figure 15.23**

The existence of $C$ in $R$ is guaranteed by the fact that $R$ is connected. You can show that $f$ is a potential function of **F** by considering two different paths between $(x_0, y_0)$ and $(x, y)$. For the *first* path, choose $(x_1, y)$ in $R$ such that $x \neq x_1$. This is possible because $R$ is open. Then choose $C_1$ and $C_2$, as shown in Figure 15.23. Using the independence of path, it follows that

$$f(x, y) = \int_C M\, dx + N\, dy$$

$$= \int_{C_1} M\, dx + N\, dy + \int_{C_2} M\, dx + N\, dy.$$

Because the first integral does not depend on $x$ and because $dy = 0$ in the second integral, you have

$$f(x, y) = g(y) + \int_{C_2} M\, dx$$

and it follows that the partial derivative of $f$ with respect to $x$ is $f_x(x, y) = M$. For the *second* path, choose a point $(x, y_1)$. Using reasoning similar to that used for the first path, you can conclude that $f_y(x, y) = N$. Therefore,

$$\nabla f(x, y) = f_x(x, y)\mathbf{i} + f_y(x, y)\mathbf{j}$$
$$= M\mathbf{i} + N\mathbf{j}$$
$$= \mathbf{F}(x, y)$$

and it follows that **F** is conservative. ∎

**EXAMPLE 4**   **Finding Work in a Conservative Force Field**

For the force field given by

$$\mathbf{F}(x, y, z) = e^x \cos y\mathbf{i} - e^x \sin y\mathbf{j} + 2\mathbf{k}$$

show that $\int_C \mathbf{F} \cdot d\mathbf{r}$ is independent of path, and calculate the work done by $\mathbf{F}$ on an object moving along a curve $C$ from $(0, \pi/2, 1)$ to $(1, \pi, 3)$.

**Solution**   Writing the force field in the form $\mathbf{F}(x, y, z) = M\mathbf{i} + N\mathbf{j} + P\mathbf{k}$, you have $M = e^x \cos y$, $N = -e^x \sin y$, and $P = 2$, and it follows that

$$\frac{\partial P}{\partial y} = 0 = \frac{\partial N}{\partial z}$$

$$\frac{\partial P}{\partial x} = 0 = \frac{\partial M}{\partial z}$$

and

$$\frac{\partial N}{\partial x} = -e^x \sin y = \frac{\partial M}{\partial y}.$$

So, $\mathbf{F}$ is conservative. If $f$ is a potential function of $\mathbf{F}$, then

$$f_x(x, y, z) = e^x \cos y$$

$$f_y(x, y, z) = -e^x \sin y$$

and

$$f_z(x, y, z) = 2.$$

By integrating with respect to $x$, $y$, and $z$ separately, you obtain

$$f(x, y, z) = \int f_x(x, y, z) \, dx = \int e^x \cos y \, dx = e^x \cos y + g(y, z)$$

$$f(x, y, z) = \int f_y(x, y, z) \, dy = \int -e^x \sin y \, dy = e^x \cos y + h(x, z)$$

and

$$f(x, y, z) = \int f_z(x, y, z) \, dz = \int 2 \, dz = 2z + k(x, y).$$

By comparing these three versions of $f(x, y, z)$, you can conclude that

$$f(x, y, z) = e^x \cos y + 2z + K.$$

Therefore, the work done by $\mathbf{F}$ along *any* curve $C$ from $(0, \pi/2, 1)$ to $(1, \pi, 3)$ is

$$W = \int_C \mathbf{F} \cdot d\mathbf{r}$$

$$= f(1, \pi, 3) - f\left(0, \frac{\pi}{2}, 1\right)$$

$$= (-e + 6) - (0 + 2)$$

$$= 4 - e. \qquad \blacksquare$$

For the object in Example 4, how much work is done when the object moves on a curve from $(0, \pi/2, 1)$ to $(1, \pi, 3)$ and then back to the starting point $(0, \pi/2, 1)$? The Fundamental Theorem of Line Integrals states that there is zero work done. Remember that, by definition, work can be negative. So, by the time the object gets back to its starting point, the amount of work that registers positively is canceled out by the amount of work that registers negatively.

A curve $C$ given by $\mathbf{r}(t)$ for $a \leq t \leq b$ is **closed** when $\mathbf{r}(a) = \mathbf{r}(b)$. By the Fundamental Theorem of Line Integrals, you can conclude that if $\mathbf{F}$ is continuous and conservative on an open region $R$, then the line integral over every closed curve $C$ is 0.

........................▷

**•• REMARK** Theorem 15.7 gives you options for evaluating a line integral involving a conservative vector field. You can use a potential function, or it might be more convenient to choose a particularly simple path, such as a straight line.

---

**THEOREM 15.7 Equivalent Conditions**

Let $\mathbf{F}(x, y, z) = M\mathbf{i} + N\mathbf{j} + P\mathbf{k}$ have continuous first partial derivatives in an open connected region $R$, and let $C$ be a piecewise smooth curve in $R$. The conditions listed below are equivalent.

1. $\mathbf{F}$ is conservative. That is, $\mathbf{F} = \nabla f$ for some function $f$.

2. $\displaystyle\int_C \mathbf{F} \cdot d\mathbf{r}$ is independent of path.

3. $\displaystyle\int_C \mathbf{F} \cdot d\mathbf{r} = 0$ for every *closed* curve $C$ in $R$.

---

**EXAMPLE 5** **Evaluating a Line Integral**

••••▷ See LarsonCalculus.com for an interactive version of this type of example.

Evaluate $\displaystyle\int_{C_1} \mathbf{F} \cdot d\mathbf{r}$, where

$$\mathbf{F}(x, y) = (y^3 + 1)\mathbf{i} + (3xy^2 + 1)\mathbf{j}$$

and $C_1$ is the semicircular path from $(0, 0)$ to $(2, 0)$, as shown in Figure 15.24.

**Solution** You have the following three options.

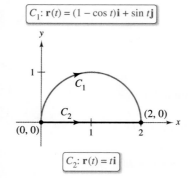

$C_1: \mathbf{r}(t) = (1 - \cos t)\mathbf{i} + \sin t\,\mathbf{j}$

$C_2: \mathbf{r}(t) = t\mathbf{i}$

**Figure 15.24**

a. You can use the method presented in Section 15.2 to evaluate the line integral along the *given curve*. To do this, you can use the parametrization $\mathbf{r}(t) = (1 - \cos t)\mathbf{i} + \sin t\,\mathbf{j}$, where $0 \leq t \leq \pi$. For this parametrization, it follows that

$$d\mathbf{r} = \mathbf{r}'(t)\,dt = (\sin t\,\mathbf{i} + \cos t\,\mathbf{j})\,dt$$

and

$$\int_{C_1} \mathbf{F} \cdot d\mathbf{r} = \int_0^\pi (\sin t + \sin^4 t + \cos t + 3\sin^2 t \cos t - 3\sin^2 t \cos^2 t)\,dt.$$

This integral should dampen your enthusiasm for this option.

b. You can try to find a *potential function* and evaluate the line integral by the Fundamental Theorem of Line Integrals. Using the technique demonstrated in Example 4, you can find the potential function to be $f(x, y) = xy^3 + x + y + K$, and, by the Fundamental Theorem,

$$W = \int_{C_1} \mathbf{F} \cdot d\mathbf{r} = f(2, 0) - f(0, 0) = 2.$$

c. Knowing that $\mathbf{F}$ is conservative, you have a third option. Because the value of the line integral is independent of path, you can replace the semicircular path with a *simpler path*. Choose the straight-line path $C_2$ from $(0, 0)$ to $(2, 0)$. Let $\mathbf{r}(t) = t\mathbf{i}$ for $0 \leq t \leq 2$, so that

$$d\mathbf{r} = \mathbf{i}\,dt \quad \text{and} \quad \mathbf{F}(x, y) = \mathbf{i} + \mathbf{j}.$$

Then the integral is

$$\int_{C_1} \mathbf{F} \cdot d\mathbf{r} = \int_{C_2} \mathbf{F} \cdot d\mathbf{r} = \int_0^2 1\,dt = t\,\Big]_0^2 = 2.$$

Of the three options, the third one is obviously the easiest. ∎

## Conservation of Energy

In 1840, the English physicist Michael Faraday wrote, "Nowhere is there a pure creation or production of power without a corresponding exhaustion of something to supply it." This statement represents the first formulation of one of the most important laws of physics—the **Law of Conservation of Energy.** In modern terminology, the law is stated as follows: *In a conservative force field, the sum of the potential and kinetic energies of an object remains constant from point to point.*

You can use the Fundamental Theorem of Line Integrals to derive this law. From physics, the **kinetic energy** of a particle of mass $m$ and speed $v$ is

$$k = \frac{1}{2}mv^2. \qquad \text{Kinetic energy}$$

The **potential energy** $p$ of a particle at point $(x, y, z)$ in a conservative vector field $\mathbf{F}$ is defined as $p(x, y, z) = -f(x, y, z)$, where $f$ is the potential function for $\mathbf{F}$. Consequently, the work done by $\mathbf{F}$ along a smooth curve $C$ from $A$ to $B$ is

$$W = \int_C \mathbf{F} \cdot d\mathbf{r} = f(x, y, z) \Big]_A^B = -p(x, y, z) \Big]_A^B = p(A) - p(B)$$

as shown in Figure 15.25. In other words, work $W$ is equal to the difference in the potential energies of $A$ and $B$. Now, suppose that $\mathbf{r}(t)$ is the position vector for a particle moving along $C$ from $A = \mathbf{r}(a)$ to $B = \mathbf{r}(b)$. At any time $t$, the particle's velocity, acceleration, and speed are $\mathbf{v}(t) = \mathbf{r}'(t)$, $\mathbf{a}(t) = \mathbf{r}''(t)$, and $v(t) = \|\mathbf{v}(t)\|$, respectively. So, by Newton's Second Law of Motion, $\mathbf{F} = m\mathbf{a}(t) = m(\mathbf{v}'(t))$, and the work done by $\mathbf{F}$ is

$$
\begin{aligned}
W &= \int_C \mathbf{F} \cdot d\mathbf{r} \\
&= \int_a^b \mathbf{F} \cdot \mathbf{r}'(t)\, dt \\
&= \int_a^b \mathbf{F} \cdot \mathbf{v}(t)\, dt \\
&= \int_a^b [m\mathbf{v}'(t)] \cdot \mathbf{v}(t)\, dt \\
&= \int_a^b m[\mathbf{v}'(t) \cdot \mathbf{v}(t)]\, dt \\
&= \frac{m}{2} \int_a^b \frac{d}{dt}[\mathbf{v}(t) \cdot \mathbf{v}(t)]\, dt \\
&= \frac{m}{2} \int_a^b \frac{d}{dt}[\|\mathbf{v}(t)\|^2]\, dt \\
&= \frac{m}{2} \Big[ \|\mathbf{v}(t)\|^2 \Big]_a^b \\
&= \frac{m}{2} \Big[ [v(t)]^2 \Big]_a^b \\
&= \frac{1}{2}m[v(b)]^2 - \frac{1}{2}m[v(a)]^2 \\
&= k(B) - k(A).
\end{aligned}
$$

Equating these two results for $W$ produces

$$p(A) - p(B) = k(B) - k(A)$$
$$p(A) + k(A) = p(B) + k(B)$$

which implies that the sum of the potential and kinetic energies remains constant from point to point.

**MICHAEL FARADAY (1791–1867)**

Several philosophers of science have considered Faraday's Law of Conservation of Energy to be the greatest generalization ever conceived by humankind. Many physicists have contributed to our knowledge of this law. Two early and influential ones were James Prescott Joule (1818–1889) and Hermann Ludwig Helmholtz (1821–1894).

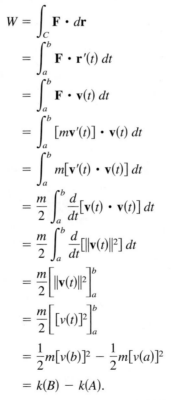

The work done by $\mathbf{F}$ along $C$ is

$$W = \int_C \mathbf{F} \cdot d\mathbf{r} = p(A) - p(B).$$

**Figure 15.25**

# 15.3 Exercises

See **CalcChat.com** for tutorial help and worked-out solutions to odd-numbered exercises.

**CONCEPT CHECK**

**1. Fundamental Theorem of Line Integrals** Explain how to evaluate a line integral using the Fundamental Theorem of Line Integrals.

**2. Independence of Path** What does it mean for a line integral to be independent of path? State the method for determining whether a line integral is independent of path.

**Line Integral of a Conservative Vector Field** In Exercises 3–8, (a) show that **F** is conservative and (b) verify that the value of

$$\int_C \mathbf{F} \cdot d\mathbf{r}$$

**is the same for each parametric representation of C.**

**3.** $\mathbf{F}(x, y) = x^2\mathbf{i} + y\mathbf{j}$

 (i) $C_1$: $\mathbf{r}_1(t) = t\mathbf{i} + t^2\mathbf{j}, \quad 0 \le t \le 1$

 (ii) $C_2$: $\mathbf{r}_2(\theta) = \sin\theta\mathbf{i} + \sin^2\theta\mathbf{j}, \quad 0 \le \theta \le \pi/2$

**4.** $\mathbf{F}(x, y) = (x^2 - y^2)\mathbf{i} - 2xy\mathbf{j}$

 (i) $C_1$: $\mathbf{r}_1(t) = t\mathbf{i} + \sqrt{t}\mathbf{j}, \quad 0 \le t \le 4$

 (ii) $C_2$: $\mathbf{r}_2(w) = w^2\mathbf{i} + w\mathbf{j}, \quad 0 \le w \le 2$

**5.** $\mathbf{F}(x, y) = 3y\mathbf{i} + 3x\mathbf{j}$

 (i) $C_1$: $\mathbf{r}_1(\theta) = \sec\theta\mathbf{i} + \tan\theta\mathbf{j}, \quad 0 \le \theta \le \pi/3$

 (ii) $C_2$: $\mathbf{r}_2(t) = \sqrt{t + 1}\mathbf{i} + \sqrt{t}\mathbf{j}, \quad 0 \le t \le 3$

**6.** $\mathbf{F}(x, y) = y\mathbf{i} + x\mathbf{j}$

 (i) $C_1$: $\mathbf{r}_1(t) = (2 + t)\mathbf{i} + (3 - t)\mathbf{j}, \quad 0 \le t \le 1$

 (ii) $C_2$: $\mathbf{r}_2(w) = (2 + \ln w)\mathbf{i} + (3 - \ln w)\mathbf{j}, \quad 1 \le w \le e$

**7.** $\mathbf{F}(x, y, z) = y^2z\mathbf{i} + 2xyz\mathbf{j} + xy^2\mathbf{k}$

 (i) $C_1$: $\mathbf{r}_1(t) = t\mathbf{i} + 2t\mathbf{j} + 4t\mathbf{k}, \quad 0 \le t \le 1$

 (ii) $C_2$: $\mathbf{r}_2(\theta) = \sin\theta\mathbf{i} + 2\sin\theta\mathbf{j} + 4\sin\theta\mathbf{k}, \quad 0 \le \theta \le \pi/2$

**8.** $\mathbf{F}(x, y, z) = 2yz\mathbf{i} + 2xz\mathbf{j} + 2xy\mathbf{k}$

 (i) $C_1$: $\mathbf{r}_1(t) = t\mathbf{i} - 4t\mathbf{j} + t^2\mathbf{k}, \quad 0 \le t \le 3$

 (ii) $C_2$: $\mathbf{r}_2(s) = s^2\mathbf{i} - \frac{4}{3}s^4\mathbf{j} + s^4\mathbf{k}, \quad 0 \le s \le \sqrt{3}$

**Using the Fundamental Theorem of Line Integrals** In Exercises 9–18, evaluate

$$\int_C \mathbf{F} \cdot d\mathbf{r}$$

**using the Fundamental Theorem of Line Integrals. Use a computer algebra system to verify your results.**

**9.** $\mathbf{F}(x, y) = 3y\mathbf{i} + 3x\mathbf{j}$

 $C$: smooth curve from $(0, 0)$ to $(3, 8)$

**10.** $\mathbf{F}(x, y) = 2(x + y)\mathbf{i} + 2(x + y)\mathbf{j}$

 $C$: smooth curve from $(-1, 1)$ to $(3, 2)$

**11.** $\mathbf{F}(x, y) = \cos x \sin y\mathbf{i} + \sin x \cos y\mathbf{j}$

 $C$: line segment from $(0, -\pi)$ to $\left(\dfrac{3\pi}{2}, \dfrac{\pi}{2}\right)$

**12.** $\mathbf{F}(x, y) = \dfrac{y}{x^2 + y^2}\mathbf{i} - \dfrac{x}{x^2 + y^2}\mathbf{j}$

 $C$: line segment from $(1, 1)$ to $(2\sqrt{3}, 2)$

**13.** $\mathbf{F}(x, y) = e^x \sin y\mathbf{i} + e^x \cos y\mathbf{j}$

 $C$: cycloid $x = \theta - \sin\theta, y = 1 - \cos\theta$ from $(0, 0)$ to $(2\pi, 0)$

**14.** $\mathbf{F}(x, y) = \dfrac{2x}{(x^2 + y^2)^2}\mathbf{i} + \dfrac{2y}{(x^2 + y^2)^2}\mathbf{j}$

 $C$: clockwise around the circle $(x - 4)^2 + (y - 5)^2 = 9$ from $(7, 5)$ to $(1, 5)$

**15.** $\mathbf{F}(x, y, z) = (z + 2y)\mathbf{i} + (2x - z)\mathbf{j} + (x - y)\mathbf{k}$

 (a) $C_1$: line segment from $(0, 0, 0)$ to $(1, 1, 1)$

 (b) $C_2$: line segments from $(0, 0, 0)$ to $(0, 0, 1)$ and $(0, 0, 1)$ to $(1, 1, 1)$

 (c) $C_3$: line segments from $(0, 0, 0)$ to $(1, 0, 0)$, from $(1, 0, 0)$ to $(1, 1, 0)$, and from $(1, 1, 0)$ to $(1, 1, 1)$

**16.** Repeat Exercise 15 using

 $\mathbf{F}(x, y, z) = zy\mathbf{i} + xz\mathbf{j} + xy\mathbf{k}.$

**17.** $\mathbf{F}(x, y, z) = -\sin x\mathbf{i} + z\mathbf{j} + y\mathbf{k}$

 $C$: smooth curve from $(0, 0, 0)$ to $\left(\dfrac{\pi}{2}, 3, 4\right)$

**18.** $\mathbf{F}(x, y, z) = 6x\mathbf{i} - 4z\mathbf{j} - (4y - 20z)\mathbf{k}$

 $C$: smooth curve from $(0, 0, 0)$ to $(3, 4, 0)$

**Finding Work in a Conservative Force Field** In Exercises 19–22, (a) show that $\int_C \mathbf{F} \cdot d\mathbf{r}$ is independent of path and (b) calculate the work done by the force field **F** on an object moving along a curve from $P$ to $Q$.

**19.** $\mathbf{F}(x, y) = 9x^2y^2\mathbf{i} + (6x^3y - 1)\mathbf{j}$

 $P(0, 0), Q(5, 9)$

**20.** $\mathbf{F}(x, y) = \dfrac{2x}{y}\mathbf{i} - \dfrac{x^2}{y^2}\mathbf{j}$

 $P(-1, 1), Q(3, 2)$

**21.** $\mathbf{F}(x, y, z) = 3\mathbf{i} + 4y\mathbf{j} - \sin z\mathbf{k}$

 $P\left(0, 1, \dfrac{\pi}{2}\right), Q(1, 4, \pi)$

**22.** $\mathbf{F}(x, y, z) = 8x^3\mathbf{i} + z^2 \cos 2y\mathbf{j} + z \sin 2y\mathbf{k}$

 $P\left(0, \dfrac{\pi}{4}, 1\right), Q(-2, 0, -1)$

**Evaluating a Line Integral** In Exercises 23–32, evaluate

$$\int_C \mathbf{F} \cdot d\mathbf{r}$$

along each path. (*Hint:* If **F** is conservative, the integration may be easier on an alternative path.)

23. $\mathbf{F}(x, y) = 2xy\mathbf{i} + x^2\mathbf{j}$
  (a) $C_1: \mathbf{r}_1(t) = t\mathbf{i} + t^2\mathbf{j},\ \ 0 \le t \le 1$
  (b) $C_2: \mathbf{r}_2(t) = t\mathbf{i} + t^3\mathbf{j},\ \ 0 \le t \le 1$

24. $\mathbf{F}(x, y) = ye^{xy}\mathbf{i} + xe^{xy}\mathbf{j}$
  (a) $C_1: \mathbf{r}_1(t) = t\mathbf{i} - (t - 3)\mathbf{j},\ \ 0 \le t \le 3$
  (b) $C_2$: The closed path consisting of line segments from $(0, 3)$ to $(0, 0)$, from $(0, 0)$ to $(3, 0)$, and then from $(3, 0)$ to $(0, 3)$

25. $\int_C y^2\,dx + 2xy\,dy$

(a)
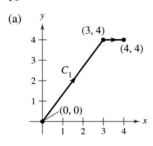

(b)
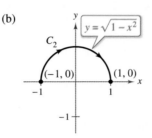

(c)
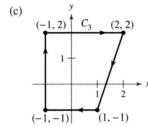

(d)
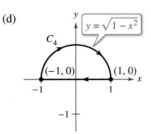

26. $\int_C (2x - 3y + 1)\,dx - (3x + y - 5)\,dy$

(a)
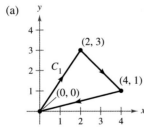

(b)
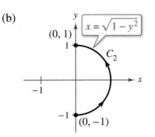

(c)
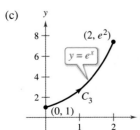

(d)
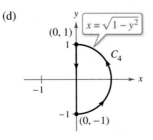

27. $\mathbf{F}(x, y, z) = yz\mathbf{i} + xz\mathbf{j} + xy\mathbf{k}$
  (a) $C_1: \mathbf{r}_1(t) = t\mathbf{i} + 2\mathbf{j} + t\mathbf{k},\ \ 0 \le t \le 4$
  (b) $C_2: \mathbf{r}_2(t) = t^2\mathbf{i} + t\mathbf{j} + t^2\mathbf{k},\ \ 0 \le t \le 2$

28. $\mathbf{F}(x, y, z) = \mathbf{i} + z\mathbf{j} + y\mathbf{k}$
  (a) $C_1: \mathbf{r}_1(t) = \cos t\mathbf{i} + \sin t\mathbf{j} + t^2\mathbf{k},\ \ 0 \le t \le \pi$
  (b) $C_2: \mathbf{r}_2(t) = (1 - 2t)\mathbf{i} + \pi^2 t\mathbf{k},\ \ 0 \le t \le 1$

29. $\mathbf{F}(x, y, z) = (2y + x)\mathbf{i} + (x^2 - z)\mathbf{j} + (2y - 4z)\mathbf{k}$
  (a) $C_1: \mathbf{r}_1(t) = t\mathbf{i} + t^2\mathbf{j} + \mathbf{k},\ \ 0 \le t \le 1$
  (b) $C_2: \mathbf{r}_2(t) = t\mathbf{i} + t\mathbf{j} + (2t - 1)^2\mathbf{k},\ \ 0 \le t \le 1$

30. $\mathbf{F}(x, y, z) = -y\mathbf{i} + x\mathbf{j} + 3xz^2\mathbf{k}$
  (a) $C_1: \mathbf{r}_1(t) = \cos t\mathbf{i} + \sin t\mathbf{j} + t\mathbf{k},\ \ 0 \le t \le \pi$
  (b) $C_2: \mathbf{r}_2(t) = (1 - 2t)\mathbf{i} + \pi t\mathbf{k},\ \ 0 \le t \le 1$

31. $\mathbf{F}(x, y, z) = e^z(y\mathbf{i} + x\mathbf{j} + xy\mathbf{k})$
  (a) $C_1: \mathbf{r}_1(t) = 4\cos t\mathbf{i} + 4\sin t\mathbf{j} + 3\mathbf{k},\ \ 0 \le t \le \pi$
  (b) $C_2: \mathbf{r}_2(t) = (4 - 8t)\mathbf{i} + 3\mathbf{k},\ \ 0 \le t \le 1$

32. $\mathbf{F}(x, y, z) = y \sin z\mathbf{i} + x \sin z\mathbf{j} + xy \cos x\mathbf{k}$
  (a) $C_1: \mathbf{r}_1(t) = t^2\mathbf{i} + t^2\mathbf{j},\ \ 0 \le t \le 2$
  (b) $C_2: \mathbf{r}_2(t) = 4t\mathbf{i} + 4t\mathbf{j},\ \ 0 \le t \le 1$

33. **Work**   A stone weighing 1 pound is attached to the end of a two-foot string and is whirled horizontally with one end held fixed. It makes 1 revolution per second. Find the work done by the force **F** that keeps the stone moving in a circular path. [*Hint:* Use Force = (mass)(centripetal acceleration).]

34. **Work**   A grappling hook weighing 1 kilogram is attached to the end of a five-meter rope and is whirled horizontally with one end held fixed. It makes 0.5 revolution per second. Find the work done by the force **F** that keeps the grappling hook moving in a circular path. [*Hint:* Use Force = (mass)(centripetal acceleration).]

35. **Work**

A zip line is installed 50 meters above ground level. It runs to a point on the ground 50 meters away from the base of the installation. Show that the work done by the gravitational force field for a 175-pound person moving the length of the zip line is the same for each path.

  (a) $C_1: \mathbf{r}_1(t) = t\mathbf{i} + (50 - t)\mathbf{j}$
  (b) $C_2: \mathbf{r}_2(t) = t\mathbf{i} + \frac{1}{50}(50 - t)^2\mathbf{j}$

36. **Work**   Can you find a path for the zip line in Exercise 35 such that the work done by the gravitational force field would differ from the amounts of work done for the two paths given? Explain why or why not.

## EXPLORING CONCEPTS

**37. Think About It** Consider

$$\mathbf{F}(x, y) = \frac{y}{x^2 + y^2}\mathbf{i} - \frac{x}{x^2 + y^2}\mathbf{j}.$$

Sketch an open connected region around the smooth curve $C$ shown in the figure such that you can use Theorem 15.7 to evaluate $\int_C \mathbf{F} \cdot d\mathbf{r}$. Explain how you created your sketch.

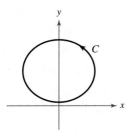

**38. Work** Let $\mathbf{F}(x, y, z) = a_1\mathbf{i} + a_2\mathbf{j} + a_3\mathbf{k}$ be a constant force vector field. Show that the work done in moving a particle along any path from $P$ to $Q$ is $W = \mathbf{F} \cdot \overrightarrow{PQ}$.

**39. Using Different Methods** Use two different methods to evaluate $\int_C \mathbf{F} \cdot d\mathbf{r}$ along the path

$$\mathbf{r}(t) = \frac{1}{t}\mathbf{i} + 3t\mathbf{j}, \quad 0.5 \le t \le 2$$

where $\mathbf{F}(x, y) = (x^2y^2 - 3x)\mathbf{i} + \frac{2}{3}x^3y\mathbf{j}.$

---

**40. HOW DO YOU SEE IT?** Consider the force field shown in the figure. To print an enlarged copy of the graph, go to *MathGraphs.com*.

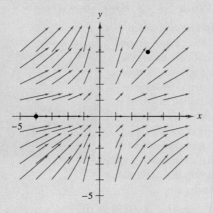

(a) Give a verbal argument that the force field is not conservative because you can identify two paths that require different amounts of work to move an object from $(-4, 0)$ to $(3, 4)$. Of the two paths, which requires the greater amount of work?

(b) Give a verbal argument that the force field is not conservative because you can find a closed curve $C$ such that $\int_C \mathbf{F} \cdot d\mathbf{r} \neq 0$.

---

**Graphical Reasoning** In Exercises 41 and 42, consider the force field shown in the figure. Is the force field conservative? Explain why or why not.

**41.**
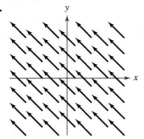

**42.**
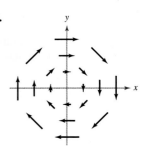

**True or False?** In Exercises 43–46, determine whether the statement is true or false. If it is false, explain why or give an example that shows it is false.

**43.** If $C_1$, $C_2$, and $C_3$ have the same initial and terminal points and $\int_{C_1} \mathbf{F} \cdot d\mathbf{r}_1 = \int_{C_2} \mathbf{F} \cdot d\mathbf{r}_2$, then $\int_{C_1} \mathbf{F} \cdot d\mathbf{r}_1 = \int_{C_3} \mathbf{F} \cdot d\mathbf{r}_3$.

**44.** If $\mathbf{F} = y\mathbf{i} + x\mathbf{j}$ and $C$ is given by $\mathbf{r}(t) = 4 \sin t\mathbf{i} + 3 \cos t\mathbf{j}$ for $0 \le t \le \pi$, then

$$\int_C \mathbf{F} \cdot d\mathbf{r} = 0.$$

**45.** If $\mathbf{F}$ is conservative in a region $R$ bounded by a simple closed path and $C$ lies within $R$, then $\int_C \mathbf{F} \cdot d\mathbf{r}$ is independent of path.

**46.** If $\mathbf{F} = M\mathbf{i} + N\mathbf{j}$ and $\dfrac{\partial M}{\partial x} = \dfrac{\partial N}{\partial y}$, then $\mathbf{F}$ is conservative.

**47. Harmonic Function** A function $f$ is called *harmonic* when

$$\frac{\partial^2 f}{\partial x^2} + \frac{\partial^2 f}{\partial y^2} = 0.$$

Prove that if $f$ is harmonic, then

$$\int_C \left(\frac{\partial f}{\partial y}\, dx - \frac{\partial f}{\partial x}\, dy\right) = 0$$

where $C$ is a smooth closed curve in the plane.

**48. Kinetic and Potential Energy** The kinetic energy of an object moving through a conservative force field is decreasing at a rate of 15 units per minute. At what rate is the potential energy changing? Explain.

**49. Investigation** Let $\mathbf{F}(x, y) = \dfrac{y}{x^2 + y^2}\mathbf{i} - \dfrac{x}{x^2 + y^2}\mathbf{j}.$

(a) Show that $\dfrac{\partial N}{\partial x} = \dfrac{\partial M}{\partial y}.$

(b) Let $\mathbf{r}(t) = \cos t\mathbf{i} + \sin t\mathbf{j}$ for $0 \le t \le \pi$. Find $\displaystyle\int_C \mathbf{F} \cdot d\mathbf{r}.$

(c) Let $\mathbf{r}(t) = \cos t\mathbf{i} - \sin t\mathbf{j}$ for $0 \le t \le \pi$. Find $\displaystyle\int_C \mathbf{F} \cdot d\mathbf{r}.$

(d) Let $\mathbf{r}(t) = \cos t\mathbf{i} + \sin t\mathbf{j}$ for $0 \le t \le 2\pi$. Find $\displaystyle\int_C \mathbf{F} \cdot d\mathbf{r}.$

(e) Do the results of parts (b)–(d) contradict Theorem 15.7? Why or why not?

(f) Show that $\nabla\left(\arctan\dfrac{x}{y}\right) = \mathbf{F}.$

# 15.4   Green's Theorem

■ Use Green's Theorem to evaluate a line integral.
■ Use alternative forms of Green's Theorem.

## Green's Theorem

In this section, you will study **Green's Theorem,** named after the English mathematician George Green (1793–1841). This theorem states that the value of a double integral over a *simply connected* plane region $R$ is determined by the value of a line integral around the boundary of $R$.

A curve $C$ given by $\mathbf{r}(t) = x(t)\mathbf{i} + y(t)\mathbf{j}$, where $a \le t \le b$, is **simple** when it does not cross itself—that is, $\mathbf{r}(c) \ne \mathbf{r}(d)$ for all $c$ and $d$ in the open interval $(a, b)$. A connected plane region $R$ is **simply connected** when every simple closed curve in $R$ encloses only points that are in $R$ (see Figure 15.26). Informally, a simply connected region cannot consist of separate parts or holes.

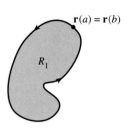
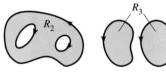

Simply connected

Not simply connected

**Figure 15.26**

> **THEOREM 15.8   Green's Theorem**
>
> Let $R$ be a simply connected region with a piecewise smooth boundary $C$, oriented counterclockwise (that is, $C$ is traversed *once* so that the region $R$ always lies to the *left*). If $M$ and $N$ have continuous first partial derivatives in an open region containing $R$, then
>
> $$\int_C M\,dx + N\,dy = \int\int_R \left( \frac{\partial N}{\partial x} - \frac{\partial M}{\partial y} \right) dA.$$

**Proof**   A proof is given only for a region that is both vertically simple and horizontally simple, as shown in Figure 15.27.

$$\int_C M\,dx = \int_{C_1} M\,dx + \int_{C_2} M\,dx$$

$$= \int_a^b M(x, f_1(x))\,dx + \int_b^a M(x, f_2(x))\,dx$$

$$= \int_a^b \left[ M(x, f_1(x)) - M(x, f_2(x)) \right] dx$$

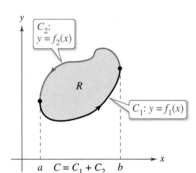

$R$ is vertically simple.

On the other hand,

$$\int\int_R \frac{\partial M}{\partial y}\,dA = \int_a^b \int_{f_1(x)}^{f_2(x)} \frac{\partial M}{\partial y}\,dy\,dx$$

$$= \int_a^b M(x, y) \Big]_{f_1(x)}^{f_2(x)}\,dx$$

$$= \int_a^b \left[ M(x, f_2(x)) - M(x, f_1(x)) \right] dx.$$

Consequently,

$$\int_C M\,dx = -\int\int_R \frac{\partial M}{\partial y}\,dA.$$

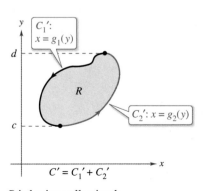

$R$ is horizontally simple.
**Figure 15.27**

Similarly, you can use $g_1(y)$ and $g_2(y)$ to show that $\int_C N\,dy = \int\int_R (\partial N/\partial x)\,dA$. By adding the integrals $\int_C M\,dx$ and $\int_C N\,dy$, you obtain the conclusion stated in the theorem.   ∎

An integral sign with a circle is sometimes used to indicate a line integral around a simple closed curve, as shown below. To indicate the orientation of the boundary, an arrow can be used. For instance, in the second integral, the arrow indicates that the boundary $C$ is oriented counterclockwise.

$$\textbf{1.} \ \oint_C M \, dx + N \, dy \qquad \textbf{2.} \ \oint_C M \, dx + N \, dy$$

## EXAMPLE 1   Using Green's Theorem

Use Green's Theorem to evaluate the line integral

$$\int_C y^3 \, dx + (x^3 + 3xy^2) \, dy$$

where $C$ is the path from $(0, 0)$ to $(1, 1)$ along the graph of $y = x^3$ and from $(1, 1)$ to $(0, 0)$ along the graph of $y = x$, as shown in Figure 15.28.

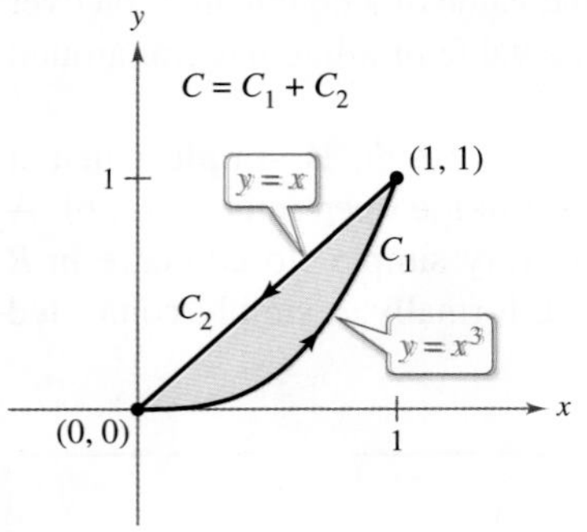

**Solution**   Because $M = y^3$ and $N = x^3 + 3xy^2$, it follows that

$$\frac{\partial N}{\partial x} = 3x^2 + 3y^2 \quad \text{and} \quad \frac{\partial M}{\partial y} = 3y^2.$$

Applying Green's Theorem, you then have

$$\int_C y^3 \, dx + (x^3 + 3xy^2) \, dy = \int\int_R \left( \frac{\partial N}{\partial x} - \frac{\partial M}{\partial y} \right) dA$$

$$= \int_0^1 \int_{x^3}^x [(3x^2 + 3y^2) - 3y^2] \, dy \, dx$$

$$= \int_0^1 \int_{x^3}^x 3x^2 \, dy \, dx$$

$$= \int_0^1 3x^2 y \Big]_{x^3}^x \, dx$$

$$= \int_0^1 (3x^3 - 3x^5) \, dx$$

$$= \left[ \frac{3x^4}{4} - \frac{x^6}{2} \right]_0^1$$

$$= \frac{1}{4}.$$

$C$ is simple and closed, and the region $R$ always lies to the left of $C$.
**Figure 15.28**

**GEORGE GREEN
(1793–1841)**

Green, a self-educated miller's son, first published the theorem that bears his name in 1828 in an essay on electricity and magnetism. At that time, there was almost no mathematical theory to explain electrical phenomena. "Considering how desirable it was that a power of universal agency, like electricity, should, as far as possible, be submitted to calculation, . . . I was induced to try whether it would be possible to discover any general relations existing between this function and the quantities of electricity in the bodies producing it."

Green's Theorem cannot be applied to every line integral. Among other restrictions stated in Theorem 15.8, the curve $C$ must be simple and closed. When Green's Theorem does apply, however, it can save time. To see this, try using the techniques described in Section 15.2 to evaluate the line integral in Example 1. To do this, you would need to write the line integral as

$$\int_C y^3 \, dx + (x^3 + 3xy^2) \, dy$$

$$= \int_{C_1} y^3 \, dx + (x^3 + 3xy^2) \, dy + \int_{C_2} y^3 \, dx + (x^3 + 3xy^2) \, dy$$

where $C_1$ is the cubic path given by

$$\mathbf{r}(t) = t\mathbf{i} + t^3\mathbf{j}$$

from $t = 0$ to $t = 1$, and $C_2$ is the line segment given by

$$\mathbf{r}(t) = (1 - t)\mathbf{i} + (1 - t)\mathbf{j}$$

from $t = 0$ to $t = 1$.

**EXAMPLE 2**    **Using Green's Theorem to Calculate Work**

While subject to the force

$$\mathbf{F}(x, y) = y^3\mathbf{i} + (x^3 + 3xy^2)\mathbf{j}$$

a particle travels once around the circle of radius 3 shown in Figure 15.29. Use Green's Theorem to find the work done by **F**.

**Solution**    From Example 1, you know by Green's Theorem that

$$\int_C y^3 \, dx + (x^3 + 3xy^2) \, dy = \iint_R 3x^2 \, dA.$$

In polar coordinates, using $x = r \cos \theta$ and $dA = r \, dr \, d\theta$, the work done is

$$
\begin{aligned}
W &= \iint_R 3x^2 \, dA \\
&= \int_0^{2\pi} \int_0^3 3(r \cos \theta)^2 r \, dr \, d\theta \\
&= 3 \int_0^{2\pi} \int_0^3 r^3 \cos^2 \theta \, dr \, d\theta \\
&= 3 \int_0^{2\pi} \frac{r^4}{4} \cos^2 \theta \Big]_0^3 \, d\theta \\
&= 3 \int_0^{2\pi} \frac{81}{4} \cos^2 \theta \, d\theta \\
&= \frac{243}{8} \int_0^{2\pi} (1 + \cos 2\theta) \, d\theta \\
&= \frac{243}{8} \left[ \theta + \frac{\sin 2\theta}{2} \right]_0^{2\pi} \\
&= \frac{243\pi}{4}.
\end{aligned}
$$

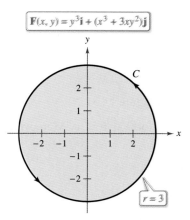

$$\boxed{\mathbf{F}(x, y) = y^3\mathbf{i} + (x^3 + 3xy^2)\mathbf{j}}$$

$r = 3$

**Figure 15.29**

When evaluating line integrals over closed curves, remember that for conservative vector fields (those for which $\partial N/\partial x = \partial M/\partial y$), the value of the line integral is 0. This is easily seen from the statement of Green's Theorem:

$$\int_C M \, dx + N \, dy = \iint_R \left( \frac{\partial N}{\partial x} - \frac{\partial M}{\partial y} \right) dA = 0.$$

**EXAMPLE 3**    **Green's Theorem and Conservative Vector Fields**

Evaluate the line integral

$$\int_C y^3 \, dx + 3xy^2 \, dy$$

where $C$ is the path shown in Figure 15.30.

**Solution**    From this line integral, $M = y^3$ and $N = 3xy^2$. So, $\partial N/\partial x = 3y^2$ and $\partial M/\partial y = 3y^2$. This implies that the vector field $\mathbf{F} = M\mathbf{i} + N\mathbf{j}$ is conservative, and because $C$ is closed, you can conclude that

$$\int_C y^3 \, dx + 3xy^2 \, dy = 0.$$

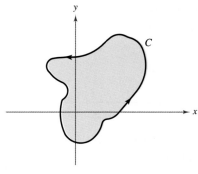

$C$ is closed.
**Figure 15.30**

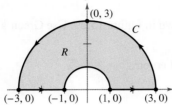

*C* is piecewise smooth.
**Figure 15.31**

**Using Green's Theorem**

$\cdots\!\vartriangleright$ *See LarsonCalculus.com for an interactive version of this type of example.*

Evaluate

$$\int_C (\arctan x + y^2)\, dx + (e^y - x^2)\, dy$$

where *C* is the path enclosing the annular region shown in Figure 15.31.

**Solution** In polar coordinates, *R* is given by $1 \le r \le 3$ for $0 \le \theta \le \pi$. Moreover,

$$\frac{\partial N}{\partial x} - \frac{\partial M}{\partial y} = -2x - 2y = -2(r\cos\theta + r\sin\theta).$$

So, by Green's Theorem,

$$
\begin{aligned}
\int_C (\arctan x + y^2)\, dx + (e^y - x^2)\, dy &= \int\!\!\int_R -2(x + y)\, dA \\
&= \int_0^\pi \int_1^3 -2r(\cos\theta + \sin\theta) r\, dr\, d\theta \\
&= \int_0^\pi -2(\cos\theta + \sin\theta)\frac{r^3}{3}\Big]_1^3 d\theta \\
&= \int_0^\pi -\frac{52}{3}(\cos\theta + \sin\theta)\, d\theta \\
&= -\frac{52}{3}\Big[\sin\theta - \cos\theta\Big]_0^\pi \\
&= -\frac{104}{3}.
\end{aligned}
$$

In Examples 1, 2, and 4, Green's Theorem was used to evaluate line integrals as double integrals. You can also use the theorem to evaluate double integrals as line integrals. One useful application occurs when $\partial N/\partial x - \partial M/\partial y = 1$.

$$
\begin{aligned}
\int_C M\, dx + N\, dy &= \int\!\!\int_R \left(\frac{\partial N}{\partial x} - \frac{\partial M}{\partial y}\right) dA \\
&= \int\!\!\int_R 1\, dA \qquad\qquad \frac{\partial N}{\partial x} - \frac{\partial M}{\partial y} = 1 \\
&= \text{area of region } R
\end{aligned}
$$

Among the many choices for *M* and *N* satisfying the stated condition, the choice of

$$M = -\frac{y}{2} \quad \text{and} \quad N = \frac{x}{2}$$

produces the following line integral for the area of region *R*.

---

**THEOREM 15.9 Line Integral for Area**

If *R* is a plane region bounded by a piecewise smooth simple closed curve *C*, oriented counterclockwise, then the area of *R* is given by

$$A = \frac{1}{2}\int_C x\, dy - y\, dx.$$

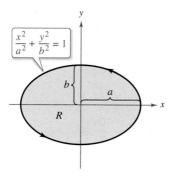

Figure 15.32

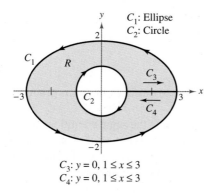

$C_1$: Ellipse
$C_2$: Circle

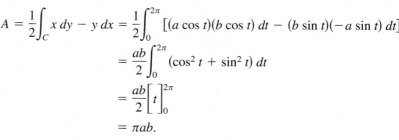

$C_3$: $y = 0, 1 \le x \le 3$
$C_4$: $y = 0, 1 \le x \le 3$

Figure 15.33

**EXAMPLE 5**    **Finding Area by a Line Integral**

Use a line integral to find the area of the ellipse $(x^2/a^2) + (y^2/b^2) = 1$.

**Solution**    Using Figure 15.32, you can induce a counterclockwise orientation to the elliptical path by letting $x = a \cos t$ and $y = b \sin t$, $0 \le t \le 2\pi$. So, the area is

$$A = \frac{1}{2} \int_C x \, dy - y \, dx = \frac{1}{2} \int_0^{2\pi} [(a \cos t)(b \cos t) \, dt - (b \sin t)(-a \sin t) \, dt]$$

$$= \frac{ab}{2} \int_0^{2\pi} (\cos^2 t + \sin^2 t) \, dt$$

$$= \frac{ab}{2} \Big[ t \Big]_0^{2\pi}$$

$$= \pi ab.$$

Green's Theorem can be extended to cover some regions that are not simply connected. This is demonstrated in the next example.

**EXAMPLE 6**    **Green's Theorem Extended to a Region with a Hole**

Let $R$ be the region inside the ellipse $(x^2/9) + (y^2/4) = 1$ and outside the circle $x^2 + y^2 = 1$. Evaluate the line integral

$$\int_C 2xy \, dx + (x^2 + 2x) \, dy$$

where $C = C_1 + C_2$ is the boundary of $R$, as shown in Figure 15.33.

**Solution**    To begin, introduce the line segments $C_3$ and $C_4$, as shown in Figure 15.33. Note that because the curves $C_3$ and $C_4$ have opposite orientations, the line integrals over them cancel. Furthermore, apply Green's Theorem to the region $R$ using the boundary $C_1 + C_4 + C_2 + C_3$ to obtain

$$\int_C 2xy \, dx + (x^2 + 2x) \, dy = \int_R \int \left( \frac{\partial N}{\partial x} - \frac{\partial M}{\partial y} \right) dA$$

$$= \int_R \int (2x + 2 - 2x) \, dA$$

$$= 2 \int_R \int dA$$

$$= 2(\text{area of } R)$$

$$= 2(\pi ab - \pi r^2)$$

$$= 2[\pi(3)(2) - \pi(1^2)]$$

$$= 10\pi.$$

In Section 15.1, a necessary and sufficient condition for conservative vector fields was listed. There, only one direction of the proof was shown. You can now outline the other direction, using Green's Theorem. Let $\mathbf{F}(x, y) = M\mathbf{i} + N\mathbf{j}$ be defined on an open disk $R$. You want to show that if $M$ and $N$ have continuous first partial derivatives and $\partial M/\partial y = \partial N/\partial x$, then $\mathbf{F}$ is conservative. Let $C$ be a closed path forming the boundary of a connected region lying in $R$. Then, using the fact that $\partial M/\partial y = \partial N/\partial x$, apply Green's Theorem to conclude that

$$\int_C \mathbf{F} \cdot d\mathbf{r} = \int_C M \, dx + N \, dy = \int_R \int \left( \frac{\partial N}{\partial x} - \frac{\partial M}{\partial y} \right) dA = 0.$$

This, in turn, is equivalent to showing that $\mathbf{F}$ is conservative (see Theorem 15.7).

## Alternative Forms of Green's Theorem

This section concludes with the derivation of two vector forms of Green's Theorem for regions in the plane. The extension of these vector forms to three dimensions is the basis for the discussion in the remaining sections of this chapter. For a vector field $\mathbf{F}$ in the plane, you can write

$$\mathbf{F}(x, y, z) = M\mathbf{i} + N\mathbf{j} + 0\mathbf{k}$$

so that the curl of $\mathbf{F}$, as described in Section 15.1, is given by

$$\operatorname{curl} \mathbf{F} = \nabla \times \mathbf{F} = \begin{vmatrix} \mathbf{i} & \mathbf{j} & \mathbf{k} \\ \dfrac{\partial}{\partial x} & \dfrac{\partial}{\partial y} & \dfrac{\partial}{\partial z} \\ M & N & 0 \end{vmatrix} = -\frac{\partial N}{\partial z}\mathbf{i} + \frac{\partial M}{\partial z}\mathbf{j} + \left(\frac{\partial N}{\partial x} - \frac{\partial M}{\partial y}\right)\mathbf{k}.$$

Consequently,

$$(\operatorname{curl} \mathbf{F}) \cdot \mathbf{k} = \left[-\frac{\partial N}{\partial z}\mathbf{i} + \frac{\partial M}{\partial z}\mathbf{j} + \left(\frac{\partial N}{\partial x} - \frac{\partial M}{\partial y}\right)\mathbf{k}\right] \cdot \mathbf{k} = \frac{\partial N}{\partial x} - \frac{\partial M}{\partial y}.$$

With appropriate conditions on $\mathbf{F}$, $C$, and $R$, you can write Green's Theorem in the vector form

$$\int_C \mathbf{F} \cdot d\mathbf{r} = \iint_R \left(\frac{\partial N}{\partial x} - \frac{\partial M}{\partial y}\right) dA$$

$$= \iint_R (\operatorname{curl} \mathbf{F}) \cdot \mathbf{k} \, dA. \qquad \text{First alternative form}$$

The extension of this vector form of Green's Theorem to surfaces in space produces **Stokes's Theorem,** discussed in Section 15.8.

For the second vector form of Green's Theorem, assume the same conditions for $\mathbf{F}$, $C$, and $R$. Using the arc length parameter $s$ for $C$, you have $\mathbf{r}(s) = x(s)\mathbf{i} + y(s)\mathbf{j}$. So, a unit tangent vector $\mathbf{T}$ to curve $C$ is given by $\mathbf{r}'(s) = \mathbf{T} = x'(s)\mathbf{i} + y'(s)\mathbf{j}$. From Figure 15.34, you can see that the *outward* unit normal vector $\mathbf{N}$ can then be written as

$$\mathbf{N} = y'(s)\mathbf{i} - x'(s)\mathbf{j}.$$

Consequently, for $\mathbf{F}(x, y) = M\mathbf{i} + N\mathbf{j}$, you can apply Green's Theorem to obtain

$$\int_C \mathbf{F} \cdot \mathbf{N} \, ds = \int_a^b (M\mathbf{i} + N\mathbf{j}) \cdot (y'(s)\mathbf{i} - x'(s)\mathbf{j}) \, ds$$

$$= \int_a^b \left(M\frac{dy}{ds} - N\frac{dx}{ds}\right) ds$$

$$= \int_C M \, dy - N \, dx$$

$$= \int_C -N \, dx + M \, dy$$

$$= \iint_R \left(\frac{\partial M}{\partial x} + \frac{\partial N}{\partial y}\right) dA \qquad \text{Green's Theorem}$$

$$= \iint_R \operatorname{div} \mathbf{F} \, dA.$$

Therefore,

$$\int_C \mathbf{F} \cdot \mathbf{N} \, ds = \iint_R \operatorname{div} \mathbf{F} \, dA. \qquad \text{Second alternative form}$$

The extension of this form to three dimensions is called the **Divergence Theorem** and will be discussed in Section 15.7. The physical interpretations of divergence and curl will be discussed in Sections 15.7 and 15.8.

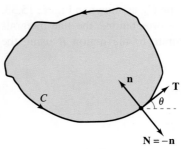

$\mathbf{T} = \cos\theta\mathbf{i} + \sin\theta\mathbf{j}$

$\mathbf{n} = \cos\left(\theta + \dfrac{\pi}{2}\right)\mathbf{i} + \sin\left(\theta + \dfrac{\pi}{2}\right)\mathbf{j}$

$\quad = -\sin\theta\mathbf{i} + \cos\theta\mathbf{j}$

$\mathbf{N} = \sin\theta\mathbf{i} - \cos\theta\mathbf{j}$

**Figure 15.34**

# 15.4  Exercises

**CONCEPT CHECK**

1. **Writing**  What does it mean for a curve to be simple? What does it mean for a plane region to be simply connected?

2. **Green's Theorem**  Explain the usefulness of Green's Theorem.

3. **Integral Sign**  What information do you learn from the integral sign $\oint_C$ ?

4. **Area**  Describe how to find the area of a plane region bounded by a piecewise smooth simple closed curve that is oriented counterclockwise.

 **Verifying Green's Theorem**  In Exercises 5–8, verify Green's Theorem by evaluating both integrals

$$\int_C y^2\, dx + x^2\, dy = \int\int_R \left(\frac{\partial N}{\partial x} - \frac{\partial M}{\partial y}\right) dA$$

**for the given path.**

5. $C$: boundary of the region lying between the graphs of $y = x$ and $y = x^2$

6. $C$: boundary of the region lying between the graphs of $y = x$ and $y = \sqrt{x}$

7. $C$: square with vertices $(0, 0)$, $(1, 0)$, $(1, 1)$, and $(0, 1)$

8. $C$: rectangle with vertices $(0, 0)$, $(3, 0)$, $(3, 4)$, and $(0, 4)$

 **Verifying Green's Theorem**  In Exercises 9 and 10, verify Green's Theorem by using a computer algebra system to evaluate both integrals $\int_C xe^y\, dx + e^x\, dy = \int\int_R \left(\frac{\partial N}{\partial x} - \frac{\partial M}{\partial y}\right) dA$ for the given path.

9. $C$: circle given by $x^2 + y^2 = 4$

10. $C$: boundary of the region lying between the graphs of $y = x$ and $y = x^3$ in the first quadrant

**Evaluating a Line Integral Using Green's Theorem**  In Exercises 11–14, use Green's Theorem to evaluate the line integral $\int_C (y - x)\, dx + (2x - y)\, dy$ for the given path.

11. $C$: boundary of the region lying between the graphs of $y = x$ and $y = x^2 - 2x$

12. $C$: $x = 2\cos\theta$, $y = \sin\theta$

13. $C$: boundary of the region lying inside the rectangle with vertices $(5, 3)$, $(-5, 3)$, $(-5, -3)$, and $(5, -3)$, and outside the square with vertices $(1, 1)$, $(-1, 1)$, $(-1, -1)$, and $(1, -1)$

14. $C$: boundary of the region lying inside the semicircle $y = \sqrt{25 - x^2}$ and outside the semicircle $y = \sqrt{9 - x^2}$

 **Evaluating a Line Integral Using Green's Theorem**  In Exercises 15–24, use Green's Theorem to evaluate the line integral.

15. $\int_C 2xy\, dx + (x + y)\, dy$

   $C$: boundary of the region lying between the graphs of $y = 0$ and $y = 1 - x^2$

16. $\int_C y^2\, dx + xy\, dy$

   $C$: boundary of the region lying between the graphs of $y = 0$, $y = \sqrt{x}$, and $x = 9$

17. $\int_C (x^2 - y^2)\, dx + 2xy\, dy$

   $C$: $x^2 + y^2 = 16$

18. $\int_C (x^2 - y^2)\, dx + 2xy\, dy$

   $C$: $r = 1 + \cos\theta$

19. $\int_C e^x \cos 2y\, dx - 2e^x \sin 2y\, dy$

   $C$: $x^2 + y^2 = a^2$

20. $\int_C 2\arctan\frac{y}{x}\, dx + \ln(x^2 + y^2)\, dy$

   $C$: $x = 4 + 2\cos\theta$, $y = 4 + \sin\theta$

21. $\int_C \cos y\, dx + (xy - x \sin y)\, dy$

   $C$: boundary of the region lying between the graphs of $y = x$ and $y = \sqrt{x}$

22. $\int_C (e^{-x^2/2} - y)\, dx + (e^{-y^2/2} + x)\, dy$

   $C$: boundary of the region lying between the graphs of the circle $x = 6\cos\theta$, $y = 6\sin\theta$ and the ellipse $x = 3\cos\theta$, $y = 2\sin\theta$

23. $\int_C (x - 3y)\, dx + (x + y)\, dy$

   $C$: boundary of the region lying between the graphs of $x^2 + y^2 = 1$ and $x^2 + y^2 = 9$

24. $\int_C 3x^2 e^y\, dx + e^y\, dy$

   $C$: boundary of the region lying between the squares with vertices $(1, 1)$, $(-1, 1)$, $(-1, -1)$, and $(1, -1)$, and $(2, 2)$, $(-2, 2)$, $(-2, -2)$, and $(2, -2)$

 **Work**  In Exercises 25–28, use Green's Theorem to calculate the work done by the force **F** on a particle that is moving counterclockwise around the closed path $C$.

25. $\mathbf{F}(x, y) = xy\mathbf{i} + (x + y)\mathbf{j}$

   $C$: $x^2 + y^2 = 1$

**26.** $\mathbf{F}(x, y) = (e^x - 3y)\mathbf{i} + (e^y + 6x)\mathbf{j}$

$C$: $r = 2 \cos \theta$

**27.** $\mathbf{F}(x, y) = (x^{3/2} - 3y)\mathbf{i} + (6x + 5\sqrt{y})\mathbf{j}$

$C$: triangle with vertices $(0, 0)$, $(5, 0)$, and $(0, 5)$

**28.** $\mathbf{F}(x, y) = (3x^2 + y)\mathbf{i} + 4xy^2\mathbf{j}$

$C$: boundary of the region lying between the graphs of $y = \sqrt{x}$, $y = 0$, and $x = 9$

 **Area** In Exercises 29–32, use a line integral to find the area of the region $R$.

**29.** $R$: region bounded by the graph of $x^2 + y^2 = 4$

**30.** $R$: triangle bounded by the graphs of $x = 0$, $3x - 2y = 0$, and $x + 2y = 8$

**31.** $R$: region bounded by the graphs of $y = 5x - 3$ and $y = x^2 + 1$

**32.** $R$: region inside the loop of the folium of Descartes bounded by the graph of

$$x = \frac{3t}{t^3 + 1}, \quad y = \frac{3t^2}{t^3 + 1}$$

**Using Green's Theorem to Verify a Formula** In Exercises 33 and 34, use Green's Theorem to verify the line integral formula(s).

**33.** The centroid of the region having area $A$ bounded by the simple closed path $C$ has coordinates

$$\bar{x} = \frac{1}{2A} \int_C x^2 \, dy \quad \text{and} \quad \bar{y} = -\frac{1}{2A} \int_C y^2 \, dx.$$

**34.** The area of a plane region bounded by the simple closed path $C$ given in polar coordinates is

$$A = \frac{1}{2} \int_C r^2 \, d\theta.$$

**Centroid** In Exercises 35–38, use the results of Exercise 33 to find the centroid of the region.

**35.** $R$: region bounded by the graphs of $y = 0$ and $4 - x^2$

**36.** $R$: region bounded by the graphs of $y = \sqrt{1 - x^2}$ and $y = 0$

**37.** $R$: region bounded by the graphs of $y = x^3$ and $y = x$, $0 \le x \le 1$

**38.** $R$: triangle with vertices $(-a, 0)$, $(a, 0)$, and $(b, c)$, where $-a \le b \le a$

**Area** In Exercises 39–42, use the result of Exercise 34 to find the area of the region bounded by the graph of the polar equation.

**39.** $r = 6(1 - \cos \theta)$

**40.** $r = a \cos 3\theta$

**41.** $r = 1 + 2 \cos \theta$ (inner loop)

**42.** $r = \dfrac{3}{2 - \cos \theta}$

**43. Maximum Value**

(a) Evaluate $\displaystyle\int_{C_1} y^3 \, dx + (27x - x^3) \, dy$, where $C_1$ is the unit circle given by $\mathbf{r}(t) = \cos t\mathbf{i} + \sin t\mathbf{j}$, for $0 \le t \le 2\pi$.

(b) Find the maximum value of $\displaystyle\int_C y^3 \, dx + (27x - x^3) \, dy$, where $C$ is any circle centered at the origin in the $xy$-plane, oriented counterclockwise.

**44.** 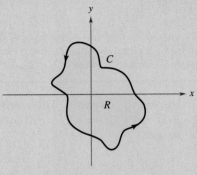 **HOW DO YOU SEE IT?** The figure shows a region $R$ bounded by a piecewise smooth simple closed path $C$.

(a) Is $R$ simply connected? Explain.

(b) Explain why $\displaystyle\int_C f(x) \, dx + g(y) \, dy = 0$, where $f$ and $g$ are differentiable functions.

**45. Green's Theorem: Region with a Hole** Let $R$ be the region inside the circle $x = 5 \cos \theta$, $y = 5 \sin \theta$ and outside the ellipse $x = 2 \cos \theta$, $y = \sin \theta$. Evaluate the line integral

$$\int_C (e^{-x^2/2} - y) \, dx + (e^{-y^2/2} + x) \, dy$$

where $C = C_1 + C_2$ is the boundary of $R$, as shown in the figure.

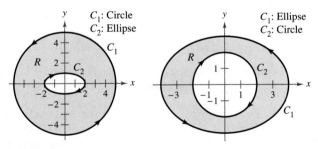

Figure for 45        Figure for 46

**46. Green's Theorem: Region with a Hole** Let $R$ be the region inside the ellipse $x = 4 \cos \theta$, $y = 3 \sin \theta$ and outside the circle $x = 2 \cos \theta$, $y = 2 \sin \theta$. Evaluate the line integral

$$\int_C (3x^2y + 1) \, dx + (x^3 + 4x) \, dy$$

where $C = C_1 + C_2$ is the boundary of $R$, as shown in the figure.

## EXPLORING CONCEPTS

**47. Think About It**  Let

$$I = \int_C \frac{y\,dx - x\,dy}{x^2 + y^2}$$

where $C$ is a circle oriented counterclockwise.

(a) Show that $I = 0$ when $C$ does not contain the origin.

(b) What is $I$ when $C$ does contain the origin?

**48. Think About It**  For each given path, verify Green's Theorem by showing that

$$\int_C y^2\,dx + x^2\,dy = \iint_R \left(\frac{\partial N}{\partial x} - \frac{\partial M}{\partial y}\right) dA.$$

For each path, which integral is easier to evaluate? Explain.

(a) $C$: triangle with vertices $(0, 0)$, $(4, 0)$, and $(4, 4)$

(b) $C$: circle given by $x^2 + y^2 = 1$

**49. Proof**

(a) Let $C$ be the line segment joining $(x_1, y_1)$ and $(x_2, y_2)$. Show that $\int_C -y\,dx + x\,dy = x_1y_2 - x_2y_1$.

(b) Let $(x_1, y_1)$, $(x_2, y_2)$, . . . , $(x_n, y_n)$ be the vertices of a polygon. Prove that the area enclosed is

$$\tfrac{1}{2}[(x_1y_2 - x_2y_1) + (x_2y_3 - x_3y_2) + \cdots +$$
$$(x_{n-1}y_n - x_ny_{n-1}) + (x_ny_1 - x_1y_n)].$$

**50. Area**  Use the result of Exercise 49(b) to find the area enclosed by the polygon with the given vertices.

(a) Pentagon: $(0, 0)$, $(2, 0)$, $(3, 2)$, $(1, 4)$, and $(-1, 1)$

(b) Hexagon: $(0, 0)$, $(2, 0)$, $(3, 2)$, $(2, 4)$, $(0, 3)$, and $(-1, 1)$

**Proof**  In Exercises 51 and 52, prove the identity, where $R$ is a simply connected region with piecewise smooth boundary $C$. Assume that the required partial derivatives of the scalar functions $f$ and $g$ are continuous. The expressions $D_N f$ and $D_N g$ are the derivatives in the direction of the outward normal vector $\mathbf{N}$ of $C$ and are defined by $D_N f = \nabla f \cdot \mathbf{N}$ and $D_N g = \nabla g \cdot \mathbf{N}$.

**51.** Green's first identity:

$$\iint_R (f\nabla^2 g + \nabla f \cdot \nabla g)\,dA = \int_C f D_N g\,ds$$

[*Hint:* Use the second alternative form of Green's Theorem and the property $\text{div}(f\mathbf{G}) = f\,\text{div}\,\mathbf{G} + \nabla f \cdot \mathbf{G}$.]

**52.** Green's second identity:

$$\iint_R (f\nabla^2 g - g\nabla^2 f)\,dA = \int_C (f D_N g - g D_N f)\,ds$$

(*Hint:* Use Green's first identity from Exercise 51 twice.)

**53. Proof**  Let $\mathbf{F} = M\mathbf{i} + N\mathbf{j}$, where $M$ and $N$ have continuous first partial derivatives in a simply connected region $R$. Prove that if $C$ is simple, smooth, and closed, and $N_x = M_y$, then $\int_C \mathbf{F} \cdot d\mathbf{r} = 0$.

### PUTNAM EXAM CHALLENGE

**54.** Find the least possible area of a convex set in the plane that intersects both branches of the hyperbola $xy = 1$ and both branches of the hyperbola $xy = -1$. (A set $S$ in the plane is called *convex* if for any two points in $S$ the line segment connecting them is contained in $S$.)

## SECTION PROJECT

# Hyperbolic and Trigonometric Functions

(a) Sketch the plane curve represented by the vector-valued function $\mathbf{r}(t) = \cosh t\,\mathbf{i} + \sinh t\,\mathbf{j}$ on the interval $0 \le t \le 5$. Show that the rectangular equation corresponding to $\mathbf{r}(t)$ is the hyperbola $x^2 - y^2 = 1$. Verify your sketch by using a graphing utility to graph the hyperbola.

(b) Let $P = (\cosh \phi, \sinh \phi)$ be the point on the hyperbola corresponding to $\mathbf{r}(\phi)$ for $\phi > 0$. Use the formula for area

$$A = \frac{1}{2}\int_C x\,dy - y\,dx$$

to verify that the area of the region shown in the figure is $\tfrac{1}{2}\phi$.

(c) Show that the area of the region shown in the figure is also given by the integral

$$A = \int_0^{\sinh \phi} \left[\sqrt{1 + y^2} - (\coth \phi)y\right] dy.$$

Confirm your answer in part (b) by evaluating this integral for $\phi = 1, 2, 4$, and $10$.

(d) Consider the unit circle given by $x^2 + y^2 = 1$. Let $\theta$ be the angle formed by the $x$-axis and the radius to $(x, y)$. The area of the corresponding sector is $\tfrac{1}{2}\theta$. That is, the trigonometric functions $f(\theta) = \cos \theta$ and $g(\theta) = \sin \theta$ could have been defined as the coordinates of the point $(\cos \theta, \sin \theta)$ on the unit circle that determines a sector of area $\tfrac{1}{2}\theta$. Write a short paragraph explaining how you could define the hyperbolic functions in a similar manner, using the "unit hyperbola" $x^2 - y^2 = 1$.

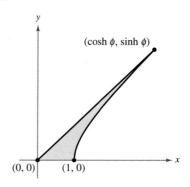

# 15.5    Parametric Surfaces

■ Understand the definition of a parametric surface, and sketch the surface.
■ Find a set of parametric equations to represent a surface.
■ Find a normal vector and a tangent plane to a parametric surface.
■ Find the area of a parametric surface.

## Parametric Surfaces

You already know how to represent a curve in the plane or in space by a set of parametric equations—or, equivalently, by a vector-valued function.

$$\mathbf{r}(t) = x(t)\mathbf{i} + y(t)\mathbf{j} \qquad \text{Plane curve}$$
$$\mathbf{r}(t) = x(t)\mathbf{i} + y(t)\mathbf{j} + z(t)\mathbf{k} \qquad \text{Space curve}$$

In this section, you will learn how to represent a surface in space by a set of parametric equations—or by a vector-valued function. For curves, note that the vector-valued function $\mathbf{r}$ is a function of a *single* parameter $t$. For surfaces, the vector-valued function is a function of *two* parameters $u$ and $v$.

---

### Definition of Parametric Surface

Let $x$, $y$, and $z$ be functions of $u$ and $v$ that are continuous on a domain $D$ in the $uv$-plane. The set of points $(x, y, z)$ given by

$$\mathbf{r}(u, v) = x(u, v)\mathbf{i} + y(u, v)\mathbf{j} + z(u, v)\mathbf{k} \qquad \text{Parametric surface}$$

is called a **parametric surface.** The equations

$$x = x(u, v), \quad y = y(u, v), \quad \text{and} \quad z = z(u, v) \qquad \text{Parametric equations}$$

are the **parametric equations** for the surface.

---

If $S$ is a parametric surface given by the vector-valued function $\mathbf{r}$, then $S$ is traced out by the position vector $\mathbf{r}(u, v)$ as the point $(u, v)$ moves throughout the domain $D$, as shown in Figure 15.35.

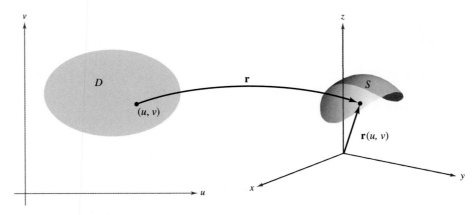

The parametric surface $S$ given by the vector-valued function $\mathbf{r}$, where $\mathbf{r}$ is a function of two variables $u$ and $v$ defined on a domain $D$
**Figure 15.35**

▷ **TECHNOLOGY** Some computer algebra systems are capable of graphing surfaces that are represented parametrically. If you have access to such software, use it to graph some of the surfaces in the examples and exercises in this section.

EXAMPLE 1    **Sketching a Parametric Surface**

Identify and sketch the parametric surface $S$ given by

$$\mathbf{r}(u, v) = 3 \cos u\mathbf{i} + 3 \sin u\mathbf{j} + v\mathbf{k}$$

where $0 \le u \le 2\pi$ and $0 \le v \le 4$.

**Solution**    Because $x = 3 \cos u$ and
$y = 3 \sin u$, you know that for each point
$(x, y, z)$ on the surface, $x$ and $y$ are related
by the equation

$$x^2 + y^2 = 3^2.$$

In other words, each cross section of $S$ taken
parallel to the $xy$-plane is a circle of radius 3,
centered on the $z$-axis. Because $z = v$, where

$$0 \le v \le 4$$

you can see that the surface is a right circular
cylinder of height 4. The radius of the cylinder
is 3, and the $z$-axis forms the axis of the cylinder,
as shown in Figure 15.36.

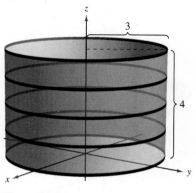

**Figure 15.36**

As with parametric representations of curves, parametric representations of
surfaces are not unique. That is, there are many other sets of parametric equations that
could be used to represent the surface shown in Figure 15.36.

EXAMPLE 2    **Sketching a Parametric Surface**

Identify and sketch the parametric surface $S$ given by

$$\mathbf{r}(u, v) = \sin u \cos v\mathbf{i} + \sin u \sin v\mathbf{j} + \cos u\mathbf{k}$$

where $0 \le u \le \pi$ and $0 \le v \le 2\pi$.

**Solution**    To identify the surface, you can try to use trigonometric identities to
eliminate the parameters. After some experimentation, you can discover that

$$\begin{aligned}
x^2 + y^2 + z^2 &= (\sin u \cos v)^2 + (\sin u \sin v)^2 + (\cos u)^2 \\
&= \sin^2 u \cos^2 v + \sin^2 u \sin^2 v + \cos^2 u \\
&= (\sin^2 u)(\cos^2 v + \sin^2 v) + \cos^2 u \\
&= \sin^2 u + \cos^2 u \\
&= 1.
\end{aligned}$$

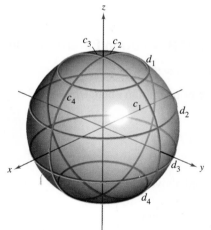

**Figure 15.37**

So, each point on $S$ lies on the unit sphere, centered at the origin, as shown in Figure 15.37.
For fixed $u = d_i$, $\mathbf{r}(u, v)$ traces out latitude circles

$$x^2 + y^2 = \sin^2 d_i, \quad 0 \le d_i \le \pi$$

that are parallel to the $xy$-plane, and for fixed $v = c_i$, $\mathbf{r}(u, v)$ traces out longitude (or
meridian) half-circles.

To convince yourself further that $\mathbf{r}(u, v)$ traces out the entire unit sphere, recall that
the parametric equations

$$x = \rho \sin \phi \cos \theta, \quad y = \rho \sin \phi \sin \theta, \quad \text{and} \quad z = \rho \cos \phi$$

where $0 \le \theta \le 2\pi$ and $0 \le \phi \le \pi$, describe the conversion from spherical to rectangular
coordinates, as discussed in Section 11.7.

## Finding Parametric Equations for Surfaces

In Examples 1 and 2, you were asked to identify the surface described by a given set of parametric equations. The reverse problem—that of writing a set of parametric equations for a given surface—is generally more difficult. One type of surface for which this problem is straightforward, however, is a surface that is given by $z = f(x, y)$. You can parametrize such a surface as

$$\mathbf{r}(x, y) = x\mathbf{i} + y\mathbf{j} + f(x, y)\mathbf{k}.$$

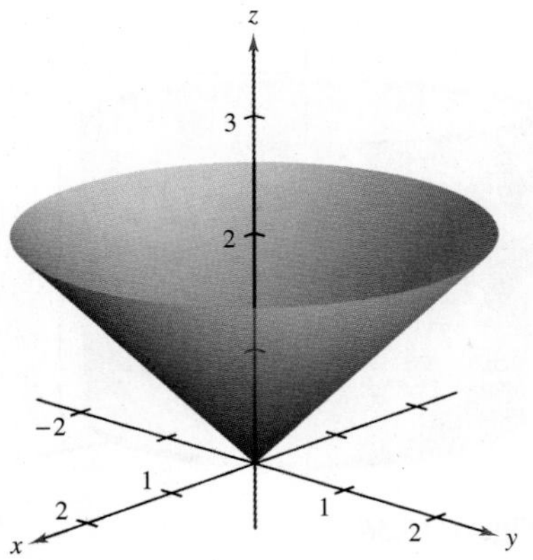

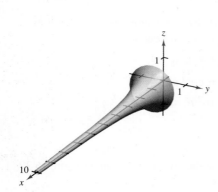

**Figure 15.38**

### EXAMPLE 3   Representing a Surface Parametrically

Write a set of parametric equations for the cone given by

$$z = \sqrt{x^2 + y^2}$$

as shown in Figure 15.38.

**Solution**   Because this surface is given in the form $z = f(x, y)$, you can let $x$ and $y$ be the parameters. Then the cone is represented by the vector-valued function

$$\mathbf{r}(x, y) = x\mathbf{i} + y\mathbf{j} + \sqrt{x^2 + y^2}\mathbf{k}$$

where $(x, y)$ varies over the entire $xy$-plane. ■

A second type of surface that is easily represented parametrically is a surface of revolution. For instance, to represent the surface formed by revolving the graph of

$$y = f(x), \quad a \le x \le b$$

about the $x$-axis, use

$$x = u, \quad y = f(u) \cos v, \quad \text{and} \quad z = f(u) \sin v$$

where $a \le u \le b$ and $0 \le v \le 2\pi$.

### EXAMPLE 4   Representing a Surface of Revolution Parametrically

∴ ▷ *See LarsonCalculus.com for an interactive version of this type of example.*

Write a set of parametric equations for the surface of revolution obtained by revolving

$$f(x) = \frac{1}{x}, \quad 1 \le x \le 10$$

about the $x$-axis.

**Solution**   Use the parameters $u$ and $v$ as described above to write

$$x = u, \quad y = f(u) \cos v = \frac{1}{u} \cos v, \quad \text{and} \quad z = f(u) \sin v = \frac{1}{u} \sin v$$

**Figure 15.39**

where

$$1 \le u \le 10 \quad \text{and} \quad 0 \le v \le 2\pi.$$

The resulting surface is a portion of *Gabriel's Horn*, as shown in Figure 15.39. ■

The surface of revolution in Example 4 is formed by revolving the graph of $y = f(x)$ about the $x$-axis. For other types of surfaces of revolution, a similar parametrization can be used. For instance, to parametrize the surface formed by revolving the graph of $x = f(z)$ about the $z$-axis, you can use

$$z = u, \quad x = f(u) \cos v, \quad \text{and} \quad y = f(u) \sin v.$$

## Normal Vectors and Tangent Planes

Let $S$ be a parametric surface given by

$$\mathbf{r}(u, v) = x(u, v)\mathbf{i} + y(u, v)\mathbf{j} + z(u, v)\mathbf{k}$$

over an open region $D$ such that $x$, $y$, and $z$ have continuous partial derivatives on $D$. The **partial derivatives of r** with respect to $u$ and $v$ are defined as

$$\mathbf{r}_u = \frac{\partial x}{\partial u}(u, v)\mathbf{i} + \frac{\partial y}{\partial u}(u, v)\mathbf{j} + \frac{\partial z}{\partial u}(u, v)\mathbf{k}$$

and

$$\mathbf{r}_v = \frac{\partial x}{\partial v}(u, v)\mathbf{i} + \frac{\partial y}{\partial v}(u, v)\mathbf{j} + \frac{\partial z}{\partial v}(u, v)\mathbf{k}.$$

Each of these partial derivatives is a vector-valued function that can be interpreted geometrically in terms of tangent vectors. For instance, if $v = v_0$ is held constant, then $\mathbf{r}(u, v_0)$ is a vector-valued function of a single parameter and defines a curve $C_1$ that lies on the surface $S$. The tangent vector to $C_1$ at the point

$$(x(u_0, v_0), y(u_0, v_0), z(u_0, v_0))$$

is given by

$$\mathbf{r}_u(u_0, v_0) = \frac{\partial x}{\partial u}(u_0, v_0)\mathbf{i} + \frac{\partial y}{\partial u}(u_0, v_0)\mathbf{j} + \frac{\partial z}{\partial u}(u_0, v_0)\mathbf{k}$$

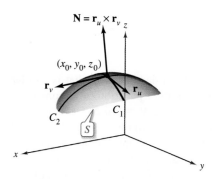

as shown in Figure 15.40. In a similar way, if $u = u_0$ is held constant, then $\mathbf{r}(u_0, v)$ is a vector-valued function of a single parameter and defines a curve $C_2$ that lies on the surface $S$. The tangent vector to $C_2$ at the point $(x(u_0, v_0), y(u_0, v_0), z(u_0, v_0))$ is given by

$$\mathbf{r}_v(u_0, v_0) = \frac{\partial x}{\partial v}(u_0, v_0)\mathbf{i} + \frac{\partial y}{\partial v}(u_0, v_0)\mathbf{j} + \frac{\partial z}{\partial v}(u_0, v_0)\mathbf{k}.$$

If the normal vector $\mathbf{r}_u \times \mathbf{r}_v$ is not $\mathbf{0}$ for any $(u, v)$ in $D$, then the surface $S$ is called **smooth** and will have a tangent plane. Informally, a smooth surface is one that has no sharp points or cusps. For instance, spheres, ellipsoids, and paraboloids are smooth, whereas the cone given in Example 3 is not smooth.

**Figure 15.40**

---

### Normal Vector to a Smooth Parametric Surface

Let $S$ be a smooth parametric surface

$$\mathbf{r}(u, v) = x(u, v)\mathbf{i} + y(u, v)\mathbf{j} + z(u, v)\mathbf{k}$$

defined over an open region $D$ in the $uv$-plane. Let $(u_0, v_0)$ be a point in $D$. A normal vector at the point

$$(x_0, y_0, z_0) = (x(u_0, v_0), y(u_0, v_0), z(u_0, v_0))$$

is given by

$$\mathbf{N} = \mathbf{r}_u(u_0, v_0) \times \mathbf{r}_v(u_0, v_0) = \begin{vmatrix} \mathbf{i} & \mathbf{j} & \mathbf{k} \\ \dfrac{\partial x}{\partial u} & \dfrac{\partial y}{\partial u} & \dfrac{\partial z}{\partial u} \\ \dfrac{\partial x}{\partial v} & \dfrac{\partial y}{\partial v} & \dfrac{\partial z}{\partial v} \end{vmatrix}.$$

---

Figure 15.40 shows the normal vector $\mathbf{r}_u \times \mathbf{r}_v$. The vector $\mathbf{r}_v \times \mathbf{r}_u$ is also normal to $S$ and points in the opposite direction.

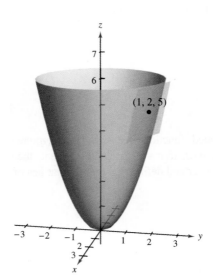

**Figure 15.41**

| EXAMPLE 5 | Finding a Tangent Plane to a Parametric Surface |

Find an equation of the tangent plane to the paraboloid

$$\mathbf{r}(u, v) = u\mathbf{i} + v\mathbf{j} + (u^2 + v^2)\mathbf{k}$$

at the point $(1, 2, 5)$.

**Solution**    The point in the $uv$-plane that is mapped to the point $(x, y, z) = (1, 2, 5)$ is $(u, v) = (1, 2)$. The partial derivatives of $\mathbf{r}$ are

$$\mathbf{r}_u = \mathbf{i} + 2u\mathbf{k} \quad \text{and} \quad \mathbf{r}_v = \mathbf{j} + 2v\mathbf{k}.$$

The normal vector is given by

$$\mathbf{r}_u \times \mathbf{r}_v = \begin{vmatrix} \mathbf{i} & \mathbf{j} & \mathbf{k} \\ 1 & 0 & 2u \\ 0 & 1 & 2v \end{vmatrix} = -2u\mathbf{i} - 2v\mathbf{j} + \mathbf{k}$$

which implies that the normal vector at $(1, 2, 5)$ is

$$\mathbf{r}_u \times \mathbf{r}_v = -2\mathbf{i} - 4\mathbf{j} + \mathbf{k}.$$

So, an equation of the tangent plane at $(1, 2, 5)$ is

$$-2(x - 1) - 4(y - 2) + (z - 5) = 0$$
$$-2x - 4y + z = -5.$$

The tangent plane is shown in Figure 15.41.    ∎

## Area of a Parametric Surface

To define the area of a parametric surface, you can use a development that is similar to that given in Section 14.5. Begin by constructing an inner partition of $D$ consisting of $n$ rectangles, where the area of the $i$th rectangle $D_i$ is $\Delta A_i = \Delta u_i \Delta v_i$, as shown in Figure 15.42. In each $D_i$, let $(u_i, v_i)$ be the point that is closest to the origin. At the point $(x_i, y_i, z_i) = (x(u_i, v_i), y(u_i, v_i), z(u_i, v_i))$ on the surface $S$, construct a tangent plane $T_i$. The area of the portion of $S$ that corresponds to $D_i$, $\Delta S_i$, can be approximated by a parallelogram $\Delta T_i$ in the tangent plane. That is, $\Delta T_i \approx \Delta S_i$. So, the surface area of $S$ is given by $\Sigma \, \Delta S_i \approx \Sigma \, \Delta T_i$. The area of the parallelogram in the tangent plane is

$$\text{Area of } \Delta T_i = \|\Delta u_i \mathbf{r}_u \times \Delta v_i \mathbf{r}_v\| = \|\mathbf{r}_u \times \mathbf{r}_v\| \, \Delta u_i \Delta v_i$$

which leads to the next definition.

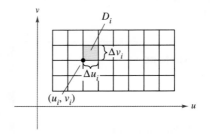

**Figure 15.42**

---

**Area of a Parametric Surface**

Let $S$ be a smooth parametric surface

$$\mathbf{r}(u, v) = x(u, v)\mathbf{i} + y(u, v)\mathbf{j} + z(u, v)\mathbf{k}$$

defined over an open region $D$ in the $uv$-plane. If each point on the surface $S$ corresponds to exactly one point in the domain $D$, then the **surface area** of $S$ is given by

$$\text{Surface area} = \iint_S dS = \iint_D \|\mathbf{r}_u \times \mathbf{r}_v\| \, dA$$

where

$$\mathbf{r}_u = \frac{\partial x}{\partial u}\mathbf{i} + \frac{\partial y}{\partial u}\mathbf{j} + \frac{\partial z}{\partial u}\mathbf{k} \quad \text{and} \quad \mathbf{r}_v = \frac{\partial x}{\partial v}\mathbf{i} + \frac{\partial y}{\partial v}\mathbf{j} + \frac{\partial z}{\partial v}\mathbf{k}.$$

---

For a surface $S$ given by $z = f(x, y)$, this formula for surface area corresponds to that given in Section 14.5. To see this, you can parametrize the surface using the vector-valued function $\mathbf{r}(x, y) = x\mathbf{i} + y\mathbf{j} + f(x, y)\mathbf{k}$ defined over the region $R$ in the $xy$-plane. Using $\mathbf{r}_x = \mathbf{i} + f_x(x, y)\mathbf{k}$ and $\mathbf{r}_y = \mathbf{j} + f_y(x, y)\mathbf{k}$, you have

$$\mathbf{r}_x \times \mathbf{r}_y = \begin{vmatrix} \mathbf{i} & \mathbf{j} & \mathbf{k} \\ 1 & 0 & f_x(x, y) \\ 0 & 1 & f_y(x, y) \end{vmatrix} = -f_x(x, y)\mathbf{i} - f_y(x, y)\mathbf{j} + \mathbf{k}$$

and

$$\|\mathbf{r}_x \times \mathbf{r}_y\| = \sqrt{[f_x(x, y)]^2 + [f_y(x, y)]^2 + 1}.$$

This implies that the surface area of $S$ is

$$\text{Surface area} = \iint_R \|\mathbf{r}_x \times \mathbf{r}_y\| \, dA$$

$$= \iint_R \sqrt{1 + [f_x(x, y)]^2 + [f_y(x, y)]^2} \, dA.$$

**EXAMPLE 6**   **Finding Surface Area**

Find the surface area of the unit sphere

$$\mathbf{r}(u, v) = \sin u \cos v\mathbf{i} + \sin u \sin v\mathbf{j} + \cos u\mathbf{k}$$

where the domain $D$ is $0 \le u \le \pi$ and $0 \le v \le 2\pi$.

**Solution**   Begin by calculating $\mathbf{r}_u$ and $\mathbf{r}_v$.

$$\mathbf{r}_u = \cos u \cos v\mathbf{i} + \cos u \sin v\mathbf{j} - \sin u\mathbf{k}$$
$$\mathbf{r}_v = -\sin u \sin v\mathbf{i} + \sin u \cos v\mathbf{j}$$

The cross product of these two vectors is

$$\mathbf{r}_u \times \mathbf{r}_v = \begin{vmatrix} \mathbf{i} & \mathbf{j} & \mathbf{k} \\ \cos u \cos v & \cos u \sin v & -\sin u \\ -\sin u \sin v & \sin u \cos v & 0 \end{vmatrix}$$

$$= \sin^2 u \cos v\mathbf{i} + \sin^2 u \sin v\mathbf{j} + \sin u \cos u\mathbf{k}$$

which implies that

$$\|\mathbf{r}_u \times \mathbf{r}_v\| = \sqrt{(\sin^2 u \cos v)^2 + (\sin^2 u \sin v)^2 + (\sin u \cos u)^2}$$

$$= \sqrt{\sin^4 u + \sin^2 u \cos^2 u}$$

$$= \sqrt{\sin^2 u}$$

$$= \sin u. \qquad \text{\small $\sin u > 0$ for $0 \le u \le \pi$}$$

Finally, the surface area of the sphere is

$$A = \iint_D \|\mathbf{r}_u \times \mathbf{r}_v\| \, dA$$

$$= \int_0^{2\pi} \int_0^{\pi} \sin u \, du \, dv$$

$$= \int_0^{2\pi} 2 \, dv$$

$$= 4\pi. \qquad \blacksquare$$

The surface in Example 6 does not quite fulfill the hypothesis that each point on the surface corresponds to exactly one point in $D$. For this surface, $\mathbf{r}(u, 0) = \mathbf{r}(u, 2\pi)$ for any fixed value of $u$. However, because the overlap consists of only a semicircle (which has no area), you can still apply the formula for the area of a parametric surface.

Because of high surface gravity, the shape of a neutron star is almost a perfect sphere. Using the surface area along with other data, scientists can estimate the mass and radius of the star.

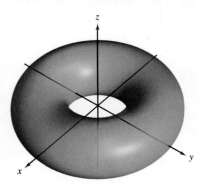

z

y

x

**Figure 15.43**

**EXAMPLE 7    Finding Surface Area**

Find the surface area of the torus given by

$$\mathbf{r}(u, v) = (2 + \cos u) \cos v\mathbf{i} + (2 + \cos u) \sin v\mathbf{j} + \sin u\mathbf{k}$$

where the domain $D$ is given by $0 \le u \le 2\pi$ and $0 \le v \le 2\pi$. (See Figure 15.43.)

**Solution**   Begin by calculating $\mathbf{r}_u$ and $\mathbf{r}_v$.

$$\mathbf{r}_u = -\sin u \cos v\mathbf{i} - \sin u \sin v\mathbf{j} + \cos u\mathbf{k}$$
$$\mathbf{r}_v = -(2 + \cos u) \sin v\mathbf{i} + (2 + \cos u) \cos v\mathbf{j}$$

The cross product of these two vectors is

$$\mathbf{r}_u \times \mathbf{r}_v = \begin{vmatrix} \mathbf{i} & \mathbf{j} & \mathbf{k} \\ -\sin u \cos v & -\sin u \sin v & \cos u \\ -(2 + \cos u) \sin v & (2 + \cos u) \cos v & 0 \end{vmatrix}$$

$$= -(2 + \cos u)(\cos v \cos u\mathbf{i} + \sin v \cos u\mathbf{j} + \sin u\mathbf{k})$$

which implies that

$$\|\mathbf{r}_u \times \mathbf{r}_v\| = (2 + \cos u)\sqrt{(\cos v \cos u)^2 + (\sin v \cos u)^2 + \sin^2 u}$$
$$= (2 + \cos u)\sqrt{\cos^2 u(\cos^2 v + \sin^2 v) + \sin^2 u}$$
$$= (2 + \cos u)\sqrt{\cos^2 u + \sin^2 u}$$
$$= 2 + \cos u.$$

Finally, the surface area of the torus is

$$A = \int\!\!\int_D \|\mathbf{r}_u \times \mathbf{r}_v\| \, dA$$
$$= \int_0^{2\pi}\!\!\int_0^{2\pi} (2 + \cos u) \, du \, dv$$
$$= \int_0^{2\pi} 4\pi \, dv$$
$$= 8\pi^2.$$

**Exploration**

For the torus in Example 7, describe the function $\mathbf{r}(u, v)$ for fixed $u$. Then describe the function $\mathbf{r}(u, v)$ for fixed $v$.

For a surface of revolution, you can show that the formula for surface area given in Section 7.4 is equivalent to the formula given in this section. For instance, suppose $f$ is a nonnegative function such that $f'$ is continuous over the interval $[a, b]$. Let $S$ be the surface of revolution formed by revolving the graph of $f$, where $a \le x \le b$, about the $x$-axis. From Section 7.4, you know that the surface area is given by

$$\text{Surface area} = 2\pi \int_a^b f(x)\sqrt{1 + [f'(x)]^2} \, dx.$$

To represent $S$ parametrically, let

$$x = u, \quad y = f(u) \cos v, \quad \text{and} \quad z = f(u) \sin v$$

where $a \le u \le b$ and $0 \le v \le 2\pi$. Then

$$\mathbf{r}(u, v) = u\mathbf{i} + f(u) \cos v\mathbf{j} + f(u) \sin v\mathbf{k}.$$

Try showing that the formula

$$\text{Surface area} = \int\!\!\int_D \|\mathbf{r}_u \times \mathbf{r}_v\| \, dA$$

is equivalent to the formula given above (see Exercise 56).

# 15.5 Exercises

See CalcChat.com for tutorial help and worked-out solutions to odd-numbered exercises.

## CONCEPT CHECK

**1. Parametric Surface** Explain how a parametric surface is represented by a vector-valued function and how the vector-valued function is used to sketch the parametric surface.

**2. Surface Area** A surface $S$ is represented by $z = f(x, y)$. What are the parametric equations for $S$?

**Matching** In Exercises 3–8, match the vector-valued function with its graph. [The graphs are labeled (a), (b), (c), (d), (e), and (f).]

(a)

(b)

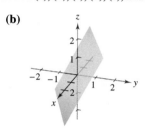

(c)

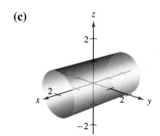

(d)

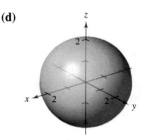

(e)

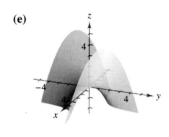

(f)

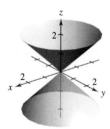

**3.** $\mathbf{r}(u, v) = u\mathbf{i} + v\mathbf{j} + uv\mathbf{k}$

**4.** $\mathbf{r}(u, v) = u \cos v\mathbf{i} + u \sin v\mathbf{j} + u\mathbf{k}$

**5.** $\mathbf{r}(u, v) = u\mathbf{i} + \frac{1}{2}(u + v)\mathbf{j} + v\mathbf{k}$

**6.** $\mathbf{r}(u, v) = v\mathbf{i} + \cos u\mathbf{j} + \sin u\mathbf{k}$

**7.** $\mathbf{r}(u, v) = 2 \cos v \cos u\mathbf{i} + 2 \cos v \sin u\mathbf{j} + 2 \sin v\mathbf{k}$

**8.** $\mathbf{r}(u, v) = u\mathbf{i} + \frac{1}{4}v^3\mathbf{j} + v\mathbf{k}$

 **Sketching a Parametric Surface** In Exercises 9–12, find the rectangular equation for the surface by eliminating the parameters from the vector-valued function. Identify the surface and sketch its graph.

**9.** $\mathbf{r}(u, v) = u\mathbf{i} + v\mathbf{j} + \frac{v}{2}\mathbf{k}$

**10.** $\mathbf{r}(u, v) = 2u \cos v\mathbf{i} + 2u \sin v\mathbf{j} + \frac{1}{2}u^2\mathbf{k}$

**11.** $\mathbf{r}(u, v) = 2 \cos u\mathbf{i} + v\mathbf{j} + 2 \sin u\mathbf{k}$

**12.** $\mathbf{r}(u, v) = 3 \cos v \cos u\mathbf{i} + 3 \cos v \sin u\mathbf{j} + 5 \sin v\mathbf{k}$

 **Graphing a Parametric Surface** In Exercises 13–16, use a computer algebra system to graph the surface represented by the vector-valued function.

**13.** $\mathbf{r}(u, v) = 2u \cos v\mathbf{i} + 2u \sin v\mathbf{j} + u^4\mathbf{k}$

$0 \le u \le 1, \quad 0 \le v \le 2\pi$

**14.** $\mathbf{r}(u, v) = 2u \cos v\mathbf{i} + 2u \sin v\mathbf{j} + v\mathbf{k}$

$0 \le u \le 1, \quad 0 \le v \le 3\pi$

**15.** $\mathbf{r}(u, v) = (u - \sin u) \cos v\mathbf{i} + (1 - \cos u) \sin v\mathbf{j} + u\mathbf{k}$

$0 \le u \le \pi, \quad 0 \le v \le 2\pi$

**16.** $\mathbf{r}(u, v) = \cos^3 u \cos v\mathbf{i} + \sin^3 u \sin v\mathbf{j} + u\mathbf{k}$

$0 \le u \le \frac{\pi}{2}, \quad 0 \le v \le 2\pi$

 **Representing a Surface Parametrically** In Exercises 17–26, find a vector-valued function whose graph is the indicated surface.

**17.** The plane $z = 3y$

**18.** The plane $x + y + z = 6$

**19.** The cone $y = \sqrt{4x^2 + 9z^2}$

**20.** The cone $x = \sqrt{16y^2 + z^2}$

**21.** The cylinder $x^2 + y^2 = 25$

**22.** The cylinder $4x^2 + y^2 = 16$

**23.** The paraboloid $x = y^2 + z^2 + 7$

**24.** The ellipsoid $\dfrac{x^2}{9} + \dfrac{y^2}{4} + \dfrac{z^2}{1} = 1$

**25.** The part of the plane $z = 4$ that lies inside the cylinder $x^2 + y^2 = 9$

**26.** The part of the paraboloid $z = x^2 + y^2$ that lies inside the cylinder $x^2 + y^2 = 9$

**Representing a Surface of Revolution Parametrically** In Exercises 27–32, write a set of parametric equations for the surface of revolution obtained by revolving the graph of the function about the given axis.

| Function | Axis of Revolution |
|---|---|
| **27.** $y = \dfrac{x}{2}, \quad 0 \le x \le 6$ | $x$-axis |
| **28.** $y = \sqrt{x}, \quad 0 \le x \le 4$ | $x$-axis |
| **29.** $x = \sin z, \quad 0 \le z \le \pi$ | $z$-axis |
| **30.** $x = z - 2, \quad 2 \le z \le 5$ | $z$-axis |
| **31.** $z = \cos^2 y, \quad \dfrac{\pi}{2} \le y \le \pi$ | $y$-axis |
| **32.** $z = y^2 + 1, \quad 0 \le y \le 2$ | $y$-axis |

**Finding a Tangent Plane** In Exercises 33–36, find an equation of the tangent plane to the surface represented by the vector-valued function at the given point.

**33.** $\mathbf{r}(u, v) = 3 \cos v \cos u\mathbf{i} + 2 \cos v \sin u\mathbf{j} + 4 \sin v\mathbf{k}$, $(0, \sqrt{3}, 2)$

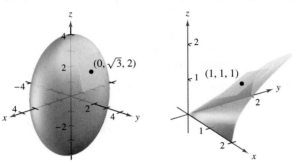

Figure for 33          Figure for 34

**34.** $\mathbf{r}(u, v) = u\mathbf{i} + v\mathbf{j} + \sqrt{uv}\,\mathbf{k}$, $(1, 1, 1)$

**35.** $\mathbf{r}(u, v) = 2u \cos v\mathbf{i} + 3u \sin v\mathbf{j} + u^2\mathbf{k}$, $(0, 6, 4)$

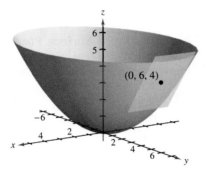

**36.** $\mathbf{r}(u, v) = 2u \cosh v\mathbf{i} + 2u \sinh v\mathbf{j} + \frac{1}{2}u^2\mathbf{k}$, $(-4, 0, 2)$

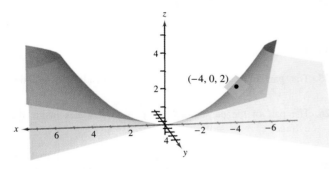

**Finding Surface Area** In Exercises 37–42, find the area of the surface over the given region. Use a computer algebra system to verify your results.

**37.** $\mathbf{r}(u, v) = 4u\mathbf{i} - v\mathbf{j} + v\mathbf{k}$, $0 \le u \le 2$, $0 \le v \le 1$

**38.** $\mathbf{r}(u, v) = 2u \cos v\mathbf{i} + 2u \sin v\mathbf{j} + u^2\mathbf{k}$, $0 \le u \le 2$, $0 \le v \le 2\pi$

**39.** $\mathbf{r}(u, v) = au \cos v\mathbf{i} + au \sin v\mathbf{j} + u\mathbf{k}$, $0 \le u \le b$, $0 \le v \le 2\pi$

**40.** $\mathbf{r}(u, v) = (a + b \cos v) \cos u\mathbf{i} + (a + b \cos v) \sin u\mathbf{j} + b \sin v\mathbf{k}$, $a > b$, $0 \le u \le 2\pi$, $0 \le v \le 2\pi$

**41.** $\mathbf{r}(u, v) = \sqrt{u} \cos v\mathbf{i} + \sqrt{u} \sin v\mathbf{j} + u\mathbf{k}$, $0 \le u \le 4$, $0 \le v \le 2\pi$

**42.** $\mathbf{r}(u, v) = \sin u \cos v\mathbf{i} + u\mathbf{j} + \sin u \sin v\mathbf{k}$, $0 \le u \le \pi$, $0 \le v \le 2\pi$

**EXPLORING CONCEPTS**

**Think About It** In Exercises 43–46, determine how the graph of the surface $s(u, v)$ differs from the graph of $\mathbf{r}(u, v) = u \cos v\mathbf{i} + u \sin v\mathbf{j} + u^2\mathbf{k}$, where $0 \le u \le 2$ and $0 \le v \le 2\pi$, as shown in the figure. (It is not necessary to graph s.)

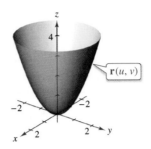

**43.** $s(u, v) = u \cos v\mathbf{i} + u \sin v\mathbf{j} - u^2\mathbf{k}$
$0 \le u \le 2$,  $0 \le v \le 2\pi$

**44.** $s(u, v) = u \cos v\mathbf{i} + u^2\mathbf{j} + u \sin v\mathbf{k}$
$0 \le u \le 2$,  $0 \le v \le 2\pi$

**45.** $s(u, v) = u \cos v\mathbf{i} + u \sin v\mathbf{j} + u^2\mathbf{k}$
$0 \le u \le 3$,  $0 \le v \le 2\pi$

**46.** $s(u, v) = 4u \cos v\mathbf{i} + 4u \sin v\mathbf{j} + u^2\mathbf{k}$
$0 \le u \le 2$,  $0 \le v \le 2\pi$

**47. Representing a Cone Parametrically** Show that the cone in Example 3 can be represented parametrically by $\mathbf{r}(u, v) = u \cos v\mathbf{i} + u \sin v\mathbf{j} + u\mathbf{k}$, where $u \ge 0$ and $0 \le v \le 2\pi$.

**48.** **HOW DO YOU SEE IT?** The figures below are graphs of $\mathbf{r}(u, v) = u\mathbf{i} + \sin u \cos v\mathbf{j} + \sin u \sin v\mathbf{k}$, where $0 \le u \le \pi/2$ and $0 \le v \le 2\pi$. Match each of the four graphs with the point in space from which the surface is viewed.

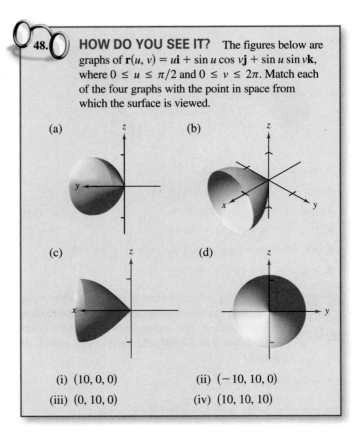

(a)          (b)

(c)          (d)

(i) $(10, 0, 0)$          (ii) $(-10, 10, 0)$

(iii) $(0, 10, 0)$          (iv) $(10, 10, 10)$

**49. Astroidal Sphere**  An equation of an **astroidal sphere** in $x$, $y$, and $z$ is

$$x^{2/3} + y^{2/3} + z^{2/3} = a^{2/3}.$$

A graph of an astroidal sphere is shown below. Show that this surface can be represented parametrically by

$$\mathbf{r}(u, v) = a \sin^3 u \cos^3 v \mathbf{i} + a \sin^3 u \sin^3 v \mathbf{j} + a \cos^3 u \mathbf{k}$$

where $0 \le u \le \pi$ and $0 \le v \le 2\pi$.

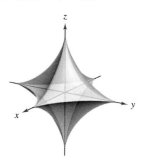

**50. Different Views of a Surface**  Use a computer algebra system to graph the vector-valued function

$$\mathbf{r}(u, v) = u \cos v \mathbf{i} + u \sin v \mathbf{j} + v \mathbf{k}, \quad 0 \le u \le \pi, \ 0 \le v \le \pi$$

from each of the points $(10, 0, 0)$, $(0, 0, 10)$, and $(10, 10, 10)$.

**51. Investigation**  Use a computer algebra system to graph the torus

$$\mathbf{r}(u, v) = (a + b \cos v) \cos u \mathbf{i} + (a + b \cos v) \sin u \mathbf{j} + b \sin v \mathbf{k}$$

for each set of values of $a$ and $b$, where $0 \le u \le 2\pi$ and $0 \le v \le 2\pi$. Use the results to describe the effects of $a$ and $b$ on the shape of the torus.

(a) $a = 4$,  $b = 1$

(b) $a = 4$,  $b = 2$

(c) $a = 8$,  $b = 1$

(d) $a = 8$,  $b = 3$

**52. Investigation**  Consider the function in Exercise 14.

(a) Sketch a graph of the function where $u$ is held constant at $u = 1$. Identify the graph.

(b) Sketch a graph of the function where $v$ is held constant at $v = 2\pi/3$. Identify the graph.

(c) Assume that a surface is represented by the vector-valued function $\mathbf{r} = \mathbf{r}(u, v)$. What generalization can you make about the graph of the function when one of the parameters is held constant?

**53. Surface Area**  The surface of the dome on a new museum is given by

$$\mathbf{r}(u, v) = 20 \sin u \cos v \mathbf{i} + 20 \sin u \sin v \mathbf{j} + 20 \cos u \mathbf{k}$$

where $0 \le u \le \pi/3$, $0 \le v \le 2\pi$, and $\mathbf{r}$ is in meters. Find the surface area of the dome.

**54. Hyperboloid**  Find a vector-valued function for the hyperboloid

$$x^2 + y^2 - z^2 = 1$$

and determine the tangent plane at $(1, 0, 0)$.

**55. Area**  Use a computer algebra system to graph one turn of the spiral ramp $\mathbf{r}(u, v) = u \cos v \mathbf{i} + u \sin v \mathbf{j} + 2v \mathbf{k}$, where $0 \le u \le 3$ and $0 \le v \le 2\pi$. Then analytically find the area of one turn of the spiral ramp.

**56. Surface Area**  Let $f$ be a nonnegative function such that $f'$ is continuous over the interval $[a, b]$. Let $S$ be the surface of revolution formed by revolving the graph of $f$, where $a \le x \le b$, about the $x$-axis. Let $x = u$, $y = f(u) \cos v$, and $z = f(u) \sin v$, where $a \le u \le b$ and $0 \le v \le 2\pi$. Then $S$ is represented parametrically by

$$\mathbf{r}(u, v) = u \mathbf{i} + f(u) \cos v \mathbf{j} + f(u) \sin v \mathbf{k}.$$

Show that the following formulas are equivalent.

$$\text{Surface area} = 2\pi \int_a^b f(x) \sqrt{1 + [f'(x)]^2}\, dx$$

$$\text{Surface area} = \int_D \int \|\mathbf{r}_u \times \mathbf{r}_v\|\, dA$$

**57. Open-Ended Project**  The parametric equations

$$x = 3 + [7 - \cos(3u - 2v) - 2\cos(3u + v)]\sin u$$
$$y = 3 + [7 - \cos(3u - 2v) - 2\cos(3u + v)]\cos u$$
$$z = \sin(3u - 2v) + 2\sin(3u + v)$$

where $-\pi \le u \le \pi$ and $-\pi \le v \le \pi$, represent the surface shown below. Try to create your own parametric surface using a computer algebra system.

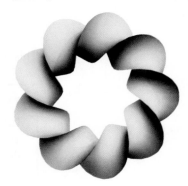

**58. Möbius Strip**  The surface shown in the figure is called a **Möbius strip** and can be represented by the parametric equations

$$x = \left(a + u \cos\frac{v}{2}\right)\cos v, \ y = \left(a + u \cos\frac{v}{2}\right)\sin v, \ z = u \sin\frac{v}{2}$$

where $-1 \le u \le 1$, $0 \le v \le 2\pi$, and $a = 3$. Try to graph other Möbius strips for different values of $a$ using a computer algebra system.

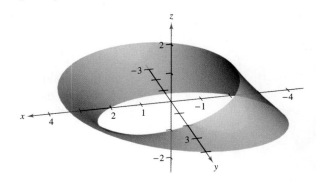

# 15.6 Surface Integrals

- Evaluate a surface integral as a double integral.
- Evaluate a surface integral for a parametric surface.
- Determine the orientation of a surface.
- Understand the concept of a flux integral.

## Surface Integrals

The remainder of this chapter deals primarily with **surface integrals.** You will first consider surfaces given by $z = g(x, y)$. Later in this section, you will consider more general surfaces given in parametric form.

Let $S$ be a surface given by $z = g(x, y)$ and let $R$ be its projection onto the $xy$-plane, as shown in Figure 15.44. Let $g$, $g_x$, and $g_y$ be continuous at all points in $R$ and let $f$ be a scalar function defined on $S$. Employing the procedure used to find surface area in Section 14.5, evaluate $f$ at $(x_i, y_i, z_i)$ and form the sum

$$\sum_{i=1}^{n} f(x_i, y_i, z_i) \, \Delta S_i$$

where

$$\Delta S_i \approx \sqrt{1 + [g_x(x_i, y_i)]^2 + [g_y(x_i, y_i)]^2} \, \Delta A_i.$$

Provided the limit of this sum as $\|\Delta\|$ approaches 0 exists, the **surface integral of $f$ over $S$** is defined as

$$\int_S \int f(x, y, z) \, dS = \lim_{\|\Delta\| \to 0} \sum_{i=1}^{n} f(x_i, y_i, z_i) \, \Delta S_i.$$

This integral can be evaluated by a double integral.

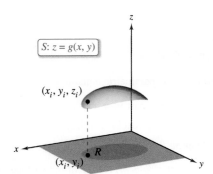
$S: z = g(x, y)$

$(x_i, y_i, z_i)$

$(x_i, y_i)$

Scalar function $f$ assigns a number to each point of $S$.
**Figure 15.44**

---

**THEOREM 15.10  Evaluating a Surface Integral**

Let $S$ be a surface given by $z = g(x, y)$ and let $R$ be its projection onto the $xy$-plane. If $g$, $g_x$, and $g_y$ are continuous on $R$ and $f$ is continuous on $S$, then the surface integral of $f$ over $S$ is

$$\int_S \int f(x, y, z) \, dS = \int_R \int f(x, y, g(x, y)) \sqrt{1 + [g_x(x, y)]^2 + [g_y(x, y)]^2} \, dA.$$

---

For surfaces described by functions of $x$ and $z$ (or $y$ and $z$), you can make the following adjustments to Theorem 15.10. If $S$ is the graph of $y = g(x, z)$ and $R$ is its projection onto the $xz$-plane, then

$$\int_S \int f(x, y, z) \, dS = \int_R \int f(x, g(x, z), z) \sqrt{1 + [g_x(x, z)]^2 + [g_z(x, z)]^2} \, dA.$$

If $S$ is the graph of $x = g(y, z)$ and $R$ is its projection onto the $yz$-plane, then

$$\int_S \int f(x, y, z) \, dS = \int_R \int f(g(y, z), y, z) \sqrt{1 + [g_y(y, z)]^2 + [g_z(y, z)]^2} \, dA.$$

If $f(x, y, z) = 1$, the surface integral over $S$ yields the surface area of $S$. For instance, suppose the surface $S$ is the plane given by $z = x$, where $0 \le x \le 1$ and $0 \le y \le 1$. The surface area of $S$ is $\sqrt{2}$ square units. Try verifying that

$$\int_S \int f(x, y, z) \, dS = \sqrt{2}.$$

> **EXAMPLE 1** **Evaluating a Surface Integral**

Evaluate the surface integral

$$\iint_S (y^2 + 2yz)\, dS$$

where $S$ is the first-octant portion of the plane

$$2x + y + 2z = 6.$$

**Solution** Begin by writing $S$ as

$$z = g(x, y) = \frac{1}{2}(6 - 2x - y).$$

Using the partial derivatives $g_x(x, y) = -1$ and $g_y(x, y) = -\frac{1}{2}$, you can write

$$\sqrt{1 + [g_x(x, y)]^2 + [g_y(x, y)]^2} = \sqrt{1 + 1 + \frac{1}{4}} = \frac{3}{2}.$$

Using Figure 15.45 and Theorem 15.10, you obtain

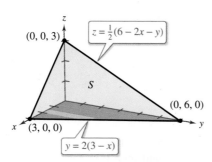

$z = \frac{1}{2}(6 - 2x - y)$

$(0, 0, 3)$

$S$

$(0, 6, 0)$

$(3, 0, 0)$

$y = 2(3 - x)$

**Figure 15.45**

$$\iint_S (y^2 + 2yz)\, dS = \iint_R f(x, y, g(x, y))\sqrt{1 + [g_x(x, y)]^2 + [g_y(x, y)]^2}\, dA$$

$$= \iint_R \left[ y^2 + 2y\left(\frac{1}{2}\right)(6 - 2x - y) \right]\left(\frac{3}{2}\right) dA$$

$$= 3\int_0^3 \int_0^{2(3-x)} y(3 - x)\, dy\, dx \qquad \text{Convert to iterated integral.}$$

$$= 3\int_0^3 \frac{y^2}{2}(3 - x)\Big]_0^{2(3-x)} dx \qquad \text{Integrate with respect to } y.$$

$$= 6\int_0^3 (3 - x)^3\, dx$$

$$= -\frac{3}{2}(3 - x)^4\Big]_0^3 \qquad \text{Integrate with respect to } x.$$

$$= \frac{243}{2}.$$

An alternative solution to Example 1 would be to project $S$ onto the $yz$-plane, as shown in Figure 15.46. Then $x = \frac{1}{2}(6 - y - 2z)$, and

$$\sqrt{1 + [g_y(y, z)]^2 + [g_z(y, z)]^2} = \sqrt{1 + \frac{1}{4} + 1} = \frac{3}{2}.$$

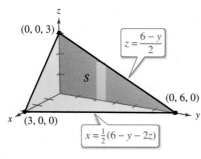

$z = \frac{6 - y}{2}$

$(0, 0, 3)$

$S$

$(0, 6, 0)$

$(3, 0, 0)$

$x = \frac{1}{2}(6 - y - 2z)$

**Figure 15.46**

So, the surface integral is

$$\iint_S (y^2 + 2yz)\, dS = \iint_R f(g(y, z), y, z)\sqrt{1 + [g_y(y, z)]^2 + [g_z(y, z)]^2}\, dA$$

$$= \int_0^6 \int_0^{(6-y)/2} (y^2 + 2yz)\left(\frac{3}{2}\right) dz\, dy$$

$$= \frac{3}{8}\int_0^6 (36y - y^3)\, dy$$

$$= \frac{243}{2}.$$

Try reworking Example 1 by projecting $S$ onto the $xz$-plane.

In Example 1, you could have projected the surface $S$ onto any one of the three coordinate planes. In Example 2, $S$ is a portion of a cylinder centered about the $x$-axis, and you can project it onto either the $xz$-plane or the $xy$-plane.

### EXAMPLE 2 Evaluating a Surface Integral

•••• ▷ *See LarsonCalculus.com for an interactive version of this type of example.*

Evaluate the surface integral

$$\iint_S (x + z)\, dS$$

where $S$ is the first-octant portion of the cylinder

$$y^2 + z^2 = 9$$

between $x = 0$ and $x = 4$, as shown in Figure 15.47.

**Solution** Project $S$ onto the $xy$-plane so that

$$z = g(x, y) = \sqrt{9 - y^2}$$

and obtain

$$\sqrt{1 + [g_x(x, y)]^2 + [g_y(x, y)]^2} = \sqrt{1 + \left(\frac{-y}{\sqrt{9 - y^2}}\right)^2}$$

$$= \frac{3}{\sqrt{9 - y^2}}.$$

Theorem 15.10 does not apply directly, because $g_y$ is not continuous when $y = 3$. However, you can apply Theorem 15.10 for $0 \le b < 3$ and then take the limit as $b$ approaches 3, as follows.

$$\iint_S (x + z)\, dS = \lim_{b \to 3^-} \int_0^b \int_0^4 (x + \sqrt{9 - y^2})\frac{3}{\sqrt{9 - y^2}}\, dx\, dy$$

$$= \lim_{b \to 3^-} 3 \int_0^b \int_0^4 \left(\frac{x}{\sqrt{9 - y^2}} + 1\right) dx\, dy$$

$$= \lim_{b \to 3^-} 3 \int_0^b \left[\frac{x^2}{2\sqrt{9 - y^2}} + x\right]_0^4 dy \qquad \text{Integrate with respect to } x.$$

$$= \lim_{b \to 3^-} 3 \int_0^b \left(\frac{8}{\sqrt{9 - y^2}} + 4\right) dy$$

$$= \lim_{b \to 3^-} 3 \left[4y + 8 \arcsin \frac{y}{3}\right]_0^b \qquad \text{Integrate with respect to } y.$$

$$= \lim_{b \to 3^-} 3 \left(4b + 8 \arcsin \frac{b}{3}\right)$$

$$= 36 + 24\left(\frac{\pi}{2}\right) \qquad \text{Evaluate limit.}$$

$$= 36 + 12\pi \qquad \blacksquare$$

▷ **TECHNOLOGY** Some computer algebra systems are capable of evaluating improper integrals. If you have access to such software, use it to evaluate the improper integral

$$\int_0^3 \int_0^4 (x + \sqrt{9 - y^2})\frac{3}{\sqrt{9 - y^2}}\, dx\, dy.$$

Do you obtain the same result as in Example 2?

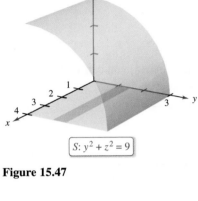

$R: 0 \le x \le 4$
$0 \le y \le 3$

$S: y^2 + z^2 = 9$

**Figure 15.47**

You have already seen that when the function $f$ defined on the surface $S$ is simply $f(x, y, z) = 1$, the surface integral yields the *surface area* of $S$.

$$\text{Area of surface} = \iint_S 1 \, dS$$

On the other hand, when $S$ is a lamina of variable density and $\rho(x, y, z)$ is the density at the point $(x, y, z)$, then the *mass* of the lamina is given by

$$\text{Mass of lamina} = \iint_S \rho(x, y, z) \, dS.$$

## EXAMPLE 3    Finding the Mass of a Surface Lamina

A cone-shaped surface lamina $S$ is given by

$$z = 4 - 2\sqrt{x^2 + y^2}, \quad 0 \le z \le 4$$

as shown in Figure 15.48. At each point on $S$, the density is proportional to the distance between the point and the $z$-axis. Find the mass $m$ of the lamina.

**Solution**   Projecting $S$ onto the $xy$-plane produces

$$S: z = 4 - 2\sqrt{x^2 + y^2} = g(x, y), \quad 0 \le z \le 4$$
$$R: x^2 + y^2 \le 4$$

with a density of $\rho(x, y, z) = k\sqrt{x^2 + y^2}$, where $k$ is the constant of proportionality. Using a surface integral, you can find the mass to be

$$m = \iint_S \rho(x, y, z) \, dS$$

$$= \iint_R k\sqrt{x^2 + y^2}\sqrt{1 + [g_x(x, y)]^2 + [g_y(x, y)]^2} \, dA$$

$$= k\iint_R \sqrt{x^2 + y^2}\sqrt{1 + \frac{4x^2}{x^2 + y^2} + \frac{4y^2}{x^2 + y^2}} \, dA$$

$$= k\iint_R \sqrt{5}\sqrt{x^2 + y^2} \, dA$$

$$= k\int_0^{2\pi}\int_0^2 \left(\sqrt{5}r\right)r \, dr \, d\theta \qquad \text{Polar coordinates}$$

$$= \frac{\sqrt{5}k}{3}\int_0^{2\pi} r^3\Big]_0^2 \, d\theta \qquad \text{Integrate with respect to } r.$$

$$= \frac{8\sqrt{5}k}{3}\int_0^{2\pi} d\theta$$

$$= \frac{8\sqrt{5}k}{3}\Big[\theta\Big]_0^{2\pi} \qquad \text{Integrate with respect to } \theta.$$

$$= \frac{16\sqrt{5}k\pi}{3}.$$

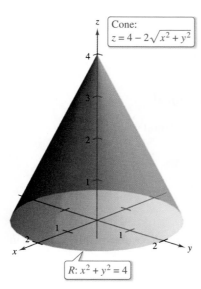

Cone: $z = 4 - 2\sqrt{x^2 + y^2}$

$R: x^2 + y^2 = 4$

**Figure 15.48**

▷ **TECHNOLOGY**   Use a computer algebra system to confirm the result shown in Example 3. The computer algebra system *Mathematica* evaluated the integral as follows.

$$k\int_{-2}^2\int_{-\sqrt{4-y^2}}^{\sqrt{4-y^2}} \sqrt{5}\sqrt{x^2 + y^2} \, dx \, dy = k\int_0^{2\pi}\int_0^2 \left(\sqrt{5}r\right)r \, dr \, d\theta = \frac{16\sqrt{5}k\pi}{3}$$

## Parametric Surfaces and Surface Integrals

For a surface $S$ given by the vector-valued function

$$\mathbf{r}(u, v) = x(u, v)\mathbf{i} + y(u, v)\mathbf{j} + z(u, v)\mathbf{k} \qquad \text{Parametric surface}$$

defined over a region $D$ in the $uv$-plane, you can show that the surface integral of $f(x, y, z)$ over $S$ is given by

$$\int_S \int f(x, y, z) \, dS = \int_D \int f(x(u, v), y(u, v), z(u, v)) \|\mathbf{r}_u(u, v) \times \mathbf{r}_v(u, v)\| \, dA.$$

Note the similarity to a line integral over a space curve $C$.

$$\int_C f(x, y, z) \, ds = \int_a^b f(x(t), y(t), z(t)) \|\mathbf{r}'(t)\| \, dt \qquad \text{Line integral}$$

Also, notice that $ds$ and $dS$ can be written as

$$ds = \|\mathbf{r}'(t)\| \, dt \quad \text{and} \quad dS = \|\mathbf{r}_u(u, v) \times \mathbf{r}_v(u, v)\| \, dA.$$

---

**EXAMPLE 4**    **Evaluating a Surface Integral**

Example 2 demonstrated an evaluation of the surface integral

$$\int_S \int (x + z) \, dS$$

where $S$ is the first-octant portion of the cylinder

$$y^2 + z^2 = 9$$

between $x = 0$ and $x = 4$, as shown in Figure 15.49. Reevaluate this integral in parametric form.

**Solution**    In parametric form, the surface is given by

$$\mathbf{r}(x, \theta) = x\mathbf{i} + 3\cos\theta\mathbf{j} + 3\sin\theta\mathbf{k}$$

where $0 \le x \le 4$ and $0 \le \theta \le \pi/2$. To evaluate the surface integral in parametric form, begin by calculating the following.

$$\mathbf{r}_x = \mathbf{i}$$
$$\mathbf{r}_\theta = -3\sin\theta\mathbf{j} + 3\cos\theta\mathbf{k}$$

$$\mathbf{r}_x \times \mathbf{r}_\theta = \begin{vmatrix} \mathbf{i} & \mathbf{j} & \mathbf{k} \\ 1 & 0 & 0 \\ 0 & -3\sin\theta & 3\cos\theta \end{vmatrix} = -3\cos\theta\mathbf{j} - 3\sin\theta\mathbf{k}$$

$$\|\mathbf{r}_x \times \mathbf{r}_\theta\| = \sqrt{9\cos^2\theta + 9\sin^2\theta} = 3$$

So, the surface integral can be evaluated as follows.

$$\int_D \int (x + 3\sin\theta)3 \, dA = \int_0^4 \int_0^{\pi/2} (3x + 9\sin\theta) \, d\theta \, dx$$

$$= \int_0^4 \left[ 3x\theta - 9\cos\theta \right]_0^{\pi/2} dx$$

$$= \int_0^4 \left( \frac{3\pi}{2}x + 9 \right) dx$$

$$= \left[ \frac{3\pi}{4}x^2 + 9x \right]_0^4$$

$$= 12\pi + 36$$

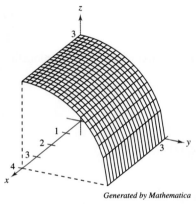

*Generated by Mathematica*

**Figure 15.49**

# Orientation of a Surface

Unit normal vectors are used to induce an orientation to a surface $S$ in space. A surface is **orientable** when a unit normal vector $\mathbf{N}$ can be defined at every nonboundary point of $S$ in such a way that the normal vectors vary continuously over the surface $S$. The surface $S$ is called an **oriented surface.**

An orientable surface $S$ has two distinct sides. So, when you orient a surface, you are selecting one of the two possible unit normal vectors. For a closed surface such as a sphere, it is customary to choose the unit normal vector $\mathbf{N}$ to be the one that points outward from the sphere.

Most common surfaces, such as spheres, paraboloids, ellipses, and planes, are orientable. (See Exercise 43 for an example of a surface that is *not* orientable.) Moreover, for an orientable surface, the gradient provides a convenient way to find a unit normal vector. That is, for an orientable surface $S$ given by

$$z = g(x, y) \qquad \text{Orientable surface}$$

let

$$G(x, y, z) = z - g(x, y).$$

Then, $S$ can be oriented by either the unit normal vector

$$\mathbf{N} = \frac{\nabla G(x, y, z)}{\|\nabla G(x, y, z)\|}$$

$$= \frac{-g_x(x, y)\mathbf{i} - g_y(x, y)\mathbf{j} + \mathbf{k}}{\sqrt{1 + [g_x(x, y)]^2 + [g_y(x, y)]^2}} \qquad \text{Upward unit normal vector}$$

or the unit normal vector

$$\mathbf{N} = \frac{-\nabla G(x, y, z)}{\|\nabla G(x, y, z)\|}$$

$$= \frac{g_x(x, y)\mathbf{i} + g_y(x, y)\mathbf{j} - \mathbf{k}}{\sqrt{1 + [g_x(x, y)]^2 + [g_y(x, y)]^2}} \qquad \text{Downward unit normal vector}$$

as shown in Figure 15.50. If the smooth orientable surface $S$ is given in parametric form by

$$\mathbf{r}(u, v) = x(u, v)\mathbf{i} + y(u, v)\mathbf{j} + z(u, v)\mathbf{k} \qquad \text{Parametric surface}$$

then the unit normal vectors are given by

$$\mathbf{N} = \frac{\mathbf{r}_u \times \mathbf{r}_v}{\|\mathbf{r}_u \times \mathbf{r}_v\|} \qquad \text{Upward unit normal vector}$$

and

$$\mathbf{N} = \frac{\mathbf{r}_v \times \mathbf{r}_u}{\|\mathbf{r}_v \times \mathbf{r}_u\|}. \qquad \text{Downward unit normal vector}$$

For an orientable surface given by

$$y = g(x, z) \quad \text{or} \quad x = g(y, z)$$

you can use the gradient

$$\nabla G(x, y, z) = -g_x(x, z)\mathbf{i} + \mathbf{j} - g_z(x, z)\mathbf{k} \qquad G(x, y, z) = y - g(x, z)$$

or

$$\nabla G(x, y, z) = \mathbf{i} - g_y(y, z)\mathbf{j} - g_z(y, z)\mathbf{k} \qquad G(x, y, z) = x - g(y, z)$$

to orient the surface.

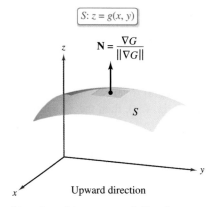

$\boxed{S: z = g(x, y)}$

$\mathbf{N} = \dfrac{\nabla G}{\|\nabla G\|}$

Upward direction

$S$ is oriented in an upward direction.

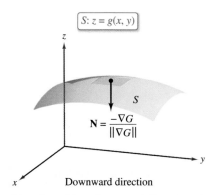

$\boxed{S: z = g(x, y)}$

$\mathbf{N} = \dfrac{-\nabla G}{\|\nabla G\|}$

Downward direction

$S$ is oriented in a downward direction.
**Figure 15.50**

## Flux Integrals

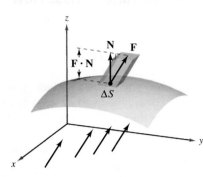

The velocity field **F** indicates the direction of the fluid flow.

**Figure 15.51**

One of the principal applications involving the vector form of a surface integral relates to the flow of a fluid through a surface. Consider an oriented surface $S$ submerged in a fluid having a continuous velocity field **F**. Let $\Delta S$ be the area of a small patch of the surface $S$ over which **F** is nearly constant. Then the amount of fluid crossing this region per unit of time is approximated by the volume of the column of height $\mathbf{F} \cdot \mathbf{N}$, as shown in Figure 15.51. That is,

$$\Delta V = (\text{height})(\text{area of base})$$
$$= (\mathbf{F} \cdot \mathbf{N})\,\Delta S.$$

Consequently, the volume of fluid crossing the surface $S$ per unit of time (called the **flux of F across** $S$) is given by the surface integral in the next definition.

---

**Definition of Flux Integral**

Let $\mathbf{F}(x, y, z) = M\mathbf{i} + N\mathbf{j} + P\mathbf{k}$, where $M$, $N$, and $P$ have continuous first partial derivatives on the surface $S$ oriented by a unit normal vector **N**. The **flux integral of F across** $S$ is given by

$$\iint_S \mathbf{F} \cdot \mathbf{N}\, dS.$$

---

Geometrically, a flux integral is the surface integral over $S$ of the *normal component* of **F**. If $\rho(x, y, z)$ is the density of the fluid at $(x, y, z)$, then the flux integral

$$\iint_S \rho\,\mathbf{F} \cdot \mathbf{N}\, dS$$

represents the *mass* of the fluid flowing across $S$ per unit of time.

To evaluate a flux integral for a surface given by $z = g(x, y)$, let

$$G(x, y, z) = z - g(x, y).$$

Then **N** $dS$ can be written as follows.

$$\mathbf{N}\, dS = \frac{\nabla G(x, y, z)}{\|\nabla G(x, y, z)\|}\, dS$$

$$= \frac{\nabla G(x, y, z)}{\sqrt{(g_x)^2 + (g_y)^2 + 1}}\sqrt{(g_x)^2 + (g_y)^2 + 1}\, dA$$

$$= \nabla G(x, y, z)\, dA$$

---

**THEOREM 15.11   Evaluating a Flux Integral**

Let $S$ be an oriented surface given by $z = g(x, y)$ and let $R$ be its projection onto the $xy$-plane.

$$\iint_S \mathbf{F} \cdot \mathbf{N}\, dS = \iint_R \mathbf{F} \cdot \left[-g_x(x, y)\mathbf{i} - g_y(x, y)\mathbf{j} + \mathbf{k}\right] dA \qquad \text{Oriented upward}$$

$$\iint_S \mathbf{F} \cdot \mathbf{N}\, dS = \iint_R \mathbf{F} \cdot \left[g_x(x, y)\mathbf{i} + g_y(x, y)\mathbf{j} - \mathbf{k}\right] dA \qquad \text{Oriented downward}$$

For the first integral, the surface is oriented upward, and for the second integral, the surface is oriented downward.

---

**EXAMPLE 5**    **Using a Flux Integral to Find the Rate of Mass Flow**

Let $S$ be the portion of the paraboloid

$$z = g(x, y) = 4 - x^2 - y^2$$

lying above the $xy$-plane, oriented by an
upward unit normal vector, as shown in
Figure 15.52. A fluid of constant density $\rho$
is flowing through the surface $S$ according to
the vector field

$$\mathbf{F}(x, y, z) = x\mathbf{i} + y\mathbf{j} + z\mathbf{k}.$$

Find the rate of mass flow through $S$.

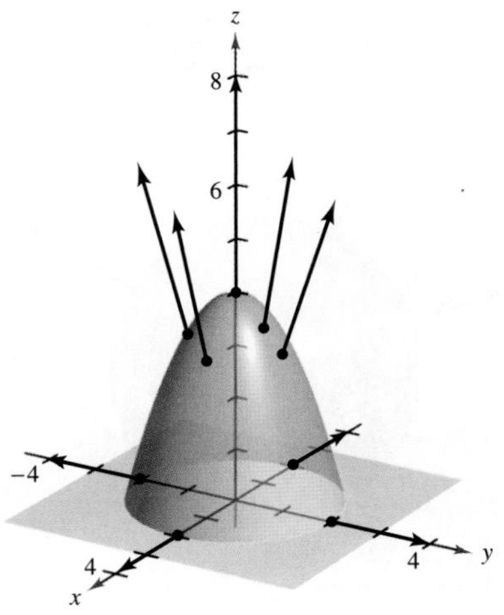

**Solution**    Note that $S$ is oriented upward and
the partial derivatives of $g$ are

$$g_x(x, y) = -2x$$

and

$$g_y(x, y) = -2y.$$

**Figure 15.52**

So, the rate of mass flow through the surface $S$ is

$$\iint_S \rho \mathbf{F} \cdot \mathbf{N} \, dS = \rho \iint_R \mathbf{F} \cdot [-g_x(x, y)\mathbf{i} - g_y(x, y)\mathbf{j} + \mathbf{k}] \, dA$$

$$= \rho \iint_R [x\mathbf{i} + y\mathbf{j} + (4 - x^2 - y^2)\mathbf{k}] \cdot (2x\mathbf{i} + 2y\mathbf{j} + \mathbf{k}) \, dA$$

$$= \rho \iint_R [2x^2 + 2y^2 + (4 - x^2 - y^2)] \, dA$$

$$= \rho \iint_R (4 + x^2 + y^2) \, dA$$

$$= \rho \int_0^{2\pi} \int_0^2 (4 + r^2) r \, dr \, d\theta \qquad \text{Polar coordinates}$$

$$= \rho \int_0^{2\pi} \left[ 2r^2 + \frac{r^4}{4} \right]_0^2 \, d\theta$$

$$= \rho \int_0^{2\pi} 12 \, d\theta$$

$$= 24\pi\rho.$$

For an oriented upward surface $S$ given by the vector-valued function

$$\mathbf{r}(u, v) = x(u, v)\mathbf{i} + y(u, v)\mathbf{j} + z(u, v)\mathbf{k} \qquad \text{Parametric surface}$$

defined over a region $D$ in the $uv$-plane, you can define the flux integral of $\mathbf{F}$ across
$S$ as

$$\iint_S \mathbf{F} \cdot \mathbf{N} \, dS = \iint_D \mathbf{F} \cdot \left( \frac{\mathbf{r}_u \times \mathbf{r}_v}{\|\mathbf{r}_u \times \mathbf{r}_v\|} \right) \|\mathbf{r}_u \times \mathbf{r}_v\| \, dA = \iint_D \mathbf{F} \cdot (\mathbf{r}_u \times \mathbf{r}_v) \, dA.$$

Note the similarity of this integral to the line integral

$$\int_C \mathbf{F} \cdot d\mathbf{r} = \int_C \mathbf{F} \cdot \mathbf{T} \, ds.$$

A summary of formulas for line and surface integrals is presented on page 1107.

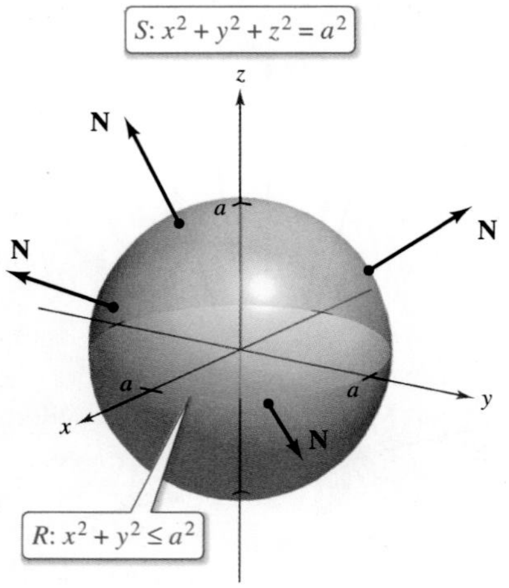

$S: x^2 + y^2 + z^2 = a^2$

$R: x^2 + y^2 \le a^2$

**Figure 15.53**

---

EXAMPLE 6 **Finding the Flux of an Inverse Square Field**

Find the flux over the sphere $S$ given by

$$x^2 + y^2 + z^2 = a^2 \qquad \text{Sphere } S$$

where $\mathbf{F}$ is an inverse square field given by

$$\mathbf{F}(x, y, z) = \frac{kq}{\|\mathbf{r}\|^2} \frac{\mathbf{r}}{\|\mathbf{r}\|} = \frac{kq\mathbf{r}}{\|\mathbf{r}\|^3} \qquad \text{Inverse square field } \mathbf{F}$$

and

$$\mathbf{r} = x\mathbf{i} + y\mathbf{j} + z\mathbf{k}.$$

Assume $S$ is oriented outward, as shown in Figure 15.53.

**Solution** The sphere is given by

$$\begin{aligned}
\mathbf{r}(u, v) &= x(u, v)\mathbf{i} + y(u, v)\mathbf{j} + z(u, v)\mathbf{k} \\
&= a \sin u \cos v\,\mathbf{i} + a \sin u \sin v\,\mathbf{j} + a \cos u\mathbf{k}
\end{aligned}$$

where $0 \le u \le \pi$ and $0 \le v \le 2\pi$. The partial derivatives of $\mathbf{r}$ are

$$\mathbf{r}_u(u, v) = a \cos u \cos v\,\mathbf{i} + a \cos u \sin v\,\mathbf{j} - a \sin u\mathbf{k}$$

and

$$\mathbf{r}_v(u, v) = -a \sin u \sin v\,\mathbf{i} + a \sin u \cos v\,\mathbf{j}$$

which implies that the normal vector $\mathbf{r}_u \times \mathbf{r}_v$ is

$$\begin{aligned}
\mathbf{r}_u \times \mathbf{r}_v &= \begin{vmatrix} \mathbf{i} & \mathbf{j} & \mathbf{k} \\ a \cos u \cos v & a \cos u \sin v & -a \sin u \\ -a \sin u \sin v & a \sin u \cos v & 0 \end{vmatrix} \\
&= a^2(\sin^2 u \cos v\,\mathbf{i} + \sin^2 u \sin v\,\mathbf{j} + \sin u \cos u\mathbf{k}).
\end{aligned}$$

Now, using

$$\begin{aligned}
\mathbf{F}(x, y, z) &= \frac{kq\mathbf{r}}{\|\mathbf{r}\|^3} \\
&= kq\frac{x\mathbf{i} + y\mathbf{j} + z\mathbf{k}}{\|x\mathbf{i} + y\mathbf{j} + z\mathbf{k}\|^3} \\
&= \frac{kq}{a^3}(a \sin u \cos v\,\mathbf{i} + a \sin u \sin v\,\mathbf{j} + a \cos u\mathbf{k})
\end{aligned}$$

it follows that

$$\begin{aligned}
\mathbf{F} \cdot (\mathbf{r}_u \times \mathbf{r}_v) &= \frac{kq}{a^3}[(a \sin u \cos v\,\mathbf{i} + a \sin u \sin v\,\mathbf{j} + a \cos u\mathbf{k}) \cdot \\
&\qquad a^2(\sin^2 u \cos v\,\mathbf{i} + \sin^2 u \sin v\,\mathbf{j} + \sin u \cos u\mathbf{k})] \\
&= kq(\sin^3 u \cos^2 v + \sin^3 u \sin^2 v + \sin u \cos^2 u) \\
&= kq \sin u.
\end{aligned}$$

Finally, the flux over the sphere $S$ is given by

$$\begin{aligned}
\iint_S \mathbf{F} \cdot \mathbf{N}\, dS &= \iint_D kq \sin u\, dA \\
&= kq \int_0^{2\pi} \int_0^\pi \sin u\, du\, dv \\
&= kq \int_0^{2\pi} 2\, dv \\
&= 4\pi kq.
\end{aligned}$$

■

The result in Example 6 shows that the flux across a sphere $S$ in an inverse square field is independent of the radius of $S$. In particular, if $\mathbf{E}$ is an electric field, then the result in Example 6, along with Coulomb's Law (see Section 15.1), yields one of the basic laws of electrostatics, known as **Gauss's Law:**

$$\iint_S \mathbf{E} \cdot \mathbf{N}\, dS = 4\pi kq \qquad \text{Gauss's Law}$$

where $q$ is a point charge located at the center of the sphere and $k$ is the Coulomb constant. Gauss's Law is valid for more general closed surfaces that enclose the origin, and relates the flux out of the surface to the total charge inside the surface.

Surface integrals are also used in the study of **heat flow.** Heat flows from areas of higher temperature to areas of lower temperature in the direction of greatest change. As a result, measuring **heat flux** involves the gradient of the temperature. The flux depends on the area of the surface. It is the normal direction to the surface that is important, because heat that flows in directions tangential to the surface will produce no heat loss. So, assume that the heat flux across a portion of the surface of area $\Delta S$ is given by $\Delta H \approx -k\nabla T \cdot \mathbf{N}\, dS$, where $T$ is the temperature, $\mathbf{N}$ is the unit normal vector to the surface in the direction of the heat flow, and $k$ is the thermal diffusivity of the material. The heat flux across the surface is given by

$$H = \iint_S -k\nabla T \cdot \mathbf{N}\, dS. \qquad \text{Heat flux across } S$$

This section concludes with a summary of different forms of line integrals and surface integrals.

---

**SUMMARY OF LINE AND SURFACE INTEGRALS**

**Line Integrals**

$$\begin{aligned} ds &= \|\mathbf{r}'(t)\|\, dt \\ &= \sqrt{[x'(t)]^2 + [y'(t)]^2 + [z'(t)]^2}\, dt \end{aligned}$$

$$\int_C f(x, y, z)\, ds = \int_a^b f(x(t), y(t), z(t))\sqrt{[x'(t)]^2 + [y'(t)]^2 + [z'(t)]^2}\, dt \qquad \text{Scalar form}$$

$$\begin{aligned} \int_C \mathbf{F} \cdot d\mathbf{r} &= \int_C \mathbf{F} \cdot \mathbf{T}\, ds \\ &= \int_a^b \mathbf{F}(x(t), y(t), z(t)) \cdot \mathbf{r}'(t)\, dt \qquad \text{Vector form} \end{aligned}$$

**Surface Integrals** $[z = g(x, y)]$

$$dS = \sqrt{1 + [g_x(x, y)]^2 + [g_y(x, y)]^2}\, dA$$

$$\iint_S f(x, y, z)\, dS = \iint_R f(x, y, g(x, y))\sqrt{1 + [g_x(x, y)]^2 + [g_y(x, y)]^2}\, dA \qquad \text{Scalar form}$$

$$\iint_S \mathbf{F} \cdot \mathbf{N}\, dS = \iint_R \mathbf{F} \cdot [-g_x(x, y)\mathbf{i} - g_y(x, y)\mathbf{j} + \mathbf{k}]\, dA \qquad \text{Vector form (upward normal)}$$

**Surface Integrals (parametric form)**

$$dS = \|\mathbf{r}_u(u, v) \times \mathbf{r}_v(u, v)\|\, dA$$

$$\iint_S f(x, y, z)\, dS = \iint_D f(x(u, v), y(u, v), z(u, v))\|\mathbf{r}_u(u, v) \times \mathbf{r}_v(u, v)\|\, dA \qquad \text{Scalar form}$$

$$\iint_S \mathbf{F} \cdot \mathbf{N}\, dS = \iint_D \mathbf{F} \cdot (\mathbf{r}_u \times \mathbf{r}_v)\, dA \qquad \text{Vector form (upward normal)}$$

## 15.6 Exercises

See **CalcChat.com** for tutorial help and worked-out solutions to odd-numbered exercises.

### CONCEPT CHECK

1. **Surface Integral** Explain how to set up a surface integral given that you will project the surface onto the $xz$-plane.

2. **Surface Integral** For what condition does the surface integral over $S$ yield the surface area of $S$?

3. **Orientation of a Surface** Describe a physical characteristic of an orientable surface.

4. **Flux** What is the physical interpretation of the flux of $\mathbf{F}$ across $S$? How do you calculate it?

 **Evaluating a Surface Integral** In Exercises 5–8, evaluate $\displaystyle\iint_S (x - 2y + z)\,dS.$

5. $S: z = 4 - x, \quad 0 \le x \le 4, \quad 0 \le y \le 3$
6. $S: z = 15 - 2x + 3y, \quad 0 \le x \le 2, \quad 0 \le y \le 4$
7. $S: z = 2, \quad x^2 + y^2 \le 1$
8. $S: z = 3y, \quad 0 \le x \le 2, \quad 0 \le y \le x$

**Evaluating a Surface Integral** In Exercises 9 and 10, evaluate $\displaystyle\iint_S xy\,dS.$

9. $S: z = 3 - x - y$, first octant
10. $S: z = \frac{1}{4}x^4, \quad 0 \le x \le 1, \quad 0 \le y \le x^2$

 **Evaluating a Surface Integral** In Exercises 11 and 12, use a computer algebra system to evaluate $\displaystyle\iint_S (x^2 - 2xy)\,dS.$

11. $S: z = 10 - x^2 - y^2, \quad 0 \le x \le 2, \quad 0 \le y \le 2$
12. $S: z = \cos x, \quad 0 \le x \le \dfrac{\pi}{2}, \quad 0 \le y \le \dfrac{1}{2}x$

**Mass** In Exercises 13 and 14, find the mass of the surface lamina $S$ of density $\rho$.

13. $S: 2x + 3y + 6z = 12$, first octant, $\rho(x, y, z) = x^2 + y^2$
14. $S: z = \sqrt{a^2 - x^2 - y^2}, \quad \rho(x, y, z) = kz$

**Evaluating a Surface Integral** In Exercises 15–18, evaluate $\displaystyle\iint_S f(x, y)\,dS.$

15. $f(x, y) = y + 5$

   $S: \mathbf{r}(u, v) = u\mathbf{i} + v\mathbf{j} + 2v\mathbf{k}$

   $0 \le u \le 1, \quad 0 \le v \le 2$

16. $f(x, y) = xy$

   $S: \mathbf{r}(u, v) = 2\cos u\mathbf{i} + 2\sin u\mathbf{j} + v\mathbf{k}$

   $0 \le u \le \dfrac{\pi}{2}, \quad 0 \le v \le 1$

17. $f(x, y) = 3y - x$

   $S: \mathbf{r}(u, v) = \cos u\mathbf{i} + \sin u\mathbf{j} + v\mathbf{k}$

   $0 \le u \le \dfrac{\pi}{3}, \quad 0 \le v \le 1$

18. $f(x, y) = x + y$

   $S: \mathbf{r}(u, v) = 4u\cos v\mathbf{i} + 4u\sin v\mathbf{j} + 3u\mathbf{k}$

   $0 \le u \le 4, \quad 0 \le v \le \pi$

**Evaluating a Surface Integral** In Exercises 19–24, evaluate $\displaystyle\iint_S f(x, y, z)\,dS.$

19. $f(x, y, z) = x^2 + y^2 + z^2$

   $S: z = x + y, \quad x^2 + y^2 \le 1$

20. $f(x, y, z) = \dfrac{xy}{z}$

   $S: z = x^2 + y^2, \quad 4 \le x^2 + y^2 \le 16$

21. $f(x, y, z) = \sqrt{x^2 + y^2 + z^2}$

   $S: z = \sqrt{x^2 + y^2}, \quad x^2 + y^2 \le 4$

22. $f(x, y, z) = \sqrt{x^2 + y^2 + z^2}$

   $S: z = \sqrt{x^2 + y^2}, \quad (x - 1)^2 + y^2 \le 1$

23. $f(x, y, z) = x^2 + y^2 + z^2$

   $S: x^2 + y^2 = 9, \quad 0 \le x \le 3, \quad 0 \le y \le 3, \quad 0 \le z \le 9$

24. $f(x, y, z) = x^2 + y^2 + z^2$

   $S: x^2 + y^2 = 9, \quad 0 \le x \le 3, \quad 0 \le z \le x$

 **Evaluating a Flux Integral** In Exercises 25–30, find the flux of F across $S$,

$$\iint_S \mathbf{F} \cdot \mathbf{N}\,dS$$

where $\mathbf{N}$ is the upward unit normal vector to $S$.

25. $\mathbf{F}(x, y, z) = 3z\mathbf{i} - 4\mathbf{j} + y\mathbf{k}; \ S: z = 1 - x - y$, first octant
26. $\mathbf{F}(x, y, z) = x\mathbf{i} + 2y\mathbf{j}; \ S: z = 6 - 3x - 2y$, first octant
27. $\mathbf{F}(x, y, z) = x\mathbf{i} + y\mathbf{j} + z\mathbf{k}; \ S: z = 1 - x^2 - y^2, \ z \ge 0$
28. $\mathbf{F}(x, y, z) = x\mathbf{i} + y\mathbf{j} + z\mathbf{k}$

   $S: x^2 + y^2 + z^2 = 36$, first octant

29. $\mathbf{F}(x, y, z) = 4\mathbf{i} - 3\mathbf{j} + 5\mathbf{k}$

   $S: z = x^2 + y^2, \quad x^2 + y^2 \le 4$

30. $\mathbf{F}(x, y, z) = x\mathbf{i} + y\mathbf{j} - 2z\mathbf{k}$

   $S: z = \sqrt{a^2 - x^2 - y^2}$

**Evaluating a Flux Integral**    In Exercises 31 and 32, find the flux of F over the closed surface. (Let N be the outward unit normal vector of the surface.)

**31.** $F(x, y, z) = (x + y)\mathbf{i} + y\mathbf{j} + z\mathbf{k}$

  S: $z = 16 - x^2 - y^2$,   $z = 0$

**32.** $F(x, y, z) = 4xy\mathbf{i} + z^2\mathbf{j} + yz\mathbf{k}$

  S: unit cube bounded by the planes $x = 0$, $x = 1$, $y = 0$, $y = 1$, $z = 0$, $z = 1$

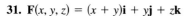**Flow Rate**    In Exercises 33 and 34, use a computer algebra system to find the rate of mass flow of a fluid of density $\rho$ through the surface $S$ oriented upward when the velocity field is given by $F(x, y, z) = 0.5z\mathbf{k}$.

**33.** S: $z = 16 - x^2 - y^2$,   $z \geq 0$

**34.** S: $z = \sqrt{16 - x^2 - y^2}$

**Gauss's Law**    In Exercises 35 and 36, evaluate $\iint_S \mathbf{E} \cdot \mathbf{N}\, dS$ to find the total charge of the electrostatic field E enclosed by the closed surface consisting of the hemisphere $z = \sqrt{1 - x^2 - y^2}$ and its circular base in the $xy$-plane.

**35.** $\mathbf{E} = yz\mathbf{i} + xz\mathbf{j} + xy\mathbf{k}$      **36.** $\mathbf{E} = x\mathbf{i} + y\mathbf{j} + 2z\mathbf{k}$

**Moments of Inertia**    In Exercises 37–40, use the following formulas for the moments of inertia about the coordinate axes of a surface lamina of density $\rho$.

$$I_x = \iint_S (y^2 + z^2)\rho(x, y, z)\, dS \qquad I_y = \iint_S (x^2 + z^2)\rho(x, y, z)\, dS$$

$$I_z = \iint_S (x^2 + y^2)\rho(x, y, z)\, dS$$

**37.** Verify that the moment of inertia of a conical shell of uniform density about its axis is $\frac{1}{2}ma^2$, where $m$ is the mass and $a$ is the radius and height.

**38.** Verify that the moment of inertia of a spherical shell of uniform density about its diameter is $\frac{2}{3}ma^2$, where $m$ is the mass and $a$ is the radius.

**39.** Find the moment of inertia about the $z$-axis for the surface lamina $x^2 + y^2 = a^2$, where $0 \leq z \leq h$, with a uniform density of 1.

**40.** Find the moment of inertia about the $z$-axis for the surface lamina $z = x^2 + y^2$, where $0 \leq z \leq h$, with a uniform density of 1.

**EXPLORING CONCEPTS**
**41. Using Different Methods**    Evaluate

$$\iint_S (x + 2y)\, dS$$

where $S$ is the first-octant portion of the plane

$2x + 2y + z = 4$

by projecting $S$ onto (a) the $xy$-plane, (b) the $xz$-plane, and (c) the $yz$-plane. Verify that all answers are the same.

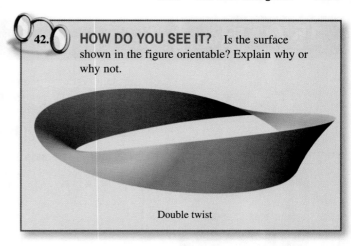

**42. HOW DO YOU SEE IT?**    Is the surface shown in the figure orientable? Explain why or why not.

Double twist

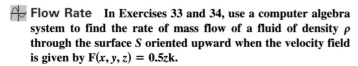**43. Investigation**

(a) Use a computer algebra system to graph the vector-valued function

$$\mathbf{r}(u, v) = (4 - v \sin u) \cos(2u)\mathbf{i} + (4 - v \sin u) \sin(2u)\mathbf{j} + v \cos u\,\mathbf{k}$$

where $0 \leq u \leq \pi$ and $-1 \leq v \leq 1$. This surface is called a Möbius strip.

(b) Is the surface orientable? Explain why or why not.

(c) Use a computer algebra system to graph the space curve represented by $\mathbf{r}(u, 0)$. Identify the curve.

(d) Cut a strip of paper and draw a line lengthwise through the center. Construct a Möbius strip by making a single twist and pasting the ends of the strip of paper together.

(e) Cut the Möbius strip along the line you drew and describe the result.

**SECTION PROJECT**

## Hyperboloid of One Sheet

Consider the parametric surface given by the function

$$\mathbf{r}(u, v) = a \cosh u \cos v\,\mathbf{i} + a \cosh u \sin v\,\mathbf{j} + b \sinh u\,\mathbf{k}.$$

(a) Use a graphing utility to graph $\mathbf{r}$ for various values of the constants $a$ and $b$. Describe the effect of the constants on the shape of the surface.

(b) Show that the surface is a hyperboloid of one sheet given by

$$\frac{x^2}{a^2} + \frac{y^2}{a^2} - \frac{z^2}{b^2} = 1.$$

(c) For fixed values $u = u_0$, describe the curves given by

$$\mathbf{r}(u_0, v) = a \cosh u_0 \cos v\,\mathbf{i} + a \cosh u_0 \sin v\,\mathbf{j} + b \sinh u_0\,\mathbf{k}.$$

(d) For fixed values $v = v_0$, describe the curves given by

$$\mathbf{r}(u, v_0) = a \cosh u \cos v_0\,\mathbf{i} + a \cosh u \sin v_0\,\mathbf{j} + b \sinh u\,\mathbf{k}.$$

(e) Find a normal vector to the surface at $(u, v) = (0, 0)$.

# 15.7 Divergence Theorem

■ Understand and use the Divergence Theorem.
■ Use the Divergence Theorem to calculate flux.

## Divergence Theorem

Recall from Section 15.4 that an alternative form of Green's Theorem is

$$\int_C \mathbf{F} \cdot \mathbf{N}\, ds = \int_R\!\!\int \left( \frac{\partial M}{\partial x} + \frac{\partial N}{\partial y} \right) dA$$

$$= \int_R\!\!\int \text{div } \mathbf{F}\, dA.$$

In an analogous way, the **Divergence Theorem** gives the relationship between a triple integral over a solid region $Q$ and a surface integral over the surface of $Q$. In the statement of the theorem, the surface $S$ is **closed** in the sense that it forms the complete boundary of the solid $Q$. Regions bounded by spheres, ellipsoids, cubes, tetrahedrons, or combinations of these surfaces are typical examples of closed surfaces. Let $Q$ be a solid region on which a triple integral can be evaluated, and let $S$ be a closed surface that is oriented by *outward* unit normal vectors, as shown in Figure 15.54. With these restrictions on $S$ and $Q$, the Divergence Theorem can be stated as shown below the figure.

**CARL FRIEDRICH GAUSS
(1777–1855)**

The *Divergence Theorem* is also called *Gauss's Theorem*, after the famous German mathematician Carl Friedrich Gauss. Gauss is recognized, with Newton and Archimedes, as one of the three greatest mathematicians in history. One of his many contributions to mathematics was made at the age of 22, when, as part of his doctoral dissertation, he proved the *Fundamental Theorem of Algebra*. *See LarsonCalculus.com to read more of this biography.*

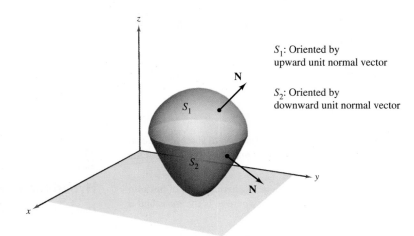

$S_1$: Oriented by upward unit normal vector

$S_2$: Oriented by downward unit normal vector

**Figure 15.54**

---

> **THEOREM 15.12 The Divergence Theorem**
>
> Let $Q$ be a solid region bounded by a closed surface $S$ oriented by a unit normal vector directed outward from $Q$. If $\mathbf{F}$ is a vector field whose component functions have continuous first partial derivatives in $Q$, then
>
> $$\int_S\!\!\int \mathbf{F} \cdot \mathbf{N}\, dS = \int\!\!\int_Q\!\!\int \text{div } \mathbf{F}\, dV.$$

**REMARK** As noted at the left above, the Divergence Theorem is sometimes called Gauss's Theorem. It is also sometimes called Ostrogradsky's Theorem, after the Russian mathematician Michel Ostrogradsky (1801–1861).

· · · · · · · · · · · · · · · · · ▷

**Proof**   For $\mathbf{F}(x, y, z) = M\mathbf{i} + N\mathbf{j} + P\mathbf{k}$, the theorem takes the form

$$\iint_S \mathbf{F} \cdot \mathbf{N} \, dS = \iint_S (M\mathbf{i} \cdot \mathbf{N} + N\mathbf{j} \cdot \mathbf{N} + P\mathbf{k} \cdot \mathbf{N}) \, dS$$

$$= \iiint_Q \left( \frac{\partial M}{\partial x} + \frac{\partial N}{\partial y} + \frac{\partial P}{\partial z} \right) dV.$$

You can prove this by verifying that the following three equations are valid.

$$\iint_S M\mathbf{i} \cdot \mathbf{N} \, dS = \iiint_Q \frac{\partial M}{\partial x} \, dV$$

$$\iint_S N\mathbf{j} \cdot \mathbf{N} \, dS = \iiint_Q \frac{\partial N}{\partial y} \, dV$$

$$\iint_S P\mathbf{k} \cdot \mathbf{N} \, dS = \iiint_Q \frac{\partial P}{\partial z} \, dV$$

Because the verifications of the three equations are similar, only the third is discussed.
Restrict the proof to a **simple solid** region with upper surface

$$z = g_2(x, y) \qquad \text{Upper surface}$$

and lower surface

$$z = g_1(x, y) \qquad \text{Lower surface}$$

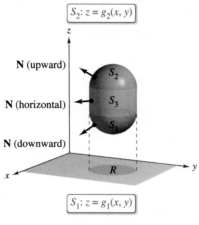

$S_2: z = g_2(x, y)$

N (upward)

N (horizontal)

N (downward)

$S_1: z = g_1(x, y)$

**Figure 15.55**

whose projections onto the $xy$-plane coincide and form region $R$. If $Q$ has a lateral
surface like $S_3$ in Figure 15.55, then a normal vector is horizontal, which implies that
$P\mathbf{k} \cdot \mathbf{N} = 0$. Consequently, you have

$$\iint_S P\mathbf{k} \cdot \mathbf{N} \, dS = \iint_{S_1} P\mathbf{k} \cdot \mathbf{N} \, dS + \iint_{S_2} P\mathbf{k} \cdot \mathbf{N} \, dS + 0.$$

On the upper surface $S_2$, the outward normal vector is upward, whereas on the lower
surface $S_1$, the outward normal vector is downward. So, by Theorem 15.11, you have

$$\iint_{S_1} P\mathbf{k} \cdot \mathbf{N} \, dS = \iint_R P(x, y, g_1(x, y))\mathbf{k} \cdot \left( \frac{\partial g_1}{\partial x}\mathbf{i} + \frac{\partial g_1}{\partial y}\mathbf{j} - \mathbf{k} \right) dA$$

$$= -\iint_R P(x, y, g_1(x, y)) \, dA$$

and

$$\iint_{S_2} P\mathbf{k} \cdot \mathbf{N} \, dS = \iint_R P(x, y, g_2(x, y))\mathbf{k} \cdot \left( -\frac{\partial g_2}{\partial x}\mathbf{i} - \frac{\partial g_2}{\partial y}\mathbf{j} + \mathbf{k} \right) dA$$

$$= \iint_R P(x, y, g_2(x, y)) \, dA.$$

Adding these results, you obtain

$$\iint_S P\mathbf{k} \cdot \mathbf{N} \, dS = \iint_R [P(x, y, g_2(x, y)) - P(x, y, g_1(x, y))] \, dA$$

$$= \iint_R \left[ \int_{g_1(x, y)}^{g_2(x, y)} \frac{\partial P}{\partial z} \, dz \right] dA$$

$$= \iiint_Q \frac{\partial P}{\partial z} \, dV.$$

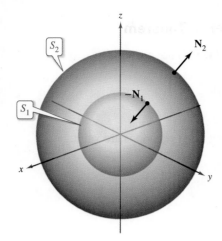

**Figure 15.56**

Even though the Divergence Theorem was stated for a simple solid region $Q$ bounded by a closed surface, the theorem is also valid for regions that are the finite unions of simple solid regions. For example, let $Q$ be the solid bounded by the closed surfaces $S_1$ and $S_2$, as shown in Figure 15.56. To apply the Divergence Theorem to this solid, let $S = S_1 \cup S_2$. The normal vector $\mathbf{N}$ to $S$ is given by $-\mathbf{N}_1$ on $S_1$ and by $\mathbf{N}_2$ on $S_2$. So, you can write

$$\iiint_Q \operatorname{div} \mathbf{F} \, dV = \iint_S \mathbf{F} \cdot \mathbf{N} \, dS$$

$$= \iint_{S_1} \mathbf{F} \cdot (-\mathbf{N}_1) \, dS + \iint_{S_2} \mathbf{F} \cdot \mathbf{N}_2 \, dS$$

$$= -\iint_{S_1} \mathbf{F} \cdot \mathbf{N}_1 \, dS + \iint_{S_2} \mathbf{F} \cdot \mathbf{N}_2 \, dS.$$

For the remainder of this section, you will apply the Divergence Theorem to simple solid regions bounded by closed surfaces.

**EXAMPLE 1** **Using the Divergence Theorem**

Let $Q$ be the solid region bounded by the coordinate planes and the plane

$$2x + 2y + z = 6$$

and let $\mathbf{F} = x\mathbf{i} + y^2\mathbf{j} + z\mathbf{k}$. Find $\iint_S \mathbf{F} \cdot \mathbf{N} \, dS$, where $S$ is the surface of $Q$.

**Solution** From Figure 15.57, you can see that $Q$ is bounded by four subsurfaces. So, you would need four *surface integrals* to evaluate

$$\iint_S \mathbf{F} \cdot \mathbf{N} \, dS.$$

However, by the Divergence Theorem, you need only one triple integral. Because

$$\operatorname{div} \mathbf{F} = \frac{\partial M}{\partial x} + \frac{\partial N}{\partial y} + \frac{\partial P}{\partial z} = 1 + 2y + 1 = 2 + 2y$$

you have

$$\iint_S \mathbf{F} \cdot \mathbf{N} \, dS = \iiint_Q \operatorname{div} \mathbf{F} \, dV$$

$$= \int_0^3 \int_0^{3-y} \int_0^{6-2x-2y} (2 + 2y) \, dz \, dx \, dy$$

$$= \int_0^3 \int_0^{3-y} (2z + 2yz)\Big]_0^{6-2x-2y} dx \, dy$$

$$= \int_0^3 \int_0^{3-y} (12 - 4x + 8y - 4xy - 4y^2) \, dx \, dy$$

$$= \int_0^3 \left[12x - 2x^2 + 8xy - 2x^2y - 4xy^2\right]_0^{3-y} dy$$

$$= \int_0^3 (18 + 6y - 10y^2 + 2y^3) \, dy$$

$$= \left[18y + 3y^2 - \frac{10y^3}{3} + \frac{y^4}{2}\right]_0^3$$

$$= \frac{63}{2}.$$

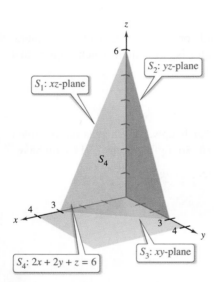

**Figure 15.57**

$S_2$: yz-plane
$S_1$: xz-plane
$S_4$
$S_3$: xy-plane
$S_4$: $2x + 2y + z = 6$

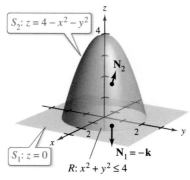

$S_2: z = 4 - x^2 - y^2$

$\mathbf{N}_2$

$S_1: z = 0$

$\mathbf{N}_1 = -\mathbf{k}$

$R: x^2 + y^2 \leq 4$

**Figure 15.58**

**EXAMPLE 2** **Verifying the Divergence Theorem**

Let $Q$ be the solid region between the paraboloid

$$z = 4 - x^2 - y^2$$

and the $xy$-plane. Verify the Divergence Theorem for

$$\mathbf{F}(x, y, z) = 2z\mathbf{i} + x\mathbf{j} + y^2\mathbf{k}.$$

**Solution** From Figure 15.58, you can see that the outward normal vector for the surface $S_1$ is $\mathbf{N}_1 = -\mathbf{k}$, whereas the outward normal vector for the surface $S_2$ is

$$\mathbf{N}_2 = \frac{2x\mathbf{i} + 2y\mathbf{j} + \mathbf{k}}{\sqrt{4x^2 + 4y^2 + 1}}.$$

So, by Theorem 15.11, you have

$$\iint_S \mathbf{F} \cdot \mathbf{N} \, dS = \iint_{S_1} \mathbf{F} \cdot \mathbf{N}_1 \, dS + \iint_{S_2} \mathbf{F} \cdot \mathbf{N}_2 \, dS$$

$$= \iint_{S_1} \mathbf{F} \cdot (-\mathbf{k}) \, dS + \iint_{S_2} \mathbf{F} \cdot \frac{(2x\mathbf{i} + 2y\mathbf{j} + \mathbf{k})}{\sqrt{4x^2 + 4y^2 + 1}} \, dS$$

$$= \iint_R -y^2 \, dA + \iint_R (4xz + 2xy + y^2) \, dA$$

$$= -\int_{-2}^{2} \int_{-\sqrt{4-y^2}}^{\sqrt{4-y^2}} y^2 \, dx \, dy + \int_{-2}^{2} \int_{-\sqrt{4-y^2}}^{\sqrt{4-y^2}} (4xz + 2xy + y^2) \, dx \, dy$$

$$= \int_{-2}^{2} \int_{-\sqrt{4-y^2}}^{\sqrt{4-y^2}} (4xz + 2xy) \, dx \, dy$$

$$= \int_{-2}^{2} \int_{-\sqrt{4-y^2}}^{\sqrt{4-y^2}} [4x(4 - x^2 - y^2) + 2xy] \, dx \, dy$$

$$= \int_{-2}^{2} \int_{-\sqrt{4-y^2}}^{\sqrt{4-y^2}} (16x - 4x^3 - 4xy^2 + 2xy) \, dx \, dy$$

$$= \int_{-2}^{2} \left[ 8x^2 - x^4 - 2x^2y^2 + x^2y \right]_{-\sqrt{4-y^2}}^{\sqrt{4-y^2}} \, dy$$

$$= \int_{-2}^{2} 0 \, dy$$

$$= 0.$$

On the other hand, because

$$\text{div } \mathbf{F} = \frac{\partial}{\partial x}[2z] + \frac{\partial}{\partial y}[x] + \frac{\partial}{\partial z}[y^2]$$

$$= 0 + 0 + 0$$

$$= 0$$

you can apply the Divergence Theorem to obtain the equivalent result

$$\iint_S \mathbf{F} \cdot \mathbf{N} \, dS = \iiint_Q \text{div } \mathbf{F} \, dV$$

$$= \iiint_Q 0 \, dV$$

$$= 0.$$

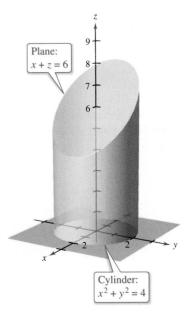

Plane:
$x + z = 6$

Cylinder:
$x^2 + y^2 = 4$

**Figure 15.59**

| EXAMPLE 3 | **Using the Divergence Theorem** |

Let $Q$ be the solid bounded by the cylinder $x^2 + y^2 = 4$, the plane $x + z = 6$, and the $xy$-plane, as shown in Figure 15.59. Find

$$\iint_S \mathbf{F} \cdot \mathbf{N} \, dS$$

where $S$ is the surface of $Q$ and

$$\mathbf{F}(x, y, z) = (x^2 + \sin z)\mathbf{i} + (xy + \cos z)\mathbf{j} + e^y\mathbf{k}.$$

**Solution**  Direct evaluation of this surface integral would be difficult. However, by the Divergence Theorem, you can evaluate the integral as follows.

$$\iint_S \mathbf{F} \cdot \mathbf{N} \, dS = \iiint_Q \text{div } \mathbf{F} \, dV = \iiint_Q (2x + x + 0) \, dV = \iiint_Q 3x \, dV$$

Next, use cylindrical coordinates with $x = r \cos \theta$ and $dV = r \, dz \, dr \, d\theta$.

$$\iiint_Q 3x \, dV = \int_0^{2\pi} \int_0^2 \int_0^{6 - r \cos \theta} (3r \cos \theta) r \, dz \, dr \, d\theta \qquad \text{Cylindrical coordinates}$$

$$= \int_0^{2\pi} \int_0^2 (18r^2 \cos \theta - 3r^3 \cos^2 \theta) \, dr \, d\theta$$

$$= \int_0^{2\pi} (48 \cos \theta - 12 \cos^2 \theta) \, d\theta$$

$$= \left[ 48 \sin \theta - 6\left(\theta + \frac{1}{2} \sin 2\theta\right) \right]_0^{2\pi}$$

$$= -12\pi$$

## Flux and the Divergence Theorem

To help understand the Divergence Theorem, consider the two sides of the equation

$$\iint_S \mathbf{F} \cdot \mathbf{N} \, dS = \iiint_Q \text{div } \mathbf{F} \, dV.$$

You know from Section 15.6 that the flux integral on the left determines the total fluid flow across the surface $S$ per unit of time. This can be approximated by summing the fluid flow across small patches of the surface. The triple integral on the right measures this same fluid flow across $S$ but from a very different perspective—namely, by calculating the flow of fluid into (or out of) small *cubes* of volume $\Delta V_i$. The flux of the $i$th cube is approximately div $\mathbf{F}(x_i, y_i, z_i) \, \Delta V_i$ for some point $(x_i, y_i, z_i)$ in the $i$th cube. Note that for a cube in the interior of $Q$, the gain (or loss) of fluid through any one of its six sides is offset by a corresponding loss (or gain) through one of the sides of an adjacent cube. After summing over all the cubes in $Q$, the only fluid flow that is not canceled by adjoining cubes is that on the outside edges of the cubes on the boundary. So, the sum

$$\sum_{i=1}^n \text{div } \mathbf{F}(x_i, y_i, z_i) \, \Delta V_i$$

approximates the total flux into (or out of) $Q$ and therefore through the surface $S$.

To see what is meant by the divergence of $\mathbf{F}$ at a point, consider $\Delta V_\alpha$ to be the volume of a small sphere $S_\alpha$ of radius $\alpha$ and center $(x_0, y_0, z_0)$ contained in region $Q$, as shown in Figure 15.60.

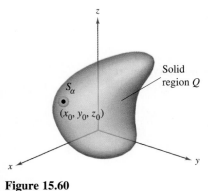

$S_\alpha$
$(x_0, y_0, z_0)$
Solid region $Q$

**Figure 15.60**

Applying the Divergence Theorem to $S_\alpha$ produces

$$\text{Flux of } \mathbf{F} \text{ across } S_\alpha = \iiint_{Q_\alpha} \text{div } \mathbf{F} \, dV \approx \text{div } \mathbf{F}(x_0, y_0, z_0) \, \Delta V_\alpha$$

where $Q_\alpha$ is the interior of $S_\alpha$. Consequently, you have

$$\text{div } \mathbf{F}(x_0, y_0, z_0) \approx \frac{\text{flux of } \mathbf{F} \text{ across } S_\alpha}{\Delta V_\alpha}.$$

By taking the limit as $\alpha \to 0$, you obtain the divergence of $\mathbf{F}$ at the point $(x_0, y_0, z_0)$.

$$\text{div } \mathbf{F}(x_0, y_0, z_0) = \lim_{\alpha \to 0} \frac{\text{flux of } \mathbf{F} \text{ across } S_\alpha}{\Delta V_\alpha} = \text{flux per unit volume at } (x_0, y_0, z_0)$$

The point $(x_0, y_0, z_0)$ in a vector field is classified as a source, a sink, or incompressible, as shown in the list below.

1. **Source,** for div $\mathbf{F} > 0$    See Figure 15.61(a).
2. **Sink,** for div $\mathbf{F} < 0$    See Figure 15.61(b).
3. **Incompressible,** for div $\mathbf{F} = 0$    See Figure 15.61(c).

• • • • • • • • • • • • • • ▷
: **REMARK** In hydrodynamics, a *source* is a point at which additional fluid is considered as being introduced to the region occupied by the fluid. A *sink* is a point at which fluid is considered as being removed.

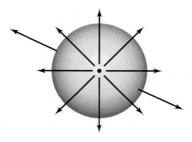 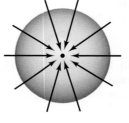 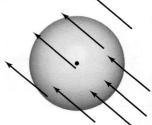

**(a)** Source: div $\mathbf{F} > 0$      **(b)** Sink: div $\mathbf{F} < 0$      **(c)** Incompressible: div $\mathbf{F} = 0$
**Figure 15.61**

---

**EXAMPLE 4**  **Calculating Flux by the Divergence Theorem**

: • • • ▷ *See LarsonCalculus.com for an interactive version of this type of example.*

Let $Q$ be the region bounded by the sphere $x^2 + y^2 + z^2 = 4$. Find the outward flux of the vector field $\mathbf{F}(x, y, z) = 2x^3\mathbf{i} + 2y^3\mathbf{j} + 2z^3\mathbf{k}$ through the sphere.

**Solution**  By the Divergence Theorem, you have

$$\text{Flux across } S = \iint_S \mathbf{F} \cdot \mathbf{N} \, dS = \iiint_Q \text{div } \mathbf{F} \, dV = \iiint_Q 6(x^2 + y^2 + z^2) \, dV.$$

Next, use spherical coordinates with $\rho^2 = x^2 + y^2 + z^2$ and $dV = \rho^2 \sin \phi \, d\theta \, d\phi \, d\rho$.

$$\iiint_Q 6(x^2 + y^2 + z^2) \, dV = 6 \int_0^2 \int_0^\pi \int_0^{2\pi} \rho^4 \sin \phi \, d\theta \, d\phi \, d\rho \qquad \text{Spherical coordinates}$$

$$= 6 \int_0^2 \int_0^\pi 2\pi \rho^4 \sin \phi \, d\phi \, d\rho$$

$$= 12\pi \int_0^2 2\rho^4 \, d\rho$$

$$= 24\pi \left( \frac{32}{5} \right)$$

$$= \frac{768\pi}{5}.$$

■

# 15.7 Exercises

See **CalcChat.com** for tutorial help and worked-out solutions to odd-numbered exercises.

## CONCEPT CHECK

**1. Using Different Methods** Suppose that a solid region $Q$ is bounded by $z = x^2 + y^2$ and $z = 2$, as shown in the figure. What methods can you use to evaluate $\int_S \int \mathbf{F} \cdot \mathbf{N} \, dS$, where $\mathbf{F} = 2x\mathbf{i} + 3y\mathbf{j} - z^2\mathbf{k}$? Which method do you prefer?

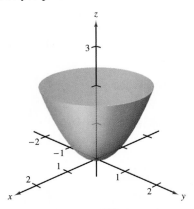

**2. Classifying a Point in a Vector Field** How do you determine whether a point $(x_0, y_0, z_0)$ in a vector field is a source, a sink, or incompressible?

 **Verifying the Divergence Theorem** In Exercises 3–8, verify the Divergence Theorem by evaluating

$$\int_S \int \mathbf{F} \cdot \mathbf{N} \, dS$$

**as a surface integral and as a triple integral.**

**3.** $\mathbf{F}(x, y, z) = 2x\mathbf{i} - 2y\mathbf{j} + z^2\mathbf{k}$

 $S$: cube bounded by the planes $x = 0$, $x = 1$, $y = 0$, $y = 1$, $z = 0$, $z = 1$

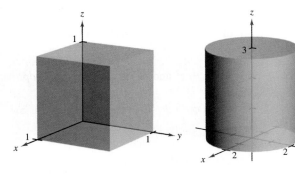

Figure for 3                 Figure for 4

**4.** $\mathbf{F}(x, y, z) = 2x\mathbf{i} - 2y\mathbf{j} + z^2\mathbf{k}$

 $S$: cylinder $x^2 + y^2 = 4$, $0 \le z \le 3$

**5.** $\mathbf{F}(x, y, z) = (2x - y)\mathbf{i} - (2y - z)\mathbf{j} + z\mathbf{k}$

 $S$: surface bounded by the plane $2x + 4y + 2z = 12$ and the coordinate planes

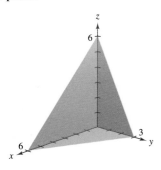

**6.** $\mathbf{F}(x, y, z) = xy\mathbf{i} + z\mathbf{j} + (x + y)\mathbf{k}$

 $S$: surface bounded by the planes $y = 4$ and $z = 4 - x$ and the coordinate planes

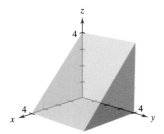

**7.** $\mathbf{F}(x, y, z) = xz\mathbf{i} + zy\mathbf{j} + 2z^2\mathbf{k}$

 $S$: surface bounded by $z = 1 - x^2 - y^2$ and $z = 0$

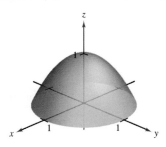

**8.** $\mathbf{F}(x, y, z) = xy^2\mathbf{i} + yx^2\mathbf{j} + e\mathbf{k}$

 $S$: surface bounded by $z = \sqrt{x^2 + y^2}$ and $z = 4$

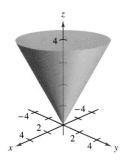

**Using the Divergence Theorem** In Exercises 9–18, use the Divergence Theorem to evaluate

$$\iint_S \mathbf{F} \cdot \mathbf{N}\, dS$$

and find the outward flux of $\mathbf{F}$ through the surface of the solid $S$ bounded by the graphs of the equations. Use a computer algebra system to verify your results.

9. $\mathbf{F}(x, y, z) = x^2\mathbf{i} + y^2\mathbf{j} + z^2\mathbf{k}$

   $S$: $x = 0, x = a, y = 0, y = a, z = 0, z = a$

10. $\mathbf{F}(x, y, z) = x^2z^2\mathbf{i} - 2y\mathbf{j} + 3xyz\mathbf{k}$

    $S$: $x = 0, x = a, y = 0, y = a, z = 0, z = a$

11. $\mathbf{F}(x, y, z) = x^2\mathbf{i} - 2xy\mathbf{j} + xyz^2\mathbf{k}$

    $S$: $z = \sqrt{a^2 - x^2 - y^2}, z = 0$

12. $\mathbf{F}(x, y, z) = xy\mathbf{i} + yz\mathbf{j} - yz\mathbf{k}$

    $S$: $z = \sqrt{a^2 - x^2 - y^2}, z = 0$

13. $\mathbf{F}(x, y, z) = x\mathbf{i} + y\mathbf{j} + z\mathbf{k}$    14. $\mathbf{F}(x, y, z) = xyz\mathbf{j}$

    $S$: $x^2 + y^2 + z^2 = 9$    $\quad$ $S$: $x^2 + y^2 = 4, z = 0, z = 5$

15. $\mathbf{F}(x, y, z) = x\mathbf{i} + y^2\mathbf{j} - z\mathbf{k}$

    $S$: $x^2 + y^2 = 25, z = 0, z = 7$

16. $\mathbf{F}(x, y, z) = (xy^2 + \cos z)\mathbf{i} + (x^2y + \sin z)\mathbf{j} + e^z\mathbf{k}$

    $S$: $z = \frac{1}{2}\sqrt{x^2 + y^2}, z = 8$

17. $\mathbf{F}(x, y, z) = xe^z\mathbf{i} + ye^z\mathbf{j} + e^z\mathbf{k}$

    $S$: $z = 4 - y, z = 0, x = 0, x = 6, y = 0$

18. $\mathbf{F}(x, y, z) = xy\mathbf{i} + 4y\mathbf{j} + xz\mathbf{k}$

    $S$: $x^2 + y^2 + z^2 = 16$

**Classifying a Point** In Exercises 19–22, a vector field and a point in the vector field are given. Determine whether the point is a source, a sink, or incompressible.

19. $\mathbf{F}(x, y, z) = 2\mathbf{i} + y\mathbf{j} + \mathbf{k}$,   $(2, 2, 1)$

20. $\mathbf{F}(x, y, z) = e^{-x}\mathbf{i} - xy^2\mathbf{j} + \ln z\mathbf{k}$,   $(0, -3, 1)$

21. $\mathbf{F}(x, y, z) = \sin x\mathbf{i} + \cos y\mathbf{j} + z^3 \sin y\mathbf{k}$,   $\left(\dfrac{\pi}{2}, \pi, 4\right)$

22. $\mathbf{F}(x, y, z) = (4xy + z^2)\mathbf{i} + (2x^2 + 6yz)\mathbf{j} + 2xz\mathbf{k}$,   $(1, -4, 2)$

23. **Source** Find a point that is a source in the vector field

    $\mathbf{F}(x, y, z) = x^2yz\mathbf{i} + x\mathbf{j} - z\mathbf{k}$.

24. **Sink** Find a point that is a sink in the vector field

    $\mathbf{F}(x, y, z) = e^{-x}\mathbf{i} + 4y\mathbf{j} + xyz^2\mathbf{k}$.

---

**EXPLORING CONCEPTS**

25. **Closed Surface** What is the value of

    $$\iint_S \text{curl } \mathbf{F} \cdot \mathbf{N}\, dS$$

    for any closed surface $S$? Explain.

---

26. **HOW DO YOU SEE IT?** The graph of a vector field $\mathbf{F}$ is shown. Does the graph suggest that the divergence of $\mathbf{F}$ at $P$ is positive, negative, or zero?

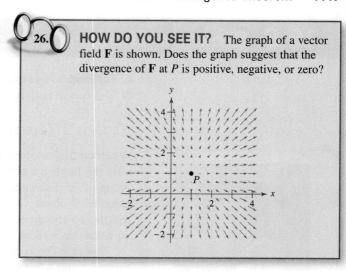

27. **Volume**

    (a) Use the Divergence Theorem to verify that the volume of the solid bounded by a surface $S$ is

    $$\iint_S x\, dy\, dz = \iint_S y\, dz\, dx = \iint_S z\, dx\, dy.$$

    (b) Verify the result of part (a) for the cube bounded by $x = 0$, $x = a, y = 0, y = a, z = 0$, and $z = a$.

28. **Constant Vector Field** For the constant vector field $\mathbf{F}(x, y, z) = a_1\mathbf{i} + a_2\mathbf{j} + a_3\mathbf{k}$, verify the following integral for any closed surface $S$.

    $$\iint_S \mathbf{F} \cdot \mathbf{N}\, dS = 0$$

29. **Volume** For the vector field $\mathbf{F}(x, y, z) = x\mathbf{i} + y\mathbf{j} + z\mathbf{k}$, verify the following integral, where $V$ is the volume of the solid bounded by the closed surface $S$.

    $$\iint_S \mathbf{F} \cdot \mathbf{N}\, dS = 3V$$

30. **Verifying an Identity** For the vector field $\mathbf{F}(x, y, z) = x\mathbf{i} + y\mathbf{j} + z\mathbf{k}$, verify that

    $$\frac{1}{\|\mathbf{F}\|}\iint_S \mathbf{F} \cdot \mathbf{N}\, dS = \frac{3}{\|\mathbf{F}\|}\iiint_Q dV.$$

**Proof** In Exercises 31 and 32, prove the identity, assuming that $Q$, $S$, and $\mathbf{N}$ meet the conditions of the Divergence Theorem and that the required partial derivatives of the scalar functions $f$ and $g$ are continuous. The expressions $D_\mathbf{N}f$ and $D_\mathbf{N}g$ are the derivatives in the direction of the vector $\mathbf{N}$ and are defined by $D_\mathbf{N}f = \nabla f \cdot \mathbf{N}$ and $D_\mathbf{N}g = \nabla g \cdot \mathbf{N}$.

31. $\displaystyle\iiint_Q (f\nabla^2 g + \nabla f \cdot \nabla g)\, dV = \iint_S f D_\mathbf{N}g\, dS$

    [*Hint:* Use div$(f\mathbf{G}) = f$ div $\mathbf{G} + \nabla f \cdot \mathbf{G}$.]

32. $\displaystyle\iiint_Q (f\nabla^2 g - g\nabla^2 f)\, dV = \iint_S (f D_\mathbf{N}g - g D_\mathbf{N}f)\, dS$

    (*Hint:* Use Exercise 31 twice.)

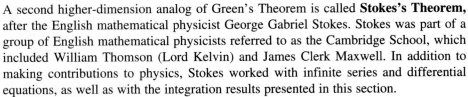

# 15.8 Stokes's Theorem

■ Understand and use Stokes's Theorem.
■ Use curl to analyze the motion of a rotating liquid.

## Stokes's Theorem

A second higher-dimension analog of Green's Theorem is called **Stokes's Theorem,** after the English mathematical physicist George Gabriel Stokes. Stokes was part of a group of English mathematical physicists referred to as the Cambridge School, which included William Thomson (Lord Kelvin) and James Clerk Maxwell. In addition to making contributions to physics, Stokes worked with infinite series and differential equations, as well as with the integration results presented in this section.

Stokes's Theorem gives the relationship between a surface integral over an oriented surface $S$ and a line integral along a closed space curve $C$ forming the boundary of $S$, as shown in Figure 15.62. The positive direction along $C$ is counterclockwise relative to the normal vector $\mathbf{N}$. That is, if you imagine grasping the normal vector $\mathbf{N}$ with your right hand, with your thumb pointing in the direction of $\mathbf{N}$, then your fingers will point in the positive direction $C$, as shown in Figure 15.63.

**GEORGE GABRIEL STOKES
(1819–1903)**

Stokes became a Lucasian professor of mathematics at Cambridge in 1849. Five years later, he published the theorem that bears his name as a prize examination question there.
*See LarsonCalculus.com to read more of this biography.*

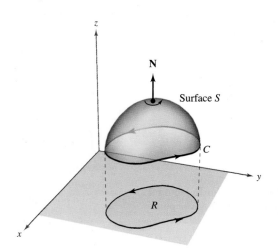

An oriented surface $S$ bounded by a closed space curve $C$
**Figure 15.62**

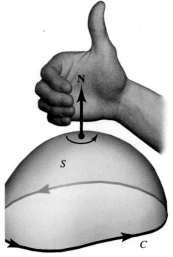

The positive direction along $C$ is counterclockwise relative to $\mathbf{N}$.
**Figure 15.63**

---

**THEOREM 15.13 Stokes's Theorem**

Let $S$ be an oriented surface with unit normal vector $\mathbf{N}$, bounded by a piecewise smooth simple closed curve $C$ with a positive orientation. If $\mathbf{F}$ is a vector field whose component functions have continuous first partial derivatives on an open region containing $S$ and $C$, then

$$\int_C \mathbf{F} \cdot d\mathbf{r} = \iint_S (\text{curl } \mathbf{F}) \cdot \mathbf{N} \, dS.$$

---

In Theorem 15.13, note that the line integral may be written in the differential form $\int_C M \, dx + N \, dy + P \, dz$ or in the vector form $\int_C \mathbf{F} \cdot \mathbf{T} \, ds$.

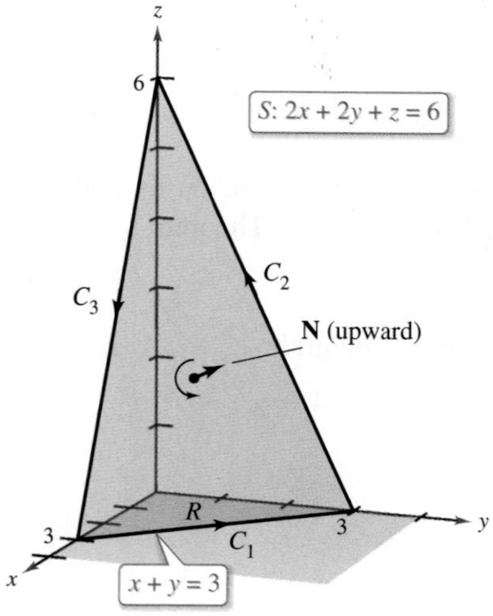

Figure 15.64

**EXAMPLE 1**   **Using Stokes's Theorem**

Let $C$ be the oriented triangle lying in the plane

$$2x + 2y + z = 6$$

as shown in Figure 15.64. Evaluate

$$\int_C \mathbf{F} \cdot d\mathbf{r}$$

where $\mathbf{F}(x, y, z) = -y^2\mathbf{i} + z\mathbf{j} + x\mathbf{k}$.

**Solution**   Using Stokes's Theorem, begin by finding the curl of $\mathbf{F}$.

$$\text{curl } \mathbf{F} = \begin{vmatrix} \mathbf{i} & \mathbf{j} & \mathbf{k} \\ \dfrac{\partial}{\partial x} & \dfrac{\partial}{\partial y} & \dfrac{\partial}{\partial z} \\ -y^2 & z & x \end{vmatrix} = -\mathbf{i} - \mathbf{j} + 2y\mathbf{k}$$

Considering

$$z = g(x, y) = 6 - 2x - 2y$$

you can use Theorem 15.11 for an upward normal vector to obtain

$$\begin{aligned}
\int_C \mathbf{F} \cdot d\mathbf{r} &= \iint_S (\text{curl } \mathbf{F}) \cdot \mathbf{N} \, dS \\
&= \iint_R (-\mathbf{i} - \mathbf{j} + 2y\mathbf{k}) \cdot [-g_x(x, y)\mathbf{i} - g_y(x, y)\mathbf{j} + \mathbf{k}] \, dA \\
&= \iint_R (-\mathbf{i} - \mathbf{j} + 2y\mathbf{k}) \cdot (2\mathbf{i} + 2\mathbf{j} + \mathbf{k}) \, dA \\
&= \int_0^3 \int_0^{3-y} (2y - 4) \, dx \, dy \\
&= \int_0^3 (-2y^2 + 10y - 12) \, dy \\
&= \left[ -\frac{2y^3}{3} + 5y^2 - 12y \right]_0^3 \\
&= -9.
\end{aligned}$$

Try evaluating the line integral in Example 1 directly, *without* using Stokes's Theorem. One way to do this would be to consider $C$ as the union of $C_1$, $C_2$, and $C_3$, as follows.

$$\begin{aligned}
C_1: \mathbf{r}_1(t) &= (3 - t)\mathbf{i} + t\mathbf{j}, \quad 0 \le t \le 3 \\
C_2: \mathbf{r}_2(t) &= (6 - t)\mathbf{j} + (2t - 6)\mathbf{k}, \quad 3 \le t \le 6 \\
C_3: \mathbf{r}_3(t) &= (t - 6)\mathbf{i} + (18 - 2t)\mathbf{k}, \quad 6 \le t \le 9
\end{aligned}$$

The value of the line integral is

$$\begin{aligned}
\int_C \mathbf{F} \cdot d\mathbf{r} &= \int_{C_1} \mathbf{F} \cdot \mathbf{r}_1'(t) \, dt + \int_{C_2} \mathbf{F} \cdot \mathbf{r}_2'(t) \, dt + \int_{C_3} \mathbf{F} \cdot \mathbf{r}_3'(t) \, dt \\
&= \int_0^3 t^2 \, dt + \int_3^6 (-2t + 6) \, dt + \int_6^9 (-2t + 12) \, dt \\
&= 9 - 9 - 9 \\
&= -9.
\end{aligned}$$

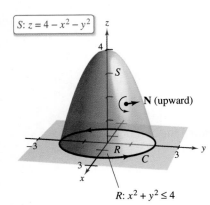

$S: z = 4 - x^2 - y^2$

$R: x^2 + y^2 \le 4$

**Figure 15.65**

### EXAMPLE 2    Verifying Stokes's Theorem

⋯▷ *See LarsonCalculus.com for an interactive version of this type of example.*

Let $S$ be the portion of the paraboloid

$$z = 4 - x^2 - y^2$$

lying above the $xy$-plane, oriented upward (see Figure 15.65). Let $C$ be its boundary curve in the $xy$-plane, oriented counterclockwise. Verify Stokes's Theorem for

$$\mathbf{F}(x, y, z) = 2z\mathbf{i} + x\mathbf{j} + y^2\mathbf{k}$$

by evaluating the surface integral and the equivalent line integral.

**Solution**    As a *surface integral,* you have $z = g(x, y) = 4 - x^2 - y^2$, $g_x = -2x$, $g_y = -2y$, and

$$\text{curl } \mathbf{F} = \begin{vmatrix} \mathbf{i} & \mathbf{j} & \mathbf{k} \\ \dfrac{\partial}{\partial x} & \dfrac{\partial}{\partial y} & \dfrac{\partial}{\partial z} \\ 2z & x & y^2 \end{vmatrix} = 2y\mathbf{i} + 2\mathbf{j} + \mathbf{k}.$$

By Theorem 15.11 (for an upward normal vector), you obtain

$$\iint_S (\text{curl } \mathbf{F}) \cdot \mathbf{N} \, dS = \iint_R (2y\mathbf{i} + 2\mathbf{j} + \mathbf{k}) \cdot (2x\mathbf{i} + 2y\mathbf{j} + \mathbf{k}) \, dA$$

$$= \int_{-2}^{2} \int_{-\sqrt{4-x^2}}^{\sqrt{4-x^2}} (4xy + 4y + 1) \, dy \, dx$$

$$= \int_{-2}^{2} \left[ 2xy^2 + 2y^2 + y \right]_{-\sqrt{4-x^2}}^{\sqrt{4-x^2}} dx$$

$$= \int_{-2}^{2} 2\sqrt{4 - x^2} \, dx$$

$$= \text{Area of circle of radius } 2$$

$$= 4\pi.$$

As a *line integral,* you can parametrize $C$ as

$$\mathbf{r}(t) = 2 \cos t\mathbf{i} + 2 \sin t\mathbf{j} + 0\mathbf{k}, \quad 0 \le t \le 2\pi.$$

For $\mathbf{F}(x, y, z) = 2z\mathbf{i} + x\mathbf{j} + y^2\mathbf{k}$, you obtain

$$\int_C \mathbf{F} \cdot d\mathbf{r} = \int_C M \, dx + N \, dy + P \, dz$$

$$= \int_C 2z \, dx + x \, dy + y^2 \, dz$$

$$= \int_0^{2\pi} [0 + (2 \cos t)(2 \cos t) + 0] \, dt$$

$$= \int_0^{2\pi} 4 \cos^2 t \, dt$$

$$= 2 \int_0^{2\pi} (1 + \cos 2t) \, dt$$

$$= 2 \left[ t + \frac{1}{2} \sin 2t \right]_0^{2\pi}$$

$$= 4\pi.$$

# Physical Interpretation of Curl

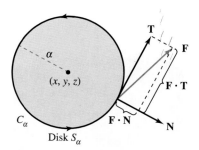

**Figure 15.66**

Stokes's Theorem provides insight into a physical interpretation of curl. In a vector field $\mathbf{F}$, let $S_\alpha$ be a *small* circular disk of radius $\alpha$, centered at $(x, y, z)$ and with boundary $C_\alpha$, as shown in Figure 15.66. At each point on the circle $C_\alpha$, $\mathbf{F}$ has a normal component $\mathbf{F} \cdot \mathbf{N}$ and a tangential component $\mathbf{F} \cdot \mathbf{T}$. The more closely $\mathbf{F}$ and $\mathbf{T}$ are aligned, the greater the value of $\mathbf{F} \cdot \mathbf{T}$. So, a fluid tends to move along the circle rather than across it. Consequently, you say that the line integral around $C_\alpha$ measures the **circulation of F around $C_\alpha$.** That is,

$$\int_{C_\alpha} \mathbf{F} \cdot \mathbf{T} \, ds = \text{circulation of } \mathbf{F} \text{ around } C_\alpha.$$

Now consider a small disk $S_\alpha$ to be centered at some point $(x, y, z)$ on the surface $S$, as shown in Figure 15.67. On such a small disk, curl $\mathbf{F}$ is nearly constant, because it varies little from its value at $(x, y, z)$. Moreover, $(\text{curl } \mathbf{F}) \cdot \mathbf{N}$ is also nearly constant on $S_\alpha$ because all unit normals to $S_\alpha$ are about the same. Consequently, Stokes's Theorem yields

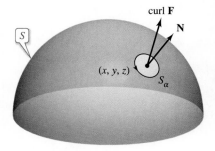

**Figure 15.67**

$$\int_{C_\alpha} \mathbf{F} \cdot \mathbf{T} \, ds = \iint_{S_\alpha} (\text{curl } \mathbf{F}) \cdot \mathbf{N} \, dS$$

$$\approx (\text{curl } \mathbf{F}) \cdot \mathbf{N} \iint_{S_\alpha} dS$$

$$\approx (\text{curl } \mathbf{F}) \cdot \mathbf{N}(\pi\alpha^2).$$

So,

$$(\text{curl } \mathbf{F}) \cdot \mathbf{N} \approx \frac{\displaystyle\int_{C_\alpha} \mathbf{F} \cdot \mathbf{T} \, ds}{\pi\alpha^2}$$

$$= \frac{\text{circulation of } \mathbf{F} \text{ around } C_\alpha}{\text{area of disk } S_\alpha}$$

$$= \text{rate of circulation.}$$

Assuming conditions are such that the approximation improves for smaller and smaller disks $(\alpha \to 0)$, it follows that

$$(\text{curl } \mathbf{F}) \cdot \mathbf{N} = \lim_{\alpha \to 0} \frac{1}{\pi\alpha^2} \int_{C_\alpha} \mathbf{F} \cdot \mathbf{T} \, ds$$

which is referred to as the **rotation of F about N.** That is,

$$\text{curl } \mathbf{F}(x, y, z) \cdot \mathbf{N} = \text{rotation of } \mathbf{F} \text{ about } \mathbf{N} \text{ at } (x, y, z).$$

In this case, the rotation of $\mathbf{F}$ is maximum when curl $\mathbf{F}$ and $\mathbf{N}$ have the same direction. Normally, this tendency to rotate will vary from point to point on the surface $S$, and Stokes's Theorem

$$\underbrace{\iint_S (\text{curl } \mathbf{F}) \cdot \mathbf{N} \, dS}_{\text{Surface integral}} = \underbrace{\int_C \mathbf{F} \cdot d\mathbf{r}}_{\text{Line integral}}$$

says that the collective measure of this *rotational* tendency taken over the entire surface $S$ (surface integral) is equal to the tendency of a fluid to *circulate* around the boundary $C$ (line integral).

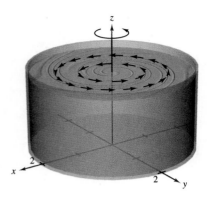

**EXAMPLE 3**    **An Application of Curl**

A liquid is swirling around in a cylindrical container of radius 2, so that its motion is described by the velocity field

$$\mathbf{F}(x, y, z) = -y\sqrt{x^2 + y^2}\,\mathbf{i} + x\sqrt{x^2 + y^2}\,\mathbf{j}$$

as shown in the figure. Find

$$\iint_S (\text{curl }\mathbf{F}) \cdot \mathbf{N}\,dS$$

where $S$ is the upper surface of the cylindrical container.

**Solution**    The curl of $\mathbf{F}$ is given by

$$\text{curl }\mathbf{F} = \begin{vmatrix} \mathbf{i} & \mathbf{j} & \mathbf{k} \\ \dfrac{\partial}{\partial x} & \dfrac{\partial}{\partial y} & \dfrac{\partial}{\partial z} \\ -y\sqrt{x^2 + y^2} & x\sqrt{x^2 + y^2} & 0 \end{vmatrix} = 3\sqrt{x^2 + y^2}\,\mathbf{k}.$$

Letting $\mathbf{N} = \mathbf{k}$, you have

$$\iint_S (\text{curl }\mathbf{F}) \cdot \mathbf{N}\,dS = \iint_R 3\sqrt{x^2 + y^2}\,dA$$

$$= \int_0^{2\pi} \int_0^2 (3r)r\,dr\,d\theta$$

$$= \int_0^{2\pi} r^3 \Big]_0^2\,d\theta$$

$$= \int_0^{2\pi} 8\,d\theta$$

$$= 16\pi.$$

If curl $\mathbf{F} = \mathbf{0}$ throughout region $Q$, then the rotation of $\mathbf{F}$ about each unit normal $\mathbf{N}$ is 0. That is, $\mathbf{F}$ is irrotational. From Section 15.1, you know that this is a characteristic of conservative vector fields.

---

## SUMMARY OF INTEGRATION FORMULAS

**Fundamental Theorem of Calculus**

$$\int_a^b F'(x)\,dx = F(b) - F(a)$$

**Fundamental Theorem of Line Integrals**

$$\int_C \mathbf{F} \cdot d\mathbf{r} = \int_C \nabla f \cdot d\mathbf{r} = f(x(b), y(b)) - f(x(a), y(a))$$

**Green's Theorem**

$$\int_C M\,dx + N\,dy = \iint_R \left( \frac{\partial N}{\partial x} - \frac{\partial M}{\partial y} \right) dA = \int_C \mathbf{F} \cdot \mathbf{T}\,ds = \int_C \mathbf{F} \cdot d\mathbf{r} = \iint_R (\text{curl }\mathbf{F}) \cdot \mathbf{k}\,dA$$

$$\int_C \mathbf{F} \cdot \mathbf{N}\,ds = \iint_R \text{div }\mathbf{F}\,dA$$

**Divergence Theorem**

$$\iint_S \mathbf{F} \cdot \mathbf{N}\,dS = \iiint_Q \text{div }\mathbf{F}\,dV$$

**Stokes's Theorem**

$$\int_C \mathbf{F} \cdot d\mathbf{r} = \iint_S (\text{curl }\mathbf{F}) \cdot \mathbf{N}\,dS$$

# 15.8 Exercises

See **CalcChat.com** for tutorial help and worked-out solutions to odd-numbered exercises.

### CONCEPT CHECK

**1. Stokes's Theorem** Explain the benefit of Stokes's Theorem when the boundary of the surface is a piecewise curve.

**2. Curl** What is the physical interpretation of curl?

 **Verifying Stokes's Theorem** In Exercises 3–6, verify Stokes's Theorem by evaluating $\int_C \mathbf{F} \cdot d\mathbf{r}$ as a line integral and as a double integral.

**3.** $\mathbf{F}(x, y, z) = (-y + z)\mathbf{i} + (x - z)\mathbf{j} + (x - y)\mathbf{k}$

$S:\ z = 9 - x^2 - y^2,\quad z \geq 0$

**4.** $\mathbf{F}(x, y, z) = (-y + z)\mathbf{i} + (x - z)\mathbf{j} + (x - y)\mathbf{k}$

$S:\ z = \sqrt{1 - x^2 - y^2}$

**5.** $\mathbf{F}(x, y, z) = xyz\mathbf{i} + y\mathbf{j} + z\mathbf{k}$

$S:\ 6x + 6y + z = 12$, first octant

**6.** $\mathbf{F}(x, y, z) = z^2\mathbf{i} + x^2\mathbf{j} + y^2\mathbf{k}$

$S:\ z = y^2,\quad 0 \leq x \leq a,\quad 0 \leq y \leq a$

 **Using Stokes's Theorem** In Exercises 7–16, use Stokes's Theorem to evaluate $\int_C \mathbf{F} \cdot d\mathbf{r}$. In each case, $C$ is oriented counterclockwise as viewed from above.

**7.** $\mathbf{F}(x, y, z) = 2y\mathbf{i} + 3z\mathbf{j} + x\mathbf{k}$

$C$: triangle with vertices $(2, 0, 0)$, $(0, 2, 0)$, and $(0, 0, 2)$

**8.** $\mathbf{F}(x, y, z) = 4z\mathbf{i} + x^2\mathbf{j} + e^y\mathbf{k}$

$C$: triangle with vertices $(4, 0, 0)$, $(0, 2, 0)$, and $(0, 0, 8)$

**9.** $\mathbf{F}(x, y, z) = z^2\mathbf{i} + 2x\mathbf{j} + y^2\mathbf{k}$

$S:\ z = 1 - x^2 - y^2,\quad z \geq 0$

**10.** $\mathbf{F}(x, y, z) = 4xz\mathbf{i} + y\mathbf{j} + 4xy\mathbf{k}$

$S:\ z = 9 - x^2 - y^2,\quad z \geq 0$

**11.** $\mathbf{F}(x, y, z) = z^2\mathbf{i} + y\mathbf{j} + z\mathbf{k}$

$S:\ z = \sqrt{4 - x^2 - y^2}$

**12.** $\mathbf{F}(x, y, z) = x^2\mathbf{i} + z^2\mathbf{j} - xyz\mathbf{k}$

$S:\ z = \sqrt{4 - x^2 - y^2}$

**13.** $\mathbf{F}(x, y, z) = -\ln\sqrt{x^2 + y^2}\,\mathbf{i} + \arctan\dfrac{x}{y}\mathbf{j} + \mathbf{k}$

$S:\ z = 9 - 2x - 3y$ over $r = 2\sin 2\theta$ in the first octant

**14.** $\mathbf{F}(x, y, z) = yz\mathbf{i} + (2 - 3y)\mathbf{j} + (x^2 + y^2)\mathbf{k},\quad x^2 + y^2 \leq 16$

$S$: the first-octant portion of $x^2 + z^2 = 16$ over $x^2 + y^2 = 16$

**15.** $\mathbf{F}(x, y, z) = xyz\mathbf{i} + y\mathbf{j} + z\mathbf{k}$

$S:\ z = x^2,\quad 0 \leq x \leq a,\quad 0 \leq y \leq a$

**16.** $\mathbf{F}(x, y, z) = xyz\mathbf{i} + y\mathbf{j} + z\mathbf{k},\quad x^2 + y^2 \leq a^2$

$S$: the first-octant portion of $z = x^2$ over $x^2 + y^2 = a^2$

 **Motion of a Liquid** In Exercises 17 and 18, the motion of a liquid in a cylindrical container of radius 3 is described by the velocity field $\mathbf{F}(x, y, z)$. Find $\int_S \int (\text{curl } \mathbf{F}) \cdot \mathbf{N}\, dS$, where $S$ is the upper surface of the cylindrical container.

**17.** $\mathbf{F}(x, y, z) = -\frac{1}{6}y^3\mathbf{i} + \frac{1}{6}x^3\mathbf{j} + 5\mathbf{k}$

**18.** $\mathbf{F}(x, y, z) = -z\mathbf{i} + y^2\mathbf{k}$

### EXPLORING CONCEPTS

**19. Think About It** Let $\mathbf{K}$ be a constant vector. Let $S$ be an oriented surface with a unit normal vector $\mathbf{N}$, bounded by a smooth curve $C$. Determine whether

$$\iint_S \mathbf{K} \cdot \mathbf{N}\, dS = \frac{1}{2}\int_C (\mathbf{K} \times \mathbf{r}) \cdot d\mathbf{r}.$$

Explain. (*Hint:* Use $\mathbf{r} = x\mathbf{i} + y\mathbf{j} + z\mathbf{k}$.)

**20. HOW DO YOU SEE IT?** Let $S_1$ be the portion of the paraboloid lying above the $xy$-plane, and let $S_2$ be the hemisphere, as shown in the figures. Both surfaces are oriented upward. For a vector field $\mathbf{F}(x, y, z)$ with continuous partial derivatives, does

$$\iint_{S_1} (\text{curl } \mathbf{F}) \cdot \mathbf{N}\, dS_1 = \iint_{S_2} (\text{curl } \mathbf{F}) \cdot \mathbf{N}\, dS_2?$$

Explain your reasoning.

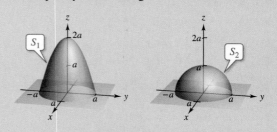

### PUTNAM EXAM CHALLENGE

**21.** Let $\mathbf{G}(x, y) = \left(\dfrac{-y}{x^2 + 4y^2}, \dfrac{x}{x^2 + 4y^2}, 0\right)$.

Prove or disprove that there is a vector-valued function $\mathbf{F}(x, y, z) = (M(x, y, z), N(x, y, z), P(x, y, z))$ with the following properties:

(i) $M, N, P$ have continuous partial derivatives for all $(x, y, z) \neq (0, 0, 0)$;

(ii) Curl $\mathbf{F} = \mathbf{0}$ for all $(x, y, z) \neq (0, 0, 0)$;

(iii) $\mathbf{F}(x, y, 0) = \mathbf{G}(x, y)$.

**Sketching a Vector Field**  In Exercises 1 and 2, find $\|\mathbf{F}\|$ and sketch several representative vectors in the vector field. Use a computer algebra system to verify your results.

1. $\mathbf{F}(x, y, z) = x\mathbf{i} + \mathbf{j} + 2\mathbf{k}$    2. $\mathbf{F}(x, y) = \mathbf{i} - 2y\mathbf{j}$

**Finding a Conservative Vector Field**  In Exercises 3–6, find the conservative vector field for the potential function by finding its gradient.

3. $f(x, y) = \sin xy - y^2$

4. $f(x, y) = \sqrt{xy}$

5. $f(x, y, z) = 2x^2 + xy + z^2$

6. $f(x, y, z) = x^2 e^{yz}$

**Testing for a Conservative Vector Field**  In Exercises 7–10, determine whether the vector field is conservative.

7. $\mathbf{F}(x, y) = \cosh y\mathbf{i} + x \sinh x\mathbf{j}$

8. $\mathbf{F}(x, y) = \dfrac{y \ln x}{x}\mathbf{i} + (\ln x)^2\mathbf{j}$

9. $\mathbf{F}(x, y, z) = y^2\mathbf{i} + 2xy\mathbf{j} + \cos z\mathbf{k}$

10. $\mathbf{F}(x, y, z) = 3e^{xy}\mathbf{i} + 3e^{x+y}\mathbf{j} + e^{3yz}\mathbf{k}$

**Finding a Potential Function**  In Exercises 11–18, determine whether the vector field is conservative. If it is, find a potential function for the vector field.

11. $\mathbf{F}(x, y) = -\dfrac{y}{x^2}\mathbf{i} + \dfrac{1}{x}\mathbf{j}$    12. $\mathbf{F}(x, y) = \dfrac{1}{y}\mathbf{i} - \dfrac{y}{x^2}\mathbf{j}$

13. $\mathbf{F}(x, y) = (xy^2 - x^2)\mathbf{i} + (x^2y + y^2)\mathbf{j}$

14. $\mathbf{F}(x, y) = (-2y^3 \sin 2x)\mathbf{i} + 3y^2(1 + \cos 2x)\mathbf{j}$

15. $\mathbf{F}(x, y, z) = 4xy^2\mathbf{i} + 2x^2\mathbf{j} + 2z\mathbf{k}$

16. $\mathbf{F}(x, y, z) = (4xy + z^2)\mathbf{i} + (2x^2 + 6yz)\mathbf{j} + 2xz\mathbf{k}$

17. $\mathbf{F}(x, y, z) = \dfrac{yz\mathbf{i} - xz\mathbf{j} - xy\mathbf{k}}{y^2 z^2}$

18. $\mathbf{F}(x, y, z) = (\sin z)(y\mathbf{i} + x\mathbf{j} + \mathbf{k})$

**Divergence and Curl**  In Exercises 19–26, find (a) the divergence of the vector field and (b) the curl of the vector field.

19. $\mathbf{F}(x, y, z) = x^2\mathbf{i} + xy^2\mathbf{j} + x^2z\mathbf{k}$

20. $\mathbf{F}(x, y, z) = y^2\mathbf{j} - z^2\mathbf{k}$

21. $\mathbf{F}(x, y, z) = (\cos y + y \cos x)\mathbf{i} + (\sin x - x \sin y)\mathbf{j} + xyz\mathbf{k}$

22. $\mathbf{F}(x, y, z) = (3x - y)\mathbf{i} + (y - 2z)\mathbf{j} + (z - 3x)\mathbf{k}$

23. $\mathbf{F}(x, y, z) = \arcsin x\mathbf{i} + xy^2\mathbf{j} + yz^2\mathbf{k}$

24. $\mathbf{F}(x, y, z) = (x^2 - y)\mathbf{i} - (x + \sin^2 y)\mathbf{j}$

25. $\mathbf{F}(x, y, z) = \ln(x^2 + y^2)\mathbf{i} + \ln(x^2 + y^2)\mathbf{j} + z\mathbf{k}$

26. $\mathbf{F}(x, y, z) = \dfrac{z}{x}\mathbf{i} + \dfrac{z}{y}\mathbf{j} + z^2\mathbf{k}$

**Evaluating a Line Integral**  In Exercises 27–30, evaluate the line integral along the given path(s).

27. $\displaystyle\int_C (x^2 + y^2)\, ds$

(a) $C$: line segment from $(0, 0)$ to $(3, 4)$

(b) $C$: one revolution counterclockwise around the circle $x^2 + y^2 = 1$, starting at $(1, 0)$

28. $\displaystyle\int_C xy\, ds$

(a) $C$: line segment from $(0, 0)$ to $(5, 4)$

(b) $C$: counterclockwise around the triangle with vertices $(0, 0)$, $(4, 0)$, and $(0, 2)$

29. $\displaystyle\int_C (x^2 + y^2)\, ds$

$C$: $\mathbf{r}(t) = (1 - \sin t)\mathbf{i} + (1 - \cos t)\mathbf{j}$,  $0 \le t \le 2\pi$

30. $\displaystyle\int_C (x^2 + y^2)\, ds$

$C$: $\mathbf{r}(t) = (\cos t + t \sin t)\mathbf{i} + (\sin t - t \cos t)\mathbf{j}$,  $0 \le t \le 2\pi$

**Evaluating a Line Integral Using Technology**  In Exercises 31 and 32, use a computer algebra system to evaluate the line integral along the given path.

31. $\displaystyle\int_C (2x + y)\, ds$

$C$: $\mathbf{r}(t) = a \cos^3 t\mathbf{i} + a \sin^3 t\mathbf{j}$,  $0 \le t \le \dfrac{\pi}{2}$

32. $\displaystyle\int_C (x^2 + y^2 + z^2)\, ds$

$C$: $\mathbf{r}(t) = t\mathbf{i} + t^2\mathbf{j} + t^{3/2}\mathbf{k}$,  $0 \le t \le 4$

**Mass**  In Exercises 33 and 34, find the total mass of the wire with density $\rho$ whose shape is modeled by $\mathbf{r}$.

33. $\mathbf{r}(t) = 3 \cos t\mathbf{i} + 3 \sin t\mathbf{j}$,  $0 \le t \le \pi$,  $\rho(x, y) = 1 + x$

34. $\mathbf{r}(t) = 3\mathbf{i} + t^2\mathbf{j} + 2t\mathbf{k}$,  $2 \le t \le 4$,  $\rho(x, y, z) = xz$

**Evaluating a Line Integral of a Vector Field**  In Exercises 35–38, evaluate $\displaystyle\int_C \mathbf{F} \cdot d\mathbf{r}$.

35. $\mathbf{F}(x, y) = xy\mathbf{i} + 2xy\mathbf{j}$

$C$: $\mathbf{r}(t) = t^2\mathbf{i} + t^2\mathbf{j}$,  $0 \le t \le 1$

36. $\mathbf{F}(x, y) = (x - y)\mathbf{i} + (x + y)\mathbf{j}$

$C$: $\mathbf{r}(t) = 4 \cos t\mathbf{i} + 3 \sin t\mathbf{j}$,  $0 \le t \le 2\pi$

37. $\mathbf{F}(x, y, z) = x\mathbf{i} + y\mathbf{j} + z\mathbf{k}$

$C$: $\mathbf{r}(t) = 2 \cos t\mathbf{i} + 2 \sin t\mathbf{j} + t\mathbf{k}$,  $0 \le t \le 2\pi$

38. $\mathbf{F}(x, y, z) = (2y - z)\mathbf{i} + (z - x)\mathbf{j} + (x - y)\mathbf{k}$

$C$: $\mathbf{r}(t) = -3t\mathbf{i} + (2t + 1)\mathbf{j} + 4\mathbf{k}$,  $0 \le t \le 2$

**Work** In Exercises 39 and 40, find the work done by the force field F on a particle moving along the given path.

**39.** $\mathbf{F}(x, y) = x\mathbf{i} - \sqrt{y}\mathbf{j}$

   $C$: $x = t$, $y = t^{3/2}$ from $(0, 0)$ to $(4, 8)$

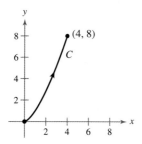

Figure for 39

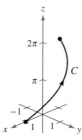

Figure for 40

**40.** $\mathbf{F}(x, y, z) = 2\mathbf{i} + y\mathbf{j} + z\mathbf{k}$

   $C$: $\mathbf{r}(t) = \cos t\mathbf{i} + \sin t\mathbf{j} + 2t\mathbf{k}$, $\quad 0 \le t \le \pi$

**Evaluating a Line Integral in Differential Form** In Exercises 41 and 42, evaluate $\displaystyle\int_C (y - x)\, dx + (2x + 5y)\, dy$.

**41.** $C$: line segments from $(0, 0)$ to $(2, -4)$ and $(2, -4)$ to $(4, -4)$

**42.** $C$: arc on $y = \sqrt{x}$ from $(0, 0)$ to $(9, 3)$

**Lateral Surface Area** In Exercises 43 and 44, find the area of the lateral surface over the curve $C$ in the $xy$-plane and under the surface $z = f(x, y)$, where

Lateral surface area $= \displaystyle\int_C f(x, y)\, ds.$

**43.** $f(x, y) = 3 + \sin(x + y)$; $C$: $y = 2x$ from $(0, 0)$ to $(2, 4)$

**44.** $f(x, y) = 12 - x - y$; $C$: $y = x^2$ from $(0, 0)$ to $(2, 4)$

**Line Integral of a Conservative Vector Field** In Exercises 45 and 46, (a) show that F is conservative and (b) verify that the value of $\displaystyle\int_C \mathbf{F} \cdot d\mathbf{r}$ is the same for each parametric representation of $C$.

**45.** $\mathbf{F}(x, y) = (3x + 4)\mathbf{i} + y^3\mathbf{j}$

   (i) $\mathbf{r}(t) = t\mathbf{i} + t\mathbf{j}$, $0 \le t \le 4$

   (ii) $\mathbf{r}(w) = w^2\mathbf{i} + w^2\mathbf{j}$, $0 \le w \le 2$

**46.** $\mathbf{F}(x, y) = xy\mathbf{i} + \frac{1}{2}x^2\mathbf{j}$

   (i) $\mathbf{r}(\theta) = \sin\theta\mathbf{i} + \cos\theta\mathbf{j}$, $0 \le \theta \le \dfrac{\pi}{2}$

   (ii) $\mathbf{r}(t) = t\mathbf{i} + (1 - t)\mathbf{j}$, $0 \le t \le 1$

**Using the Fundamental Theorem of Line Integrals** In Exercises 47–50, evaluate $\displaystyle\int_C \mathbf{F} \cdot d\mathbf{r}$ using the Fundamental Theorem of Line Integrals.

**47.** $\mathbf{F}(x, y) = e^{2x}\mathbf{i} + e^{2y}\mathbf{j}$

   $C$: line segment from $(-1, -1)$ to $(0, 0)$

**48.** $\mathbf{F}(x, y) = -\sin y\mathbf{i} - x\cos y\mathbf{j}$

   $C$: clockwise around the circle $(x + 1)^2 + y^2 = 16$ from $(-1, 4)$ to $(3, 0)$

**49.** $\mathbf{F}(x, y, z) = 2xyz\mathbf{i} + x^2z\mathbf{j} + x^2y\mathbf{k}$

   $C$: smooth curve from $(0, 0, 0)$ to $(1, 3, 2)$

**50.** $\mathbf{F}(x, y, z) = y\mathbf{i} + x\mathbf{j} + \dfrac{1}{z}\mathbf{k}$

   $C$: smooth curve from $(0, 0, 1)$ to $(4, 4, 4)$

**Finding Work in a Conservative Force Field** In Exercises 51 and 52, (a) show that $\displaystyle\int_C \mathbf{F} \cdot d\mathbf{r}$ is independent of path and (b) calculate the work done by the force field F on an object moving along a curve from $P$ to $Q$.

**51.** $\mathbf{F}(x, y) = (1 - 3xy^2)\mathbf{i} - 3x^2y\mathbf{j}$; $\quad P(4, 2)$, $Q(0, 1)$

**52.** $\mathbf{F}(x, y) = e^{2y}\mathbf{i} + 2xe^{2y}\mathbf{j}$; $\quad P(-1, 3)$, $Q(4, 5)$

**Evaluating a Line Integral Using Green's Theorem** In Exercises 53–58, use Green's Theorem to evaluate the line integral.

**53.** $\displaystyle\int_C y\, dx + 2x\, dy$

   $C$: square with vertices $(0, 0)$, $(0, 1)$, $(1, 0)$, and $(1, 1)$

**54.** $\displaystyle\int_C xy\, dx + (x^2 + y^2)\, dy$

   $C$: square with vertices $(0, 0)$, $(0, 2)$, $(2, 0)$, and $(2, 2)$

**55.** $\displaystyle\int_C xy^2\, dx + x^2y\, dy$

   $C$: $x = 4\cos t$, $y = 4\sin t$

**56.** $\displaystyle\int_C (x^2 - y^2)\, dx + 3y^2\, dy$

   $C$: $x^2 + y^2 = 9$

**57.** $\displaystyle\int_C xy\, dx + x^2\, dy$

   $C$: boundary of the region between the graphs of $y = x^2$ and $y = 1$

**58.** $\displaystyle\int_C y^2\, dx + x^{4/3}\, dy$

   $C$: $x^{2/3} + y^{2/3} = 1$

**Work** In Exercises 59 and 60, use Green's Theorem to calculate the work done by the force F on a particle that is moving counterclockwise around the closed path $C$.

**59.** $\mathbf{F}(x, y) = y^2\mathbf{i} + 2xy\mathbf{j}$

   $C$: $x^2 + y^2 = 36$

**60.** $\mathbf{F}(x, y) = 3\mathbf{i} + (x^3 + 1)\mathbf{j}$

   $C$: boundary of the region lying between the graphs of $y = x^2$ and $y = 4$

**Area** In Exercises 61 and 62, use a line integral to find the area of the region $R$.

**61.** $R$: triangle bounded by the graphs of $y = \frac{1}{2}x$, $y = 6 - x$, and $y = x$

**62.** $R$: region bounded by the graphs of $y = 3x$ and $y = 4 - x^2$

**Sketching a Parametric Surface** In Exercises 63 and 64, find the rectangular equation for the surface by eliminating the parameters from the vector-valued function. Identify the surface and sketch its graph.

**63.** $\mathbf{r}(u, v) = 3u \cos v\mathbf{i} + 3u \sin v\mathbf{j} + 18u^2\mathbf{k}$

**64.** $\mathbf{r}(u, v) = 3(u + v)\mathbf{i} + u\mathbf{j} - 6v\mathbf{k}$

**Graphing a Parametric Surface** In Exercises 65 and 66, use a computer algebra system to graph the surface represented by the vector-valued function.

**65.** $\mathbf{r}(u, v) = \sec u \cos v\mathbf{i} + (1 + 2 \tan u) \sin v\mathbf{j} + 2u\mathbf{k}$

$0 \le u \le \dfrac{\pi}{3}, \quad 0 \le v \le 2\pi$

**66.** $\mathbf{r}(u, v) = e^{-u/4} \cos v\mathbf{i} + e^{-u/4} \sin v\mathbf{j} + \dfrac{u}{6}\mathbf{k}$

$0 \le u \le 4, \quad 0 \le v \le 2\pi$

**Representing a Surface Parametrically** In Exercises 67 and 68, find a vector-valued function whose graph is the indicated surface.

**67.** The ellipsoid $\dfrac{x^2}{1} + \dfrac{y^2}{8} + \dfrac{z^2}{9} = 1$

**68.** The part of the plane $z = 2$ that lies inside the cylinder $x^2 + y^2 = 25$

**Representing a Surface of Revolution Parametrically** In Exercises 69 and 70, write a set of parametric equations for the surface of revolution obtained by revolving the graph of the function about the given axis.

| Function | Axis of Revolution |
|---|---|
| **69.** $y = 2x^3, 0 \le x \le 2$ | $x$-axis |
| **70.** $z = \sqrt{y + 1}, 0 \le y \le 3$ | $y$-axis |

**Finding Surface Area** In Exercises 71 and 72, find the area of the surface over the given region. Use a computer algebra system to verify your results.

**71.** $\mathbf{r}(u, v) = 4u\mathbf{i} + (3u - v)\mathbf{j} + v\mathbf{k}$

$0 \le u \le 3, \quad 0 \le v \le 1$

**72.** $\mathbf{r}(u, v) = 3u \cos v\mathbf{i} + 3u \sin v\mathbf{j} + u\mathbf{k}$

$0 \le u \le 2, \quad 0 \le v \le 2\pi$

**Evaluating a Surface Integral** In Exercises 73 and 74, evaluate

$$\iint_S (5x + y - 2z) \, dS.$$

**73.** $S: z = x + \dfrac{y}{2}, \quad 0 \le x \le 2, \quad 0 \le y \le 5$

**74.** $S: z = e^2 - x, \quad 0 \le x \le 4, \quad 0 \le y \le \sqrt{x}$

**Mass** In Exercises 75 and 76, find the mass of the surface lamina $S$ of density $\rho$.

**75.** $S: 2y + 6x + z = 18$, first octant, $\rho(x, y, z) = 2x$

**76.** $S: z = 20 - 4x - 5y$, first octant, $\rho(x, y, z) = ky$

**Evaluating a Surface Integral** In Exercises 77 and 78, evaluate $\displaystyle\iint_S f(x, y) \, dS.$

**77.** $f(x, y) = x + y$

$S: \mathbf{r}(u, v) = u\mathbf{i} + v\mathbf{j} + 5v\mathbf{k}, \quad 0 \le u \le 1, \quad 0 \le v \le 3$

**78.** $f(x, y) = x^2y$

$S: \mathbf{r}(u, v) = 5 \cos u\mathbf{i} + 5 \sin u\mathbf{j} + v\mathbf{k}$

$0 \le u \le \dfrac{\pi}{2}, \quad 0 \le v \le 1$

**Evaluating a Flux Integral** In Exercises 79 and 80, find the flux of F across $S$,

$$\iint_S \mathbf{F} \cdot \mathbf{N} \, dS$$

where N is the upward unit normal vector to $S$.

**79.** $\mathbf{F}(x, y, z) = -2\mathbf{i} - 2\mathbf{j} + \mathbf{k}$

$S: z = 25 - x^2 - y^2, \quad z \ge 0$

**80.** $\mathbf{F}(x, y, z) = x\mathbf{i} + 2y\mathbf{j} + 2z\mathbf{k}$

$S: x + y + 3z = 3$, first octant

**Using the Divergence Theorem** In Exercises 81 and 82, use the Divergence Theorem to evaluate

$$\iint_S \mathbf{F} \cdot \mathbf{N} \, dS$$

and find the outward flux of F through the surface of the solid bounded by the graphs of the equations.

**81.** $\mathbf{F}(x, y, z) = x^2\mathbf{i} + xy\mathbf{j} + z\mathbf{k}$

$Q$: solid region bounded by the coordinate planes and the plane $2x + 3y + 4z = 12$

**82.** $\mathbf{F}(x, y, z) = x\mathbf{i} + y\mathbf{j} + z\mathbf{k}$

$Q$: solid region bounded by the coordinate planes and the plane $2x + 3y + 4z = 12$

**Using Stokes's Theorem** In Exercises 83 and 84, use Stokes's Theorem to evaluate

$$\int_C \mathbf{F} \cdot d\mathbf{r}.$$

In each case, $C$ is oriented counterclockwise as viewed from above.

**83.** $\mathbf{F}(x, y, z) = (\cos y + y \cos x)\mathbf{i} + (\sin x - x \sin y)\mathbf{j} + xyz\mathbf{k}$

$S$: portion of $z = y^2$ over the square in the $xy$-plane with vertices $(0, 0)$, $(a, 0)$, $(a, a)$, and $(0, a)$

**84.** $\mathbf{F}(x, y, z) = (x - z)\mathbf{i} + (y - z)\mathbf{j} + x^2\mathbf{k}$

$S$: first-octant portion of the plane $3x + y + 2z = 12$

**Motion of a Liquid** In Exercises 85 and 86, the motion of a liquid in a cylindrical container of radius 4 is described by the velocity field $\mathbf{F}(x, y, z)$. Find $\displaystyle\iint_S (\text{curl } \mathbf{F}) \cdot \mathbf{N} \, dS$, where $S$ is the upper surface of the cylindrical container.

**85.** $\mathbf{F}(x, y, z) = \mathbf{i} + x\mathbf{j} - \mathbf{k}$ **86.** $\mathbf{F}(x, y, z) = y^2\mathbf{i} + 3z\mathbf{j} + \mathbf{k}$

# P.S. Problem Solving

See **CalcChat.com** for tutorial help and worked-out solutions to odd-numbered exercises.

**1. Heat Flux**   Consider a single heat source located at the origin with temperature

$$T(x, y, z) = \frac{25}{\sqrt{x^2 + y^2 + z^2}}.$$

(a) Calculate the heat flux across the surface

$$S = \left\{ (x, y, z): z = \sqrt{1 - x^2}, -\frac{1}{2} \le x \le \frac{1}{2}, 0 \le y \le 1 \right\}$$

as shown in the figure.

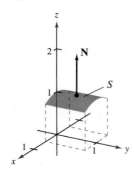

(b) Repeat the calculation in part (a) using the parametrization

$$x = \cos u, \quad y = v, \quad z = \sin u$$

where

$$\frac{\pi}{3} \le u \le \frac{2\pi}{3} \quad \text{and} \quad 0 \le v \le 1.$$

**2. Heat Flux**   Consider a single heat source located at the origin with temperature

$$T(x, y, z) = \frac{25}{\sqrt{x^2 + y^2 + z^2}}.$$

(a) Calculate the heat flux across the surface

$$S = \left\{ (x, y, z): z = \sqrt{1 - x^2 - y^2}, x^2 + y^2 \le 1 \right\}$$

as shown in the figure.

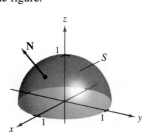

(b) Repeat the calculation in part (a) using the parametrization

$$x = \sin u \cos v, \quad y = \sin u \sin v, \quad z = \cos u$$

where

$$0 \le u \le \frac{\pi}{2} \quad \text{and} \quad 0 \le v \le 2\pi.$$

**3. Moments of Inertia**   Consider a wire of density $\rho(x, y, z)$ given by the space curve

$$C: \mathbf{r}(t) = x(t)\mathbf{i} + y(t)\mathbf{j} + z(t)\mathbf{k}, \quad a \le t \le b.$$

The **moments of inertia** about the $x$-, $y$-, and $z$-axes are given by

$$I_x = \int_C (y^2 + z^2)\rho(x, y, z) \, ds$$

$$I_y = \int_C (x^2 + z^2)\rho(x, y, z) \, ds$$

$$I_z = \int_C (x^2 + y^2)\rho(x, y, z) \, ds.$$

Find the moments of inertia for a wire of uniform density $\rho = 1$ in the shape of the helix

$$\mathbf{r}(t) = 3 \cos t\mathbf{i} + 3 \sin t\mathbf{j} + 2t\mathbf{k}, \quad 0 \le t \le 2\pi \text{ (see figure)}.$$

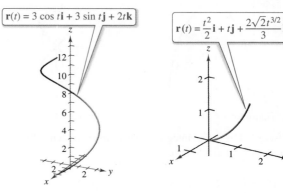

Figure for 3          Figure for 4

**4. Moments of Inertia**   Using the formulas from Exercise 3, find the moments of inertia for a wire of density $\rho = \dfrac{1}{1 + t}$ given by the curve

$$C: \mathbf{r}(t) = \frac{t^2}{2}\mathbf{i} + t\mathbf{j} + \frac{2\sqrt{2}t^{3/2}}{3}\mathbf{k}, \quad 0 \le t \le 1 \text{ (see figure)}.$$

**5. Laplace's Equation**   Let $\mathbf{F}(x, y, z) = x\mathbf{i} + y\mathbf{j} + z\mathbf{k}$, and let $f(x, y, z) = \|\mathbf{F}(x, y, z)\|$.

(a) Show that $\nabla(\ln f) = \dfrac{\mathbf{F}}{f^2}$.

(b) Show that $\nabla\left(\dfrac{1}{f}\right) = -\dfrac{\mathbf{F}}{f^3}$.

(c) Show that $\nabla f^n = nf^{n-2}\mathbf{F}$.

(d) The **Laplacian** is the differential operator

$$\nabla^2 = \nabla \cdot \nabla = \frac{\partial^2}{\partial x^2} + \frac{\partial^2}{\partial y^2} + \frac{\partial^2}{\partial z^2}$$

and **Laplace's equation** is

$$\nabla^2 w = \frac{\partial^2 w}{\partial x^2} + \frac{\partial^2 w}{\partial y^2} + \frac{\partial^2 w}{\partial z^2} = 0.$$

Any function that satisfies this equation is called **harmonic**. Show that the function $w = 1/f$ is harmonic.

**6. Green's Theorem** Consider the line integral

$$\int_C y^n \, dx + x^n \, dy$$

where $C$ is the boundary of the region lying between the graphs of $y = \sqrt{a^2 - x^2}$, $a > 0$, and $y = 0$.

(a) Use a computer algebra system to verify Green's Theorem for $n$, an odd integer from 1 through 7.

(b) Use a computer algebra system to verify Green's Theorem for $n$, an even integer from 2 through 8.

(c) For $n$ an odd integer, make a conjecture about the value of the integral.

**7. Area** Use a line integral to find the area bounded by one arch of the cycloid $x(\theta) = a(\theta - \sin \theta)$, $y(\theta) = a(1 - \cos \theta)$, $0 \le \theta \le 2\pi$, as shown in the figure.

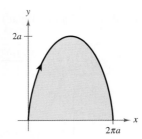

Figure for 7          Figure for 8

**8. Area** Use a line integral to find the area bounded by the two loops of the eight curve

$$x(t) = \frac{1}{2}\sin 2t, \quad y(t) = \sin t, \quad 0 \le t \le 2\pi$$

as shown in the figure.

**9. Work** The force field $\mathbf{F}(x, y) = (x + y)\mathbf{i} + (x^2 + 1)\mathbf{j}$ acts on an object moving from the point $(0, 0)$ to the point $(0, 1)$, as shown in the figure.

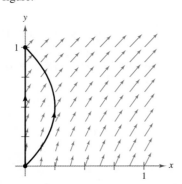

(a) Find the work done when the object moves along the path $x = 0, 0 \le y \le 1$.

(b) Find the work done when the object moves along the path $x = y - y^2, 0 \le y \le 1$.

(c) The object moves along the path $x = c(y - y^2), 0 \le y \le 1$, $c > 0$. Find the value of the constant $c$ that minimizes the work.

**10. Work** The force field $\mathbf{F}(x, y) = 3x^2y^2\mathbf{i} + 2x^3y\mathbf{j}$ is shown in the figure below. Three particles move from the point $(1, 1)$ to the point $(2, 4)$ along different paths. Explain why the work done is the same for each particle and find the value of the work.

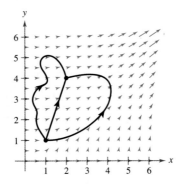

**11. Area and Work** How does the area of the ellipse

$$\frac{x^2}{a^2} + \frac{y^2}{b^2} = 1$$

compare with the magnitude of the work done by the force field

$$\mathbf{F}(x, y) = -\frac{1}{2}y\mathbf{i} + \frac{1}{2}x\mathbf{j}$$

on a particle that moves once around the ellipse (see figure)?

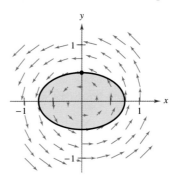

**12. Verifying Identities**

(a) Let $f$ and $g$ be scalar functions with continuous partial derivatives, and let $C$ and $S$ satisfy the conditions of Stokes's Theorem. Verify each identity.

(i) $\displaystyle\int_C (f\nabla g) \cdot d\mathbf{r} = \iint_S (\nabla f \times \nabla g) \cdot \mathbf{N} \, dS$

(ii) $\displaystyle\int_C (f\nabla f) \cdot d\mathbf{r} = 0$

(iii) $\displaystyle\int_C (f\nabla g + g\nabla f) \cdot d\mathbf{r} = 0$

(b) Demonstrate the results of part (a) for the functions

$$f(x, y, z) = xyz \quad \text{and} \quad g(x, y, z) = z.$$

Let $S$ be the hemisphere $z = \sqrt{4 - x^2 - y^2}$.

# 16 Additional Topics in Differential Equations

■ ■ ■ ■ ■ ■ ■ ■ ■ ■ ■ ■ ■ ■ ■

Parachute Jump
*(Section Project, p. 1152)*

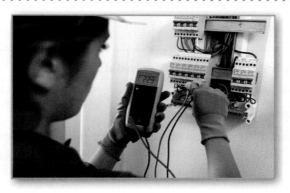

Electrical Circuits *(Exercises 29 and 30, p. 1151)*

Undamped or Damped Motion? *(Exercise 53, p. 1144)*

Motion of a Spring
*(Example 8, p. 1142)*

Cost *(Exercise 45, p. 1136)*

# 16.1    Exact First-Order Equations

■ Solve an exact differential equation.
■ Use an integrating factor to make a differential equation exact.

## Exact Differential Equations

In Chapter 6, you studied applications of differential equations to growth and decay problems. You also learned more about the basic ideas of differential equations and studied the solution technique known as separation of variables. In this chapter, you will learn more about solving differential equations and using them in real-life applications. This section introduces you to a method for solving the first-order differential equation

$$M(x, y) \, dx + N(x, y) \, dy = 0$$

for the special case in which this equation represents the exact differential of a function $z = f(x, y)$.

---

**Definition of an Exact Differential Equation**

The equation

$$M(x, y) \, dx + N(x, y) \, dy = 0$$

is an **exact differential equation** when there exists a function $f$ of two variables $x$ and $y$ having continuous partial derivatives such that

$$f_x(x, y) = M(x, y) \quad \text{and} \quad f_y(x, y) = N(x, y).$$

The general solution of the equation is $f(x, y) = C$.

---

From Section 13.3, you know that if $f$ has continuous second partials, then

$$\frac{\partial M}{\partial y} = \frac{\partial^2 f}{\partial y \partial x} = \frac{\partial^2 f}{\partial x \partial y} = \frac{\partial N}{\partial x}.$$

This suggests the following test for exactness.

---

**THEOREM 16.1    Test for Exactness**

Let $M$ and $N$ have continuous partial derivatives on an open disk $R$. The differential equation

$$M(x, y) \, dx + N(x, y) \, dy = 0$$

is exact if and only if

$$\frac{\partial M}{\partial y} = \frac{\partial N}{\partial x}.$$

---

Every differential equation of the form

$$M(x) \, dx + N(y) \, dy = 0$$

is exact. In other words, a separable differential equation is actually a special type of an exact equation.

Exactness is a fragile condition in the sense that seemingly minor alterations in an exact equation can destroy its exactness. This is demonstrated in the next example.

EXAMPLE 1　Testing for Exactness

Determine whether each differential equation is exact.

**a.** $(xy^2 + x) dx + x^2y dy = 0$　　**b.** $\cos y\, dx + (y^2 - x \sin y) dy = 0$

**Solution**

**a.** This differential equation is exact because

$$\frac{\partial M}{\partial y} = \frac{\partial}{\partial y}[xy^2 + x] = 2xy \quad \text{and} \quad \frac{\partial N}{\partial x} = \frac{\partial}{\partial x}[x^2y] = 2xy.$$

Notice that the equation $(y^2 + 1) dx + xy\, dy = 0$ is not exact, even though it is obtained by dividing each side of the first equation by $x$.

**b.** This differential equation is exact because

$$\frac{\partial M}{\partial y} = \frac{\partial}{\partial y}[\cos y] = -\sin y \quad \text{and} \quad \frac{\partial N}{\partial x} = \frac{\partial}{\partial x}[y^2 - x \sin y] = -\sin y.$$

Notice that the equation $\cos y\, dx + (y^2 + x \sin y) dy = 0$ is not exact, even though it differs from the first equation only by a single sign. ∎

Note that the test for exactness of $M(x, y) dx + N(x, y) dy = 0$ is the same as the test for determining whether $\mathbf{F}(x, y) = M(x, y)\mathbf{i} + N(x, y)\mathbf{j}$ is the gradient of a potential function (Theorem 15.1). This means that a general solution $f(x, y) = C$ to an exact differential equation can be found by the method used to find a potential function for a conservative vector field.

EXAMPLE 2　Solving an Exact Differential Equation

⋮
•••▷ *See LarsonCalculus.com for an interactive version of this type of example.*

Solve the differential equation $(2xy - 3x^2) dx + (x^2 - 2y) dy = 0$

**Solution**　This differential equation is exact because

$$\frac{\partial M}{\partial y} = \frac{\partial}{\partial y}[2xy - 3x^2] = 2x \quad \text{and} \quad \frac{\partial N}{\partial x} = \frac{\partial}{\partial x}[x^2 - 2y] = 2x.$$

The general solution, $f(x, y) = C$, is

$$f(x, y) = \int M(x, y)\, dx = \int (2xy - 3x^2)\, dx = x^2y - x^3 + g(y).$$

In Section 15.1, you determined $g(y)$ by integrating $N(x, y)$ with respect to $y$ and reconciling the two expressions for $f(x, y)$. An alternative method is to partially differentiate this version of $f(x, y)$ with respect to $y$ and compare the result with $N(x, y)$. In other words,

$$f_y(x, y) = \frac{\partial}{\partial y}[x^2y - x^3 + g(y)] = x^2 + g'(y) = \overbrace{x^2 - 2y}^{N(x, y)}.$$

$$\boxed{g'(y) = -2y}$$

So, $g'(y) = -2y$, and it follows that $g(y) = -y^2 + C_1$. Therefore,

$$f(x, y) = x^2y - x^3 - y^2 + C_1$$

and the general solution is $x^2y - x^3 - y^2 = C$. ∎

EXAMPLE 3  **Solving an Exact Differential Equation**

Find the particular solution of $(\cos x - x \sin x + y^2)\, dx + 2xy\, dy = 0$ that satisfies the initial condition $y = 1$ when $x = \pi$.

**Solution**  The differential equation is exact because

$$\underbrace{\frac{\partial}{\partial y}[\cos x - x \sin x + y^2]}_{\frac{\partial M}{\partial y}} = 2y = \underbrace{\frac{\partial}{\partial x}[2xy]}_{\frac{\partial N}{\partial x}}.$$

Because $N(x, y)$ is simpler than $M(x, y)$, it is better to begin by integrating $N(x, y)$.

$$f(x, y) = \int N(x, y)\, dy = \int 2xy\, dy = xy^2 + g(x)$$

Next, find $f_x(x, y)$ and compare the result with $M(x, y)$.

$$f_x(x, y) = \frac{\partial}{\partial x}[xy^2 + g(x)] = y^2 + g'(x) = \overbrace{\cos x - x \sin x + y^2}^{M(x, y)}$$

$$\boxed{g'(x) = \cos x - x \sin x}$$

So, $g'(x) = \cos x - x \sin x$ and it follows that

$$g(x) = \int (\cos x - x \sin x)\, dx$$
$$= x \cos x + C_1.$$

This implies that $f(x, y) = xy^2 + x \cos x + C_1$, and the general solution is

$$xy^2 + x \cos x = C. \qquad \text{General solution}$$

Applying the given initial condition produces

$$\pi(1)^2 + \pi \cos \pi = C$$

which implies that $C = 0$. So, the particular solution is

$$xy^2 + x \cos x = 0.$$

The graph of the particular solution is shown in Figure 16.2. Notice that the graph consists of two parts: the ovals are given by $y^2 + \cos x = 0$, and the $y$-axis is given by $x = 0$.

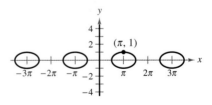

**Figure 16.2**

▷ **TECHNOLOGY**  A graphing utility can be used to graph a particular solution that satisfies the initial condition of a differential equation. In Example 3, the differential equation and initial condition are satisfied when $xy^2 + x \cos x = 0$, which implies that the particular solution can be written as $x = 0$ or $y = \pm\sqrt{-\cos x}$. On a graphing utility screen, the solution would be represented by Figure 16.1 together with the $y$-axis.

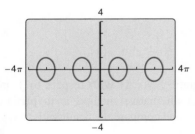

**Figure 16.1**

In Example 3, note that for $z = f(x, y) = xy^2 + x \cos x$, the total differential of $z$ is given by

$$dz = f_x(x, y)\, dx + f_y(x, y)\, dy$$
$$= (\cos x - x \sin x + y^2)\, dx + 2xy\, dy$$
$$= M(x, y)\, dx + N(x, y)\, dy.$$

In other words, $M\, dx + N\, dy = 0$ is called an *exact* differential equation because $M\, dx + N\, dy$ is exactly the differential of $f(x, y)$.

## Integrating Factors

When the differential equation $M(x, y) \, dx + N(x, y) \, dy = 0$ is not exact, it may be possible to make it exact by multiplying by an appropriate factor $u(x, y)$, which is called an **integrating factor** for the differential equation.

**EXAMPLE 4** **Multiplying by an Integrating Factor**

**a.** When the differential equation

$$2y \, dx + x \, dy = 0 \qquad \text{Not an exact equation}$$

is multiplied by the integrating factor $u(x, y) = x$, the resulting equation

$$2xy \, dx + x^2 \, dy = 0 \qquad \text{Exact equation}$$

is exact—the left side is the total differential of $x^2 y$.

**b.** When the equation

$$y \, dx - x \, dy = 0 \qquad \text{Not an exact equation}$$

is multiplied by the integrating factor $u(x, y) = 1/y^2$, the resulting equation

$$\frac{1}{y} \, dx - \frac{x}{y^2} \, dy = 0 \qquad \text{Exact equation}$$

is exact—the left side is the total differential of $x/y$. ∎

Finding an integrating factor can be difficult. There are two classes of differential equations, however, whose integrating factors can be found routinely—namely, those that possess integrating factors that are functions of either $x$ alone or $y$ alone. The next theorem, which is presented without proof, outlines a procedure for finding these two special categories of integrating factors.

• • • • • • • • • • • • • • • • ▷

• • **REMARK** When either $h(x)$ or $k(y)$ is constant, Theorem 16.2 still applies. As an aid to remembering these formulas, note that the subtracted partial derivative identifies both the denominator and the variable for the integrating factor.

**THEOREM 16.2 Integrating Factors**

Consider the differential equation $M(x, y) \, dx + N(x, y) \, dy = 0$.

**1.** If

$$\frac{1}{N(x, y)}[M_y(x, y) - N_x(x, y)] = h(x)$$

is a function of $x$ alone, then $e^{\int h(x) \, dx}$ is an integrating factor.

**2.** If

$$\frac{1}{M(x, y)}[N_x(x, y) - M_y(x, y)] = k(y)$$

is a function of $y$ alone, then $e^{\int k(y) \, dy}$ is an integrating factor.

**Exploration**

In Chapter 6, you solved the first-order linear differential equation

$$\frac{dy}{dx} + P(x)y = Q(x)$$

by using the integrating factor $u(x) = e^{\int P(x) \, dx}$. Show that you can obtain this integrating factor by using the methods of this section.

EXAMPLE 5    **Finding an Integrating Factor**

Solve the differential equation $(y^2 - x) \, dx + 2y \, dy = 0$.

**Solution**    This equation is not exact because

$$M_y(x, y) = 2y \quad \text{and} \quad N_x(x, y) = 0.$$

However, because

$$\frac{M_y(x, y) - N_x(x, y)}{N(x, y)} = \frac{2y - 0}{2y} = 1 = h(x)$$

it follows that $e^{\int h(x) \, dx} = e^{\int dx} = e^x$ is an integrating factor. Multiplying the differential equation by $e^x$ produces the exact differential equation

$$(y^2 e^x - xe^x) \, dx + 2ye^x \, dy = 0.$$

Next, integrate $N(x, y)$, as shown.

$$f(x, y) = \int N(x, y) \, dy = \int 2ye^x \, dy = y^2 e^x + g(x)$$

Now, find $f_x(x, y)$ and compare the result with $M(x, y)$.

$$f_x(x, y) = y^2 e^x + g'(x) = \overbrace{y^2 e^x - xe^x}^{M(x, y)}$$

$$g'(x) = -xe^x$$

Therefore, $g'(x) = -xe^x$ and $g(x) = -xe^x + e^x + C_1$, which implies that

$$f(x, y) = y^2 e^x - xe^x + e^x + C_1.$$

The general solution is $y^2 e^x - xe^x + e^x = C$, or

$$y^2 - x + 1 = Ce^{-x}. \qquad \text{General solution}$$

The next example shows how a differential equation can help in sketching a force field given by $\mathbf{F}(x, y) = M(x, y)\mathbf{i} + N(x, y)\mathbf{j}$.

EXAMPLE 6    **An Application to Force Fields**

Sketch the force field

$$\mathbf{F}(x, y) = \frac{2y}{\sqrt{x^2 + y^2}}\mathbf{i} - \frac{y^2 - x}{\sqrt{x^2 + y^2}}\mathbf{j}$$

by finding and sketching the family of curves tangent to $\mathbf{F}$.

**Solution**    At the point $(x, y)$ in the plane, the vector $\mathbf{F}(x, y)$ has a slope of

$$\frac{dy}{dx} = \frac{-(y^2 - x)/\sqrt{x^2 + y^2}}{2y/\sqrt{x^2 + y^2}} = \frac{-(y^2 - x)}{2y}$$

which, in differential form, is

$$2y \, dy = -(y^2 - x) \, dx$$

$$(y^2 - x) \, dx + 2y \, dy = 0.$$

From Example 5, you know that the general solution of this differential equation is $y^2 = x - 1 + Ce^{-x}$. Figure 16.3 shows several representative curves from this family. Note that the force vector at $(x, y)$ is tangent to the curve passing through $(x, y)$.

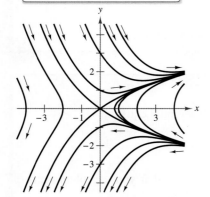

Force field:
$$\mathbf{F}(x, y) = \frac{2y}{\sqrt{x^2 + y^2}}\mathbf{i} - \frac{y^2 - x}{\sqrt{x^2 + y^2}}\mathbf{j}$$
Family of curves tangent to $\mathbf{F}$:
$$y^2 = x - 1 + Ce^{-x}$$

**Figure 16.3**

# 16.1  Exercises

See **CalcChat.com** for tutorial help and worked-out solutions to odd-numbered exercises.

## CONCEPT CHECK

**1. Exactness**  What does it mean for the differential equation $M(x, y)\,dx + N(x, y)\,dy = 0$ to be exact? Explain how to determine whether this differential equation is exact.

**2. Integrating Factor**  When is it beneficial to use an integrating factor to find the solution of the differential equation $M(x, y)\,dx + N(x, y)\,dy = 0$?

 **Testing for Exactness**  In Exercises 3–6, determine whether the differential equation is exact.

**3.** $(2x + xy^2)\,dx + (3 + x^2y)\,dy = 0$

**4.** $(2xy - y)\,dx + (x^2 - xy)\,dy = 0$

**5.** $x \sin y\,dx + x \cos y\,dy = 0$    **6.** $ye^{xy}\,dx + xe^{xy}\,dy = 0$

 **Solving an Exact Differential Equation**  In Exercises 7–14, verify that the differential equation is exact. Then find the general solution.

**7.** $(2x - 3y)\,dx + (2y - 3x)\,dy = 0$

**8.** $ye^x\,dx + e^x\,dy = 0$

**9.** $(3y^2 + 10xy^2)\,dx + (6xy - 2 + 10x^2y)\,dy = 0$

**10.** $2 \cos(2x - y)\,dx - \cos(2x - y)\,dy = 0$

**11.** $\dfrac{1}{x^2 + y^2}(x\,dy - y\,dx) = 0$    **12.** $e^{-(x^2+y^2)}(x\,dx + y\,dy) = 0$

**13.** $\dfrac{x}{y^2}\,dx - \dfrac{x^2}{y^3}\,dy = 0$

**14.** $(e^y \cos xy)[y\,dx + (x + \tan xy)\,dy] = 0$

**Graphical and Analytic Analysis**  In Exercises 15 and 16, (a) sketch an approximate solution of the differential equation satisfying the initial condition on the slope field, (b) find the particular solution that satisfies the initial condition, and (c) use a graphing utility to graph the particular solution. Compare the graph with the sketch in part (a).

| Differential Equation | Initial Condition |
|---|---|
| **15.** $(2x \tan y + 5)\,dx + (x^2 \sec^2 y)\,dy = 0$ | $y\left(\dfrac{1}{2}\right) = \dfrac{\pi}{4}$ |
| **16.** $\dfrac{1}{\sqrt{x^2 + y^2}}(x\,dx + y\,dy) = 0$ | $y(4) = 3$ |

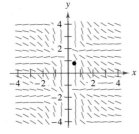

Figure for 15

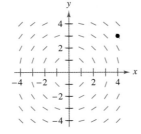

Figure for 16

 **Finding a Particular Solution**  In Exercises 17–22, find the particular solution of the differential equation that satisfies the initial condition.

**17.** $(2xy - 9x^2)\,dx + (2y + x^2 + 1)\,dy = 0,\quad y(0) = -3$

**18.** $(2xy^2 + 4)\,dx + (2x^2y - 6)\,dy = 0,\quad y(-1) = 8$

**19.** $e^{3x}(\sin 3y\,dx + \cos 3y\,dy) = 0,\quad y(0) = \pi$

**20.** $(x^2 + y^2)\,dx + 2xy\,dy = 0,\quad y(3) = 1$

**21.** $\dfrac{y}{x - 1}\,dx + [\ln(x - 1) + 2y]\,dy = 0,\quad y(2) = 4$

**22.** $\dfrac{1}{x^2 + y^2}(x\,dx + y\,dy) = 0,\quad y(0) = 4$

 **Finding an Integrating Factor**  In Exercises 23–32, find the integrating factor that is a function of $x$ or $y$ alone and use it to find the general solution of the differential equation.

**23.** $y^2\,dx + 5xy\,dy = 0$

**24.** $(2x^3 + y)\,dx - x\,dy = 0$

**25.** $y\,dx - (x + 6y^2)\,dy = 0$

**26.** $(5x^2 - y^2)\,dx + 2y\,dy = 0$

**27.** $(x + y)\,dx + \tan x\,dy = 0$

**28.** $(2x^2y - 1)\,dx + x^3\,dy = 0$

**29.** $y^2\,dx + (xy - 1)\,dy = 0$

**30.** $(x^2 + 2x + y)\,dx + 2\,dy = 0$

**31.** $2y\,dx + (x - \sin\sqrt{y})\,dy = 0$

**32.** $(-2y^3 + 1)\,dx + (3xy^2 + x^3)\,dy = 0$

**Using an Integrating Factor**  In Exercises 33–36, use the integrating factor to find the general solution of the differential equation.

| | Integrating Factor | Differential Equation |
|---|---|---|
| **33.** | $u(x, y) = xy^2$ | $(4x^2y + 2y^2)\,dx + (3x^3 + 4xy)\,dy = 0$ |
| **34.** | $u(x, y) = x^2y$ | $(3y^2 + 5x^2y)\,dx + (3xy + 2x^3)\,dy = 0$ |
| **35.** | $u(x, y) = x^{-2}y^{-3}$ | $(-y^5 + x^2y)\,dx + (2xy^4 - 2x^3)\,dy = 0$ |
| **36.** | $u(x, y) = x^{-2}y^{-2}$ | $-y^3\,dx + (xy^2 - x^2)\,dy = 0$ |

**37. Integrating Factor**  Show that each expression is an integrating factor for the differential equation $y\,dx - x\,dy = 0$.

(a) $\dfrac{1}{x^2}$  (b) $\dfrac{1}{y^2}$  (c) $\dfrac{1}{xy}$  (d) $\dfrac{1}{x^2 + y^2}$

**38. Integrating Factor**  Show that the differential equation $(axy^2 + by)\,dx + (bx^2y + ax)\,dy = 0$ is exact only when $a = b$. For $a \neq b$, show that $x^m y^n$ is an integrating factor, where

$$m = -\frac{2b + a}{a + b},\quad n = -\frac{2a + b}{a + b}.$$

**Tangent Curves** In Exercises 39–42, use a graphing utility to graph the family of curves tangent to the force field.

39. $\mathbf{F}(x, y) = \dfrac{y}{\sqrt{x^2 + y^2}}\mathbf{i} - \dfrac{x}{\sqrt{x^2 + y^2}}\mathbf{j}$

40. $\mathbf{F}(x, y) = \dfrac{x}{\sqrt{x^2 + y^2}}\mathbf{i} - \dfrac{y}{\sqrt{x^2 + y^2}}\mathbf{j}$

41. $\mathbf{F}(x, y) = 4x^2 y\mathbf{i} - \left(2xy^2 + \dfrac{x}{y^2}\right)\mathbf{j}$

42. $\mathbf{F}(x, y) = (1 + x^2)\mathbf{i} - 2xy\mathbf{j}$

**Finding an Equation of a Curve** In Exercises 43 and 44, find an equation of the curve with the specified slope passing through the given point.

| Slope | Point |
|---|---|
| 43. $\dfrac{dy}{dx} = \dfrac{y - x}{3y - x}$ | $(2, 1)$ |
| 44. $\dfrac{dy}{dx} = \dfrac{-2xy}{x^2 + y^2}$ | $(-1, 1)$ |

• • **45. Cost** • • • • • • • • • • • • • • • • • • •

In a manufacturing process where $y = C(x)$ represents the cost of producing $x$ units, the **elasticity of cost** is defined as

$E(x) = \dfrac{\text{marginal cost}}{\text{average cost}} = \dfrac{C'(x)}{C(x)/x} = \dfrac{x}{y}\dfrac{dy}{dx}.$

Find the cost function when the elasticity function is

$E(x) = \dfrac{20x - y}{2y - 10x}$

where

$C(100) = 500$

and $x \geq 100.$

**46.** **HOW DO YOU SEE IT?** The graph shows several representative curves from the family of curves tangent to a force field $\mathbf{F}$. Which is the equation of the force field? Explain your reasoning.

(a) $\mathbf{F}(x, y) = -\mathbf{i} + 2\mathbf{j}$  (b) $\mathbf{F}(x, y) = -3x\mathbf{i} + y\mathbf{j}$

(c) $\mathbf{F}(x, y) = e^x\mathbf{i} - \mathbf{j}$  (d) $\mathbf{F}(x, y) = 2\mathbf{i} + e^{-y}\mathbf{j}$

**Euler's Method** In Exercises 47 and 48, (a) use Euler's Method and a graphing utility to graph the particular solution of the differential equation over the indicated interval with the specified value of $h$ and initial condition, (b) find the particular solution of the differential equation analytically, and (c) use a graphing utility to graph the particular solution and compare the result with the graph in part (a).

| Differential Equation | Interval | $h$ | Initial Condition |
|---|---|---|---|
| 47. $y' = \dfrac{-xy}{x^2 + y^2}$ | $[2, 4]$ | 0.05 | $y(2) = 1$ |
| 48. $y' = \dfrac{6x + y^2}{y(3y - 2x)}$ | $[0, 5]$ | 0.2 | $y(0) = 1$ |

49. **Euler's Method** Repeat Exercise 47 for $h = 1$ and discuss how the accuracy of the result changes.

50. **Euler's Method** Repeat Exercise 48 for $h = 0.5$ and discuss how the accuracy of the result changes.

**EXPLORING CONCEPTS**

**Exact Differential Equation** In Exercises 51 and 52, find all values of $k$ such that the differential equation is exact.

51. $(xy^2 + kx^2 y + x^3)\, dx + (x^3 + x^2 y + y^2)\, dy = 0$

52. $(ye^{2xy} + 2x)\, dx + (kxe^{2xy} - 2y)\, dy = 0$

53. **Exact Differential Equation** Find all nonzero functions $f$ and $g$ such that

$g(y) \sin x\, dx + y^2 f(x)\, dy = 0$

is exact.

54. **Exact Differential Equation** Find all nonzero functions $g$ such that

$g(y)e^y\, dx + xy\, dy = 0$

is exact.

**True or False?** In Exercises 55–58, determine whether the statement is true or false. If it is false, explain why or give an example that shows it is false.

55. Every separable equation is an exact equation.

56. Every exact equation is a separable equation.

57. If $M\, dx + N\, dy = 0$ is exact, then

$[f(x) + M]\, dx + [g(y) + N]\, dy = 0$

is also exact.

58. If $M\, dx + N\, dy = 0$ is exact, then

$xM\, dx + xN\, dy = 0$

is also exact.

# 16.2 Second-Order Homogeneous Linear Equations

- Solve a second-order linear differential equation.
- Solve a higher-order linear differential equation.
- Use a second-order linear differential equation to solve an applied problem.

## Second-Order Linear Differential Equations

In this section and the next section, you will learn methods for solving higher-order linear differential equations.

---

**Definition of Linear Differential Equation of Order *n***

Let $g_1, g_2, \ldots, g_n$ and $f$ be functions of $x$ with a common (interval) domain. An equation of the form

$$y^{(n)} + g_1(x)y^{(n-1)} + g_2(x)y^{(n-2)} + \cdots + g_{n-1}(x)y' + g_n(x)y = f(x)$$

is a **linear differential equation of order *n*.** If $f(x) = 0$, then the equation is **homogeneous;** otherwise, it is **nonhomogeneous.**

---

• • **REMARK** Notice that this use of the term *homogeneous* differs from that in Section 6.3.

Homogeneous equations are discussed in this section, and the nonhomogeneous case is discussed in the next section.

The functions $y_1, y_2, \ldots, y_n$ are **linearly independent** when the *only* solution of the equation

$$C_1y_1 + C_2y_2 + \cdots + C_ny_n = 0$$

is the trivial one, $C_1 = C_2 = \cdots = C_n = 0$. Otherwise, this set of functions is **linearly dependent.**

---

**EXAMPLE 1** **Linearly Independent and Dependent Functions**

Determine whether the functions are linearly independent or linearly dependent.

**a.** $y_1(x) = \sin x, y_2(x) = x$      **b.** $y_1(x) = x, y_2(x) = 3x$

**Solution**

**a.** The functions $y_1(x) = \sin x$ and $y_2(x) = x$ are linearly independent because the only values of $C_1$ and $C_2$ for which

$$C_1 \sin x + C_2 x = 0$$

for all $x$ are $C_1 = 0$ and $C_2 = 0$.

**b.** It can be shown that two functions form a linearly dependent set if and only if one is a constant multiple of the other. For example, $y_1(x) = x$ and $y_2(x) = 3x$ are linearly dependent because

$$C_1 x + C_2(3x) = 0$$

has the nonzero solutions $C_1 = -3$ and $C_2 = 1$.

The theorem on the next page points out the importance of linear independence in constructing the general solution of a second-order linear homogeneous differential equation with constant coefficients.

---

### THEOREM 16.3    Linear Combinations of Solutions

If $y_1$ and $y_2$ are linearly independent solutions of the differential equation $y'' + ay' + by = 0$, then the general solution is

$$y = C_1 y_1 + C_2 y_2 \qquad \text{General solution}$$

where $C_1$ and $C_2$ are constants.

---

**Proof**    Letting $y_1$ and $y_2$ be solutions of $y'' + ay' + by = 0$, you obtain the following system of equations.

$$y_1''(x) + ay_1'(x) + by_1(x) = 0$$
$$y_2''(x) + ay_2'(x) + by_2(x) = 0$$

Multiplying the first equation by $C_1$, multiplying the second by $C_2$, and adding the resulting equations together, you obtain

$$[C_1 y_1''(x) + C_2 y_2''(x)] + a[C_1 y_1'(x) + C_2 y_2'(x)] + b[C_1 y_1(x) + C_2 y_2(x)] = 0$$

which means that $y = C_1 y_1 + C_2 y_2$ is a solution, as desired. The proof that all solutions are of this form is best left to a full course on differential equations. ∎

Theorem 16.3 states that when you can find two linearly independent solutions, you can obtain the general solution by forming a **linear combination** of the two solutions.

To find two linearly independent solutions, note that the nature of the equation $y'' + ay' + by = 0$ suggests that it may have solutions of the form $y = e^{mx}$. If so, then

$$y' = me^{mx} \quad \text{and} \quad y'' = m^2 e^{mx}.$$

So, by substitution, $y = e^{mx}$ is a solution if and only if

$$y'' + ay' + by = 0$$
$$m^2 e^{mx} + ame^{mx} + be^{mx} = 0$$
$$e^{mx}(m^2 + am + b) = 0.$$

Because $e^{mx}$ is never 0, $y = e^{mx}$ is a solution if and only if

$$m^2 + am + b = 0. \qquad \text{Characteristic equation}$$

This is the **characteristic equation** of the differential equation $y'' + ay' + by = 0$. Note that the characteristic equation can be determined from its differential equation simply by replacing $y''$ with $m^2$, $y'$ with $m$, and $y$ with 1.

### EXAMPLE 2    Characteristic Equation: Distinct Real Zeros

Solve the differential equation $y'' - 4y = 0$.

**Solution**    In this case, the characteristic equation is

$$m^2 - 4 = 0. \qquad \text{Characteristic equation}$$

So, $m = \pm 2$. Therefore, $y_1 = e^{m_1 x} = e^{2x}$ and $y_2 = e^{m_2 x} = e^{-2x}$ are particular solutions of the differential equation. Furthermore, because these two solutions are linearly independent, you can apply Theorem 16.3 to conclude that the general solution is

$$y = C_1 e^{2x} + C_2 e^{-2x}. \qquad \text{General solution} \quad ∎$$

---

### Exploration

For each differential equation below, find the characteristic equation. Solve the characteristic equation for $m$, and use the values of $m$ to find a general solution of the differential equation. Using your results, develop a general solution of differential equations with characteristic equations that have distinct real roots.

(a) $y'' - 9y = 0$

(b) $y'' - 6y' + 8y = 0$

The characteristic equation in Example 2 has two distinct real zeros. From algebra, you know that this is only one of *three* possibilities for quadratic equations. In general, the quadratic equation $m^2 + am + b = 0$ has zeros

$$m_1 = \frac{-a + \sqrt{a^2 - 4b}}{2} \quad \text{and} \quad m_2 = \frac{-a - \sqrt{a^2 - 4b}}{2}$$

which fall into one of three cases.

1. Two distinct real zeros, $m_1 \neq m_2$
2. Two equal real zeros, $m_1 = m_2$
3. Two complex conjugate zeros, $m_1 = \alpha + \beta i$ and $m_2 = \alpha - \beta i$

In terms of the differential equation $y'' + ay' + by = 0$, these three cases correspond to three different types of general solutions.

■ **FOR FURTHER INFORMATION**
For more information on Theorem 16.4, see the article "A Note on a Differential Equation" by Russell Euler in the 1989 winter issue of the *Missouri Journal of Mathematical Sciences*.

> **THEOREM 16.4   Solutions of $y'' + ay' + by = 0$**
>
> The solutions of $y'' + ay' + by = 0$ fall into one of three cases, depending on the solutions of the characteristic equation, $m^2 + am + b = 0$.
>
> 1. *Distinct Real Zeros*   If $m_1 \neq m_2$ are distinct real zeros of the characteristic equation, then the general solution is
> $$y = C_1 e^{m_1 x} + C_2 e^{m_2 x}.$$
> 2. *Equal Real Zeros*   If $m_1 = m_2$ are equal real zeros of the characteristic equation, then the general solution is
> $$y = C_1 e^{m_1 x} + C_2 x e^{m_1 x} = (C_1 + C_2 x) e^{m_1 x}.$$
> 3. *Complex Zeros*   If $m_1 = \alpha + \beta i$ and $m_2 = \alpha - \beta i$ are complex zeros of the characteristic equation, then the general solution is
> $$y = C_1 e^{\alpha x} \cos \beta x + C_2 e^{\alpha x} \sin \beta x.$$

**EXAMPLE 3**   **Characteristic Equation: Complex Zeros**

Find the general solution of the differential equation $y'' + 6y' + 12y = 0$.

**Solution**   The characteristic equation $m^2 + 6m + 12 = 0$ has two complex zeros, as follows.

$$m = \frac{-6 \pm \sqrt{36 - 48}}{2} \qquad \text{Use Quadratic Formula with } a = 1, b = 6, \text{ and } c = 12.$$

$$= \frac{-6 \pm \sqrt{-12}}{2}$$

$$= \frac{-6 \pm 2\sqrt{-3}}{2}$$

$$= -3 \pm \sqrt{-3}$$

$$= -3 \pm \sqrt{3}\,i$$

So, $\alpha = -3$ and $\beta = \sqrt{3}$, and the general solution is

$$y = C_1 e^{-3x} \cos \sqrt{3}\,x + C_2 e^{-3x} \sin \sqrt{3}\,x.$$

Several members of the family of solutions, including $f(x) = e^{-3x} \cos \sqrt{3}\,x$ and $g(x) = e^{-3x} \sin \sqrt{3}\,x$, are shown in Figure 16.4. (Note that although the characteristic equation has two *complex* zeros, the solution of the differential equation is *real*.) ■

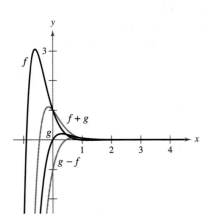

Several members of the family of solutions to Example 3, including $f(x) = e^{-3x} \cos \sqrt{3}\,x$ and $g(x) = e^{-3x} \sin \sqrt{3}\,x$, are shown in the graph. Notice that as $x \to \infty$, all of these solutions approach 0.

**Figure 16.4**

**EXAMPLE 4**    **Characteristic Equation: Repeated Zeros**

Solve the differential equation

$$y'' + 4y' + 4y = 0$$

subject to the initial conditions $y(0) = 2$ and $y'(0) = 1$.

**Solution**    The characteristic equation and its zeros are

$$m^2 + 4m + 4 = 0$$
$$(m + 2)^2 = 0$$
$$m = -2.$$

So, the characteristic equation has two equal zeros given by $m = -2$, and the general solution is

$$y = C_1 e^{-2x} + C_2 x e^{-2x}. \qquad \text{General solution}$$

Now, because $y = 2$ when $x = 0$, you have

$$2 = C_1(1) + C_2(0)(1)$$
$$2 = C_1.$$

Furthermore, because $y' = 1$ when $x = 0$, you have

$$y' = -2C_1 e^{-2x} + C_2(-2x e^{-2x} + e^{-2x}) \qquad \text{Derivative of } y \text{ with respect to } x$$
$$1 = -2(2)(1) + C_2[-2(0)(1) + 1] \qquad \text{Substitute.}$$
$$1 = -4 + C_2$$
$$5 = C_2.$$

Therefore, the particular solution is

$$y = 2e^{-2x} + 5x e^{-2x}. \qquad \text{Particular solution}$$

Try checking this solution in the original differential equation.    ■

## Higher-Order Linear Differential Equations

For higher-order homogeneous linear differential equations, you can find the general solution in much the same way as you do for second-order equations. That is, you begin by determining the $n$ zeros of the characteristic equation. Then, based on these $n$ zeros, you form a linearly independent collection of $n$ solutions. The major difference is that with equations of third or higher order, zeros of the characteristic equation may occur more than twice. When this happens, the linearly independent solutions are formed by multiplying by increasing powers of $x$, as demonstrated in Examples 6 and 7.

**EXAMPLE 5**    **Solving a Third-Order Equation**

Find the general solution of

$$y''' - y' = 0.$$

**Solution**    The characteristic equation and its zeros are

$$m^3 - m = 0$$
$$m(m - 1)(m + 1) = 0$$
$$m = 0, 1, -1.$$

Because the characteristic equation has three distinct zeros, the general solution is

$$y = C_1 + C_2 e^{-x} + C_3 e^x. \qquad \text{General solution}$$

EXAMPLE 6    **Solving a Third-Order Equation**

Find the general solution of

$$y''' + 3y'' + 3y' + y = 0.$$

**Solution**   The characteristic equation and its zeros are

$$m^3 + 3m^2 + 3m + 1 = 0$$
$$(m + 1)^3 = 0$$
$$m = -1.$$

Because the zero $m = -1$ occurs three times, the general solution is

$$y = C_1 e^{-x} + C_2 x e^{-x} + C_3 x^2 e^{-x}.$$    General solution

EXAMPLE 7    **Solving a Fourth-Order Equation**

⋯▷ *See LarsonCalculus.com for an interactive version of this type of example.*

Find the general solution of

$$y^{(4)} + 2y'' + y = 0.$$

**Solution**   The characteristic equation and its zeros are

$$m^4 + 2m^2 + 1 = 0$$
$$(m^2 + 1)^2 = 0$$
$$m = \pm i.$$

Because each of the zeros

$$m_1 = \alpha + \beta i = 0 + i \quad \text{and} \quad m_2 = \alpha - \beta i = 0 - i$$

occurs twice, the general solution is

$$y = C_1 \cos x + C_2 \sin x + C_3 x \cos x + C_4 x \sin x.$$    General solution    ■

## Application

One of the many applications of linear differential equations is describing the motion of an oscillating spring. According to Hooke's Law, a spring that is stretched (or compressed) $y$ units from its natural length $L$ tends to *restore* itself to its natural length by a force $F$ that is proportional to $y$. That is, $F(y) = -ky$, where $k$ is the **spring constant** and indicates the stiffness of the spring.

A rigid object of mass $m$ is attached to the end of a spring and causes a displacement, as shown in Figure 16.5. Assume that the mass of the spring is negligible compared with $m$. When the object is pulled downward and released, the resulting oscillations are a product of two opposing forces—the spring force $F(y) = -ky$ and the weight $mg$ of the object. Under such conditions, you can use a differential equation to find the position $y$ of the object as a function of time $t$. According to Newton's Second Law of Motion, the force acting on the weight is $F = ma$, where $a = d^2y/dt^2$ is the acceleration. Assuming that the motion is **undamped**—that is, there are no other external forces acting on the object—it follows that

$$m\frac{d^2y}{dt^2} = -ky$$

and you have

$$\frac{d^2y}{dt^2} + \frac{k}{m}y = 0.$$    Undamped motion of a spring

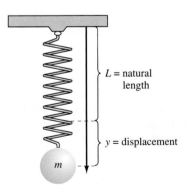

$L$ = natural length

$y$ = displacement

$m$

A rigid object of mass $m$ attached to the end of the spring causes a displacement of $y$.

**Figure 16.5**

A common type of spring is a *coil spring*, also called a *helical spring* because the shape of the spring is a helix. A *tension* coil spring resists being stretched (see photo above and Example 8). A *compression* coil spring resists being compressed, such as the spring in a car suspension.

EXAMPLE 8 **Undamped Motion of a Spring**

A 4-pound weight stretches a spring 8 inches from its natural length. The weight is pulled downward an additional 6 inches and released with an initial upward velocity of 8 feet per second. Find a formula for the position of the weight as a function of time $t$.

**Solution** The 4-pound weight stretches the spring 8 inches $= \frac{2}{3}$ foot from its natural length, so by Hooke's Law

$$4 = k\left(\frac{2}{3}\right) \implies k = 6.$$

Moreover, because the weight $w$ is given by $mg$, it follows that

$$m = \frac{w}{g} = \frac{4}{32} = \frac{1}{8}.$$

So, the resulting differential equation for this undamped motion is

$$\frac{d^2y}{dt^2} + \left(\frac{6}{1/8}\right)y = 0 \implies \frac{d^2y}{dt^2} + 48y = 0.$$

The characteristic equation $m^2 + 48 = 0$ has complex zeros $m = 0 \pm 4\sqrt{3}\,i$, so the general solution is

$$y = C_1 e^0 \cos 4\sqrt{3}\,t + C_2 e^0 \sin 4\sqrt{3}\,t$$
$$= C_1 \cos 4\sqrt{3}\,t + C_2 \sin 4\sqrt{3}\,t.$$

When $t = 0$ seconds, $y = 6$ inches $= \frac{1}{2}$ foot. Using this initial condition, you have

$$\frac{1}{2} = C_1(1) + C_2(0) \implies C_1 = \frac{1}{2}. \qquad y(0) = \tfrac{1}{2}$$

To determine $C_2$, note that $y' = 8$ feet per second when $t = 0$ seconds.

$$y' = -4\sqrt{3}\,C_1 \sin 4\sqrt{3}\,t + 4\sqrt{3}\,C_2 \cos 4\sqrt{3}\,t \qquad \text{Derivative of } y \text{ with respect to } t$$

$$8 = -4\sqrt{3}\left(\frac{1}{2}\right)(0) + 4\sqrt{3}\,C_2(1) \qquad \text{Substitute.}$$

$$\frac{2\sqrt{3}}{3} = C_2$$

Consequently, the position at time $t$ is given by

$$y = \frac{1}{2} \cos 4\sqrt{3}\,t + \frac{2\sqrt{3}}{3} \sin 4\sqrt{3}\,t. \qquad \blacksquare$$

The object in Figure 16.6 undergoes an additional damping or frictional force that is proportional to its velocity. A case in point would be the damping force resulting from friction and movement through a fluid. Considering this damping force

$$-p\frac{dy}{dt} \qquad \text{Damping force}$$

the differential equation for the oscillation is

$$m\frac{d^2y}{dt^2} = -ky - p\frac{dy}{dt}$$

or, in standard linear form,

A damped vibration could be caused by friction and movement through a liquid.

**Figure 16.6**

$$\frac{d^2y}{dt^2} + \frac{p}{m}\left(\frac{dy}{dt}\right) + \frac{k}{m}y = 0. \qquad \text{Damped motion of a spring}$$

# 16.2  Exercises

See **CalcChat.com** for tutorial help and worked-out solutions to odd-numbered exercises.

## CONCEPT CHECK

**1. Linear Differential Equation**   Determine the order of each linear differential equation and decide whether each equation is homogeneous.

(a) $y^{(5)} + x^6 y' + xy = 0$   (b) $y'' + 3e^x y + 2x = 0$

**2. Linearly Independent**   Describe what it means for the functions $y_1$ and $y_2$ to be linearly independent.

**3. Using Zeros**   The zeros of the characteristic equation for two differential equations of the form $y'' + ay' + by = 0$ are given. Write the corresponding general solution for each set of zeros.

(a) $m = -1, 3$   (b) $m = 2, 2$

**4. Finding a General Solution**   Explain how to find the general solution of a higher-order homogenous linear differential equation.

**Verifying a Solution**   In Exercises 5–8, verify the solution of the differential equation. Then use a graphing utility to graph the particular solutions for several different values of $C_1$ and $C_2$.

| Solution | Differential Equation |
|---|---|
| 5. $y = (C_1 + C_2 x)e^{-3x}$ | $y'' + 6y' + 9y = 0$ |
| 6. $y = C_1 + C_2 e^{3x}$ | $y'' - 3y' = 0$ |
| 7. $y = C_1 \cos 2x + C_2 \sin 2x$ | $y'' + 4y = 0$ |
| 8. $y = C_1 e^{-x} \cos 3x + C_2 e^{-x} \sin 3x$ | $y'' + 2y' + 10y = 0$ |

**Finding a General Solution**   In Exercises 9–36, find the general solution of the linear differential equation.

9. $y'' - y' = 0$        10. $y'' + 2y' = 0$

11. $y'' - y' - 6y = 0$        12. $y'' + 6y' + 5y = 0$

13. $2y'' + 3y' - 2y = 0$        14. $16y'' - 16y' + 3y = 0$

15. $y'' + 6y' + 9y = 0$        16. $y'' - 10y' + 25y = 0$

17. $16y'' - 8y' + y = 0$        18. $9y'' - 12y' + 4y = 0$

19. $y'' + y = 0$        20. $y'' + 4y = 0$

21. $4y'' - 5y = 0$        22. $y'' - 2y = 0$

23. $y'' - 2y' + 4y = 0$        24. $y'' - 4y' + 21y = 0$

25. $y'' - 3y' + y = 0$        26. $3y'' + 4y' - y = 0$

27. $9y'' - 12y' + 11y = 0$        28. $2y'' - 6y' + 7y = 0$

29. $y^{(4)} - y = 0$        30. $y^{(4)} - y'' = 0$

31. $y''' - 6y'' + 11y' - 6y = 0$        32. $y''' - y'' - y' + y = 0$

33. $y''' - 3y'' + 7y' - 5y = 0$

34. $y''' - 3y'' + 3y' - y = 0$

35. $y^{(4)} - 2y'' + y = 0$

36. $y^{(4)} - 2y''' + y'' = 0$

**37. Finding a Particular Solution**   Consider the differential equation $y'' + 100y = 0$ and the solution $y = C_1 \cos 10x + C_2 \sin 10x$. Find the particular solution satisfying each initial condition.

(a) $y(0) = 2$,   $y'(0) = 0$     (b) $y(0) = 0$,   $y'(0) = 2$

(c) $y(0) = -1$,   $y'(0) = 3$

**38. Finding a Particular Solution**   Determine $C$ and $\omega$ such that $y = C \sin \sqrt{3}\,t$ is a particular solution of the differential equation $y'' + \omega y = 0$, where $y'(0) = -5$.

**Finding a Particular Solution: Initial Conditions**   In Exercises 39–44, find the particular solution of the linear differential equation that satisfies the initial conditions.

39. $y'' - y' - 30y = 0$        40. $y'' - 7y' + 12y = 0$

    $y(0) = 1$,   $y'(0) = -4$        $y(0) = 3$,   $y'(0) = 3$

41. $y'' + 16y = 0$        42. $9y'' - 6y' + y = 0$

    $y(0) = 0$,   $y'(0) = 2$        $y(0) = 2$,   $y'(0) = 1$

43. $y'' + 2y' + 3y = 0$        44. $4y'' + 4y' + y = 0$

    $y(0) = 2$,   $y'(0) = 1$        $y(0) = 3$,   $y'(0) = -1$

**Finding a Particular Solution: Boundary Conditions**   In Exercises 45–50, find the particular solution of the linear differential equation that satisfies the boundary conditions.

45. $y'' - 4y' + 3y = 0$

    $y(0) = 1$,   $y(1) = 3$

46. $4y'' + y = 0$

    $y(0) = 2$,   $y(\pi) = -5$

47. $y'' + 9y = 0$

    $y(0) = 3$,   $y\left(\dfrac{\pi}{2}\right) = 4$

48. $4y'' + 20y' + 21y = 0$

    $y(0) = 3$,   $y(2) = 0$

49. $4y'' - 28y' + 49y = 0$

    $y(0) = 2$,   $y(1) = -1$

50. $y'' + 6y' + 45y = 0$

    $y(0) = 4$,   $y\left(\dfrac{\pi}{12}\right) = 2$

## EXPLORING CONCEPTS

**51. Finding Another Solution**   Show that the equation $y = C_1 \sinh x + C_2 \cosh x$ is a solution of the homogeneous linear differential equation $y'' - y = 0$. Then use hyperbolic definitions to find another solution of the differential equation.

**52. General Solution of a Differential Equation**   What is the general solution of $y^{(n)} = 0$? Explain.

• • 53. **Undamped or Damped Motion?** • • • • • • • • •

Several shock absorbers are shown at the right. Do you think the motion of the spring in a shock absorber is undamped or damped?

 **Motion of a Spring** In Exercises 59–64, a 32-pound weight stretches a spring $\frac{2}{3}$ foot from its natural length. Use the given information to find a formula for the position of the weight as a function of time.

59. The weight is pulled $\frac{1}{2}$ foot below equilibrium and released.

60. The weight is raised $\frac{2}{3}$ foot above equilibrium and released.

61. The weight is raised $\frac{2}{3}$ foot above equilibrium and released with an initial downward velocity of $\frac{1}{2}$ foot per second.

62. The weight is pulled $\frac{1}{2}$ foot below equilibrium and released with an initial upward velocity of $\frac{1}{2}$ foot per second.

63. The weight is pulled $\frac{1}{2}$ foot below equilibrium and released. The motion takes place in a medium that furnishes a damping force of magnitude $\frac{1}{8}|v|$ at all times.

64. The weight is pulled $\frac{1}{2}$ foot below equilibrium and released. The motion takes place in a medium that furnishes a damping force of magnitude $\frac{1}{4}|v|$ at all times.

65. **Real Zeros** The characteristic equation of the differential equation $y'' + ay' + by = 0$ has two equal real zeros given by $m = r$. Show that $y = C_1 e^{rx} + C_2 x e^{rx}$ is a solution.

66. **Complex Zeros** The characteristic equation of the differential equation

$$y'' + ay' + by = 0$$

has complex zeros given by $m_1 = \alpha + \beta i$ and $m_2 = \alpha - \beta i$. Show that $y = C_1 e^{\alpha x} \cos \beta x + C_2 e^{\alpha x} \sin \beta x$ is a solution.

54. **HOW DO YOU SEE IT?** Give a geometric argument to explain why the graph cannot be a solution of the differential equation. (It is not necessary to solve the differential equation.)

(a) $y'' = y'$          (b) $y'' = -\frac{1}{2}y'$

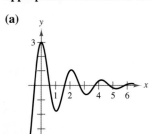

**True or False?** In Exercises 67–70, determine whether the statement is true or false. If it is false, explain why or give an example that shows it is false.

67. $y_1 = e^x$ and $y_2 = 3e^x$ are linearly dependent.

68. $y_1 = x$ and $y_2 = x^2$ are linearly dependent.

69. $y = x$ is a solution of

$$a_n y^{(n)} + a_{n-1} y^{(n-1)} + \cdots + a_1 y' + a_0 y = 0$$

if and only if $a_1 = a_0 = 0$.

70. It is possible to choose $a$ and $b$ such that $y = x^2 e^x$ is a solution of $y'' + ay' + by = 0$.

**Motion of a Spring** In Exercises 55–58, match the differential equation with the graph of a particular solution. [The graphs are labeled (a), (b), (c), and (d).] The correct match can be made by comparing the frequency of the oscillations or the rate at which the oscillations are being damped with the appropriate coefficient in the differential equation.

(a) 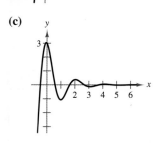          (b)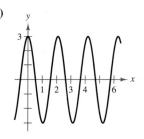

(c) 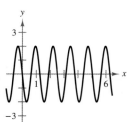          (d)

**Wronskian** The *Wronskian* of two differentiable functions $f$ and $g$, denoted by $W(f, g)$, is defined as the function given by the determinant

$$W(f, g) = \begin{vmatrix} f & g \\ f' & g' \end{vmatrix}.$$

The functions $f$ and $g$ are linearly independent when there exists at least one value of $x$ for which $W(f, g) \neq 0$. In Exercises 71–74, use the Wronskian to verify that the two functions are linearly independent.

55. $y'' + 9y = 0$

56. $y'' + 25y = 0$

57. $y'' + 2y' + 10y = 0$

58. $y'' + y' + \frac{37}{4}y = 0$

71. $y_1 = e^{ax}$          72. $y_1 = e^{ax}$
    $y_2 = e^{bx}, \quad a \neq b$          $y_2 = xe^{ax}$

73. $y_1 = e^{ax} \sin bx$          74. $y_1 = x$
    $y_2 = e^{ax} \cos bx, \quad b \neq 0$          $y_2 = x^2$

# 16.3 Second-Order Nonhomogeneous Linear Equations

■ Recognize the general solution of a second-order nonhomogeneous linear differential equation.
■ Use the method of undetermined coefficients to solve a second-order nonhomogeneous linear differential equation.
■ Use the method of variation of parameters to solve a second-order nonhomogeneous linear differential equation.

## Nonhomogeneous Equations

In the preceding section, damped oscillations of a spring were represented by the *homogeneous* second-order linear equation

$$\frac{d^2y}{dt^2} + \frac{p}{m}\left(\frac{dy}{dt}\right) + \frac{k}{m}y = 0. \qquad \text{Free motion}$$

This type of oscillation is called **free** because it is determined solely by the spring and gravity and is free of the action of other external forces. If such a system is also subject to an external periodic force, such as $a \sin bt$, caused by vibrations at the opposite end of the spring, then the motion is called **forced,** and it is characterized by the *nonhomogeneous* equation

$$\frac{d^2y}{dt^2} + \frac{p}{m}\left(\frac{dy}{dt}\right) + \frac{k}{m}y = a \sin bt. \qquad \text{Forced motion}$$

In this section, you will study two methods for finding the general solution of a nonhomogeneous linear differential equation. In both methods, the first step is to find the general solution of the corresponding homogeneous equation.

$$y = y_h \qquad \qquad \text{General solution of homogeneous equation}$$

Having done this, you try to find a particular solution of the nonhomogeneous equation.

$$y = y_p \qquad \qquad \text{Particular solution of nonhomogeneous equation}$$

By combining these two results, you can conclude that the general solution of the nonhomogeneous equation is

$$y = y_h + y_p \qquad \qquad \text{General solution of nonhomogeneous equation}$$

as stated in the next theorem.

**SOPHIE GERMAIN**
**(1776–1831)**

Many of the early contributors to calculus were interested in forming mathematical models for vibrating strings and membranes, oscillating springs, and elasticity. One of these was the French mathematician Sophie Germain, who in 1816 was awarded a prize by the French Academy for a paper entitled "Memoir on the Vibrations of Elastic Plates."
*See LarsonCalculus.com to read more of this biography.*

---

**THEOREM 16.5   Solution of Nonhomogeneous Linear Equation**

Let

$$y'' + ay' + by = F(x)$$

be a second-order nonhomogeneous linear differential equation. If $y_p$ is a particular solution of this equation and $y_h$ is the general solution of the corresponding homogeneous equation, then

$$y = y_h + y_p$$

is the general solution of the nonhomogeneous equation.

---

## Method of Undetermined Coefficients

You already know how to find the solution $y_h$ of a linear *homogeneous* differential equation. The remainder of this section looks at ways to find the particular solution $y_p$. When $F(x)$ in

$$y'' + ay' + by = F(x)$$

consists of sums or products of $x^n$, $e^{mx}$, $\cos \beta x$, or $\sin \beta x$, you can find a particular solution $y_p$ by the method of **undetermined coefficients.** The object of this method is to guess that the solution $y_p$ is a generalized form of $F(x)$. Here are some examples.

1. For $F(x) = 3x^2$, choose $y_p = Ax^2 + Bx + C$.
2. For $F(x) = 4xe^x$, choose $y_p = Axe^x + Be^x$.
3. For $F(x) = x + \sin 2x$, choose $y_p = (Ax + B) + C \sin 2x + D \cos 2x$.

Then, by substitution, determine the coefficients for the generalized solution.

---

**EXAMPLE 1**    **Method of Undetermined Coefficients**

Find the general solution of the equation $y'' - 2y' - 3y = 2 \sin x$.

**Solution**    Note that the corresponding homogeneous equation is $y'' - 2y' - 3y = 0$. To find the general solution $y_h$, solve the characteristic equation as shown.

$$m^2 - 2m - 3 = 0$$
$$(m + 1)(m - 3) = 0$$
$$m = -1, 3$$

So, the general solution of the homogeneous equation is $y_h = C_1 e^{-x} + C_2 e^{3x}$. Next, let $y_p$ be a generalized form of $2 \sin x$.

$$y_p = A \cos x + B \sin x$$
$$y_p{}' = -A \sin x + B \cos x$$
$$y_p{}'' = -A \cos x - B \sin x$$

Substitution into the original differential equation yields

$$y'' - 2y' - 3y = 2 \sin x$$
$$-A \cos x - B \sin x - 2(-A \sin x + B \cos x) - 3(A \cos x + B \sin x) = 2 \sin x$$
$$-A \cos x - B \sin x + 2A \sin x - 2B \cos x - 3A \cos x - 3B \sin x = 2 \sin x$$
$$(-4A - 2B) \cos x + (2A - 4B) \sin x = 2 \sin x.$$

By equating coefficients of like terms, you obtain the system of two equations

$$-4A - 2B = 0 \quad \text{and} \quad 2A - 4B = 2$$

with solutions

$$A = \frac{1}{5} \quad \text{and} \quad B = -\frac{2}{5}.$$

Therefore, a particular solution of the original nonhomogeneous equation is

$$y_p = \frac{1}{5} \cos x - \frac{2}{5} \sin x \qquad \text{Particular solution of nonhomogeneous equation}$$

and the general solution is

$$y = y_h + y_p$$
$$= C_1 e^{-x} + C_2 e^{3x} + \frac{1}{5} \cos x - \frac{2}{5} \sin x. \qquad \text{General solution of nonhomogeneous equation}$$

In Example 1, the form of the homogeneous solution $y_h = C_1e^{-x} + C_2e^{3x}$ has no overlap with the function $F(x)$ in the equation $y'' + ay' + by = F(x)$. However, suppose the given differential equation in Example 1 were of the form

$$y'' - 2y' - 3y = e^{-x}.$$

Now it would make no sense to guess that the particular solution was

$$y = Ae^{-x}$$

because in the equation $y'' - 2y' - 3y = e^{-x}$, this solution yields

$$Ae^{-x} - 2(-Ae^{-x}) - 3Ae^{-x} = 0 \neq e^{-x}. \qquad y = Ae^{-x}, y' = -Ae^{-x}, y'' = Ae^{-x}$$

In such cases, you should alter your guess by multiplying by the lowest power of $x$ that removes the duplication. For this particular problem, you would guess

$$y_p = Axe^{-x}.$$

Another case where you need to alter your guess for the particular solution $y_p$ is shown in the next example.

---

**EXAMPLE 2**    **Method of Undetermined Coefficients**

Find the general solution of

$$y'' - 2y' = x + 2e^x.$$

**Solution**    The corresponding homogeneous equation is $y'' - 2y' = 0$, and the characteristic equation

$$m^2 - 2m = 0$$
$$m(m - 2) = 0$$

has solutions $m = 0$ and $m = 2$. So,

$$y_h = C_1 + C_2e^{2x}. \qquad \text{General solution of homogeneous equation}$$

Because $F(x) = x + 2e^x$, your first choice for $y_p$ would be $(A + Bx) + Ce^x$. However, because $y_h$ *already* contains a constant term $C_1$, you should multiply the *polynomial part* $(A + Bx)$ by $x$ and use

$$y_p = Ax + Bx^2 + Ce^x$$
$$y_p' = A + 2Bx + Ce^x$$
$$y_p'' = 2B + Ce^x.$$

Substitution into the original differential equation produces

$$y'' - 2y' = x + 2e^x$$
$$2B + Ce^x - 2(A + 2Bx + Ce^x) = x + 2e^x$$
$$(2B - 2A) - 4Bx - Ce^x = x + 2e^x.$$

Equating coefficients of like terms yields the system of three equations

$$2B - 2A = 0, \quad -4B = 1, \quad \text{and} \quad -C = 2$$

with solutions $A = B = -\frac{1}{4}$ and $C = -2$. Therefore,

$$y_p = -\frac{1}{4}x - \frac{1}{4}x^2 - 2e^x \qquad \text{Particular solution of nonhomogeneous equation}$$

and the general solution is

$$y = y_h + y_p$$
$$= C_1 + C_2e^{2x} - \frac{1}{4}x - \frac{1}{4}x^2 - 2e^x. \qquad \text{General solution of nonhomogeneous equation}$$

In Example 2, the polynomial part of the initial guess $(A + Bx) + Ce^x$ for $y_p$ overlapped by a constant term with

$$y_h = C_1 + C_2e^{2x}$$

and it was necessary to multiply the polynomial part by a power of $x$ that removed the overlap. The next example further illustrates some choices for $y_p$ that eliminate overlap with $y_h$. Remember that in all cases, the first guess for $y_p$ should match the types of functions occurring in $F(x)$.

**EXAMPLE 3    Choosing the Form of the Particular Solution**

Determine a suitable choice for $y_p$ for each differential equation, given its general solution of the homogeneous equation.

| $y'' + ay' + by = F(x)$ | $y_h$ |
|---|---|
| **a.** $y'' = x^2$ | $C_1 + C_2x$ |
| **b.** $y'' + 2y' + 10y = 4 \sin 3x$ | $C_1e^{-x} \cos 3x + C_2e^{-x} \sin 3x$ |
| **c.** $y'' - 4y' + 4 = e^{2x}$ | $C_1e^{2x} + C_2xe^{2x}$ |

**Solution**

**a.** Because $F(x) = x^2$, the normal choice for $y_p$ would be $A + Bx + Cx^2$. However, because $y_h = C_1 + C_2x$ already contains a constant term and a linear term, you should multiply by $x^2$ to obtain

$$y_p = Ax^2 + Bx^3 + Cx^4.$$

**b.** Because $F(x) = 4 \sin 3x$ and each term in $y_h$ contains a factor of $e^{-x}$, you can simply let

$$y_p = A \cos 3x + B \sin 3x.$$

**c.** Because $F(x) = e^{2x}$, the normal choice for $y_p$ would be $Ae^{2x}$. However, because $y_h = C_1e^{2x} + C_2xe^{2x}$ already contains an $e^{2x}$ term and an $xe^{2x}$ term, you should multiply by $x^2$ to get

$$y_p = Ax^2e^{2x}.$$

**EXAMPLE 4    Solving a Third-Order Equation**

⋮⋯▷ *See LarsonCalculus.com for an interactive version of this type of example.*

Find the general solution of $y''' + 3y'' + 3y' + y = x$.

**Solution**    From Example 6 in Section 16.2, you know that the homogeneous solution is

$$y_h = C_1e^{-x} + C_2xe^{-x} + C_3x^2e^{-x}.$$

Because $F(x) = x$, let $y_p = A + Bx$ and obtain $y_p' = B$ and $y_p'' = 0$. So, by substitution into the original differential equation, you have

$$0 + 3(0) + 3(B) + A + Bx = x$$
$$(3B + A) + Bx = x.$$

Equating coefficients of like terms yields the system of two equations

$$3B + A = 0 \quad \text{and} \quad B = 1.$$

So, $B = 1$ and $A = -3$, which implies that $y_p = -3 + x$. Therefore, the general solution is

$$y = y_h + y_p$$
$$= C_1e^{-x} + C_2xe^{-x} + C_3x^2e^{-x} - 3 + x.$$

∎

## Variation of Parameters

The method of undetermined coefficients works well when $F(x)$ is made up of polynomials or functions whose successive derivatives have a cyclical pattern. For functions such as $1/x$ and $\tan x$, which do not have such characteristics, it is better to use a more general method called **variation of parameters.** In this method, you assume that $y_p$ has the same *form* as $y_h$, except that the constants in $y_h$ are replaced by variables.

---

**Variation of Parameters**

To find the general solution of the equation $y'' + ay' + by = F(x)$, use these steps.

1. Find $y_h = C_1 y_1 + C_2 y_2$.
2. Replace the constants by variables to form $y_p = u_1 y_1 + u_2 y_2$.
3. Solve the following system for $u_1'$ and $u_2'$.

$$u_1' y_1 + u_2' y_2 = 0$$
$$u_1' y_1' + u_2' y_2' = F(x)$$

4. Integrate to find $u_1$ and $u_2$. The general solution is $y = y_h + y_p$.

---

**EXAMPLE 5**  **Variation of Parameters**

Solve the differential equation

$$y'' - 2y' + y = \frac{e^x}{2x}, \quad x > 0.$$

**Solution**  The characteristic equation

$$m^2 - 2m + 1 = 0 \quad \Longrightarrow \quad (m - 1)^2 = 0$$

has one repeated solution, $m = 1$. So, the homogeneous solution is

$$y_h = C_1 y_1 + C_2 y_2 = C_1 e^x + C_2 x e^x.$$

Replacing $C_1$ and $C_2$ by $u_1$ and $u_2$ produces

$$y_p = u_1 y_1 + u_2 y_2 = u_1 e^x + u_2 x e^x.$$

The resulting system of equations is

$$u_1' e^x + u_2' x e^x = 0$$
$$u_1' e^x + u_2'(x e^x + e^x) = \frac{e^x}{2x}.$$

Subtracting the second equation from the first produces $u_2' = 1/(2x)$. Then, by substitution in the first equation, you have $u_1' = -\frac{1}{2}$. Finally, integration yields

$$u_1 = -\int \frac{1}{2}\, dx = -\frac{x}{2} \quad \text{and} \quad u_2 = \frac{1}{2} \int \frac{1}{x}\, dx = \frac{1}{2} \ln x = \ln \sqrt{x}.$$

From this result, it follows that a particular solution is

$$y_p = -\frac{1}{2} x e^x + \left( \ln \sqrt{x} \right) x e^x$$

and the general solution is

$$y = C_1 e^x + C_2 x e^x - \frac{1}{2} x e^x + x e^x \ln \sqrt{x}.$$

■

**Exploration**

Notice in Example 5 that the constants of integration were not introduced when finding $u_1$ and $u_2$. Show that for

$$u_1 = -\frac{x}{2} + a_1 \quad \text{and} \quad u_2 = \ln \sqrt{x} + a_2$$

the general solution is equivalent to

$$y = y_h + y_p = C_1 e^x + C_2 e^x - \frac{1}{2} x e^x + x e^x \ln \sqrt{x}$$

which is the same result as the solution obtained in the example.

**EXAMPLE 6**    **Variation of Parameters**

Solve the differential equation $y'' + y = \tan x$.

**Solution**    Because the characteristic equation $m^2 + 1 = 0$ has solutions $m = \pm i$, the homogeneous solution is $y_h = C_1 \cos x + C_2 \sin x$. Replacing $C_1$ and $C_2$ by $u_1$ and $u_2$ produces $y_p = u_1 \cos x + u_2 \sin x$. The resulting system of equations is

$$u_1' \cos x + u_2' \sin x = 0$$
$$-u_1' \sin x + u_2' \cos x = \tan x.$$

Multiplying the first equation by $\sin x$ and the second by $\cos x$ produces

$$u_1' \sin x \cos x + u_2' \sin^2 x = 0$$
$$-u_1' \sin x \cos x + u_2' \cos^2 x = \sin x.$$

Adding these two equations produces $u_2' = \sin x$, which implies that

$$u_1' \sin x \cos x + (\sin x) \sin^2 x = 0$$
$$u_1' \sin x \cos x = -\sin^3 x$$
$$u_1' = -\frac{\sin^2 x}{\cos x}$$
$$u_1' = \frac{\cos^2 x - 1}{\cos x}$$
$$u_1' = \cos x - \sec x.$$

Integration yields

$$u_1 = \int (\cos x - \sec x)\, dx = \sin x - \ln |\sec x + \tan x|$$

and

$$u_2 = \int \sin x\, dx = -\cos x$$

so that the particular solution is

$$y_p = \sin x \cos x - \cos x \ln |\sec x + \tan x| - \sin x \cos x$$
$$= -\cos x \ln |\sec x + \tan x|$$

and the general solution is

$$y = y_h + y_p$$
$$= C_1 \cos x + C_2 \sin x - \cos x \ln |\sec x + \tan x|.$$

## 16.3 Exercises

See CalcChat.com for tutorial help and worked-out solutions to odd-numbered exercises.

**CONCEPT CHECK**

**1. Writing** What is the form of the general solution of a second-order nonhomogeneous linear differential equation?

**2. Choosing a Method** Determine whether you would use the method of undetermined coefficients or the method of variation of parameters to find the general solution of each differential equation. Explain your reasoning. (Do not solve the equations.)

(a) $y'' + 3y' + y = x^2$  (b) $y'' + y = \csc x$

(c) $y''' - 6y = 3x - e^{-2x}$

 **Method of Undetermined Coefficients** In Exercises 3–6, solve the differential equation by the method of undetermined coefficients.

3. $y'' + 7y' + 12y = 3x + 1$  4. $y'' - y' - 6y = 4$

5. $y'' - 8y' + 16y = e^{3x}$  6. $y'' - 2y' - 15y = \sin x$

 **Choosing the Form of the Particular Solution** In Exercises 7–10, determine a suitable choice for $y_p$ for the differential equation, given its general solution of the homogeneous equation. Explain your reasoning. (Do not solve the equation.)

$$y'' + ay' + by = F(x) \qquad y_h$$

7. $y'' + y' = 4x + 6$  $\qquad C_1 + C_2e^{-x}$

8. $y'' - 9y = x + 2e^{-3x}$  $\qquad C_1e^{-3x} + C_2e^{3x}$

9. $3y'' + 6y' = 4 + \sin x$  $\qquad C_1 + C_2e^{-2x}$

10. $y'' + y = 8\cos x$  $\qquad C_1\cos x + C_2\sin x$

 **Method of Undetermined Coefficients** In Exercises 11–16, solve the differential equation by the method of undetermined coefficients.

11. $y'' + 2y' = e^{-2x}$  12. $y'' - 9y = 5e^{3x}$

13. $y'' + 9y = \sin 3x$

14. $16y'' - 8y' + y = 4(x + e^{x/4})$

15. $y''' - 3y'' + 4y = 2 + e^{2x}$  16. $y''' - 3y' + 2y = 2e^{-2x}$

**Using Initial Conditions** In Exercises 17–22, solve the differential equation by the method of undetermined coefficients subject to the initial condition(s).

17. $y'' + y = x^3$

$\quad y(0) = 1, y'(0) = 0$

18. $y'' + 4y = 4$

$\quad y(0) = 1, y'(0) = 6$

19. $y'' + y' = 2\sin x$

$\quad y(0) = 0, y'(0) = -3$

20. $y'' + y' - 2y = 3\cos 2x$

$\quad y(0) = -1, y'(0) = 2$

21. $y' - 4y = xe^x - xe^{4x}$

$\quad y(0) = \dfrac{1}{3}$

22. $y' + 2y = \sin x$

$\quad y\left(\dfrac{\pi}{2}\right) = \dfrac{2}{5}$

 **Method of Variation of Parameters** In Exercises 23–28, solve the differential equation by the method of variation of parameters.

23. $y'' + y = \sec x$  24. $y'' + y = \sec x \tan x$

25. $y'' + 4y = \csc 2x$  26. $y'' - 4y' + 4y = x^2e^{2x}$

27. $y'' - 2y' + y = e^x \ln x$  28. $y'' - 4y' + 4y = \dfrac{e^{2x}}{x}$

•• **Electrical Circuits** ••••••••••••••

In Exercises 29 and 30, use the electrical circuit differential equation

$$\frac{d^2q}{dt^2} + \left(\frac{R}{L}\right)\frac{dq}{dt} + \left(\frac{1}{LC}\right)q = \left(\frac{1}{L}\right)E(t)$$

where $R$ is the resistance (in ohms), $C$ is the capacitance (in farads), $L$ is the inductance (in henrys), $E(t)$ is the electromotive force (in volts), and $q$ is the charge on the capacitor (in coulombs). Find the charge $q$ as a function of time $t$ for the electrical circuit described. Assume that $q(0) = 0$ and $q'(0) = 0$.

29. $R = 20, C = 0.02, L = 2, E(t) = 12\sin 5t$

30. $R = 20, C = 0.02, L = 1, E(t) = 10\sin 5t$

**Motion of a Spring** In Exercises 31–34, use the differential equation

$$\frac{w}{g}y''(t) + by'(t) + ky(t) = \frac{w}{g}F(t)$$

which models the oscillating motion of an object on the end of a spring (see figure). In the equation, $y$ is the displacement from equilibrium (positive direction is downward), measured in feet, $t$ is time in seconds, $w$ is the weight of the object, $g$ is the acceleration due to gravity, $b$ is the magnitude of the resistance to the motion, $k$ is the spring constant from Hooke's Law, and $F(t)$ is the acceleration imposed on the system. Find the displacement $y$ as a function of time $t$ for the oscillating motion described subject to the initial conditions. Use a graphing utility to graph the displacement function.

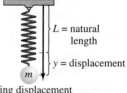

Spring displacement

$L$ = natural length

$y$ = displacement

31. $w = 24, g = 32, b = 0, k = 48, F(t) = 48\sin 4t$

$\quad y(0) = \frac{1}{4}, y'(0) = 0$

32. $w = 2$, $g = 32$, $b = 0$, $k = 4$, $F(t) = 4 \sin 8t$

$y(0) = \frac{1}{4}$, $y'(0) = 0$

33. $w = 2$, $g = 32$, $b = 1$, $k = 4$, $F(t) = 4 \sin 8t$

$y(0) = \frac{1}{4}$, $y'(0) = -3$

34. $w = 4$, $g = 32$, $b = \frac{1}{2}$, $k = \frac{25}{2}$, $F(t) = 0$

$y(0) = \frac{1}{2}$, $y'(0) = -4$

---

**EXPLORING CONCEPTS**

35. **Motion of a Spring** Rewrite $y_h$ in the solution to Exercise 31 by using the identity

$$a \cos \omega t + b \sin \omega t = \sqrt{a^2 + b^2}\, \sin(\omega t + \phi)$$

where $\phi = \arctan a/b$.

---

36. **HOW DO YOU SEE IT?** The figure shows the particular solution of the differential equation

$$\frac{4}{32}y'' + by' + \frac{25}{2}y = 0$$

that models the oscillating motion of an object on the end of a spring and satisfies the initial conditions $y(0) = \frac{1}{2}$ and $y'(0) = -4$ for values of the resistance component $b$ in the interval $[0, 1]$. According to the figure, is the motion damped or undamped when $b = 0$? When $b > 0$? (You do not need to solve the differential equation.)

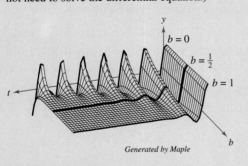

*Generated by Maple*

37. **Motion of a Spring** Refer to the differential equation and the initial conditions given in Exercise 36.

(a) When there is no resistance to the motion $(b = 0)$, describe the motion.

(b) For $b > 0$, what is the ultimate effect of the retarding force?

(c) Is there a real number $M$ such that there will be no oscillations of the spring for $b > M$? Explain your answer.

38. **Solving a Differential Equation** Solve the differential equation given that $y_1$ and $y_2$ are solutions of the corresponding homogeneous equation.

(a) $x^2 y'' - xy' + y = 4x \ln x$

$y_1 = x$, $y_2 = x \ln x$

(b) $x^2 y'' + xy' + 4y = \sin(\ln x)$

$y_1 = \sin(\ln x^2)$, $y_2 = \cos(\ln x^2)$

**True or False?** In Exercises 39 and 40, determine whether the statement is true or false. If it is false, explain why or give an example that shows it is false.

39. $y_p = -e^{2x} \cos e^{-x}$ is a particular solution of the differential equation

$$y'' - 3y' + 2y = \cos e^{-x}$$

40. $y_p = -\frac{1}{8}e^{2x}$ is a particular solution of the differential equation

$$y'' - 6y' = e^{2x}.$$

---

**PUTNAM EXAM CHALLENGE**

41. For all real $x$, the real-valued function $y = f(x)$ satisfies $y'' - 2y' + y = 2e^x$.

(a) If $f(x) > 0$ for all real $x$, must $f'(x) > 0$ for all real $x$? Explain.

(b) If $f'(x) > 0$ for all real $x$, must $f(x) > 0$ for all real $x$? Explain.

This problem was composed by the Committee on the Putnam Prize Competition.
© The Mathematical Association of America. All rights reserved.

---

**SECTION PROJECT**

## Parachute Jump

The fall of a parachutist is described by the second-order linear differential equation

$$\left(-\frac{w}{g}\right)\frac{d^2y}{dt^2} - k\frac{dy}{dt} = w$$

where $w$ is the weight of the parachutist, $y$ is the height at time $t$, $g$ is the acceleration due to gravity, and $k$ is the drag factor of the parachute.

(a) The parachute is opened at 2000 feet, so

$$y(0) = 2000.$$

At that time, the velocity is

$$y'(0) = -100 \text{ feet per second.}$$

For a 160-pound parachutist who has a parachute with a drag factor of $k = 8$, the differential equation is

$$-5y'' - 8y' = 160.$$

Using the initial conditions, verify that the solution of the differential equation is

$$y = 1950 + 50e^{-1.6t} - 20t.$$

(b) Consider a 192-pound parachutist who has a parachute with a drag factor of $k = 9$. Using the initial conditions given in part (a), write and solve a differential equation that describes the fall of the parachutist.

# 16.4 Series Solutions of Differential Equations

■ Use a power series to solve a differential equation.
■ Use a Taylor series to find the series solution of a differential equation.

## Power Series Solution of a Differential Equation

Power series can be used to solve certain types of differential equations. This section begins with the general **power series solution** method.

Recall from Chapter 9 that a power series represents a function $f$ on an interval of convergence and that you can successively differentiate the power series to obtain a series for $f'$, $f''$, and so on. These properties are used in the power series solution method demonstrated in the first two examples.

### EXAMPLE 1   Power Series Solution

Use a power series to solve the differential equation $y' - 2y = 0$.

**Solution**   Assume that $y = \sum_{n=0}^{\infty} a_n x^n$ is a solution. Then,

$$y' = \sum_{n=1}^{\infty} n a_n x^{n-1}.$$

Substituting for $y'$ and $-2y$, you obtain the following series form of the differential equation. (Note that, from the third step to the fourth, the index of summation is changed to ensure that $x^n$ occurs in both sums.)

$$y' - 2y = 0$$

$$\sum_{n=1}^{\infty} n a_n x^{n-1} - 2\sum_{n=0}^{\infty} a_n x^n = 0$$

$$\sum_{n=1}^{\infty} n a_n x^{n-1} = \sum_{n=0}^{\infty} 2 a_n x^n$$

$$\sum_{n=0}^{\infty} (n+1) a_{n+1} x^n = \sum_{n=0}^{\infty} 2 a_n x^n$$

Now, by equating coefficients of like terms, you obtain the **recursion formula**

$$(n+1) a_{n+1} = 2 a_n$$

which implies that

$$a_{n+1} = \frac{2 a_n}{n+1}, \quad n \geq 0.$$

This formula generates the following results.

| $a_0$ | $a_1$ | $a_2$ | $a_3$ | $a_4$ | $a_5$ | $\cdots$ |
|---|---|---|---|---|---|---|
| $a_0$ | $2a_0$ | $\dfrac{2^2 a_0}{2}$ | $\dfrac{2^3 a_0}{3!}$ | $\dfrac{2^4 a_0}{4!}$ | $\dfrac{2^5 a_0}{5!}$ | $\cdots$ |

Using these values as the coefficients for the *solution* series, you have

$$y = \sum_{n=0}^{\infty} \frac{2^n a_0}{n!} x^n$$

$$= a_0 \sum_{n=0}^{\infty} \frac{(2x)^n}{n!}$$

$$= a_0 e^{2x}. \qquad \blacksquare$$

### Exploration

In Example 1, the differential equation could be solved easily without using a series. Determine which method should be used to solve the differential equation

$$y' - 2y = 0$$

and show that the result is the same as that obtained in the example.

In Example 1, the differential equation could be solved easily without using a series. The differential equation in Example 2 cannot be solved by any of the methods discussed in previous sections.

### EXAMPLE 2  Power Series Solution

Use a power series to solve the differential equation

$$y'' + xy' + y = 0.$$

**Solution**  Assume that $y = \sum_{n=0}^{\infty} a_n x^n$ is a solution. Then you have

$$y' = \sum_{n=1}^{\infty} na_n x^{n-1}, \quad xy' = \sum_{n=1}^{\infty} na_n x^n, \quad y'' = \sum_{n=2}^{\infty} n(n-1)a_n x^{n-2}.$$

Substituting for $y''$, $xy'$, and $y$ in the given differential equation, you obtain the following series.

$$\sum_{n=2}^{\infty} n(n-1)a_n x^{n-2} + \sum_{n=0}^{\infty} na_n x^n + \sum_{n=0}^{\infty} a_n x^n = 0$$

$$\sum_{n=2}^{\infty} n(n-1)a_n x^{n-2} = -\sum_{n=0}^{\infty} (n+1)a_n x^n$$

To obtain equal powers of $x$, adjust the summation indices by replacing $n$ by $n+2$ in the left-hand sum, to obtain

$$\sum_{n=0}^{\infty} (n+2)(n+1)a_{n+2} x^n = -\sum_{n=0}^{\infty} (n+1)a_n x^n.$$

By equating coefficients, you have

$$(n+2)(n+1)a_{n+2} = -(n+1)a_n$$

from which you obtain the recursion formula

$$a_{n+2} = -\frac{(n+1)}{(n+2)(n+1)}a_n = -\frac{a_n}{n+2}, \quad n \ge 0,$$

and the coefficients of the solution series are as follows.

$$a_2 = -\frac{a_0}{2} \qquad\qquad a_3 = -\frac{a_1}{3}$$

$$a_4 = -\frac{a_2}{4} = \frac{a_0}{2 \cdot 4} \qquad\qquad a_5 = -\frac{a_3}{5} = \frac{a_1}{3 \cdot 5}$$

$$a_6 = -\frac{a_4}{6} = -\frac{a_0}{2 \cdot 4 \cdot 6} \qquad\qquad a_7 = -\frac{a_5}{7} = -\frac{a_1}{3 \cdot 5 \cdot 7}$$

$$\vdots \qquad\qquad\qquad\qquad \vdots$$

$$a_{2k} = \frac{(-1)^k a_0}{2 \cdot 4 \cdot 6 \cdots (2k)} = \frac{(-1)^k a_0}{2^k(k!)} \qquad a_{2k+1} = \frac{(-1)^k a_1}{3 \cdot 5 \cdot 7 \cdots (2k+1)}$$

So, you can represent the general solution as the sum of two series—one for the even-powered terms with coefficients in terms of $a_0$, and one for the odd-powered terms with coefficients in terms of $a_1$.

$$y = a_0\left(1 - \frac{x^2}{2} + \frac{x^4}{2 \cdot 4} - \cdots\right) + a_1\left(x - \frac{x^3}{3} + \frac{x^5}{3 \cdot 5} - \cdots\right)$$

$$= a_0 \sum_{k=0}^{\infty} \frac{(-1)^k x^{2k}}{2^k(k!)} + a_1 \sum_{k=0}^{\infty} \frac{(-1)^k x^{2k+1}}{3 \cdot 5 \cdot 7 \cdots (2k+1)}$$

The solution has two arbitrary constants, $a_0$ and $a_1$, as you would expect in the general solution of a second-order differential equation.

# Approximation by Taylor Series

A second type of series solution method involves a differential equation with *initial conditions* and makes use of Taylor series, as given in Section 9.10.

**EXAMPLE 3** Approximation by Taylor Series

⋯▷ *See LarsonCalculus.com for an interactive version of this type of example.*

Use a Taylor series to find the first six terms of the series solution of

$$y' = y^2 - x$$

for the initial condition $y = 1$ when $x = 0$. Then use this polynomial to approximate values of $y$ for $0 \le x \le 1$.

**Solution**  Recall from Section 9.10 that, for $c = 0$,

$$y = y(0) + y'(0)x + \frac{y''(0)}{2!}x^2 + \frac{y'''(0)}{3!}x^3 + \cdots .$$

Because $y(0) = 1$ and $y' = y^2 - x$, you obtain the following.

$$y(0) = 1$$

$$y' = y^2 - x \qquad\qquad y'(0) = 1$$
$$y'' = 2yy' - 1 \qquad\qquad y''(0) = 2 - 1 = 1$$
$$y''' = 2yy'' + 2(y')^2 \qquad\qquad y'''(0) = 2 + 2 = 4$$
$$y^{(4)} = 2yy''' + 6y'y'' \qquad\qquad y^{(4)}(0) = 8 + 6 = 14$$
$$y^{(5)} = 2yy^{(4)} + 8y'y''' + 6(y'')^2 \qquad\qquad y^{(5)}(0) = 28 + 32 + 6 = 66$$

So, $y$ can be approximated by the first six terms of the series solution shown below.

$$y \approx y(0) + y'(0)x + \frac{y''(0)}{2!}x^2 + \frac{y'''(0)}{3!}x^3 + \frac{y^{(4)}(0)}{4!}x^4 + \frac{y^{(5)}(0)}{5!}x^5$$

$$= 1 + x + \frac{1}{2}x^2 + \frac{4}{3!}x^3 + \frac{14}{4!}x^4 + \frac{66}{5!}x^5$$

Using this polynomial, you can approximate values for $y$ in the interval $0 \le x \le 1$, as shown in the table below.

| $x$ | 0.0 | 0.1 | 0.2 | 0.3 | 0.4 | 0.5 | 0.6 | 0.7 | 0.8 | 0.9 | 1.0 |
|---|---|---|---|---|---|---|---|---|---|---|---|
| $y$ | 1.0000 | 1.1057 | 1.2264 | 1.3691 | 1.5432 | 1.7620 | 2.0424 | 2.4062 | 2.8805 | 3.4985 | 4.3000 |

In addition to approximating values of a function, you can also use a series solution to sketch a graph. In Figure 16.7, the series solutions of $y' = y^2 - x$ using the first two, four, and six terms are shown, along with an approximation found using a computer algebra system. The approximations are nearly the same for values of $x$ close to 0. As $x$ approaches 1, however, there is a noticeable difference among the approximations. For a series solution that is more accurate near $x = 1$, repeat Example 3 using $c = 1$.

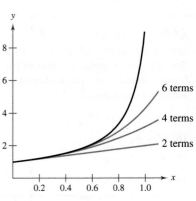

**Figure 16.7**

# 16.4 Exercises

See **CalcChat.com** for tutorial help and worked-out solutions to odd-numbered exercises.

## CONCEPT CHECK

**1. Power Series Solution Method**   Describe how to use power series to solve a differential equation.

**2. Recursion Formula**   What is a recursion formula? Give an example.

**Power Series Solution**   In Exercises 3–6, use a power series to solve the differential equation.

**3.** $5y' + y = 0$

**4.** $(x + 2)y' + y = 0$

**5.** $y' + 3xy = 0$

**6.** $y' - 2xy = 0$

**Finding Terms of a Power Series Solution** In Exercises 7 and 8, the solution of the differential equation is a sum of two power series. Find the first three terms of each power series. (See Example 2.)

**7.** $(x^2 + 4)y'' + y = 0$

**8.** $y'' + x^2 y = 0$

**Approximation by Taylor Series** In Exercises 9–14, use a Taylor series to find the first $n$ terms of the series solution of the differential equation that satisfies the initial condition(s). Use this polynomial to approximate $y$ for the given value of $x$.

**9.** $y' + (2x - 1)y = 0,\ y(0) = 2,\ n = 5,\ x = \frac{1}{2}$

**10.** $y' - 2xy = 0,\ y(0) = 1,\ n = 4,\ x = 1$

**11.** $y'' - 2xy = 0,\ y(0) = 1,\ y'(0) = -3,\ n = 6,\ x = \frac{1}{4}$

**12.** $y'' - 2xy' + y = 0,\ y(0) = 1,\ y'(0) = 2,\ n = 8,\ x = \frac{1}{2}$

**13.** $y'' + x^2 y' - (\cos x)y = 0,\ y(0) = 3,\ y'(0) = 2,\ n = 4,\ x = \frac{1}{3}$

**14.** $y'' + e^x y' - (\sin x)y = 0,\ y(0) = -2,\ y'(0) = 1,\ n = 4,\ x = \frac{1}{5}$

## EXPLORING CONCEPTS

**Using Different Methods**   In Exercises 15–18, verify that the power series solution of the differential equation is equivalent to the solution found using previously learned solution techniques.

**15.** $y' - ky = 0$

**16.** $y' + ky = 0$

**17.** $y'' - k^2 y = 0$

**18.** $y'' + k^2 y = 0$

**19. Investigation**   Consider the differential equation

$$y'' - xy' = 0$$

with the initial conditions

$$y(0) = 0 \quad \text{and} \quad y'(0) = 2.$$

(a) Find the series solution satisfying the initial conditions.

(b) Use a graphing utility to graph the third-degree and fifth-degree series approximations of the solution. Identify the approximations.

(c) Identify the symmetry of the solution.

**20. HOW DO YOU SEE IT?**   Consider the differential equation

$$y'' + 9y = 0$$

with initial conditions $y(0) = 2$ and $y'(0) = 6$. The figure shows the graph of the solution of the differential equation and the third-degree and fifth-degree polynomial approximations of the solution. Identify each.

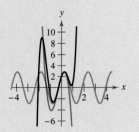

**Verifying that a Series Converges**   In Exercises 21–24, use the power series solution of the differential equation to verify that the series converges to the given function on the indicated interval.

**21.** $\displaystyle\sum_{n=0}^{\infty} \frac{x^n}{n!} = e^x,\ (-\infty, \infty)$

Differential equation: $y' - y = 0$

**22.** $\displaystyle\sum_{n=0}^{\infty} \frac{(-1)^n x^{2n}}{(2n)!} = \cos x,\ (-\infty, \infty)$

Differential equation: $y'' + y = 0$

**23.** $\displaystyle\sum_{n=0}^{\infty} \frac{(-1)^n x^{2n+1}}{2n + 1} = \arctan x,\ (-1, 1)$

Differential equation: $(x^2 + 1)y'' + 2xy' = 0$

**24.** $\displaystyle\sum_{n=0}^{\infty} \frac{(2n)! x^{2n+1}}{(2^n n!)^2 (2n + 1)} = \arcsin x,\ (-1, 1)$

Differential equation: $(1 - x^2)y'' - xy' = 0$

**25. Airy's Equation**   Find the first six terms of the series solution of Airy's equation, $y'' - xy = 0$.

# Review Exercises   See CalcChat.com for tutorial help and worked-out solutions to odd-numbered exercises.

**Testing for Exactness** In Exercises 1 and 2, determine whether the differential equation is exact.

**1.** $(y + x^3 + xy^2)\,dx - x\,dy = 0$

**2.** $(5x - y)\,dx + (5y - x)\,dy = 0$

**Solving an Exact Differential Equation** In Exercises 3–6, verify that the differential equation is exact. Then find the general solution.

**3.** $(10x + 8y + 2)\,dx + (8x + 5y + 2)\,dy = 0$

**4.** $(2x - 2y^3 + y)\,dx + (x - 6xy^2)\,dy = 0$

**5.** $(x - y - 5)\,dx - (x + 3y - 2)\,dy = 0$

**6.** $y\sin(xy)\,dx + [x\sin(xy) + y]\,dy = 0$

**Graphical and Analytic Analysis** In Exercises 7 and 8, (a) sketch an approximate solution of the differential equation satisfying the initial condition on the slope field, (b) find the particular solution that satisfies the initial condition, and (c) use a graphing utility to graph the particular solution. Compare the graph with the sketch in part (a).

**7.** $(2x - y)\,dx + (2y - x)\,dy = 0,\ y(2) = 2$

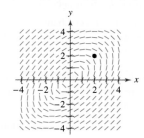

**8.** $(6xy - y^3)\,dx + (4y + 3x^2 - 3xy^2)\,dy = 0,\ y(0) = 1$

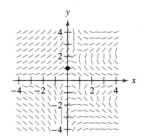

**Finding a Particular Solution** In Exercises 9–12, find the particular solution of the differential equation that satisfies the initial condition.

**9.** $(2x + y - 3)\,dx + (x - 3y + 1)\,dy = 0,\ y(2) = 0$

**10.** $3x^2y^2\,dx + (2x^3y - 3y^2)\,dy = 0,\ y(1) = 2$

**11.** $-\cos 2y\,dx + 2x\sin 2y\,dy = 0,\ y(3) = \pi$

**12.** $[9 + \ln(xy^3)]\,dx + \dfrac{3x}{y}\,dy = 0,\ y(1) = 1$

**Finding an Integrating Factor** In Exercises 13–16, find the integrating factor that is a function of $x$ or $y$ alone and use it to find the general solution of the differential equation.

**13.** $(3x^2 - y^2)\,dx + 2xy\,dy = 0$

**14.** $2xy\,dx + (y^2 - x^2)\,dy = 0$

**15.** $dx + (3x - e^{-2y})\,dy = 0$

**16.** $\cos y\,dx - [2(x - y)\sin y + \cos y]\,dy = 0$

**Verifying a Solution** In Exercises 17 and 18, verify the solution of the differential equation. Then use a graphing utility to graph the particular solutions for several different values of $C_1$ and $C_2$.

| Solution | Differential Equation |
|---|---|
| **17.** $y = C_1 e^{-3x} + C_2 e^{2x}$ | $y'' + y' - 6y = 0$ |
| **18.** $y = C_1 \cos 3x + C_2 \sin 3x$ | $y'' + 9y = 0$ |

**Finding a General Solution** In Exercises 19–28, use a characteristic equation to find the general solution of the linear differential equation.

**19.** $2y'' + 5y' + 3y = 0$

**20.** $y'' - 4y' - 2y = 0$

**21.** $y'' - 6y' = 0$

**22.** $25y'' + 30y' + 9y = 0$

**23.** $y'' + 8y = 0$

**24.** $y'' + y' + 3y = 0$

**25.** $y''' - 2y'' - 3y' = 0$

**26.** $y''' - 6y'' + 12y' - 8y = 0$

**27.** $y^{(4)} - 5y'' = 0$

**28.** $y^{(4)} + 6y'' + 9y = 0$

**Finding a Particular Solution: Initial Conditions** In Exercises 29–32, use a characteristic equation to find the particular solution of the linear differential equation that satisfies the initial conditions.

**29.** $y'' - y' - 2y = 0$
$y(0) = 0, y'(0) = 3$

**30.** $y'' + 4y' + 5y = 0$
$y(0) = 2, y'(0) = -7$

**31.** $y'' + 2y' - 3y = 0$
$y(0) = 2, y'(0) = 0$

**32.** $y'' + 12y' + 36y = 0$
$y(0) = 2, y'(0) = 1$

**Finding a Particular Solution: Boundary Conditions** In Exercises 33 and 34, use a charcteristic equation to find the particular solution of the linear differential equation that satisfies the boundary conditions.

**33.** $y'' + 2y' + 5y = 0$
$y(1) = 4, y(2) = 0$

**34.** $y'' + y = 0$
$y(0) = 2, y(\pi/2) = 1$

**Motion of a Spring** In Exercises 35 and 36, a 64-pound weight stretches a spring $\frac{4}{3}$ feet from its natural length. Use the given information to find a formula for the position of the weight as a function of time.

**35.** The weight is pulled $\frac{1}{2}$ foot below equilibrium and released.

**36.** The weight is pulled $\frac{3}{4}$ foot below equilibrium and released. The motion takes place in a medium that furnishes a damping force of magnitude $\frac{1}{8}|v|$ at all times.

**Method of Undetermined Coefficients** In Exercises 37–40, solve the differential equation by the method of undetermined coefficients.

**37.** $y'' + y = x^3 + x$

**38.** $y'' + 2y = e^{2x} + x$

**39.** $y'' - 8y' - 9y = 9x - 10$

**40.** $y'' + 5y' + 4y = x^2 + \sin 2x$

**Choosing the Form of the Particular Solution** In Exercises 41 and 42, determine a suitable choice for $y_p$ for the differential equation, given its general solution of the homogeneous equation. Explain your reasoning. (Do not solve the equation.)

$$y'' + ay' + by = F(x) \qquad\qquad y_h$$

**41.** $y'' - 4y' + 3y = e^x + 8e^{3x} \qquad C_1 e^x + C_2 e^{3x}$

**42.** $y'' = 2x + 1 \qquad\qquad\qquad C_1 + C_2 x$

**Method of Undetermined Coefficients** In Exercises 43 and 44, solve the differential equation by the method of undetermined coefficients.

**43.** $y'' + y = 2 \cos x$

**44.** $2y'' - y' = 4x$

**Using Initial Conditions** In Exercises 45–50, solve the differential equation by the method of undetermined coefficients subject to the initial conditions.

**45.** $y'' - y' - 6y = 54$

   $y(0) = 2, y'(0) = 0$

**46.** $y'' + 25y = e^x$

   $y(0) = 0, y'(0) = 0$

**47.** $y'' + 4y = \cos x$

   $y(0) = 6, y'(0) = -6$

**48.** $y'' + 3y' = 6x$

   $y(0) = 2, y'(0) = \frac{10}{3}$

**49.** $y'' - y' - 2y = 1 + xe^{-x}$

   $y(0) = 1, y'(0) = 3$

**50.** $y''' - y'' = 4x^2$

   $y(0) = 1, y'(0) = 1, y''(0) = 1$

**Method of Variation of Parameters** In Exercises 51–54, solve the differential equation by the method of variation of parameters.

**51.** $y'' + 9y = \csc 3x$

**52.** $4y'' + y = \sec \dfrac{x}{2} \tan \dfrac{x}{2}$

**53.** $y'' - 2y' + y = 2xe^x$

**54.** $y'' + 2y' + y = \dfrac{1}{x^2 e^x}$

**55. Electrical Circuit** The differential equation

$$\frac{d^2 q}{dt^2} + 4\frac{dq}{dt} + 8q = 3 \sin 4t$$

models the charge $q$ on a capacitor of an electrical circuit. Find the charge $q$ as a function of time $t$. Assume that $q(0) = 0$ and $q'(0) = 0$.

**56. Investigation** The differential equation

$$\frac{8}{32}y'' + by' + ky = \frac{8}{32}F(t), \quad y(0) = \frac{1}{2}, y'(0) = 0$$

models the oscillating motion of an object on the end of a spring, where $y$ is the displacement from equilibrium (positive direction is downward), measured in feet, $t$ is time in seconds, $b$ is the magnitude of the resistance to the motion, $k$ is the spring constant from Hooke's Law, and $F(t)$ is the acceleration imposed on the system.

(a) Solve the differential equation and use a graphing utility to graph the solution for each of the assigned quantities for $b$, $k$, and $F(t)$.

  (i)   $b = 0, k = 1, F(t) = 24 \sin \pi t$

  (ii)  $b = 0, k = 2, F(t) = 24 \sin(2\sqrt{2}t)$

  (iii) $b = 0.1, k = 2, F(t) = 0$

  (iv)  $b = 1, k = 2, F(t) = 0$

(b) Describe the effect of increasing the resistance to motion $b$.

(c) Explain how the motion of the object changes when a stiffer spring (greater value of $k$) is used.

**57. Think About It**

(a) Explain how, by observation, you know that a form of a particular solution of the differential equation $y'' + 3y = 12 \sin x$ is $y_p = A \sin x$.

(b) Use your explanation in part (a) to find a particular solution of the differential equation $y'' + 5y = 10 \cos x$.

(c) Compare the algebra required to find particular solutions in parts (a) and (b) with that required when the form of the particular solution is $y_p = A \cos x + B \sin x$.

**58. Think About It** Explain how you can find a particular solution of the differential equation $y'' + 4y' + 6y = 30$ by observation.

**Power Series Solution** In Exercises 59 and 60, use a power series to solve the differential equation.

**59.** $(x - 4)y' + y = 0$

**60.** $y'' + 3xy' - 3y = 0$

**Approximation by Taylor Series** In Exercises 61 and 62, use a Taylor series to find the first $n$ terms of the series solution of the differential equation that satisfies the initial conditions. Use this polynomial to approximate $y$ for the given value of $x$.

**61.** $y'' + y' - e^x y = 0, y(0) = 2, y'(0) = 0, n = 4, x = \dfrac{1}{4}$

**62.** $y'' + xy = 0, y(0) = 1, y'(0) = 1, n = 6, x = \dfrac{1}{2}$

# P.S. Problem Solving

See **CalcChat.com** for tutorial help and
worked-out solutions to odd-numbered exercises.

1. **Finding a General Solution** Find the value of $k$ that makes the differential equation

$$(3x^2 + kxy^2)\, dx - (5x^2y + ky^2)\, dy = 0$$

exact. Using this value of $k$, find the general solution.

2. **Using an Integrating Factor** The differential equation $(kx^2 + y^2)\, dx - kxy\, dy = 0$ is not exact, but the integrating factor $1/x^2$ makes it exact.

   (a) Use this information to find the value of $k$.

   (b) Using this value of $k$, find the general solution.

3. **Finding a General Solution** Find the general solution of the differential equation $y'' - a^2y = 0$, $a > 0$. Show that the general solution can be written in the form

$$y = C_1 \cosh ax + C_2 \sinh ax.$$

4. **Finding a General Solution** Find the general solution of the differential equation $y'' + \beta^2 y = 0$. Show that the general solution can be written in the form

$$y = C\sin(\beta x + \phi),\ 0 \le \phi < 2\pi.$$

5. **Distinct Real Zeros** Given that the characteristic equation of the differential equation $y'' + ay' + by = 0$ has two distinct real zeros, $m_1 = r + s$ and $m_2 = r - s$, where $r$ and $s$ are real numbers, show that the general solution of the differential equation can be written in the form

$$y = e^{rx}(C_1 \cosh sx + C_2 \sinh sx).$$

6. **Limit of a Solution** Given that $a$ and $b$ are positive and that $y(x)$ is a solution of the differential equation

$$y'' + ay' + by = 0$$

show that $\displaystyle\lim_{x\to\infty} y(x) = 0$.

7. **Trivial and Nontrivial Solutions** Consider the differential equation $y'' + ay = 0$ with boundary conditions $y(0) = 0$ and $y(L) = 0$ for some nonzero real number $L$.

   (a) For $a = 0$, show that the differential equation has only the trivial solution $y = 0$.

   (b) For $a < 0$, show that the differential equation has only the trivial solution $y = 0$.

   (c) For $a > 0$, find the value(s) of $a$ for which the solution is nontrivial. Then find the corresponding solution(s).

8. **Euler's Differential Equation** **Euler's differential equation** is of the form

$$x^2y'' + axy' + by = 0,\quad x > 0$$

where $a$ and $b$ are constants.

   (a) Show that this equation can be transformed into a second-order linear differential equation with constant coefficients by using the substitution $x = e^t$.

   (b) Solve $x^2y'' + 6xy' + 6y = 0$.

9. **Pendulum** Consider a pendulum of length $L$ that swings by the force of gravity only.

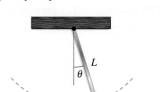

   For small values of $\theta = \theta(t)$, the motion of the pendulum can be approximated by the differential equation

$$\frac{d^2\theta}{dt^2} + \frac{g}{L}\theta = 0$$

   where $g$ is the acceleration due to gravity.

   (a) Find the general solution of the differential equation and show that it can be written in the form

$$\theta(t) = A\cos\left[\sqrt{\frac{g}{L}}(t + \phi)\right].$$

   (b) Find the particular solution for a pendulum of length 0.25 meter when the initial conditions are $\theta(0) = 0.1$ radian and $\theta'(0) = 0.5$ radian per second. (Use $g = 9.8$ meters per second per second.)

   (c) Determine the period of the pendulum.

   (d) Determine the maximum value of $\theta$.

   (e) How much time from $t = 0$ does it take for $\theta$ to be 0 the first time? the second time?

   (f) What is the angular velocity $\theta'$ when $\theta = 0$ the first time? the second time?

10. **Deflection of a Beam** A horizontal beam with a length of 2 meters rests on supports located at the ends of the beam.

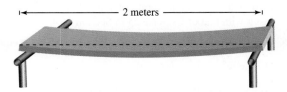

   The beam is supporting a load of $W$ kilograms per meter. The resulting deflection $y$ of the beam at a horizontal distance of $x$ meters from the left end can be modeled by

$$A\frac{d^2y}{dx^2} = 2Wx - \frac{1}{2}Wx^2$$

   where $A$ is a positive constant.

   (a) Solve the differential equation to find the deflection $y$ as a function of the horizontal distance $x$.

   (b) Use a graphing utility to determine the location and value of the maximum deflection.

**Damped Motion**  In Exercises 11–14, consider a damped mass-spring system whose motion is described by the differential equation

$$\frac{d^2y}{dt^2} + 2\lambda \frac{dy}{dt} + \omega^2 y = 0.$$

**The zeros of its characteristic equation are**

$$m_1 = -\lambda + \sqrt{\lambda^2 - \omega^2}$$

**and**

$$m_2 = -\lambda - \sqrt{\lambda^2 - \omega^2}.$$

**For $\lambda^2 - \omega^2 > 0$, the system is *overdamped*; for $\lambda^2 - \omega^2 = 0$, it is *critically damped*; and for $\lambda^2 - \omega^2 < 0$, it is *underdamped*.**

(a) **Determine whether the differential equation represents an overdamped, critically damped, or underdamped system.**

(b) **Find the particular solution that satisfies the initial conditions.**

(c) **Use a graphing utility to graph the particular solution found in part (b). Explain how the graph illustrates the type of damping in the system.**

11. $\dfrac{d^2y}{dt^2} + 8\dfrac{dy}{dt} + 16y = 0$    12. $\dfrac{d^2y}{dt^2} + 2\dfrac{dy}{dt} + 26y = 0$

$y(0) = 1, \ y'(0) = 1$        $y(0) = 1, \ y'(0) = 4$

13. $\dfrac{d^2y}{dt^2} + 20\dfrac{dy}{dt} + 64y = 0$    14. $\dfrac{d^2y}{dt^2} + 2\dfrac{dy}{dt} + y = 0$

$y(0) = 2, \ y'(0) = -20$      $y(0) = 2, \ y'(0) = -1$

15. **Airy's Equation**  Consider Airy's equation given in Section 16.4, Exercise 25. Rewrite the equation as

$$y'' - (x - 1)y - y = 0.$$

Then use a power series of the form

$$y = \sum_{n=0}^{\infty} a_n(x - 1)^n$$

to find the first *eight* terms of the solution. Compare your result with that of Exercise 25 in Section 16.4.

16. **Chebyshev's Equation**  Consider **Chebyshev's equation**

$$(1 - x^2)y'' - xy' + k^2 y = 0.$$

Polynomial solutions of this differential equation are called *Chebyshev polynomials* and are denoted by $T_k(x)$. They satisfy the recursion equation

$$T_{n+1}(x) = 2xT_n(x) - T_{n-1}(x).$$

(a) Given that $T_0(x) = 1$ and $T_1(x) = x$, determine the Chebyshev polynomials $T_2(x)$, $T_3(x)$, and $T_4(x)$.

(b) Verify that $T_0(x)$, $T_1(x)$, $T_2(x)$, $T_3(x)$, and $T_4(x)$ are solutions of the given differential equation.

(c) Verify the following Chebyshev polynomials.

$T_5(x) = 16x^5 - 20x^3 + 5x$

$T_6(x) = 32x^6 - 48x^4 + 18x^2 - 1$

$T_7(x) = 64x^7 - 112x^5 + 56x^3 - 7x$

17. **Bessel's Equation: Order Zero**  The differential equation $x^2y'' + xy' + x^2y = 0$ is known as **Bessel's equation of order zero.**

(a) Use a power series of the form

$$y = \sum_{n=0}^{\infty} a_n x^n$$

to find the solution.

(b) Compare your result with that of the function $J_0(x)$ given in Section 9.8, Exercise 65.

18. **Bessel's Equation: Order One**  The differential equation

$$x^2y'' + xy' + (x^2 - 1)y = 0$$

is known as **Bessel's equation of order one.**

(a) Use a power series of the form

$$y = \sum_{n=0}^{\infty} a_n x^n$$

to find the solution.

(b) Compare your result with that of the function $J_1(x)$ given in Section 9.8, Exercise 66.

19. **Hermite's Equation**  Consider **Hermite's equation**

$$y'' - 2xy' + 2ky = 0.$$

(a) Use a power series of the form

$$y = \sum_{n=0}^{\infty} a_n x^n$$

to find the solution when $k = 4$. [*Hint:* Choose the arbitrary constants such that the leading term is $(2x)^k$.]

(b) Polynomial solutions of Hermite's equation are called *Hermite polynomials* and are denoted by $H_k(x)$. The general form for $H_k(x)$ can be written as

$$H_k(x) = \sum_{n=0}^{P} \frac{(-1)^n k! (2x)^{k-2n}}{n!(k - 2n)!}$$

where $P$ is the greatest integer less than or equal to $k/2$. Use this formula to determine the Hermite polynomials $H_0(x)$, $H_1(x)$, $H_2(x)$, $H_3(x)$, and $H_4(x)$.

20. **Laguerre's Equation**  Consider **Laguerre's equation**

$$xy'' + (1 - x)y' + ky = 0.$$

(a) Polynomial solutions of Laguerre's equation are called *Laguerre polynomials* and are denoted by $L_k(x)$. Use a power series of the form

$$y = \sum_{n=0}^{\infty} a_n x^n$$

to show that

$$L_k(x) = \sum_{n=0}^{k} \frac{(-1)^n k! x^n}{(k - n)!(n!)^2}.$$

Assume that $a_0 = 1$.

(b) Determine the Laguerre polynomials $L_0(x)$, $L_1(x)$, $L_2(x)$, $L_3(x)$, and $L_4(x)$.

# Appendices

# A  Proofs of Selected Theorems

The text version of Appendix A, Proofs of Selected Theorems, is available at *CengageBrain.com*. Also, to enhance your study of calculus, each proof is available in video format at *LarsonCalculus.com*. At this website, you can watch videos of Bruce Edwards explaining each proof in the text and in Appendix A. To access a video, visit the website at *LarsonCalculus.com* or scan the code near the proof or the proof's reference.

**Sample Video:  Bruce Edwards's Proof of the
Power Rule at *LarsonCalculus.com***

---

## The Power Rule

Before proving the next rule, it is important to review the procedure for expanding a binomial.

$$(x + \Delta x)^2 = x^2 + 2x\Delta x + (\Delta x)^2$$
$$(x + \Delta x)^3 = x^3 + 3x^2\Delta x + 3x(\Delta x)^2 + (\Delta x)^3$$
$$(x + \Delta x)^4 = x^4 + 4x^3\Delta x + 6x^2(\Delta x)^2 + 4x(\Delta x)^3 + (\Delta x)^4$$
$$(x + \Delta x)^5 = x^5 + 5x^4\Delta x + 10x^3(\Delta x)^2 + 10x^2(\Delta x)^3 + 5x(\Delta x)^4 + (\Delta x)^5$$

The general binomial expansion for a positive integer $n$ is

$$(x + \Delta x)^n = x^n + nx^{n-1}(\Delta x) + \underbrace{\frac{n(n-1)x^{n-2}}{2}(\Delta x)^2 + \cdots + (\Delta x)^n}.$$

$(\Delta x)^2$ is a factor of these terms.

This binomial expansion is used in proving a special case of the Power Rule.

> **THEOREM 2.3   The Power Rule**
>
> If $n$ is a rational number, then the function $f(x) = x^n$ is differentiable and
>
> $$\frac{d}{dx}[x^n] = nx^{n-1}.$$
>
> For $f$ to be differentiable at $x = 0$, $n$ must be a number such that $x^{n-1}$ is defined on an interval containing 0.

**••REMARK** From Example 7 in Section 2.1, you know that the function $f(x) = x^{1/3}$ is defined at $x = 0$ but is not differentiable at $x = 0$. This is because $x^{-2/3}$ is not defined on an interval containing 0.

**Proof**  If $n$ is a positive integer greater than 1, then the binomial expansion produces

$$\frac{d}{dx}[x^n] = \lim_{\Delta x \to 0} \frac{(x + \Delta x)^n - x^n}{\Delta x}$$

$$= \lim_{\Delta x \to 0} \frac{x^n + nx^{n-1}(\Delta x) + \frac{n(n-\ }{ }}{ }$$

$$= \lim_{\Delta x \to 0} \left[ nx^{n-1} + \frac{n(n-1)x^{n-2}}{2}(\ \right.$$

$$= nx^{n-1} + 0 + \cdots + 0$$

$$= nx^{n-1}.$$

This proves the case for which $n$ is a positive inte the case for $n = 1$. Example 7 in Section 2.3 integer. In Exercise 73 in Section 2.5, you are rational. (In Section 5.5, the Power Rule will be

When using the Power Rule, the case f separate differentiation rule. That is,

$$\frac{d}{dx}[x] = 1.$$

This rule is consistent with the fact that the s Figure 2.15.

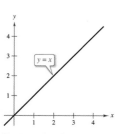

The slope of the line $y = x$ is 1.
**Figure 2.15**

Proof: The Power Rule

02:21

IIII :: vimeo

## B    Integration Tables

**Forms Involving $u^n$**

**1.** $\displaystyle\int u^n \, du = \frac{u^{n+1}}{n+1} + C, \quad n \neq -1$

**2.** $\displaystyle\int \frac{1}{u} \, du = \ln|u| + C$

**Forms Involving $a + bu$**

**3.** $\displaystyle\int \frac{u}{a+bu} \, du = \frac{1}{b^2}(bu - a\ln|a+bu|) + C$

**4.** $\displaystyle\int \frac{u}{(a+bu)^2} \, du = \frac{1}{b^2}\left(\frac{a}{a+bu} + \ln|a+bu|\right) + C$

**5.** $\displaystyle\int \frac{u}{(a+bu)^n} \, du = \frac{1}{b^2}\left[\frac{-1}{(n-2)(a+bu)^{n-2}} + \frac{a}{(n-1)(a+bu)^{n-1}}\right] + C, \quad n \neq 1, 2$

**6.** $\displaystyle\int \frac{u^2}{a+bu} \, du = \frac{1}{b^3}\left[-\frac{bu}{2}(2a - bu) + a^2\ln|a+bu|\right] + C$

**7.** $\displaystyle\int \frac{u^2}{(a+bu)^2} \, du = \frac{1}{b^3}\left(bu - \frac{a^2}{a+bu} - 2a\ln|a+bu|\right) + C$

**8.** $\displaystyle\int \frac{u^2}{(a+bu)^3} \, du = \frac{1}{b^3}\left[\frac{2a}{a+bu} - \frac{a^2}{2(a+bu)^2} + \ln|a+bu|\right] + C$

**9.** $\displaystyle\int \frac{u^2}{(a+bu)^n} \, du = \frac{1}{b^3}\left[\frac{-1}{(n-3)(a+bu)^{n-3}} + \frac{2a}{(n-2)(a+bu)^{n-2}} - \frac{a^2}{(n-1)(a+bu)^{n-1}}\right] + C, \quad n \neq 1, 2, 3$

**10.** $\displaystyle\int \frac{1}{u(a+bu)} \, du = \frac{1}{a}\ln\left|\frac{u}{a+bu}\right| + C$

**11.** $\displaystyle\int \frac{1}{u(a+bu)^2} \, du = \frac{1}{a}\left(\frac{1}{a+bu} + \frac{1}{a}\ln\left|\frac{u}{a+bu}\right|\right) + C$

**12.** $\displaystyle\int \frac{1}{u^2(a+bu)} \, du = -\frac{1}{a}\left(\frac{1}{u} + \frac{b}{a}\ln\left|\frac{u}{a+bu}\right|\right) + C$

**13.** $\displaystyle\int \frac{1}{u^2(a+bu)^2} \, du = -\frac{1}{a^2}\left[\frac{a+2bu}{u(a+bu)} + \frac{2b}{a}\ln\left|\frac{u}{a+bu}\right|\right] + C$

**Forms Involving $a + bu + cu^2, \ b^2 \neq 4ac$**

**14.** $\displaystyle\int \frac{1}{a+bu+cu^2} \, du = \begin{cases} \dfrac{2}{\sqrt{4ac-b^2}}\arctan\dfrac{2cu+b}{\sqrt{4ac-b^2}} + C, & b^2 < 4ac \\[3mm] \dfrac{1}{\sqrt{b^2-4ac}}\ln\left|\dfrac{2cu+b-\sqrt{b^2-4ac}}{2cu+b+\sqrt{b^2-4ac}}\right| + C, & b^2 > 4ac \end{cases}$

**15.** $\displaystyle\int \frac{u}{a+bu+cu^2} \, du = \frac{1}{2c}\left(\ln|a+bu+cu^2| - b\int \frac{1}{a+bu+cu^2} \, du\right)$

**Forms Involving $\sqrt{a+bu}$**

**16.** $\displaystyle\int u^n\sqrt{a+bu} \, du = \frac{2}{b(2n+3)}\left[u^n(a+bu)^{3/2} - na\int u^{n-1}\sqrt{a+bu} \, du\right]$

**17.** $\displaystyle\int \frac{1}{u\sqrt{a+bu}} \, du = \begin{cases} \dfrac{1}{\sqrt{a}}\ln\left|\dfrac{\sqrt{a+bu} - \sqrt{a}}{\sqrt{a+bu} + \sqrt{a}}\right| + C, & a > 0 \\[3mm] \dfrac{2}{\sqrt{-a}}\arctan\sqrt{\dfrac{a+bu}{-a}} + C, & a < 0 \end{cases}$

**18.** $\displaystyle\int \frac{1}{u^n\sqrt{a+bu}} \, du = \frac{-1}{a(n-1)}\left[\frac{\sqrt{a+bu}}{u^{n-1}} + \frac{(2n-3)b}{2}\int \frac{1}{u^{n-1}\sqrt{a+bu}} \, du\right], \quad n \neq 1$

**19.** $\displaystyle\int \frac{\sqrt{a + bu}}{u}\, du = 2\sqrt{a + bu} + a\int \frac{1}{u\sqrt{a + bu}}\, du$

**20.** $\displaystyle\int \frac{\sqrt{a + bu}}{u^n}\, du = \frac{-1}{a(n - 1)}\left[\frac{(a + bu)^{3/2}}{u^{n-1}} + \frac{(2n - 5)b}{2}\int \frac{\sqrt{a + bu}}{u^{n-1}}\, du\right],\ n \neq 1$

**21.** $\displaystyle\int \frac{u}{\sqrt{a + bu}}\, du = \frac{-2(2a - bu)}{3b^2}\sqrt{a + bu} + C$

**22.** $\displaystyle\int \frac{u^n}{\sqrt{a + bu}}\, du = \frac{2}{(2n + 1)b}\left(u^n\sqrt{a + bu} - na\int \frac{u^{n-1}}{\sqrt{a + bu}}\, du\right)$

## Forms Involving $a^2 \pm u^2$, $a > 0$

**23.** $\displaystyle\int \frac{1}{a^2 + u^2}\, du = \frac{1}{a}\arctan \frac{u}{a} + C$

**24.** $\displaystyle\int \frac{1}{u^2 - a^2}\, du = -\int \frac{1}{a^2 - u^2}\, du = \frac{1}{2a}\ln\left|\frac{u - a}{u + a}\right| + C$

**25.** $\displaystyle\int \frac{1}{(a^2 \pm u^2)^n}\, du = \frac{1}{2a^2(n - 1)}\left[\frac{u}{(a^2 \pm u^2)^{n-1}} + (2n - 3)\int \frac{1}{(a^2 \pm u^2)^{n-1}}\, du\right],\ n \neq 1$

## Forms Involving $\sqrt{u^2 \pm a^2}$, $a > 0$

**26.** $\displaystyle\int \sqrt{u^2 \pm a^2}\, du = \frac{1}{2}\left(u\sqrt{u^2 \pm a^2} \pm a^2 \ln\left|u + \sqrt{u^2 \pm a^2}\right|\right) + C$

**27.** $\displaystyle\int u^2\sqrt{u^2 \pm a^2}\, du = \frac{1}{8}\left[u(2u^2 \pm a^2)\sqrt{u^2 \pm a^2} - a^4 \ln\left|u + \sqrt{u^2 \pm a^2}\right|\right] + C$

**28.** $\displaystyle\int \frac{\sqrt{u^2 + a^2}}{u}\, du = \sqrt{u^2 + a^2} - a \ln\left|\frac{a + \sqrt{u^2 + a^2}}{u}\right| + C$

**29.** $\displaystyle\int \frac{\sqrt{u^2 - a^2}}{u}\, du = \sqrt{u^2 - a^2} - a\, \text{arcsec}\, \frac{|u|}{a} + C$

**30.** $\displaystyle\int \frac{\sqrt{u^2 \pm a^2}}{u^2}\, du = \frac{-\sqrt{u^2 \pm a^2}}{u} + \ln\left|u + \sqrt{u^2 \pm a^2}\right| + C$

**31.** $\displaystyle\int \frac{1}{\sqrt{u^2 \pm a^2}}\, du = \ln\left|u + \sqrt{u^2 \pm a^2}\right| + C$

**32.** $\displaystyle\int \frac{1}{u\sqrt{u^2 + a^2}}\, du = \frac{-1}{a}\ln\left|\frac{a + \sqrt{u^2 + a^2}}{u}\right| + C$

**33.** $\displaystyle\int \frac{1}{u\sqrt{u^2 - a^2}}\, du = \frac{1}{a}\,\text{arcsec}\, \frac{|u|}{a} + C$

**34.** $\displaystyle\int \frac{u^2}{\sqrt{u^2 \pm a^2}}\, du = \frac{1}{2}\left(u\sqrt{u^2 \pm a^2} \mp a^2 \ln\left|u + \sqrt{u^2 \pm a^2}\right|\right) + C$

**35.** $\displaystyle\int \frac{1}{u^2\sqrt{u^2 \pm a^2}}\, du = \mp\frac{\sqrt{u^2 \pm a^2}}{a^2 u} + C$

**36.** $\displaystyle\int \frac{1}{(u^2 \pm a^2)^{3/2}}\, du = \frac{\pm u}{a^2\sqrt{u^2 \pm a^2}} + C$

## Forms Involving $\sqrt{a^2 - u^2}$, $a > 0$

**37.** $\displaystyle\int \sqrt{a^2 - u^2}\, du = \frac{1}{2}\left(u\sqrt{a^2 - u^2} + a^2 \arcsin \frac{u}{a}\right) + C$

**38.** $\displaystyle\int u^2\sqrt{a^2 - u^2}\, du = \frac{1}{8}\left[u(2u^2 - a^2)\sqrt{a^2 - u^2} + a^4 \arcsin \frac{u}{a}\right] + C$

**39.** $\displaystyle\int \frac{\sqrt{a^2 - u^2}}{u}\, du = \sqrt{a^2 - u^2} - a \ln\left|\frac{a + \sqrt{a^2 - u^2}}{u}\right| + C$    **40.** $\displaystyle\int \frac{\sqrt{a^2 - u^2}}{u^2}\, du = \frac{-\sqrt{a^2 - u^2}}{u} - \arcsin\frac{u}{a} + C$

**41.** $\displaystyle\int \frac{1}{\sqrt{a^2 - u^2}}\, du = \arcsin\frac{u}{a} + C$    **42.** $\displaystyle\int \frac{1}{u\sqrt{a^2 - u^2}}\, du = \frac{-1}{a} \ln\left|\frac{a + \sqrt{a^2 - u^2}}{u}\right| + C$

**43.** $\displaystyle\int \frac{u^2}{\sqrt{a^2 - u^2}}\, du = \frac{1}{2}\left(-u\sqrt{a^2 - u^2} + a^2 \arcsin\frac{u}{a}\right) + C$    **44.** $\displaystyle\int \frac{1}{u^2 \sqrt{a^2 - u^2}}\, du = \frac{-\sqrt{a^2 - u^2}}{a^2 u} + C$

**45.** $\displaystyle\int \frac{1}{(a^2 - u^2)^{3/2}}\, du = \frac{u}{a^2 \sqrt{a^2 - u^2}} + C$

### Forms Involving sin *u* or cos *u*

**46.** $\displaystyle\int \sin u\, du = -\cos u + C$    **47.** $\displaystyle\int \cos u\, du = \sin u + C$

**48.** $\displaystyle\int \sin^2 u\, du = \frac{1}{2}(u - \sin u \cos u) + C$    **49.** $\displaystyle\int \cos^2 u\, du = \frac{1}{2}(u + \sin u \cos u) + C$

**50.** $\displaystyle\int \sin^n u\, du = -\frac{\sin^{n-1} u \cos u}{n} + \frac{n-1}{n}\int \sin^{n-2} u\, du$    **51.** $\displaystyle\int \cos^n u\, du = \frac{\cos^{n-1} u \sin u}{n} + \frac{n-1}{n}\int \cos^{n-2} u\, du$

**52.** $\displaystyle\int u \sin u\, du = \sin u - u \cos u + C$    **53.** $\displaystyle\int u \cos u\, du = \cos u + u \sin u + C$

**54.** $\displaystyle\int u^n \sin u\, du = -u^n \cos u + n\int u^{n-1} \cos u\, du$    **55.** $\displaystyle\int u^n \cos u\, du = u^n \sin u - n\int u^{n-1} \sin u\, du$

**56.** $\displaystyle\int \frac{1}{1 \pm \sin u}\, du = \tan u \mp \sec u + C$    **57.** $\displaystyle\int \frac{1}{1 \pm \cos u}\, du = -\cot u \pm \csc u + C$

**58.** $\displaystyle\int \frac{1}{\sin u \cos u}\, du = \ln|\tan u| + C$

### Forms Involving tan *u*, cot *u*, sec *u*, or csc *u*

**59.** $\displaystyle\int \tan u\, du = -\ln|\cos u| + C$    **60.** $\displaystyle\int \cot u\, du = \ln|\sin u| + C$

**61.** $\displaystyle\int \sec u\, du = \ln|\sec u + \tan u| + C$

**62.** $\displaystyle\int \csc u\, du = \ln|\csc u - \cot u| + C$    or    $\displaystyle\int \csc u\, du = -\ln|\csc u + \cot u| + C$

**63.** $\displaystyle\int \tan^2 u\, du = -u + \tan u + C$    **64.** $\displaystyle\int \cot^2 u\, du = -u - \cot u + C$

**65.** $\displaystyle\int \sec^2 u\, du = \tan u + C$    **66.** $\displaystyle\int \csc^2 u\, du = -\cot u + C$

**67.** $\displaystyle\int \tan^n u\, du = \frac{\tan^{n-1} u}{n-1} - \int \tan^{n-2} u\, du,\ n \neq 1$    **68.** $\displaystyle\int \cot^n u\, du = -\frac{\cot^{n-1} u}{n-1} - \int \cot^{n-2} u\, du,\ n \neq 1$

**69.** $\displaystyle\int \sec^n u\, du = \frac{\sec^{n-2} u \tan u}{n-1} + \frac{n-2}{n-1}\int \sec^{n-2} u\, du,\ n \neq 1$

**70.** $\displaystyle\int \csc^n u\, du = -\frac{\csc^{n-2} u \cot u}{n-1} + \frac{n-2}{n-1}\int \csc^{n-2} u\, du,\ n \neq 1$

**71.** $\displaystyle\int \frac{1}{1 \pm \tan u}\, du = \frac{1}{2}(u \pm \ln|\cos u \pm \sin u|) + C$

**72.** $\displaystyle\int \frac{1}{1 \pm \cot u}\, du = \frac{1}{2}(u \mp \ln|\sin u \pm \cos u|) + C$

**73.** $\displaystyle\int \frac{1}{1 \pm \sec u}\, du = u + \cot u \mp \csc u + C$

**74.** $\displaystyle\int \frac{1}{1 \pm \csc u}\, du = u - \tan u \pm \sec u + C$

**Forms Involving Inverse Trigonometric Functions**

**75.** $\displaystyle\int \arcsin u\, du = u \arcsin u + \sqrt{1 - u^2} + C$

**76.** $\displaystyle\int \arccos u\, du = u \arccos u - \sqrt{1 - u^2} + C$

**77.** $\displaystyle\int \arctan u\, du = u \arctan u - \ln\sqrt{1 + u^2} + C$

**78.** $\displaystyle\int \operatorname{arccot} u\, du = u \operatorname{arccot} u + \ln\sqrt{1 + u^2} + C$

**79.** $\displaystyle\int \operatorname{arcsec} u\, du = u \operatorname{arcsec} u - \ln\left|u + \sqrt{u^2 - 1}\right| + C$

**80.** $\displaystyle\int \operatorname{arccsc} u\, du = u \operatorname{arccsc} u + \ln\left|u + \sqrt{u^2 - 1}\right| + C$

**Forms Involving $e^u$**

**81.** $\displaystyle\int e^u\, du = e^u + C$

**82.** $\displaystyle\int u e^u\, du = (u - 1)e^u + C$

**83.** $\displaystyle\int u^n e^u\, du = u^n e^u - n\int u^{n-1} e^u\, du$

**84.** $\displaystyle\int \frac{1}{1 + e^u}\, du = u - \ln(1 + e^u) + C$

**85.** $\displaystyle\int e^{au} \sin bu\, du = \frac{e^{au}}{a^2 + b^2}(a \sin bu - b \cos bu) + C$

**86.** $\displaystyle\int e^{au} \cos bu\, du = \frac{e^{au}}{a^2 + b^2}(a \cos bu + b \sin bu) + C$

**Forms Involving ln $u$**

**87.** $\displaystyle\int \ln u\, du = u(-1 + \ln u) + C$

**88.** $\displaystyle\int u \ln u\, du = \frac{u^2}{4}(-1 + 2 \ln u) + C$

**89.** $\displaystyle\int u^n \ln u\, du = \frac{u^{n+1}}{(n+1)^2}[-1 + (n+1)\ln u] + C,\ n \neq -1$

**90.** $\displaystyle\int (\ln u)^2\, du = u[2 - 2\ln u + (\ln u)^2] + C$

**91.** $\displaystyle\int (\ln u)^n\, du = u(\ln u)^n - n\int (\ln u)^{n-1}\, du$

**Forms Involving Hyperbolic Functions**

**92.** $\displaystyle\int \cosh u\, du = \sinh u + C$

**93.** $\displaystyle\int \sinh u\, du = \cosh u + C$

**94.** $\displaystyle\int \operatorname{sech}^2 u\, du = \tanh u + C$

**95.** $\displaystyle\int \operatorname{csch}^2 u\, du = -\coth u + C$

**96.** $\displaystyle\int \operatorname{sech} u \tanh u\, du = -\operatorname{sech} u + C$

**97.** $\displaystyle\int \operatorname{csch} u \coth u\, du = -\operatorname{csch} u + C$

**Forms Involving Inverse Hyperbolic Functions (in logarithmic form)**

**98.** $\displaystyle\int \frac{du}{\sqrt{u^2 \pm a^2}} = \ln\left(u + \sqrt{u^2 \pm a^2}\right) + C$

**99.** $\displaystyle\int \frac{du}{a^2 - u^2} = \frac{1}{2a} \ln\left|\frac{a + u}{a - u}\right| + C$

**100.** $\displaystyle\int \frac{du}{u\sqrt{a^2 \pm u^2}} = -\frac{1}{a} \ln\frac{a + \sqrt{a^2 \pm u^2}}{|u|} + C$

# Chapter 11

## Section 11.1 *(page 759)*

1. Answers will vary. Sample answer: A scalar is a single real number, such as 2. A vector is a line segment having both direction and magnitude. The vector $\langle \sqrt{3}, 1 \rangle$, given in component form, has a direction of $\pi/6$ and a magnitude of 2.

3. (a) $\langle 4, 2 \rangle$

   (b)
   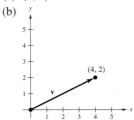

5. $\mathbf{u} = \mathbf{v} = \langle 2, 4 \rangle$    7. $\mathbf{u} = \mathbf{v} = \langle 6, -5 \rangle$

9. (a) and (d)    11. (a) and (d)

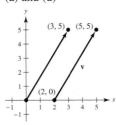

   (b) $\langle 3, 5 \rangle$    (b) $\langle -2, -4 \rangle$
   (c) $\mathbf{v} = 3\mathbf{i} + 5\mathbf{j}$    (c) $\mathbf{v} = -2\mathbf{i} - 4\mathbf{j}$

13. (a) and (d)    15. (a) and (d)

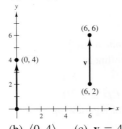

   (b) $\langle 0, 4 \rangle$    (c) $\mathbf{v} = 4\mathbf{j}$    (b) $\langle -1, \frac{5}{3} \rangle$
                                      (c) $\mathbf{v} = -\mathbf{i} + \frac{5}{3}\mathbf{j}$

17. $\langle 3, 5 \rangle$    19. 4    21. 17    23. $\sqrt{26}$

25. (a) $\langle 6, 10 \rangle$    (b) $\langle -9, -15 \rangle$

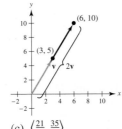

   (c) $\langle \frac{21}{2}, \frac{35}{2} \rangle$    (d) $\langle 2, \frac{10}{3} \rangle$

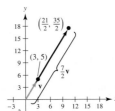    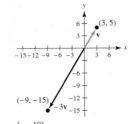

27. (a) $\langle \frac{8}{3}, 6 \rangle$    (b) $\langle 6, -15 \rangle$
    (c) $\langle -2, -14 \rangle$    (d) $\langle 18, -7 \rangle$

29.    31.

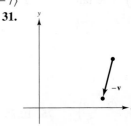

33.

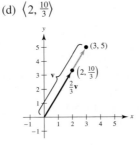

35. $\left\langle \dfrac{\sqrt{17}}{17}, \dfrac{4\sqrt{17}}{17} \right\rangle$    37. $\left\langle \dfrac{3\sqrt{34}}{34}, \dfrac{5\sqrt{34}}{34} \right\rangle$

39. (a) $\sqrt{2}$    (b) $\sqrt{5}$    (c) 1    (d) 1    (e) 1    (f) 1

41. (a) $\dfrac{\sqrt{5}}{2}$    (b) $\sqrt{13}$    (c) $\dfrac{\sqrt{85}}{2}$    (d) 1    (e) 1    (f) 1

43.

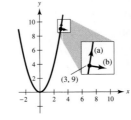

   $\|\mathbf{u}\| + \|\mathbf{v}\| = \sqrt{5} + \sqrt{41}$ and $\|\mathbf{u} + \mathbf{v}\| = \sqrt{74}$
   $\sqrt{74} \le \sqrt{5} + \sqrt{41}$

45. $\langle 0, 6 \rangle$    47. $\langle -\sqrt{5}, 2\sqrt{5} \rangle$    49. $\langle 3, 0 \rangle$

51. $\langle -\sqrt{3}, 1 \rangle$    53. $\left\langle \dfrac{2 + 3\sqrt{2}}{2}, \dfrac{3\sqrt{2}}{2} \right\rangle$

55. $\langle 2\cos 4 + \cos 2, 2\sin 4 + \sin 2 \rangle$

57. $\theta = 0°$

59. 0; Vectors that start and end at the same point have a magnitude of 0.

61. $a = 3, b = 1$    63. $a = -2, b = -4$

65. $a = -\frac{2}{3}, b = \frac{5}{3}$

67. (a) $\pm\dfrac{1}{\sqrt{37}}\langle 1, 6 \rangle$    69. (a) $\pm\dfrac{1}{\sqrt{10}}\langle 1, 3 \rangle$

    (b) $\pm\dfrac{1}{\sqrt{37}}\langle 6, -1 \rangle$    (b) $\pm\dfrac{1}{\sqrt{10}}\langle 3, -1 \rangle$

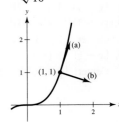

**71.** (a) $\pm\dfrac{1}{5}\langle -4, 3\rangle$

(b) $\pm\dfrac{1}{5}\langle 3, 4\rangle$

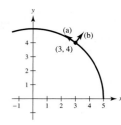

**73.** $\left\langle -\dfrac{\sqrt{2}}{2}, \dfrac{\sqrt{2}}{2}\right\rangle$    **75.** 10.7°, 584.6 lb    **77.** 71.3°, 228.5 lb

**79.** Tension in cable $CB$: 1958.1 lb
Tension in cable $CA$: 2638.2 lb

**81.** Horizontal: 1193.43 ft/sec
Vertical: 125.43 ft/sec

**83.** 38.3° north of west, 882.9 km/h

**85.** False. Weight has direction.    **87.** True

**89.** True    **91.** True    **93.** False. $\|a\mathbf{i} + b\mathbf{j}\| = \sqrt{2}|a|$

**95–97.** Proofs    **99.** $x^2 + y^2 = 25$

# Section 11.2 *(page 767)*

**1.** $x_0$ is directed distance to $yz$-plane.
$y_0$ is directed distance to $xz$-plane.
$z_0$ is directed distance to $xy$-plane.

**3.** (a) Point    (b) Vertical line    (c) Plane

**5.**

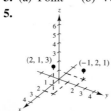

**7.**

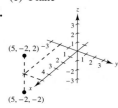

**9.** $(-3, 4, 5)$    **11.** $(12, 0, 0)$

**13.** One unit above the $xy$-plane

**15.** Three units behind the $yz$-plane

**17.** To the left of the $xz$-plane

**19.** Within three units of the $xz$-plane

**21.** Three units below the $xy$-plane and below either Quadrant I or Quadrant III

**23.** Above the $xy$-plane and above Quadrants II or IV *or* below the $xy$-plane and below Quadrants I or III

**25.** $3\sqrt{2}$    **27.** 5    **29.** $7, 7\sqrt{5}, 14$; Right triangle

**31.** $\sqrt{41}, \sqrt{41}, \sqrt{14}$; Isosceles triangle    **33.** $(6, 4, 7)$

**35.** $(2, 6, 3)$

**37.** $(x - 7)^2 + (y - 1)^2 + (z + 2)^2 = 1$

**39.** $\left(x - \dfrac{3}{2}\right)^2 + (y - 2)^2 + (z - 1)^2 = \dfrac{21}{4}$

**41.** $(x + 7)^2 + (y - 7)^2 + (z - 6)^2 = 36$

**43.** $(x - 1)^2 + (y + 3)^2 + (z + 4)^2 = 25$
Center: $(1, -3, -4)$    Radius: 5

**45.** $\left(x - \dfrac{1}{3}\right)^2 + (y + 1)^2 + z^2 = 1$
Center: $\left(\dfrac{1}{3}, -1, 0\right)$    Radius: 1

**47.** (a) $\langle -2, 2, 2\rangle$
(b) $\mathbf{v} = -2\mathbf{i} + 2\mathbf{j} + 2\mathbf{k}$
(c)

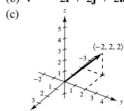

**49.** (a) and (d)

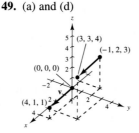

(b) $\langle 4, 1, 1\rangle$
(c) $\mathbf{v} = 4\mathbf{i} + \mathbf{j} + \mathbf{k}$

**51.** $\mathbf{v} = \langle 1, -1, 6\rangle$
$\|\mathbf{v}\| = \sqrt{38}$
$\mathbf{u} = \dfrac{1}{\sqrt{38}}\langle 1, -1, 6\rangle$

**53.** $\mathbf{v} = \langle -4, 3, 2\rangle$
$\|\mathbf{v}\| = \sqrt{29}$
$\mathbf{u} = \dfrac{1}{\sqrt{29}}\langle -4, 3, 2\rangle$

**55.** $(3, 1, 8)$

**57.** (a)

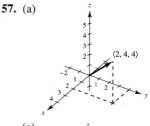

(b)

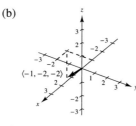

(c)

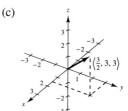

(d)

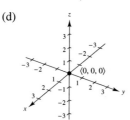

**59.** $\langle 3, 0, 0\rangle$    **61.** $\langle 21, 18, 15\rangle$    **63.** a and b

**65.** a    **67.** Collinear    **69.** Not collinear

**71.** $\overrightarrow{AB} = \langle 1, 2, 3\rangle$, $\overrightarrow{CD} = \langle 1, 2, 3\rangle$, $\overrightarrow{BD} = \langle -2, 1, 1\rangle$,
$\overrightarrow{AC} = \langle -2, 1, 1\rangle$; Because $\overrightarrow{AB} = \overrightarrow{CD}$ and $\overrightarrow{BD} = \overrightarrow{AC}$,
the given points form the vertices of a parallelogram.

**73.** $\sqrt{2}$    **75.** $\sqrt{34}$    **77.** $\sqrt{14}$

**79.** (a) $\dfrac{1}{3}\langle 2, -1, 2\rangle$    (b) $-\dfrac{1}{3}\langle 2, -1, 2\rangle$

**81.** (a) $\dfrac{2\sqrt{2}}{5}\mathbf{i} - \dfrac{\sqrt{2}}{2}\mathbf{j} + \dfrac{3\sqrt{2}}{10}\mathbf{k}$

(b) $-\dfrac{2\sqrt{2}}{5}\mathbf{i} + \dfrac{\sqrt{2}}{2}\mathbf{j} - \dfrac{3\sqrt{2}}{10}\mathbf{k}$

**83.** $\left\langle 0, \dfrac{10}{\sqrt{2}}, \dfrac{10}{\sqrt{2}}\right\rangle$    **85.** $\left\langle 1, -1, \dfrac{1}{2}\right\rangle$

**87.**    $\langle 0, \sqrt{3}, \pm 1\rangle$    **89.** $(2, -1, 2)$

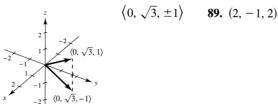

**91.** A sphere of radius 4 centered at $(x_1, y_1, z_1)$:
$(x - x_1)^2 + (y - y_1)^2 + (z - z_1)^2 = 16$

**93.** The set of points outside a sphere of radius 1 centered at the origin

**95.** The terminal points of the vectors $t\mathbf{u}$, $\mathbf{u} + t\mathbf{v}$, and $s\mathbf{u} + t\mathbf{v}$ are collinear.

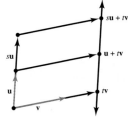

**97.** $\dfrac{\sqrt{3}}{3}\langle 1, 1, 1\rangle$

**99.** (a) $T = \dfrac{8L}{\sqrt{L^2 - 18^2}}$, $L > 18$

(b)

| $L$ | 20 | 25 | 30 | 35 | 40 | 45 | 50 |
|---|---|---|---|---|---|---|---|
| $T$ | 18.4 | 11.5 | 10 | 9.3 | 9.0 | 8.7 | 8.6 |

(c) 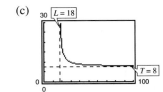    (d) Proof    (e) 30 in.

**101.** Tension in cable $AB$: 202.919 N
Tension in cable $AC$: 157.909 N
Tension in cable $AD$: 226.521 N

**103.** $\left(x - \tfrac{4}{3}\right)^2 + (y - 3)^2 + \left(z + \tfrac{1}{3}\right)^2 = \tfrac{44}{9}$

Sphere; center: $\left(\dfrac{4}{3}, 3, -\dfrac{1}{3}\right)$, radius: $\dfrac{2\sqrt{11}}{3}$

## Section 11.3    (page 777)

**1.** The vectors are orthogonal.

**3.** (a) 17    (b) 25    (c) 26    (d) $\langle -17, 85\rangle$    (e) 51

**5.** (a) $-26$    (b) 52    (c) 13    (d) $\langle 78, -52\rangle$    (e) $-78$

**7.** (a) 2    (b) 29    (c) 61    (d) $\langle 0, 12, 10\rangle$    (e) 6

**9.** (a) 1    (b) 6    (c) 2    (d) $\mathbf{i} - \mathbf{k}$    (e) 3

**11.** (a) $\dfrac{\pi}{2}$    (b) $90°$    **13.** (a) 1.7127    (b) $98.1°$

**15.** (a) 1.0799    (b) $61.9°$    **17.** (a) 2.0306    (b) $116.3°$

**19.** 20    **21.** Orthogonal    **23.** Neither    **25.** Orthogonal

**27.** Right triangle; Answers will vary.

**29.** Acute triangle; Answers will vary.

**31.** $\cos\alpha = \dfrac{1}{3}$, $\alpha \approx 70.5°$    **33.** $\cos\alpha = \dfrac{7}{\sqrt{51}}$, $\alpha \approx 11.4°$

$\cos\beta = \dfrac{2}{3}$, $\beta \approx 48.2°$    $\cos\beta = \dfrac{1}{\sqrt{51}}$, $\beta \approx 82.0°$

$\cos\gamma = \dfrac{2}{3}$, $\gamma \approx 48.2°$    $\cos\gamma = -\dfrac{1}{\sqrt{51}}$, $\gamma \approx 98.0°$

**35.** $\cos\alpha = 0$, $\alpha \approx 90°$

$\cos\beta = \dfrac{3}{\sqrt{13}}$, $\beta \approx 33.7°$

$\cos\gamma = -\dfrac{2}{\sqrt{13}}$, $\gamma \approx 123.7°$

**37.** (a) $\langle 2, 8\rangle$    (b) $\langle 4, -1\rangle$    **39.** (a) $\left\langle \tfrac{5}{2}, \tfrac{1}{2}\right\rangle$    (b) $\left\langle -\tfrac{1}{2}, \tfrac{5}{2}\right\rangle$

**41.** (a) $\langle -2, 2, 2\rangle$    (b) $\langle 2, 1, 1\rangle$

**43.** (a) $\langle 0, -3, -3\rangle$    (b) $\langle -9, 1, -1\rangle$

**45.** You cannot add a vector to a scalar.

**47.** Yes.    **49.** \$17,490.25; Total revenue

$$\left\|\frac{\mathbf{u} \cdot \mathbf{v}}{\|\mathbf{v}\|^2}\mathbf{v}\right\| = \left\|\frac{\mathbf{v} \cdot \mathbf{u}}{\|\mathbf{u}\|^2}\mathbf{u}\right\|$$

$$|\mathbf{u} \cdot \mathbf{v}|\frac{\|\mathbf{v}\|}{\|\mathbf{v}\|^2} = |\mathbf{v} \cdot \mathbf{u}|\frac{\|\mathbf{u}\|}{\|\mathbf{u}\|^2}$$

$$\frac{1}{\|\mathbf{v}\|} = \frac{1}{\|\mathbf{u}\|}$$

$$\|\mathbf{u}\| = \|\mathbf{v}\|$$

**51.** Answers will vary. Sample answer: $\langle 12, 2\rangle$ and $\langle -12, -2\rangle$

**53.** Answers will vary. Sample answer: $\langle 2, 0, 3\rangle$ and $\langle -2, 0, -3\rangle$

**55.** $\arccos\dfrac{1}{\sqrt{3}} \approx 54.7°$

**57.** (a) 8335.1 lb    (b) 47,270.8 lb

**59.** 425 ft-lb    **61.** 2900.2 km-N

**63.** False. For example, $\langle 1, 1\rangle \cdot \langle 2, 3\rangle = 5$ and $\langle 1, 1\rangle \cdot \langle 1, 4\rangle = 5$, but $\langle 2, 3\rangle \neq \langle 1, 4\rangle$.

**65.** (a) $(0, 0), (1, 1)$

(b) To $y = x^2$ at $(1, 1)$: $\left\langle \pm\dfrac{\sqrt{5}}{5}, \pm\dfrac{2\sqrt{5}}{5}\right\rangle$

To $y = x^{1/3}$ at $(1, 1)$: $\left\langle \pm\dfrac{3\sqrt{10}}{10}, \pm\dfrac{\sqrt{10}}{10}\right\rangle$

To $y = x^2$ at $(0, 0)$: $\langle \pm 1, 0\rangle$

To $y = x^{1/3}$ at $(0, 0)$: $\langle 0, \pm 1\rangle$

(c) At $(1, 1)$: $\theta = 45°$

At $(0, 0)$: $\theta = 90°$

**67.** (a) $(-1, 0), (1, 0)$

(b) To $y = 1 - x^2$ at $(1, 0)$: $\left\langle \pm\dfrac{\sqrt{5}}{5}, \mp\dfrac{2\sqrt{5}}{5}\right\rangle$

To $y = x^2 - 1$ at $(1, 0)$: $\left\langle \pm\dfrac{\sqrt{5}}{5}, \pm\dfrac{2\sqrt{5}}{5}\right\rangle$

To $y = 1 - x^2$ at $(-1, 0)$: $\left\langle \pm\dfrac{\sqrt{5}}{5}, \pm\dfrac{2\sqrt{5}}{5}\right\rangle$

To $y = x^2 - 1$ at $(-1, 0)$: $\left\langle \pm\dfrac{\sqrt{5}}{5}, \mp\dfrac{2\sqrt{5}}{5}\right\rangle$

(c) At $(1, 0)$: $\theta = 53.13°$

At $(-1, 0)$: $\theta = 53.13°$

**69.** Proof

**71.** (a)

(b) $k\sqrt{2}$    (c) $60°$    (d) $109.5°$

**73–75.** Proofs

## Section 11.4    (page 785)

**1.** $\mathbf{u} \times \mathbf{v}$ is a vector that is perpendicular to both $\mathbf{u}$ and $\mathbf{v}$.

**3.** $-\mathbf{k}$

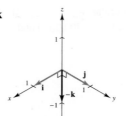

**5.** $-\mathbf{j}$

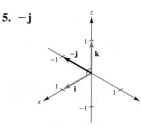

**7.** (a) $20\mathbf{i} + 10\mathbf{j} - 16\mathbf{k}$   (b) $-20\mathbf{i} - 10\mathbf{j} + 16\mathbf{k}$   (c) $\mathbf{0}$

**9.** (a) $17\mathbf{i} - 33\mathbf{j} - 10\mathbf{k}$   (b) $-17\mathbf{i} + 33\mathbf{j} + 10\mathbf{k}$   (c) $\mathbf{0}$

**11.** $\langle 0, 0, 6 \rangle$     **13.** $\langle -2, 3, -1 \rangle$

**15.** $\left\langle -\dfrac{7}{9\sqrt{3}}, -\dfrac{5}{9\sqrt{3}}, \dfrac{13}{9\sqrt{3}} \right\rangle$ or $\left\langle \dfrac{7}{9\sqrt{3}}, \dfrac{5}{9\sqrt{3}}, -\dfrac{13}{9\sqrt{3}} \right\rangle$

**17.** $\left\langle \dfrac{3}{\sqrt{59}}, \dfrac{7}{\sqrt{59}}, \dfrac{1}{\sqrt{59}} \right\rangle$ or $\left\langle -\dfrac{3}{\sqrt{59}}, -\dfrac{7}{\sqrt{59}}, -\dfrac{1}{\sqrt{59}} \right\rangle$

**19.** $1$     **21.** $6\sqrt{5}$     **23.** $9\sqrt{5}$     **25.** $\dfrac{11}{2}$

**27.** $10 \cos 40° \approx 7.66$ ft-lb

**29.** (a) $\mathbf{F} = -180(\cos\theta\,\mathbf{j} + \sin\theta\,\mathbf{k})$

  (b) $\|\overrightarrow{AB} \times \mathbf{F}\| = |225 \sin\theta + 180 \cos\theta|$

  (c) $\|\overrightarrow{AB} \times \mathbf{F}\| = 225\left(\dfrac{1}{2}\right) + 180\left(\dfrac{\sqrt{3}}{2}\right) \approx 268.38$

  (d) $\theta = 141.34°$
    $\overrightarrow{AB}$ and $\mathbf{F}$ are perpendicular.

  (e)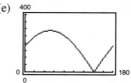

    From part (d), the zero is $\theta \approx 141.34°$ when the vectors are parallel.

**31.** $1$     **33.** $6$     **35.** $2$     **37.** $75$

**39.** $a = b = c = h$ and $e = f = g$

**41.** On the $x$-axis; The cross product has the form $\langle k, 0, 0 \rangle$.

**43.** False. The cross product of two vectors is not defined in a two-dimensional coordinate system.

**45.** False. Let $\mathbf{u} = \langle 1, 0, 0 \rangle$, $\mathbf{v} = \langle 1, 0, 0 \rangle$, and $\mathbf{w} = \langle -1, 0, 0 \rangle$. Then $\mathbf{u} \times \mathbf{v} = \mathbf{u} \times \mathbf{w} = \mathbf{0}$, but $\mathbf{v} \neq \mathbf{w}$.

**47–55.** Proofs

## Section 11.5 *(page 794)*

**1.** Parametric equations: $x = x_1 + at$, $y = y_1 + bt$, $z = z_1 + ct$

  Symmetric equations: $\dfrac{x - x_1}{a} = \dfrac{y - y_1}{b} = \dfrac{z - z_1}{c}$

  You need a vector $\mathbf{v} = \langle a, b, c \rangle$ parallel to the line and a point $P(x_1, y_1, z_1)$ on the line.

**3.** Answers will vary. Sample answer: $3y - z = 5$

**5.** (a) Yes   (b) No   (c) Yes

| Parametric Equations (a) | Symmetric Equations (b) | Direction Numbers |
|---|---|---|
| **7.** $x = 3t$ $y = t$ $z = 5t$ | $\dfrac{x}{3} = y = \dfrac{x}{5}$ | $3, 1, 5$ |
| **9.** $x = -2 + 2t$ $y = 4t$ $z = 3 - 2t$ | $\dfrac{x + 2}{2} = \dfrac{y}{4} = \dfrac{z - 3}{-2}$ | $2, 4, -2$ |

| Parametric Equations (a) | Symmetric Equations (b) | Direction Numbers |
|---|---|---|
| **11.** $x = 1 + 3t$ $y = -2t$ $z = 1 + t$ | $\dfrac{x - 1}{3} = \dfrac{y}{-2} = \dfrac{z - 1}{1}$ | $3, -2, 1$ |
| **13.** $x = 5 + 17t$ $y = -3 - 11t$ $z = -2 - 9t$ | $\dfrac{x - 5}{17} = \dfrac{y + 3}{-11} = \dfrac{z + 2}{-9}$ | $17, -11, -9$ |
| **15.** $x = 7 - 10t$ $y = -2 + 2t$ $z = 6$ | Not possible | $-10, 2, 0$ |

**17.** $x = 2$     **19.** $x = 2 + 3t$     **21.** $x = 5 + 2t$
  $y = 3$        $y = 3 + 2t$         $y = -3 - t$
  $z = 4 + t$    $z = 4 - t$          $z = -4 + 3t$

**23.** $x = 2 - t$     **25.** $P(3, -1, -2)$     **27.** $P(7, -6, -2)$
  $y = 1 + t$         $\mathbf{v} = \langle -1, 2, 0 \rangle$     $\mathbf{v} = \langle 4, 2, 1 \rangle$
  $z = 2 + t$

**29.** Identical     **31.** Identical     **33.** $(2, 3, 1)$, $55.5°$

**35.** Not intersecting     **37.** (a) Yes   (b) Yes   (c) No

**39.** $y - 3 = 0$     **41.** $2x + 3y - z = 10$

**43.** $2x - y - 2z + 6 = 0$     **45.** $3x - 19y - 2z = 0$

**47.** $4x - 3y + 4z = 10$     **49.** $z = 3$     **51.** $x + y + z = 5$

**53.** $7x + y - 11z = 5$     **55.** $y - z = -1$     **57.** $x - z = 0$

**59.** $9x - 3y + 2z - 21 = 0$     **61.** Parallel     **63.** Identical

**65.** (a) $\theta \approx 65.91°$       **67.** (a) $\theta \approx 69.67°$
  (b) $x = 2$              (b) $x = 2 - 9t$
    $y = 1 + t$              $y = -1 - 5t$
    $z = 1 + 2t$             $z = 22t$

**69.** Orthogonal     **71.** Neither; $83.5°$     **73.** Parallel

**75.**                                **77.**

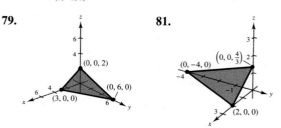

**79.**          **81.**

**83.** The line lies in the plane.     **85.** Not intersecting

**87.** $\dfrac{6\sqrt{14}}{7}$     **89.** $\dfrac{11\sqrt{6}}{6}$     **91.** $\dfrac{2\sqrt{26}}{13}$     **93.** $\dfrac{27\sqrt{94}}{188}$

**95.** $\dfrac{\sqrt{2533}}{17}$     **97.** $\dfrac{7\sqrt{3}}{3}$     **99.** $\dfrac{\sqrt{66}}{3}$     **101.** Exactly 1

**103.** Yes. Consider three points, two on one line and one on the second line. A unique plane contains all three points.

**105.** (a)

| Year | 2009 | 2010 | 2011 | 2012 | 2013 | 2014 |
|------|------|------|------|------|------|------|
| $z$ (approx.) | 18.93 | 19.46 | 20.31 | 21.10 | 21.58 | 22.62 |

    The approximations are close to the actual values.

  (b) An increase

**107.** (a) $\sqrt{70}$ in.

  (b)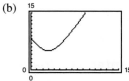

  (c) The distance is never zero.

  (d) 5 in.

**109.** $\left(\frac{77}{13}, \frac{48}{13}, -\frac{23}{13}\right)$    **111.** $x = 21t,\ y = 1 + 11t,\ z = 4 + 13t$

**113.** True    **115.** True

**117.** False. Plane $7x + y - 11z = 5$ and plane $5x + 2y - 4z = 1$ are both perpendicular to plane $2x - 3y + z = 3$ but are not parallel.

## Section 11.6  *(page 806)*

**1.** Quadric surfaces are the three-dimensional analogs of conic sections.

**3.** The trace of a surface is the intersection of the surface with a plane. You find a trace by setting one variable equal to a constant, such as $x = 0$ or $z = 2$.

**5.** c    **6.** e    **7.** f    **8.** b    **9.** d    **10.** a

**11.** Right circular cylinder    **13.** Elliptic cylinder

 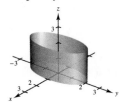

**15.** Hyperboloid of two sheets    **17.** Hyperboloid of one sheet

 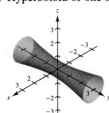

**19.** Ellipsoid    **21.** Elliptic cone

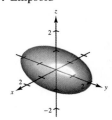 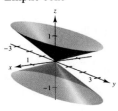

**23.** Hyperbolic paraboloid    **25.** Elliptic paraboloid

 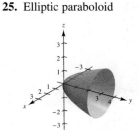

**27.** There have to be two minus signs to have a hyperboloid of two sheets. The number of sheets is the same as the number of minus signs.

**29.** No. See table on pages 800 and 801.

**31.** $x^2 + z^2 = 25y^2$    **33.** $x^2 + y^2 = 4z^{2/3}$

**35.** $y^2 + z^2 = \dfrac{4}{x^2}$    **37.** $y = \sqrt{2z}\ \left(\text{or } x = \sqrt{2z}\right)$

**39.** $y = \sqrt{5 - 8x^2}\ \left(\text{or } z = \sqrt{5 - 8x^2}\right)$    **41.** $\dfrac{128\pi}{3}$

**43.** (a) Major axis: $4\sqrt{2}$      (b) Major axis: $8\sqrt{2}$

       Minor axis: 4            Minor axis: 8

       Foci: $(0, \pm 2, 2)$       Foci: $(0, \pm 4, 8)$

**45.** $x^2 + z^2 = 8y$, elliptic paraboloid

**47.** $\dfrac{x^2}{3963^2} + \dfrac{y^2}{3963^2} + \dfrac{z^2}{3950^2} = 1$

**49.** $x = at,\ y = -bt,\ z = 0;$

    $x = at,\ y = bt + ab^2,\ z = 2abt + a^2b^2$

**51.** The Klein bottle does not have both an "inside" and an "outside." It is formed by inserting the small open end through the side of the bottle and making it contiguous with the top of the bottle.

## Section 11.7  *(page 813)*

**1.** The cylindrical coordinate system is an extension of the polar coordinate system. In this system, a point $P$ in space is represented by an ordered triple $(r, \theta, z)$. $(r, \theta)$ is a polar representation of the projection of $P$ in the $xy$-plane, and $z$ is the directed distance from $(r, \theta)$ to $P$.

**3.** $(-7, 0, 5)$    **5.** $\left(\dfrac{3\sqrt{2}}{2}, \dfrac{3\sqrt{2}}{2}, 1\right)$    **7.** $\left(-2\sqrt{3}, -2, -3\right)$

**9.** $\left(5, \dfrac{\pi}{2}, 1\right)$    **11.** $\left(2\sqrt{2}, -\dfrac{\pi}{4}, -4\right)$    **13.** $\left(2, \dfrac{\pi}{3}, 4\right)$

**15.** $z = 4$    **17.** $r^2 - 2z^2 = 5$    **19.** $r = \sec\theta\tan\theta$

**21.** $r^2\sin^2\theta = 10 - z^2$

**23.** $x^2 + y^2 = 9$           **25.** $x - \sqrt{3}y = 0$

 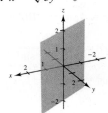

**27.** $x^2 + y^2 + z^2 = 5$

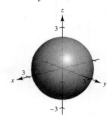

**29.** $x^2 + (y - 2)^2 = 4$

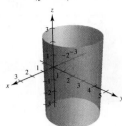

**31.** $\left(4, 0, \dfrac{\pi}{2}\right)$    **33.** $\left(4\sqrt{2}, \dfrac{2\pi}{3}, \dfrac{\pi}{4}\right)$    **35.** $\left(4, \dfrac{\pi}{6}, \dfrac{\pi}{6}\right)$

**37.** $\left(\sqrt{6}, \sqrt{2}, 2\sqrt{2}\right)$    **39.** $(0, 0, 12)$

**41.** $(0.915, 0.915, 4.830)$    **43.** $\rho = 2 \csc \phi \csc \theta$

**45.** $\rho = 7$    **47.** $\rho = 4 \csc \phi$    **49.** $\tan^2 \phi = 2$

**51.** $x^2 + y^2 + z^2 = 1$    **53.** $3x^2 + 3y^2 - z^2 = 0$

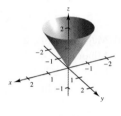

**55.** $x^2 + y^2 + (z - 2)^2 = 4$    **57.** $x^2 + y^2 = 1$

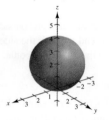

**59.** $\left(4, \dfrac{\pi}{4}, \dfrac{\pi}{2}\right)$    **61.** $\left(6\sqrt{2}, \dfrac{\pi}{2}, \dfrac{3\pi}{4}\right)$

**63.** $\left(13, \pi, \arccos \dfrac{5}{13}\right)$    **65.** $\left(10, \dfrac{\pi}{6}, 0\right)$

**67.** $\left(3\sqrt{3}, -\dfrac{\pi}{6}, 3\right)$    **69.** $\left(4, \dfrac{7\pi}{6}, 4\sqrt{3}\right)$

**71.** d    **72.** e    **73.** c    **74.** a    **75.** f    **76.** b

**77.** Because of the restriction $r \geq 0$

**79.** (a) $r^2 + z^2 = 27$    (b) $\rho = 3\sqrt{3}$

**81.** (a) $r^2 + (z - 1)^2 = 1$    (b) $\rho = 2 \cos \phi$

**83.** (a) $r = 4 \sin \theta$    (b) $\rho = \dfrac{4 \sin \theta}{\sin \phi} = 4 \sin \theta \csc \phi$

**85.** (a) $r^2 = \dfrac{9}{\cos^2 \theta - \sin^2 \theta}$

(b) $\rho^2 = \dfrac{9 \csc^2 \phi}{\cos^2 \theta - \sin^2 \theta}$

**87.**

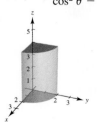

**89.**

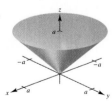

**91.**

**93.**

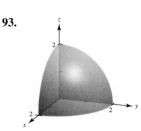

**95.** Rectangular: $0 \leq x \leq 10, 0 \leq y \leq 10, 0 \leq z \leq 10$

**97.** Spherical: $4 \leq \rho \leq 6$

**99.** Cylindrical: $r^2 + z^2 \leq 9, r \leq 3 \cos \theta, 0 \leq \theta \leq \pi$

**101.** False. See page 809.    **103.** Ellipse

## Review Exercises for Chapter 11    *(page 815)*

**1.** (a) $\mathbf{u} = \langle 3, -1 \rangle, \mathbf{v} = \langle 4, 2 \rangle$    (b) $\mathbf{u} = 3\mathbf{i} - \mathbf{j}, \mathbf{v} = 4\mathbf{i} + 2\mathbf{j}$

(c) $\|\mathbf{u}\| = \sqrt{10}, \|\mathbf{v}\| = 2\sqrt{5}$    (d) $\langle -5, 5 \rangle$

**3.** $\mathbf{v} = \langle 4, 4\sqrt{3} \rangle$    **5.** $(-5, 4, 0)$    **7.** $\sqrt{22}$

**9.** $(x - 3)^2 + (y + 2)^2 + (z - 6)^2 = 16$

**11.** $(x - 2)^2 + (y - 3)^2 + z^2 = 9$

Center: $(2, 3, 0)$    Radius: 3

**13.** (a) and (d)

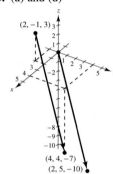

(b) $\mathbf{u} = \langle 2, 5, -10 \rangle$    (c) $\mathbf{u} = 2\mathbf{i} + 5\mathbf{j} - 10\mathbf{k}$

**15.** $\langle -8, 5, 1 \rangle$    **17.** Collinear    **19.** $\dfrac{1}{\sqrt{38}} \langle 2, 3, 5 \rangle$

**21.** (a) $\mathbf{u} = \langle -1, 4, 0 \rangle$

$\mathbf{v} = \langle -3, 0, 6 \rangle$

(b) 3    (c) 45

**23.** (a) $\dfrac{\pi}{12}$    (b) $15°$    **25.** Orthogonal

**27.** (a) $\langle \frac{12}{5}, \frac{16}{5} \rangle$    (b) $\langle \frac{8}{5}, -\frac{6}{5} \rangle$

**29.** Answers will vary. Sample answer: $\langle -6, 5, 0 \rangle, \langle 6, -5, 0 \rangle$

**31.** (a) $-9\mathbf{i} + 26\mathbf{j} - 7\mathbf{k}$    (b) $9\mathbf{i} - 26\mathbf{j} + 7\mathbf{k}$    (c) $\mathbf{0}$

**33.** $\left\langle \dfrac{8}{\sqrt{377}}, \dfrac{12}{\sqrt{377}}, \dfrac{13}{\sqrt{377}} \right\rangle$ or $\left\langle -\dfrac{8}{\sqrt{377}}, -\dfrac{12}{\sqrt{377}}, -\dfrac{13}{\sqrt{377}} \right\rangle$

**35.** 15 ft-lb

**37.** (a) $x = 3 + 6t, y = 11t, z = 2 + 4t$

(b) $\dfrac{x - 3}{6} = \dfrac{y}{11} = \dfrac{z - 2}{4}$

**39.** $x = -6, y = -8 + t, z = 2$

**41.** $27x + 4y + 32z + 33 = 0$    **43.** $x + 2y = 1$    **45.** $\frac{8}{7}$

**47.** $\dfrac{\sqrt{35}}{7}$

**49.** Plane

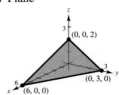

**51.** Plane

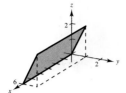

**53.** Ellipsoid

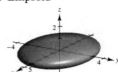

**55.** Hyperboloid of two sheets

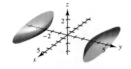

**57.** Cylinder

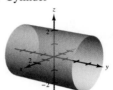

**59.** $x^2 + z^2 = 2y$

**61.** (a) $\left(2\sqrt{3}, -\dfrac{\pi}{3}, -5\right)$ (b) $\left(\sqrt{37}, -\dfrac{\pi}{3}, \arccos\left(-\dfrac{5\sqrt{37}}{37}\right)\right)$

**63.** $(-5, 0, 1)$    **65.** $\left(-2\sqrt{2}, 0, 2\sqrt{2}\right)$

**67.** (a) $r^2 \cos 2\theta = 2z$ (b) $\rho = 2 \sec 2\theta \cos \phi \csc^2 \phi$

**69.** $z = y^2 + 3x$

**71.** $x^2 + y^2 - z^2 = 0$

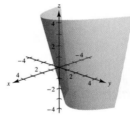
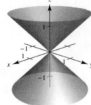

## P.S. Problem Solving (page 817)

**1–3.** Proofs     **5.** (a) $\dfrac{3\sqrt{2}}{2} \approx 2.12$ (b) $\sqrt{5} \approx 2.24$

**7.** (a) $\dfrac{\pi}{2}$ (b) $\dfrac{1}{2}(\pi abk)\mathbf{k}$

   (c) $V = \dfrac{1}{2}(\pi ab)k^2$

   $V = \dfrac{1}{2}(\text{area of base})\text{height}$

**9.** Proof

**11.** (a)

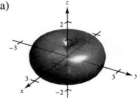

(b)

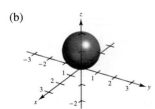

**13.** (a) Tension: $\dfrac{2\sqrt{3}}{3} \approx 1.1547$ lb

   Magnitude of $\mathbf{u}$: $\dfrac{\sqrt{3}}{3} \approx 0.5774$ lb

   (b) $T = \sec \theta$, $\|\mathbf{u}\| = \tan \theta$; Domain: $0° \le \theta \le 90°$

(c)

| $\theta$ | 0° | 10° | 20° | 30° |
|---|---|---|---|---|
| $T$ | 1 | 1.0154 | 1.0642 | 1.1547 |
| $\|\mathbf{u}\|$ | 0 | 0.1763 | 0.3640 | 0.5774 |

| $\theta$ | 40° | 50° | 60° |
|---|---|---|---|
| $T$ | 1.3054 | 1.5557 | 2 |
| $\|\mathbf{u}\|$ | 0.8391 | 1.1918 | 1.7321 |

(d)

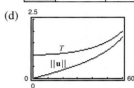

(e) Both are increasing functions.

(f) $\displaystyle\lim_{\theta \to \pi/2^-} T = \infty$ and $\displaystyle\lim_{\theta \to \pi/2^-} \|\mathbf{u}\| = \infty$

   Yes. As $\theta$ increases, both $T$ and $\|\mathbf{u}\|$ increase.

**15.** $\langle 0, 0, \cos \alpha \sin \beta - \cos \beta \sin \alpha \rangle$; Proof

**17.** $D = \dfrac{|\overrightarrow{PQ} \cdot \mathbf{n}|}{\|\mathbf{n}\|}$

   $= \dfrac{|\mathbf{w} \cdot (\mathbf{u} \times \mathbf{v})|}{\|\mathbf{u} \times \mathbf{v}\|} = \dfrac{|(\mathbf{u} \times \mathbf{v}) \cdot \mathbf{w}|}{\|\mathbf{u} \times \mathbf{v}\|} = \dfrac{|\mathbf{u} \cdot (\mathbf{v} \times \mathbf{w})|}{\|\mathbf{u} \times \mathbf{v}\|}$

**19.** Proof

# Chapter 12

## Section 12.1 *(page 825)*

**1.** You can use a vector-valued function to trace the graph of a curve. Recall that the terminal point of the position vector $\mathbf{r}(t)$ coincides with a point on the curve.

**3.** $(-\infty, -1) \cup (-1, \infty)$     **5.** $(0, \infty)$

**7.** $[0, \infty)$     **9.** $(-\infty, \infty)$

**11.** (a) $\dfrac{1}{2}\mathbf{i}$ (b) $\mathbf{j}$ (c) $\dfrac{1}{2}(s + 1)^2\mathbf{i} - s\mathbf{j}$

   (d) $\dfrac{1}{2}\Delta t(\Delta t + 4)\mathbf{i} - \Delta t\mathbf{j}$

**13.** $\mathbf{r}(t) = 5t\mathbf{i} + 2t\mathbf{j} + 2t\mathbf{k}, 0 \le t \le 1$

   $x = 5t, y = 2t, z = 2t, 0 \le t \le 1$

**15.** $\mathbf{r}(t) = (-3 + 2t)\mathbf{i} + (-6 - 3t)\mathbf{j} + (-1 - 7t)\mathbf{k}, 0 \le t \le 1$

   $x = -3 + 2t, y = -6 - 3t, z = -1 - 7t, 0 \le t \le 1$

**17.** $t^2(5t - 1)$; No, the dot product is a scalar.

**19.** b     **20.** c     **21.** d     **22.** a

**23.**

**25.**

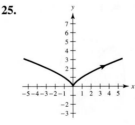

**27.**

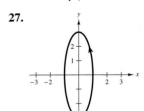

**29.**

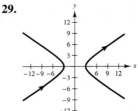

**31.**

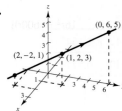

**33.**

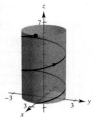

**35.**

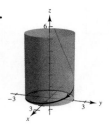

**37.**

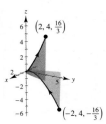

**39.**

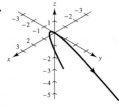

Parabola

**41.**

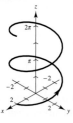

(a) The helix is translated two units back on the $x$-axis.
(b) The height of the helix increases at a greater rate.
(c) The orientation of the graph is reversed.
(d) The radius of the helix is increased from 2 to 6.

**43.** $\mathbf{u}(t) = 3t^2\mathbf{i} + (t - 1)\mathbf{j} + (t + 2)\mathbf{k}$
**45.** $\mathbf{u}(t) = 3t^2\mathbf{i} + 2(t - 1)\mathbf{j} + t\mathbf{k}$
**47–53.** Answers will vary.
**55.**

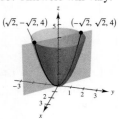

$\mathbf{r}(t) = t\mathbf{i} - t\mathbf{j} + 2t^2\mathbf{k}$
**57.**

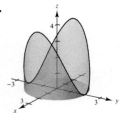

$\mathbf{r}(t) = 2\sin t\mathbf{i} + 2\cos t\mathbf{j} + 4\sin^2 t\mathbf{k}$

**59.**

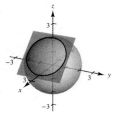

$\mathbf{r}(t) = (1 + \sin t)\mathbf{i} + \sqrt{2}\cos t\mathbf{j} + (1 - \sin t)\mathbf{k}$ and
$\mathbf{r}(t) = (1 + \sin t)\mathbf{i} - \sqrt{2}\cos t\mathbf{j} + (1 - \sin t)\mathbf{k}$
**61.**
$$\mathbf{r}(t) = t\mathbf{i} + t\mathbf{j} + \sqrt{4 - t^2}\mathbf{k}$$

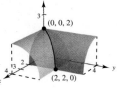

**63.** Let $x = t$, $y = 2t\cos t$, and $z = 2t\sin t$. Then
$$y^2 + z^2 = (2t\cos t)^2 + (2t\sin t)^2$$
$$= 4t^2\cos^2 t + 4t^2\sin^2 t$$
$$= 4t^2(\cos^2 t + \sin^2 t)$$
$$= 4t^2.$$
Because $x = t$, $y^2 + z^2 = 4x^2$.

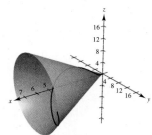

**65.** $\pi\mathbf{i} - \mathbf{j}$     **67.** 0     **69.** $\mathbf{i} + \mathbf{j} + \mathbf{k}$
**71.** $\left(-\infty, -\frac{1}{2}\right), \left(-\frac{1}{2}, 0\right), (0, \infty)$     **73.** $[-1, 1]$
**75.** $\left(-\dfrac{\pi}{2} + n\pi, \dfrac{\pi}{2} + n\pi\right)$, $n$ is an integer.
**77.** It is a line; Answers will vary.
**79.** $\mathbf{r}(t) = \begin{cases} \mathbf{i} + \mathbf{j}, & t \geq 3 \\ -\mathbf{i} + \mathbf{j}, & t < 3 \end{cases}$
**81.** $\mathbf{r}(t) = \cos t\mathbf{i} + \sin t\mathbf{j} + \dfrac{1}{\pi}t\mathbf{k}$, $0 \leq t \leq 4\pi$

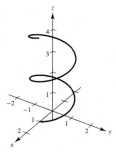

**83–85.** Proofs     **87.** Not necessarily     **89.** Yes; yes

## Section 12.2   *(page 834)*

**1.** $\mathbf{r}'(t_0)$ represents the vector that is tangent to the curve represented by $\mathbf{r}(t)$ at the point $t_0$.

**3.** $\mathbf{r}'(t) = -2t\mathbf{i} + \mathbf{j}$
$\mathbf{r}(3) = -8\mathbf{i} + 3\mathbf{j}$
$\mathbf{r}'(3) = -6\mathbf{i} + \mathbf{j}$

**5.** $\mathbf{r}'(t) = -\sin t\mathbf{i} + \cos t\mathbf{j}$
$\mathbf{r}\left(\dfrac{\pi}{2}\right) = \mathbf{j}$
$\mathbf{r}'\left(\dfrac{\pi}{2}\right) = -\mathbf{i}$

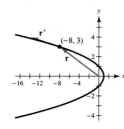

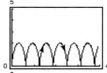

**7.** $\mathbf{r}'(t) = \langle e^t, 2e^{2t} \rangle$
$\mathbf{r}(0) = \mathbf{i} + \mathbf{j}$
$\mathbf{r}'(0) = \mathbf{i} + 2\mathbf{j}$

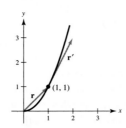

**9.** $\mathbf{r}'(t) = -2\sin t\mathbf{i} + 2\cos t\mathbf{j} + \mathbf{k}$
$\mathbf{r}\left(\dfrac{3\pi}{2}\right) = -2\mathbf{j} + \left(\dfrac{3\pi}{2}\right)\mathbf{k}$
$\mathbf{r}'\left(\dfrac{3\pi}{2}\right) = 2\mathbf{i} + \mathbf{k}$

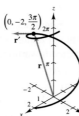

**11.** $4t^3\mathbf{i} - 5\mathbf{j}$    **13.** $-9\sin t\cos^2 t\mathbf{i} + 6\sin^2 t\cos t\mathbf{j}$

**15.** $-e^{-t}\mathbf{i} + (5te^t + 5e^t)\mathbf{k}$

**17.** $\langle \sin t + t\cos t, \cos t - t\sin t, 1 \rangle$

**19.** (a) $3t^2\mathbf{i} + t\mathbf{j}$    (b) $6t\mathbf{i} + \mathbf{j}$    (c) $18t^3 + t$

**21.** (a) $-4\sin t\mathbf{i} + 4\cos t\mathbf{j}$    (b) $-4\cos t\mathbf{i} - 4\sin t\mathbf{j}$    (c) $0$

**23.** (a) $t\mathbf{i} - \mathbf{j} + \frac{1}{2}t^2\mathbf{k}$    (b) $\mathbf{i} + t\mathbf{k}$    (c) $\dfrac{t^3}{2} + t$
(d) $-t\mathbf{i} - \frac{1}{2}t^2\mathbf{j} + \mathbf{k}$

**25.** (a) $\langle t\cos t, t\sin t, 1 \rangle$
(b) $\langle \cos t - t\sin t, \sin t + t\cos t, 0 \rangle$    (c) $t$
(d) $\langle -\sin t - t\cos t, \cos t - t\sin t, t^2 \rangle$

**27.** $(-\infty, 0), (0, \infty)$    **29.** $\left(\dfrac{\pi}{2}, 2\pi\right)$

**31.** $(-\infty, -2), (-2, \infty)$

**33.** $\left(-\dfrac{\pi}{2} + n\pi, \dfrac{\pi}{2} + n\pi\right)$, $n$ is an integer

**35.** (a) $\mathbf{i} + 3\mathbf{j} + 2t\mathbf{k}$    (b) $-\mathbf{i} + (9 - 2t)\mathbf{j} + (6t - 3t^2)\mathbf{k}$
(c) $40t\mathbf{i} + 15t^2\mathbf{j} + 20t^3\mathbf{k}$    (d) $8t + 9t^2 + 5t^4$
(e) $8t^3\mathbf{i} + (12t^2 - 4t^3)\mathbf{j} + (3t^2 - 24t)\mathbf{k}$
(f) $2\mathbf{i} + 6\mathbf{j} + 8t\mathbf{k}$

**37.** (a) $7t^6$    (b) $12t^5\mathbf{i} - 5t^4\mathbf{j}$    **39.** $t^2\mathbf{i} + t\mathbf{j} + 9t\mathbf{k} + \mathbf{C}$

**41.** $\ln|t|\mathbf{i} + t\mathbf{j} - \frac{2}{5}t^{5/2}\mathbf{k} + \mathbf{C}$    **43.** $t\mathbf{i} + t^4\mathbf{j} + \dfrac{5^t}{\ln 5}\mathbf{k} + \mathbf{C}$

**45.** $e^t\mathbf{i} + t\mathbf{j} + (t\sin t + \cos t)\mathbf{k} + \mathbf{C}$

**47.** $4\mathbf{i} + \dfrac{1}{2}\mathbf{j} - \mathbf{k}$    **49.** $5\mathbf{i} + 6\mathbf{j} + \dfrac{\pi}{2}\mathbf{k}$

**51.** $2\mathbf{i} + (e^2 - 1)\mathbf{j} - (e^2 + 1)\mathbf{k}$

**53.** $2e^{2t}\mathbf{i} + 3(e^t - 1)\mathbf{j}$    **55.** $600\sqrt{3}t\mathbf{i} + (-16t^2 + 600t)\mathbf{j}$

**57.** $\dfrac{2 - e^{-t^2}}{2}\mathbf{i} + (e^{-t} - 2)\mathbf{j} + (t + 1)\mathbf{k}$

**59.** The three components of $\mathbf{u}$ are increasing functions of $t$ at $t = t_0$.

**61–67.** Proofs

**69.** (a) 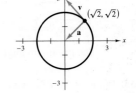    The curve is a cycloid.

(b) The maximum of $\|\mathbf{r}'\|$ is 2 and the minimum of $\|\mathbf{r}'\|$ is 0. The maximum and the minimum of $\|\mathbf{r}'\|$ are 1.

**71.** Proof    **73.** True

**75.** False. Let $\mathbf{r}(t) = \cos t\mathbf{i} + \sin t\mathbf{j} + \mathbf{k}$, then $\dfrac{d}{dt}[\|\mathbf{r}(t)\|] = 0$, but $\|\mathbf{r}'(t)\| = 1$.

## Section 12.3    *(page 842)*

**1.** The direction of the velocity vector provides the direction of motion at time $t$ and the magnitude of the velocity vector provides the speed of the object.

**3.** (a) $\mathbf{v}(t) = 3\mathbf{i} + \mathbf{j}$
$\|\mathbf{v}(t)\| = \sqrt{10}$
$\mathbf{a}(t) = \mathbf{0}$
(b) $\mathbf{v}(1) = 3\mathbf{i} + \mathbf{j}$
$\mathbf{a}(1) = \mathbf{0}$
(c)

**5.** (a) $\mathbf{v}(t) = 2t\mathbf{i} + \mathbf{j}$
$\|\mathbf{v}(t)\| = \sqrt{4t^2 + 1}$
$\mathbf{a}(t) = 2\mathbf{i}$
(b) $\mathbf{v}(2) = 4\mathbf{i} + \mathbf{j}$
$\mathbf{a}(2) = 2\mathbf{i}$
(c)

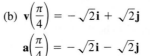

**7.** (a) $\mathbf{v}(t) = -2\sin t\mathbf{i} + 2\cos t\mathbf{j}$    (c)
$\|\mathbf{v}(t)\| = 2$
$\mathbf{a}(t) = -2\cos t\mathbf{i} - 2\sin t\mathbf{j}$
(b) $\mathbf{v}\left(\dfrac{\pi}{4}\right) = -\sqrt{2}\mathbf{i} + \sqrt{2}\mathbf{j}$
$\mathbf{a}\left(\dfrac{\pi}{4}\right) = -\sqrt{2}\mathbf{i} - \sqrt{2}\mathbf{j}$

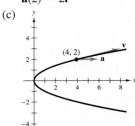

**9.** (a) $\mathbf{v}(t) = \langle 1 - \cos t, \sin t \rangle$    (c)
$\|\mathbf{v}(t)\| = \sqrt{2 - 2\cos t}$
$\mathbf{a}(t) = \langle \sin t, \cos t \rangle$
(b) $\mathbf{v}(\pi) = \langle 2, 0 \rangle$
$\mathbf{a}(\pi) = \langle 0, -1 \rangle$

**11.** (a) $\mathbf{v}(t) = \mathbf{i} + 5\mathbf{j} + 3\mathbf{k}$
$\|\mathbf{v}(t)\| = \sqrt{35}$
$\mathbf{a}(t) = \mathbf{0}$
(b) $\mathbf{v}(1) = \mathbf{i} + 5\mathbf{j} + 3\mathbf{k}$
$\mathbf{a}(1) = \mathbf{0}$

**13.** (a) $\mathbf{v}(t) = \mathbf{i} + 2t\mathbf{j} + t\mathbf{k}$
$\|\mathbf{v}(t)\| = \sqrt{1 + 5t^2}$
$\mathbf{a}(t) = 2\mathbf{j} + \mathbf{k}$
(b) $\mathbf{v}(4) = \mathbf{i} + 8\mathbf{j} + 4\mathbf{k}$
$\mathbf{a}(4) = 2\mathbf{j} + \mathbf{k}$

**15.** (a) $\mathbf{v}(t) = \mathbf{i} - \mathbf{j} - \dfrac{t}{\sqrt{9 - t^2}}\mathbf{k}$

$\|\mathbf{v}(t)\| = \sqrt{\dfrac{18 - t^2}{9 - t^2}}$

$\mathbf{a}(t) = -\dfrac{9}{(9 - t^2)^{3/2}}\mathbf{k}$

(b) $\mathbf{v}(0) = \mathbf{i} - \mathbf{j}$

$\mathbf{a}(0) = -\frac{1}{3}\mathbf{k}$

**17.** (a) $\mathbf{v}(t) = 4\mathbf{i} - 3 \sin t\mathbf{j} + 3 \cos t\mathbf{k}$

$\|\mathbf{v}(t)\| = 5$

$\mathbf{a}(t) = -3 \cos t\mathbf{j} - 3 \sin t\mathbf{k}$

(b) $\mathbf{v}(\pi) = \langle 4, 0, -3 \rangle$

$\mathbf{a}(\pi) = \langle 0, 3, 0 \rangle$

**19.** (a) $\mathbf{v}(t) = (e^t \cos t - e^t \sin t)\mathbf{i} + (e^t \sin t + e^t \cos t)\mathbf{j} + e^t\mathbf{k}$

$\|\mathbf{v}(t)\| = e^t\sqrt{3}$

$\mathbf{a}(t) = -2e^t \sin t\mathbf{i} + 2e^t \cos t\mathbf{j} + e^t\mathbf{k}$

(b) $\mathbf{v}(0) = \langle 1, 1, 1 \rangle$

$\mathbf{a}(0) = \langle 0, 2, 1 \rangle$

**21.** $\mathbf{v}(t) = t(\mathbf{i} + \mathbf{j} + \mathbf{k})$

$\mathbf{r}(t) = \dfrac{t^2}{2}(\mathbf{i} + \mathbf{j} + \mathbf{k})$

$\mathbf{r}(2) = 2(\mathbf{i} + \mathbf{j} + \mathbf{k})$

**23.** $\mathbf{v}(t) = \left(\dfrac{t^2}{2} + \dfrac{9}{2}\right)\mathbf{j} + \left(\dfrac{t^2}{2} - \dfrac{1}{2}\right)\mathbf{k}$

$\mathbf{r}(t) = \left(\dfrac{t^3}{6} + \dfrac{9}{2}t - \dfrac{14}{3}\right)\mathbf{j} + \left(\dfrac{t^3}{6} - \dfrac{1}{2}t + \dfrac{1}{3}\right)\mathbf{k}$

$\mathbf{r}(2) = \dfrac{17}{3}\mathbf{j} + \dfrac{2}{3}\mathbf{k}$

**25.** $\mathbf{v}(t) = -\sin t\mathbf{i} + \cos t\mathbf{j} + \mathbf{k}$

$\mathbf{r}(t) = \cos t\mathbf{i} + \sin t\mathbf{j} + t\mathbf{k}$

$\mathbf{r}(2) = (\cos 2)\mathbf{i} + (\sin 2)\mathbf{j} + 2\mathbf{k}$

**27.** 45.5 ft; The ball will clear the fence.

**29.** $v_0 = 40\sqrt{6}$ ft/sec; 78 ft    **31.** Proof

**33.** (a) $\mathbf{r}(t) = \left(\frac{440}{3} \cos \theta_0\right)t\mathbf{i} + \left[3 + \left(\frac{440}{3} \sin \theta_0\right)t - 16t^2\right]\mathbf{j}$

(b)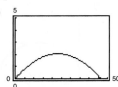

$\theta_0 = 20 \quad \theta_0 = 25$

$\theta_0 = 10 \quad \theta_0 = 15$

The minimum angle appears to be $\theta_0 = 20°$.

(c) $\theta_0 \approx 19.38°$

**35.** (a) $v_0 = 28.78$ ft/sec, $\theta = 58.28°$    (b) $v_0 \approx 32$ ft/sec

**37.** 1.91°

**39.** (a) 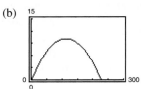    (b)

Maximum height: 2.1 ft    Maximum height: 10.0 ft
Range: 46.6 ft    Range: 227.8 ft

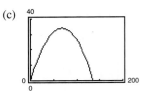    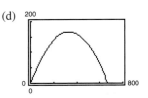

(c)    (d)

Maximum height: 34.0 ft    Maximum height: 166.5 ft
Range: 136.1 ft    Range: 666.1 ft

(e) 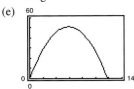    (f)

Maximum height: 51.0 ft    Maximum height: 249.8 ft
Range: 117.9 ft    Range: 576.9 ft

**41.** Maximum height: 129.1 m; Range: 886.3 m    **43.** Proof

**45.** $\mathbf{v}(t) = b\omega[(1 - \cos \omega t)\mathbf{i} + \sin \omega t\mathbf{j}]$

$\mathbf{a}(t) = b\omega^2(\sin \omega t\mathbf{i} + \cos \omega t\mathbf{j})$

(a) $\|\mathbf{v}(t)\| = 0$ when $\omega t = 0, 2\pi, 4\pi, \ldots$

(b) $\|\mathbf{v}(t)\|$ is maximum when $\omega t = \pi, 3\pi, \ldots$

**47.** $\mathbf{v}(t) = -b\omega \sin \omega t\mathbf{i} + b\omega \cos \omega t\mathbf{j}$

$\mathbf{v}(t) \cdot \mathbf{r}(t) = 0$

**49.** $\mathbf{a}(t) = -b\omega^2(\cos \omega t\mathbf{i} + \sin \omega t\mathbf{j}) = -\omega^2\mathbf{r}(t)$; $\mathbf{a}(t)$ is a negative multiple of a unit vector from $(0, 0)$ to $(\cos \omega t, \sin \omega t)$, so $\mathbf{a}(t)$ is directed toward the origin.

**51.** $8\sqrt{2}$ ft/sec

**53.** The particle could be changing direction.

**55.** This is true for uniform circular motion but not true for non-uniform circular motion.

**57–59.** Proofs    **61.** True

**63.** False. Consider $\mathbf{r}(t) = \langle t^2, -t^2 \rangle$. Then $\mathbf{v}(t) = \langle 2t, -2t \rangle$ and $\|\mathbf{v}(t)\| = \sqrt{8t^2}$.

## Section 12.4    *(page 852)*

**1.** The unit tangent vector points in the direction of motion.

**3.** $\mathbf{T}(1) = \dfrac{\sqrt{2}}{2}(\mathbf{i} + \mathbf{j})$    **5.** $\mathbf{T}\left(\dfrac{\pi}{3}\right) = -\dfrac{\sqrt{3}}{2}\mathbf{i} + \dfrac{1}{2}\mathbf{j}$

**7.** $\mathbf{T}(e) = \dfrac{3e\mathbf{i} - \mathbf{j}}{\sqrt{9e^2 + 1}} \approx 0.9926\mathbf{i} - 0.1217\mathbf{j}$

**9.** $\mathbf{T}(0) = \dfrac{\sqrt{2}}{2}(\mathbf{i} + \mathbf{k})$    **11.** $\mathbf{T}(0) = \dfrac{\sqrt{2}}{2}(\mathbf{j} + \mathbf{k})$

$x = t$        $x = 1$

$y = 0$        $y = 3t$

$z = t$        $z = -4 + 3t$

**13.** $\mathbf{T}\left(\dfrac{\pi}{4}\right) = \dfrac{1}{2}\langle -\sqrt{2}, \sqrt{2}, 0 \rangle$

$x = \sqrt{2} - \sqrt{2}t$

$y = \sqrt{2} + \sqrt{2}t$

$z = 4$

**15.** $\mathbf{N}(2) = \dfrac{\sqrt{5}}{5}(-2\mathbf{i} + \mathbf{j})$

**17.** $\mathbf{N}(1) = -\dfrac{\sqrt{14}}{14}(\mathbf{i} - 2\mathbf{j} + 3\mathbf{k})$

**19.** $\mathbf{N}\left(\dfrac{3\pi}{4}\right) = \dfrac{\sqrt{2}}{2}(\mathbf{i} - \mathbf{j})$

**21.** $\mathbf{r}(2) = 2\mathbf{i} + \frac{1}{2}\mathbf{j}$

$\mathbf{T}(2) = \dfrac{\sqrt{17}}{17}(4\mathbf{i} - \mathbf{j})$

$\mathbf{N}(2) = \dfrac{\sqrt{17}}{17}(\mathbf{i} + 4\mathbf{j})$

**23.** $\mathbf{r}(2) = 5\mathbf{i} - 4\mathbf{j}$

$\mathbf{T}(2) = \dfrac{\mathbf{i} - 2\mathbf{j}}{\sqrt{5}}$

$\mathbf{N}(2) = \dfrac{-2\mathbf{i} - \mathbf{j}}{\sqrt{5}}$,

perpendicular to $\mathbf{T}(2)$

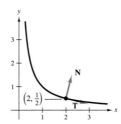

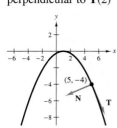

**25.** $a_\mathbf{T} = -\sqrt{2}$

$a_\mathbf{N} = \sqrt{2}$

**27.** $a_\mathbf{T} = -\dfrac{7\sqrt{5}}{5}$

$a_\mathbf{N} = \dfrac{6\sqrt{5}}{5}$

**29.** $a_\mathbf{T} = \sqrt{2}e^{\pi/2}$

$a_\mathbf{N} = \sqrt{2}e^{\pi/2}$

**31.** $\mathbf{T}(t) = -\sin\omega t\mathbf{i} + \cos\omega t\mathbf{j}$

$\mathbf{N}(t) = -\cos\omega t\mathbf{i} - \sin\omega t\mathbf{j}$

$a_\mathbf{T} = 0$

$a_\mathbf{N} = a\omega^2$

**33.** $\|\mathbf{v}(t)\| = a\omega$; The speed is constant because $a_\mathbf{T} = 0$.

**35.** $a_\mathbf{T}$ is undefined.

$a_\mathbf{N}$ is undefined.

**37.** $a_\mathbf{T} = \dfrac{5\sqrt{6}}{6}$

$a_\mathbf{N} = \dfrac{\sqrt{30}}{6}$

**39.** $a_\mathbf{T} = \sqrt{3}$

$a_\mathbf{N} = \sqrt{2}$

**41.** The particle's motion is in a straight line.

**43.** $\mathbf{v}(t) = \mathbf{r}'(t) = 3\mathbf{i} + 4\mathbf{j}$

$\|\mathbf{v}(t)\| = \sqrt{9 + 16} = 5$

$\mathbf{a}(t) = \mathbf{v}'(t) = \mathbf{0}$

$\mathbf{T}(t) = \dfrac{\mathbf{v}(t)}{\|\mathbf{v}(t)\|} = \dfrac{3}{5}\mathbf{i} + \dfrac{4}{5}\mathbf{j}$

$\mathbf{T}'(t) = 0 \implies \mathbf{N}(t)$ does not exist.

The path is a line. The speed is constant (5).

**45.** (a) $t = \frac{1}{2}$: $a_\mathbf{T} = \dfrac{\sqrt{2}\pi^2}{2}$, $a_\mathbf{N} = \dfrac{\sqrt{2}\pi^2}{2}$

$t = 1$: $a_\mathbf{T} = 0$, $a_\mathbf{N} = \pi^2$

$t = \frac{3}{2}$: $a_\mathbf{T} = -\dfrac{\sqrt{2}\pi^2}{2}$, $a_\mathbf{N} = \dfrac{\sqrt{2}\pi^2}{2}$

(b) $t = \frac{1}{2}$: Increasing because $a_\mathbf{T} > 0$.

$t = 1$: Maximum because $a_\mathbf{T} = 0$.

$t = \frac{3}{2}$: Decreasing because $a_\mathbf{T} < 0$.

**47.** $\mathbf{T}\left(\dfrac{\pi}{2}\right) = \dfrac{\sqrt{17}}{17}(-4\mathbf{i} + \mathbf{k})$

$\mathbf{N}\left(\dfrac{\pi}{2}\right) = -\mathbf{j}$

$\mathbf{B}\left(\dfrac{\pi}{2}\right) = \dfrac{\sqrt{17}}{17}(\mathbf{i} + 4\mathbf{k})$

**49.** $\mathbf{T}\left(\dfrac{\pi}{4}\right) = \dfrac{\sqrt{2}}{2}(\mathbf{j} - \mathbf{k})$

$\mathbf{N}\left(\dfrac{\pi}{4}\right) = -\dfrac{\sqrt{2}}{2}(\mathbf{j} + \mathbf{k})$

$\mathbf{B}\left(\dfrac{\pi}{4}\right) = -\mathbf{i}$

**51.** $\mathbf{T}\left(\dfrac{\pi}{3}\right) = \dfrac{\sqrt{5}}{5}(\mathbf{i} - \sqrt{3}\mathbf{j} + \mathbf{k})$

$\mathbf{N}\left(\dfrac{\pi}{3}\right) = -\dfrac{1}{2}(\sqrt{3}\mathbf{i} + \mathbf{j})$

$\mathbf{B}\left(\dfrac{\pi}{3}\right) = \dfrac{\sqrt{5}}{10}(\mathbf{i} - \sqrt{3}\mathbf{j} - 4\mathbf{k})$

**53.** $\mathbf{N}(t) = \dfrac{1}{\sqrt{16t^2 + 9}}(-4t\mathbf{i} + 3\mathbf{j})$

**55.** $\mathbf{N}(t) = \dfrac{1}{\sqrt{5t^2 + 25}}(-t\mathbf{i} - 2t\mathbf{j} + 5\mathbf{k})$

**57.** $a_\mathbf{T} = \dfrac{-32(v_0\sin\theta - 32t)}{\sqrt{v_0^2\cos^2\theta + (v_0\sin\theta - 32t)^2}}$

$a_\mathbf{N} = \dfrac{32v_0\cos\theta}{\sqrt{v_0^2\cos^2\theta + (v_0\sin\theta - 32t)^2}}$

At maximum height, $a_\mathbf{T} = 0$ and $a_\mathbf{N} = 32$.

**59.** (a) $\mathbf{r}(t) = 60\sqrt{3}t\mathbf{i} + (5 + 60t - 16t^2)\mathbf{j}$

(b)

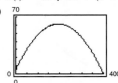

Maximum height $\approx 61.245$ ft

Range $\approx 398.186$ ft

(c) $\mathbf{v}(t) = 60\sqrt{3}\mathbf{i} + (60 - 32t)\mathbf{j}$

$\|\mathbf{v}(t)\| = 8\sqrt{16t^2 - 60t + 225}$

$\mathbf{a}(t) = -32\mathbf{j}$

(d)

| $t$ | 0.5 | 1.0 | 1.5 |
|---|---|---|---|
| Speed | 112.85 | 107.63 | 104.61 |

| $t$ | 2.0 | 2.5 | 3.0 |
|---|---|---|---|
| Speed | 104 | 105.83 | 109.98 |

(e)

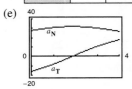

The speed is decreasing when $a_\mathbf{T}$ and $a_\mathbf{N}$ have opposite signs.

**61.** (a) $4\sqrt{625\pi^2 + 1} \approx 314$ mi/h

(b) $a_\mathbf{T} = 0$, $a_\mathbf{N} = 1000\pi^2$

$a_\mathbf{T} = 0$ because the speed is constant.

**63.** (a) The centripetal component is quadrupled.

(b) The centripetal component is halved.

**65.** 4.74 mi/sec     **67.** 4.67 mi/sec

**69.** False. These vectors are perpendicular for an object traveling at a constant speed but not for an object traveling at a variable speed.

**71.** (a) and (b) Proofs     **73–75.** Proofs

## Section 12.5   (page 864)

**1.** The curve bends more sharply at $Q$ than at $P$.

**3.**

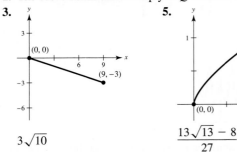

$3\sqrt{10}$

**5.**

$\dfrac{13\sqrt{13} - 8}{27}$

**7.**

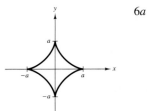

6a

**9.** 362.9 ft

**11.**

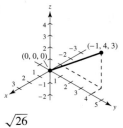

$\sqrt{26}$

**13.**

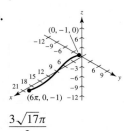

$\dfrac{3\sqrt{17}\pi}{2}$

$2\pi\sqrt{a^2+b^2}$

**15.**

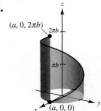

**17.** (a) $2\sqrt{21} \approx 9.165$    (b) 9.529
(c) Increase the number of line segments.    (d) 9.571

**19.** 0    **21.** $\dfrac{1}{4}$    **23.** 0    **25.** $\dfrac{\sqrt{2}}{2}$    **27.** 1    **29.** $\dfrac{1}{4}$

**31.** $\dfrac{1}{a}$    **33.** $\dfrac{\sqrt{5}}{(1+5t^2)^{3/2}}$    **35.** $\dfrac{3}{25}$    **37.** $\dfrac{12}{125}$

**39.** $\dfrac{7\sqrt{26}}{676}$    **41.** $K=0$, $\dfrac{1}{K}$ is undefined.

**43.** $K=\dfrac{10}{101^{3/2}}$, $\dfrac{1}{K}=\dfrac{101^{3/2}}{10}$    **45.** $K=4$, $\dfrac{1}{K}=\dfrac{1}{4}$

**47.** $K=\dfrac{12}{145^{3/2}}$, $\dfrac{1}{K}=\dfrac{145^{3/2}}{12}$    **49.** (a) $(1,3)$    (b) 0

**51.** (a) $K\to\infty$ as $x\to 0$  (No maximum)    (b) 0

**53.** (a) $\left(\dfrac{1}{\sqrt{2}}, -\dfrac{\ln 2}{2}\right)$    (b) 0    **55.** $(0,1)$

**57.** $(\pi+2n\pi, 0)$    **59.** $c=\pm\sqrt{2}$

**61.** (a) $K=\dfrac{2|6x^2-1|}{(16x^6-16x^4+4x^2+1)^{3/2}}$

(b) $x=0$: $x^2+\left(y+\dfrac{1}{2}\right)^2=\dfrac{1}{4}$

$x=1$: $x^2+\left(y-\dfrac{1}{2}\right)^2=\dfrac{5}{4}$

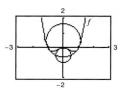

(c)

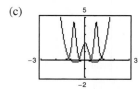

The curvature tends to be greatest near the extrema of the function and decreases as $x\to\pm\infty$. However, $f$ and $K$ do not have the same critical numbers.

Critical numbers of $f$: $x=0, \pm\dfrac{\sqrt{2}}{2}\approx\pm 0.7071$

Critical numbers of $K$: $x=0, \pm 0.7647, \pm 0.4082$

**63.** $a=\dfrac{1}{4}$, $b=2$

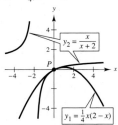

**65.** (a) 12.25 units    (b) $\dfrac{1}{2}$    **67–69.** Proofs

**71.** (a) 0    (b) 0    **73.** $\dfrac{1}{4}$    **75.** Proof

**77.** $K=\dfrac{1}{4a}\left|\csc\dfrac{\theta}{2}\right|$

Minimum: $K=\dfrac{1}{4a}$

There is no maximum.

**79.** 3327.5 lb    **81.** Proof

**83.** False. See Exploration on page 855.

**85.** True    **87–93.** Proofs

## Review Exercises for Chapter 12   *(page 867)*

**1.** (a) All reals except $\dfrac{\pi}{2}+n\pi$, $n$ is an integer.

(b) Continuous except at $t=\dfrac{\pi}{2}+n\pi$, $n$ is an integer.

**3.** (a) $[3,\infty)$    (b) Continuous for all $t\ge 3$

**5.** (a) $\mathbf{i}-\sqrt{2}\mathbf{k}$    (b) $-3\mathbf{i}+4\mathbf{j}$
(c) $(2c-1)\mathbf{i}+(c-1)^2\mathbf{j}-\sqrt{c+1}\,\mathbf{k}$
(d) $2\Delta t\mathbf{i}+\Delta t(\Delta t+2)\mathbf{j}-\left(\sqrt{\Delta t+3}-\sqrt{3}\right)\mathbf{k}$

**7.** $\mathbf{r}(t)=(3-t)\mathbf{i}-2t\mathbf{j}+(5-2t)\mathbf{k}$, $0\le t\le 1$
$x=3-t$, $y=-2t$, $z=5-2t$, $0\le t\le 1$

**9.**

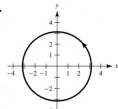

**11.**

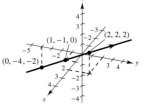

**13.** $\mathbf{r}(t)=t\mathbf{i}+\left(-\dfrac{3}{4}t+3\right)\mathbf{j}$    **15.**

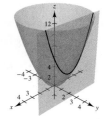

$x=t$, $y=2$, $z=t^2+4$

**17.** $\ln 3\mathbf{j}-\dfrac{1}{3}\mathbf{k}$

**19.** (a) $(2t+4)\mathbf{i}-6t\mathbf{j}$
(b) $2\mathbf{i}-6\mathbf{j}$
(c) $40t+8$

**21.** (a) $6t^2\mathbf{i}+4\mathbf{j}-2t\mathbf{k}$
(b) $12t\mathbf{i}-2\mathbf{k}$
(c) $72t^3+4t$
(d) $-8\mathbf{i}-12t^2\mathbf{j}-48t\mathbf{k}$

**23.** $(-\infty,1)$, $(1,\infty)$

**25.** (a) $3\mathbf{i} + \mathbf{j}$     (b) $-5\mathbf{i} + (2t - 2)\mathbf{j} + 2t^2\mathbf{k}$
(c) $18t\mathbf{i} + (6t - 3)\mathbf{j}$     (d) $4t + 3t^2$
(e) $\left(\frac{8}{3}t^3 - 2t^2\right)\mathbf{i} - 8t^3\mathbf{j} + (9t^2 - 2t + 1)\mathbf{k}$
(f) $2\mathbf{i} + 8t\mathbf{j} + 16t^2\mathbf{k}$

**27.** $\frac{1}{3}t^3\mathbf{i} + \frac{5}{2}t^2\mathbf{j} + 2t^4\mathbf{k} + \mathbf{C}$     **29.** $2t^{3/2}\mathbf{i} + 2\ln|t|\mathbf{j} + t\mathbf{k} + \mathbf{C}$

**31.** $\frac{32}{3}\mathbf{j}$     **33.** $2(e - 1)\mathbf{i} - 8\mathbf{j} - 2\mathbf{k}$

**35.** $\mathbf{r}(t) = (t^2 + 1)\mathbf{i} + (e^t + 2)\mathbf{j} - (e^{-t} + 4)\mathbf{k}$

**37.** (a) $\mathbf{v}(t) = 4\mathbf{i} + 3t^2\mathbf{j} - \mathbf{k}$
$\|\mathbf{v}(t)\| = \sqrt{17 + 9t^4}$
$\mathbf{a}(t) = 6t\mathbf{j}$
(b) $\mathbf{v}(1) = 4\mathbf{i} + 3\mathbf{j} - \mathbf{k}$
$\mathbf{a}(1) = 6\mathbf{j}$

**39.** (a) $\mathbf{v}(t) = \langle -3\cos^2 t \sin t, 3\sin^2 t \cos t, 3\rangle$
$\|\mathbf{v}(t)\| = 3\sqrt{\sin^2 t \cos^2 t + 1}$
$\mathbf{a}(t) = \langle 3\cos t(2\sin^2 t - \cos^2 t), 3\sin t(2\cos^2 t - \sin^2 t), 0\rangle$
(b) $\mathbf{v}(\pi) = \langle 0, 0.3\rangle$
$\mathbf{a}(\pi) = \langle 3, 0, 0\rangle$

**41.** 11.67 ft; The ball will clear the fence.

**43.** $\mathbf{T}(2) = \dfrac{3\mathbf{i} - 2\mathbf{j}}{\sqrt{13}}$

**45.** $\mathbf{T}(0) = \dfrac{2\mathbf{i} - 3\mathbf{k}}{\sqrt{13}}$; $x = 1 + 2t, y = 1, z = -3t$

**47.** $\mathbf{N}(1) = -\dfrac{3\sqrt{10}}{10}\mathbf{i} + \dfrac{\sqrt{10}}{10}\mathbf{j}$     **49.** $\mathbf{N}\left(\dfrac{\pi}{4}\right) = -\mathbf{j}$

**51.** $a_{\mathbf{T}} = -\dfrac{2\sqrt{13}}{585}$     **53.** $a_{\mathbf{T}} = 0$
$a_{\mathbf{N}} = \dfrac{4\sqrt{13}}{65}$          $a_{\mathbf{N}} = 1$

**55.**
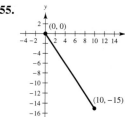
$5\sqrt{13}$

**57.**
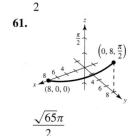
$2$

**59.**
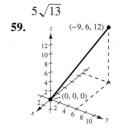
$3\sqrt{29}$

**61.**
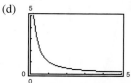
$\dfrac{\sqrt{65}\pi}{2}$

**63.** 0     **65.** $\dfrac{2\sqrt{5}}{(4 + 5t^2)^{3/2}}$     **67.** $\dfrac{\sqrt{2}}{3}$

**69.** $K = \dfrac{1}{26^{3/2}}, \dfrac{1}{K} = 26\sqrt{26}$     **71.** $K = \dfrac{\sqrt{2}}{4}, r = 2\sqrt{2}$

**73.** 2016.7 lb

## P.S. Problem Solving     *(page 869)*

**1.** (a) $a$     (b) $\pi a$     (c) $K = \pi a$

**3.** Initial speed: 447.21 ft/sec; $\theta \approx 63.43°$     **5–7.** Proofs

**9.** Unit tangent: $\langle -\frac{4}{5}, 0, \frac{3}{5}\rangle$
Principal unit normal: $\langle 0, -1, 0\rangle$
Binormal: $\langle \frac{3}{5}, 0, \frac{4}{5}\rangle$

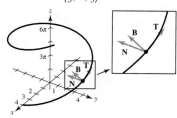

**11.** (a) and (b)  Proofs
**13.** (a)

(b) 6.766

(c) $K = \dfrac{\pi(\pi^2 t^2 + 2)}{(\pi^2 t^2 + 1)^{3/2}}$
$K(0) = 2\pi$
$K(1) = \dfrac{\pi(\pi^2 + 2)}{(\pi^2 + 1)^{3/2}} \approx 1.04$
$K(2) \approx 0.51$

(d)

(e) $\lim_{t \to \infty} K = 0$

(f) As $t \to \infty$, the graph spirals outward and the curvature decreases.

# Chapter 13

## Section 13.1     *(page 880)*

**1.** There is not a unique value of $z$ for each ordered pair.
**3.** $z$ is a function of $x$ and $y$.     **5.** $z$ is a function of $x$ and $y$.
**7.** $z$ is not a function of $x$ and $y$.
**9.** (a) 1     (b) 1     (c) $-17$
(d) $9 - y$     (e) $2x - 1$     (f) $13 - t$
**11.** (a) $-1$     (b) 0     (c) $xe^3$     (d) $te^{-y}$
**13.** (a) 3     (b) 2     (c) $\dfrac{16}{t}$     (d) $-\dfrac{6}{5}$
**15.** (a) $\sqrt{2}$     (b) $3\sin 1$     (c) 0     (d) 4
**17.** (a) $-4$     (b) $-6$     (c) $-\frac{25}{4}$     (d) $\frac{9}{4}$
**19.** (a) $2, \Delta x \neq 0$     (b) $2y + \Delta y, \Delta y \neq 0$
**21.** Domain: $\{(x, y): x$ is any real number, $y$ is any real number$\}$
Range: all real numbers
**23.** Domain: $\{(x, y): y \geq 0\}$
Range: all real numbers
**25.** Domain: $\{(x, y): x \neq 0, y \neq 0\}$
Range: all real numbers
**27.** Domain: $\{(x, y): x^2 + y^2 \leq 4\}$
Range: $0 \leq z \leq 2$
**29.** Domain: $\{(x, y): -1 \leq x + y \leq 1\}$
Range: $0 \leq z \leq \pi$

**31.** Domain: $\{(x, y): y < -x + 5\}$
Range: all real numbers

**33.** (a) $(20, 0, 0)$    (b) $(-15, 10, 20)$
(c) $(20, 15, 25)$    (d) $(20, 20, 0)$

**35.** Plane

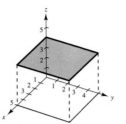

**37.** Cylinder with rulings parallel to the $x$-axis

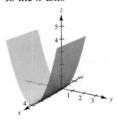

**39.** Paraboloid

**41.** Cylinder with rulings parallel to the $y$-axis

**43.**

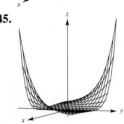

**45.**

**47.** c    **48.** d    **49.** b    **50.** a

**51.** Lines: $x + y = c$

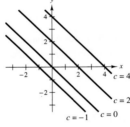

**53.** Ellipses: $x^2 + 4y^2 = c$
[except $x^2 + 4y^2 = 0$ is the point $(0, 0)$]

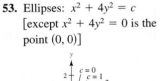

**55.** Hyperbolas: $xy = c$

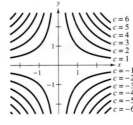

**57.** Circles passing through $(0, 0)$

Centered at $\left(\dfrac{1}{2c}, 0\right)$

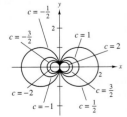

**59.**

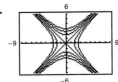

**61.**

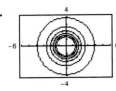

**63.** Yes; The definition of a function of two variables requires that $z$ be unique for each ordered pair $(x, y)$ in the domain.

**65.** $f(x, y) = \dfrac{x}{y}$ $\left(\text{The level curves are the lines } y = \dfrac{x}{c}.\right)$

**67.** The surface may be shaped like a saddle. For example, let $f(x, y) = xy$. The graph is not unique because any vertical translation will produce the same level curves.

**69.**

| Tax Rate | Inflation Rate | | |
|---|---|---|---|
| | 0 | 0.03 | 0.05 |
| 0 | $1790.85 | $1332.56 | $1099.43 |
| 0.28 | $1526.43 | $1135.80 | $937.09 |
| 0.35 | $1466.07 | $1090.90 | $900.04 |

**71.** Plane

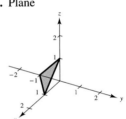

**73.** Sphere

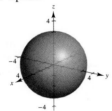

**75.** Elliptic cone

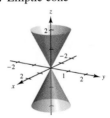

**77.** (a) 243 board-ft    (b) 507 board-ft

**79.**

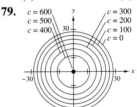

**81.** 36,661 units

**83.** Proof

**85.** (a) $k = \dfrac{520}{3}$

(b) $P = \dfrac{520T}{3V}$

The level curves are lines.

**87.** (a) $C$    (b) $A$    (c) $B$

**89.** $C = 4.50xy + 5.00(xz + yz)$    **91.** False. Let $f(x, y) = 4$.

**93.** False. The equation of a sphere is not a function.

**95.** Putnam Problem A1, 2008

## Section 13.2  (page 891)

**1.** As $x$ approaches $-1$ and $y$ approaches 3, $z$ approaches 1.

**3–5.** Proofs    **7.** 9    **9.** $-20$    **11.** 7, continuous

**13.** $e^2$, continuous    **15.** 0, continuous for $y \neq 0$

**17.** $\frac{1}{2}$, continuous except at $(0, 0)$    **19.** $-1$, continuous

**21.** 0, continuous for $xy \neq 1$, $|xy| \leq 1$

**23.** $2\sqrt{2}$, continuous for $x + y + z \geq 0$    **25.** 0

**27.** Limit does not exist.    **29.** Limit does not exist.

**31.** Limit does not exist.    **33.** 0

**35.** Limit does not exist.

**37.** No. The existence of $f(2, 3)$ has no bearing on the existence of the limit as $(x, y) \to (2, 3)$.

**39.** $\lim_{x \to 0} f(x, 0) = 0$ if $f(x, 0)$ exists.    **41.** Continuous, 1

**43.**

| $(x, y)$ | $(1, 0)$ | $(0.5, 0)$ | $(0.1, 0)$ | $(0.01, 0)$ | $(0.001, 0)$ |
|---|---|---|---|---|---|
| $f(x, y)$ | 0 | 0 | 0 | 0 | 0 |

$y = 0$: 0

| $(x, y)$ | $(1, 1)$ | $(0.5, 0.5)$ | $(0.1, 0.1)$ |
|---|---|---|---|
| $f(x, y)$ | $\frac{1}{2}$ | $\frac{1}{2}$ | $\frac{1}{2}$ |

| $(x, y)$ | $(0.01, 0.01)$ | $(0.001, 0.001)$ |
|---|---|---|
| $f(x, y)$ | $\frac{1}{2}$ | $\frac{1}{2}$ |

$y = x$: $\frac{1}{2}$

Limit does not exist.

Continuous except at $(0, 0)$

**45.**

| $(x, y)$ | $(1, 0)$ | $(0.5, 0)$ | $(0.1, 0)$ | $(0.01, 0)$ | $(0.001, 0)$ |
|---|---|---|---|---|---|
| $f(x, y)$ | 0 | 0 | 0 | 0 | 0 |

$y = 0$: 0

| $(x, y)$ | $(1, 1)$ | $(0.5, 0.5)$ | $(0.1, 0.1)$ |
|---|---|---|---|
| $f(x, y)$ | $\frac{1}{2}$ | 1 | 5 |

| $(x, y)$ | $(0.01, 0.01)$ | $(0.001, 0.001)$ |
|---|---|---|
| $f(x, y)$ | 50 | 500 |

$y = x$: $\infty$

The limit does not exist.

Continuous except at $(0, 0)$

**47.** (a) $\dfrac{1 + a^2}{a}$, $a \neq 0$    (b) Limit does not exist.

(c) No; Different paths result in different limits.

**49.** $f$ is continuous. $g$ is continuous except at $(0, 0)$. $g$ has a removable discontinuity at $(0, 0)$.

**51.** 0    **53.** 0    **55.** 1    **57.** 1    **59.** 0

**61.** Continuous except at $(0, 0, 0)$    **63.** Continuous

**65.** Continuous    **67.** Continuous

**69.** Continuous for $y \neq \dfrac{2x}{3}$    **71.** (a) $2x$    (b) $-4$

**73.** (a) $\dfrac{1}{y}$    (b) $-\dfrac{x}{y^2}$    **75.** (a) $3 + y$    (b) $x - 2$

**77.** 0

**79.** True

**81.** False. Let $f(x, y) = \begin{cases} \ln(x^2 + y^2), & x \neq 0, y \neq 0 \\ 0, & x = 0, y = 0 \end{cases}$.

**83.** $\dfrac{\pi}{2}$    **85.** Proof

## Section 13.3  (page 900)

**1.** $z_x$, $f_x(x, y)$, $\dfrac{\partial z}{\partial x}$

**3.** (a) Differentiate first with respect to $y$, then with respect to $x$, and last with respect to $z$.

(b) Differentiate first with respect to $z$ and then with respect to $x$.

**5.** No. Because you are finding the partial derivative with respect to $x$, you consider $y$ to be constant. So, the denominator is considered a constant and does not contain any variables.

**7.** No. Because you are finding the partial derivative with respect to $y$, you consider $x$ to be constant. So, the denominator is considered a constant and does not contain any variables.

**9.** Yes. Because you are finding the partial derivative with respect to $x$, you consider $y$ to be constant. So, both the numerator and denominator contain variables.

**11.** $f_x(x, y) = 2$
$f_y(x, y) = -5$

**13.** $\dfrac{\partial z}{\partial x} = 6 - 2xy$
$\dfrac{\partial z}{\partial y} = -x^2 + 16y$

**15.** $\dfrac{\partial z}{\partial x} = \sqrt{y}$
$\dfrac{\partial z}{\partial y} = \dfrac{x}{2\sqrt{y}}$

**17.** $\dfrac{\partial z}{\partial x} = ye^{xy}$
$\dfrac{\partial z}{\partial y} = xe^{xy}$

**19.** $\dfrac{\partial z}{\partial x} = 2xe^{2y}$
$\dfrac{\partial z}{\partial y} = 2x^2e^{2y}$

**21.** $\dfrac{\partial z}{\partial x} = \dfrac{1}{x}$
$\dfrac{\partial z}{\partial y} = -\dfrac{1}{y}$

**23.** $\dfrac{\partial z}{\partial x} = \dfrac{2x}{x^2 + y^2}$
$\dfrac{\partial z}{\partial y} = \dfrac{2y}{x^2 + y^2}$

**25.** $\dfrac{\partial z}{\partial x} = \dfrac{x^3 - 3y^3}{x^2y}$
$\dfrac{\partial z}{\partial y} = \dfrac{-x^3 + 12y^3}{2xy^2}$

**27.** $h_x(x, y) = -2xe^{-(x^2+y^2)}$
$h_y(x, y) = -2ye^{-(x^2+y^2)}$

**29.** $f_x(x, y) = \dfrac{x}{\sqrt{x^2 + y^2}}$
$f_y(x, y) = \dfrac{y}{\sqrt{x^2 + y^2}}$

**31.** $\dfrac{\partial z}{\partial x} = -y \sin xy$
$\dfrac{\partial z}{\partial y} = -x \sin xy$

**33.** $\dfrac{\partial z}{\partial x} = 2 \sec^2(2x - y)$
$\dfrac{\partial z}{\partial y} = -\sec^2(2x - y)$

**35.** $\dfrac{\partial z}{\partial x} = 8ye^y \cos 8xy$
$\dfrac{\partial z}{\partial y} = e^y(8x \cos 8xy + \sin 8xy)$

**37.** $\dfrac{\partial z}{\partial x} = 2 \cosh(2x + 3y)$
$\dfrac{\partial z}{\partial y} = 3 \cosh(2x + 3y)$

**39.** $f_x(x, y) = 1 - x^2$
$f_y(x, y) = y^2 - 1$

**41.** $f_x(x, y) = 3$
$f_y(x, y) = 2$

**43.** $f_x(x, y) = \dfrac{1}{2\sqrt{x + y}}$

$f_y(x, y) = \dfrac{1}{2\sqrt{x + y}}$

**45.** $f_x = 12$

$f_y = 12$

**47.** $f_x = -1$

$f_y = \frac{1}{2}$

**49.** $f_x = \frac{1}{4}$

$f_y = \frac{1}{4}$

**51.** $f_x = -\dfrac{1}{4}$

$f_y = \dfrac{1}{4}$

**53.** $\dfrac{\partial z}{\partial x}(1, 2) = 2$

$\dfrac{\partial z}{\partial y}(1, 2) = 1$

**55.** $g_x(1, 1) = -2$

$g_y(1, 1) = -2$

**57.** $H_x(x, y, z) = \cos(x + 2y + 3z)$

$H_y(x, y, z) = 2\cos(x + 2y + 3z)$

$H_z(x, y, z) = 3\cos(x + 2y + 3z)$

**59.** $\dfrac{\partial w}{\partial x} = \dfrac{x}{\sqrt{x^2 + y^2 + z^2}}$

$\dfrac{\partial w}{\partial y} = \dfrac{y}{\sqrt{x^2 + y^2 + z^2}}$

$\dfrac{\partial w}{\partial z} = \dfrac{z}{\sqrt{x^2 + y^2 + z^2}}$

**61.** $F_x(x, y, z) = \dfrac{x}{x^2 + y^2 + z^2}$

$F_y(x, y, z) = \dfrac{y}{x^2 + y^2 + z^2}$

$F_z(x, y, z) = \dfrac{z}{x^2 + y^2 + z^2}$

**63.** $f_x = 3,\ f_y = 1,\ f_z = 2$    **65.** $f_x = 1,\ f_y = 0,\ f_z = 0$

**67.** $f_x = 4,\ f_y = 24,\ f_z = 0$    **69.** $x = 2,\ y = -2$

**71.** $x = -6,\ y = 4$    **73.** $x = 1,\ y = 1$    **75.** $x = 0,\ y = 0$

**77.** $\dfrac{\partial^2 z}{\partial x^2} = 0$

$\dfrac{\partial^2 z}{\partial y^2} = 6x$

$\dfrac{\partial^2 z}{\partial y \partial x} = \dfrac{\partial^2 z}{\partial x \partial y} = 6y$

**79.** $\dfrac{\partial^2 z}{\partial x^2} = 12x^2$

$\dfrac{\partial^2 z}{\partial y^2} = 18y$

$\dfrac{\partial^2 z}{\partial y \partial x} = \dfrac{\partial^2 z}{\partial x \partial y} = -2$

**81.** $\dfrac{\partial^2 z}{\partial x^2} = \dfrac{y^2}{(x^2 + y^2)^{3/2}}$

$\dfrac{\partial^2 z}{\partial y^2} = \dfrac{x^2}{(x^2 + y^2)^{3/2}}$

$\dfrac{\partial^2 z}{\partial y \partial x} = \dfrac{\partial^2 z}{\partial x \partial y} = \dfrac{-xy}{(x^2 + y^2)^{3/2}}$

**83.** $\dfrac{\partial^2 z}{\partial x^2} = e^x \tan y$

$\dfrac{\partial^2 z}{\partial y^2} = 2e^x \sec^2 y \tan y$

$\dfrac{\partial^2 z}{\partial y \partial x} = \dfrac{\partial^2 z}{\partial x \partial y} = e^x \sec^2 y$

**85.** $\dfrac{\partial^2 z}{\partial x^2} = -y^2 \cos xy$

$\dfrac{\partial^2 z}{\partial y^2} = -x^2 \cos xy$

$\dfrac{\partial^2 z}{\partial y \partial x} = \dfrac{\partial^2 z}{\partial x \partial y} = -xy \cos xy - \sin xy$

**87.** $\dfrac{\partial z}{\partial x} = \sec y$

$\dfrac{\partial z}{\partial y} = x \sec y \tan y$

$\dfrac{\partial^2 z}{\partial x^2} = 0$

$\dfrac{\partial^2 z}{\partial y^2} = x \sec y(\sec^2 y + \tan^2 y)$

$\dfrac{\partial^2 z}{\partial y \partial x} = \dfrac{\partial^2 z}{\partial x \partial y} = \sec y \tan y$

No values of $x$ and $y$ exist such that $f_x(x, y) = f_y(x, y) = 0$.

**89.** $\dfrac{\partial z}{\partial x} = \dfrac{y^2 - x^2}{x(x^2 + y^2)}$

$\dfrac{\partial z}{\partial y} = \dfrac{-2y}{x^2 + y^2}$

$\dfrac{\partial^2 z}{\partial x^2} = \dfrac{x^4 - 4x^2y^2 - y^4}{x^2(x^2 + y^2)^2}$

$\dfrac{\partial^2 z}{\partial y^2} = \dfrac{2(y^2 - x^2)}{(x^2 + y^2)^2}$

$\dfrac{\partial^2 z}{\partial y \partial x} = \dfrac{\partial^2 z}{\partial x \partial y} = \dfrac{4xy}{(x^2 + y^2)^2}$

No values of $x$ and $y$ exist such that $f_x(x, y) = f_y(x, y) = 0$.

**91.** $f_{xyy}(x, y, z) = f_{yxy}(x, y, z) = f_{yyx}(x, y, z) = 0$

**93.** $f_{xyy}(x, y, z) = f_{yxy}(x, y, z) = f_{yyx}(x, y, z) = z^2 e^{-x} \sin yz$

**95.** $\dfrac{\partial^2 z}{\partial x^2} + \dfrac{\partial^2 z}{\partial y^2} = 0 + 0 = 0$

**97.** $\dfrac{\partial^2 z}{\partial x^2} + \dfrac{\partial^2 z}{\partial y^2} = e^x \sin y - e^x \sin y = 0$

**99.** $\dfrac{\partial^2 z}{\partial t^2} = -c^2 \sin(x - ct) = c^2\left(\dfrac{\partial^2 z}{\partial x^2}\right)$

**101.** $\dfrac{\partial^2 z}{\partial t^2} = \dfrac{-c^2}{(x + ct)^2} = c^2\left(\dfrac{\partial^2 z}{\partial x^2}\right)$

**103.** $\dfrac{\partial z}{\partial t} = \dfrac{-e^{-t} \cos x}{c} = c^2\left(\dfrac{\partial^2 z}{\partial x^2}\right)$    **105.** Proof

**107.** Yes; $f(x, y) = \cos(3x - 2y)$

**109.** No. Let $z = x + y + 1$.

**111.**

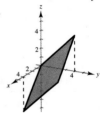

**113.** Dollars/yr; negative; You expect the influence that age has on the cost of the car to be negative.

**115.** (a) $\sqrt{2}$    (b) $\frac{5}{2}$    **117.** (a) 72    (b) 72

**119.** $IQ_M = \dfrac{100}{C},\ IQ_M(12, 10) = 10$

$IQ$ increases at a rate of 10 points per year of mental age when the mental age is 12 and the chronological age is 10.

$IQ_C = -\dfrac{100M}{C^2},\ IQ_C(12, 10) = -12$

$IQ$ decreases at a rate of 12 points per year of chronological age when the mental age is 12 and the chronological age is 10.

**121.** An increase in either the charge for food and housing or the tuition will cause a decrease in the number of applicants.

**123.** $\dfrac{\partial T}{\partial x} = -2.4°/\text{m},\ \dfrac{\partial T}{\partial y} = -9°/\text{m}$

**125.** $T = \dfrac{PV}{nR} \implies \dfrac{\partial T}{\partial P} = \dfrac{v}{nR}$

$P = \dfrac{nRT}{V} \implies \dfrac{\partial P}{\partial V} = \dfrac{-nRT}{V^2}$

$V = \dfrac{nRT}{P} \implies \dfrac{\partial V}{\partial T} = \dfrac{nR}{P}$

$\dfrac{\partial T}{\partial P} \cdot \dfrac{\partial P}{\partial V} \cdot \dfrac{\partial V}{\partial T} = -\dfrac{nRT}{VP} = -\dfrac{nRT}{nRT} = -1$

**127.** (a) $\dfrac{\partial z}{\partial x} = 0.23, \dfrac{\partial z}{\partial y} = 0.14$

(b) As the expenditures on amusement parks and campgrounds $(x)$ increase, the expenditures on spectator sports $(z)$ increase. As the expenditures on live entertainment $(y)$ increase, the expenditures on spectator sports $(z)$ also increase.

**129.** (a) $f_x(x, y) = \dfrac{y(x^4 + 4x^2y^2 - y^4)}{(x^2 + y^2)^2}$

$f_y(x, y) = \dfrac{x(x^4 - 4x^2y^2 - y^4)}{(x^2 + y^2)^2}$

(b) $f_x(0, 0) = 0, f_y(0, 0) = 0$

(c) $f_{xy}(0, 0) = -1, f_{yx}(0, 0) = 1$

(d) $f_{xy}$ or $f_{yx}$ or both are not continuous at $(0, 0)$.

**131.** Proof

## Section 13.4   *(page 909)*

**1.** In general, the accuracy worsens as $\Delta x$ and $\Delta y$ increase.

**3.** $dz = 15x^2y^2\, dx + 10x^3y\, dy$

**5.** $dz = (e^{x^2+y^2} + e^{-x^2-y^2})(x\, dx + y\, dy)$

**7.** $dw = 2xyz^2\, dx + (x^2z^2 + z\cos yz)\, dy + (2x^2yz + y\cos yz)\, dz$

**9.** (a) $f(2, 1) = 1, f(2.1, 1.05) = 1.05, \Delta z = 0.05$

(b) $dz = 0.05$

**11.** (a) $f(2, 1) = 11, f(2.1, 1.05) = 10.4875, \Delta z = -0.5125$

(b) $dz = -0.5$

**13.** (a) $f(2, 1) = e^2 \approx 7.3891, f(2.1, 1.05) = 1.05e^{2.1} \approx 8.5745,$

$\Delta z \approx 1.1854$

(b) $dz \approx 1.1084$

**15.** 0.44     **17.** 0

**19.** Yes. Because $f_x$ and $f_y$ are continuous on $R$, you know that $f$ is differentiable on $R$. Because $f$ is differentiable on $R$, you know that $f$ is continuous on $R$.

**21.** $dA = h\, dl + l\, dh$

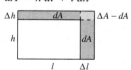

$\Delta A - dA = dl\, dh$

**23.** $dV = \pm 3.92$ in.$^3$, $\dfrac{dV}{V} = 0.82\%$

**25.**

| $\Delta r$ | $\Delta h$ | $dV$ | $\Delta V$ | $\Delta V - dV$ |
|---|---|---|---|---|
| 0.1 | 0.1 | 8.3776 | 8.5462 | 0.1686 |
| 0.1 | $-0.1$ | 5.0265 | 5.0255 | $-0.0010$ |
| 0.001 | 0.002 | 0.1005 | 0.1006 | 0.0001 |
| $-0.0001$ | 0.0002 | $-0.0034$ | $-0.0034$ | 0.0000 |

**27.** $dC = \pm 2.4418, \dfrac{dC}{C} = 19\%$     **29.** 10%

**31.** (a) $V = 18\sin\theta$ ft$^3$, $\theta = \dfrac{\pi}{2}$   (b) 1.047 ft$^3$

**33.** $L \approx 8.096 \times 10^{-4} \pm 6.6 \times 10^{-6}$ microhenrys

**35.** Answers will vary.
Sample answer:
$\varepsilon_1 = \Delta x$
$\varepsilon_2 = 0$

**37.** Answers will vary.
Sample answer:
$\varepsilon_1 = y\,\Delta x$
$\varepsilon_2 = 2x\,\Delta x + (\Delta x)^2$

**39.** Proof

## Section 13.5   *(page 917)*

**1.** You can convert $w$ into a function of $s$ and $t$, or you can use the Chain Rule given in Theorem 13.7.

**3.** $8t + 5$; 21     **5.** $e^t(\sin t + \cos t)$; 1

**7.** (a) and (b) $2e^{2t} + \dfrac{3}{t^4}$   **9.** (a) and (b) $2e^{2t}$

**11.** (a) and (b) $3(2t^2 - 1)$   **13.** $\dfrac{-11\sqrt{29}}{29} \approx -2.04$

**15.** $\dfrac{\partial w}{\partial s} = 4s, 4$

$\dfrac{\partial w}{\partial t} = 4t, 12$

**17.** $\dfrac{\partial w}{\partial s} = 5\cos(5s - t), 0$

$\dfrac{\partial w}{\partial t} = -\cos(5s - t), 0$

**19.** (a) and (b)

$\dfrac{\partial w}{\partial s} = t^2(3s^2 - t^2)$

$\dfrac{\partial w}{\partial t} = 2st(s^2 - 2t^2)$

**21.** (a) and (b)

$\dfrac{\partial w}{\partial s} = te^{s^2 - t^2}(2s^2 + 1)$

$\dfrac{\partial w}{\partial t} = se^{s^2 - t^2}(1 - 2t^2)$

**23.** $\dfrac{y - 2x + 1}{2y - x + 1}$

**25.** $\dfrac{x^2 + y^2 + x}{x^2 + y^2 + y}$

**27.** $\dfrac{\partial z}{\partial x} = -\dfrac{x}{z}$

$\dfrac{\partial z}{\partial y} = -\dfrac{y}{z}$

**29.** $\dfrac{\partial z}{\partial x} = -\dfrac{x}{y + z}$

$\dfrac{\partial z}{\partial y} = -\dfrac{z}{y + z}$

**31.** $\dfrac{\partial z}{\partial x} = \dfrac{\partial z}{\partial y} = \dfrac{\sec^2(x + y)}{\sin z}$

**33.** $\dfrac{\partial z}{\partial x} = -\dfrac{(ze^{xz} + y)}{xe^{xz}}$

$\dfrac{\partial z}{\partial y} = -e^{-xz}$

**35.** $\dfrac{\partial w}{\partial x} = \dfrac{7y + w^2}{4z - 2wz - 2wx}$

$\dfrac{\partial w}{\partial y} = \dfrac{7x + z^2}{4z - 2wz - 2wx}$

$\dfrac{\partial w}{\partial z} = \dfrac{2yz - 4w + w^2}{4z - 2wz - 2wx}$

**37.** $\dfrac{\partial w}{\partial x} = \dfrac{y\sin xy}{z}$

$\dfrac{\partial w}{\partial y} = \dfrac{x\sin xy - z\cos yz}{z}$

$\dfrac{\partial w}{\partial z} = -\dfrac{y\cos yz + w}{z}$

**39.** (a) $f(tx, ty) = 2(tx)^2 - 5(tx)(ty)$

$= t^2(2x^2 - 5xy) = t^2f(x, y); n = 2$

(b) $xf_x(x, y) + yf_y(x, y) = 4x^2 - 10xy = 2f(x, y)$

**41.** (a) $f(tx, ty) = e^{tx/ty} = e^{x/y} = f(x, y); n = 0$

(b) $xf_x(x, y) + yf_y(x, y) = \dfrac{xe^{x/y}}{y} - \dfrac{xe^{x/y}}{y} = 0$

**43.** 47     **45.** Proof

**47.** (a) $\dfrac{\partial F}{\partial u}\dfrac{\partial u}{\partial x} + \dfrac{\partial F}{\partial v}\dfrac{\partial v}{\partial x} = 4\dfrac{\partial F}{\partial u}$

(b) $\dfrac{\partial F}{\partial u}\dfrac{\partial u}{\partial x} + \dfrac{\partial F}{\partial v}\dfrac{\partial v}{\partial x} = -2\dfrac{\partial F}{\partial u} + 2x\dfrac{\partial F}{\partial v}$

**49.** $4608\pi$ in.$^3$/min, $624\pi$ in.$^2$/min     **51.** $28m$ cm$^2$/sec

**53–55.** Proofs

## Section 13.6  (page 928)

1. The partial derivative with respect to $x$ is the directional derivative in the direction of the positive $x$-axis. That is, the directional derivative for $\theta = 0$.

3. $-\sqrt{2}$    5. $\frac{1}{2} + \sqrt{3}$    7. 1    9. $-\frac{7}{25}$    11. 6

13. $\dfrac{2\sqrt{5}}{5}$    15. $3\mathbf{i} + 10\mathbf{j}$    17. $2\mathbf{i} - \dfrac{1}{2}\mathbf{j}$

19. $20\mathbf{i} - 14\mathbf{j} - 30\mathbf{k}$    21. $-1$    23. $\dfrac{2\sqrt{3}}{3}$    25. $3\sqrt{2}$

27. $-\dfrac{8}{\sqrt{5}}$    29. $-\sqrt{y}\,\mathbf{i} + \left(2y - \dfrac{x}{2\sqrt{y}}\right)\mathbf{j};\ \sqrt{39}$

31. $\tan y\,\mathbf{i} + x\sec^2 y\,\mathbf{j};\ \sqrt{17}$

33. $\cos x^2 y^3 (2x\mathbf{i} + 3y^2\mathbf{j});\ \dfrac{1}{\pi}\sqrt{4 + 9\pi^6}$

35. $\dfrac{x\mathbf{i} + y\mathbf{j} + z\mathbf{k}}{\sqrt{x^2 + y^2 + z^2}};\ 1$    37. $yz(yz\mathbf{i} + 2xz\mathbf{j} + 2xy\mathbf{k});\ \sqrt{33}$

39. $-2\mathbf{i} - 3\mathbf{j}$    41. $3\mathbf{i} - \mathbf{j}$

43. (a) $16\mathbf{i} - \mathbf{j}$    (b) $\dfrac{\sqrt{257}}{257}(16\mathbf{i} - \mathbf{j})$    (c) $y = 16x - 22$

(d)

45. (a) $6\mathbf{i} - 4\mathbf{j}$    (b) $\dfrac{\sqrt{13}}{13}(3\mathbf{i} - 2\mathbf{j})$    (c) $y = \dfrac{3}{2}x - \dfrac{1}{2}$

(d)

47. (a)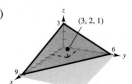

(b) (i) $-\dfrac{5\sqrt{2}}{12}$    (ii) $\dfrac{2 - 3\sqrt{3}}{12}$

(iii) $\dfrac{2 + 3\sqrt{3}}{12}$    (iv) $\dfrac{3 - 2\sqrt{3}}{12}$

(c) (i) $-\dfrac{5\sqrt{2}}{12}$    (ii) $\dfrac{3}{5}$    (iii) $-\dfrac{1}{5}$    (iv) $-\dfrac{11\sqrt{10}}{60}$

(d) $-\dfrac{1}{3}\mathbf{i} - \dfrac{1}{2}\mathbf{j}$    (e) $\dfrac{\sqrt{13}}{6}$

(f) $\mathbf{u} = \dfrac{1}{\sqrt{13}}(3\mathbf{i} - 2\mathbf{j})$

$D_\mathbf{u}f(3, 2) = \nabla f \cdot \mathbf{u} = 0$

$\nabla f$ is the direction of the greatest rate of change of $f$. So, in a direction orthogonal to $\nabla f$, the rate of change of $f$ is 0.

49. (a)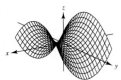

(b) $D_\mathbf{u}f(4, -3) = 8\cos\theta + 6\sin\theta$

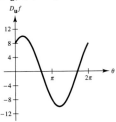

Generated by Mathematica

(c) $\theta \approx 2.21,\ \theta \approx 5.36$
Directions in which there is no change in $f$

(d) $\theta \approx 0.64,\ \theta \approx 3.79$
Directions of greatest rate of change in $f$

(e) 10; Magnitude of the greatest rate of change

(f)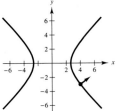

Generated by Mathematica

Orthogonal to the level curve

51. No; Answers will vary.    53. $5\nabla h = -(5\mathbf{i} + 12\mathbf{j})$

55. $\dfrac{1}{625}(7\mathbf{i} - 24\mathbf{j})$    57. $6\mathbf{i} - 10\mathbf{j};\ 11.66°/\text{cm}$    59. $y^2 = 10x$

61. True    63. True    65. $f(x, y, z) = e^x\cos y + \dfrac{1}{2}z^2 + C$

67. (a) and (b)  Proofs

(c)

## Section 13.7  (page 937)

1. $\nabla F(x_0, y_0, z_0)$ and any tangent vector $\mathbf{v}$ at $(x_0, y_0, z_0)$ are orthogonal. So, $\nabla F(x_0, y_0, z_0) \cdot \mathbf{v} = 0$.

3. The level surface can be written as $3x - 5y + 3z = 15$, which is an equation of a plane in space.

5. The level surface can be written as $4x^2 + 9y^2 - 4z^2 = 0$, which is an elliptic cone that lies on the $z$-axis.

7. $4x + 2y - z = 2$    9. $3x + 4y - 5z = 0$

11. $2x - 2y - z = 2$    13. $3x + 4y - 25z = 25(1 - \ln 5)$

15. $4x + 2y + 5z = -15$

17. (a) $x + y + z = 9$    (b) $x - 3 = y - 3 = z - 3$

**19.** (a) $x - 2y + 2z = 7$   (b) $x - 1 = \dfrac{y + 1}{-2} = \dfrac{z - 2}{2}$

**21.** (a) $6x - 4y - z = 5$   (b) $\dfrac{x - 3}{6} = \dfrac{y - 2}{-4} = \dfrac{z - 5}{-1}$

**23.** (a) $10x + 5y + 2z = 30$   (b) $\dfrac{x - 1}{10} = \dfrac{y - 2}{5} = \dfrac{z - 5}{2}$

**25.** (a) $8x + y - z = 0$   (b) $\dfrac{x}{8} = \dfrac{y - 2}{1} = \dfrac{z - 2}{-1}$

**27.** $x = t + 1, y = 1 - t, z = t + 1$

**29.** $x = 4t + 3, y = 4t + 3, z = 4 - 3t$

**31.** $x = t + 3, y = 5t + 1, z = 2 - 4t$

**33.** $86.0°$   **35.** $77.4°$   **37.** $(0, 3, 12)$   **39.** $(2, 2, -4)$

**41.** $(0, 0, 0)$   **43.** Proof   **45.** (a) and (b) Proofs

**47.** Not necessarily; They only need to be parallel.

**49.** $\left(-\frac{1}{2}, \frac{1}{4}, \frac{1}{4}\right)$ or $\left(\frac{1}{2}, -\frac{1}{4}, -\frac{1}{4}\right)$   **51.** $(-2, 1, -1)$ or $(2, -1, 1)$

**53.** (a) Line: $x = 1, y = 1, z = 1 - t$

    Plane: $z = 1$

   (b) Line: $x = -1, y = 2 + \frac{6}{25}t, z = -\frac{4}{5} - t$

    Plane: $6y - 25z - 32 = 0$

   (c)

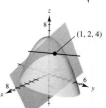

**55.** (a) $x = 1 + t$      (b)

    $y = 2 - 2t$

    $z = 4$

    $\theta \approx 48.2°$

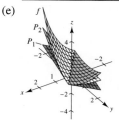

**57.** $F(x, y, z) = \dfrac{x^2}{a^2} + \dfrac{y^2}{b^2} + \dfrac{z^2}{c^2} - 1$

  $F_x(x, y, z) = \dfrac{2x}{a^2}$

  $F_y(x, y, z) = \dfrac{2y}{b^2}$

  $F_z(x, y, z) = \dfrac{2z}{c^2}$

  Plane: $\dfrac{2x_0}{a^2}(x - x_0) + \dfrac{2y_0}{b^2}(y - y_0) + \dfrac{2z_0}{c^2}(z - z_0) = 0$

    $\dfrac{x_0 x}{a^2} + \dfrac{y_0 y}{b^2} + \dfrac{z_0 z}{c^2} = 1$

**59.** $F(x, y, z) = a^2 x^2 + b^2 y^2 - z^2$

  $F_x(x, y, z) = 2a^2 x$

  $F_y(x, y, z) = 2b^2 y$

  $F_z(x, y, z) = -2z$

  Plane: $2a^2 x_0(x - x_0) + 2b^2 y_0(y - y_0) - 2z_0(z - z_0) = 0$

    $a^2 x_0 x + b^2 y_0 y - z_0 z = 0$

  Therefore, the plane passes through the origin.

**61.** (a) $P_1(x, y) = 1 + x - y$

   (b) $P_2(x, y) = 1 + x - y + \frac{1}{2}x^2 - xy + \frac{1}{2}y^2$

   (c) If $x = 0, P_2(0, y) = 1 - y + \frac{1}{2}y^2$.

    This is the second-degree Taylor polynomial for $e^{-y}$.

    If $y = 0, P_2(x, 0) = 1 + x + \frac{1}{2}x^2$.

    This is the second-degree Taylor polynomial for $e^x$.

   (d)

| $x$ | $y$ | $f(x, y)$ | $P_1(x, y)$ | $P_2(x, y)$ |
|---|---|---|---|---|
| 0 | 0 | 1 | 1 | 1 |
| 0 | 0.1 | 0.9048 | 0.9000 | 0.9050 |
| 0.2 | 0.1 | 1.1052 | 1.1000 | 1.1050 |
| 0.2 | 0.5 | 0.7408 | 0.7000 | 0.7450 |
| 1 | 0.5 | 1.6487 | 1.5000 | 1.6250 |

   (e)

**63.** Proof

## Section 13.8  *(page 946)*

**1.** (a) To say that $f$ has a relative minimum at $(x_0, y_0)$ means that the point $(x_0, y_0, z_0)$ is at least as low as all nearby points on the graph of $z = f(x, y)$.

   (b) To say that $f$ has a relative maximum at $(x_0, y_0)$ means that the point $(x_0, y_0, z_0)$ is at least as high as all nearby points in the graph of $z = f(x, y)$.

   (c) Critical points of $f$ are the points at which the gradient of $f$ is 0 or the points at which one of the partial derivatives does not exist.

   (d) A critical point is a saddle point if it is neither a relative minimum nor a relative maximum.

**3.** Relative minimum:    **5.** Relative minimum:

  $(1, 3, 0)$               $(0, 0, 1)$

**7.** Relative minimum:    **9.** Relative minimum:

  $(-1, 3, -4)$          $(-4, 6, -55)$

**11.** Every point along the $x$- or $y$-axis is a critical point. Each of the critical points yields an absolute maximum.

**13.** Relative maximum:    **15.** Relative minimum:

  $\left(\frac{1}{2}, -1, \frac{31}{4}\right)$       $\left(\frac{1}{2}, -4, -\frac{187}{4}\right)$

**17.** Relative minimum:    **19.** Relative maximum:

  $(3, -4, -5)$          $(0, 0, -12)$

**21.** Saddle point:    **23.** No critical numbers

  $(1, -1, -1)$

**25.**                 **27.**

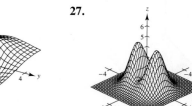

Relative maximum: $(-1, 0, 2)$   Relative minimum: $(0, 0, 0)$

Relative minimum: $(1, 0, -2)$   Relative maxima: $(0, \pm 1, 4)$

                                Saddle points: $(\pm 1, 0, 1)$

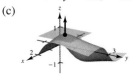

**29.** $z$ is never negative. Minimum: $z = 0$ when $x = y \neq 0$.

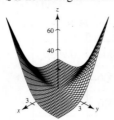

**31.** Insufficient information    **33.** Saddle point

**35.** (a) $(0, 0)$    (b) Saddle point: $(0, 0, 0)$    (c) $(0, 0)$

(d)

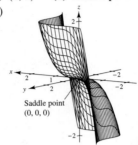

Saddle point
$(0, 0, 0)$

**37.** (a) $(1, a), (b, -4)$

(b) Absolute minima: $(1, a, 0), (b, -4, 0)$

(c) $(1, a), (b, -4)$

(d)

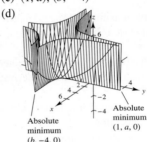

Absolute
minimum
$(b, -4, 0)$

Absolute
minimum
$(1, a, 0)$

**39.** Absolute maximum:    **41.** Absolute maximum:
$(4, 0, 21)$    $(0, 1, 10)$
Absolute minimum:    Absolute minimum:
$(4, 2, -11)$    $(1, 2, 5)$

**43.** Absolute maxima:    **45.** Absolute maxima:
$(\pm 2, 4, 28)$    $(-2, -1, 9), (2, 1, 9)$
Absolute minimum:    Absolute minima:
$(0, 1, -2)$    $(x, -x, 0), |x| \leq 1$

**47.** Relative minimum: $(0, 3, -1)$    **49.** $-4 < f_{xy}(3, 7) < 4$

**51.**

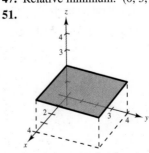

Extrema at all $(x, y)$

**53.** (a) $f_x = 2x = 0, f_y = -2y = 0 \Rightarrow (0, 0)$ is a critical point.

$g_x = 2x = 0, g_y = 2y = 0 \Rightarrow (0, 0)$ is a critical point.

(b) $d = 2(-2) - 0 < 0 \Rightarrow (0, 0)$ is a saddle point.

$d = 2(2) - 0 > 0 \Rightarrow (0, 0)$ is a relative minimum.

**55.** False. Let $f(x, y) = 1 - |x| - |y|$ at the point $(0, 0, 1)$.

**57.** False. Let $f(x, y) = x^2 y^2$ (see Example 4 on page 944).

## Section 13.9    *(page 953)*

**1.** Write the equation to be maximized or minimized as a function of two variables. Take the partial derivatives and set them equal to zero or undefined to obtain the critical points. Use the Second Partials Test to test for relative extrema using the critical points. Check the boundary points.

**3.** $\sqrt{3}$    **5.** $\sqrt{7}$    **7.** $x = y = z = 3$

**9.** $x = y = z = 10$    **11.** 9 ft $\times$ 9 ft $\times$ 8.25 ft; \$26.73

**13.** Let $x$, $y$, and $z$ be the length, width, and height, respectively, and let $V_0$ be the given volume. Then $V_0 = xyz$ and $z = \dfrac{V_0}{xy}$.

The surface area is

$$S = 2xy + 2yz + 2xz = 2\left(xy + \frac{V_0}{x} + \frac{V_0}{y}\right).$$

$$\left.\begin{array}{l} S_x = 2\left(y - \dfrac{V_0}{x^2}\right) = 0 \\[2mm] S_y = 2\left(x - \dfrac{V_0}{y^2}\right) = 0 \end{array}\right\} \begin{array}{l} x^2 y - V_0 = 0 \\[2mm] xy^2 - V_0 = 0 \end{array}$$

So, $x = \sqrt[3]{V_0}$, $y = \sqrt[3]{V_0}$, and $z = \sqrt[3]{V_0}$.

**15.** $x_1 = 3, x_2 = 6$    **17.** Proof

**19.** $x = \dfrac{\sqrt{2}}{2} \approx 0.707$ km

$y = \dfrac{3\sqrt{2} + 2\sqrt{3}}{6} \approx 1.284$ km

**21.** (a) $y = \frac{3}{4}x + \frac{4}{3}$    (b) $\frac{1}{6}$    **23.** (a) $y = -2x + 4$    (b) 2

**25.** $y = \frac{84}{43}x - \frac{12}{43}$    **27.** $y = -\frac{175}{148}x + \frac{945}{148}$

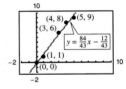

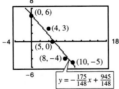

**29.** (a) $y = 0.23x + 2.38$    (b) \$301.4 billion

(c) The new model is $y = 0.23x + 5.09$, so the constant increases.

**31.** $a\displaystyle\sum_{i=1}^{n} x_i^4 + b\sum_{i=1}^{n} x_i^3 + c\sum_{i=1}^{n} x_i^2 = \sum_{i=1}^{n} x_i^2 y_i$

$a\displaystyle\sum_{i=1}^{n} x_i^3 + b\sum_{i=1}^{n} x_i^2 + c\sum_{i=1}^{n} x_i = \sum_{i=1}^{n} x_i y_i$

$a\displaystyle\sum_{i=1}^{n} x_i^2 + b\sum_{i=1}^{n} x_i + cn = \sum_{i=1}^{n} y_i$

**33.** $y = \frac{3}{7}x^2 + \frac{6}{5}x + \frac{26}{35}$    **35.** $y = x^2 - x$

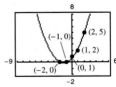

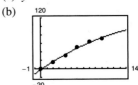

**37.** (a) $y = -0.22x^2 + 9.66x - 1.79$

(b)

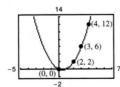

**39.** (a) $\ln P = -0.1499h + 9.3018$    (b) $P = 10,957.7e^{-0.1499h}$

(c)

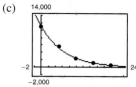

**41.** Proof

## Section 13.10  *(page 962)*

**1.** Optimization problems that have restrictions or constraints on the values that can be used to produce the optimal solutions are called constrained optimization problems.

**3.** $f(5, 5) = 25$    **5.** $f(1, 2) = 5$    **7.** $f(25, 50) = 2600$

**9.** $f(1, 1) = 2$    **11.** $f(3, 3, 3) = 27$    **13.** $f\left(\frac{1}{3}, \frac{1}{3}, \frac{1}{3}\right) = \frac{1}{3}$

**15.** Maxima: $f\left(\frac{\sqrt{2}}{2}, \frac{\sqrt{2}}{2}\right) = \frac{5}{2}$

$f\left(-\frac{\sqrt{2}}{2}, -\frac{\sqrt{2}}{2}\right) = \frac{5}{2}$

Minima: $f\left(-\frac{\sqrt{2}}{2}, \frac{\sqrt{2}}{2}\right) = -\frac{1}{2}$

$f\left(\frac{\sqrt{2}}{2}, -\frac{\sqrt{2}}{2}\right) = -\frac{1}{2}$

**17.** $f(8, 16, 8) = 1024$    **19.** $\frac{\sqrt{2}}{2}$    **21.** $3\sqrt{2}$    **23.** $\frac{\sqrt{11}}{2}$

**25.** 2    **27.** $\sqrt{3}$    **29.** $(-4, 0, 4)$    **31.** $\sqrt{3}$

**33.** $x = y = z = 3$    **35.** 9 ft $\times$ 9 ft $\times$ 8.25 ft; $26.73

**37.** Proof    **39.** $\frac{2\sqrt{3}a}{3} \times \frac{2\sqrt{3}b}{3} \times \frac{2\sqrt{3}c}{3}$

**41.** At $(0, 0)$, the Lagrange equations are inconsistent.

**43.** $\sqrt[3]{360} \times \sqrt[3]{360} \times \frac{4}{3}\sqrt[3]{360}$ ft

**45.** $r = \sqrt[3]{\dfrac{v_0}{2\pi}}$ and $h = 2\sqrt[3]{\dfrac{v_0}{2\pi}}$    **47.** Proof

**49.** $P\left(\dfrac{15,625}{28}, 3125\right) \approx 203,144$

**51.** $x \approx 237.4$

$y \approx 640.9$

Cost $\approx$ $68,364.80

**53.** Putnam Problem 2, morning session, 1938

## Review Exercises for Chapter 13  *(page 964)*

**1.** (a) $-3$    (b) $-7$    (c) 15    (d) $7x^2 - 3$

**3.** Domain: $\{(x, y): x \geq 0 \text{ and } y \neq 0\}$

Range: all real numbers

**5.**

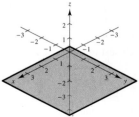

Plane

**7.** Lines: $y = 2x - 3 + c$

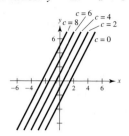

**9.** (a)

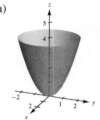

(b) $g$ is a vertical translation of $f$ two units upward.

(c) $g$ is a horizontal translation of $f$ two units to the right.

(d)

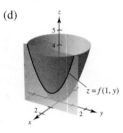

$z = f(1, y)$

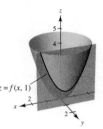

$z = f(x, 1)$

**11.** Elliptic paraboloid

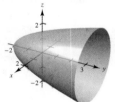

**13.** Limit: $\frac{1}{2}$

Continuous except at $(0, 0)$

**15.** Limit: 0

Continuous

**17.** Limit: $-\dfrac{\ln 2}{5}$

Continuous for $x \neq \dfrac{z}{y}$

**19.** $f_x(x, y) = 15x^2$

$f_y(x, y) = 7$

**21.** $f_x(x, y) = e^x \cos y$

$f_y(x, y) = -e^x \sin y$

**23.** $f_x(x, y) = -\dfrac{y^4}{x^2}e^{y/x}$

$f_y(x, y) = \dfrac{y^3}{x}e^{y/x} + 3y^2 e^{y/x}$

**25.** $f_x(x, y, z) = 2z^2 + 6yz$

$f_y(x, y, z) = 6xz$

$f_z(x, y, z) = 4xz + 6xy$

**27.** $f_x(0, 2) = 0$

$f_y(0, 2) = -1$

**29.** $f_x\left(2, 3, -\dfrac{\pi}{3}\right) = -\sqrt{3}\pi - \dfrac{3}{2}$

$f_y\left(2, 3, -\dfrac{\pi}{3}\right) = -1$

$f_z\left(2, 3, -\dfrac{\pi}{3}\right) = 6\sqrt{3}$

**31.** $f_{xx}(x, y) = 6$

$f_{yy}(x, y) = 12y$

$f_{xy}(x, y) = f_{yx}(x, y) = -1$

**33.** $h_{xx}(x, y) = -y \cos x$
$h_{yy}(x, y) = -x \sin y$
$h_{xy}(x, y) = h_{yx}(x, y) = \cos y - \sin x$

**35.** Slope in $x$-direction: $0$
Slope in $y$-direction: $4$

**37.** $(xy \cos xy + \sin xy)\, dx + (x^2 \cos xy)\, dy$

**39.** $dw = (3y^2 - 6x^2yz^2)\, dx + (6xy - 2x^3z^2)\, dy + (-4x^3yz)\, dz$

**41.** (a) $f(2, 1) = 10$          (b) $dz = 0.5$
$f(2.1, 1.05) = 10.5$
$\Delta z = 0.5$

**43.** $dV = \pm\pi$ in.$^3$, $\dfrac{dV}{V} = 15\%$     **45.** Proof

**47.** (a) and (b) $\dfrac{dw}{dt} = \dfrac{8t - 1}{4t^2 - t + 4}$

**49.** (a) and (b) $\dfrac{dw}{dt} = 2t^2e^{2t} + 2te^{2t} + 2t + 1$

**51.** (a) and (b) $\dfrac{\partial w}{\partial r} = \dfrac{4r^2t - 4rt^2 - t^3}{(2r - t)^2}$
$\dfrac{\partial w}{\partial t} = \dfrac{4r^2t - rt^2 - 4r^3}{(2r - t)^2}$

**53.** $\dfrac{-3x^2 + y}{-x + 5}$

**55.** $\dfrac{\partial z}{\partial x} = \dfrac{-2x - y}{y + 2z}$
$\dfrac{\partial z}{\partial y} = \dfrac{-x - 2y - z}{y + 2z}$

**57.** $-50$     **59.** $\frac{2}{3}$     **61.** $\langle 4, 4 \rangle$, $4\sqrt{2}$     **63.** $\left\langle -\frac{1}{2}, 0 \right\rangle$, $\frac{1}{2}$

**65.** $\langle -2, -3, -1 \rangle$, $\sqrt{14}$

**67.** (a) $54\mathbf{i} - 16\mathbf{j}$   (b) $\dfrac{27}{\sqrt{793}}\mathbf{i} - \dfrac{8}{\sqrt{793}}\mathbf{j}$   (c) $y = \dfrac{27}{8}x - \dfrac{65}{8}$

(d)
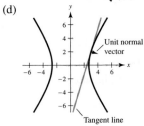

**69.** $2x + 6y - z = 8$     **71.** $z = 4$

**73.** (a) $4x + 4y - z = 8$
(b) $x = 2 + 4t$, $y = 1 + 4t$, $z = 4 - t$

**75.** $36.7°$     **77.** $(0, 0, 9)$

**79.** Relative maximum: $(4, -1, 9)$

**81.** Relative minimum: $\left(-4, \frac{4}{3}, -2\right)$

**83.** Relative minimum: $(1, 1, 3)$     **85.** $\sqrt{3}$

**87.** $x_1 = 2$, $x_2 = 4$

**89.** $y = \frac{161}{226}x + \frac{456}{113}$

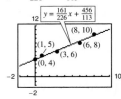

**91.** (a) $y = 0.138x + 22.1$     (b) $46.25$ bushels/acre

**93.** $f(4, 4) = 32$     **95.** $f(15, 7) = 352$     **97.** $f(3, 6) = 36$

**99.** $x = \dfrac{\sqrt{2}}{2} \approx 0.707$ km, $y = \dfrac{\sqrt{3}}{3} \approx 0.577$ km,
$z = \left(60 - 3\sqrt{2} - 2\sqrt{3}\right)6 \approx 8.716$ km

## P.S. Problem Solving   *(page 967)*

**1.** (a) $12$ square units     (b) and (c) Proofs

**3.** (a) $y_0z_0(x - x_0) + x_0z_0(y - y_0) + x_0y_0(z - z_0) = 0$

(b) $x_0y_0z_0 = 1 \implies z_0 = \dfrac{1}{x_0y_0}$
Then the tangent plane is
$$y_0\left(\frac{1}{x_0y_0}\right)(x - x_0) + x_0\left(\frac{1}{x_0y_0}\right)(y - y_0) + x_0y_0\left(z - \frac{1}{x_0y_0}\right) = 0.$$
Intercepts: $(3x_0, 0, 0)$, $(0, 3y_0, 0)$, $\left(0, 0, \dfrac{3}{x_0y_0}\right)$

**5.** (a) 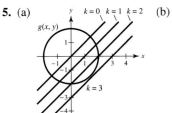    (b)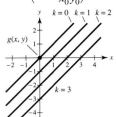

Maximum value: $2\sqrt{2}$     Maximum and minimum value: $0$
The method of Lagrange multipliers does not work because $\nabla g(x_0, y_0) = \mathbf{0}$.

**7.** $2\sqrt[3]{150}$ ft $\times$ $2\sqrt[3]{150}$ ft $\times$ $\dfrac{5\sqrt[3]{150}}{3}$ ft

**9.** (a) $x\dfrac{\partial f}{\partial x} + y\dfrac{\partial f}{\partial y} = xCy^{1-a}ax^{a-1} + yCx^a(1 - a)y^{1-a-1}$
$= ax^aCy^{1-a} + (1 - a)x^aC(y^{1-a})$
$= Cx^ay^{1-a}[a + (1 - a)]$
$= Cx^ay^{1-a}$
$= f(x, y)$

(b) $f(tx, ty) = C(tx)^a(ty)^{1-a}$
$= Ctx^ay^{1-a}$
$= tCx^ay^{1-a}$
$= tf(x, y)$

**11.** (a) $x = 32\sqrt{2}t$
$y = 32\sqrt{2}t - 16t^2$

(b) $\alpha = \arctan\left(\dfrac{y}{x + 50}\right) = \arctan\left(\dfrac{32\sqrt{2}t - 16t^2}{32\sqrt{2}t + 50}\right)$

(c) $\dfrac{d\alpha}{dt} = \dfrac{-16\left(8\sqrt{2}t^2 + 25t - 25\sqrt{2}\right)}{64t^4 - 256\sqrt{2}t^3 + 1024t^2 + 800\sqrt{2}t + 625}$

(d)
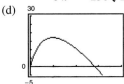
No; The rate of change of $\alpha$ is greatest when the projectile is closest to the camera.

(e) $\alpha$ is maximum when $t = 0.98$ second.
No, the projectile is at its maximum height when $t = \sqrt{2} \approx 1.41$ seconds.

**13.** (a)  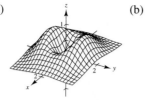  (b)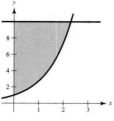

Minimum: $(0, 0, 0)$    Minima: $(\pm 1, 0, -e^{-1})$
Maxima: $(0, \pm 1, 2e^{-1})$    Maxima: $(0, \pm 1, 2e^{-1})$
Saddle points: $(\pm 1, 0, e^{-1})$    Saddle point: $(0, 0, 0)$
(c) $\alpha > 0$             $\alpha < 0$
Minimum: $(0, 0, 0)$    Minima: $(\pm 1, 0, \alpha e^{-1})$
Maxima: $(0, \pm 1, \beta e^{-1})$    Maxima: $(0, \pm 1, \beta e^{-1})$
Saddle points:        Saddle point: $(0, 0, 0)$
$(\pm 1, 0, \alpha e^{-1})$

**15.** (a)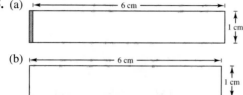

(b)

(c) Height
(d) $dl = 0.01,\ dh = 0:\ dA = 0.01$
      $dl = 0,\ dh = 0.01:\ dA = 0.06$

**17–21.** Proofs

# Chapter 14

## Section 14.1 *(page 976)*

**1.** An iterated integral is an integral of a function of several variables. Integrate with respect to one variable while holding the other variables constant.

**3.** $\dfrac{3x^2}{2}$    **5.** $\dfrac{4x^2 - x^4}{2}$    **7.** $\dfrac{y}{2}[(\ln y)^2 - y^2]$

**9.** $x^2(1 - e^{-x^2} - x^2 e^{-x^2})$    **11.** 3    **13.** $\dfrac{\sqrt{2}}{4}$    **15.** 64

**17.** $\dfrac{3}{2}$   **19.** $\dfrac{1}{3}$   **21.** $\dfrac{2}{3}$   **23.** 4   **25.** $\dfrac{\pi}{2}$   **27.** $\dfrac{\pi^2}{32} + \dfrac{1}{8}$

**29.** $\dfrac{1}{2}$   **31.** Diverges   **33.** 8   **35.** $\dfrac{16}{3}$   **37.** 36

**39.** $\dfrac{8}{3}$   **41.** $\dfrac{9}{2}$

**43.**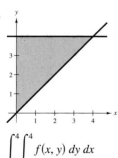

$$\int_0^4 \int_x^4 f(x, y)\, dy\, dx$$

**45.**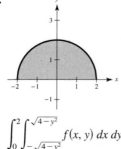

$$\int_0^2 \int_{-\sqrt{4-y^2}}^{\sqrt{4-y^2}} f(x, y)\, dx\, dy$$

**47.**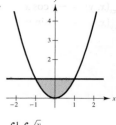

$$\int_0^{\ln 10} \int_{e^x}^{10} f(x, y)\, dy\, dx$$

**49.**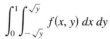

$$\int_0^1 \int_{-\sqrt{y}}^{\sqrt{y}} f(x, y)\, dx\, dy$$

**51.**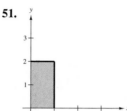

$$\int_0^1 \int_0^2 dy\, dx = \int_0^2 \int_0^1 dx\, dy = 2$$

**53.**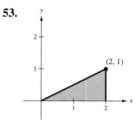

$$\int_0^1 \int_{2y}^2 dx\, dy = \int_0^2 \int_0^{x/2} dy\, dx = 1$$

**55.**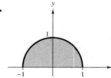

$$\int_0^1 \int_{-\sqrt{1-y^2}}^{\sqrt{1-y^2}} dx\, dy = \int_{-1}^1 \int_0^{\sqrt{1-x^2}} dy\, dx = \frac{\pi}{2}$$

**57.**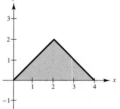

$$\int_0^2 \int_0^x dy\, dx + \int_2^4 \int_0^{4-x} dy\, dx = \int_0^2 \int_y^{4-y} dx\, dy = 4$$

**59.**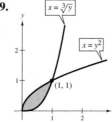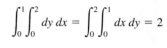

$$\int_0^1 \int_{y^2}^{\sqrt[3]{y}} dx\, dy = \int_0^1 \int_{x^3}^{\sqrt{x}} dy\, dx = \frac{5}{12}$$

**61.**

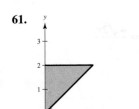

$$\int_0^2 \int_x^2 x\sqrt{1 + y^3} \, dy \, dx = \frac{26}{9}$$

**63.**

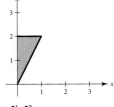

$$\int_0^1 \int_{2x}^2 4e^{y^2} \, dy \, dx = e^4 - 1 \approx 53.598$$

**65.**

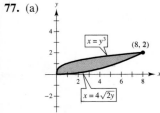

$$\int_0^1 \int_y^1 \sin x^2 \, dx \, dy = \frac{1}{2}(1 - \cos 1) \approx 0.230$$

**67.** $4\int_0^5 \int_0^{\sqrt{25-x^2}} dy \, dx = 25\pi$ square units

**69.** (a) No    (b) Yes    (c) Yes    **71.** $\dfrac{\sin 2}{2} - \dfrac{\sin 3}{3}$

**73.** $(\ln 5)^2$    **75.** $\dfrac{15\pi}{2}$

**77.** (a)

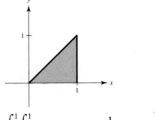

(b) $\displaystyle\int_0^8 \int_{x^2/32}^{\sqrt[3]{x}} (x^2y - xy^2) \, dy \, dx$    (c) $\dfrac{67,520}{693}$

**79.** True

## Section 14.2    *(page 987)*

1. Use rectangular prisms to approximate the volume, where $f(x_i, y_i)$ is the height of prism $i$ and $\Delta A_i$ is the area of the rectangular base of the prism. You can improve the approximation by using more rectangular prisms of smaller rectangular bases.

3. 24 (approximation is exact)

5. Approximation: 52; Exact: $\frac{160}{3}$

**7.**

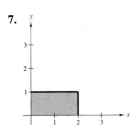

2

**9.**

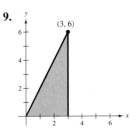

(3, 6)

36

**11.**

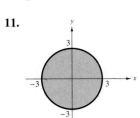

0

**13.** $\displaystyle\int_0^3 \int_0^5 xy \, dy \, dx = \frac{225}{4}$

$\displaystyle\int_0^5 \int_0^3 xy \, dx \, dy = \frac{225}{4}$

**15.** $\displaystyle\int_1^2 \int_x^{2x} \frac{y}{x^2 + y^2} \, dy \, dx = \frac{1}{2} \ln \frac{5}{2}$

$\displaystyle\int_1^2 \int_1^y \frac{y}{x^2 + y^2} \, dx \, dy + \int_2^4 \int_{y/2}^2 \frac{y}{x^2 + y^2} \, dx \, dy = \frac{1}{2} \ln \frac{5}{2}$

**17.** $\displaystyle\int_0^1 \int_{4-x}^{4-x^2} -2y \, dy \, dx = -\frac{6}{5}$

$\displaystyle\int_3^4 \int_{4-y}^{\sqrt{4-y}} -2y \, dx \, dy = -\frac{6}{5}$

**19.** $\displaystyle\int_0^3 \int_{4y/3}^{\sqrt{25-y^2}} x \, dx \, dy = 25$

$\displaystyle\int_0^4 \int_0^{3x/4} x \, dy \, dx + \int_4^5 \int_0^{\sqrt{25-x^2}} x \, dy \, dx = 25$

**21.** 4    **23.** 12    **25.** $\frac{3}{8}$    **27.** 1

**29.** $\displaystyle\int_0^1 \int_0^{x^3} xy \, dy \, dx = \frac{1}{16}$    **31.** $\displaystyle\int_0^2 \int_0^{\sqrt{4-x^2}} (x + y) \, dy \, dx = \frac{16}{3}$

**33.** $\displaystyle\int_0^2 \int_0^{4-x^2} (4 - x^2) \, dy \, dx = \frac{256}{15}$

**35.** $2\displaystyle\int_0^2 \int_0^{\sqrt{1-(x-1)^2}} (2x - x^2 + y^2) \, dy \, dx$

**37.** $4\displaystyle\int_0^2 \int_0^{\sqrt{4-x^2}} (x^2 + y^2) \, dy \, dx$

**39.** $\displaystyle\int_0^2 \int_{-\sqrt{2-2(y-1)^2}}^{\sqrt{2-2(y-1)^2}} (4y - x^2 - 2y^2) \, dx \, dy$

**41.** $\dfrac{81\pi}{2}$    **43.** 1.2315

**45.**

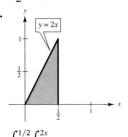

$$\int_0^{1/2} \int_0^{2x} e^{-x^2} \, dy \, dx = 1 - e^{-1/4} \approx 0.221$$

**47.**

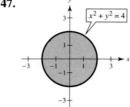

$$\int_{-2}^{2}\int_{-\sqrt{4-y^2}}^{\sqrt{4-y^2}}\sqrt{4-y^2}\,dx\,dy = \frac{64}{3}$$

**49.**

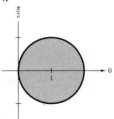

$$\int_{0}^{4}\int_{0}^{y/2}\sin y^2\,dx\,dy = \frac{1-\cos 16}{4} \approx 0.489$$

**51.** 2    **53.** $\frac{8}{3}$    **55.** $(e-1)^2$    **57.** 25,645.24

**59.** $kB$; Answers will vary.    **61.** Proof; $\frac{2}{3}$    **63.** Proof; $\frac{4}{9}$

**65.** Proof    **67.** 400; 272

**69.** False. $V = 8\int_{0}^{1}\int_{0}^{\sqrt{1-y^2}}\sqrt{1-x^2-y^2}\,dx\,dy$

**71.** $R$: $x^2 + y^2 \le 9$    **73.** $\frac{1}{2}(1-e)$

**75.** Putnam Problem A2, 1989

## Section 14.3 *(page 995)*

**1.** Rectangular

**3.** $r$-simple regions have fixed bounds for $\theta$ and variable bounds for $r$. $\theta$-simple regions have variable bounds for $\theta$ and fixed bounds for $r$.

**5.** $R = \{(r, \theta): 0 \le r \le 8, 0 \le \theta \le \pi\}$

**7.** $R = \left\{(r, \theta): 4 \le r \le 8, 0 \le \theta \le \frac{\pi}{2}\right\}$

**9.** $\pi$            **11.** 0

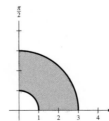

**13.** $\frac{8\sqrt{2}\pi}{3}$       **15.** $\frac{9}{8} + \frac{3\pi^2}{32}$

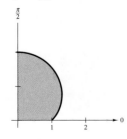

**17.** 9    **19.** $4\pi$    **21.** $\frac{\pi}{10}$    **23.** $\frac{2}{3}$

**25.** $\frac{\pi}{2}\sin 1$    **27.** $\int_{0}^{\pi/4}\int_{0}^{2\sqrt{2}}r^2\,dr\,d\theta = \frac{4\sqrt{2}\pi}{3}$

**29.** $\int_{0}^{\pi/2}\int_{0}^{6}(\cos\theta + \sin\theta)r^2\,dr\,d\theta = 144$

**31.** $\int_{0}^{\pi/4}\int_{1}^{2}r\theta\,dr\,d\theta = \frac{3\pi^2}{64}$    **33.** $\frac{1}{8}$    **35.** $\frac{250\pi}{3}$

**37.** $\frac{64}{9}(3\pi - 4)$    **39.** $2\sqrt{4 - 2\sqrt[3]{2}}$    **41.** $9\pi$

**43.** $\frac{3\pi}{2}$    **45.** $\pi$

**47.**                  **49.**

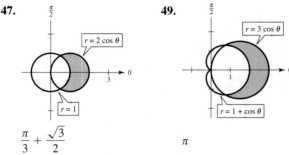

$\frac{\pi}{3} + \frac{\sqrt{3}}{2}$                    $\pi$

**51.**

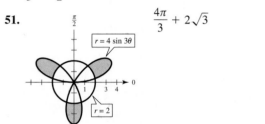

$\frac{4\pi}{3} + 2\sqrt{3}$

**53.** $\int_{0}^{\pi/6}\int_{1}^{\sqrt{3}\sec\theta}r\,dr\,d\theta + \int_{\pi/6}^{\pi/2}\int_{1}^{\csc\theta}r\,dr\,d\theta = \sqrt{3} - \frac{\pi}{4}$

**55.** 486,788    **57.** 1.2858    **59.** 56.051

**61.** False. Let $f(r, \theta) = r - 1$ and let $R$ be a sector where $0 \le r \le 6$ and $0 \le \theta \le \pi$.

**63.** (a) $2\pi$    (b) $\sqrt{2\pi}$

**65.** (a) $\int_{2}^{4}\int_{y/\sqrt{3}}^{y}f\,dx\,dy$

(b) $\int_{2/\sqrt{3}}^{2}\int_{2}^{\sqrt{3}x}f\,dy\,dx + \int_{2}^{4/\sqrt{3}}\int_{x}^{\sqrt{3}x}f\,dy\,dx + \int_{4/\sqrt{3}}^{4}\int_{x}^{4}f\,dy\,dx$

(c) $\int_{\pi/4}^{\pi/3}\int_{2\csc\theta}^{4\csc\theta}fr\,dr\,d\theta$

**67.** $\frac{4}{\pi}$

## Section 14.4 *(page 1004)*

**1.** Use a double integral when the density of the lamina is not constant.

**3.** $m = 4$    **5.** $m = \frac{1}{8}$

**7.** (a) $m = ka^2, \left(\frac{a}{2}, \frac{a}{2}\right)$    (b) $m = \frac{ka^3}{2}, \left(\frac{a}{2}, \frac{2a}{3}\right)$

(c) $m = \frac{ka^3}{2}, \left(\frac{2a}{3}, \frac{a}{2}\right)$

**9.** (a) $m = \frac{ka^2}{2}, \left(\frac{a}{3}, \frac{2a}{3}\right)$    (b) $m = \frac{ka^3}{3}, \left(\frac{3a}{8}, \frac{3a}{4}\right)$

(c) $m = \frac{ka^3}{6}, \left(\frac{a}{2}, \frac{3a}{4}\right)$

**11.** (a) $\left(\dfrac{a}{2}+5,\dfrac{a}{2}\right)$   (b) $\left(\dfrac{a}{2}+5,\dfrac{2a}{3}\right)$

  (c) $\left(\dfrac{2(a^2+15a+75)}{3(a+10)},\dfrac{a}{2}\right)$

**13.** $m=\dfrac{k}{4},\left(\dfrac{2}{3},\dfrac{8}{15}\right)$   **15.** $m=30k,\left(\dfrac{14}{5},\dfrac{4}{5}\right)$

**17.** $m=k(e-1),\left(\dfrac{1}{e-1},\dfrac{e+1}{4}\right)$

**19.** $m=\dfrac{256k}{15},\left(0,\dfrac{16}{7}\right)$   **21.** $m=\dfrac{6k}{\pi},\left(\dfrac{3}{2},\dfrac{\pi}{8}\right)$

**23.** $m=\dfrac{9\pi k}{2},\left(\dfrac{8\sqrt{2}}{\pi},\dfrac{8(2-\sqrt{2})}{\pi}\right)$

**25.** $m=\dfrac{k}{8}(1-5e^{-4}),\left(\dfrac{e^4-13}{e^4-5},\dfrac{8}{27}\left[\dfrac{e^6-7}{e^6-5e^2}\right]\right)$

**27.** $m=\dfrac{k\pi}{3},\left(\dfrac{81\sqrt{3}}{40\pi},0\right)$

**29.** $\bar{\bar{x}}=\dfrac{\sqrt{3}b}{3}$   **31.** $\bar{\bar{x}}=\dfrac{a}{2}$   **33.** $\bar{\bar{x}}=\dfrac{a}{2}$

   $\bar{\bar{y}}=\dfrac{\sqrt{3}h}{3}$      $\bar{\bar{y}}=\dfrac{a}{2}$      $\bar{\bar{y}}=\dfrac{a}{2}$

**35.** $I_x=\dfrac{32k}{3}$          **37.** $I_x=16k$

   $I_y=\dfrac{16k}{3}$           $I_y=\dfrac{512k}{5}$

   $I_0=16k$             $I_0=\dfrac{592k}{5}$

   $\bar{\bar{x}}=\dfrac{2\sqrt{3}}{3}$           $\bar{\bar{x}}=\dfrac{4\sqrt{15}}{5}$

   $\bar{\bar{y}}=\dfrac{2\sqrt{6}}{3}$           $\bar{\bar{y}}=\dfrac{\sqrt{6}}{2}$

**39.** $2k\displaystyle\int_{-b}^{b}\int_{0}^{\sqrt{b^2-x^2}}(x-a)^2\,dy\,dx=\dfrac{k\pi b^2}{4}(b^2+4a^2)$

**41.** $\displaystyle\int_{-a}^{a}\int_{0}^{\sqrt{a^2-x^2}}ky(y-a)^2\,dy\,dx=ka^5\left(\dfrac{56-15\pi}{60}\right)$

**43.** $\dfrac{L}{3}$   **45.** $\dfrac{L}{2}$

**47.** The object with a greater polar moment of inertia has more resistance, so more torque is required to twist the object.

**49.** Proof

## Section 14.5   *(page 1011)*

**1.** If $f$ and its first partial derivatives are continuous on the closed region $R$ in the $xy$-plane, then the differential of the surface area given by $z=f(x,y)$ over $R$ is $dS=\sqrt{1+[f_x(x,y)]^2+[f_y(x,y)]^2}\,dA.$

**3.** 24   **5.** $4\pi\sqrt{62}$   **7.** $\frac{1}{2}\left[4\sqrt{17}+\ln\left(4+\sqrt{17}\right)\right]$

**9.** $\frac{8}{27}\left(10\sqrt{10}-1\right)$   **11.** $\sqrt{2}-1$   **13.** $\sqrt{2}\pi$

**15.** $2\pi a\left(a-\sqrt{a^2-b^2}\right)$   **17.** $12\sqrt{14}$   **19.** $20\pi$

**21.** $\displaystyle\int_{0}^{1}\int_{0}^{x}\sqrt{5+4x^2}\,dy\,dx=\dfrac{27-5\sqrt{5}}{12}\approx 1.3183$

**23.** $\displaystyle\int_{-3}^{3}\int_{-\sqrt{9-x^2}}^{\sqrt{9-x^2}}\sqrt{1+4x^2+4y^2}\,dy\,dx$

   $=\dfrac{\pi}{6}\left(37\sqrt{37}-1\right)\approx 117.3187$

**25.** $\displaystyle\int_{0}^{1}\int_{0}^{1}\sqrt{1+4x^2+4y^2}\,dy\,dx\approx 1.8616$

**27.** $\displaystyle\int_{0}^{4}\int_{0}^{10}\sqrt{1+e^{2xy}(x^2+y^2)}\,dy\,dx$

**29.** $\displaystyle\int_{-2}^{2}\int_{-\sqrt{4-x^2}}^{\sqrt{4-x^2}}\sqrt{1+e^{-2x}}\,dy\,dx$

**31.** No. The size and shape of the graph stay the same, just the position is changed. So, the surface area does not increase.

**33.** (a) Yes. For example, let $R$ be the square given by $0\le x\le 1$ and $0\le y\le 1$, and let $S$ be the square parallel to $R$ given by $0\le x\le 1,0\le y\le 1$, and $z=1$.

  (b) Yes. Let $R$ be the region in part (a) and let $S$ be the surface given by $f(x,y)=xy$.

  (c) No

**35.** (a) $812\pi\sqrt{609}$ cm$^3$   (b) $100\pi\sqrt{609}$ cm$^2$   **37.** 16

## Section 14.6   *(page 1021)*

**1.** The volume of the solid region $Q$   **3.** 18   **5.** $\frac{1}{9}$

**7.** $\dfrac{15}{2}\left(1-\dfrac{1}{e}\right)$   **9.** $\dfrac{189}{2}$   **11.** $\dfrac{324}{5}$

**13.** $V=\displaystyle\int_{0}^{7}\int_{0}^{(7-x)/2}\int_{0}^{7-x-2y}dz\,dy\,dx$

**15.** $V=\displaystyle\int_{-\sqrt{6}}^{\sqrt{6}}\int_{-\sqrt{6-y^2}}^{\sqrt{6-y^2}}\int_{0}^{6-x^2-y^2}dz\,dx\,dy$

**17.** $V=\displaystyle\int_{-4}^{4}\int_{-\sqrt{16-x^2}}^{\sqrt{16-x^2}}\int_{(x^2+y^2)/2}^{\sqrt{80-x^2-y^2}}dz\,dy\,dx$   **19.** $\dfrac{256}{15}$

**21.** $\frac{3}{2}$   **23.** 10

**25.**

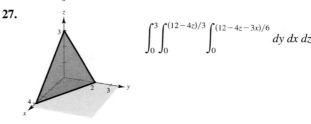

$\displaystyle\int_{0}^{1}\int_{0}^{1}\int_{-1}^{-\sqrt{z}}dy\,dz\,dx$

**27.**

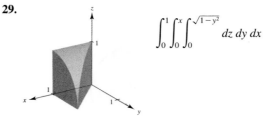

$\displaystyle\int_{0}^{3}\int_{0}^{(12-4z)/3}\int_{0}^{(12-4z-3x)/6}dy\,dx\,dz$

**29.**

$\displaystyle\int_{0}^{1}\int_{0}^{x}\int_{0}^{\sqrt{1-y^2}}dz\,dy\,dx$

**31.** $\displaystyle\int_0^3\int_0^5\int_{y/5}^1 xyz\,dx\,dy\,dz,\quad \int_0^3\int_0^1\int_0^{5x} xyz\,dy\,dx\,dz,$

$\displaystyle\int_0^5\int_0^3\int_{y/5}^1 xyz\,dx\,dz\,dy,\quad \int_0^1\int_0^3\int_0^{5x} xyz\,dy\,dz\,dx,$

$\displaystyle\int_0^5\int_{y/5}^1\int_0^3 xyz\,dz\,dx\,dy,\quad \int_0^1\int_0^{5x}\int_0^3 xyz\,dz\,dy\,dx;\ \dfrac{225}{16}$

**33.** $\displaystyle\int_{-3}^3\int_{-\sqrt{9-x^2}}^{\sqrt{9-x^2}}\int_0^4 xyz\,dz\,dy\,dx,\quad \int_{-3}^3\int_{-\sqrt{9-y^2}}^{\sqrt{9-y^2}}\int_0^4 xyz\,dz\,dx\,dy,$

$\displaystyle\int_{-3}^3\int_0^4\int_{-\sqrt{9-x^2}}^{\sqrt{9-x^2}} xyz\,dy\,dz\,dx,\quad \int_0^4\int_{-3}^3\int_{-\sqrt{9-x^2}}^{\sqrt{9-x^2}} xyz\,dy\,dx\,dz,$

$\displaystyle\int_0^4\int_{-3}^3\int_{-\sqrt{9-y^2}}^{\sqrt{9-y^2}} xyz\,dx\,dy\,dz,\quad \int_{-3}^3\int_0^4\int_{-\sqrt{9-y^2}}^{\sqrt{9-y^2}} xyz\,dx\,dz\,dy;\ 0$

**35.** $\displaystyle\int_0^1\int_0^{1-z}\int_0^{1-y^2} dx\,dy\,dz,\quad \int_0^1\int_0^{1-y}\int_0^{1-y^2} dx\,dz\,dy,$

$\displaystyle\int_0^1\int_0^{2z-z^2}\int_0^{1-z} 1\,dy\,dx\,dz + \int_0^1\int_{2z-z^2}^1\int_0^{\sqrt{1-x}} 1\,dy\,dx\,dz,$

$\displaystyle\int_0^1\int_{1-\sqrt{1-x}}^1\int_0^{1-z} 1\,dy\,dz\,dx + \int_0^1\int_0^{1-\sqrt{1-x}}\int_0^{\sqrt{1-x}} 1\,dy\,dz\,dx,$

$\displaystyle\int_0^1\int_0^{\sqrt{1-x}}\int_0^{1-y} dz\,dy\,dx$

**37.** $m = 8k,\ \bar{x} = \dfrac{3}{2}$     **39.** $m = \dfrac{128k}{3},\ \bar{z} = 1$

**41.** $m = k\displaystyle\int_0^b\int_0^b\int_0^b xy\,dz\,dy\,dx$

$M_{yz} = k\displaystyle\int_0^b\int_0^b\int_0^b x^2y\,dz\,dy\,dx$

$M_{xz} = k\displaystyle\int_0^b\int_0^b\int_0^b xy^2\,dz\,dy\,dx$

$M_{xy} = k\displaystyle\int_0^b\int_0^b\int_0^b xyz\,dz\,dy\,dx$

**43.** $\bar{x}$ will be greater than 2, and $\bar{y}$ and $\bar{z}$ will be unchanged.

**45.** $\bar{x}$ and $\bar{z}$ will be unchanged, and $\bar{y}$ will be greater than 0.

**47.** $\left(0, 0, \dfrac{3h}{4}\right)$     **49.** $\left(0, 0, \dfrac{3}{2}\right)$     **51.** $\left(5, 6, \dfrac{5}{4}\right)$

**53.** (a) $I_x = \dfrac{2ka^5}{3}$     **55.** (a) $I_x = 256k$

$I_y = \dfrac{2ka^5}{3}$     $I_y = \dfrac{512k}{3}$

$I_z = \dfrac{2ka^5}{3}$     $I_z = 256k$

(b) $I_x = \dfrac{ka^8}{8}$     (b) $I_x = \dfrac{2048k}{3}$

$I_y = \dfrac{ka^8}{8}$     $I_y = \dfrac{1024k}{3}$

$I_z = \dfrac{ka^8}{8}$     $I_z = \dfrac{2048k}{3}$

**57.** Proof

**59.** $\displaystyle\int_{-1}^1\int_{-1}^1\int_0^{1-x} (x^2 + y^2)\sqrt{x^2 + y^2 + z^2}\,dz\,dy\,dx$

**61.** (a) $m = \displaystyle\int_{-2}^2\int_{-\sqrt{4-x^2}}^{\sqrt{4-x^2}}\int_0^{4-x^2-y^2} kz\,dz\,dy\,dx$

(b) $\bar{x} = \bar{y} = 0$ by symmetry.

$\bar{z} = \dfrac{1}{m}\displaystyle\int_{-2}^2\int_{-\sqrt{4-x^2}}^{\sqrt{4-x^2}}\int_0^{4-x^2-y^2} kz^2\,dz\,dy\,dx$

(c) $I_z = \displaystyle\int_{-2}^2\int_{-\sqrt{4-x^2}}^{\sqrt{4-x^2}}\int_0^{4-x^2-y^2} kz(x^2 + y^2)\,dz\,dy\,dx$

**63.** $\dfrac{13}{3}$     **65.** $\dfrac{3}{2}$     **67.** Increase

**69.** b     **71.** $Q: 2x^2 + y^2 + 3z^2 \le 1;\ 0.684;\ \dfrac{4\sqrt{6}\pi}{45}$

**73.** Putnam Problem B1, 1965

# Section 14.7   (page 1029)

**1.** Some solids are represented by equations involving $x^2$ and $y^2$. Often, converting these equations to cylindrical or spherical coordinates yields equations you can work with more easily.

**3.** 27     **5.** $\dfrac{11}{10}$     **7.** $\dfrac{\pi}{3}$     **9.** $\pi(e^4 + 3)$

**11.** 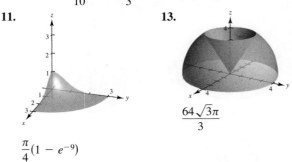     **13.**

$\dfrac{64\sqrt{3}\pi}{3}$

$\dfrac{\pi}{4}(1 - e^{-9})$

**15.** $48(3\pi - 4)$     **17.** $\dfrac{\pi}{6}$     **19.** $\dfrac{250}{9}(3\pi - 4)$     **21.** $48k\pi$

**23.** $\dfrac{\pi r_0^2 h}{3}$     **25.** $\left(0, 0, \dfrac{h}{5}\right)$

**27.** $I_z = 4k\displaystyle\int_0^{\pi/2}\int_0^{r_0}\int_0^{h(r_0-r)/r_0} r^3\,dz\,dr\,d\theta = \dfrac{3mr_0^2}{10}$

**29.** Proof     **31.** $9\pi\sqrt{2}$     **33.** $16\pi^2$     **35.** $k\pi a^4$

**37.** $\left(0, 0, \dfrac{3r}{8}\right)$     **39.** $\dfrac{k\pi}{192}$

**41.** Cylindrical: $\displaystyle\int_0^{2\pi}\int_0^2\int_{r^2}^4 r^2\cos\theta\,dz\,dr\,d\theta = 0$

Spherical: $\displaystyle\int_0^{2\pi}\int_0^{\arctan(1/2)}\int_0^{4\sec\phi} \rho^3\sin^2\phi\cos\theta\,d\rho\,d\phi\,d\theta$

$+ \displaystyle\int_0^{2\pi}\int_{\arctan(1/2)}^{\pi/2}\int_0^{\cot\phi\csc\phi} \rho^3\sin^2\phi\cos\phi\,d\rho\,d\phi\,d\theta = 0$

**43.** Cylindrical: $\displaystyle\int_0^{2\pi}\int_0^1\int_1^{1+\sqrt{1-r^2}} r^2\cos\theta\,dz\,dr\,d\theta = 0$

Spherical: $\displaystyle\int_0^{\pi/4}\int_0^{2\pi}\int_{\sec\phi}^{2\cos\phi} \rho^3\sin^2\phi\cos\theta\,d\rho\,d\theta\,d\phi = 0$

**45.** (a) $r$ constant: right circular cylinder about $z$-axis

$\theta$ constant: plane parallel to $z$-axis

$z$ constant: plane parallel to $xy$-plane

(b) $\rho$ constant: sphere

$\theta$ constant: plane parallel to $z$-axis

$\phi$ constant: cone

**47.** Putnam Problem A1, 2006

## Section 14.8  *(page 1036)*

**1.** $\dfrac{\partial x}{\partial u}\dfrac{\partial y}{\partial v} - \dfrac{\partial y}{\partial u}\dfrac{\partial x}{\partial v}$    **3.** $-\frac{1}{2}$    **5.** $1 + 2v$

**7.** 1    **9.** $-e^{2u}$

**11.** 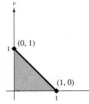    **13.**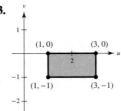

**15.** $\displaystyle\iint_R 3xy\,dA = \int_{-2/3}^{2/3}\int_{1-x}^{(1/2)x+2} 3xy\,dy\,dx$

$\displaystyle + \int_{2/3}^{4/3}\int_{(1/2)x}^{(1/2)x+2} 3xy\,dy\,dx + \int_{4/3}^{8/3}\int_{(1/2)x}^{4-x} 3xy\,dy\,dx = \dfrac{164}{9}$

**17.** $\frac{8}{3}$    **19.** 36    **21.** $(e^{-1/2} - e^{-2})\ln 8 \approx 0.9798$    **23.** 18

**25.** $12(e^4 - 1)$    **27.** $\frac{100}{9}$    **29.** $\frac{2}{5}a^{5/2}$    **31.** One

**33.** (a)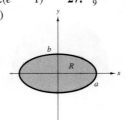

(b) $ab$    (c) $\pi ab$

**35.** $u^2v$    **37.** $-uv$    **39.** $-\rho^2 \sin\phi$

**41.** Putnam Problem A2, 1994

## Review Exercises for Chapter 14  *(page 1038)*

**1.** $\dfrac{1 - \cos 3x^2}{x}$    **3.** $\dfrac{29}{6}$    **5.** $\dfrac{1}{6}$    **7.** $\dfrac{3}{2}$    **9.** 16

**11.**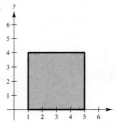

$\displaystyle\int_1^5\int_0^4 dy\,dx = \int_0^4\int_1^5 dx\,dy = 16$

**13.**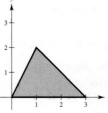

$\displaystyle\int_0^2\int_{y/2}^{3-y} dx\,dy = \int_0^1\int_0^{2x} dy\,dx + \int_1^3\int_0^{3-x} dy\,dx = 3$

**15.** $\displaystyle\int_0^2\int_0^4 4xy\,dy\,dx = \int_0^4\int_0^2 4xy\,dx\,dy = 64$    **17.** 21

**19.** $\dfrac{40}{3}$    **21.** $\dfrac{40}{3}$    **23.** 13.67°C    **25.** $\dfrac{5\sqrt{5}\pi}{6}$

**27.** $\dfrac{81}{5}$    **29.** $\dfrac{3\pi}{2}$

**31.**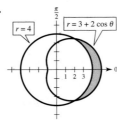

$\dfrac{13\sqrt{3}}{2} - \dfrac{5\pi}{3}$

**33.** (a) $r = 3\sqrt{\cos 2\theta}$

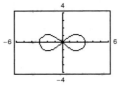

(b) 9    (c) $3(3\pi - 16\sqrt{2} + 20) \approx 20.392$

**35.** 7    **37.** $m = \dfrac{32k}{5}, \left(\dfrac{5}{3}, \dfrac{5}{2}\right)$    **39.** $m = \dfrac{k}{4}, \left(\dfrac{32}{45}, \dfrac{64}{55}\right)$

**41.** $I_x = 12k$

$I_y = \dfrac{81k}{2}$

$I_0 = \dfrac{105k}{2}$

$\bar{\bar{x}} = \dfrac{3\sqrt{2}}{2}$

$\bar{\bar{y}} = \dfrac{2\sqrt{3}}{3}$

**43.** $\dfrac{\pi}{6}(101\sqrt{101} - 1)$    **45.** $\dfrac{1}{6}(37\sqrt{37} - 1)$

**47.** (a) 30,415.74 ft³    (b) 2081.53 ft²    **49.** 56

**51.** $\dfrac{16}{3} + 2e$    **53.** $\dfrac{8\pi}{5}$    **55.** 36

**57.**

$\displaystyle\int_0^1\int_x^1\int_0^{\sqrt{1-x^2}} dz\,dy\,dx$

**59.** $m = \dfrac{500k}{3}, \bar{x} = \dfrac{5}{2}$    **61.** $12(\sqrt{3} - 1)$    **63.** $\dfrac{\pi}{15}$

**65.** $\pi\left(3\sqrt{13} + 4\ln\dfrac{3 + \sqrt{13}}{2}\right) \approx 48.995$    **67.** $16\pi$

**69.** $\dfrac{8\pi}{3}(2 - \sqrt{3})$    **71.** $-6(v + u)$    **73.** $\sin^2\theta - \cos^2\theta$

**75.** $5\ln 5 - 3\ln 3 - 2 \approx 2.751$    **77.** 81

## P.S. Problem Solving   *(page 1041)*

**1.** $8(2 - \sqrt{2})$    **3.** $\frac{1}{3}$    **5.** (a)–(g) Proofs

**7.** $-\frac{1}{2}$; $\frac{1}{2}$; No; Fubini's Theorem is not valid because $f$ is not continuous on the region $0 \le x \le 1, 0 \le y \le 1$.

**9.** $\dfrac{\sqrt{\pi}}{4}$    **11.** If $a, k > 0$, then $1 = ka^2$ or $a = \dfrac{1}{\sqrt{k}}$.

**13.** Answers will vary.

**15.** The greater the angle between the given plane and the $xy$-plane, the greater the surface area. So, $z_2 < z_1 < z_4 < z_3$.

**17.**

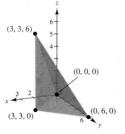

$$\int_0^3 \int_0^{2x} \int_x^{6-x} dy\, dz\, dx = 18$$

# Chapter 15

## Section 15.1   *(page 1053)*

**1.** See "Definition of Vector Field" on page 1044. Some physical examples of vector fields include velocity fields, gravitational fields, and electric force fields.

**3.** Reconstruct a function from its partial derivatives by integrating and comparing versions of the function to determine constants.

**5.** d    **6.** c    **7.** a    **8.** b

**9.** $\sqrt{2}$    **11.** $\sqrt{1 + 9y^2}$

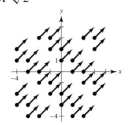

**13.** $\sqrt{3}$    **15.**

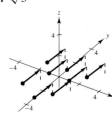

**17.**

**19.** $2x\mathbf{i} + 4y\mathbf{j}$    **21.** $(10x + 3y)\mathbf{i} + (3x + 2y)\mathbf{j}$

**23.** $6yz\mathbf{i} + 6xz\mathbf{j} + 6xy\mathbf{k}$    **25.** $2xye^{x^2}\mathbf{i} + e^{x^2}\mathbf{j} + \mathbf{k}$

**27.** $\left[\dfrac{xy}{x + y} + y\ln(x + y)\right]\mathbf{i} + \left[\dfrac{xy}{x + y} + x\ln(x + y)\right]\mathbf{j}$

**29.** Conservative    **31.** Not conservative    **33.** Conservative

**35.** Not conservative

**37.** Conservative; $f(x, y) = 3xy - \dfrac{x^3}{3} + \dfrac{y^2}{2} + K$

**39.** Conservative; $f(x, y) = e^{x^2y} + K$    **41.** Not conservative

**43.** Conservative; $f(x, y) = x \sin y + K$    **45.** $4\mathbf{i} - \mathbf{j} - 3\mathbf{k}$

**47.** $-2\mathbf{k}$    **49.** $\dfrac{2x}{x^2 + y^2}\mathbf{k}$

**51.** Conservative; $f(x, y, z) = x^3 + y^3 + z^3 + xyz + K$

**53.** Not conservative

**55.** Conservative; $f(x, y, z) = \dfrac{xz}{y} - z + K$    **57.** $2x + 4y$

**59.** $2 \sin x \cos x + 3z^2$    **61.** 28    **63.** 0

**65.** Vector field; The curl of a vector field is a vector field.

**67.** Neither; The expression is meaningless because you can only take the curl of a vector field.

**69.** $9x\mathbf{j} - 2y\mathbf{k}$    **71.** $z\mathbf{j} + y\mathbf{k}$    **73.** $3z + 2x$    **75.** 0

**77.** (a)–(h) Proofs

## Section 15.2   *(page 1065)*

**1.** (a) The arc length of $C$    (b) The mass of the string

**3.** $\mathbf{r}(t) = \begin{cases} t\mathbf{i} + t\mathbf{j}, & 0 \le t \le 1 \\ (2 - t)\mathbf{i} + \sqrt{2 - t}\,\mathbf{j}, & 1 \le t \le 2 \end{cases}$

**5.** $\mathbf{r}(t) = \begin{cases} t\mathbf{i}, & 0 \le t \le 3 \\ 3\mathbf{i} + (t - 3)\mathbf{j}, & 3 \le t \le 6 \\ (9 - t)\mathbf{i} + 3\mathbf{j}, & 6 \le t \le 9 \\ (12 - t)\mathbf{j}, & 9 \le t \le 12 \end{cases}$

**7.** $\mathbf{r}(t) = 3 \cos t\mathbf{i} + 3 \sin t\mathbf{j},\ 0 \le t \le 2\pi$

**9.** (a) $C$: $\mathbf{r}(t) = t\mathbf{i} + t\mathbf{j},\ 0 \le t \le 1$    (b) $\dfrac{2\sqrt{2}}{3}$

**11.** (a) $C$: $\mathbf{r}(t) = \cos t\mathbf{i} + \sin t\mathbf{j},\ 0 \le t \le \dfrac{\pi}{2}$    (b) $\dfrac{\pi}{2}$

**13.** (a) $C$: $\mathbf{r}(t) = \begin{cases} t\mathbf{i}, & 0 \le t \le 1 \\ t\mathbf{i} + (4t - 4)\mathbf{j}, & 1 \le t \le 2 \end{cases}$

(b) $1 + 7\sqrt{17}$

**15.** (a) $C$: $\mathbf{r}(t) = \begin{cases} t\mathbf{i}, & 0 \le t \le 1 \\ (2 - t)\mathbf{i} + (t - 1)\mathbf{j}, & 1 \le t \le 2 \\ (3 - t)\mathbf{j}, & 2 \le t \le 3 \end{cases}$

(b) $3 + 3\sqrt{2}$

**17.** (a) $C$: $\mathbf{r}(t) = \begin{cases} t\mathbf{i}, & 0 \le t \le 1 \\ \mathbf{i} + (t - 1)\mathbf{k}, & 1 \le t \le 2 \\ \mathbf{i} + (t - 2)\mathbf{j} + \mathbf{k}, & 2 \le t \le 3 \end{cases}$    (b) $\dfrac{23}{6}$

**19.** 20    **21.** $\dfrac{5\pi}{2}$    **23.** $8\sqrt{5}\pi\left(1 + \dfrac{4\pi^2}{3}\right) \approx 795.7$

**25.** $2\pi + 2$    **27.** $\dfrac{k}{12}(41\sqrt{41} - 27)$    **29.** 8

**31.** $\frac{1}{3}e^6 + \frac{95}{3}$    **33.** $\frac{9}{4}$    **35.** About 249.49    **37.** 66

**39.** 0    **41.** $-10\pi^2$

**43.** Positive; The vector field determined by $\mathbf{F}$ points in the general direction of the path $C$, so $\mathbf{F} \cdot \mathbf{T} > 0$.

**45.** Zero; The vector field determined by $\mathbf{F}$ is perpendicular to the path $C$.

**47.** (a) $\frac{236}{3}$; Orientation is from left to right, so the value is positive.

(b) $-\frac{236}{3}$; Orientation is from right to left, so the value is negative.

**49.** $\mathbf{F}(t) = -2t\mathbf{i} - t\mathbf{j}$
$\mathbf{r}'(t) = \mathbf{i} - 2\mathbf{j}$
$\mathbf{F}(t) \cdot \mathbf{r}'(t) = -2t + 2t = 0$
$\int_C \mathbf{F} \cdot d\mathbf{r} = 0$

**51.** $\mathbf{F}(t) = (t^3 - 2t^2)\mathbf{i} + \left(t - \frac{t^2}{2}\right)\mathbf{j}$
$\mathbf{r}'(t) = \mathbf{i} + 2t\mathbf{j}$
$\mathbf{F}(t) \cdot \mathbf{r}'(t) = t^3 - 2t^2 + 2t^2 - t^3 = 0$
$\int_C \mathbf{F} \cdot d\mathbf{r} = 0$

**53.** 68    **55.** $\frac{40}{3}$    **57.** 25    **59.** $\frac{63}{2}$    **61.** $-\frac{11}{6}$

**63.** $\frac{316}{3}$    **65.** $5h$    **67.** $\frac{1}{2}$    **69.** $\frac{h}{4}\left[2\sqrt{5} + \ln(2 + \sqrt{5})\right]$

**71.** $\frac{1}{120}\left(25\sqrt{5} - 11\right)$

**73.** (a) $12\pi \approx 37.70 \text{ cm}^2$    (b) $\frac{12\pi}{5} \approx 7.54 \text{ cm}^3$

(c)

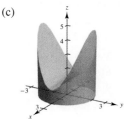

**75.** $I_x = I_y = a^3\pi$

**77.** (a)

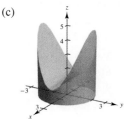

(b) $9\pi \text{ cm}^2 \approx 28.274 \text{ cm}^2$    (c) $\frac{27\pi}{2} \text{ cm}^2 \approx 42.412 \text{ cm}^3$

**79.** 1750 ft-lb    **81.** No. $y = 2x$, so $dy = 2\ dx$.

**83.** $z_3, z_1, z_2, z_4$; The greater the height of the surface over the curve $y = \sqrt{x}$, the greater the lateral surface area.

**85.** False. $\int_C xy\ ds = \sqrt{2}\int_0^1 t^2\ dt$    **87.** $-12$

## Section 15.3   *(page 1076)*

**1.** Verify that the vector field is conservative. Find a potential function. Calculate the difference of the values of the function evaluated at the endpoints.

**3.** (a) Proof

(b) $\int_{C_1} \mathbf{F} \cdot d\mathbf{r} = \int_0^1 (t^2 + 2t^3)\ dt = \frac{5}{6}$

$\int_{C_2} \mathbf{F} \cdot d\mathbf{r} = \int_0^{\pi/2} (\sin^2 \theta \cos \theta + 2\sin^3 \theta \cos \theta)\ d\theta = \frac{5}{6}$

**5.** (a) Proof

(b) $\int_{C_1} \mathbf{F} \cdot d\mathbf{r} = \int_0^{\pi/3} (3\tan^2 \theta \sec \theta + 3\sec^3 \theta)\ d\theta$
$\approx 10.392$

$\int_{C_2} \mathbf{F} \cdot d\mathbf{r} = \int_0^3 \left(\frac{3\sqrt{t}}{2\sqrt{t+1}} + \frac{3\sqrt{t+1}}{2\sqrt{t}}\right)\ dt \approx 10.392$

**7.** (a) Proof

(b) $\int_{C_1} \mathbf{F} \cdot d\mathbf{r} = \int_0^1 64t^3\ dt = 16$

$\int_{C_2} \mathbf{F} \cdot d\mathbf{r} = \int_0^{\pi/2} 64\sin^3 \theta \cos \theta\ d\theta = 16$

**9.** 72    **11.** $-1$    **13.** 0    **15.** (a) 2    (b) 2    (c) 2

**17.** 11    **19.** (a) Proof    (b) 30,366

**21.** (a) Proof    (b) 32    **23.** (a) 1    (b) 1

**25.** (a) 64    (b) 0    (c) 0    (d) 0

**27.** (a) 32    (b) 32    **29.** (a) $\frac{2}{3}$    (b) $\frac{17}{6}$    **31.** (a) 0    (b) 0

**33.** 0

**35.** (a) $d\mathbf{r} = (\mathbf{i} - \mathbf{j})\ dt \implies \int_0^{50} 175\ dt = 8750$ ft-lb

(b) $d\mathbf{r} = \left(\mathbf{i} - \frac{1}{25}(50 - t)\mathbf{j}\right)\ dt$

$7\int_0^{50} (50 - t)\ dt = 8750$ ft-lb

**37.**

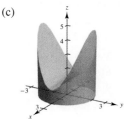

The partial derivatives of $\mathbf{F}$ are not continuous at $(0, 0)$. Draw an open connected region that excludes that point.

**39.** 1.125

**41.** Yes, because the work required to get from point to point is independent of the path taken.

**43.** False. It would be true if $\mathbf{F}$ were conservative.

**45.** True    **47.** Proof

**49.** (a) Proof    (b) $-\pi$    (c) $\pi$    (d) $-2\pi$

(e) No, because $\mathbf{F}$ is not continuous at $(0, 0)$ in $R$ enclosed by $C$.

(f) $\nabla\left(\arctan \frac{x}{y}\right) = \frac{1/y}{1 + (x/y)^2}\mathbf{i} + \frac{-x/y^2}{1 + (x/y)^2}\mathbf{j}$

## Section 15.4   *(page 1085)*

**1.** A curve is simple when it does not cross itself. A connected plane region is simply connected when every simple closed curve in the region encloses only points that are in the region. For example, a region with a hole is not simply connected.

**3.** You are working with a simple closed curve with a boundary whose orientation is counterclockwise.

**5.** $\frac{1}{30}$    **7.** 0    **9.** About 19.99    **11.** $\frac{9}{2}$    **13.** 56

**15.** $\frac{4}{3}$  **17.** 0  **19.** 0  **21.** $\frac{1}{12}$  **23.** $32\pi$

**25.** $\pi$  **27.** $\frac{225}{2}$  **29.** $4\pi$  **31.** $\frac{9}{2}$  **33.** Proof

**35.** $\left(0, \frac{8}{5}\right)$  **37.** $\left(\frac{8}{15}, \frac{8}{21}\right)$  **39.** $54\pi$  **41.** $\pi - \frac{3\sqrt{3}}{2}$

**43.** (a) $\frac{51\pi}{2}$  (b) $\frac{243\pi}{2}$  **45.** $46\pi$

**47.** (a) $\displaystyle\int_C \mathbf{F} \cdot d\mathbf{r} = \int_C M\,dx + N\,dy = \iint_R \left(\frac{\partial N}{\partial x} - \frac{\partial M}{\partial y}\right) dA = 0$
(b) $I = -2\pi$ when $C$ is a circle that contains the origin.
**49–53.** Proofs

## Section 15.5  *(page 1095)*

**1.** $S$ is traced out by the position vector $\mathbf{r}(u, v)$ as the point $(u, v)$ moves throughout the domain. To sketch the surface, it is helpful to relate $x$, $y$, and $z$, where $x$, $y$, and $z$ are functions of $u$ and $v$.

**3.** e  **4.** f  **5.** b  **6.** c  **7.** d  **8.** a

**9.** $y - 2z = 0$        **11.** $x^2 + z^2 = 4$
Plane            Cylinder

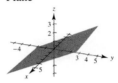  

**13.**                **15.**

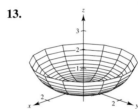        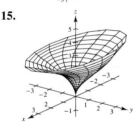

**17.** $\mathbf{r}(u, v) = u\mathbf{i} + v\mathbf{j} + 3v\mathbf{k}$
**19.** $\mathbf{r}(u, v) = \frac{1}{2}u \cos v\mathbf{i} + u\mathbf{j} + \frac{1}{3}u \sin v\mathbf{k}$, $u \geq 0, 0 \leq v \leq 2\pi$ or
$\mathbf{r}(x, y) = x\mathbf{i} + \sqrt{4x^2 + 9y^2}\,\mathbf{j} + z\mathbf{k}$
**21.** $\mathbf{r}(u, v) = 5 \cos u\mathbf{i} + 5 \sin u\mathbf{j} + v\mathbf{k}$
**23.** $\mathbf{r}(u, v) = u\mathbf{i} + \sqrt{u - 7} \cos v\mathbf{j} + \sqrt{u - 7} \sin v\mathbf{k}$ or
$\mathbf{r}(y, z) = (y^2 + z^2 + 7)\mathbf{i} + y\mathbf{j} + z\mathbf{k}$
**25.** $\mathbf{r}(u, v) = v \cos u\mathbf{i} + v \sin u\mathbf{j} + 4\mathbf{k}$,  $0 \leq v \leq 3$

**27.** $x = u, y = \frac{u}{2} \cos v, z = \frac{u}{2} \sin v, 0 \leq u \leq 6, 0 \leq v \leq 2\pi$

**29.** $x = \sin u \cos v, y = \sin u \sin v, z = u$
$0 \leq u \leq \pi, 0 \leq v \leq 2\pi$
**31.** $x = \cos^2 u \cos v, y = u, z = \cos^2 u \sin v$

$\frac{\pi}{2} \leq u \leq \pi, 0 \leq v \leq 2\pi$

**33.** $9y + \frac{3\sqrt{3}}{2}z = 12\sqrt{3}$    **35.** $4y - 3z = 12$    **37.** $8\sqrt{2}$

**39.** $\pi ab^2\sqrt{a^2 + 1}$    **41.** $\frac{\pi}{6}\left(17\sqrt{17} - 1\right) \approx 36.177$

**43.** The paraboloid is reflected (inverted) through the $xy$-plane.
**45.** The height of the paraboloid is increased from 4 to 9.
**47–49.** Proofs

**51.** (a)        (b)
(c)        (d)

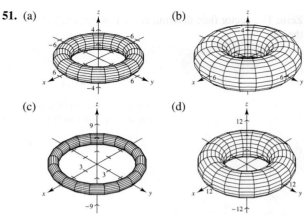

The radius of the generating circle that is revolved about the $z$-axis is $b$, and its center is $a$ units from the axis of revolution.
**53.** $400\pi$ m$^2$
**55.**

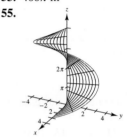

$2\pi\left[\frac{3}{2}\sqrt{13} + 2\ln\left(3 + \sqrt{13}\right) - 2\ln 2\right]$
**57.** Answers will vary. Sample answer: Let
$x = (2 - u)(5 + \cos v) \cos 3\pi u$
$y = (2 - u)(5 + \cos v) \sin 3\pi u$
$z = 5u + (2 - u) \sin v$
where $-\pi \leq u \leq \pi$ and $-\pi \leq v \leq \pi$.

## Section 15.6  *(page 1108)*

**1.** Solve for $y$ in the equation of the surface. Then use the integral
$\displaystyle\iint_S f(x, g(x, z), z)\sqrt{1 + [g_x(x, z)]^2 + [g_z(x, z)]^2}\,dA$.

**3.** An orientable surface has two distinct sides.

**5.** $12\sqrt{2}$  **7.** $2\pi$  **9.** $\frac{27\sqrt{3}}{8}$  **11.** About $-11.47$

**13.** $\frac{364}{3}$  **15.** $12\sqrt{5}$  **17.** $\frac{3 - \sqrt{3}}{2}$  **19.** $\sqrt{3}\pi$

**21.** $\frac{32\pi}{3}$  **23.** $486\pi$  **25.** $-\frac{4}{3}$  **27.** $\frac{3\pi}{2}$  **29.** $20\pi$

**31.** $384\pi$  **33.** $64\pi\rho$  **35.** 0  **37.** Proof  **39.** $2\pi a^3 h$
**41.** (a) 12  (b) 12  (c) 12

**43.** (a)

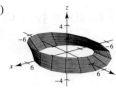

(b) No. If a normal vector at a point $P$ on the surface is moved around the Möbius strip once, it will point in the opposite direction.

(c)

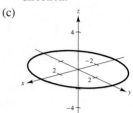

Circle

(d) Construction

(e) You obtain a strip with a double twist that is twice as long as the Möbius strip.

## Section 15.7 *(page 1116)*

**1.** Divergence Theorem or two surface integrals; In this case, it is easier to use the Divergence Theorem.

**3.** 1　　**5.** 18　　**7.** $\pi$　　**9.** $3a^4$　　**11.** 0　　**13.** $108\pi$

**15.** 0　　**17.** $18(e^4 - 5)$　　**19.** Source

**21.** Incompressible　　**23.** Any point that satisfies $xyz > \frac{1}{2}$

**25.** 0; Proof　　**27–31.** Proofs

## Section 15.8 *(page 1123)*

**1.** Stokes's Theorem allows you to evaluate a line integral using a single double integral.

**3.** $18\pi$　　**5.** 0　　**7.** $-12$　　**9.** $2\pi$　　**11.** 0　　**13.** $\frac{8}{3}$

**15.** $-\dfrac{a^5}{4}$　　**17.** $\dfrac{81\pi}{4}$　　**19.** Yes; Proof

**21.** Putnam Problem A5, 1987

## Review Exercises for Chapter 15 *(page 1124)*

**1.** $\sqrt{x^2 + 5}$

**3.** $y \cos xy\mathbf{i} + (x \cos xy - 2y)\mathbf{j}$　　**5.** $(4x + y)\mathbf{i} + x\mathbf{j} + 2z\mathbf{k}$

**7.** Not conservative　　**9.** Conservative

**11.** Conservative; $f(x, y) = \dfrac{y}{x} + K$

**13.** Conservative; $f(x, y) = \frac{1}{2}x^2y^2 - \frac{1}{3}x^3 + \frac{1}{3}y^3 + K$

**15.** Not conservative　　**17.** Conservative; $f(x, y, z) = \dfrac{x}{yz} + K$

**19.** (a) div $\mathbf{F} = 2x + 2xy + x^2$　　(b) curl $\mathbf{F} = -2xz\mathbf{j} + y^2\mathbf{k}$

**21.** (a) div $\mathbf{F} = -y \sin x - x \cos y + xy$

(b) curl $\mathbf{F} = xz\mathbf{i} - yz\mathbf{j}$

**23.** (a) div $\mathbf{F} = \dfrac{1}{\sqrt{1 - x^2}} + 2xy + 2yz$

(b) curl $\mathbf{F} = z^2\mathbf{i} + y^2\mathbf{k}$

**25.** (a) div $\mathbf{F} = \dfrac{2x + 2y}{x^2 + y^2} + 1$　　(b) curl $\mathbf{F} = \dfrac{2x - 2y}{x^2 + y^2}\mathbf{k}$

**27.** (a) $\frac{125}{3}$　　(b) $2\pi$　　**29.** $6\pi$　　**31.** $\dfrac{9a^2}{5}$　　**33.** $3\pi$

**35.** 1　　**37.** $2\pi^2$　　**39.** $\frac{8}{3}(3 - 4\sqrt{2}) \approx -7.085$　　**41.** 12

**43.** $\dfrac{\sqrt{5}}{3}(19 - \cos 6) \approx 13.446$

**45.** (a) Proof

(b) (i) $\displaystyle\int_C \mathbf{F} \cdot d\mathbf{r} = \int_0^4 (3t + 4 + t^3)\, dt = 104$

(ii) $\displaystyle\int_C \mathbf{F} \cdot d\mathbf{r} = \int_0^2 \left[(3w^2 + 4)(2w) + w^6(2w)\right] dw = 104$

**47.** $1 - \dfrac{1}{e^2}$　　**49.** 6　　**51.** (a) Proof　　(b) 92　　**53.** 1

**55.** 0　　**57.** 0　　**59.** 0　　**61.** 3

**63.** $z = 2(x^2 + y^2)$　　　　**65.**

Paraboloid

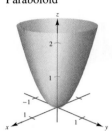

**67.** $\mathbf{r}(u, v) = \cos v \cos u\mathbf{i} + 2\sqrt{2}\cos v \sin u\mathbf{j} + 3 \sin v\mathbf{k}$

**69.** $x = u,\ y = 2u^3 \cos v,\ z = 2u^3 \sin v,\ 0 \le u \le 2,\ 0 \le v \le 2\pi$

**71.** $3\sqrt{41}$　　**73.** 45　　**75.** $27\sqrt{41}$　　**77.** $6\sqrt{26}$

**79.** $25\pi$　　**81.** 66　　**83.** $\dfrac{2a^6}{5}$　　**85.** $16\pi$

## P.S. Problem Solving *(page 1127)*

**1.** (a) and (b) $\dfrac{25\sqrt{2}k\pi}{6}$

**3.** $I_x = \dfrac{\sqrt{13}\pi}{3}(27 + 32\pi^2)$

$I_y = \dfrac{\sqrt{13}\pi}{3}(27 + 32\pi^2)$

$I_z = 18\sqrt{13}\pi$

**5.** (a)–(d) Proofs　　**7.** $3a^2\pi$　　**9.** (a) 1　　(b) $\frac{13}{15}$　　(c) $\frac{5}{2}$

**11.** The area is the same as the magnitude.

# Chapter 16

## Section 16.1    *(page 1135)*

1. When there exists a function $f$ of two variables $x$ and $y$ having continuous partial derivatives such that $f_x(x, y) = M(x, y)$ and $f_y(x, y) = N(x, y)$; To test for exactness, determine whether $\dfrac{\partial M}{\partial y} = \dfrac{\partial N}{\partial x}$.

3. Exact    5. Not exact    7. $x^2 - 3xy + y^2 = C$

9. $3xy^2 + 5x^2y^2 - 2y = C$

11. $\arctan \dfrac{x}{y} = C$    13. $\dfrac{x^2}{2y^2} = C$

15. (a) Answers will vary.    (b) $x^2 \tan y + 5x = \dfrac{11}{4}$

(c)
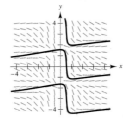

17. $x^2y - 3x^3 + y^2 + y = 6$    19. $e^{3x} \sin 3y = 0$

21. $y \ln(x - 1) + y^2 = 16$

23. Integrating factor: $y^3$

$xy^5 = C$

25. Integrating factor: $\dfrac{1}{y^2}$

$\dfrac{x}{y} - 6y = C$

27. Integrating factor: $\cos x$

$y \sin x + x \sin x + \cos x = C$

29. Integrating factor: $\dfrac{1}{y}$

$xy - \ln |y| = C$

31. Integrating factor: $\dfrac{1}{\sqrt{y}}$

$x\sqrt{y} + \cos \sqrt{y} = C$

33. $x^4y^3 + x^2y^4 = C$    35. $\dfrac{y^2}{x} + \dfrac{x}{y^2} = C$

37. (a)–(d) Proofs

39. $x^2 + y^2 = C$

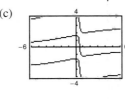

41. $2x^2y^4 + x^2 = C$

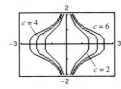

43. $x^2 - 2xy + 3y^2 = 3$    45. $C = \dfrac{5(x^2 + \sqrt{x^4 - 1,000,000x})}{x}$

47. (a)

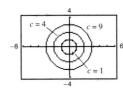

(c)

(b) $y^2(2x^2 + y^2) = 9$

49. (a)

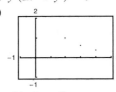

(c)

(b) $y^2(2x^2 + y^2) = 9$    Less accurate

51. $k = 3$    53. $f(x) = -\cos x + C_1$, $g(y) = \frac{1}{3}y^3 + C_2$

55. True    57. True

## Section 16.2    *(page 1143)*

1. (a) Order 5; homogeneous    (b) Order 2; nonhomogeneous

3. (a) $y = C_1e^{-x} + C_2e^{3x}$    (b) $y = C_1e^{2x} + C_2xe^{2x}$

5.
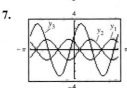
$y_1$: $C_1 = 0$, $C_2 = 1$
$y_2$: $C_1 = 1$, $C_2 = 1$
$y_3$: $C_1 = -1$, $C_2 = -2$

7.
$y_1$: $C_1 = 1$, $C_2 = -1$
$y_2$: $C_1 = -1$, $C_2 = 1$
$y_3$: $C_1 = 2$, $C_2 = 3$

9. $y = C_1 + C_2e^x$    11. $y = C_1e^{3x} + C_2e^{-2x}$

13. $y = C_1e^{x/2} + C_2e^{-2x}$    15. $y = C_1e^{-3x} + C_2xe^{-3x}$

17. $y = C_1e^{x/4} + C_2xe^{x/4}$    19. $y = C_1 \sin x + C_2 \cos x$

21. $y = C_1e^{(\sqrt{5}/2)x} + C_2e^{(-\sqrt{5}/2)x}$

23. $y = e^x\left(C_1 \sin \sqrt{3}x + C_2 \cos \sqrt{3}x\right)$

25. $y = C_1e^{[(3+\sqrt{5})/2]x} + C_2e^{[(3-\sqrt{5})/2]x}$

27. $y = e^{2x/3}\left(C_1 \sin \dfrac{\sqrt{7}x}{3} + C_2 \cos \dfrac{\sqrt{7}x}{3}\right)$

29. $y = C_1e^x + C_2e^{-x} + C_3 \sin x + C_4 \cos x$

31. $y = C_1e^x + C_2e^{2x} + C_3e^{3x}$

33. $y = C_1e^x + e^x(C_2 \sin 2x + C_3 \cos 2x)$

35. $y = C_1e^{-x} + C_2xe^{-x} + C_3e^x + C_4xe^x$

37. (a) $y = 2 \cos 10x$    (b) $y = \frac{1}{5} \sin 10x$

(c) $y = -\cos 10x + \frac{3}{10} \sin 10x$

39. $y = \frac{1}{11}(e^{6x} + 10e^{-5x})$    41. $y = \frac{1}{2} \sin 4x$

43. $y = 2e^{x/3} + \frac{1}{3}xe^{x/3}$    45. $y = \left(\dfrac{e - 3}{e - e^3}\right)e^{3x} + \left(\dfrac{3 - e^3}{e - e^3}\right)e^x$

47. $y = 3 \cos 3x - 4 \sin 3x$

49. $y = 2e^{7x/2} + \left(-\dfrac{1}{e^{7/2}} - 2\right)xe^{7x/2}$

51. Proof; $y = C_3e^x + C_4e^{-x}$    53. Damped    55. b

56. d    57. c    58. a    59. $y = \dfrac{1}{2} \cos 4\sqrt{3}t$

61. $y = -\dfrac{2}{3} \cos 4\sqrt{3}t + \dfrac{\sqrt{3}}{24} \sin 4\sqrt{3}t$

63. $y = \dfrac{e^{-t/16}}{2}\left(\cos \dfrac{\sqrt{12,287}t}{16} + \dfrac{\sqrt{12,287}}{12,287} \sin \dfrac{\sqrt{12,287}t}{16}\right)$

65. Proof    67. True    69. True    71–73. Proofs

## Section 16.3    *(page 1151)*

1. The sum of the general solution of the corresponding homogeneous equation and a particular solution of the nonhomogeneous equation

3. $y_h = C_1e^{-3x} + C_2e^{-4x}$
$y_p = -\frac{1}{16} + \frac{1}{4}x$
$y = C_1e^{-3x} + C_2e^{-4x} - \frac{1}{16} + \frac{1}{4}x$

**5.** $y_h = C_1 e^{4x} + C_2 x e^{4x}$

$y_p = e^{3x}$

$y = C_1 e^{4x} + C_2 x e^{4x} + e^{3x}$

**7.** The initial guess of $y_p = A + Bx$ requires modification because $y_h$ already has a constant term. Use $y_p = Ax + Bx^2$.

**9.** The initial guess of $y_p = A + B \sin x + C \cos x$ requires modification because $y_h$ already has a constant term. Use $y_p = Ax + B \sin x + C \cos x$.

**11.** $y_h = C_1 + C_2 e^{-2x}$

$y_p = -\frac{1}{2} x e^{-2x}$

$y = C_1 + C_2 e^{-2x} - \frac{1}{2} x e^{-2x}$

**13.** $y_h = C_1 \cos 3x + C_2 \sin 3x$

$y_p = -\frac{x}{6} \cos 3x$

$y = \left( C_1 - \frac{x}{6} \right) \cos 3x + C_2 \sin 3x$

**15.** $y_h = C_1 e^{2x} + C_2 x e^{2x} + C_3 e^{-x}$

$y_p = \frac{1}{2} + \frac{1}{6} x^2 e^{2x}$

$y = C_1 e^{2x} + C_2 x e^{2x} + C_3 e^{-x} + \frac{1}{2} + \frac{1}{6} x^2 e^{2x}$

**17.** $y = 6 \sin x + \cos x + x^3 - 6x$

**19.** $y = -1 + 2e^{-x} - \sin x - \cos x$

**21.** $y = \left( \frac{4}{9} - \frac{1}{2} x^2 \right) e^{4x} - \frac{1}{9}(1 + 3x) e^x$

**23.** $y = (C_1 + \ln|\cos x|)\cos x + (C_2 + x)\sin x$

**25.** $y = \left( C_1 - \frac{x}{2} \right) \cos 2x + \left( C_2 + \frac{1}{4} \ln|\sin 2x| \right) \sin 2x$

**27.** $y = (C_1 + C_2 x)e^x + \frac{x^2 e^x}{4}(\ln x^2 - 3)$

**29.** $q = \frac{3}{25}(e^{-5t} + 5t e^{-5t} - \cos 5t)$

**31.** $y = \frac{1}{4} \cos 8t - \frac{1}{2} \sin 8t + \sin 4t$

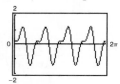

**33.** $y = \left( \frac{9}{32} - \frac{3}{4}t \right) e^{-8t} - \frac{1}{32} \cos 8t$

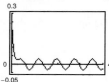

**35.** $y = \frac{\sqrt{5}}{4} \sin\left( 8t + \pi - \arctan \frac{1}{2} \right) \approx \frac{\sqrt{5}}{4} \sin(8t + 2.6779)$

**37.** (a) Undamped   (b) Damped

(c) $b > \frac{5}{2}$; No oscillations in this case.

**39.** True     **41.** Putnam Problem A3, 1987

## Section 16.4 *(page 1156)*

**1.** Given a differential equation, assume that the solution is of the form $y = \sum a_n x^n$. Then substitute $y$ and its derivatives into the differential equation. You should then be able to determine the coefficients $(a_0, a_1, \dots)$ for the solution series.

**3.** $y = a_0 \sum_{k=0}^{\infty} \left( -\frac{x}{5} \right)^k \frac{1}{k!}$     **5.** $y = a_0 \sum_{k=0}^{\infty} \frac{(-3)^k}{2^k k!} x^{2k}$

**7.** $y = a_0 \left( 1 - \frac{x^2}{8} + \frac{x^4}{128} - \cdots \right)$

$+ a_1 \left( x - \frac{x^3}{24} + \frac{7x^5}{1920} - \cdots \right)$

**9.** $y = 2 + \frac{2x}{1!} - \frac{2x^2}{2!} - \frac{10x^3}{3!} + \frac{2x^4}{4!}; y\left( \frac{1}{2} \right) \approx 2.547$

**11.** $y \approx 1 - \frac{3x}{1!} + \frac{2x^3}{3!} - \frac{12x^4}{4!} + \frac{16x^6}{6!} - \frac{120x^7}{7!}; y\left( \frac{1}{4} \right) \approx 0.253$

**13.** $y \approx 3 + \frac{2x}{1!} + \frac{3x^2}{2!} + \frac{2x^3}{3!}; y\left( \frac{1}{3} \right) \approx 3.846$

**15–17.** Proofs

**19.** (a) $y = 2x + \frac{x^3}{3} + \frac{x^5}{20} + \cdots$

(b)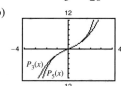

(c) The solution is symmetric about the origin.

**21–23.** Proofs

**25.** $y = a_0 + a_1 x + \frac{a_0}{6} x^3 + \frac{a_1}{12} x^4 + \frac{a_0}{180} x^6 + \frac{a_1}{504} x^7$

## Review Exercises for Chapter 16   *(page 1157)*

**1.** Not exact   **3.** $16xy + 10x^2 + 4x + 5y^2 + 4y = C$

**5.** $-2xy - 3y^2 + 4y + x^2 - 10x = C$

**7.** (a) Answers will vary.   (b) $x^2 + y^2 - xy = 4$

(c)

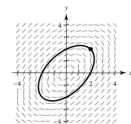   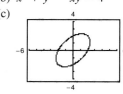

**9.** $2xy + 2x^2 - 6x - 3y^2 + 2y = -4$   **11.** $x \cos 2y = 3$

**13.** Integrating factor: $\dfrac{1}{x^2}$   **15.** Integrating factor: $e^{3y}$

$3x + \dfrac{y^2}{x} = C$     $xe^{3y} - e^y = C$

**17.**

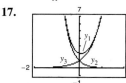

$y_1: C_1 = C_2 = 1$

$y_2: C_1 = 1, C_2 = 0$

$y_3: C_1 = 0, C_2 = 1$

**19.** $y = C_1 e^{-x} + C_2 e^{(-3/2)x}$   **21.** $y = C_1 + C_2 e^{6x}$

**23.** $y = C_1 \sin(2\sqrt{2}x) + C_2 \cos(2\sqrt{2}x)$

**25.** $y = C_1 e^{-x} + C_2 + C_3 e^{3x}$

**27.** $y = C_1 + C_2 x + C_3 e^{-\sqrt{5}x} + C_4 e^{\sqrt{5}x}$

**29.** $y = e^{2x} - e^{-x}$   **31.** $y = \frac{1}{2} e^{-3x} + \frac{3}{2} e^x$

**33.** $y = \left( \dfrac{4e}{\sin 2 - \tan 4 \cos 2} \right)(e^{-x} \sin 2x - \tan 4e^{-x} \cos 2x)$

**35.** $y = \frac{1}{2} \cos(2\sqrt{6}t)$

**37.** $y_h = C_1 \sin x + C_2 \cos x$
$y_p = -5x + x^3$
$y = C_1 \sin x + C_2 \cos x - 5x + x^3$

**39.** $y_h = C_1 e^{-x} + C_2 e^{9x}$
$y_p = -x + 2$
$y = C_1 e^{-x} + C_2 e^{9x} - x + 2$

**41.** The initial guess of $y_p = Ae^x + Be^{3x}$ requires modification because $y_h$ already contains an $e^x$ and $e^{3x}$ term. Use $y_p = Axe^x + Bxe^{3x}$.

**43.** $y_h = C_1 \sin x + C_2 \cos x$
$y_p = x \sin x$
$y = (C_1 + x) \sin x + C_2 \cos x$

**45.** $y = \frac{11}{5}(2e^{3x} + 3e^{-2x}) - 9$

**47.** $y = \frac{17}{3} \cos 2x - 3 \sin 2x + \frac{1}{3} \cos x$

**49.** $y = -\frac{1}{2} - \frac{1}{27}e^{-x} - \frac{1}{9}xe^{-x} - \frac{1}{6}x^2 e^{-x} + \frac{83}{54}e^{2x}$

**51.** $y = C_1 \cos 3x + C_2 \sin 3x - \frac{1}{3}x \cos 3x + \frac{1}{9} \sin 3x \ln|\sin 3x|$

**53.** $y = \left(C_1 + C_2 x + \frac{1}{3}x^3\right)e^x$

**55.** $q(t) = \frac{3}{10}e^{-2t} \sin 2t + \frac{3}{20}e^{-2t} \cos 2t - \frac{3}{40} \sin 4t - \frac{3}{20} \cos 4t$

**57.** (a) Only a second derivative is used, so a cosine is unnecessary.
(b) $y_p = \frac{5}{2} \cos x$
(c) If $y_p = A \cos x + B \sin x$, then $y_p'' = -A \cos x - B \sin x$. So it would be more difficult to solve for $A$ and $B$.

**59.** $y = a_0 \sum_{n=0}^{\infty} \frac{x^n}{4^n}$

**61.** $y \approx 2 + \frac{2x^2}{2!} + \frac{4x^4}{4!} + \frac{4x^5}{5!}; \ y\left(\frac{1}{4}\right) \approx 2.063$

## P.S. Problem Solving   (page 1159)

**1.** $k = -5; 6x^3 + 10y^3 - 15x^2y^2 = C$

**3.** $y = B_1 e^{ax} + B_2 e^{-ax};$ Proof    **5.** Proof

**7.** (a) and (b) Proofs    (c) $a = \left(\frac{n\pi}{L}\right)^2$, $n$ is an integer

**9.** (a) $\theta(t) = C_1 \cos\left(\sqrt{\frac{g}{L}}\,t\right) + C_2 \sin\left(\sqrt{\frac{g}{L}}\,t\right)$; Proof
(b) $\theta(t) = 0.128 \cos\left[\sqrt{39.2}(t - 0.108)\right]$
(c) About 1 sec    (d) 0.128    (e) 0.358 sec; 0.860 sec
(f) $\theta'(0.358) = -0.8012; \theta'(0.860) = 0.8012$

**11.** (a) Critically damped    (b) $y = e^{-4t} + 5te^{-4t}$
(c)
Answers will vary.

**13.** (a) Overdamped    (b) $y = e^{-16t} + e^{-4t}$
(c)
Answers will vary.

**15.** $y = a_0 + a_1(x - 1) + \frac{a_0}{2}(x - 1)^2 + \frac{a_0 + a_1}{6}(x - 1)^3$
$+ \frac{a_0 + 2a_1}{24}(x - 1)^4 + \frac{4a_0 + a_1}{120}(x - 1)^5$
$+ \frac{5a_0 + 6a_1}{720}(x - 1)^6 + \frac{9a_0 + 11a_1}{5040}(x - 1)^7$
Answers will vary.

**17.** (a) $y = a_0 \sum_{k=0}^{\infty} \frac{(-1)^k x^{2k}}{2^{2k}(k!)^2}$    (b) $y = a_0 J_0(x)$

**19.** (a) $y = 16x^4 - 48x^2 + 12$
(b) $H_0(x) = 1$
$H_1(x) = 2x$
$H_2(x) = 4x^2 - 2$
$H_3(x) = 8x^3 - 12x$
$H_4(x) = 16x^4 - 48x^2 + 12$

# Index

# ALGEBRA

## Factors and Zeros of Polynomials

Let $p(x) = a_n x^n + a_{n-1} x^{n-1} + \cdots + a_1 x + a_0$ be a polynomial. If $p(a) = 0$, then $a$ is a *zero* of the polynomial and a solution of the equation $p(x) = 0$. Furthermore, $(x - a)$ is a *factor* of the polynomial.

## Fundamental Theorem of Algebra

An *n*th degree polynomial has $n$ (not necessarily distinct) zeros. Although all of these zeros may be imaginary, a real polynomial of odd degree must have at least one real zero.

## Quadratic Formula

If $p(x) = ax^2 + bx + c$, and $0 \le b^2 - 4ac$, then the real zeros of $p$ are $x = \left(-b \pm \sqrt{b^2 - 4ac}\right)/2a$.

## Special Factors

$x^2 - a^2 = (x - a)(x + a)$

$x^3 + a^3 = (x + a)(x^2 - ax + a^2)$

$x^3 - a^3 = (x - a)(x^2 + ax + a^2)$

$x^4 - a^4 = (x - a)(x + a)(x^2 + a^2)$

## Binomial Theorem

$(x + y)^2 = x^2 + 2xy + y^2$

$(x + y)^3 = x^3 + 3x^2y + 3xy^2 + y^3$

$(x + y)^4 = x^4 + 4x^3y + 6x^2y^2 + 4xy^3 + y^4$

$(x - y)^2 = x^2 - 2xy + y^2$

$(x - y)^3 = x^3 - 3x^2y + 3xy^2 - y^3$

$(x - y)^4 = x^4 - 4x^3y + 6x^2y^2 - 4xy^3 + y^4$

$(x + y)^n = x^n + nx^{n-1}y + \dfrac{n(n-1)}{2!}x^{n-2}y^2 + \cdots + nxy^{n-1} + y^n$

$(x - y)^n = x^n - nx^{n-1}y + \dfrac{n(n-1)}{2!}x^{n-2}y^2 - \cdots \pm nxy^{n-1} \mp y^n$

## Rational Zero Theorem

If $p(x) = a_n x^n + a_{n-1} x^{n-1} + \cdots + a_1 x + a_0$ has integer coefficients, then every *rational zero* of $p$ is of the form $x = r/s$, where $r$ is a factor of $a_0$ and $s$ is a factor of $a_n$.

## Factoring by Grouping

$acx^3 + adx^2 + bcx + bd = ax^2(cx + d) + b(cx + d) = (ax^2 + b)(cx + d)$

## Arithmetic Operations

$ab + ac = a(b + c)$

$\dfrac{a}{b} + \dfrac{c}{d} = \dfrac{ad + bc}{bd}$

$\dfrac{a + b}{c} = \dfrac{a}{c} + \dfrac{b}{c}$

$\dfrac{\left(\dfrac{a}{b}\right)}{\left(\dfrac{c}{d}\right)} = \left(\dfrac{a}{b}\right)\left(\dfrac{d}{c}\right) = \dfrac{ad}{bc}$

$\dfrac{\left(\dfrac{a}{b}\right)}{c} = \dfrac{a}{bc}$

$\dfrac{a}{\left(\dfrac{b}{c}\right)} = \dfrac{ac}{b}$

$a\left(\dfrac{b}{c}\right) = \dfrac{ab}{c}$

$\dfrac{a - b}{c - d} = \dfrac{b - a}{d - c}$

$\dfrac{ab + ac}{a} = b + c$

## Exponents and Radicals

$a^0 = 1, \quad a \ne 0$ 
$(ab)^x = a^x b^x$ 
$a^x a^y = a^{x+y}$ 
$\sqrt{a} = a^{1/2}$ 
$\dfrac{a^x}{a^y} = a^{x-y}$ 
$\sqrt[n]{a} = a^{1/n}$

$\left(\dfrac{a}{b}\right)^x = \dfrac{a^x}{b^x}$ 
$\sqrt[n]{a^m} = a^{m/n}$ 
$a^{-x} = \dfrac{1}{a^x}$ 
$\sqrt[n]{ab} = \sqrt[n]{a}\sqrt[n]{b}$ 
$(a^x)^y = a^{xy}$ 
$\sqrt[n]{\dfrac{a}{b}} = \dfrac{\sqrt[n]{a}}{\sqrt[n]{b}}$

# FORMULAS FROM GEOMETRY

## Triangle

$h = a \sin \theta$

Area $= \dfrac{1}{2}bh$

(Law of Cosines)

$c^2 = a^2 + b^2 - 2ab \cos \theta$

## Right Triangle

(Pythagorean Theorem)

$c^2 = a^2 + b^2$

## Equilateral Triangle

$h = \dfrac{\sqrt{3}s}{2}$

Area $= \dfrac{\sqrt{3}s^2}{4}$

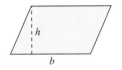

## Parallelogram

Area $= bh$

## Trapezoid

Area $= \dfrac{h}{2}(a + b)$

## Circle

Area $= \pi r^2$

Circumference $= 2\pi r$

## Sector of Circle

($\theta$ in radians)

Area $= \dfrac{\theta r^2}{2}$

$s = r\theta$

## Circular Ring

($p$ = average radius,

$w$ = width of ring)

Area $= \pi(R^2 - r^2)$

$\quad = 2\pi pw$

## Sector of Circular Ring

($p$ = average radius,

$w$ = width of ring,

$\theta$ in radians)

Area $= \theta pw$

## Ellipse

Area $= \pi ab$

Circumference $\approx 2\pi \sqrt{\dfrac{a^2 + b^2}{2}}$

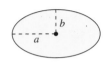

## Cone

($A$ = area of base)

Volume $= \dfrac{Ah}{3}$

## Right Circular Cone

Volume $= \dfrac{\pi r^2 h}{3}$

Lateral Surface Area $= \pi r \sqrt{r^2 + h^2}$

## Frustum of Right Circular Cone

Volume $= \dfrac{\pi(r^2 + rR + R^2)h}{3}$

Lateral Surface Area $= \pi s(R + r)$

## Right Circular Cylinder

Volume $= \pi r^2 h$

Lateral Surface Area $= 2\pi rh$

## Sphere

Volume $= \dfrac{4}{3}\pi r^3$

Surface Area $= 4\pi r^2$

## Wedge

($A$ = area of upper face,

$B$ = area of base)

$A = B \sec \theta$

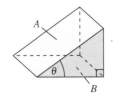